## ARITHMETIC SEQUENCE

$$a_{k+1} - a_k = d$$

$$a_n = a_1 + (n-1)d$$

$$\sum_{k=1}^{n} a_k = \frac{n}{2}(a_1 + a_n)$$

## GEOMETRIC SEQUENCE

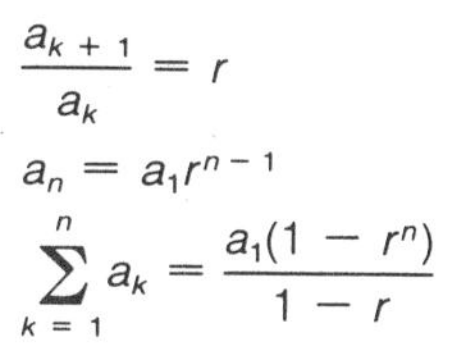

$$\frac{a_{k+1}}{a_k} = r$$

$$a_n = a_1 r^{n-1}$$

$$\sum_{k=1}^{n} a_k = \frac{a_1(1 - r^n)}{1 - r}$$

## DISTANCE FORMULA

$$d = \sqrt{(x_2 - x_1)^2 + (y_2 - y_1)^2}$$

## GEOMETRIC SERIES

$$\sum_{k=1}^{\infty} a_k = \frac{a_1}{1 - r} \qquad \text{if } |r| < 1$$

## CONIC SECTIONS

circle: $\begin{cases} x^2 + y^2 = r^2 \\ (x - h)^2 + (y - k)^2 = r^2 \end{cases}$

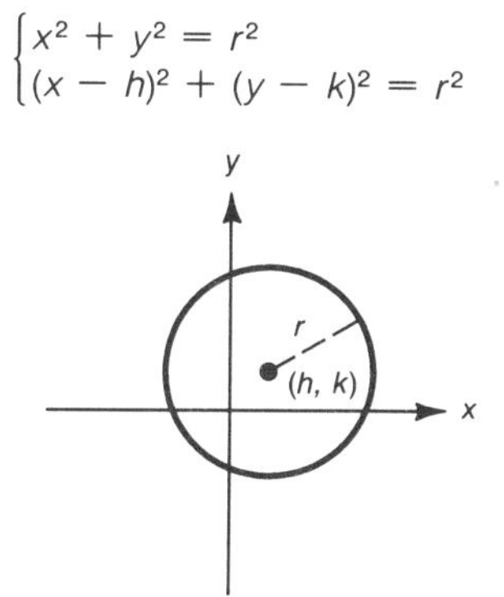


parabola: $\begin{cases} y = ax^2 + bx + c \\ x = ay^2 + by + c \end{cases}$

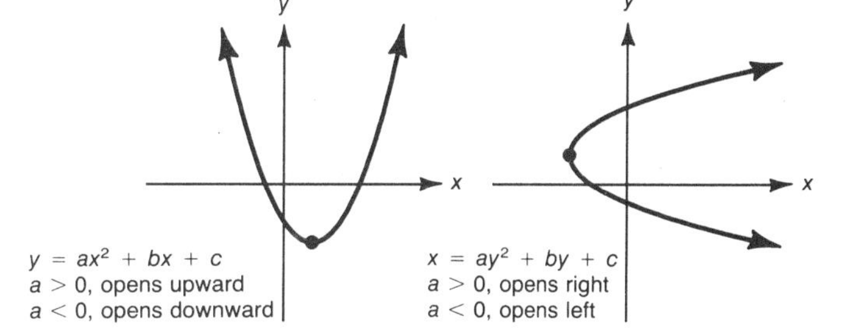


ellipse: $\left\{ \frac{x^2}{a^2} + \frac{y^2}{b^2} = 1 \right.$

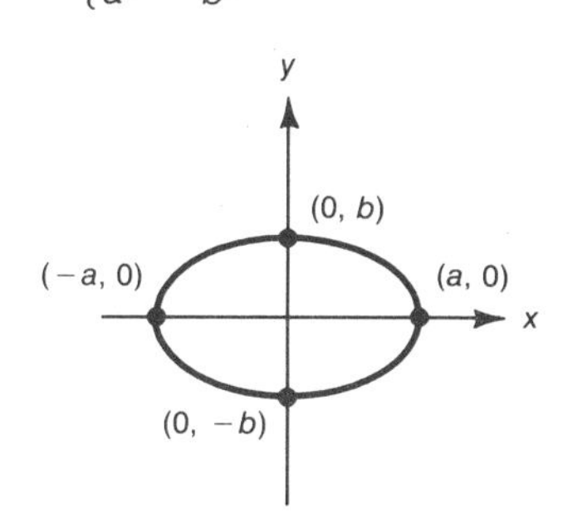


hyperbola:

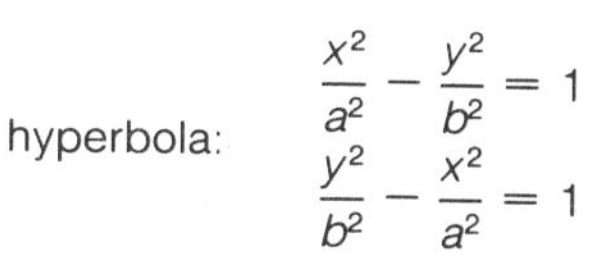

$$\frac{x^2}{a^2} - \frac{y^2}{b^2} = 1$$

$$\frac{y^2}{b^2} - \frac{x^2}{a^2} = 1$$

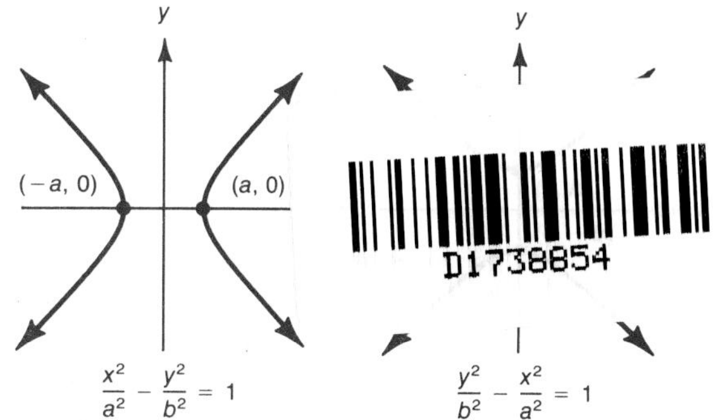


## BINOMIAL THEOREM

$$(a + b)^n = \binom{n}{0}a^n + \binom{n}{1}a^{n-1}b^1 + \cdots + \binom{n}{k}a^{n-k}b^k + \cdots + \binom{n}{n}b^n$$

$$\binom{n}{r} = \frac{n!}{r!(n - r)!}$$

# Algebra for College Students

# Algebra for College Students

D. Franklin Wright
*Cerritos College*

Bill D. New
*Cerritos College*

Wm. C. Brown Publishers

**Book Team**

Editor *Earl McPeek*
Developmental Editor *Theresa Grutz*
Designer *K. Wayne Harms*
Art Editor *Gayle A. Salow*
Visuals Processor *Jodi Wagner*

**Wm. C. Brown Publishers**



A study guide and solutions manual are available in your college bookstore. The titles are: *Student's Study Guide to accompany Algebra for College Students* and *Student's Solutions Manual to accompany Algebra for College Students*. They have been written to help you study, review, and master the course material. Ask the bookstore manager to order a copy for you if it is not in stock.

Cover photo by Mike Mitchell, Cedar Rapids, Iowa

Chapter Opening Art by Mary Coons

The credits section for this book begins on page 574, and is considered an extension of the copyright page.



Library of Congress Catalog Card Number: 89-83367

ISBN 0-697-07870-1



Printed in the United States of America by Wm. C. Brown Publishers, 2460 Kerper Boulevard, Dubuque, IA 52001

10 9 8 7 6 5 4 3 2 1

# CONTENTS

## CHAPTER 13 SEQUENCES AND SERIES 447

# PREFACE

## The Purpose

*Algebra for College Students* provides a solid review of intermediate algebra and a transition into topics traditionally presented in courses called college algebra. The quantity and level of the topics provided allows for a great deal of flexibility in the content of the course and provides for the possibility of a one-semester or a two-semester course. Business and social science majors who will continue their studies in courses such as statistics and short calculus will be well prepared for success in those courses. Students continuing in the science areas will have a solid foundation for trigonometry, precalculus, and courses in computer science.

We have provided very little overlap with material covered in a beginning algebra course. While Chapter 1 provides a review of topics from beginning algebra, students will find that the review is comprehensive and that the pace of coverage is somewhat faster and in more depth than they have seen in previous courses. As with any text in mathematics, students should read the text carefully and thoroughly. With insightful comments from many reviewers and an active editorial staff, we have confidence that students and instructors alike will find that this text is indeed a superior teaching and learning tool.

## The Style

The style of the text is informal and nontechnical while maintaining mathematical accuracy. Each topic is developed in a straightforward step-by-step manner. Each section contains many carefully developed and worked out examples to lead the students successfully through the exercises and prepare them for examinations. Whenever appropriate, information is presented in list form for organized learning and easy reference. Common Errors are highlighted and explained so that students can avoid such pitfalls and better understand the correct corresponding techniques. Practice Problems with answers are provided in almost every section to allow the students to "warm up" and to provide the instructor with immediate classroom feedback.

## Special Features

### Chapter Introductions

Each chapter begins with an introduction previewing the chapter coverage and preparing students for the new topics.

### Section Objectives

Each section begins with a listing of objectives, clearly identifying the skills to be learned.

**Practice Problems** Practice Problems can be found in almost every section.

**Calculator Problems** Many sections also contain Calculator Problems.

**Chapter Summaries** The chapter summaries are organized for easy reference and review. They are subdivided into Key Terms and Formulas, Properties and Rules, and Procedures.

**Cumulative Reviews** Beginning with Chapter 2, each chapter contains a Cumulative Review of topics chosen from previous chapters to continually refresh students' memories, maintain skill levels, and help the students understand the nature of the development of algebraic concepts.

**Examples** Examples have been carefully chosen and developed with a step-by-step analysis and commentary.

**Applications** Applications are a particularly important part of the text and are carefully chosen to be meaningful to the students. They appear in Chapters 1, 2, 3, 5, 7, 9, 11, and 13.

**The Content** **Chapter 1,** Real Numbers and Solving Equations, is a review of topics from beginning algebra. Real numbers and their properties, exponents, absolute value, solving equations, and solving for terms in formulas are all revealed. Other topics include a variety of word problems, interval notation (possibly new for some students), and solving and graphing inequalities by testing points in intervals.

**Chapter 2,** Exponents and Polynomials, discusses properties of exponents and scientific notation, operations with polynomials, and factoring polynomials. A special section on factoring includes factoring with negative exponents, a skill particularly useful in calculus.

**Chapter 3,** Rational Expressions, applies the factoring skills from Chapter 2 in operations with rational expressions. Synthetic division is now included along with long division and complex fractions. The interval notation introduced in Chapter 1 is used in discussing the solutions and graphs of rational inequalities.

**Chapter 4,** Radicals and Complex Numbers, develops the algebraic skills and related understanding of radicals and complex numbers needed for Chapter 5.

**Chapter 5,** Quadratic Equations and Inequalities, contains three techniques for solving quadratic equations: factoring, completing the square, and the quadratic formula. Quadratic equations are then related to applications, equations involving radicals, equations in quadratic form, and solutions to quadratic inequalities. In a manner similar to that in Chapter 1, inequalities are solved by using the concepts of intervals and testing points within key intervals.

**Chapter 6,** Linear Equations and Inequalities, includes complete discussions on the three basic forms for equations of straight lines in a plane: the standard form, the slope-intercept form, and the point-slope form. Linear inequalities are discussed in terms of half-planes and solved by testing single points in one half-plane.

**Chapter 7,** Systems of Equations and Inequalities, develops the basic methods of substitution and addition for solving systems of linear equations. Graphical methods are discussed for systems of two equations in two variables, and there are many related word problems for systems in two variables and in three variables. The concept of linear programming and the graphs of systems of linear inequalities are also developed.

**Chapter 8,** Functions, introduces functions and functional notation and goes into some detail with quadratic functions and the graphs of parabolas. The value of functional notation is emphasized through the concepts of horizontal and vertical translations, a unique approach at this level. The development of the composition of functions lays the groundwork for understanding the relationship between exponential and logarithmic functions in Chapter 9.

**Chapter 9,** Exponential and Logarithmic Functions, makes considerable use of the hand-held calculator. While the properties of real exponents and logarithms are presented completely, most numerical calculations are performed with the aid of a calculator and with the assumption that the student has such a calculator. Special emphasis is placed on the number $e$ and applications with natural logarithms. Students who plan to take a course in calculus should be aware that many of the applications found in calculus involve exponential and logarithmic expressions in some form. For those students, we recommend that a special effort should be made to grasp these rather abstract concepts.

**Chapter 10,** Conic Sections, is designed to provide a basic understanding of conic sections and their graphs. The distance formula is introduced to motivate the discussion of circles, and parabolas are discussed in some detail. However, as is appropriate for this level, circles, ellipses, and hyperbolas are restricted to those cases centered at the origin.

**Chapter 11,** Matrices and Determinants, shows the student two more techniques for solving systems of linear equations: Cramer's Rule with the use of determinants and the Gaussian elimination method with the use of matrices. Students should be made aware that matrices and the operations with matrices are particularly valuable tools for applications in business and in computer science.

**Chapter 12,** Roots and Polynomial Equations, analyzes the general properties of the roots (solutions) of polynomial equations: how many roots to expect, how to find possible rational roots, how to approximate irrational roots, and how to restrict the search for roots. These ideas come under the heading of theory of equations. The intended effect is to create an awareness of the problems related to solving polynomial equations other than quadratic equations and first-degree equations.

**Chapter 13,** Sequences and Series, provides flexibility for the instructor and reference material for the students. The topics presented here, including permutations and combinations, are likely to appear in courses in probability and statistics, finite mathematics, and higher level courses in mathematics. Any of these topics covered at this time will give students additional mathematical insights and a head start on future studies.

## Special Tools for Learning Provided in Every Chapter

**"Mathematical Challenges"** To develop interest and enthusiasm for mathematics

**Introduction** Comments on the nature of and reasons for the topics to be covered

**Highlighted Definitions and Theorems** Provide concise meanings in an easily identifiable format

**Procedure Boxes** Summarize step-by-step techniques for easy reference and review

**Section Exercises** Paired and graded problems reinforce section concepts and skills

**Chapter Summary** Coded for reference to corresponding sections and formatted according to definitions, lists of rules, and procedures

**Chapter Review** Problems for extra practice

**Chapter Test** Practice under test conditions

**Cumulative Review** Provides for a continual reinforcement

Each section begins with a list of Objectives.
Most sections contain Practice Problems with Solutions.

## Additional Features

1. Answers to the odd-numbered exercises and to all the Chapter Review and Chapter Test and Cumulative Review questions are in the back of the book.
2. There are over 555 numbered examples completely worked out with detailed analysis. There are also over 295 Practice Problems.
3. Key ideas, procedures, and rules are in list form and boldface print for easy identification and reference.
4. The wide format gives an "open" look to the text, and the design allows for easy identification of the various parts of the text: text material, examples, definitions, theorems, practice problems, rules, procedures, common errors, and exercises.

**Calculators**

Chapter 9 has been written to encourage the use of calculators. Also, calculator exercises are provided in many sections throughout the text. We believe that the calculator is now part of every student's life and it is important that they understand both the capabilities and the limitations of calculators. Having learned the necessary skills in the mathematics classroom, they will be able to apply this knowledge in their daily lives as well as in other classes.

The Exercises

More than 3,200 carefully selected and graded section exercises are provided. They proceed from relatively easy problems to more difficult ones.

**Ancillary Package**

For the Instructor

The **Instructor's Manual** has been expanded to include chapter overviews, objectives, Points for Emphasis, chapter summaries, chapter quizzes, challenge problems, and all solutions for the quizzes and challenge problems, and concise answer key for all text exercises.

The ***Instructor's Solutions Manual*** provides complete solutions for all problems in the section exercises, chapter reviews, chapter tests, and cumulative reviews.

**Wm. C. Brown Publishers Math TestPak** is a free, computerized testing service with two convenient options. You may use your own Apple® IIe, IIc, or IBM PC to produce your test. It is menu-driven with on-line help screens to guide you through the test-making process. You may select items from the bank, edit the existing items, add new items of your own, or have the system randomly select items for you by chapter, section, or objective.

**Wm. C. Brown Publishers Call-in Testing Service** offers customized student test masters and answer keys based on your selections from the *Test Item File*. Within two working days of your request, the master and answer key will be mailed to you.

The printed ***Test Item File*** in an 8½″ × 11″ format contains all of the questions on the Wm. C. Brown Publishers Math TestPak. It will serve as a ready-reference if you use your own computer to generate tests or use the Wm. C. Brown Publishers Call-in Testing Service.

**Wm. C. Brown Publishers Computerized Gradebook Software** is available for the IBM PC and Apple II family of computers.

For the Student

The ***Student Study Guide*** provides alternative explanations and examples, additional problems, self-testing materials, and built-in answer keys. It is available for student purchase.

The ***Student's Solutions Manual*** contains helpful hints, summaries, and detailed solutions to every other odd-numbered exercise problem. It is available for student purchase.

**Algebra Problem Solver,** Interactive Student Tutorial Software, by Michael Hoban and Kathirgama Nathan, is available to help your students master algebra topics. Contact your Wm. C. Brown Publishers sales representative for details.

**Math Lab Software** helps students master algebraic concepts in an interactive format. This program generates problems at random on a given topic, provides immediate feedback, and channels the student into a correct problem-solving procedure.

On the **Videotapes,** the instructor introduces a concept, provides detailed explanations of example problems that illustrate the concept, including applications, and concludes with a summary. The tapes are available in ½″ VHS format and are free to qualified adopters.

The **Audiotapes** start with a brief synopsis of the section and worked out sample problems, explaining each step and offering helpful hints and useful warnings. Important concepts and procedures are stressed. These tapes are free to qualified adopters.

## Acknowledgments

The following reviewers have been helpful with their constructive and critical comments:

**James Blackburn**
Tulsa Junior College

**Bill Chatfield**
University of Wisconsin-Platteville

**Sally Copeland**
Johnson County Community College

**Larry Curnutt**
Bellevue Community College

**Thomas F. Davis**
Daytona Beach Community College

**Richard De Tar**
Sacramento City College

**Elizabeth Hodes**
Santa Barbara City College

**Daniel A. Hogan**
Hinds Junior College

**Bill E. Jordan**
Seminole Community College

**Rahim G. Karimpour**
Southern Illinois University

**Gerald R. Krusinski**
College of DuPage

**James Newsom**
Tidewater Community College

**Sandra V. Rumore**
Liberty University

**Ronald P. Shassberger**
Grand Rapids Junior College

**Terry R. Tiballi**
North Harris County College

**Gerry C. Vidrine**
Louisiana State University

**Tom Worthington**
Grand Rapids Junior College

We would particularly like to thank the staff at Wm. C. Brown Publishers—Earl McPeek (editor), Theresa Grutz (developmental editor and mother to us all), and Gene Collins (production editor who understands authors)—for their outstanding work and enthusiasm for the entire project. Pat Wright types and prepares manuscripts in the best form possible. Thanks again, Pat, for such outstanding work and putting up with two of your best friends. Thanks also to our students and colleagues at Cerritos College for their interest and comments over the years.

Thank You All

Frank Wright
Bill New

CHAPTER

# 1 REAL NUMBERS AND SOLVING EQUATIONS

## MATHEMATICAL CHALLENGES

A freight train one kilometer long goes through a tunnel that is one kilometer long. If the train is traveling at a speed of fifteen kilometers per hour, how long does it take to pass through the tunnel?

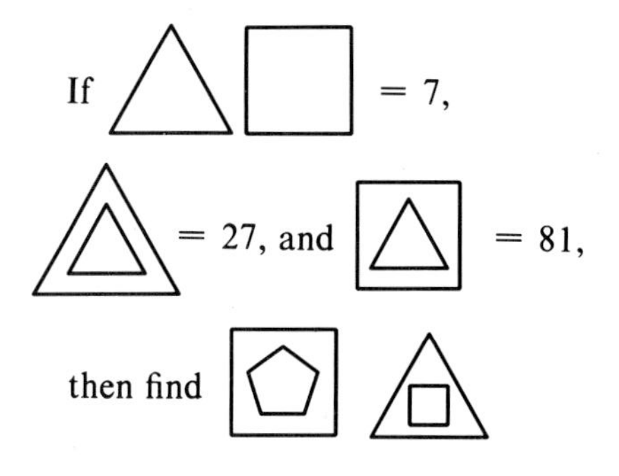


This drawing of four sticks and a pebble represents a stone on a shovel. Move exactly two sticks to remove the stone from the shovel.

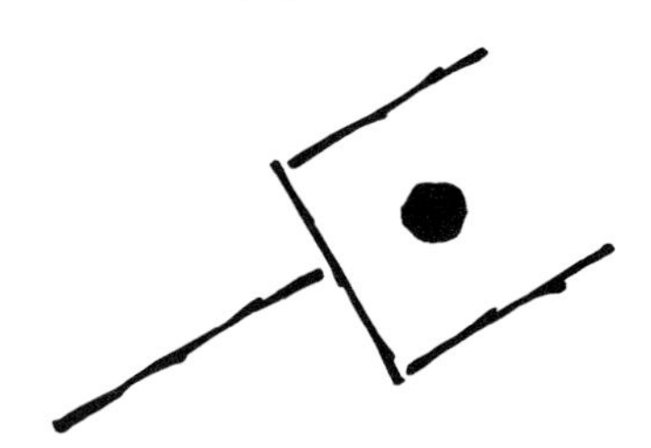

## CHAPTER OUTLINE



Welcome to your second course in algebra. You will find that many of the topics presented, such as solving equations, factoring, working with fractional expressions, and solving word problems, are familiar because they were discussed in beginning algebra. However, you will also find that the coverage of each topic, including many new topics, is in greater depth and that the pace is somewhat faster. The intent is to prepare you for success with the algebraic content in your future studies in mathematics, science, business, and economics. You should be prepared to expend considerable time and effort throughout the semester.

We begin in Chapter 1 with a review of the properties and operations with real numbers. These ideas form the foundation for the study of algebra and need to be thoroughly understood. As an example of a deeper analysis of familiar topics, we will discuss solving first-degree equations, then expand the techniques to include solving absolute value equations and solving formulas for specified variables. You will also be introduced to the new topic of intervals of real numbers. Intervals and interval notation are related to graphs and are common in higher level mathematics courses.

There is much to look forward to in this course. With hard work and perseverance, you should have a very rewarding experience.

## 1.1 Properties of Real Numbers

**OBJECTIVES**

In this section, you will be learning to:

1. Identify given numbers as members of one or more of the following sets:
   natural numbers,
   whole numbers,
   integers,
   rational numbers,
   irrational numbers, and
   real numbers.
2. Write rational numbers as infinite repeating decimals.
3. Graph sets of numbers on real number lines.
4. Describe sets of numbers using set-builder notation given their graphs.
5. Name the properties of real numbers that justify given statements.
6. Complete statements using the real number properties.

We begin with a development of the terminology and properties of numbers that form the foundation for the study of algebra. The following kinds of numbers are studied in some detail in beginning algebra courses.

### Types of Numbers

***Natural Numbers*** (or ***Counting Numbers***)

$N = \{1, 2, 3, 4, 5, 6, \ldots\}$  The three dots ( . . . ) indicate that the pattern is to continue without end.

***Whole Numbers*** (The number 0 is included with $N$.)

$W = \{0, 1, 2, 3, 4, 5, 6, \ldots\}$

***Integers***

$J = \{\ldots, -4, -3, -2, -1, 0, 1, 2, 3, 4, \ldots\}$

The integers are one of the important stepping stones from arithmetic to algebra since the concept of positive and negative numbers is basic to algebra. We can represent the integers on a number line by marking 0 at some point and then marking the **positive integers** to the right of 0 and their **opposites** or **negative integers** to the left of 0 (Figure 1.1).

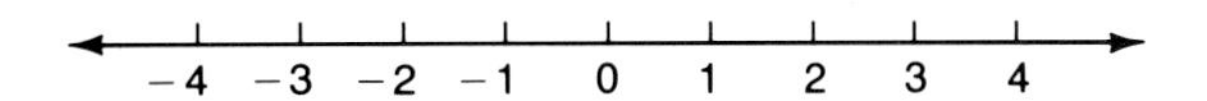


**Figure 1.1**

We also need to know about **rational numbers, irrational numbers,** and **real numbers.** We define these numbers first, then discuss their properties.

***Rational Number***

A **rational number** is any number that can be written in the form $\frac{a}{b}$ where $a$ and $b$ are integers and $b \neq 0$.
($Q$ represents all rational numbers.)

**EXAMPLES**

**1.** $\frac{2}{3}, \frac{7}{1}, \frac{-5}{3}, \frac{27}{10}$, and $\frac{3}{-10}$ are all rational numbers. Each is in the form $\frac{a}{b}$ where $a$ and $b$ are integers and $b \neq 0$.

**2.** $1\frac{3}{4}$ and 2.33 are also rational numbers. They are not in the form $\frac{a}{b}$, but they can be written in that form:

$$1\frac{3}{4} = \frac{7}{4} \qquad \text{and} \qquad 2.33 = 2\frac{33}{100} = \frac{233}{100}$$ ■

You may generally think of rational numbers as **fractions.** However, be aware that this is technically incorrect because there are fractions, such as $\frac{\pi}{6}$ and $\frac{\sqrt{2}}{3}$, that are **not** rational numbers.

Using the definition of a rational number, it can be proved that **any rational number can be written as an infinite repeating decimal.** Thus, there are two basic forms for rational numbers. Consider the three rational numbers

$\frac{2}{3}$, $\frac{1}{7}$, and $\frac{4}{11}$. Long division shows the repeating decimal pattern for each:

$$\begin{array}{r} 0.6666\ldots \\ 3\overline{)2.0000\ldots} \\ \underline{1\,8}\phantom{000} \\ 20\phantom{00} \\ \underline{18}\phantom{00} \\ 20\phantom{0} \\ \underline{18}\phantom{0} \\ 20 \\ \underline{18} \\ 2 \end{array} \qquad \begin{array}{r} 0.14285714\ldots \\ 7\overline{)1.00000000\ldots} \\ \underline{7}\phantom{0000000} \\ 30\phantom{000000} \\ \underline{28}\phantom{000000} \\ 20\phantom{00000} \\ \underline{14}\phantom{00000} \\ 60\phantom{0000} \\ \underline{56}\phantom{0000} \\ 40\phantom{000} \\ \underline{35}\phantom{000} \\ 50\phantom{00} \\ \underline{49}\phantom{00} \\ 10\phantom{0} \\ \underline{7}\phantom{0} \\ 30 \\ \underline{28} \\ 2 \end{array} \qquad \begin{array}{r} 0.3636\ldots \\ 11\overline{)4.0000\ldots} \\ \underline{3\,3}\phantom{000} \\ 70\phantom{00} \\ \underline{66}\phantom{00} \\ 40\phantom{0} \\ \underline{33}\phantom{0} \\ 70 \\ \underline{66} \\ 4 \end{array}$$

Thus, $\frac{2}{3} = 0.6666\ldots$, $\frac{1}{7} = 0.14285714\ldots$, and $\frac{4}{11} = 0.3636\ldots$.

Or we can write $\frac{2}{3} = 0.\overline{6}$, $\frac{1}{7} = 0.\overline{142857}$, and $\frac{4}{11} = 0.\overline{36}$. The bar is written over the repeating pattern of digits.

In addition, fractions that can be written as terminating decimals, such as $\frac{1}{4} = 0.25$ and $\frac{3}{8} = 0.375$, are considered to be repeating decimals with a repeating pattern of 0s.

$$\frac{1}{4} = 0.25 = 0.25\overline{0} \qquad \text{and} \qquad \frac{3}{8} = 0.375 = 0.375\overline{0}$$

---

***Irrational Number***

An **irrational number** is any number that can be written as an infinite nonrepeating decimal.
($H$ represents all irrational numbers.)

---

**EXAMPLES** The following numbers are irrational numbers and have no repeating pattern in their decimal representation. The three dots . . . indicate no end to the digits, but, in these cases, there is no pattern involved.

**3.** $\pi = 3.14159265358979\ldots$ — $\pi$ has no repeating pattern in its decimal form.

**4.** $\sqrt{2} = 1.414213\ldots$ — The square root of 2 has no repeating pattern in its decimal form.

**5.** $e = 2.718281828359045 \ldots$ — $e$ is a number used in higher mathematics and engineering courses.

**6.** $0.01001000100001 \ldots$ — Even though there is a pattern to the digits, the pattern is not repeating. ∎

Together, the rational numbers ($Q$) and irrational numbers ($H$) form the set of **real numbers** ($R$). The relationships between the various types of real numbers can be seen in Figure 1.2.

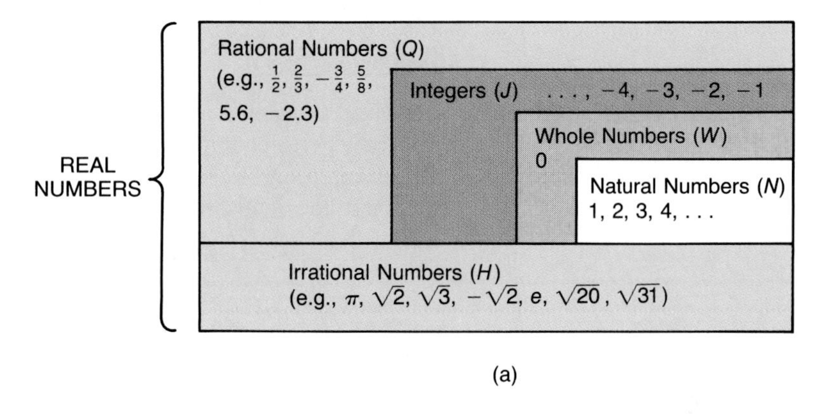


(a)

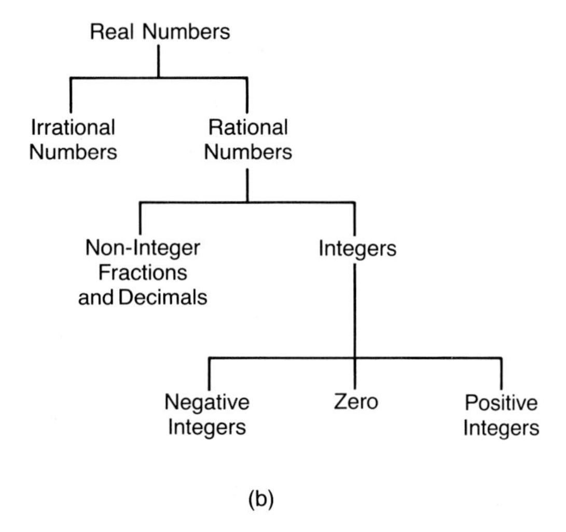


(b)

**Figure 1.2**

Understanding these relationships is **critical** for success in algebra. For example, if a problem calls for integer solutions and your solution is $\frac{3}{5}$, then you need to understand that you have not found an **integer** solution.

In summary:

1. Every natural number is also a whole number, an integer, a rational number, and a real number.
2. Every whole number is also an integer, a rational number, and a real number.
3. Every integer is also a rational number and a real number.
4. Every rational number is also a real number.
5. Every irrational number is also a real number.

In Chapter 4, we will expand our knowledge of number types to include **imaginary numbers,** such as $\sqrt{-1}$, and **complex numbers.**

### Number Lines

When you draw a number line, you should know that every real number (every rational and irrational number) has a corresponding point on that line. We say that **there is a one-to-one correspondence between the real numbers and the points on a line.** Thus, number lines are also called **real number lines** (Figure 1.3).

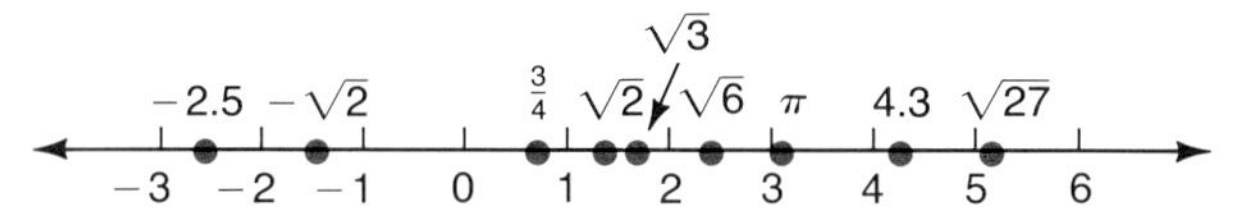


**Figure 1.3**

We will thoroughly discuss irrational numbers in the form of various roots, such as $\sqrt{3}$, in Chapter 4. For now, to estimate the placement of numbers such as $\sqrt{3}$, $\sqrt{6}$, or $\sqrt{27}$ on a number line, you can note their relationships to the square roots of perfect square integers such as 1, 4, 9, 16, 25, and so on. Thus,

$$\sqrt{3} \text{ is slightly less than } \sqrt{4} = 2 \qquad (\sqrt{3} = 1.7320508\ldots)$$
$$\sqrt{6} \text{ is slightly more than } \sqrt{4} = 2 \qquad (\sqrt{6} = 2.4494897\ldots)$$
$$\sqrt{27} \text{ is slightly more than } \sqrt{25} = 5 \qquad (\sqrt{27} = 5.1961524\ldots)$$

To understand how an infinite nonrepeating decimal corresponds to a single point on a line, we will illustrate how $\pi = 3.14159265\ldots$ can be marked. If a circle has a diameter of 1 unit, then its circumference is $\pi$. By rolling such a circle along a line, the number $\pi$ can be located (Figure 1.4).

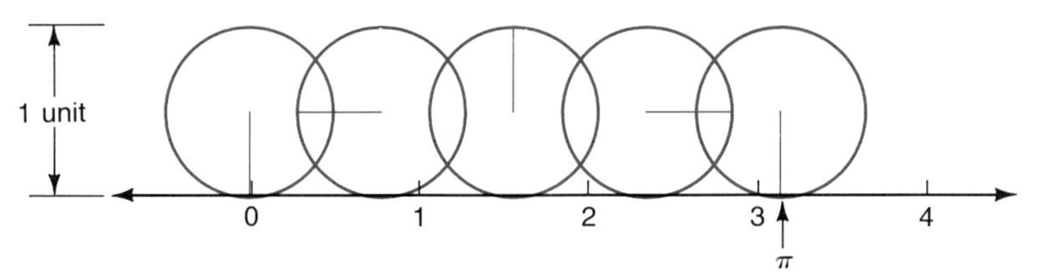


**Figure 1.4**

The notation $\{x \mid \ \}$ is read "the set of all $x$ such that . . ." and is called **set-builder notation.** The bar ( | ) is read "such that." This notation, along with the inequalities

$<$ "less than"
$\leq$ "less than or equal to"
$>$ "greater than"
$\geq$ "greater than or equal to"

is used to indicate sets of real numbers that are to be graphed.

| Notation | Meaning | Graph |
|---|---|---|
| $\{x \mid x \leq a\}$ | "the set of all $x$ such that $x$ is less than or equal to $a$" | a |
| $\{x \mid x \geq b\}$ | "the set of all $x$ such that $x$ is greater than or equal to $b$" | b |
| $\{x \mid x < a$ **or** $x > b\}$ (This is also known as the **union** of two sets of numbers.) | "the set of all $x$ such that $x$ is less than $a$ **or** $x$ is greater than $b$" | a b |
| $\{x \mid x > a$ **and** $x < b\}$ or $\{x \mid a < x < b\}$ (This is also known as the **intersection** of two sets of numbers.) | "the set of all $x$ such that $x$ is greater than $a$ **and** $x$ is less than $b$" | a b |

The following examples show how various sets of real numbers can be graphed. Note that an open circle means that the point is not included and a shaded circle means that the point is included.

**EXAMPLES**

**7.** Graph the set of real numbers $\{x \mid -1 \leq x < 2\}$.

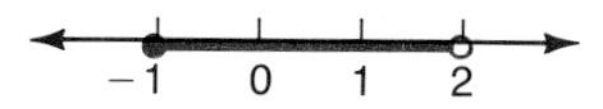


**8.** Graph the set $\{x \mid x > 3\}$.

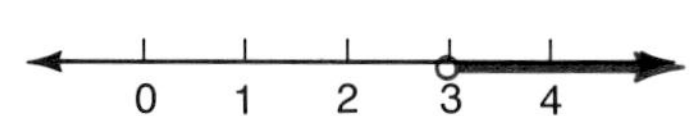


■

In Examples 9 and 10, note carefully the use of the two key words **or** and **and.**

## EXAMPLES

**9.** Graph the set $\{x \mid x \leq 2$ **and** $x \geq 0\}$. The word **and** implies that we want those values of $x$ that satisfy **both** inequalities.

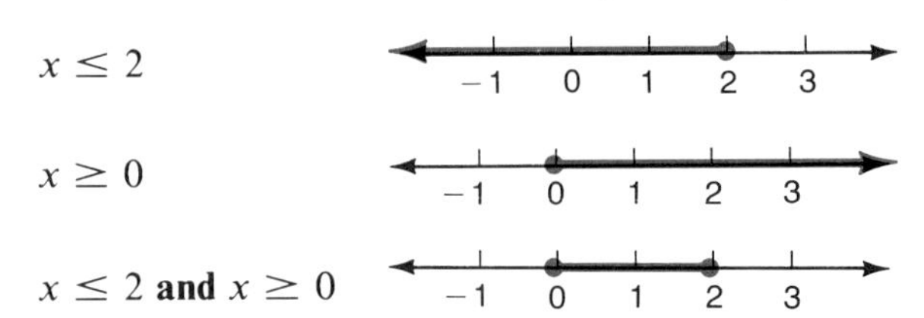


This set can also be indicated as $\{x \mid 0 \leq x \leq 2\}$.

**10.** Graph the set $\{x \mid x > 5$ **or** $x \leq 4\}$. The word **or** implies that we want any $x$ that satisfies either inequality (or both).

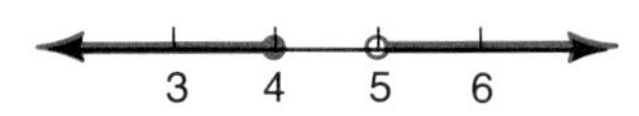


**11.** Given the set of numbers $\left\{-2, -\sqrt{3}, -1.1, -\frac{1}{2}, 0, \frac{5}{8}, \sqrt{1.7}, 1.7\right\}$,

**a.** graph the set of numbers on a number line;
**b.** tell which numbers are integers;
**c.** tell which numbers are rational numbers;
**d.** tell which numbers are irrational numbers.

**Solution**

**a.**

**b.** $-2$ and $0$ are integers.

**c.** $-2, -1.1, -\frac{1}{2}, 0, \frac{5}{8}$, and $1.7$ are rational numbers.

**d.** $-\sqrt{3}$ and $\sqrt{1.7}$ are irrational numbers. As decimals, $-\sqrt{3}$ is approximately $-1.732$ and $\sqrt{1.7}$ is approximately $1.304$.

(**Note:** Such decimal approximations to square roots can be found using a hand-held calculator.) ■

---

***Substitution Property*** If $a$ and $b$ are real numbers and $a = b$, then $a$ and $b$ may be substituted for each other in any expression.

---

The **operations** of addition and multiplication with real numbers have five properties each, and there is one property that combines both operations. These eleven properties are called the **field properties of real numbers.**

***Field Properties of the Real Numbers***

For real numbers $a$, $b$, and $c$:

| ***Addition*** | | ***Name of Property*** | ***Multiplication*** | |
|---|---|---|---|---|
| A1 | $a + b$ is a real number. | Closure | M1 | $a \cdot b$ is a real number. |
| A2 | $a + b = b + a$ | Commutative | M2 | $a \cdot b = b \cdot a$ |
| A3 | $(a + b) + c = a + (b + c)$ | Associative | M3 | $(a \cdot b) \cdot c = a \cdot (b \cdot c)$ |
| A4 | $a + 0 = 0 + a = a$ | Identity | M4 | $a \cdot 1 = 1 \cdot a = a$ |
| A5 | $a + (-a) = 0$ | Inverse | M5 | $a \cdot \frac{1}{a} = 1$ $(a \neq 0)$ |

D Distributive Property of Multiplication over Addition: $a(b + c) = a \cdot b + a \cdot c$

**EXAMPLES** Tell which field property justifies each statement.

**12.** $-3 \cdot 1 = -3$ — Identity of multiplication

**13.** $\frac{1}{2} + 0 = \frac{1}{2}$ — Identity of addition

**14.** $7 + 16 = 16 + 7$ — Commutative property of addition

**15.** $(2 \cdot 3) \cdot 8 = 2 \cdot (3 \cdot 8)$ — Associative property of multiplication

**16.** $6 + (-6) = 0$ — Inverse property of addition

**17.** $9 \cdot \frac{1}{9} = 1$ — Inverse property of multiplication

**18.** $4(x + 3) = 4 \cdot x + 12$ — Distributive property ■

You should be aware that neither subtraction nor division is commutative or associative. For example, with subtraction,

$$6 - 2 = 4 \quad \text{and} \quad 2 - 6 = -4 \quad \text{so} \quad 6 - 2 \neq 2 - 6.$$

Also,

$$10 - (5 - 3) = 10 - (2) = 8 \quad \text{and} \quad (10 - 5) - 3 = 5 - 3 = 2$$
$$\text{so} \quad 10 - (5 - 3) \neq (10 - 5) - 3.$$

With division,

$$6 \div 2 = 3 \quad \text{and} \quad 2 \div 6 = \frac{1}{3} \quad \text{so} \quad 6 \div 2 \neq 2 \div 6.$$

Also,

$24 \div (4 \div 2) = 24 \div (2) = 12$ and $(24 \div 4) \div 2 = 6 \div 2 = 3$
so $24 \div (4 \div 2) \neq (24 \div 4) \div 2$.

For $a \neq 0$,

$\frac{1}{a}$ is the **multiplicative inverse** (or **reciprocal**) of $a$ and $\boldsymbol{a \cdot \frac{1}{a} = 1}$.

For 0,

$$\frac{1}{0} = \text{undefined (0 does not have a reciprocal.)}$$

(That is, if $\frac{1}{0} = x$, then $1 = 0 \cdot x$. But this is not possible, so we say that $\frac{1}{0}$ is undefined.)

The following three statements are theorems about real numbers and can be proved. They are reasonably intuitive, and their proofs are not given here. We will use these theorems in solving equations throughout the text.

***Zero Factor Law*** For any real number $a$,

$$a \cdot 0 = 0$$

***Addition Property of Equality*** For real numbers $a$, $b$, and $c$,

If $a = b$, then $a + c = b + c$.

***Multiplication Property of Equality*** For real numbers $a$, $b$, and $c$,

If $a = b$, then $a \cdot c = b \cdot c$.

There are two basic properties related to inequalities (or order) with real numbers. You have probably used these before, but you may not have known their names.

***Properties of Inequality (Order)*** For real numbers $a$, $b$, and $c$:

O1 Exactly one of the following is true: — Trichotomy property

$a < b$, $a = b$, or $a > b$

O2 If $a < b$ and $b < c$, then $a < c$. — Transitive property

**EXAMPLES** State which property of equality or order is illustrated.

**19.** $-5 \cdot 0 = 0$ — Zero factor law

**20.** If $x < 3$ and $3 < y$, then $x < y$. — Transitive property of order

| | |
|---|---|
| **21.** If $a = 7$, then $a + 6 = 13$. | Addition property of equality |
| **22.** If $x = \frac{1}{2}$, then $2x = 1$. | Multiplication property of equality |
| **23.** If $a$ is a real number, then either $a = 0.5$ or $a > 0.5$ or $a < 0.5$. | Trichotomy property of order ■ |

***Practice Problems***

1. List the integers.
2. What type of number is $\pi$?
3. Graph the set $\{x \mid 0 < x \leq 1\}$.
4. $a + 3 = 3 + a$ illustrates which property of addition?
5. $2(x + y) = 2x + 2y$ illustrates which field property?

## EXERCISES 1.1

Given the set of numbers $\left\{-8, -\sqrt{5}, -\sqrt{4}, -\frac{4}{3}, -1.2, -\frac{\sqrt{3}}{2}, 0, \frac{4}{5}, \sqrt{3}, \sqrt{11}, \sqrt{16}, 4.2, 6\right\}$, list those numbers in the set that are described in Exercises 1–6.

**1.** $\{x \mid x \text{ is a whole number}\}$

**2.** $\{x \mid x \text{ is a natural number}\}$

**3.** $\{x \mid x \text{ is an integer}\}$

**4.** $\{x \mid x \text{ is an irrational number}\}$

**5.** $\{x \mid x \text{ is a rational number}\}$

**6.** $\{x \mid x \text{ is a real number}\}$

In Exercises 7–12, choose the word that correctly completes each statement.

**7.** If $x$ is a rational number, then $x$ is (never, sometimes, always) a real number.

**8.** If $x$ is a rational number, then $x$ is (never, sometimes, always) an irrational number.

**9.** If $x$ is an integer, then $x$ is (never, sometimes, always) a whole number.

**10.** If $x$ is a real number, then $x$ is (never, sometimes, always) a rational number.

**11.** If $x$ is a rational number, then $x$ is (never, sometimes, always) an integer.

**12.** If $x$ is a natural number, then $x$ is (never, sometimes, always) a whole number.

Write each of the rational numbers in Exercises 13–18 as an infinite repeating decimal.

**13.** $\frac{5}{8}$ **14.** $\frac{9}{16}$ **15.** $-\frac{7}{3}$ **16.** $-\frac{8}{9}$ **17.** $\frac{71}{20}$ **18.** $\frac{5}{7}$

Graph each of the sets in Exercises 19–26 on a number line.

**19.** $\{x \mid x < 7, x \text{ is a whole number}\}$

**20.** $\{x \mid x < -12, x \text{ is an integer}\}$

**21.** $\{x \mid x \geq -4, x \text{ is an integer}\}$

**22.** $\{x \mid -9 < x \leq 2, x \text{ is an integer}\}$

**Answers to Practice Problems** **1.** $\{\ldots, -4, -3, -2, -1, 0, 1, 2, \ldots\}$
**2.** Irrational (or real) **3.** [number line graph from 0 (open) to 1 (closed)] 0 1 **4.** Commutative property of addition
**5.** Distributive property of multiplication over addition

**23.** $\{x \mid -5 < x < 6, x \text{ is a natural number}\}$

**24.** $\{x \mid -8 < x < 0, x \text{ is a whole number}\}$

**25.** $\{x \mid x < 12 \text{ and } x > 0, x \text{ is an integer}\}$

**26.** $\{x \mid x > 0 \text{ and } x \leq 8, x \text{ is an integer}\}$

Graph each of the sets of real numbers in Exercises 27–38 on a number line.

**27.** $\{x \mid 3 \leq x < 5\}$

**28.** $\{x \mid -4 < x < 4\}$

**29.** $\{x \mid -1.8 < x \leq 3\}$

**30.** $\{x \mid x \geq -2.5\}$

**31.** $\{x \mid x < 2 \text{ or } x > 8\}$

**32.** $\left\{x \mid x \leq -5 \text{ or } x \geq \frac{9}{5}\right\}$

**33.** $\{x \mid -\sqrt{2} < x < 0\}$

**34.** $\{x \mid -4 \leq x < \sqrt{5}\}$

**35.** $\{x \mid x \geq -1 \text{ and } -3 \leq x \leq 0\}$

**36.** $\left\{x \mid -\frac{7}{4} < x \leq 2 \text{ or } 2 < x < 3\right\}$

**37.** $\{x \mid -1.6 < x < 0 \text{ or } 2 \leq x \leq 3.7\}$

**38.** $\left\{x \mid -\frac{3}{5} < x < 0 \text{ or } 0 \leq x < \pi\right\}$

Name the property of real numbers that justifies each statement in Exercises 39–62. All variables represent real numbers.

**39.** $a + b$ is a real number.

**40.** $y + (-y) = 0$

**41.** $x < y$, $x = y$, or $x > y$

**42.** $4x + 0 = 4x$

**43.** $5 + (a + b) = (5 + a) + b$

**44.** $9(x + 5) = 9x + 45$

**45.** $(3x)y = y(3x)$

**46.** $7x \cdot \frac{1}{7x} = 1$

**47.** If $4x = 11$, then $\frac{1}{4}(4x) = \frac{1}{4}(11)$.

**48.** $x < 5$, $x = 5$, or $x > 5$

**49.** If $x = y$, then $x + 5 = y + 5$.

**50.** If $x < 11$ and $x = y$, then $y < 11$.

**51.** $\sqrt{5} \cdot x$ is a real number.

**52.** $(\sqrt{2}x)y = \sqrt{2}(xy)$

**53.** $(x + 1)(y + 4) = (x + 1)y + (x + 1)4$

**54.** $\sqrt{7} + (-\sqrt{7}) = 0$

**55.** $(y + z) \cdot 1 = y + z$

**56.** $x + (y + 7) = (x + y) + 7$

**57.** If $x < -2$ and $-2 < y$, then $x < y$.

**58.** $8x + 3x = (8 + 3)x$

**59.** $11 + (4 + y) = 11 + (y + 4)$

**60.** $(x + y) \cdot \frac{1}{x + y} = 1$

**61.** $\frac{1}{2y} \cdot 0 = 0$

**62.** If $a < b$ and $b < (x - 2)$, then $a < (x - 2)$.

Complete the expressions in Exercises 63–70 using the given property.

**63.** $2(x + 7) =$ ________________ Commutative property of addition

**64.** $x(y + 3) =$ ________________ Commutative property of multiplication

**65.** $x(6 + y) =$ ________________ Distributive property

**66.** $x + (3 + y) =$ ________________ Associative property of addition

**67.** $3(xz) =$ ________________ Associative property of multiplication

**68.** If $x + 3 = 21$ and $x = y + 1$, then ________________ Substitution property

**69.** $(x + 2)(y + 3) =$ ________________ Distributive property

**70.** If $x + 2 = y - 4$, then $5(x + 2) =$ ________________ Multiplication property of equality

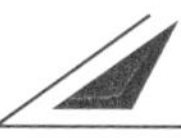

# 1.2 Operations with Real Numbers

**OBJECTIVES**

In this section, you will be learning to:

1. Evaluate absolute value expressions.
2. Determine the values, if any, that make absolute value equations true statements.
3. Add and subtract real numbers.
4. Multiply and divide real numbers.
5. Evaluate real number expressions using the rules for order of operations.

## Absolute Value

To understand how to actually perform the operations of addition, subtraction, multiplication, and division, we need the concept of **absolute value** of a number, symbolized $|x|$. Geometrically, the absolute value of a number is its distance from 0 on a number line. Thus, $|+3| = 3$ and $|-3| = 3$, since both $+3$ and $-3$ are 3 units from 0 (Figure 1.5).

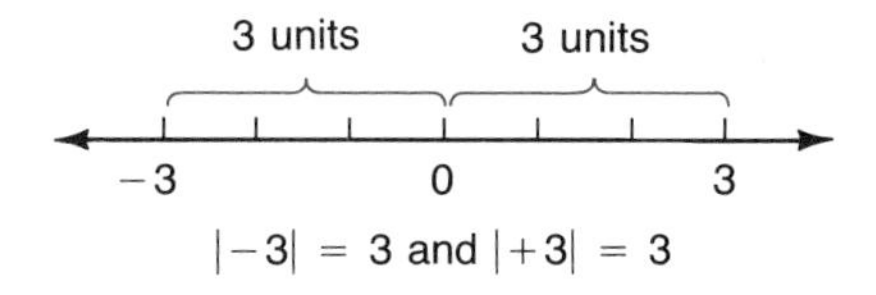


**Figure 1.5**

To help in understanding the definition of absolute value, answer the following questions:

Is $x$ positive or negative?
Is $-x$ positive or negative?

The correct answers are that you do not know because you do not know the value for $x$. In fact, if $x = 0$, then $x$ is neither positive nor negative. Suppose that $x = -7$. Then, obviously, $x$ is negative. However, what is not so obvious is that this means

$$-x = -(-7) = +7$$

and $-x$ is positive. Think of $-x$ as being the **opposite of $x$** rather than being negative. Thus,

$$\text{for } x = -7,$$
$$|x| = |-7| = +7 = -(-7) = -x$$

and again we see that the symbol $-x$ can represent a positive number. This approach will be a big help in understanding and using the following definition.

**Absolute Value**

For any real number $x$,

If $x$ is positive or 0, $|x| = x$.
If $x$ is negative, $|x| = -x$.

Another form of this same definition is the following:

$$|x| = \begin{cases} x & \text{if } x \geq 0 \\ -x & \text{if } x < 0 \end{cases}$$

In any case, remember that the absolute value of any real number is **nonnegative.**

EXAMPLES

1. $|5| = 5$
2. $|-6| = -(-6) = 6$
3. $|-\sqrt{2}| = -(-\sqrt{2}) = \sqrt{2}$
4. $|\pi| = \pi$
5. $\left|-\frac{3}{4}\right| = -\left(-\frac{3}{4}\right) = \frac{3}{4}$
6. $|0| = 0$
7. If $|x| = 6$, what are the possible values for $x$?

   **Solution** $x = 6$ or $x = -6$

   since both

   $$|6| = 6 \quad \text{and} \quad |-6| = 6$$

8. If $|x| = -5$, what are the possible values for $x$?

   **Solution** There is no value for $x$ for which $|x| = -5$ because absolute value is nonnegative. ■

## Addition and Subtraction

The rules for addition with positive and negative real numbers are stated here for easy reference. They illustrate the need for understanding absolute value. They also illustrate how mathematical symbols can simplify an idea that is rather difficult to state in words.

---

***Rules for Adding Real Numbers***

1. To add two real numbers with like signs, add their absolute values and use the common sign.

$$(+7) + (+3) = +(|+7| + |+3|) = +(7 + 3) = +10$$
$$(-7) + (-3) = -(|-7| + |-3|) = -(7 + 3) = -10$$

2. To add two real numbers with unlike signs, subtract their absolute values (the smaller from the larger) and use the sign of the number with the larger absolute value.

$$(-15) + (+9) = -(|-15| - |+9|) = -(15 - 9) = -6$$
$$(+15) + (-9) = +(|+15| - |-9|) = +(15 - 9) = +6$$

---

EXAMPLES

9. Sums with like signs:

   **a.** $(+10) + (+2) = +(|+10| + |+2|) = +(10 + 2) = +12$

   **b.** $\left(\frac{1}{5}\right) + \left(\frac{3}{5}\right) = \frac{4}{5}$

   **c.** $(-10) + (-2) = -(|-10| + |-2|) = -(10 + 2) = -12$

**10.** Sums with unlike signs:

**a.** $(-10) + (+2) = -(|-10| - |+2|) = -(10 - 2) = -8$

**b.** $\left(-\frac{7}{11}\right) + \left(\frac{5}{11}\right) = -\frac{2}{11}$

**c.** $(+10) + (-2) = +(|+10| - |-2|) = +(10 - 2) = +8$ ∎

The difference between two real numbers is defined to be the sum of the first number and the opposite of the second number, as follows.

***Subtraction***

For real numbers $a$ and $b$,

$$a - b = a + (-b)$$

We see that subtraction is defined in terms of addition. This means that any subtraction problem can be thought of as an addition problem. Thus, to find the difference between $-20$ and $+13$, we can write

$$(-20) - (+13) = (-20) + (-13) = -33$$

To subtract $-7$ from $-18$, we have

$$(-18) - (-7) = (-18) + (+7) = -11$$

**EXAMPLES**

**11.** $18 - 13 = 18 + (-13) = 5$

**12.** $-20 - 9 = -20 + (-9) = -29$

**13.** $14 - (-6) = 14 + (+6) = 20$

**14.** $-3 - 4 = -3 + (-4) = -7$

**15.** $\frac{17}{3} - \frac{25}{3} = \frac{17}{3} + \left(-\frac{25}{3}\right) = -\frac{8}{3}$ ∎

If more than two numbers are involved, add or subtract from left to right.

**EXAMPLES**

**16.** $8 - 12 - 21 = 8 + (-12) + (-21) = -4 + (-21) = -25$

**17.** $\frac{3}{5} + \frac{4}{5} - \frac{7}{5} = \frac{7}{5} - \frac{7}{5} = 0$

**18.** $\frac{2}{3} - \frac{1}{12} - \frac{3}{4} = \frac{8}{12} - \frac{1}{12} - \frac{9}{12} = \frac{7}{12} - \frac{9}{12} = -\frac{2}{12} = -\frac{1}{6}$

[**Note:** 12 is the least common multiple (LCM) of the denominators. We will discuss the LCM in detail in Chapter 3.]

**19.** $8.2 - 3.1 - 0.6 = 5.1 - 0.6 = 4.5$ ∎

## Multiplication and Division

If two real numbers are multiplied together, the result is called the **product** and the two numbers are called **factors** of the product. For example, since $5 \cdot 3 = 15$, the numbers 5 and 3 are factors of the product 15.

From arithmetic, we know that **the product of two positive real numbers is positive.** For the product of a positive integer and a negative integer, consider the product $5(-3)$. We can think of this as repeated addition, as

$$(-3) + (-3) + (-3) + (-3) + (-3) = 5(-3) = -15$$

More generally, by using the commutative and associative properties and the relationship $-a = -1 \cdot a$, we can prove that **the product of a positive real number and a negative real number is negative.** The theorem is stated without the proof.

**Theorem**

For positive real numbers $a$ and $b$,

$$a(-b) = -ab$$

**EXAMPLES**

**20.** $8(-5) = -40$

**21.** $(-6)\left(\frac{1}{2}\right) = -3$

**22.** $4(-20.1) = -80.4$

**23.** $9(-4)(2) = -36(2) = -72$ ■

The fact that the product of two negative numbers is positive is stated in the following theorem without proof.

**Theorem**

For positive real numbers $a$ and $b$,

$$(-a)(-b) = ab$$

**EXAMPLES**

**24.** $(-3)(-4) = 12$

**25.** $\left(-\frac{3}{4}\right)\left(-\frac{1}{2}\right) = \frac{3}{8}$

**26.** $(-2.1)(-0.03) = 0.063$

**27.** $(-2)(-5)(-7) = (10)(-7) = -70$ ■

Since the product of two negative numbers is positive, we can make the following useful observations:

If a product of nonzero factors contains an even number of negative factors, the product will be positive.

If a product of nonzero factors contains an odd number of negative factors, the product will be negative.

Examples 28 and 29 illustrate these two ideas.

**EXAMPLES**

**28.** Find the product $(-6)(-2)(-1)(-8)$.

**Solution** The product will be positive since there are four (an even number) negative factors.

$$\begin{aligned}(-6)(-2)(-1)(-8) &= [(-6)(-2)][(-1)(-8)] \\ &= [+12][+8] \\ &= +96\end{aligned}$$

**29.** Find the product $(-7)(5)(-2)(-9)$.

**Solution** The product will be negative since there are three (an odd number) negative factors.

$$\begin{aligned}(-7)(5)(-2)(-9) &= [(-7)(5)][(-2)(-9)] \\ &= [-35][+18] \\ &= -630\end{aligned}$$

■

The rules for multiplying with positive and negative real numbers are also related to the rules for division because division is defined in terms of multiplication. Thus, we know that

$$\frac{15}{3} = 5 \quad \text{because} \quad 15 = 3 \cdot 5.$$

***Division***

For real numbers $a$ and $b$ $(b \neq 0)$,

$$\frac{a}{b} = x \quad \text{if and only if} \quad a = b \cdot x$$

With this definition and the rules for multiplication, we have results as follows:

$$\frac{-24}{6} = -4 \quad \text{because} \quad -24 = 6(-4)$$

$$\frac{14}{-2} = -7 \quad \text{because} \quad 14 = (-2)(-7)$$

$$\frac{-36}{-12} = 3 \quad \text{because} \quad -36 = (-12)(3)$$

**EXAMPLES**

**30.** $\dfrac{30.6}{-2} = -15.3$ **31.** $\dfrac{-18}{-6} = 3$ **32.** $-\dfrac{51}{3} = -17$

■

| ***Rules for Multiplication and Division with Real Numbers*** | 1. The product (or quotient) of two nonzero real numbers with like signs is a positive real number.<br>2. The product (or quotient) of two nonzero real numbers with unlike signs is a negative real number. |
|---|---|

We have been discussing positive and negative numbers. What about 0? We know (from Section 1.1) that $a \cdot 0 = 0$ for any real number $a$. But **division by 0 is undefined.** The reasoning is based on multiplication.

Suppose that $a \neq 0$ and $\frac{a}{0} = x$. Then, by definition of division, $a = 0 \cdot x$. But this is impossible because $0 \cdot x = 0$ and $a \neq 0$.

Suppose that $a = 0$. Then $a = 0 \cdot x$ is true for any value of $x$ and there is no unique value for $x$. Therefore,

$$\frac{a}{0} \text{ is undefined for any real } a.$$

However, if $b \neq 0$, then $\frac{0}{b} = 0$ since $0 = b \cdot 0$ for any real $b$.

**EXAMPLES**

**33.** $\frac{7}{0}$ is undefined.

**34.** $\frac{0}{-6} = 0$ ■

To indicate repeated multiplication by the same factor, we use **exponents.** The factor is called the **base** of the exponent.

$$3 \cdot 3 \cdot 3 \cdot 3 = 3^4$$ (exponent 4, base 3) read "3 to the fourth power"

The **power** is the value of the expression:

$$3^4 = 81$$ (exponent, base, power)

We say that the base is "squared" if the exponent is 2 and "cubed" if the exponent is 3. Thus,

$5^2 = 25$ is read "5 squared equals 25."

and $2^3 = 8$ is read "2 cubed equals 8."

**EXAMPLES**

**35.** $2^5 = 2 \cdot 2 \cdot 2 \cdot 2 \cdot 2 = 32$

**36.** $(-6)^3 = (-6)(-6)(-6) = -216$ ■

## *Order of Operations*

To evaluate an expression that has more than one operation in it, we need to agree on a set of rules. If you evaluate $5 + 2 \cdot 3$ from left to right, you get $5 + 2 \cdot 3 = 7 \cdot 3 = 21$. If you multiply first, you get $5 + 2 \cdot 3 = 5 + 6 = 11$. Only one answer can be right, and mathematicians have agreed on the following **rules for order of operations.**

***Rules for Order of Operations***

1. Simplify within symbols of inclusion (parentheses, brackets, or braces) beginning with the innermost symbols.
2. Find any powers indicated by exponents.
3. Perform any multiplications or divisions as they occur from **left to right.**
4. Perform any additions or subtractions as they occur from **left to right.**

Using these rules, we find that the correct value for $5 + 2 \cdot 3$ is $5 + 6 = 11$.

**EXAMPLES** Using the rules for order of operations, find the value of each of the following expressions.

**37.** $10 - 21 \div 3 + 2$

**Solution**

$$\begin{aligned} 10 - 21 \div 3 + 2 &= 10 - 7 + 2 && \text{Perform the division first.} \\ &= 3 + 2 = 5 && \text{Add or subtract from left to right.} \end{aligned}$$

**38.** $5(-2) + 6 \cdot 4 - 2$

**Solution**

$$\begin{aligned} 5(-2) + 6 \cdot 4 - 2 &= -10 + 24 - 2 && \text{Do the multiplication from left to right first.} \\ &= 14 - 2 && \text{Add or subtract from left to right.} \\ &= 12 \end{aligned}$$

**39.** $2(-5 - 3) + 16 \div (+2) + (3 - 6 \cdot 8)$

**Solution**

$$\begin{aligned} &2(-5 - 3) + 16 \div 2 + (3 - 6 \cdot 8) \\ &= 2(-8) + 16 \div 2 + (3 - 48) && \text{Simplify within the parentheses first.} \\ &= 2(-8) + 16 \div 2 + (-45) \\ &= -16 + 8 - 45 && \text{Multiply or divide from left to right.} \\ &= -8 - 45 && \text{Add or subtract from left to right.} \\ &= -53 \end{aligned}$$

**40.** $16 \cdot 3 \div 2^3 - (2 \cdot 3^2 + 20)$

**Solution** $16 \cdot 3 \div 2^3 - (2 \cdot 3^2 + 20)$

$= 16 \cdot 3 \div 8 - (2 \cdot 9 + 20)$ Find the powers using exponents.

$= 16 \cdot 3 \div 8 - (18 + 20)$ Simplify within the parentheses.

$= 16 \cdot 3 \div 8 - 38$

$= 48 \div 8 - 38$ Multiply or divide from left to right.

$= 6 - 38$ Add or subtract from left to right.

$= -32$ ■

**Practice Problems** Find the value of each expression.

1. $-8.6 - 4.1 - 0.2$
2. $(-3)(-5)(-6)$
3. $\frac{8}{0}$
4. $4(-2) + 6 \cdot 3$
5. $2(-5 + 3) + 16 \div 8 \cdot 2$
6. $3(-8 + 2^3) + 6^2 \div 2^2 - 4$

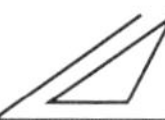

## EXERCISES 1.2

Find the value of each expression in Exercises 1–5.

**1.** $|7|$ **2.** $\left|-\frac{3}{4}\right|$ **3.** $|-\sqrt{5}|$ **4.** $|0|$ **5.** $-|-8|$

Find those values for $x$ that make Exercises 6–15 true statements. If a statement is never true, indicate by saying "no value."

**6.** $|x| = 4$ **7.** $|x| = 7$ **8.** $|x| = 0$ **9.** $|x| = 2$ **10.** $|x| = -3$

**11.** $|x| = \frac{4}{5}$ **12.** $|x| = 2.6$ **13.** $|x| = -2.8$ **14.** $|x| = -x$ **15.** $|x| = x$

Perform the indicated operations in Exercises 16–65.

**16.** $(-7) + 20$ **17.** $(-2) + (-9)$ **18.** $-5 + |-3|$ **19.** $(-8) + (-6) + 5$

**20.** $-3 + |7| + (-2)$ **21.** $\left(-\frac{3}{8}\right) + \frac{7}{8}$ **22.** $\frac{9}{16} + \left|-\frac{5}{16}\right|$ **23.** $12 - 15$

**24.** $-4 - (-8)$ **25.** $(-9) - (-9)$ **26.** $0 - (-12)$ **27.** $17 - |-4|$

**28.** $-\frac{4}{13} - \frac{3}{13}$ **29.** $\frac{3}{5} - \frac{9}{5}$ **30.** $\left|-\frac{8}{3}\right| - \left(-\frac{2}{3}\right)$ **31.** $(-1.7) + (-5.2)$

**32.** $(8.5) + (-7.9)$ **33.** $-7 - (-2) + 6$ **34.** $-18 - 22 - 41$ **35.** $-8 + (-7) - (-15)$

**Answers to Practice Problems** 1. $-12.9$ 2. $-90$ 3. Undefined 4. 10 5. 0 6. 5

**36.** $9 - (-3) + (-2)$

**37.** $21 + |-3| - |-4|$

**38.** $|13| - |-9| + |-3|$

**39.** $-\frac{7}{6} + \left(-\frac{5}{6}\right) - \frac{1}{6}$

**40.** $\frac{4}{15} + \left|-\frac{7}{15}\right| - \left|\frac{16}{15}\right|$

**41.** $\left(-\frac{9}{16}\right) + \left(-\frac{7}{8}\right)$

**42.** $\frac{1}{8} - \left(-\frac{1}{2}\right) + \frac{1}{4}$

**43.** $\frac{4}{5} + \left(-\frac{2}{3}\right) - \frac{1}{6}$

**44.** $-\frac{3}{8} - \frac{5}{6} + \left(-\frac{1}{2}\right)$

**45.** $(-8)(-7)$

**46.** $(-3)(17)$

**47.** $(-8)(-1)(-5)(6)(-2)$

**48.** $(12)\left(-\frac{5}{6}\right)$

**49.** $\frac{3}{8} \cdot \frac{5}{2}$

**50.** $-\frac{5}{16} \cdot \frac{3}{4}$

**51.** $\left(-\frac{3}{10}\right)\left(\frac{5}{6}\right)\left(-\frac{8}{7}\right)\left(\frac{1}{2}\right)\left(-\frac{1}{4}\right)$

**52.** $6(5.3)$

**53.** $(-0.8)(4.9)$

**54.** $(11.7)(-2.06)(-1.3)$

**55.** $(-20) \div (-10)$

**56.** $\frac{-39}{-13}$

**57.** $\frac{-91}{-7}$

**58.** $\frac{52}{13}$

**59.** $\frac{9}{16} \div 0$

**60.** $60 \div (-15)$

**61.** $0 \div \frac{11}{12}$

**62.** $\frac{28.4}{-7}$

**63.** $-68.05 \div 5$

**64.** $-88.64 \div (-8)$

**65.** $-6.084 \div (-9)$

Find the value of the expression in Exercises 66–80 using the rules for order of operations.

**66.** $18 \div 3 \cdot 6 - 3$

**67.** $7(4 - 2) \div 7 + 3$

**68.** $2^2 - 15 \div 3$

**69.** $2^2 \cdot 3 \div 3 + 6 \div 3$

**70.** $-6 \cdot 3 \div (-1) + 4 - 2$

**71.** $5(-2) \div (-5) + 5 - 3$

**72.** $(4^2 + 6) - 2 \cdot 19$

**73.** $(12 \cdot 4 \div 2^3) - [(3 \cdot 2^3) \div (4 \cdot 6)]$

**74.** $[(3 \cdot 0) \div (2 \cdot 1)] - (24 - 6^2) \div (4^2 - 3 \cdot 4)$

**75.** $(3 \cdot 2^3) \div (3 \cdot 4) + (2 \cdot 3 + 4) \div (6 - 1)$

**76.** $-6 + (-2)(12 \cdot 2 \div 3)4$

**77.** $14 - [11 \cdot 4 - (2 \cdot 3^2 + 1)]$

**78.** $6 + 3[-4 - 2(3 - 1)]$

**79.** $7 - [4 \cdot 3 - (4 - 3 \cdot 2)]$

**80.** $-2[6 + 4(1 + 7)] \div 4$

### Calculator Problems

**81.** $3.6841 + (-2.9431) + (0.1895)$

**82.** $-7.9431 + (-8.7925) - (-6.8392)$

**83.** $18.6454 - 22.5267 + (-2.5973)$

**84.** $(3.692)(-4.714)(0.658)$

**85.** $(59.787) \div (-73)$

**86.** $(9.94)(-2.52) \div 0.63$

**87.** $23.9412 - (9.63)^2(-4.72)$

**88.** $(-14.52)^2 - (5.93)^2$

# 1.3 Solving Equations

**OBJECTIVES**

In this section, you will be learning to:

1. Simplify expressions by removing grouping symbols and combining like terms.
2. Solve first-degree equations.
3. Solve first-degree absolute value equations.

## Combining Like Terms

An expression that involves only multiplication and/or division with constants and/or variables is called a **term.** Exponents can be used to indicate repeated factors. For example,

$$2x^5, \qquad \frac{1}{3}x^2y, \qquad -14, \qquad \text{and} \qquad 6x$$

are all terms.

**Like terms** (or **similar terms**) are those terms that have the same variable factors with the same exponents. Constants are also terms. For example,

**a.** $4x^2y$, $9x^2y$, and $-3x^2y$ are like terms.

**b.** $8xy$, $13x$, and $17y$ are not like terms.

The numerical factor of a term is called the **coefficient** of the variables. Thus, 4 is the coefficient of $x^2y$ in the term $4x^2y$.

To **combine like terms,** such as $4x + 11x$, we use the distributive property: $a(b + c) = ab + ac$. By applying the commutative property of multiplication, we use the distributive property in the form

$$ba + ca = (b + c)a$$

EXAMPLES Combine like terms in the following expressions.

**1.** $4x + 11x = (4 + 11)x$
$= 15x$

**2.** $-6y + 4y = (-6 + 4)y$
$= -2y$

**3.** $-x + 3 - (x^2 - 5x + 1)$

The sign of each term in parentheses is changed. $-1$ is understood to be the coefficient of $(x^2 - 5x + 1)$.

$= -x + 3 - x^2 + 5x - 1$

$-(x^2 - 5x + 1)$
$= -1(x^2 - 5x + 1)$
$= -x^2 + 5x - 1.$

$= -x^2 + (-1 + 5)x + 3 - 1$

$-1$ is the coefficient of $-x$.

$= -x^2 + 4x + 2$

**4.** $\dfrac{3x + 5x}{4} + 9x = \dfrac{8x}{4} + 9x$
$= 2x + 9x$
$= 11x$

The fraction bar is treated as a symbol of inclusion, and $3x$ and $5x$ are added first.

**5.** $4x^2 - [3x - (x^2 + x)]$ — Remove the innermost symbol of inclusion first.

$= 4x^2 - [3x - x^2 - x]$

$= 4x^2 - 3x + x^2 + x$ — The coefficient of $x$ (and the coefficient of $x^2$) is understood to be 1.

$= (4 + 1)x^2 + (-3 + 1)x$ — This step should be done mentally.

$= 5x^2 - 2x$ ■

## First-Degree Equations

If $x$ is replaced with 5 in the equation $3x + 4 = 10$, we get $3 \cdot 5 + 4 = 10$, a false statement. If 2 replaces $x$, we get $3 \cdot 2 + 4 = 10$, a true statement. The number 2 is a **solution** of the equation. The **solution set** of an equation consists of all values of the variable that make a true statement when substituted for the variable.

***First-Degree Equation***

An equation of the form $ax + b = 0$, where $a$ and $b$ are real numbers and $a \neq 0$, is called a **first-degree equation** in $x$.

To **solve** an equation or **find the solution set** of a first-degree equation, we need the following two properties of equations.

***Addition Property of Equality***

If the same algebraic expression is added to both sides of an equation, the new equation has the same solutions as the original equation. Symbolically, if $A$, $B$, and $C$ are algebraic expressions, then the equations

$$A = B$$

and

$$A + C = B + C$$

have the same solutions.

***Multiplication Property of Equality***

If both sides of an equation are multiplied by the same nonzero algebraic expression, the new equation has the same solutions as the original equation. Symbolically, if $A$, $B$, and $C$ are algebraic expressions, then the equations

$$A = B$$

and

$$AC = BC \qquad (\text{where } C \neq 0)$$

have the same solutions.

Two equations that have the same solution set are **equivalent.** The numbers in the solution set are said to **satisfy** the equation, and solving the equation is the process used to find the solution set. For example, the equations

$$5x + 3 = 13$$

and

$$5x = 10$$

and

$$x = 2$$

are all equivalent since the solution to all three equations is the same number, 2.

The basic strategy in solving first-degree equations is to find equivalent equations until one is found with only variable terms on one side and constants on the other. Then an equation such as $x = 5$ or $x = 7$ gives the solution to the original equation.

***To Solve a First-Degree Equation***

1. Simplify each side of the equation by removing any grouping symbols and combining like terms. (In some cases, you may want to multiply each term by a constant to clear fraction or decimal coefficients.)
2. Use the addition property of equality to add the opposites of constants and/or variables so that variables are on one side and constants on the other.
3. Use the multiplication property of equality to multiply both sides by the reciprocal of the coefficient of the variable (or divide both sides by the coefficient).
4. Check your answer by substituting it into the original equation.

**EXAMPLES** Solve the following first-degree equations.

**6.** $3x - 5 = 4x + 7 + x$

**Solution**

| | |
|---|---|
| $3x - 5 = 4x + 7 + x$ | Write the equation. |
| $3x - 5 = 5x + 7$ | Combine like terms. |
| $-5 = 2x + 7$ | Add $-3x$ to both sides of the equation. |
| $-12 = 2x$ | Add $-7$ to both sides of the equation. |
| $-6 = x$ | Divide both sides of the equation by 2. |

*Check* $3(-6) - 5 = 4(-6) + 7 + (-6)$

$$-18 - 5 = -24 + 7 - 6$$

$$-23 = -23$$

The solution is $-6$. We usually write just $x = -6$ (or $-6 = x$) to indicate the solution to the original equation.

**7.** $7(x - 3) = x + 3(x + 5)$

**Solution**

| | |
|---|---|
| $7(x - 3) = x + 3(x + 5)$ | Write the equation. |
| $7x - 21 = x + 3x + 15$ | Use the distributive property. |
| $7x - 21 = 4x + 15$ | Combine like terms. |
| $3x - 21 = 15$ | Add $-4x$ to both sides. |
| $3x = 36$ | Add 21 to both sides. |
| $x = 12$ | Divide both sides by 3. |

*Check* $7(12 - 3) = 12 + 3(12 + 5)$

$$7(9) = 12 + 3(17)$$
$$63 = 12 + 51$$
$$63 = 63$$

**8.** $\dfrac{x - 5}{4} + \dfrac{3}{2} = \dfrac{x + 2}{3}$

**Solution**

| | |
|---|---|
| $\dfrac{x - 5}{4} + \dfrac{3}{2} = \dfrac{x + 2}{3}$ | Write the equation. |
| $12\left(\dfrac{x - 5}{4}\right) + 12\left(\dfrac{3}{2}\right) = 12\left(\dfrac{x + 2}{3}\right)$ | Multiply both sides by 12, the LCM of the denominators. The LCM of the denominators also is called the LCD or least common denominator. |
| $3(x - 5) + 6(3) = 4(x + 2)$ | |
| $3x - 15 + 18 = 4x + 8$ | Use the distributive property. |
| $3x + 3 = 4x + 8$ | Combine like terms. |
| $3 = x + 8$ | Add $-3x$ to both sides. |
| $-5 = x$ | Add $-8$ to both sides. |

*Check* $\dfrac{-5 - 5}{4} + \dfrac{3}{2} = \dfrac{-5 + 2}{3}$

$$\frac{-10}{4} + \frac{6}{4} = \frac{-3}{3}$$

$$\frac{-4}{4} = \frac{-3}{3}$$

$$-1 = -1$$

**9.** $3.1x + 7.6 = 2.7x + 7.8$

**Solution**

| | |
|---|---|
| $3.1x + 7.6 = 2.7x + 7.8$ | Write the equation. |
| $31x + 76 = 27x + 78$ | Multiply each term by 10 so that all the numbers will be integers. |
| $4x + 76 = 78$ | Add $-27x$ to both sides. |
| $4x = 2$ | Add $-76$ to both sides. |
| $x = 0.5 \quad \left(\text{or } x = \frac{1}{2}\right)$ | Divide both sides by 4. |

*Check* $3.1(0.5) + 7.6 = 2.7(0.5) + 7.8$

$1.55 + 7.6 = 1.35 + 7.8$

$9.15 = 9.15$ ■

## Absolute Value Equations

The definition of **absolute value** was given in Section 1.2 and is stated again here for easy reference.

***Absolute Value***

For any real number $x$,

$$|x| = \begin{cases} x & \text{if } x \geq 0 \\ -x & \text{if } x < 0 \end{cases}$$

Equations involving absolute value may have more than one solution. For example, suppose that $|x| = 3$. Since $|+3| = 3$ and $|-3| = -(-3) = 3$, we can have either $x = 3$ or $x = -3$. In general, **any positive number and its opposite have the same absolute value.**

For $c > 0$:

**a.** If $|x| = c$, then $x = c$ or $x = -c$.

**b.** If $|ax + b| = c$, then $ax + b = c$ or $ax + b = -c$.

**EXAMPLES** Solve the following equations involving absolute value.

**10.** $|x| = 5$

**Solution** $x = 5 \quad$ or $\quad x = -5$

**11.** $|3x - 4| = 5$

**Solution**

| | | |
|---|---|---|
| $3x - 4 = 5$ | or | $3x - 4 = -5$ |
| $3x = 9$ | | $3x = -1$ |
| $x = 3$ | | $x = -\frac{1}{3}$ |

**12.** $|4x + 7| = -8$

**Solution** There is no number that has a negative absolute value.

**13.** $|3x + 7| - 4 = 2$

**Solution** $|3x + 7| - 4 = 2$

$|3x + 7| = 6$ Add 4 to both sides so that the absolute value expression is by itself.

$3x + 7 = 6$ or $3x + 7 = -6$

$3x = -1$ $\qquad$ $3x = -13$

$x = -\frac{1}{3}$ $\qquad$ $x = -\frac{13}{3}$ ∎

If two numbers have the same absolute value, then either they are equal or they are opposites of each other. This fact can be used to solve equations that involve two absolute values.

If $|a| = |b|$, then either $a = b$ or $a = -b$.

More generally,

$$\text{if } |ax + b| = |cx + d|,$$

then either

$$ax + b = cx + d \quad \text{or} \quad ax + b = -(cx + d)$$

**EXAMPLE**

**14.** Solve $|x + 6| = |3x - 1|$.

In this case, the two expressions $(x + 6)$ and $(3x - 1)$ either are equal to each other or are opposites of each other.

**Solution** $|x + 6| = |3x - 1|$

$x + 6 = 3x - 1$ or $x + 6 = -(3x - 1)$ Note the use of parentheses. We want the opposite of the entire expression $(3x - 1)$.

$6 = 2x - 1$ $\qquad$ $x + 6 = -3x + 1$

$7 = 2x$ $\qquad$ $4x + 6 = 1$

$\frac{7}{2} = x$ $\qquad$ $4x = -5$

$x = -\frac{5}{4}$ ∎

***Practice Problems***

**1.** Combine like terms: $3x^2 - [x^2 + (3x - 2x^2)]$

Solve the following equations.

**2.** $3x - 4 = 2x + 6 - x$

**3.** $6(x - 4) + x = 4(1 - x) + 4x$

**4.** $|2x - 1| = 8.2$

**Answers to Practice Problems** **1.** $4x^2 - 3x$ **2.** $x = 5$ **3.** $x = 4$ **4.** $x = 4.6$ or $x = -3.6$

## EXERCISES 1.3

In each of the expressions in Exercises 1–16, combine like terms.

**1.** $-2x + 5y + 6x - 2y$
**2.** $4x + 2x - 3y - x$
**3.** $4x - 3y + 2(x + 2y)$
**4.** $5(x - y) + 2x - 3y$
**5.** $(3x^2 + x) - (7x^2 - 2x)$
**6.** $-(4x^2 - 2x) - (5x^2 + 2x)$
**7.** $4x - [5x + 3 - (7x - 4)]$
**8.** $3x - [2y - (3x + 4y)]$
**9.** $\frac{-4x - 2x}{3} + 7x$
**10.** $\frac{2(4x - x)}{3} - \frac{3(6x - x)}{3}$
**11.** $\frac{8(5x + 2x)}{7} - \frac{2(3x + x)}{4}$
**12.** $\frac{6(5x - x)}{8} + \frac{5(x + 5x)}{3}$
**13.** $2x + [9x - 4(3x + 2) - 7]$
**14.** $7x - 3[4 - (6x - 1) + x]$
**15.** $6x^2 - [9 - 2(3x^2 - 1) + 7x^2]$
**16.** $2x^2 + [4x^2 - (8x^2 - 3x) + (2x^2 + 7x)]$

Solve each of the equations in Exercises 17–60.

**17.** $7x - 4 = 17$
**18.** $9x + 6 = -21$
**19.** $4 - 3x = 19$
**20.** $18 + 11x = 23$
**21.** $6x - 2.5 = 1.1$
**22.** $5x + 4.06 = 2.31$
**23.** $7x + 3.4 = -1.5$
**24.** $8.2 = 2.6 + 8x$
**25.** $7x - 6 = 2x + 9$
**26.** $x - 7 = 4x + 11$
**27.** $\frac{x}{5} - 1 = -6$
**28.** $\frac{3x}{4} + 11 = 20$
**29.** $\frac{2x}{3} - 4 = 8$
**30.** $\frac{5x}{4} + 1 = 11$
**31.** $2(x - 2) = 2x - 4$
**32.** $3x - 7 = 4(x + 3)$
**33.** $3(2x - 3) = 4x + 5$
**34.** $3(x - 1) = 4x + 6$
**35.** $4(3 - 2x) = 2(x - 4)$
**36.** $-3(x + 5) = 6(x + 2)$
**37.** $4x + 3 - x = 3x - 9$
**38.** $5x + 13 = x - 8 - 3x$
**39.** $x + 7 = 6x + 4 - x$
**40.** $x - 9 + 5x = 2x - 3$
**41.** $x - 2 = \frac{x + 2}{4} + 5$
**42.** $x + 8 = \frac{x}{3} - 2$
**43.** $\frac{4x}{5} + 2 = 2x - 4$
**44.** $\frac{3x}{2} + 1 = x - 1$
**45.** $0.8x + 6.2 = 0.2x - 1.0$
**46.** $2.4x - 8.5 = 1.1x + 0.6$
**47.** $2.5x + 2.0 = 0.7x + 5.6$
**48.** $3.2x + 9.5 = 1.8x - 1.7$
**49.** $12x - (4x - 6) = 3x - (9x - 27)$
**50.** $5(x - 3) - 3 = 2x - 6(2 - x)$
**51.** $\frac{x}{4} + 2 = \frac{3x}{2} - 3$
**52.** $\frac{2x + 1}{8} - \frac{1}{4} = \frac{x - 3}{2} + 1$
**53.** $\frac{x}{2} - \frac{2x}{3} = \frac{3}{4} + \frac{x}{3}$
**54.** $\frac{1}{3}x + \frac{1}{4} = \frac{1}{5}x + \frac{1}{6}$
**55.** $5(x - 4) = 3(4x - 7) - 2(3x + 4)$
**56.** $5(x + 1) - 4(3 - x) = 2x - 7(1 - x)$
**57.** $2(4 - x) - (3x + 2) = 7 + 4(x - 7)$
**58.** $3(x - 1) - (4x + 2) = 3[(2x - 1) - 2(x + 3)]$
**59.** $3 - x = \frac{1}{4}(7 - x) - \frac{1}{3}(2x - 3)$
**60.** $\frac{3}{2}(x - 1) = \frac{1}{2}(x - 3) - \frac{7}{10}$

Solve each of the absolute value equations in Exercises 61–80.

**61.** $|x| = 8$ **62.** $|x| = 6$ **63.** $|z| = -\frac{3}{7}$ **64.** $|y| = \frac{1}{5}$

**65.** $|x + 3| = 2$ **66.** $|y + 5| = -7$ **67.** $|x - 4| = \frac{1}{2}$ **68.** $|3x + 1| = 8$

**69.** $|5x - 2| + 4 = 7$ **70.** $|2x - 7| - 1 = 0$ **71.** $\left|\frac{x}{4} + 1\right| - 2 = 1$

**72.** $\left|\frac{x}{3} - 2\right| + 3 = 4$ **73.** $|2x - 1| = |x + 2|$ **74.** $|2x - 5| = |x - 3|$

**75.** $|3x + 1| = |4 - x|$ **76.** $|5x + 4| = |1 - 3x|$ **77.** $\left|\frac{3x}{2} + 2\right| = \left|\frac{x}{4} + 3\right|$

**78.** $\left|\frac{x}{3} - 4\right| = \left|\frac{5x}{6} + 1\right|$ **79.** $\left|\frac{2x}{5} - 3\right| = \left|\frac{x}{2} - 1\right|$ **80.** $\left|\frac{4x}{3} + 7\right| = \left|\frac{x}{4} + 2\right|$

**Calculator Problems**

Round off answers to the nearest hundredth.

**81.** $4.632x - 0.8144 = 1.583x + 9.647$ **82.** $11.473x + 6.9511 = 1.5731 + 2.614x$

**83.** $-5.8342 + 8.5178x = 10.4759 - 0.0163x$ **84.** $0.86(x + 1.47) = 0.91(2x - 3.42)$

**85.** $0.59(3x - 4.11) = 1.93(x + 8.37)$

## 1.4 Solving for Any Term in a Formula

**OBJECTIVES**

In this section, you will be learning to:

1. Solve applied problems by using known formulas.
2. Solve formulas for specified variables in terms of the other variables.

A **formula** is an equation that represents a general relationship between two or more quantities or measurements. Several variables may appear in a formula, and the formula is not always in the most convenient form for application in some word problems. In such situations, we may want to "solve" a formula for a particular variable and use the formula in a different form. For example, the formula $d = rt$ (distance equals rate times time) is "solved" for $d$. Solving for $r$ or $t$ gives

$$r = \frac{d}{t} \quad \text{or} \quad t = \frac{d}{r}$$

Formulas are useful in many fields of study, such as business, economics, medicine, physics, technology, and chemistry, as well as mathematics. Some formulas and their meanings are shown here and will be used with others in the exercises.

| Formula | Meaning |
|---|---|
| **1.** $I = Prt$ | The simple interest ($I$) earned by investing money is equal to the product of the principal ($P$) times the rate of interest ($r$) times the time ($t$) in years. |

| Formula | Meaning |
|---|---|
| **2.** $C = \frac{5}{9}(F - 32)$ | Temperature in degrees Celsius ($C$) equals $\frac{5}{9}$ the difference between the Fahrenheit temperature ($F$) and 32. |
| **3.** $\text{IQ} = \frac{100M}{C}$ | Intelligence Quotient (IQ) is calculated by multiplying 100 times mental age ($M$) as measured by some test and dividing by chronological age ($C$). |
| **4.** $\alpha + \beta + \gamma = 180$ | The sum of the angles ($\alpha$, $\beta$, $\gamma$) of a triangle is 180°. (**Note:** $\alpha$, $\beta$, and $\gamma$ are the Greek letters alpha, beta, and gamma, respectively.) |

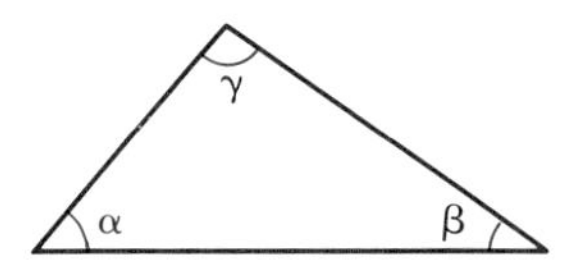


Formulas for the perimeter ($P$) and the area ($A$) of geometric figures are as follows:

**Formula** **Figure**

**5.** $P = 4s$
$A = s^2$

SQUARE

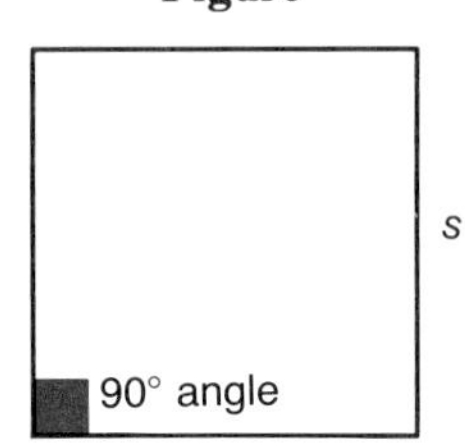


**6.** $P = 2\ell + 2w$
$A = \ell w$

RECTANGLE

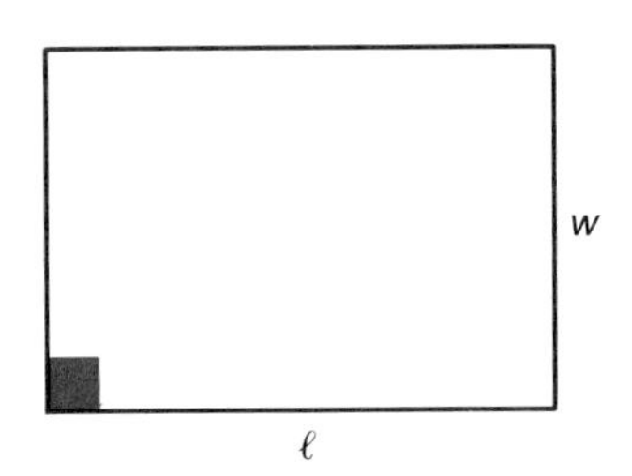


**7.** $P = 2b + 2a$
$A = bh$

PARALLELOGRAM

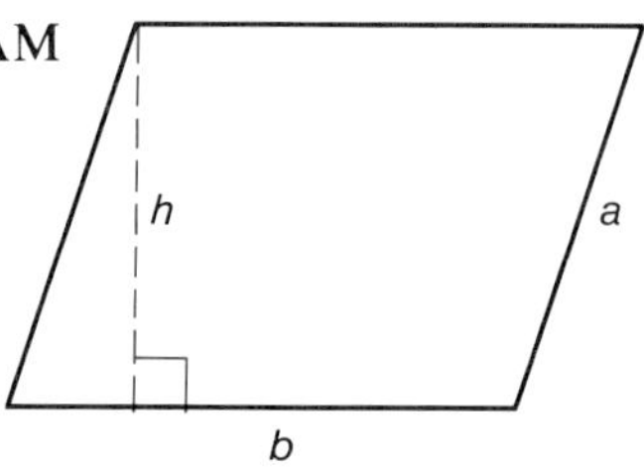

**8.** $P = a + b + c$ TRIANGLE

$A = \frac{1}{2}bh$

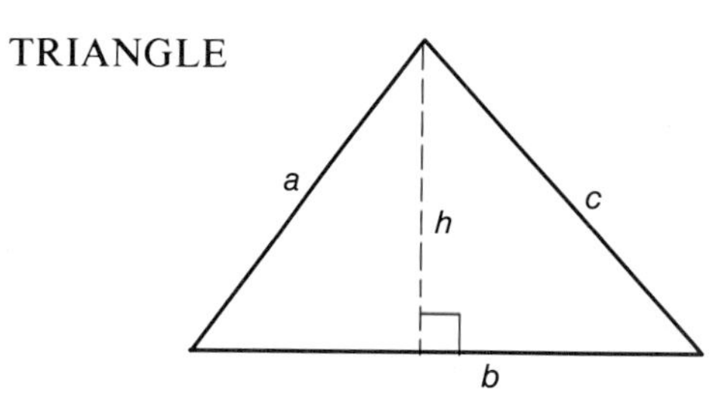


**9.** $C = 2\pi r$ CIRCLE

$C = \pi d$

$A = \pi r^2$

$r$ = radius
$d$ = diameter
$C$ = circumference or perimeter of a circle.

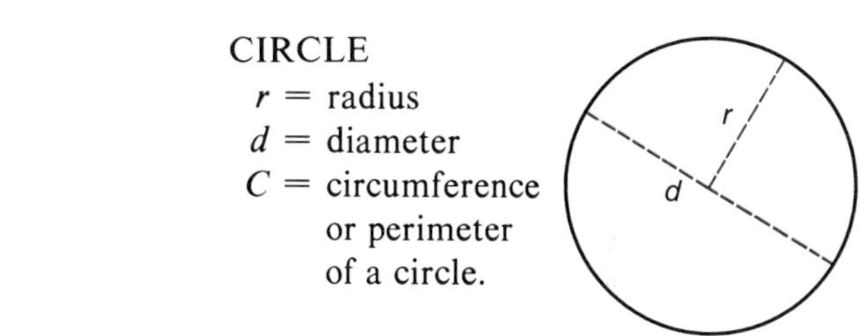


**10.** $P = a + b + c + d$ TRAPEZOID

$A = \frac{1}{2}h(b + c)$

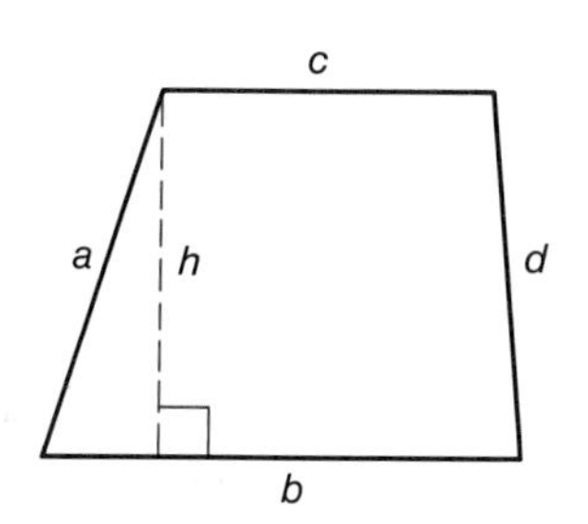


If the values for all but one variable in a formula are known, they can be substituted into the formula, and the unknown value can be found by solving the equation as we did in Section 1.3.

## EXAMPLE

**1.** The perimeter of a triangle is 38 feet, and one side is 5 feet long and a second side is 18 feet long. How long is the third side?

**Solution 1** Using the formula $P = a + b + c$, substitute $P = 38$, $a = 5$, and $b = 18$. Then solve for the third side.

$$38 = 5 + 18 + c$$
$$38 = 23 + c$$
$$15 = c$$

P = 38
a = 5
c = ?
b = 18

The third side is 15 feet long.

**Solution 2** First solve for $c$ in terms of $P$, $a$, and $b$. Then substitute for $P$, $a$, and $b$.

$$P = a + b + c$$

$$P - a - b = c$$ Treat $a$ and $b$ as constants, and add $-a - b$ to both sides.

or $$c = P - a - b$$

So $$c = 38 - 5 - 18$$
$$= 38 - 23 = 15$$ ■

Many times, we simply want the formula solved for one of the variables without substituting in particular values. In this situation, we treat all the other variables as if they were constants and follow the same procedures for solving an equation as we did in Section 2.2. Thus, in Solution 2, we solved the formula $P = a + b + c$ for $c$ and found $c = P - a - b$.

## EXAMPLES

**2.** Given $P = a + b + c$, solve for $a$ in terms of $P$, $b$, and $c$.

**Solution** $$P = a + b + c$$ Treat $P$, $b$, and $c$ as if they are constants.

$$P - b - c = a + b + c - b - c$$ Add $-b - c$ to both sides.

$$P - b - c = a$$ Simplify.

or $$a = P - b - c$$

**3.** Given $C = \frac{5}{9}(F - 32)$, solve for $F$ in terms of $C$.

**Solution** $$C = \frac{5}{9}(F - 32)$$ Treat $C$ as a constant.

$$\frac{9}{5}C = \frac{9}{5} \cdot \frac{5}{9}(F - 32)$$ Multiply both sides by $\frac{9}{5}$. (Or multiply by 9 and then divide by 5.)

$$\frac{9}{5}C = F - 32$$ Simplify.

$$\frac{9}{5}C + 32 = F - 32 + 32$$ Add 32 to both sides.

$$\frac{9}{5}C + 32 = F$$ Simplify.

Thus, the formula

$$C = \frac{5}{9}(F - 32)$$

is solved for $C$ and

$$F = \frac{9}{5}C + 32$$

is solved for $F$. These are two forms of the same formula.

**4.** Solve for $\ell$ given $P = 2\ell + 2w$.

**Solution**

$$P = 2\ell + 2w$$

$$P - 2w = 2\ell + 2w - 2w \qquad \text{Add } -2w \text{ to both sides.}$$

$$P - 2w = 2\ell \qquad \text{Simplify.}$$

$$\frac{P - 2w}{2} = \frac{\cancel{2}\ell}{\cancel{2}} \qquad \text{Divide both sides by 2.}$$

$$\frac{P - 2w}{2} = \ell \qquad \left(\text{Or multiply by } \frac{1}{2}.\right)$$

**5.** Solve the formula $y = mx + b$ for $x$.

**Solution**

$$y = mx +$$

$$y - b = mx \qquad \text{Add } -b \text{ to both sides.}$$

$$\frac{y - b}{m} = x \qquad \text{Divide both sides by } m.$$

**6.** Solve for $R$ given $F = \dfrac{1}{R + r}$.

**Solution**

$$F = \frac{1}{R + r}$$

$$(R + r)F = (R + r)\frac{1}{R + r} \qquad \text{Multiply both sides by the denominator } R + r.$$

$$RF + rF = 1 \qquad \text{Use the distributive property.}$$

$$RF = 1 - rF \qquad \text{Add } -rF \text{ to both sides.}$$

$$\frac{R\cancel{F}}{\cancel{F}} = \frac{1 - rF}{F} \qquad \text{Divide both sides by } F.$$

$$R = \frac{1 - rF}{F}$$

■

## Practice Problems

Solve for the indicated variable.

1. $P = a + 2b$; solve for $b$.
2. $y = mx + b$; solve for $m$.
3. $\alpha + \beta + \gamma = 180$; solve for $\alpha$.
4. $F = \dfrac{1}{R + r}$; solve for $r$.

**Answers to Practice Problems** **1.** $b = \dfrac{P - a}{2}$ **2.** $m = \dfrac{y - b}{x}$ **3.** $\alpha = 180 - \beta - \gamma$
**4.** $r = \dfrac{1 - FR}{F}$

## EXERCISES 1.4

1. The interest earned in 2 years on an investment is \$297. If the rate of interest is 9%, find the amount invested.
2. The Celsius temperature is 45°. Find the Fahrenheit temperature.
3. Two angles of a triangle measure 72° and 65°. Find the measure of the third angle.
4. The perimeter of a square is $10\frac{2}{3}$ meters. Find the length of the sides.
5. The perimeter of a rectangle is 88 feet. If the length is 31 feet, find the width.
6. The area of a parallelogram is 1081 square inches. If the height is 23 inches, find the length of the base.
7. The perimeter of a triangle is 147 inches. Two of the sides measure 38 inches and 48 inches. Find the length of the third side.
8. The circumference of a circle is $26\pi$ centimeters. Find the radius.
9. The radius of a circle is 14 feet. Find the area.
10. The area of a trapezoid is 51 square meters. One base is 7 meters long and the other is 10 meters long. Find the height of the trapezoid.

Solve for the indicated variable in Exercises 11–55.

11. $P = a + b + c$; solve for $b$.
12. $P = 3s$; solve for $s$.
13. $f = ma$; solve for $m$.
14. $C = \pi d$; solve for $d$.
15. $A = \ell w$; solve for $w$.
16. $P = R - C$; solve for $C$.
17. $R = np$; solve for $n$.
18. $v = k + gt$; solve for $k$.
19. $I = A - p$; solve for $p$.
20. $L = 2\pi rh$; solve for $h$.
21. $A = \dfrac{m + n}{2}$; solve for $m$.
22. $W = RI^2t$; solve for $R$.
23. $P = 4s$; solve for $s$.
24. $C = 2\pi r$; solve for $r$.
25. $d = rt$; solve for $t$.
26. $P = a + 2b$; solve for $a$.
27. $I = Prt$; solve for $t$.
28. $R = \dfrac{E}{I}$; solve for $E$.
29. $P = a + 2b$; solve for $b$.
30. $c^2 = a^2 + b^2$; solve for $b^2$.
31. $S = \dfrac{a}{1 - r}$; solve for $a$.
32. $A = \dfrac{h}{2}(a + b)$; solve for $h$.
33. $y = mx + b$; solve for $x$.
34. $V = \ell wh$; solve for $h$.
35. $A = 4\pi r^2$; solve for $r^2$.
36. $V = \pi r^2 h$; solve for $h$.
37. $\text{IQ} = \dfrac{100M}{C}$; solve for $M$.
38. $A = \dfrac{R}{2L}$; solve for $R$.
39. $V = \dfrac{1}{3}\pi r^2 h$; solve for $h$.
40. $A = \dfrac{1}{2}bh$; solve for $b$.
41. $R = \dfrac{E}{I}$; solve for $I$.
42. $\text{IQ} = \dfrac{100M}{C}$; solve for $C$.
43. $A = \dfrac{R}{2L}$; solve for $L$.
44. $K = \dfrac{mv^2}{2g}$; solve for $g$.
45. $A = \dfrac{h}{2}(a + b)$; solve for $b$.
46. $L = a + (n - 1)d$; solve for $d$.
47. $L = 2\pi rh$; solve for $h$.
48. $S = 2\pi rh + 2\pi r^2$; solve for $h$.
49. $S = \dfrac{a}{1 - r}$; solve for $r$.

**50.** $P = \dfrac{A}{1 + ni}$; solve for $n$.

**51.** $W = \dfrac{2PR}{R - r}$; solve for $P$.

**52.** $V^2 = v^2 + 2gh$; solve for $g$.

**53.** $I = \dfrac{nE}{R + nr}$; solve for $R$.

**54.** $A = P + Prt$; solve for $P$.

**55.** $S = \dfrac{rL - a}{b - a}$; solve for $a$.

## 1.5 Applications

**OBJECTIVE**

In this section, you will be learning to use first-degree equations to solve applied problems of the following types:

a. numerical expressions,
b. distance-rate-time,
c. cost-profit, and
d. simple interest.

Word problems are designed to teach you to read carefully, to organize, and to think clearly. Whether or not a particular problem is easy for you depends a great deal on your personal experiences and general reasoning abilities. The problems generally do not give specific directions to add, subtract, multiply, or divide. You must decide what relationships are indicated by your careful analysis of the problem. The following attack plan is recommended for all word problems involving one variable and one equation.

***Attack Plan for Word Problems***

1. Read the problem carefully. (Reread it several times if necessary.)
2. Decide what is asked for, and assign a variable to the unknown quantity. Label this variable so you know exactly what it represents.
3. Draw a diagram or set up a chart whenever possible.
4. Form an equation (or inequality) that relates the information provided.
5. Solve the equation (or inequality).
6. Check your solution with the wording of the problem to be sure it makes sense.

Problems involving numerical expressions will usually contain some key words indicating the operations to be performed. Learn to look for words such as those in the following list.

| **Addition** | **Subtraction** | **Multiplication** | **Division** |
|---|---|---|---|
| add | subtract | multiply | divide |
| sum | difference | product | quotient |
| plus | minus | times | |
| more than | less than | twice | |
| increased by | decreased by | of (used with fractions) | |

### EXAMPLE: Numerical Expressions

**1.** The sum of two numbers is 36. If $\frac{1}{2}$ of the smaller number is equal to $\frac{1}{4}$ of the larger number, find the two numbers.

**Solution** **Analyze the problem and identify the key words.** The key words are **sum** (indicating addition) and **of** (indicating multiplication when used with fractions).
**Assign variables to the unknown quantities.**
Let $x$ = smaller number
Now, since $x$ + (larger number) = 36, $36 - x$ = larger number.
**Form an equation relating the given information.**

$$\underbrace{\tfrac{1}{2}\text{ of the smaller number}}_{\frac{1}{2}x} \quad \underbrace{\text{is equal to}}_{=} \quad \underbrace{\tfrac{1}{4}\text{ of the larger number}}_{\frac{1}{4}(36-x)}$$

**Solve the equation.**

$$4 \cdot \frac{1}{2}x = 4 \cdot \frac{1}{4}(36 - x)$$ Multiply both sides of the equation by 4 to give integer coefficients.

$$2x = 1(36 - x)$$
$$2x = 36 - x$$
$$3x = 36$$
$$x = 12$$ Smaller number
$$36 - x = 24$$ Larger number

*Check* $12 + 24 = 36$

$$\frac{1}{2}(12) = \frac{1}{4}(24)$$

The two numbers are 12 and 24. ■

Problems involving distance usually make use of the relationship indicated by the formula $r \cdot t = d$, where $r$ = rate, $t$ = time, and $d$ = distance. A chart or table showing the known and unknown values is quite helpful and is illustrated in the next example.

**EXAMPLE: Distance**

**2.** A motorist averaged 45 mph for the first part of a trip and 54 mph for the last part of the trip. If the total trip of 303 miles took 6 hours, what was the time for each part of the trip?

**Solution**

**Analysis of Attack Plan**

Let $t$ = time for 1st part of trip
$6 - t$ = time for 2nd part of trip ← What is being asked for?

(Total time minus time for 1st part of trip gives time for 2nd part of trip.)

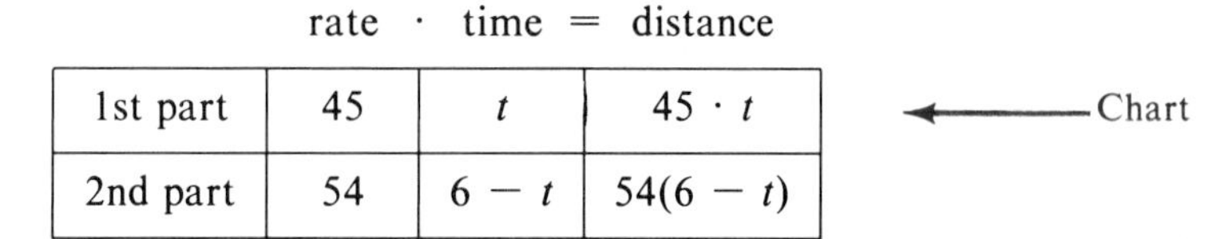

| | rate · | time = | distance |
|---|---|---|---|
| 1st part | 45 | $t$ | $45 \cdot t$ |
| 2nd part | 54 | $6 - t$ | $54(6 - t)$ |

← Chart

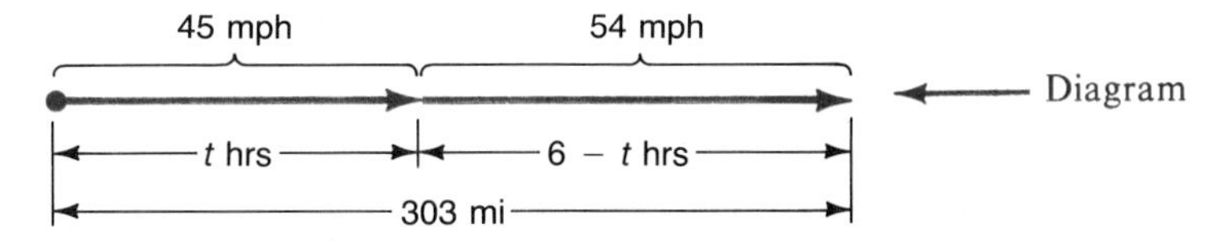


← Diagram

$$\underbrace{\begin{matrix}\text{1st part}\\ \text{distance}\end{matrix}}_{45t} + \underbrace{\begin{matrix}\text{2nd part}\\ \text{distance}\end{matrix}}_{54(6-t)} = \underbrace{\begin{matrix}\text{total}\\ \text{distance}\end{matrix}}_{303}$$

← Form the equation relating the given information.

← Solve the equation.

$$45t + 324 - 54t = 303$$
$$-9t = -21$$
$$t = \frac{21}{9} = \frac{7}{3}$$
$$6 - t = 6 - \frac{7}{3} = \frac{11}{3}$$

*Check*

$$45 \cdot \frac{7}{3} = 15 \cdot 7 = 105 \text{ mi (1st part)}$$

← Check the results.

$$54 \cdot \frac{11}{3} = 18 \cdot 11 = 198 \text{ mi (2nd part)}$$

$105 + 198 = 303$ mi total

The first part took $\frac{7}{3}$ hr or $2\frac{1}{3}$ hr. The second part took $\frac{11}{3}$ hr or $3\frac{2}{3}$ hr. ■

Problems involving cost come in a variety of forms. The next two examples illustrate these types of problems in the exercises.

**EXAMPLES: Cost**

**3.** The Berrys sold their house. After paying the real estate agent a commission of 6% of the selling price and then paying \$486 in other costs and \$30,000 on the mortgage, they received \$18,864. What was the selling price of the house?

**Solution** Use the relationship $SP - C = P$, that is, selling price − cost = profit.

Let $s$ = selling price.

$$\underbrace{\text{selling price}}\ -\ \overbrace{\text{cost}}\ =\ \overbrace{\text{profit}}$$

$$\begin{aligned} s - (0.06s + 486 + 30{,}000) &= 18{,}864 \\ s - 0.06s - 486 - 30{,}000 &= 18{,}864 \\ 0.94s &= 49{,}350 \\ 94s &= 4{,}935{,}000 \\ s &= 52{,}500 \end{aligned}$$

*Check*

| | | |
|---|---|---|
| \$52,500 selling price | \$3,150 commission | \$52,500 selling price |
| ×0.06 commission % | 486 costs | −33,636 expenses |
| \$3,150 commission | +30,000 mortgage | \$18,864 profit received |
| | \$33,636 total expenses | |

The selling price was \$52,500.

**4.** A jeweler paid \$175 for a ring. He prices the ring so that he can give a 30% discount on the selling price (or marked price) and still make a profit of 20% on his cost. What price should he mark the ring?

**Solution** Again, we make use of the relationship $SP - C = P$.

Let $x$ = the selling price,
then $0.30x + 175$ = the cost (This is the discount plus what he paid for the ring.)

$$\underbrace{\text{selling price}}\ -\ \overbrace{\text{cost}}\ =\ \overbrace{\text{profit}}$$

$$\begin{aligned} x - (0.30x + 175) &= 0.20(175) \\ x - 0.30x - 175 &= 35 \\ 0.70x &= 210 \\ 70x &= 21{,}000 \\ x &= 300 \end{aligned}$$

Here the cost is the discount plus \$175, and the profit is 20% of what he paid originally.

*Check*

| | | |
|---|---|---|
| \$300 selling price | \$90 discount | \$300 selling price |
| ×0.30 discount % | +175 cost | −265 total expense |
| \$90 discount | \$265 total expense | \$35 profit |

Also,

$$\begin{array}{r l} \$175 & \text{cost} \\ \underline{\times 0.20} & \text{profit } \% \\ \$35 & \text{profit} \end{array}$$

The jeweler should set the selling price at $300. ■

To work problems related to interest on money invested for one year, you need to know the basic relationship between the principal $P$ (amount invested), the annual rate of interest $r$, and the amount of interest $I$ (money earned). This relationship is described in the formula $P \cdot r = I$. We use this relationship in Example 5.

**EXAMPLE: Interest**

**5.** A savings and loan company pays 7% interest on long-term savings, and a high-risk stock indicates it should yield 12% interest. If a woman has $40,000 to invest and wants an annual income of $3550 from her investments, how much should she put in the savings and loan and how much in the stock?

**Solution**

Let $x$ = amount invested at 7%

$40{,}000 - x$ = amount invested at 12%

Total amount invested minus amount invested at 7% represents amount invested at 12%.

| | principal | · rate = | interest |
|---|---|---|---|
| Savings and loan | $x$ | 0.07 | $0.07(x)$ |
| Stock | $40{,}000 - x$ | 0.12 | $0.12(40{,}000 - x)$ |

$$\underbrace{\text{interest @ 7\%}} + \underbrace{\text{interest @ 12\%}} = \underbrace{\text{total interest}}$$

$$\begin{aligned} 0.07(x) + 0.12(40{,}000 - x) &= 3550 \\ 7x + 12(40{,}000 - x) &= 355{,}000 \\ 7x + 480{,}000 - 12x &= 355{,}000 \\ -5x &= -125{,}000 \\ x &= 25{,}000 \\ 40{,}000 - x &= 15{,}000 \end{aligned}$$

*Check* $25{,}000(0.07) = 1{,}750$ and $15{,}000(0.12) = 1{,}800$

and $\$1{,}750 + \$1{,}800 = \$3550$

The woman should put $25,000 in the savings and loan at 7% and $15,000 in the stock at 12%. ■

## EXERCISES 1.5

1. If 15 is added to a number, the result is 56. Find the number.
2. A number subtracted from 20 is equal to three times the number. Find the number.
3. Nine less than twice a number is equal to the number. What is the number?
4. Fifty-six is 12 more than four times a number. Find the number.
5. Five less than the quotient of a number and 4 is $-2$. Find the number.
6. If 6 is added to the quotient of a number and 3, the result is 1. What is the number?
7. Seven times a certain number is equal to the sum of three times a number and 28. What is the number?
8. Twelve more than five times a number is equal to the difference between 5 and twice the number. Find the number.
9. Four added to the quotient of a number and 6 is equal to 11 less than the number. What is the number?
10. The quotient of twice a number and 8 is equal to 3 more than the number. What is the number?
11. One number is 6 more than three times another. If their sum is 38, find the two numbers.
12. The sum of two numbers is 98 and their difference is 20. Find the two numbers.
13. The length of a backyard is 8 feet less than twice the width. If 260 feet of fencing is needed to enclose the yard, find the dimensions of the yard.
14. A pair of trousers is on sale for a discount of 15%. The sale price is \$39.95. Find the original price.
15. After a raise of 8%, Juan's salary is \$1620 per month. What was his salary before the increase?
16. The U-Drive Company charges \$20 per day plus 22¢ per mile driven. For a one-day trip, Louis paid a rental charge of \$66.20. How many miles did he drive?
17. For a long-distance call, the telephone company charges 35¢ for the first three minutes and 15¢ for each additional minute. If the cost of a call was \$1.55, how many minutes long was it?
18. Willis bought a shirt and necktie for \$30. The shirt cost \$15 more than the tie. Find the cost of each.
19. Two planes, which are 2475 miles apart, fly toward each other. Their speeds differ by 75 mph. If they pass each other in 3 hours, what is the speed of each?

| | rate · | time = | distance |
|---|---|---|---|
| 1st plane | $r$ | 3 | |
| 2nd plane | $r + 75$ | 3 | |

20. Jane rides her bike to Blue Lake. Going to the lake, she averages 12 mph. On the return trip, she averages 10 mph. If the round trip takes a total of $5\frac{1}{2}$ hours, how long does the return trip take?

| | rate · | time = | distance |
|---|---|---|---|
| Going | 12 | $\frac{11}{2} - t$ | |
| Returning | 10 | $t$ | |

21. A car travels from one town to another in 6 hours. On the return trip, the speed is increased by 10 mph and the trip takes 5 hours. Find the rate on the return trip. How far apart are the towns?

| | rate · | time = | distance |
|---|---|---|---|
| Going | $r$ | 6 | |
| Returning | $r + 10$ | 5 | |

**22.** The Reeds are moving across the state. Mr. Reed, driving a truck, leaves $3\frac{1}{2}$ hours before Mrs. Reed. If he averages 40 mph and she averages 60 mph, how long will it take her to overtake Mr. Reed?

**23.** Carol has 8 hours to spend on a hike up a mountain and back again. She can walk up the trail at an average of 2 mph and can walk down at an average of 3 mph. How long should she plan to spend on the uphill part of the hike?

**24.** After traveling for 40 minutes, Mr. Koole had to slow to $\frac{2}{3}$ his original speed for the rest of the trip, due to heavy traffic. The total trip of 84 miles took 2 hours. Find his original speed.

**25.** A train leaves Los Angeles at 2:00 P.M. A second train leaves the same station in the same direction at 4:00 P.M. The second train travels 24 mph faster than the first. If the second train overtakes the first at 7:00 P.M., what is the speed of each of the two trains?

**26.** Maria jogs to the country at a rate of 10 mph. She returns along the same route at 6 mph. If the total trip took 1 hour 36 minutes, how far did she jog?

**27.** A particular style of shoe costs the dealer $28.80 per pair. At what price should the dealer mark them so he can sell them at a 10% discount of the selling price and still make a profit of 25% of the cost?

**28.** A grocery store bought ice cream for 59¢ a half gallon and stored it in two freezers. During the night, one freezer "defrosted" and ruined 14 half gallons. If the remaining ice cream is sold for 98¢ a half gallon, how many half gallons did the store buy if it made a profit of $42.44?

**29.** A farmer raises strawberries. They cost him 18¢ a basket to produce. He is able to sell only 85% of those he produces. If he sells his strawberries at 40¢ a basket, how many must he produce to make a profit of $720? (**Hint:** He produces 100% but sells only 85%.)

**30.** Mary builds cabinets in her spare time. Good-quality cabinet plywood costs $1.00 per square foot. There is approximately a 10% waste of material (due to cutting and fitting). She also figures $60 per month for finishing material, glue, tools, etc. If she charges $2.30 per square foot of finished cabinet, how many square feet of plywood would she use if her profit is $282.40 in one month?

**31.** A citrus farmer figures that the fruit costs 16¢ a pound to grow. If he lost 20% of the crop he produced due to a frost and he sold the remaining 80% at 30¢ a pound, how many pounds did he produce to make a profit of $5000?

**32.** Last summer, Ernie sold surfboards. One style sold for $70 and the other sold for $50. He sold a total of 48 surfboards. How many of each style did he sell if the receipts from each style were equal?

**33.** The pro shop at the Divots Country Club ordered two brands of golf balls, Brand X and another brand. Brand X balls cost 90¢ each and the other brand costs 75¢ each. The total cost of the Brand X balls exceeded the total cost of the other brand of balls by $14.40. If an equal number of each brand was ordered, what was the total cost of each brand?

**34.** Sellit Realty Company gets a 6% fee for selling improved properties and 10% for selling unimproved land. Last week, the total sales were $220,000 and their total fees were $16,400. What were the sales from each of the two types of properties?

**35.** Mr. Cheep bought $2250 worth of stock, some at $3.00 per share and some at $4.50 per share. If he bought a total of 550 shares of stock, how many of each did he buy?

**36.** A total of $25,000 is invested at two rates, 5% and 6%. The annual return on the 5% investment exceeds the annual return on the 6% investment by $40. How much is invested at each rate?

**37.** The annual interest earned by \$6000 is \$120 less than the interest earned by \$10,000 invested at 1% less interest per year. What is the rate of interest on each amount?

**38.** The annual interest on a \$4000 investment exceeds the interest on a \$3000 investment by \$80. The \$4000 is invested at a $\frac{1}{2}$% higher rate of interest than the \$3000. What is the rate of each investment's interest?

**39.** Ten thousand dollars is invested, part at 5.5% and part at 6%. The interest from the 5.5% investment exceeds the interest from the 6% investment by \$251. How much is invested at each rate?

**40.** Two investments totaling \$16,000 produce an annual income of \$1140. One investment yields 6% a year, while the other yields 8% per year. How much is invested at each rate?

## 1.6 Solving Inequalities

**OBJECTIVES**

In this section, you will be learning to:

1. Solve first-degree inequalities.
2. Solve first-degree absolute value inequalities
3. Write the solutions for inequalities using interval notation.
4. Graph the solutions for inequalities on real number lines.

### First-Degree Inequalities

In Section 1.1, we graphed the sets of real numbers on number lines. Some of those graphs represent **intervals** of real numbers. Various types of intervals and the corresponding **interval notation** are listed in the following table.

**Types of Intervals**

| Name of Interval | Algebraic Notation | Interval Notation | Graph |
|---|---|---|---|
| Open interval | $a < x < b$ | $(a,b)$ | a b |
| Closed interval | $a \leq x \leq b$ | $[a,b]$ | a b |
| Half-open interval | $a \leq x < b$ | $[a,b)$ | a b |
| | $a < x \leq b$ | $(a,b]$ | a b |
| Open interval | $x > a$ | $(a,+\infty)$ | a |
| | $x < b$ | $(-\infty,b)$ | b |
| Half-open interval | $x \geq a$ | $[a,+\infty)$ | a |
| | $x \leq b$ | $(-\infty,b]$ | b |

In this section, we will solve **first-degree inequalities** such as $6x + 5 \leq -7$ and write the solution in interval notation as $(-\infty, -2]$. We say that "$x$ is in $(-\infty, -2]$."

***First-Degree Inequality***

For any real numbers $a$ and $b$, $a \neq 0$, $ax + b < 0$ is called a **first-degree inequality** in $x$. (The definition is valid if $\leq$, $>$, or $\geq$ is used instead of $<$.)

Solving first-degree inequalities is similar to solving first-degree equations, with one important difference: **Multiplying or dividing both sides of an inequality by a negative number reverses the sense of the inequality.** "Is less than" becomes "is greater than," and vice versa. For example,

**a.**
$$3 < 6$$
$$-1(3) > -1(6)$$
$$-3 > -6$$

**b.**
$$-2 \leq 5$$
$$-4(-2) \geq -4(5)$$
$$8 \geq -20$$

**c.**
$$-6x > 12$$
$$\frac{-6x}{-6} < \frac{12}{-6}$$
$$x < -2$$

Solving first-degree inequalities depends on the following procedures.

***To Solve a First-Degree Inequality***

1. Simplify each side of the inequality by removing any grouping symbols and combining like terms.
2. Add the opposites of constants and/or variables to both sides so that variables are on one side and constants on the other.
3. Divide both sides by the coefficient of the variable and
   **a.** leave the sense of the inequality unchanged if the coefficient is positive; or
   **b.** reverse the sense of the inequality if the coefficient is negative.

**Note:** Unless otherwise stated, we will assume that the replacement set for $x$ is the set of real numbers.

**EXAMPLES** Solve the following inequalities and graph their solution sets. Write the solutions in interval notation.

**1.** $6x + 5 \leq -1$

**Solution**

| | |
|---|---|
| $6x + 5 \leq -1$ | |
| $6x \leq -6$ | Add $-5$ to both sides. |
| $x \leq -1$ | Divide both sides by 6. |
| $x$ is in $(-\infty, -1]$ | In interval notation |

The circle about $-1$ is shaded to indicate that $-1$ is included in the solution.

**2.** $x - 3 > 3x + 4$

**Solution** $x - 3 > 3x + 4$

$-3 > 2x + 4$ Add $-x$ to both sides.

$-7 > 2x$ Add $-4$ to both sides.

$-\frac{7}{2} > x$ Divide both sides by 2.

$x$ is in $\left(-\infty, -\frac{7}{2}\right)$ Using interval notation

The open circle about $-\frac{7}{2}$ indicates that $-\frac{7}{2}$ is not included in the solution.

**3.** $6 - 4x \leq x + 1$

**Solution** $6 - 4x \leq x + 1$

$6 - 5x \leq 1$ Add $-x$ to both sides.

$-5x \leq -5$ Add $-6$ to both sides.

$x \geq 1$ Divide both sides by $-5$.

Note that the sense of the inequality is changed from $\leq$ to $\geq$ because we divided by a negative number.

$x$ is in $[1, +\infty)$ Using interval notation

**4.** $2x + 5 < 3x - (7 - x)$

**Solution** $2x + 5 < 3x - (7 - x)$

$2x + 5 < 3x - 7 + x$ Remove parentheses first.

$2x + 5 < 4x - 7$ Combine like terms.

$5 < 2x - 7$ Add $-2x$ to both sides.

$12 < 2x$ Add 7 to both sides.

$6 < x$ Divide both sides by 2.

$x$ is in $(6, +\infty)$ Using interval notation

■

## *Absolute Value Inequalities*

Now consider an inequality with absolute value such as $|x| < 3$. For a number to have an absolute value less than 3, it must be within 3 units of 0. That is, the numbers between $-3$ and 3 have their absolute values less than 3 because they are within 3 units of 0. Thus, for $|x| < 3$,

| Algebraic Notation | Graph | Interval Notation |
|---|---|---|
| $-3 < x < 3$ | | $x$ is in $(-3,3)$ |

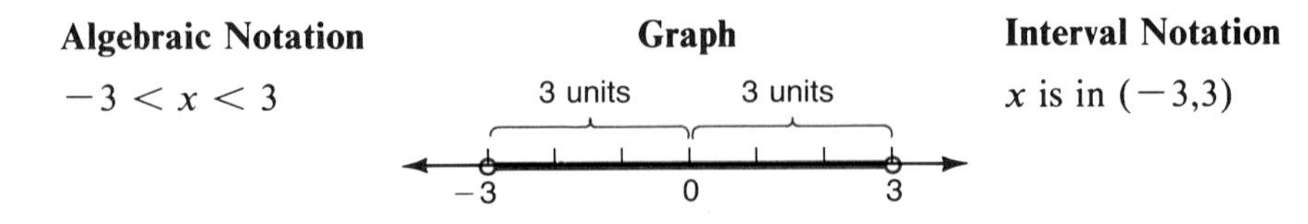

**Note:** The notation $-3 < x < 3$ is shorthand for the statement "$-3 < x$ **and** $x < 3$." The word **and** implies that both inequalities must be true. For example, we can say that $-3 < 7$ but $7 \nless 3$, so 7 is **not** a solution to $|x| < 3$. We can say that $-3 < 2$ **and** $2 < 3$, so 2 **is** a solution to $|x| < 3$.

Now consider the inequality $|x - 5| < 3$. This means that $x - 5$ is between $-3$ and 3. We can solve first-degree inequalities to find the values for $x$. For example,

| | |
|---|---|
| $\lvert x - 5 \rvert < 3$ | |
| $-3 < x - 5 < 3$ | $x - 5$ is between $-3$ and 3. |
| $-3 + 5 < x - 5 + 5 < 3 + 5$ | Add $+5$ to each part of the expression, just as in solving first-degree inequalities. |
| $2 < x < 8$ | Simplify. |
| $x$ is in $(2,8)$ | Using interval notation |

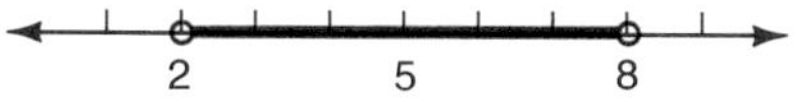


The values for $x$ are between 2 and 8 and are within 3 units of 5.

For $c > 0$:

**a.** If $|x| < c$, then $-c < x < c$.

**b.** If $|ax + b| < c$, then $-c < ax + b < c$.

**EXAMPLES** Solve the following absolute value inequalities and graph the solution sets.

**5.** $|x| \le 6$

**Solution** $-6 \le x \le 6$

or $x$ is in $[6,6]$

**6.** $|x + 3| < 2$

**Solution**

$$-2 < x + 3 < 2$$
$$-2 - 3 < x + 3 - 3 < 2 - 3$$
$$-5 < x < -1$$

or $x$ is in $(-5,-1)$

**7.** $|2x - 7| < 1$

**Solution**

$$-1 < 2x - 7 < 1$$
$$6 < 2x < 8$$
$$3 < x < 4$$

or $x$ is in $(3,4)$ ■

Now consider the inequality $|x| > 3$. For a number to have an absolute value greater than 3, its distance from 0 must be greater than 3. That is, the numbers that are greater than 3 **or** less than $-3$ will have absolute values greater than 3. Thus, for $|x| > 3$,

| **Algebraic Notation** | **Graph** | **Interval Notation** |
|---|---|---|
| $x > 3$ **or** $x < -3$ | 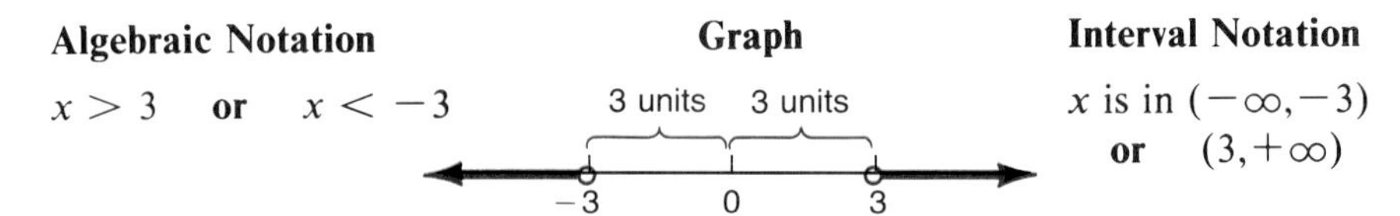 | $x$ is in $(-\infty, -3)$ **or** $(3, +\infty)$ |



**Note:** The expression $x > 3$ or $x < -3$ **cannot** be combined into one inequality expression. The word **or** must separate the inequalities since any number that satisfies one **or** the other is a solution to the absolute value inequality. There are **no** numbers that satisfy **both** inequalities.

Now consider the inequality $|x - 5| > 6$. This means that the distance of $x - 5$ from 0 must be greater than 6. That is, $x - 5$ is to be greater than 6 or less than $-6$. Thus,

$$|x - 5| > 6$$

| | | | |
|---|---|---|---|
| $x - 5 > 6$ | **or** | $x - 5 < -6$ | $x - 5$ is greater than 6 or less than $-6$. |
| $x - 5 + 5 > 6 + 5$ | **or** | $x - 5 + 5 < -6 + 5$ | Add $+5$ to each side, just as in solving first-degree inequalities. |
| $x > 11$ | **or** | $x < -1$ | Simplify. |

(**Note:** The values for $x$ are less than $-1$ or greater than 11 and are more than 6 units from 5. Thus, we can interpret the inequality $|x - 5| > 6$ to mean that the distance from $x$ to 5 is to be greater than 6.)

For $c > 0$:

**a.** If $|x| > c$, then $x > c$ or $x < -c$.

**b.** If $|ax + b| > c$, then $ax + b > c$ or $ax + b < -c$.

**EXAMPLES** Solve the following absolute value inequalities and graph the solution sets.

**8.** $|x| \geq 5$

**Solution** $x \geq 5$ or $x \leq -5$

or $x$ is in $(-\infty,-5]$ or $[5,+\infty)$

$-5$ $\quad$ $5$

**9.** $|4x - 3| > 2$

**Solution**

$$4x - 3 > 2 \quad \text{or} \quad 4x - 3 < -2$$
$$4x > 5 \quad \text{or} \quad 4x < 1$$
$$x > \frac{5}{4} \quad \text{or} \quad x < \frac{1}{4}$$

or $x$ is in $\left(-\infty,\frac{1}{4}\right)$ or $\left(\frac{5}{4},+\infty\right)$

$\frac{1}{4}$ $\quad$ $\frac{5}{4}$

**10.** $|3x - 8| > -6$

**Solution** There is nothing to do here except observe that no matter what is substituted for $x$, the absolute value will be greater than $-6$. Absolute value is always nonnegative. The solution to the inequality is all real numbers, so shade the entire number line. In interval notation, $x$ is in $(-\infty,+\infty)$.

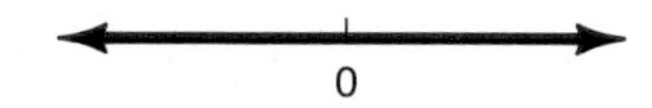


**11.** $|x + 9| < -\frac{1}{2}$

**Solution** Since absolute value is always nonnegative (greater than or equal to 0), there is no solution to this inequality. No number has an absolute value less than $-\frac{1}{2}$.

**12.** $|2x + 6| + 4 < 9$

**Solution**

$$|2x + 6| < 5$$

Add $-4$ to both sides to get the absolute value expression by itself on one side.

$$-5 < 2x + 6 < 5$$
$$-11 < 2x < -1$$
$$-\frac{11}{2} < x < -\frac{1}{2}$$

or $x$ is in $\left(-\frac{11}{2},-\frac{1}{2}\right)$

■

## EXERCISES 1.6

Solve the inequalities in Exercises 1–45 and graph the solution sets. Write the solution in interval notation. Assume that $x$ is a real number.

**1.** $2x + 3 < 5$
**2.** $4x - 7 \geq 9$
**3.** $14 - 5x < 4$
**4.** $23 < 7x - 5$
**5.** $6x - 15 > 1$
**6.** $9 - 2x < 8$
**7.** $5.6 + 3x \geq 4.4$
**8.** $12x - 8.3 < 6.1$
**9.** $1.5x + 9.6 < 12.6$
**10.** $0.8x - 2.1 \geq 1.1$
**11.** $2x - 1 > 3x + 2$
**12.** $6x + 6 < 4x - 2$
**13.** $3x - 5 > 3 - x$
**14.** $2 + 3x \geq x + 8$
**15.** $x - 6 \leq 4 - x$
**16.** $3x - 1 \leq 11 - 3x$
**17.** $5x + 6 \geq 2x - 2$
**18.** $4 - 2x < 5 + x$
**19.** $4 + x > 1 - x$
**20.** $x - 6 > 3x + 5$
**21.** $\frac{x}{2} - 1 \leq \frac{5x}{2} - 3$
**22.** $\frac{x}{4} + 1 \leq 5 - \frac{x}{4}$
**23.** $\frac{4x}{5} - 2 > x + 1$
**24.** $\frac{x}{3} - 2 > 1 - \frac{x}{3}$
**25.** $\frac{5x}{3} + 2 > \frac{x}{3} - 1$
**26.** $3.5 + 2x < 4x - 2.5$
**27.** $6x + 5.91 < 1.11 - 2x$
**28.** $4.3x + 21.5 \geq 1.7x + 0.7$
**29.** $6.2x - 5.7 > 4.8x + 3.1$
**30.** $0.9x - 11.3 < 3.1 - 0.7x$
**31.** $4(6 - x) < -2(3x + 1)$
**32.** $-3(2x - 5) \leq 3(x - 1)$
**33.** $3x + 8 \leq -3(2x - 3)$
**34.** $6(3x + 1) < 5(1 - 2x)$
**35.** $4 + 7x \leq 3x - 8 + x$
**36.** $11x + 8 - 5x \geq 2x - (4 - x)$
**37.** $1 - (2x + 8) < (9 + x) - 4x$
**38.** $5 - 3(4 - x) + x \leq -2(3 - 2x) - x$
**39.** $x - (2x + 5) \geq 7 - (4 - x) + 10$
**40.** $\frac{2(x - 1)}{3} < \frac{3(x + 1)}{4}$
**41.** $\frac{x + 2}{2} \geq \frac{2x}{3}$
**42.** $\frac{x - 2}{4} > \frac{x + 2}{2} + 6$
**43.** $\frac{x + 4}{9} \leq \frac{x}{3} - 2$
**44.** $\frac{2x + 7}{4} > \frac{x + 1}{3} - 1$
**45.** $\frac{4x}{7} - 3 > \frac{x - 6}{2} - 4$

Solve each of the absolute value problems in Exercises 46–65 and graph the solution sets. Write the solutions in interval notation.

**46.** $|x| \geq -2$
**47.** $|x| \geq 3$
**48.** $|x| \leq \frac{4}{5}$
**49.** $|x| \geq \frac{7}{2}$
**50.** $|x - 3| > 2$
**51.** $|y - 4| \leq 5$
**52.** $|x + 2| \leq 4$
**53.** $|x - 7| > -8$
**54.** $|x + 6| \leq 4$
**55.** $|2x - 1| \geq 2$
**56.** $|3 - 2x| < -2$
**57.** $|3x + 4| - 1 < 0$
**58.** $\left|\frac{3x}{2} - 4\right| \geq 5$
**59.** $\left|\frac{3}{7}y + \frac{1}{2}\right| > 2$
**60.** $|5x + 2| + 3 > 2$
**61.** $|7x - 3| + 4 \geq 6$
**62.** $|2x - 9| - 7 \leq 4$
**63.** $5 > |4 - 2x| + 2$
**64.** $-4 < |6x - 1| + 4$
**65.** $7 > |8 - 5x| + 3$

**Calculator Problems**

Solve the inequalities in Exercises 66–70. Round off answers to the nearest hundredth.

**66.** $8.31x + 14.915 < 4.73x + 8.592$
**67.** $17.431 - 6.957x \leq 11.516 - 3.476x$
**68.** $27.659 + 10.803x > 5.982x + 15.631$
**69.** $|14.3x - 10.5105| \leq 2.0735$
**70.** $|8.62x + 4.591| > 7.312$

# CHAPTER 1 SUMMARY

## Key Terms and Formulas

**Natural Numbers** (or **Counting Numbers**) [1.1]
$N = \{1, 2, 3, 4, 5, 6, \ldots\}$

**Whole Numbers** [1.1]
$W = \{0, 1, 2, 3, 4, 5, 6, \ldots\}$

**Integers** [1.1]
$J = \{\ldots, -4, -3, -2, -1, 0, 1, 2, 3, 4, \ldots\}$

**Rational Numbers** [1.1]

$Q = \{$numbers of the form $\frac{a}{b}$ where $a$ and $b$ are integers, $b \neq 0\}$

$Q = \{$infinite repeating decimals$\}$

**Irrational Numbers** [1.1]
$H = \{$infinite nonrepeating decimals$\}$

**Real Numbers** [1.1]
$R = \{$any rational or irrational number$\}$

**Real Number Line** [1.1]
There is a **one-to-one correspondence** between the real numbers and the points on a number line.

**Absolute Value** [1.2]
For any real number $x$,

$$|x| = \begin{cases} x & \text{if } x \geq 0 \\ -x & \text{if } x < 0 \end{cases}$$

**Difference** [1.2]
For real numbers $a$ and $b$, $a - b = a + (-b)$.

**Like terms** (or **similar terms**) are those terms that have the same variable factors with the same exponents. [1.3]

**First-Degree Equation** [1.3]
An equation of the form $ax + b = 0$, where $a$ and $b$ are real numbers and $a \neq 0$, is called a **first-degree equation** in $x$.

For $c > 0$:

**a.** If $|x| = c$, then $x = c$ or $x = -c$.

**b.** If $|ax + b| = c$, then $ax + b = c$ or $ax + b = -c$. [1.3]

If $|a| = |b|$, then either $a = b$ or $a = -b$.

More generally,

$$\text{if } |ax + b| = |cx + d|,$$

then either

$$ax + b = cx + d \quad \text{or}$$
$$ax + b = -(cx + d) \quad [1.3]$$

A **formula** is an equation that represents a general relationship between two or more quantities or measurements. [1.4]

**Types of Intervals** [1.6]

| Name of Interval | Algebraic Notation | Interval Notation | Graph |
|---|---|---|---|
| Open interval | $a < x < b$ | $(a,b)$ | a b |
| Closed interval | $a \leq x \leq b$ | $[a,b]$ | a b |
| Half-open interval | $a \leq x < b$ | $[a,b)$ | a b |
| | $a < x \leq b$ | $(a,b]$ | a b |
| Open interval | $x > a$ | $(a,+\infty)$ | a |
| | $x < b$ | $(-\infty,b)$ | b |
| Half-open interval | $x \geq a$ | $[a,+\infty)$ | a |
| | $x \leq b$ | $(-\infty,b]$ | b |

**First-Degree Inequality** [1.6]
For real numbers $a$ and $b$, $a \neq 0$, $ax + b < 0$ is called a **first-degree inequality** in $x$. (The definition is valid if $\leq$, $>$, or $\geq$ is used in place of $<$.)

For $c > 0$

**a.** If $|x| < c$, then $-c < x < c$.
**b.** If $|ax + b| < c$, then $-c < ax + b < c$. [1.6]

For $c > 0$

**a.** If $|x| > c$, then $x > c$ or $x < -c$.
**b.** If $|ax + b| > c$, then $ax + b > c$ or $ax + b < -c$. [1.6]

## *Properties and Rules*

**Zero Factor Law** [1.1]
For any real number $a$, $a \cdot 0 = 0$.

**Addition Property of Equality** [1.1]
For real numbers $a$, $b$, and $c$,

$$\text{If } a = b, \text{ then } a + c = b + c.$$

**Multiplication Property of Equality** [1.1]
For real numbers $a$, $b$, and $c$,

$$\text{If } a = b, \text{ then } a \cdot c = b \cdot c.$$

**Substitution Property** [1.1]
If $a$ and $b$ are real numbers and $a = b$, then $a$ and $b$ may be substituted for each other in any expression.

**Field Properties of the Real Numbers** [1.1]
For real numbers $a$, $b$, and $c$:

| Addition | | Name of Property | Multiplication | |
|---|---|---|---|---|
| A1 | $a + b$ is a real number. | Closure | M1 | $a \cdot b$ is a real number. |
| A2 | $a + b = b + a$ | Commutative | M2 | $a \cdot b = b \cdot a$ |
| A3 | $(a + b) + c = a + (b + c)$ | Associative | M3 | $(a \cdot b) \cdot c = a \cdot (b \cdot c)$ |
| A4 | $a + 0 = 0 + a = a$ | Identity | M4 | $a \cdot 1 = 1 \cdot a = a$ |
| A5 | $a + (-a) = 0$ | Inverse | M5 | $a \cdot \dfrac{1}{a} = 1$ $(a \neq 0)$ |

D Distributive Property of Multiplication over Addition: $a(b + c) = a \cdot b + a \cdot c$

**Properties of Inequality (Order)** [1.1]
For real numbers $a$, $b$, and $c$:

O1 Exactly one of the following is true: Trichotomy property

$$a < b, \quad a = b, \quad \text{or} \quad a > b$$

O2 If $a < b$ and $b < c$, then $a < c$. Transitive property

**Rules for Adding Real Numbers** [1.2]

1. To add two real numbers with like signs, add their absolute values and use the common sign.
2. To add two real numbers with unlike signs, subtract their absolute values (the smaller from the larger) and use the sign of the number with the larger absolute value.

**Rules for Multiplication and Division with Real Numbers** [1.2]

1. The product (or quotient) of two real numbers with like signs is a positive real number.
2. The product (or quotient) of two real numbers with unlike signs is a negative real number.

**Division by 0 is undefined.** [1.2]

**Rules for Order of Operations** [1.2]

1. Simplify within symbols of inclusion (parentheses, brackets, or braces) beginning with the innermost symbols.
2. Find any powers indicated by exponents.
3. Perform any multiplications or divisions as they occur from left to right.
4. Perform any additions or subtractions as they occur from left to right.

**Addition Property of Equality** [1.3]
If the same algebraic expression is added to both sides of an equation, the new equation has the same solutions as the original equation. Symbolically, if $A$, $B$, and $C$ are algebraic expressions, then the equations

$$A = B$$

and

$$A + C = B + C$$

have the same solutions.

**Multiplication Property of Equality** [1.3]
If both sides of an equation are multiplied by the same nonzero algebraic expression, the new equation has the same solutions as the original equation. Symbolically, if $A$, $B$, and $C$ are algebraic expressions, then the equations

$$A = B$$

and

$$AC = BC \text{ (where } C \neq 0)$$

have the same solutions.

## Procedures

**To Solve a First-Degree Equation** [1.3]

1. Simplify each side of the equation by removing any grouping symbols and combining like terms. (In some cases, you may want to multiply each term by a constant to clear fraction or decimal coefficients.)
2. Use the addition property of equality to add the opposites of constants and/or variables so that variables are on one side and constants on the other.
3. Use the multiplication property of equality to multiply both sides by the reciprocal of the coefficient of the variable (or divide both sides by the coefficient).
4. Check your answer by substituting it into the original equation.

**Attack Plan for Word Problems** [1.5]

1. Read the problem carefully. (Reread it several times if necessary.)
2. Decide what is asked for, and assign a variable to the unknown quantity. Label this variable so you know exactly what it represents.
3. Draw a diagram or set up a chart whenever possible.
4. Form an equation (or inequality) that relates the information provided.
5. Solve the equation (or inequality).
6. Check your solution with the wording of the problem to be sure it makes sense.

**To Solve a First-Degree Inequality** [1.6]

1. Simplify each side of the inequality by removing any grouping symbols and combining like terms.
2. Add the opposites of constants and/or variables to both sides so that variables are on one side and constants on the other.
3. Divide both sides by the coefficient of the variable and
   a. leave the sense of the inequality unchanged if the coefficient is positive; or
   b. reverse the sense of the inequality if the coefficient is negative.

## CHAPTER 1 REVIEW

Given the set of numbers $\left\{-10, -\sqrt{25}, -1.6, -\sqrt{7}, 0, \frac{1}{5}, \sqrt{9}, \pi, \sqrt{12}\right\}$, list those numbers in the set that are described in Exercises 1–6. [1.1]

1. $\{x \mid x \text{ is a natural number}\}$
2. $\{x \mid x \text{ is a whole number}\}$
3. $\{x \mid x \text{ is a rational number}\}$
4. $\{x \mid x \text{ is an integer}\}$
5. $\{x \mid x \text{ is an irrational number}\}$
6. $\{x \mid x \text{ is a real number}\}$

Write each of the rational numbers in Exercises 7–9 as an infinite repeating decimal. [1.1]

7. $\frac{5}{6}$
8. $\frac{11}{8}$
9. $\frac{13}{15}$

Graph each of the sets in Exercises 10–13 on a number line. [1.1]

10. $\{x \mid -2 < x < 10, x \text{ is a whole number}\}$
11. $\{x \mid x > -7, x \text{ is an integer}\}$
12. $\{x \mid x \leq 9, x \text{ is a natural number}\}$
13. $\{x \mid -5 \leq x < 8, x \text{ is an integer}\}$

Graph each of the sets of real numbers in Exercises 14–17 on a number line ($x$ is a real number). [1.1]

14. $\left\{x \mid -3 < x \leq \frac{7}{5}\right\}$
15. $\left\{x \mid -\frac{2}{3} < x < \sqrt{3}\right\}$
16. $\{x \mid -0.8 < x < 5 \text{ and } x \leq 3\}$
17. $\{x \mid -5 < x \leq 4 \text{ or } 2 < x \leq 5.3\}$

Name the property of real numbers that justifies each statement in Exercises 18–25. All variables represent real numbers. [1.1]

**18.** $7 + (x + 3) = 7 + (3 + x)$

**19.** $z(y + 5) = zy + z(5)$

**20.** $\sqrt{7}(xy) = (\sqrt{7}x)y$

**21.** If $x = 9$, then $3 \cdot x = 3 \cdot 9$.

**22.** $\frac{4}{5} + \left(-\frac{4}{5}\right) = 0$

**23.** $(x + 10) \cdot 1 = x + 10$

**24.** If $x < 10$ and $10 < y$, then $x < y$.

**25.** $x < -9$, $x = -9$, or $x > -9$

Complete the expressions in Exercises 26–30 using the given property. [1.1]

**26.** $(x + 5) + y =$ ______ Associative property for addition

**27.** $(x - 0.3) + 0 =$ ______ Identity for addition

**28.** $5(x + 8z) =$ ______ Distributive property

**29.** If $x + y = 15$ and $y = x - 7$, then ______ Substitution property

**30.** If $2x + 3 = y + 4$, then $(2x + 3) + 7 =$ ______ Addition property for equality

Find those values for $x$ that make Exercises 31–33 true statements. If statements are never true, indicate by saying "no value." [1.2]

**31.** $|x| = 3.9$

**32.** $|x| = 8$

**33.** $|x| = -\frac{3}{5}$

Perform the indicated operations in Exercises 34–55. [1.2]

**34.** $(-8) + 12$

**35.** $(-13) + (-7)$

**36.** $|-9| + 2$

**37.** $23 - 31$

**38.** $17 - (-5)$

**39.** $|9| - |-10|$

**40.** $\frac{3}{5} + \left(-\frac{1}{2}\right)$

**41.** $10 + (-7) + (-2)$

**42.** $6.5 + (-4.2) - 3.1$

**43.** $\frac{3}{4} - \frac{2}{3} + \left(-\frac{1}{6}\right)$

**44.** $(-7)(-12)$

**45.** $8(-5)$

**46.** $\left(-\frac{1}{3}\right)(6)$

**47.** $(1.8)(-4)$

**48.** $22 \div (-11)$

**49.** $(-4) \div (-6)$

**50.** $8 \div 0$

**51.** $28 \div |-4|$

**52.** $0 \div \frac{3}{5}$

**53.** $(-7)(-9)(-2)$

**54.** $3(-6)\left(\frac{1}{2}\right)(-5)$

**55.** $\left(\frac{7}{16}\right)\left(-\frac{2}{3}\right)\left(\frac{3}{5}\right)\left(\frac{5}{2}\right)$

Find the value of the expressions in Exercises 56–60 using the rules for order of operations. [1.2]

**56.** $12 - 6 \div 2 \cdot 3 - 5$

**57.** $5 - (13 \cdot 5 - 5) \div 3 \cdot 2$

**58.** $3^2 + 5 \cdot 4 - 10 + |7|$

**59.** $6 \cdot 4 - 2^3 - (5 \cdot 10) - 5^2$

**60.** $\left(\frac{3}{4} + \frac{5}{8}\right)\left[\left(\frac{1}{2}\right)\left(-\frac{4}{5}\right) - \frac{2}{3}\right]$

In each of the expressions in Exercises 61–66, combine like terms. [1.3]

**61.** $-4(x + 3) + 2x$

**62.** $x + \frac{x - 5x}{4}$

**63.** $\frac{2(x + 3x)}{4} + 2x$

**64.** $(x^3 + 4x - 1) - (-2x^3 + x^2)$

**65.** $-2[7x - (2x + 5) + 3]$

**66.** $2x + [4x - (8y - 3) + (2y + 7)]$

Solve each of the equations in Exercises 67–74. [1.3]

**67.** $7x + 4 = -17$

**68.** $9x - 11 = x + 5$

**69.** $5(1 - 2x) = 3x + 57$

**70.** $x + 1 = 2(x - 1) - x + 3$

**71.** $5(2x + 3) = 3(x - 4) - 1$

**72.** $\frac{7x}{8} + 5 = \frac{x}{4}$

**73.** $\frac{5x}{2} - \frac{3(x - 1)}{2} = x + 2$

**74.** $\frac{2x + 5}{3} = \frac{-3x + 2}{6} - 1$

Solve each of the absolute value equations in Exercises 75–80. [1.3]

**75.** $|2x + 1| = 5.6$

**76.** $|3x - 2| = -6$

**77.** $\left|\frac{2x}{3} + 4\right| = 2$

**78.** $\left|\frac{x}{2} - 3\right| = 5$

**79.** $|3(x + 1) + 2| = -1$

**80.** $|2(x - 4) + x| = 2$

Solve for the indicated variable in Exercises 81–83. [1.4]

**81.** $A = \frac{m + n}{2}$; solve for $n$.

**82.** $K = \frac{mv^2}{2g}$; solve for $m$.

**83.** $W = \frac{2PR}{R - r}$; solve for $r$.

Set up equations and solve the word problems in Exercises 84–88. [1.5]

**84.** The difference between twice a number and 3 is equal to the difference between five times the number and 2. Find the number.

**85.** The local supermarket charges a flat rate of \$5, plus \$3 per hour, for rental of a carpet cleaner. If it cost Ron \$26 to rent the machine, how many hours did he keep it?

**86.** Stephanie rode her new moped to Rod's house. Traveling the side streets, she averaged 20 mph. To save time on the return trip, they loaded the bike into Rod's truck and took the freeway, averaging 50 mph. The freeway distance is 2 miles less than the distance on the side streets and saves 24 minutes. Find the distance traveled on the return trip.

**87.** Julie has two part-time jobs. One job pays \$3.80 per hour and the other pays \$4.20 per hour. Last week she worked a total of 44 hours and received \$177.60. How many hours did she work at each rate?

**88.** A rancher bought some calves for \$170 each. "Jet-age" rustlers stole 6 of them. Four months later, the rancher sold the remaining calves for \$240 each. He paid a feed bill of \$400 and still made a profit of \$4110. How many calves did he buy?

Solve the inequalities in Exercises 89–96 and graph the solutions. Write the solution set in interval notation. Assume that $x$ is a real number. [1.6]

**89.** $5x - 7 > x + 9$

**90.** $\frac{x}{3} + 1 > 2$

**91.** $5x + 10 \leq 6(x + 3.8)$

**92.** $x + 8 - 5x \geq 2(x - 2)$

**93.** $\frac{2x + 1}{3} \leq \frac{3x}{5}$

**94.** $\frac{x}{3} - 2.1 > \frac{x}{2} + 1.3$

**95.** $\frac{2x + 7}{5} > x - 1$

**96.** $\frac{3(x - 2)}{4} < \frac{2(x + 2)}{3} - 2$

Solve each of the absolute value problems in Exercises 97–100 and graph the solutions. Write the solutions in interval notation. [1.6]

**97.** $|5x + 2| > 7$

**98.** $\left|\frac{4x}{3} - 1\right| < 3$

**99.** $|3x + 2| + 4 < 10$

**100.** $|4 - 3x| + 5 > 2$

## CHAPTER 1 TEST

**1.** Given the set of numbers

$\left\{-\sqrt{11}, -2, -\frac{5}{3}, 0, \frac{1}{2}, \sqrt{3}, 2, \pi\right\}$,

list these numbers in the sets that are described.

**a.** $\{x \mid x \text{ is an integer}\}$

**b.** $\{x \mid x \text{ is a rational number}\}$

**2.** Write the rational number $\frac{5}{13}$ as an infinite repeating decimal.

Graph each of the sets in Exercises 3 and 4 on a number line.

**3.** $\{x \mid x \leq 7, x \text{ is a whole number}\}$

**4.** $\{x \mid -8 < x \leq 0, x \text{ is an integer}\}$

Graph each of the sets of real numbers in Exercises 5 and 6 on a number line.

**5.** $\left\{x \mid -2 \leq x < \frac{5}{8}\right\}$

**6.** $\{x \mid -1.5 < x < 3 \text{ or } x \geq \sqrt{17}\}$

Name the property of real numbers that justifies each statement in Exercises 7 and 8. All variables represent real numbers.

**7.** $3x + 15y = 3(x + 5y)$

**8.** $3 + (x + y) = (3 + x) + y$

Complete the expressions in Exercises 9 and 10 using the given property.

**9.** $7y \cdot 0 =$ _____ Zero factor law

**10.** If $2x + y = 17$ and $y = 3x$, then _____ Substitution property

Perform the indicated operations in Exercises 11–15.

**11.** $|-19| + 43 - (-8)$

**12.** $(-96) \div (-12)$

**13.** $\left(\frac{5}{8}\right)(-3)$

**14.** $\frac{3}{5} + \left(-\frac{7}{10}\right) - \frac{1}{6}$

**15.** $4(-6)(-2)(-3)$

Find the value of the expressions in Exercises 16 and 17 using the rules for order of operations.

**16.** $6 + (7^2 - 3^2) \div 5 \cdot 4$

**17.** $\left(-\frac{1}{3}\right) + \frac{3}{4} \div \frac{1}{6}$

Simplify the expressions in Exercises 18 and 19 by combining like terms.

**18.** $4(x + 6) - (8 - 2x)$

**19.** $-2x + 3[5x - (2x + 4)]$

Solve each of the equations in Exercises 20–24.

**20.** $5x - 3(x - 2) = 4$

**21.** $4x + (5 - x) = 3(x + 2)$

**22.** $\frac{3x - 2}{8} = \frac{x}{4} - 1$

**23.** $|2x + 1| = 2.8$

**24.** $|5 - 3x| + 1 = 4$

Solve for the indicated variable in Exercises 25 and 26.

**25.** $A = P + Prt$; solve for $t$.

**26.** $I = \frac{nE}{R + nr}$; solve for $r$.

Solve the inequalities in Exercises 27 and 28, then graph the solution. Write the solution in interval notation.

**27.** $3x + 7 < 4(x + 3)$

**28.** $\dfrac{2x + 5}{4} > x + 3$

Solve the inequalities in Exercises 29 and 30 and graph the solution. Write the solution in interval notation.

**29.** $|7 - 2x| < 3$

**30.** $|2(x - 3) + 5| > 2.7$

Set up equations and solve the word problems in Exercises 31 and 32.

**31.** A bus leaves Kansas City headed for Phoenix traveling at a rate of 48 mph. Thirty minutes later, a second bus follows, traveling at 54 mph. How long will it take the second bus to overtake the first?

**32.** The Candy Shack sells a particular candy in two different sized packages. One size sells for $1.25 each and the other sells for $1.75 each. If the store received $65.50 for 42 packages of candy, how many of each size were sold?

C H A P T E R

# 2 EXPONENTS AND POLYNOMIALS

## MATHEMATICAL CHALLENGES

Which is greater, $2^{100}$ or $3^{75}$?

Express the number 96 as the difference of two squares in four different ways.

What percent of the perfect squares between 0 and 1000 are odd numbers?

## CHAPTER OUTLINE



Whole number exponents are used to indicate repeated multiplication. In this chapter we will consider negative exponents as well as positive exponents and their properties. Later, in Chapters 4 and 9, we will discuss the meanings of fractional exponents and, more generally, real exponents. While the basic properties for exponents are developed only for integer exponents in this chapter, you should be aware of the fact that these properties are valid for all real exponents and will be used even though the meanings of the exponents may be different from simple repeated multiplication.

Equally important are the factoring skills we will develop in this chapter. These skills will be applied extensively in Chapter 3 when we operate with rational expressions (fractions with polynomials in the numerator and denominator) and solve equations. Your skill in factoring polynomials may be one of the determining factors in your success in this course, so you should work particularly hard at this early stage.

Factoring with negative exponents, discussed in Section 2.5, deserves special mention. This skill gives an added dimension to your equation-solving skills and is particularly useful in simplifying expressions in calculus.

## 2.1 Properties of Exponents

### OBJECTIVE

In this section, you will be learning to simplify expressions with constant or single-variable bases using the properties of integer exponents.

In Section 1.2, whole number exponents were introduced so we could write algebraic terms, such as $3x^2$ and $-5xy^2$, and combine like terms. Now we will develop properties of exponents that are useful in simplifying algebraic expressions. These properties will include the exponent 0 and negative exponents, so we will have properties of **integer exponents.** (Fractional exponents will be discussed in Chapter 4.)

To multiply $4^2 \cdot 4^3$ or $x^3 \cdot x^5$ or $(-2)^3(-2)$, we could write

$$4^2 \cdot 4^3 = (4 \cdot 4) \cdot (4 \cdot 4 \cdot 4) = 4^5$$
$$x^3 \cdot x^5 = (x \cdot x \cdot x)(x \cdot x \cdot x \cdot x \cdot x) = x^8$$
$$(-2)^3(-2) = (-2)(-2)(-2)(-2) = (-2)^4$$

In general, to **multiply** terms with the same base, we **add** the exponents. This rule is stated as property 1 of exponents.

***Property 1 of Exponents*** If $a$ is a nonzero number and $m$ and $n$ are integers, then

$$a^m \cdot a^n = a^{m+n}$$

To clarify the difference between two expressions with negative signs and exponents such as $(-2)^4$ and $-2^4$, consider that

$$-2^4 = (-1) \cdot 2^4$$

Thus,

$$-2^4 = -1(2)(2)(2)(2) = -16$$

and the exponent 4 has base 2, not $-2$. In contrast,

$$(-2)^4 = (-2)(-2)(-2)(-2) = 16$$

and the exponent 4 has base $-2$.

Another point to clarify is that if no exponent is written for a variable or constant, it is understood to be 1. For example, $y = y^1$ and $7 = 7^1$.

Property 1 is stated for integer exponents and, applying this property with the exponent 0, we get

$$2^0 \cdot 2^3 = 2^{0+3} = 2^3 \qquad (\text{Also, } 1 \cdot 2^3 = 2^3.)$$
$$3^0 \cdot 3^4 = 3^{0+4} = 3^4 \qquad (\text{Also, } 1 \cdot 3^4 = 3^4.)$$
$$5^0 \cdot 5^2 = 5^{0+2} = 5^2 \qquad (\text{Also, } 1 \cdot 5^2 = 5^2.)$$
$$6^0 \cdot 6^2 = 6^{0+2} = 6^2 \qquad (\text{Also, } 1 \cdot 6^2 = 6^2.)$$

Analysis of this discussion indicates that $2^0 = 1$, $3^0 = 1$, $5^0 = 1$, and $6^0 = 1$. Thus, Property 2 seems reasonable.

**Property 2 of Exponents**

If $a$ is a nonzero number, then

$$a^0 = 1$$

**The expression $0^0$ is undefined.**

Throughout this text, unless specifically stated otherwise, we will assume that the bases of exponents are nonzero.

Now we want to determine the meaning of negative exponents. Consider the following examples using Property 1 and Property 2.

| **With Exponents** | **With Fractions** |
|---|---|
| $2 \cdot 2^{-1} = 2^{1+(-1)} = 2^0 = 1$ | $2 \cdot \frac{1}{2} = 1$ |
| $2^2 \cdot 2^{-2} = 2^{2+(-2)} = 2^0 = 1$ | $2^2 \cdot \frac{1}{2^2} = 4 \cdot \frac{1}{4} = 1$ |
| $2^3 \cdot 2^{-3} = 2^{3+(-3)} = 2^0 = 1$ | $2^3 \cdot \frac{1}{2^3} = 8 \cdot \frac{1}{8} = 1$ |
| $3^2 \cdot 3^{-2} = 3^{2+(-2)} = 3^0 = 1$ | $3^2 \cdot \frac{1}{3^2} = 9 \cdot \frac{1}{9} = 1$ |

These examples lead to the conclusion that

$$2^{-1} = \frac{1}{2}, \quad 2^{-2} = \frac{1}{2^2}, \quad 2^{-3} = \frac{1}{2^3}, \quad \text{and} \quad 3^{-2} = \frac{1}{3^2}$$

Similarly, we can show that

$$6^{-1} = \frac{1}{6}, \quad 6^{-2} = \frac{1}{6^2}, \quad 5^{-3} = \frac{1}{5^3}, \quad \text{and} \quad 2^{-5} = \frac{1}{2^5}$$

We now have Property 3.

**Property 3 of Exponents** If $a$ is a nonzero number and $n$ is an integer, then

$$a^{-n} = \frac{1}{a^n}$$

**EXAMPLES** Using Properties 1, 2, and 3 of exponents, simplify the following expressions so that they contain only positive exponents.

**1.** $x^2 \cdot x^5$

**Solution** $x^2 \cdot x^5 = x^{2+5} = x^7$

**2.** $3^2 \cdot 3^5$

**Solution** $3^2 \cdot 3^5 = 3^{2+5} = 3^7 = 2187$

**3.** $2y^3 \cdot 3y^4$

**Solution** $2y^3 \cdot 3y^4 = 2 \cdot 3 \cdot y^{3+4} = 6y^7$

**4.** $11^0$

**Solution** $11^0 = 1$

**5.** $x^0 \cdot x^4$

**Solution** $x^0 \cdot x^4 = x^{0+4} = x^4$

**6.** $7^{-1}$

**Solution** $7^{-1} = \frac{1}{7}$

**7.** $2^{-5} \cdot 2^8$

**Solution** $2^{-5} \cdot 2^8 = 2^{-5+8} = 2^3 = 8$

**8.** $x^{-10} \cdot x^7$

**Solution** $x^{-10} \cdot x^7 = x^{-10+7} = x^{-3} = \frac{1}{x^3}$ (Here, Property 1 is used first, then Property 3.)

**9.** $2x^{-3} \cdot 5x^{-4}$

**Solution** $2x^{-3} \cdot 5x^{-4} = 2 \cdot 5 \cdot x^{-3} \cdot x^{-4} = 10x^{-7} = 10 \cdot \frac{1}{x^7} = \frac{10}{x^7}$ ■

Next, we want to consider division. For example, what is the value of $\frac{2^5}{2^2}$ or $\frac{6}{6^3}$? We can write both expressions as multiplication and then use property 1:

$$\frac{2^5}{2^2} = 2^5 \cdot \frac{1}{2^2} = 2^5 \cdot 2^{-2} = 2^{5-2} = 2^3 = 8$$

and

$$\frac{6}{6^3} = 6 \cdot \frac{1}{6^3} = 6^1 \cdot 6^{-3} = 6^{1-3} = 6^{-2} = \frac{1}{6^2} = \frac{1}{36}$$

Or we can reduce as fractions:

$$\frac{y^2}{y^5} = \frac{\overset{1}{\cancel{y}} \cdot \overset{1}{\cancel{y}}}{\underset{1}{\cancel{y}} \cdot \underset{1}{\cancel{y}} \cdot y \cdot y \cdot y} = \frac{1}{y^3}$$

This leads us to Property 4.

**Property 4 of Exponents** If $a$ is a nonzero number and $m$ and $n$ are integers, then

$$\frac{a^m}{a^n} = a^{m-n}$$

**EXAMPLES** Using Property 4, simplify the following expressions so that they contain only positive exponents.

**10.** $\frac{x^7}{x}$

**Solution** $\frac{x^7}{x} = x^{7-1} = x^6$

**11.** $\frac{y^{-3}}{y^{-8}}$

**Solution** $\frac{y^{-3}}{y^{-8}} = y^{-3-(-8)} = y^{-3+8} = y^5$

Having negative exponents in the denominator can be confusing. Remember to **subtract** the exponent in the denominator even if it is negative.

**12.** $\frac{a^{-10}}{a^{-6}}$

**Solution** $\frac{a^{-10}}{a^{-6}} = a^{-10-(-6)} = a^{-10+6} = a^{-4} = \frac{1}{a^4}$ ■

The last property in this section involves raising a power to a power. Consider an expression such as $(2^2)^3$ or $(x^4)^5$. We can write

$$(2^2)^3 = 2^2 \cdot 2^2 \cdot 2^2 = 2^{2+2+2} = 2^6$$

and $$(x^4)^5 = x^4 \cdot x^4 \cdot x^4 \cdot x^4 \cdot x^4 = x^{4+4+4+4+4} = x^{20}$$

However, repeatedly adding the same number is the same as multiplying with that number. Thus,

$$(2^2)^3 = 2^{2 \cdot 3} = 2^6 \qquad \text{and} \qquad (x^4)^5 = x^{4 \cdot 5} = x^{20}$$

Thus we have Property 5.

**Property 5 of Exponents**

If $a$ is a nonzero number and $m$ and $n$ are integers, then

$$(a^m)^n = a^{mn}$$

**EXAMPLES** Using Property 5, simplify the following expressions so that they contain only positive exponents.

**13.** $(x^3)^2$

**Solution** $(x^3)^2 = x^{3 \cdot 2} = x^6$

**14.** $(a^{-3})^7$

**Solution** $(a^{-3})^7 = a^{-3 \cdot 7} = a^{-21} = \dfrac{1}{a^{21}}$ ■

**Special Note: An expression can be considered simplified with negative exponents in the numerator.** In Examples 12 and 14, the answers $a^{-4}$ and $a^{-21}$ are considered as simplified as $\dfrac{1}{a^4}$ and $\dfrac{1}{a^{21}}$. You may use either form in your answers.

In the following examples, simplify each of the expressions using any of the five properties of exponents that apply. You might try each example on scratch paper first and then look at the solution shown in the example. There may be more than one correct procedure, and **you should apply whichever property you "see" first.**

**EXAMPLES**

**15.** $\dfrac{x^{10}x^2}{x^3}$

**Solution** $\dfrac{x^{10}x^2}{x^3} = \dfrac{x^{10+2}}{x^3} = \dfrac{x^{12}}{x^3} = x^{12-3} = x^9$

**16.** $\dfrac{x^6x^{-2}}{x^7}$

**Solution** $\dfrac{x^6x^{-2}}{x^7} = \dfrac{x^{6-2}}{x^7} = \dfrac{x^4}{x^7} = x^{4-7} = x^{-3}$ or $\dfrac{1}{x^3}$

**17.** $\dfrac{3^{-5} \cdot 3^9}{3^3 \cdot 3}$

**Solution** $\dfrac{3^{-5} \cdot 3^9}{3^3 \cdot 3} = \dfrac{3^{-5+9}}{3^{3+1}} = \dfrac{3^4}{3^4} = 3^{4-4} = 3^0 = 1$

**18.** $\left(\dfrac{y^{-2}}{y^{-5}}\right)^{-2}$

**Solution** $\left(\dfrac{y^{-2}}{y^{-5}}\right)^{-2} = (y^{-2-(-5)})^{-2} = (y^{-2+5})^{-2} = (y^3)^{-2}$

$= y^{3(-2)} = y^{-6}$ or $\dfrac{1}{y^6}$

**19.** $\dfrac{2y^3 \cdot 6y}{4y^{-1}}$

**Solution** $\dfrac{2y^3 \cdot 6y}{4y^{-1}} = \dfrac{12 \cdot y^{3+1}}{4y^{-1}} = \dfrac{\overset{3}{\cancel{12}}y^4}{\cancel{4}y^{-1}} = 3y^{4-(-1)}$

$= 3y^{4+1} = 3y^5$

**20.** $\left(\dfrac{2x^3 \cdot 3x^2}{17x^{-5}}\right)^0$

**Solution** $\left(\dfrac{2x^3 \cdot 3x^2}{17x^{-5}}\right)^0 = 1$ ■

In Examples 21 and 22, assume that $k$ represents a whole number, and follow the appropriate rules for exponents.

**EXAMPLES**

**21.** $x^{2k} \cdot x^{2k}$

**Solution** $x^{2k} \cdot x^{2k} = x^{2k+2k} = x^{4k}$

**22.** $\dfrac{x^3 \cdot x^k}{(x^2)^k}$

**Solution** $\dfrac{x^3 \cdot x^k}{(x^2)^k} = \dfrac{x^{k+3}}{x^{2k}} = x^{k+3-2k}$

$= x^{3-k}$ or $\dfrac{1}{x^{k-3}}$ ■

## EXERCISES 2.1

Simplify the expressions in Exercises 1–70.

**1.** $(-2)^4(-2)$ **2.** $(7^2)(7^0)$ **3.** $8^0$ **4.** $7^{-2}$ **5.** $-6^2$

**6.** $3 \cdot 2^2$ **7.** $5^{-1}$ **8.** $-5^{-2}$ **9.** $-4^{-3}$ **10.** $8^{-2}$

**11.** $x^5 \cdot x^7$ **12.** $x^3 \cdot x^5$ **13.** $x^4 \cdot x^0 \cdot x$ **14.** $x^3 \cdot x^{-1}$ **15.** $y^{-2} \cdot y^{-1}$

**16.** $x^{-2} \cdot x^3 \cdot x^5$ **17.** $x^3 \cdot x^{-7} \cdot x^2$ **18.** $y^{-3} \cdot y^{-2} \cdot y^0$ **19.** $\frac{x^3}{x^7}$ **20.** $\frac{x^{12}}{x^4}$

**21.** $\frac{y^5}{y^0}$ **22.** $\frac{x^2}{x^{-1}}$ **23.** $\frac{y^{-2}}{y^2}$ **24.** $\frac{y^2}{y^{-5}}$ **25.** $\frac{x^2x^4}{x}$

**26.** $\frac{x^3x^5}{x^4}$ **27.** $\frac{x^0x^3}{x^6}$ **28.** $\frac{x \cdot x^3}{x^5}$ **29.** $\frac{x^2x^4}{x^{-2}}$ **30.** $\frac{x^{-1}x^3}{x^{-4}}$

**31.** $\frac{x \cdot x^{-2}}{x^2x^{-3}}$ **32.** $\frac{x^{16}}{x^{-2}x^{-8}}$ **33.** $(x^2)^3$ **34.** $(x^4)^2$ **35.** $(-x^4)^3$

**36.** $(-x^3)^4$ **37.** $(x^5)^{-2}$ **38.** $(x^{-2})^{-1}$ **39.** $(x^3)^{-3}$ **40.** $(x^2)^{-2}$

**41.** $(x^2x^{-3})^4$ **42.** $(y^4y^{-1})^5$ **43.** $(x^3x^{-2})^{-1}$ **44.** $(x^3x^{-3})^3$ **45.** $\frac{y^3y^3}{y^2}$

**46.** $\frac{y^2y^4}{y}$ **47.** $\frac{y^{10}y^4}{y^6}$ **48.** $\frac{y \cdot y^4}{y^5}$ **49.** $\frac{x^5x^2}{(x^2)^2}$ **50.** $\frac{x^{10}x^{-3}}{x^3x^{\ 1}}$

**51.** $\frac{x^8x^{-2}}{(x^2)^3}$ **52.** $\frac{(x^{-2})^3}{x \cdot x^{-3}}$ **53.** $\frac{(y^2)^4}{y^{-2}y^{-1}}$ **54.** $\left(\frac{y^2y^{-1}}{y^5y^2}\right)^{-2}$ **55.** $\left(\frac{x^2x^0}{x^4x^{-1}}\right)^{-3}$

**56.** $\left(\frac{x^{-3}x^0}{x^2x}\right)^3$ **57.** $\left(\frac{x^5x^{-2}}{x \cdot x^{-3}}\right)^2$ **58.** $x^k \cdot x$ **59.** $x^k \cdot x^3$ **60.** $x^k \cdot x^{2k}$

**61.** $x^{3k} \cdot x^4$ **62.** $\frac{x^k}{x^2}$ **63.** $\frac{x^{2k}}{x^k}$ **64.** $\frac{x^{k+1}}{x^3}$ **65.** $(x^k)^2$

**66.** $(x^5)^k$ **67.** $x(x^2)^k$ **68.** $\frac{x^2x^k}{(x^2)^k}$ **69.** $\frac{x^{k+1}x^{-2}}{x^4}$ **70.** $\frac{x^{k+3}x}{x^{-2}}$

## 2.2 More on Exponents and Scientific Notation

**OBJECTIVES**

In this section, you will be learning to:

1. Simplify any expressions using the properties of integer exponents.
2. Write numbers given in scientific notation as decimal numbers.
3. Write decimal numbers in scientific notation.

In Section 2.1, the base of each exponent was one constant or one variable. Two more properties of exponents will help in simplifying expressions such as $(4x)^3$, $(-3x^2y)^5$, and $\left(\frac{a^2b}{b^7}\right)^3$, where the base may be a product or a quotient or a combination of these. Using the associative and commutative properties of multiplication, we can write

$$\begin{aligned}(4x)^3 &= (4x)\cdot(4x)\cdot(4x)\\ &= 4\cdot4\cdot4\cdot x\cdot x\cdot x\\ &= 4^3\cdot x^3\\ &= 64x^3\end{aligned}$$

$$\begin{aligned}(-3x^2y)^5 &= (-3x^2y)\cdot(-3x^2y)\cdot(-3x^2y)\cdot(-3x^2y)\cdot(-3x^2y)\\ &= (-3)(-3)(-3)(-3)(-3)\cdot x^2\cdot x^2\cdot x^2\cdot x^2\cdot x^2\cdot y\cdot y\cdot y\cdot y\cdot y\\ &= (-3)^5\cdot(x^2)^5\cdot y^5\\ &= -243x^{10}y^5\end{aligned}$$

Note carefully that in the expression $(-3x^2y)^5$, the negative sign goes with the 3, and $-3$ is treated as a factor.

These examples lead to Property 6.

**Property 6 of Exponents** If $a$ and $b$ are nonzero numbers and $n$ is an integer, then

$$(ab)^n = a^nb^n$$

**EXAMPLES** Simplify the following expressions using Property 6.

**1.** $(2x)^5$

**Solution** $(2x)^5 = 2^5\cdot x^5 = 32x^5$

**2.** $(-5y)^2$

**Solution** $(-5y)^2 = (-5)^2y^2 = 25y^2$

**3.** $(-6x^5)^3$

**Solution** $(-6x^5)^3 = (-6)^3(x^5)^3 = -216x^{15}$

**4.** $(3x^{-2}y^3)^2$

**Solution** $(3x^{-2}y^3)^2 = 3^2(x^{-2})^2(y^3)^2 = 9x^{-4}y^6$ or $\frac{9y^6}{x^4}$

**5.** $(-a^2b^4c)^3$

**Solution**
$$\begin{aligned}(-a^2b^4c)^3 &= (-1\cdot a^2b^4c)^3\\ &= (-1)^3(a^2)^3(b^4)^3c^3\\ &= -a^6b^{12}c^3\end{aligned}$$

This example illustrates the important relationship concerning the coefficient $-1$ and exponents. In the expression $-a^2$, $-1$ is the coefficient and $a$ is the base of the exponent 2. That is, $-a^2 = -1 \cdot a^2$. The same is true with constant bases. Thus, $-5^2 = -1 \cdot 5^2 = -25$. [$-5^2 \neq (-5)^2$ and $-a^2 \neq (-a)^2$.] ■

Expressions with quotients are treated much the same as expressions with products. For example,

$$\left(\frac{y}{3}\right)^4 = \frac{y}{3} \cdot \frac{y}{3} \cdot \frac{y}{3} \cdot \frac{y}{3} = \frac{y \cdot y \cdot y \cdot y}{3 \cdot 3 \cdot 3 \cdot 3} = \frac{y^4}{3^4} = \frac{y^4}{81}$$

and
$$\left(\frac{a}{x}\right)^3 = \frac{a}{x} \cdot \frac{a}{x} \cdot \frac{a}{x} = \frac{a \cdot a \cdot a}{x \cdot x \cdot x} = \frac{a^3}{x^3}$$

We now have Property 7.

**Property 7 of Exponents** If $a$ and $b$ are nonzero numbers and $n$ is an integer, then

$$\left(\frac{a}{b}\right)^n = \frac{a^n}{b^n}$$

**EXAMPLES** Simplify the following expressions using any of the seven properties of exponents that apply.

**6.** $\left(\frac{5x}{3b}\right)^3$

**Solution** $\left(\frac{5x}{3b}\right)^3 = \frac{(5x)^3}{(3b)^3} = \frac{5^3x^3}{3^3b^3} = \frac{125x^3}{27b^3}$

**7.** $\left(\frac{a^3b^{-3}}{4}\right)^2$

**Solution** $\left(\frac{a^3b^{-3}}{4}\right)^2 = \frac{(a^3b^{-3})^2}{4^2} = \frac{(a^3)^2(b^{-3})^2}{4^2} = \frac{a^6b^{-6}}{16}$ or $\frac{a^6}{16b^6}$

**8.** $\frac{(3^{-2}x^{-3})^{-1}}{(x^{-2}y^3)^3(2x^{-1}y^2)^{-1}}$

**Solution** $\frac{(3^{-2}x^{-3})^{-1}}{(x^{-2}y^3)^3(2x^{-1}y^2)^{-1}} = \frac{3^2x^3}{x^{-6}y^9 \cdot 2^{-1}x^1y^{-2}} = \frac{2^1 \cdot 9x^3}{x^{-5}y^7}$

$= \frac{18x^{3-(-5)}}{y^7} = \frac{18x^8}{y^7}$ or $18x^8y^{-7}$

**9.** $\left(\dfrac{2x^2y^{-3}}{x^{-3}y}\right)^{-2}$

**Solution** $\left(\dfrac{2x^2y^{-3}}{x^{-3}y}\right)^{-2} = (2x^{2+3}y^{-3-1})^{-2} = (2x^5y^{-4})^{-2}$

$= 2^{-2}x^{-10}y^8 \quad \text{or} \quad \dfrac{1}{4}x^{-10}y^8 \quad \text{or} \quad \dfrac{y^8}{4x^{10}}$

**10.** $\left(\dfrac{x^{2k}y^k}{x^ky}\right)^2$

**Solution** $\left(\dfrac{x^{2k}y^k}{x^ky}\right)^2 = (x^{2k-k}y^{k-1})^2$

$= (x^ky^{k-1})^2 = x^{2k}y^{2(k-1)} = x^{2k}y^{2k-2}$

**11.** $\dfrac{1}{5x^{-2}}$

**Solution** $\dfrac{1}{5x^{-2}} = \dfrac{1}{5 \cdot \dfrac{1}{x^2}} = \dfrac{1}{5} \cdot \dfrac{x^2}{1} = \dfrac{x^2}{5}$

Notice that the exponent $-2$ is for $x$ only and has no effect on the coefficient 5. ■

As a practical note, **any factor of a single fraction may be moved from numerator to denominator or vice versa by changing the sign of the exponent of that factor.** Thus, as a first step we could write

$$\left(\frac{y^{-4}}{3x^{-1}}\right)^{-2} = \left(\frac{x}{3y^4}\right)^{-2}$$

Note that the 3 is not affected because the exponent $-1$ refers only to $x$.

Also, **if a single fraction is raised to a power, we may write the reciprocal of that fraction with the sign of the exponent changed.** Thus, we could write

$$\left(\frac{x}{3y^4}\right)^{-2} = \left[\left(\frac{x}{3y^4}\right)^{-1}\right]^2 = \left[\frac{3y^4}{x}\right]^2$$

In general,

$$\frac{1}{a^{-n}} = a^n \quad \text{and} \quad \frac{a^{-n}}{b^{-m}} = \frac{b^m}{a^n} \quad \text{and} \quad \left(\frac{a}{b}\right)^{-n} = \left(\frac{b}{a}\right)^n.$$

Remember that the choice of steps is yours and that as long as you follow the properties of exponents, the answer will be the same regardless of the order of the steps.

A basic application of integer exponents occurs in scientific disciplines when very large and very small numbers are involved. For example, the distance from the earth to the sun is approximately 93,000,000 miles, and 0.0000000077 centimeter is the approximate radius of a carbon atom.

In **scientific notation,** decimal numbers are written as the product of a number between 1 and 10 and an integer power of 10. Thus,

$$93{,}000{,}000 = 9.3 \times 10^7 \qquad \text{and} \qquad 0.0000000077 = 7.7 \times 10^{-9}$$

The exponent tells how many places the decimal point is to be moved and in what direction. If the exponent is positive, the decimal point is moved to the right:

$$2.7 \times 10^3 = 2\,7\,0\,0. \qquad \text{3 places right}$$

A negative exponent indicates that the decimal point should go to the left:

$$3.92 \times 10^{-6} = 0.\,0\,0\,0\,0\,0\,3\,9\,2 \qquad \text{6 places left}$$

**EXAMPLES** Write the following decimals in scientific notation.

**12.** $867{,}000{,}000{,}000 = 8.67 \times 10^{11}$

**13.** $420{,}000 = 4.2 \times 10^5$

**14.** $0.0036 = 3.6 \times 10^{-3}$

**15.** $0.000000025 = 2.5 \times 10^{-8}$

Simplify the following expressions by first writing the decimal numbers in scientific notation and then using the properties of exponents.

**16.** $\dfrac{0.0023 \times 560{,}000}{0.00014}$

**Solution**

$$\frac{0.0023 \times 560{,}000}{0.00014} = \frac{2.3 \times 10^{-3} \times 5.6 \times 10^5}{1.4 \times 10^{-4}}$$

$$= \frac{2.3 \times \overset{4}{\cancel{5.6}}}{\cancel{1.4}} \times \frac{10^{-3} \times 10^5}{10^{-4}}$$

$$= 9.2 \times \frac{10^2}{10^{-4}} = 9.2 \times 10^{2-(-4)}$$

$$= 9.2 \times 10^6$$

**17.** $\dfrac{8.1 \times 8200}{9{,}000{,}000 \times 4.1}$

**Solution**

$$\frac{8.1 \times 8200}{9{,}000{,}000 \times 4.1} = \frac{8.1 \times 8.2 \times 10^3}{9.0 \times 10^6 \times 4.1}$$

$$= \frac{\overset{0.9}{\cancel{8.1}} \times \overset{2}{\cancel{8.2}}}{\cancel{9.0} \times \cancel{4.1}} \times \frac{10^3}{10^6}$$

$$= 1.8 \times 10^{3-6} = 1.8 \times 10^{-3}$$

**18.** Light travels approximately $3 \times 10^8$ meters per second. How many meters per minute does light travel?

**Solution** Since there are 60 seconds in one minute, multiply by 60.

$$3 \times 10^8 \times 60 = 180 \times 10^8 = 1.8 \times 10^{10}$$

Thus, light travels $1.8 \times 10^{10}$ meters per minute. ■

## EXERCISES 2.2

Simplify the expressions in Exercises 1–32.

**1.** $(x^{-2})^2$ **2.** $(x^2)^{-3}$ **3.** $(-2x^4)^2$ **4.** $(3x^{-2})^{-1}$

**5.** $(x^2y^{-3})^2$ **6.** $(xy^3)^{-2}$ **7.** $\left(\frac{a}{b^2}\right)^4$ **8.** $\left(\frac{2a}{b^2}\right)^3$

**9.** $\left(\frac{2^{-1}a}{3b^2}\right)^{-2}$ **10.** $\left(\frac{-3x^{-2}}{y^3}\right)^{-1}$ **11.** $\left(\frac{6x}{x^2y^{-3}}\right)^{-2}$ **12.** $\left(\frac{2ab^4}{3b^2}\right)^{-3}$

**13.** $\left(\frac{5^{-1}x^3y^{-2}}{xy^{-1}}\right)^2$ **14.** $\left(\frac{x^2y^{-3}}{3x^{-1}y}\right)^{-1}$ **15.** $(x^2y)^k$ **16.** $(x^ky^m)^2$

**17.** $(x^{2k}y^2)(xy^n)$ **18.** $(x^{4n}y^3)(x^ny^{-k})$ **19.** $(x^{n+2}y^k)(x^{-2}y)$ **20.** $(x^{k+1}y^{3k})(x^2y^{-k})$

**21.** $\left(\frac{x^4y}{y^2}\right)\left(\frac{x^{-1}y^2}{y^{-1}}\right)^0$ **22.** $\left(\frac{a^2b}{ab^{-2}}\right)\left(\frac{a^{-3}b}{b^{-3}}\right)$ **23.** $\left(\frac{x^2y}{y^2}\right)^{-1}\left(\frac{3x^{-2}y}{y^{-2}}\right)^2$ **24.** $\left(\frac{x^2y^{-3}}{y^{-1}}\right)^2\left(\frac{xy^2}{2y}\right)^{-1}$

**25.** $\left(\frac{a^3b^{-1}}{ab^{-2}}\right)\left(\frac{a^0b^3}{a^4b^{-1}}\right)^{-1}$ **26.** $\left(\frac{5x^3y}{x^{-2}y^3}\right)^{-1}\left(\frac{4x^{-2}y^{-1}}{15xy^4}\right)^{-1}$ **27.** $\left(\frac{7x^{-2}y}{xy^4}\right)^2\left(\frac{14xy^{-3}}{x^4y^2}\right)^{-1}$ **28.** $\frac{(4^{-2}x^{-3}y)^{-1}}{(x^{-2}y^2)^3(5xy^{-2})^{-1}}$

**29.** $\frac{(xy^{-2})^4(x^{-2}y^3)^{-2}}{(xy^2)^{-1}(xy^{-2})^{-4}}$ **30.** $\frac{(6x^2y)(x^{-1}y^3)^2}{(x^{-1}y)^2(3x^2y)^3}$ **31.** $\frac{(x^3y^4)^{-1}(x^{-2}y)^2}{(xy^2)^{-3}(xy^{-1})^2}$ **32.** $\frac{(x^{-3}y^{-5})^{-2}(x^2y^{-3})^3}{(x^3y^{-4})^2(x^{-1}y^{-2})^{-2}}$

Write Exercises 33–38 in decimal notation.

**33.** $4.72 \times 10^5$ **34.** $6.91 \times 10^{-4}$ **35.** $1.28 \times 10^{-7}$

**36.** $1.63 \times 10^8$ **37.** $9.23 \times 10^{-3}$ **38.** $5.88 \times 10^6$

Write Exercises 39–54 in scientific notation.

**39.** 479,000 **40.** 0.000367 **41.** 0.000000871

**42.** $52{,}800 \times 1{,}000$ **43.** $143{,}000 \times 0.0003$ **44.** $0.007 \times 0.00012$

**45.** $0.036 \times 4{,}000{,}000$ **46.** $\frac{27{,}000}{0.0009}$ **47.** $\frac{1800 \times 0.00045}{1350}$

**48.** $\frac{0.0032 \times 120}{0.0096}$ **49.** $\frac{0.084 \times 0.0093}{0.21 \times 0.031}$ **50.** $\frac{0.0070 \times 50 \times 0.55}{1.4 \times 0.0011 \times 0.25}$

**51.** $\frac{0.36 \times 5200}{0.00052 \times 720}$ **52.** $\frac{0.0016 \times 0.09 \times 460}{0.00012 \times 0.023}$ **53.** $\frac{760 \times 84 \times 0.063}{900 \times 0.38 \times 210}$

**54.** $\frac{420 \times 0.016 \times 80}{0.028 \times 120 \times 0.2}$

**55.** Light travels approximately $3 \times 10^{10}$ centimeters per second. How many centimeters would this be per minute? per hour?

**56.** An atom of gold weighs approximately $3.25 \times 10^{-22}$ grams. What would be the weight of 3000 atoms of gold?

**57.** One light-year is approximately $9.46 \times 10^{15}$ meters. The distance to a certain star is about 4.3 light-years. How many meters is this?

**58.** One light-year is about $6 \times 10^{12}$ miles. The mean distance from the sun to Pluto is $3.675 \times 10^{9}$ miles. How many light-years is this?

**59.** The weight of an atom is measured in atomic weight units (amu), where 1 amu = $1.6605 \times 10^{-27}$ kilograms. The atomic weight of carbon 12 is 12 amu. Express the weight in kilograms.

**60.** The atomic weight of argon is about 40 amu. Express this weight in kilograms. (See Exercise 59.)

## 2.3 Operations with Polynomials

**OBJECTIVES**

In this section, you will be learning to:

1. Identify polynomial expressions.
2. Classify certain polynomials as monomials, binomials, or trinomials.
3. Add and subtract polynomials.
4. Multiply polynomials.
5. Evaluate polynomials for given values of the variables.

A **monomial** is a term that has no variable in its denominator and has only whole number exponents on its variables. Thus, a monomial does **not** have variables with negative exponents in the numerator or positive exponents in the denominator or fractional exponents.

The following terms **are** monomials: $5x$, $-7y^2$, $4$, $\frac{1}{8}x^2y^3$.

The following terms **are not** monomials: $\frac{2x^2}{y^2}$, $-6x^{-1}$, $3y^{1/2}$.

The general form of a **monomial in $x$** is

$$kx^n \qquad \textbf{where } n \textbf{ is a whole number and } k \textbf{ is any real number.}$$

$n$ is called the **degree** of the monomial and $k$ is the **coefficient.**

If a term has more than one variable, then its degree is the sum of the degrees of its variables. For example, $7x^2y$ is 3rd degree $(2 + 1 = 3)$ and $8a^2b^3$ is 5th degree $(2 + 3 = 5)$.

In the case of a constant monomial, such as 5, we can write $5 = 5 \cdot 1 = 5 \cdot x^0$. So we say that a constant is a monomial of **0 degree.** In the special case of 0, we can write $0 = 0x = 0x^3 = 0x^{16}$. Because of all the possible ways of writing 0, we say that 0 is a monomial of **no degree.**

Any monomial or algebraic sum of monomials is a **polynomial.** All of the following expressions are polynomials:

$$18, \qquad 2 + 3x^2, \qquad 5x^2y + 4xy^2, \qquad \text{and} \qquad 3x - 4w^2 + 5xy$$

The **degree of a polynomial** is the largest of the degrees of its terms after like terms have been combined. Generally, the terms are written in order of degree from left to right. For example,

$$x^3 - 2x + 4x^2 - x^3 + 5x + 1 = 4x^2 + 3x + 1$$

is **second degree** in $x$.

Some polynomial forms are used so frequently in algebra that they have been given special names, as follows:

**Classification of Polynomials**

Monomial: polynomial with one term
Binomial: polynomial with two terms
Trinomial: polynomial with three terms

## Addition

The **sum** of two or more polynomials can be found by combining like terms.

**EXAMPLES** Simplify each of the following expressions.

**1.** $(x^3 - 2x + 1) + (3x^2 + 4x + 5)$

**Solution** $(x^3 - 2x + 1) + (3x^2 + 4x + 5)$

$$= x^3 - 2x + 1 + 3x^2 + 4x + 5$$
$$= x^3 + 3x^2 - 2x + 4x + 1 + 5$$
$$= x^3 + 3x^2 + 2x + 6$$

**2.** $(2x^2 + 4x - 7) + (x^2 + 6x + 8)$

**Solution** $(2x^2 + 4x - 7) + (x^2 + 6x + 8)$

$$= 2x^2 + 4x - 7 + x^2 + 6x + 8$$
$$= 2x^2 + x^2 + 4x + 6x - 7 + 8$$
$$= 3x^2 + 10x + 1$$ ■

## Subtraction

To find the **difference** of two polynomials, we can proceed in either of the following ways:

**a.** add the opposite of each term being subtracted, **or**
**b.** multiply each term being subtracted by $-1$, then add.

**EXAMPLE**

**3.** Find the difference in simplest form: $(x^2y + 3y - 4x) - (2x^2y - 7x)$

**Solution**

**a.** Add the opposites of the terms being subtracted.

$(x^2y + 3y - 4x) - (2x^2y - 7x)$

$$= x^2y + 3y - 4x - 2x^2y + 7x$$
$$= x^2y - 2x^2y + 3y - 4x + 7x$$
$$= -x^2y + 3y + 3x$$

**b.** Multiply each term being subtracted by $-1$.

$(x^2y + 3y - 4x) - (2x^2y - 7x)$

$$= (x^2y + 3y - 4x) + (-1)(2x^2y - 7x)$$
$$= x^2y + 3y - 4x + (-2x^2y + 7x)$$
$$= x^2y + 3y - 4x - 2x^2y + 7x$$
$$= -x^2y + 3y + 3x$$ ■

## Multiplication

The **product** of two or more polynomials can be found by using the distributive property, $a(b + c) = ab + ac$. We show first the product of a monomial with a polynomial of two or more terms.

$$\begin{aligned} 5x(2x^3 + 3) &= 5x \cdot 2x^3 + 5x \cdot 3 \\ &= 10x^4 + 15x \end{aligned}$$

and

$$\begin{aligned} 3x^2(x^2 - 5x + 1) &= 3x^2 \cdot x^2 + 3x^2(-5x) + 3x^2 \cdot 1 \\ &= 3x^4 - 15x^3 + 3x^2 \end{aligned}$$

Now, for a product such as $(x + 3)(x + 8)$, we consider the distributive property in the form $(b + c)a = ba + ca$ and treat the binomial $(x + 8)$ as $a$ in the formula.

$$(b + c)a = ba + ca$$

$$\begin{aligned} (x + 3)(x + 8) &= x(x + 8) + 3(x + 8) \\ &= x \cdot x + x \cdot 8 + 3 \cdot x + 3 \cdot 8 \\ &= x^2 + 8x + 3x + 24 \\ &= x^2 + 11x + 24 \end{aligned}$$

Similarly,

$$\begin{aligned} &(2x - 1)(x^2 + x - 4) \\ &\quad = 2x(x^2 + x - 4) - 1(x^2 + x - 4) \\ &\quad = 2x \cdot x^2 + 2x \cdot x + 2x(-4) - 1 \cdot x^2 - 1 \cdot x - 1(-4) \\ &\quad = 2x^3 + 2x^2 - 8x - x^2 - x + 4 \\ &\quad = 2x^3 + x^2 - 9x + 4 \end{aligned}$$

**EXAMPLES**

**4.** $x^2y(x^2 + 3y^2) = x^2y \cdot x^2 + x^2y \cdot 3y^2$ Use the distributive property.
$= x^4y + 3x^2y^3$

**5.** $3x(2x + 1)(x - 5) = 3x[(2x + 1)(x - 5)]$ Multiply the two binomials first.
$= 3x[2x(x - 5) + 1(x - 5)]$ Use the distributive property.
$= 3x[2x^2 - 10x + x - 5]$
$= 3x[2x^2 - 9x - 5]$ Simplify.
$= 3x \cdot 2x^2 + 3x(-9x) + 3x(-5)$ Use the distributive property.
$= 6x^3 - 27x^2 - 15x$ ■

In the case of the product of two binomials such as $(2x + 5)(3x - 7)$, the FOIL method is useful. F-O-I-L is a mnemonic (memory) aid to help you remember which terms of the binomials to multiply together.

**F:** First **O:** Outside **I:** Inside **L:** Last

$$(2x + 5)(3x - 7) = \overbrace{2x \cdot 3x}^{\text{F}} + \overbrace{2x(-7)}^{\text{O}} + \overbrace{5 \cdot 3x}^{\text{I}} + \overbrace{5(-7)}^{\text{L}}$$
$$= 6x^2 - 14x + 15x - 35$$
$$= 6x^2 - x - 35$$

Some products occur so frequently that they deserve special mention and are given in the form of formulas with names. The first three are usually discussed in beginning algebra. All can be developed using the distributive property and collecting like terms.

**Special Products of Polynomials**

| | | |
|---|---|---|
| **I.** | $(x + a)(x - a) = x^2 - a^2$ | Difference of two squares |
| **II.** | $(x + a)^2 = x^2 + 2ax + a^2$ | Perfect square trinomial |
| **III.** | $(x - a)^2 = x^2 - 2ax + a^2$ | Perfect square trinomial |
| **IV.** | $(x - a)(x^2 + ax + a^2) = x^3 - a^3$ | Difference of two cubes |
| **V.** | $(x + a)(x^2 - ax + a^2) = x^3 + a^3$ | Sum of two cubes |

These forms should be memorized. Along with the FOIL method, they are part of the foundation for our work with factoring in the next section and with algebraic fractions in the next chapter.

**EXAMPLES**

**6.** $(x + 5)(x - 7)$

**Solution** Using the FOIL method,

(F: $x^2$, O: $-7x$, I: $+5x$, L: $-35$)

$$(x + 5)(x - 7) = x^2 - 7x + 5x - 35$$
$$= x^2 - 2x - 35$$

**7.** $(3y + 2)(3y - 2)$

**Solution** Using form I, the result is the difference of two squares.

Form I $\quad (x + a)(x - a) = x^2 - a^2$

$$(3y + 2)(3y - 2) = (3y)^2 - (2)^2$$
$$= 9y^2 - 4 \qquad \text{Difference of two squares}$$

**8.** $(5x + 7)^2$

**Solution** Using form II, the result is a perfect square trinomial.

Form II

$$\begin{aligned}(x + a)^2 &= x^2 + 2ax + a^2 \\ (5x + 7)^2 &= (5x)^2 + 2 \cdot 7 \cdot 5x + 7^2 \\ &= 25x^2 + 70x + 49 \end{aligned}$$

Perfect square trinomial

**9.** $(x - 3)(x^2 + 3x + 9)$

**Solution** Using form IV, the result is the difference of two cubes.

Form IV

$$\begin{aligned}(x - a)(x^2 + ax + a^2) &= x^3 - a^3 \\ (x - 3)(x^2 + 3x + 9) &= (x)^3 - (3)^3 \\ &= x^3 - 27 \end{aligned}$$

Difference of two cubes

To verify this product, multiply and combine like terms.

$$\begin{aligned}(x - 3)(x^2 + 3x + 9) &= x(x^2 + 3x + 9) - 3(x^2 + 3x + 9) \\ &= x^3 + 3x^2 + 9x - 3x^2 - 9x - 27 \\ &= x^3 - 27 \end{aligned}$$

Note that the expression $x^2 + 3x + 9$ is **not** a perfect square trinomial. ■

The squares of binomials represented in forms II and III sometimes lead to a common error.

**Common Error**

For **products** $(ab)^n = a^n b^n$ and $(ab)^2 = a^2 b^2$. But this rule does **not** apply to binomials.

$$\begin{aligned}(a + b)^2 &\neq a^2 + b^2 \\ (a - b)^2 &\neq a^2 - b^2\end{aligned}$$

Remember,

$$(a + b)^2 = (a + b)(a + b) = a^2 + 2ab + b^2$$

and

$$(a - b)^2 = (a - b)(a - b) = a^2 - 2ab + b^2$$

The following examples illustrate a few more abstract applications of the general forms.

**EXAMPLES**

**10.** $[(x + 3) - y]^2$

**Solution** First treat $(x + 3)$ as a single term:

$$\begin{aligned}[(x + 3) - y]^2 &= (x + 3)^2 - 2(x + 3)y + y^2 \\ &= x^2 + 6x + 9 - 2xy - 6y + y^2\end{aligned}$$

**11.** $(x^k + 3)(x^k - 3)$ Assume that $k$ represents a whole number.

**Solution** Using form I and the fact that $(x^k)^2 = x^{2k}$,

$$(x^k + 3)(x^k - 3) = (x^k)^2 - 3^2 = x^{2k} - 9$$ Difference of two squares

**12.** $(3y^{2k} - 5)(y^k - 4)$

**Solution** Using FOIL,

$$\begin{aligned}(3y^{2k} - 5)(y^k - 4) &= 3y^{2k} \cdot y^k - 4 \cdot 3y^{2k} - 5y^k - 5(-4)\\ &= 3y^{3k} - 12y^{2k} - 5y^k + 20\end{aligned}$$ ■

***Practice Problems***

Perform the following operations and simplify.

1. $(3x^2 - 2x + 5) + (2x^2 - x + 3)$
2. $[(x^3 - 2x^2) - (x^2 - 1)] - (1 - 3x + x^2)$
3. $(x + 5)(x^2 - 5x + 25)$
4. $x^k(x^{2k} - 1)$
5. $(7x - 1)(2x + 3)$
6. $(x - 8)^2$

## EXERCISES 2.3

State which of the expressions in Exercises 1–12 are polynomials; then classify each polynomial as a monomial, binomial, or trinomial.

**1.** $x^3 - x^2$ **2.** $9$ **3.** $x^2 - 3\sqrt{x}$ **4.** $x^2 + 8x^2y - y^2$

**5.** $\dfrac{5x^2 - 7y + 2y^2}{4}$ **6.** $x^2 + xy - \dfrac{1}{y}$ **7.** $0$ **8.** $-\sqrt{2}$

**9.** $\sqrt{x^2y^3 - xy^2}$ **10.** $\dfrac{3}{2}x^2 - \sqrt{3}x + 7$ **11.** $7x^2 - 6x + 9\sqrt{x}$ **12.** $\dfrac{x^2y - 3y^2}{x}$

Perform the indicated operations and simplify in Exercises 13–35.

**13.** $(3x^2 - 5x + 1) + (x^2 + 2x - 7)$
**14.** $(5x^2 + 8x - 3) + (-2x^2 + 6x - 4)$
**15.** $(x^2 - 9x + 2) + (-x^2 + 2x - 8)$
**16.** $(7x^2 - 4x + 6) + (4x^2 - 2x + 5)$
**17.** $(x^2 + y^2) + (2x^2 - 5y^2)$
**18.** $(x^2 - 3xy + y^2) + (2x^2 - 5xy - y^2)$
**19.** $(2x^2 + 3x + 8) - (x^2 + 4x - 2)$
**20.** $(6x^3 - 5x + 1) - (2x^3 + 3x - 4)$
**21.** $(4x^3 - 7x^2 + 3x + 2) - (-2x^3 - 5x - 1)$
**22.** $(2x^4 + 3x) - (5x^3 + 4x + 3)$
**23.** $(2x^3 - 3x^2 + 6) - (x^4 + x + 1)$
**24.** $(5x^2 + 6x - 1) + (x^4 - 3x^2 + 2x)$
**25.** $(7x^2 - 2xy + 3y^2) + (-3x^2 - 2xy + 5y^2)$
**26.** $(4x^2 - 8xy + 2y^2) + (-9x^2 + 5xy - 6y^2)$
**27.** $(3x^2 - 2y^2) + (7xy + 4y^2) - (-6x^2 - 6xy + 8y^2)$
**28.** $(9xy + 8y^2) - (6x^2 - 8xy) + (5x^2 - 3xy + 7y^2)$

**Answers to Practice Problems** 1. $5x^2 - 3x + 8$ 2. $x^3 - 4x^2 + 3x$ 3. $x^3 + 125$ 4. $x^{3k} - x^k$ 5. $14x^2 + 19x - 3$ 6. $x^2 - 16x + 64$

**29.** $(5x^3 - 14x^2) - (5x^2 + 2x + 1) - (-7x^3 + 2x^2 - 13)$
**30.** $(7x^3 + 4x^2 - x) + (3x^3 - 4x + 5) - (8x^3 + x^2 - x + 3)$
**31.** $x^3 - [3x^2 - 1 - (x^3 + 4x^2 + 1)] + (3x^3 - 3x^2 - 2)$
**32.** $3x - 4xy + [6y + (4x + 3xy + 2y)] - [-6x - (xy - 4y)]$
**33.** $x^2 - 2xy + [y^2 - (3xy + 2y^2) - (3x^2 - xy - 2y^2)]$
**34.** $[(4x^2 - 3x) - (2x^2 + 5x)] + [(x^2 - 6x) + (-3x^2 + x)]$
**35.** $[(2x + xy - y) + (x - 2xy + 4y)] - [(-3x + 5xy + y) - (2x + 3xy - 2y)]$

Find the indicated products and simplify in Exercises 36–90.

**36.** $5x(x^2 - 2x + 3)$
**37.** $2x^2(3x^2 + 5x - 1)$
**38.** $xy^2(x^2 + 4y)$
**39.** $x^2z(x + 4y - z)$
**40.** $(x + 3)(x - 6)$
**41.** $(x - 2)(x - 5)$
**42.** $(x - 8)(x - 1)$
**43.** $(x + 2)(x + 4)$
**44.** $(2y + 1)(y - 6)$
**45.** $(y + 5)(3y + 2)$
**46.** $(3x - 4)(x - 5)$
**47.** $(2x - 1)(x - 2)$
**48.** $(2y + 3)(3y + 2)$
**49.** $(5y - 2)(3y + 1)$
**50.** $(8x + 3)(x - 5)$
**51.** $(7x + 6)(2x - 3)$
**52.** $(9x + 1)(3x - 2)$
**53.** $(5x - 11)(3x + 4)$
**54.** $(3x + 1)^2$
**55.** $(4x - 3)^2$
**56.** $(5x - 2y)^2$
**57.** $(7x + 4y)^2$
**58.** $(4x + 7)(4x - 7)$
**59.** $(3x + 5)(3x - 5)$
**60.** $(2x - 3y)(2x + 3y)$
**61.** $(6x - y)(6x + y)$
**62.** $x(3x^2 - 4)(3x^2 + 4)$
**63.** $3x(7x^2 + 8)(7x^2 - 8)$
**64.** $(x - 1)(x^2 + x + 1)$
**65.** $(y + 4)(y^2 - 4y + 16)$
**66.** $(x + 3)(x^2 + 6x + 9)$
**67.** $(y - 5)(y^2 + 3y + 2)$
**68.** $(x - 3y)(x^2 + 3xy + 9y^2)$
**69.** $(x + 2y)(x^2 - 2xy + 4y^2)$
**70.** $(2x^3 - 3)^2$
**71.** $(x^3 + 2)^2$
**72.** $(2x^3 - 7)(2x^3 + 7)$
**73.** $(8y^2 - 7)(3y^2 + 2)$
**74.** $4x^2y(x^2 + 6y^2)(x^2 - 6y^2)$
**75.** $3xy(x^2 - 6y^2)(x^2 + 3y^2)$
**76.** $x^2y(5x^2 + y^2)(2x^2 - 3y^2)$
**77.** $x^3(x - 2y)(x^2 + 2xy + 4y^2)$
**78.** $[(x + y) + 2][(x + y) - 2]$
**79.** $[(x + 1) + y][(x + 1) - y]$
**80.** $[(5x - y) + 3]^2$
**81.** $[(2x + 1) - y]^2$
**82.** $[(x + 4) - 2y]^2$
**83.** $[(x - 3y) + 5]^2$
**84.** $x^2(x^k + 3)$
**85.** $x^3(x^{2k} + x)$
**86.** $(x^k + 3)(x^k - 5)$
**87.** $(x^k + 6)(x^k - 6)$
**88.** $(x^k + 1)(x^k + 4)$
**89.** $(2x^k - 3)(x^k + 2)$
**90.** $(3x^k + 2)(x^k + 5)$

Evaluate each polynomial in Exercises 91–100 for the specified values of the variable.

**91.** $2x^2 - x + 3; x = 1$
**92.** $3x^2 - 2x + 5; x = 2$
**93.** $3 - x^2; x = -2$
**94.** $x^3 - 2x^2 + x - 1; x = 2$
**95.** $x^3 + x^2 - 4; x = -3$
**96.** $4x^3 - 2x^2 - 1; x = 4$
**97.** $2x^2 - 3xy + y^2; x = 2, y = -2$
**98.** $4x - 2xy + 5y; x = 1, y = 1$
**99.** $3x + 4xy - 2yz + z; x = 1, y = 0, z = 2$
**100.** $2xyz - 3x + yz - xz; x = 2, y = -1, z = 2$

# 2.4 Factoring

**OBJECTIVES**

In this section, you will be learning to:

1. Factor out the greatest common factors of polynomials.
2. Factor trinomials (including perfect squares).
3. Factor the differences of squares.

In Section 2.3, we discussed various types of products of polynomials. For example, using the distributive property,

$$2x(x^2 - 8x + 3) = 2x \cdot x^2 + 2x(-8x) + 2x \cdot 3 = 2x^3 - 16x^2 + 6x$$

The expressions $2x$ and $x^2 - 8x + 3$ are called **factors** of the product.

$$\underset{\text{factor}}{2x}\underbrace{(x^2 - 8x + 3)}_{\text{factor}} = \underbrace{2x^3 - 16x^2 + 6x}_{\text{product}}$$

In this section, we want to start with the product and find the factors. We will rely heavily on the multiplication skills developed in Section 2.3 since factoring is the reverse of multiplication.

Furthermore, we will find that the skill of factoring polynomials is very useful when simplifying expressions (Chapter 3) and when solving equations. In other words, Sections 2.4 and 2.5 are particularly important, so study them with extra care.

## Greatest Common Factor

First we look for the **greatest common factor.** We want to find the expression of highest degree and largest integer coefficient that is a factor of each term. Then, this factor is to be "factored out" or divided into each term, resulting in another factor. For example,

$$\begin{aligned} 6x^4 + 3x^3 - 21x^2 &= 3x \cdot 2x^3 + 3x \cdot x^2 + 3x(-7x) \\ &= 3x(2x^3 + x^2 - 7x) \end{aligned}$$

But the expression $2x^3 + x^2 - 7x$ has $x$ as a common factor. Therefore, we have not factored completely. **An expression is factored completely if none of its factors can be factored.** Thus, $3x$ is not the greatest common factor. The greatest common factor is $3x^2$.

$$\begin{aligned} 6x^4 + 3x^3 - 21x^2 &= 3x^2 \cdot 2x^2 + 3x^2 \cdot x + 3x^2(-7) \\ &= 3x^2(2x^2 + x - 7) \end{aligned}$$

A common factor may be a nonmonomial, a binomial, or any polynomial. In the following example, the binomial $x^2 + 1$ is factored out.

$$x^2(x^2 + 1) + 5x(x^2 + 1) + 2(x^2 + 1) = (x^2 + 1)(x^2 + 5x + 2)$$

## Basic Forms

Finding the greatest common factor is just the first step in factoring. Certain products should be recognized quickly, while others may involve the FOIL method and some experimentation.

We begin with the first three forms from Section 2.3:

**I.** Difference of two squares $x^2 - a^2 = (x + a)(x - a)$

**II.** Perfect square trinomial $x^2 + 2ax + a^2 = (x + a)^2$

**III.** Perfect square trinomial $x^2 - 2ax + a^2 = (x - a)^2$

**EXAMPLES** Factor completely.

**1.** $x^2 - 25$

**Solution** Form: $x^2 - a^2 = (x + a)(x - a)$

$$x^2 - 25 = x^2 - 5^2 = (x + 5)(x - 5)$$

**2.** $4x^2 - 121$

**Solution** Form: $x^2 - a^2 = (x + a)(x - a)$

$$4x^2 - 121 = (2x)^2 - 11^2 = (2x + 11)(2x - 11)$$

**3.** $4y^2 + 12y + 9$

**Solution** Form: $x^2 + 2ax + a^2 = (x + a)^2$

$$4y^2 + 12y + 9 = (2y)^2 + 2(2y)(3) + 3^2 = (2y + 3)^2$$

**4.** $(z + 2)^2 - 12(z + 2) + 36$

**Solution** Form: $x^2 - 2ax + a^2 = (x - a)^2$

$$(z + 2)^2 - 12(z + 2) + 36 = [(z + 2) - 6]^2$$

**5.** $2x^2 - 128$

**Solution** **The greatest common factor should always be factored out first.**

$$\begin{aligned} 2x^2 - 128 &= 2(x^2 - 64) && \text{Difference of two squares} \\ &= 2(x + 8)(x - 8) \end{aligned}$$

**6.** $x(x - 3) - (3x + 1)(x - 3)$

**Solution** Factor out the greatest common factor, in this case, the binomial $(x - 3)$.

$$\begin{aligned} x(x - 3) - (3x + 1)(x - 3) & \\ &= (x - 3)[x - (3x + 1)] \\ &= (x - 3)[x - 3x - 1] \\ &= (x - 3)(-2x - 1) \quad \text{or} \quad -1(x - 3)(2x + 1) \end{aligned}$$

**7.** $2a^3 - 8a^2b + 8ab^2$

**Solution** Factor out the greatest common factor.

$$\begin{aligned} 2a^3 - 8a^2b + 8ab^2 &= 2a(a^2 - 4ab + 4b^2) && \text{Perfect square trinomial} \\ &= 2a(a - 2b)^2 \end{aligned}$$

■

There are two general methods for factoring trinomials in the form

$ax^2 + bx + c$ where $a$, $b$, and $c$ are integers.

The first method is called the ***ac*-method** (or **grouping**) and the second method is called the **FOIL method** (or **trial-and-error**).

## ac-Method (Grouping)

The *ac*-method is very systematic. It also involves a technique called factoring by grouping that we will develop here on a limited basis and discuss in more detail in Section 2.5.

Consider the problem of factoring $2x^2 + 9x + 10$ where $a = 2$, $b = 9$, and $c = 10$.

### Analysis of Factoring by the ac-Method

| | $2x^2 + 9x + 10$ | $ax^2 + bx + c$ |
|---|---|---|
| **Step 1** | Multiply $2 \cdot 10 = 20$. | Multiply $a \cdot c$. |
| **Step 2** | Find two integers whose product is 20 and whose sum is 9. (In this case, $4 \cdot 5 = 20$ and $4 + 5 = 9$.) | Find two integers whose product is $ac$ and whose sum is $b$. If this is not possible, then the trinomial is **not factorable.** |
| **Step 3** | Rewrite the middle term ($9x$) using 4 and 5 as coefficients.<br>$2x^2 + 9x + 10$<br>$= 2x^2 + 4x + 5x + 10$ | Rewrite the middle term ($bx$) using the two numbers found in Step 2 as coefficients. |
| **Step 4** | Factor by grouping the first two terms and the last two terms.<br>$2x^2 + 4x + 5x + 10$<br>$= 2x(x + 2) + 5(x + 2)$ | Factor by grouping the first two terms and the last two terms. |
| **Step 5** | Factor the common binomial factor $(x + 2)$. Thus,<br>$2x^2 + 9x + 10$<br>$= 2x^2 + 4x + 5x + 10$<br>$= 2x(x + 2) + 5(x + 2)$<br>$= (x + 2)(2x + 5)$ | Factor the common binomial factor to find two binomial factors of the trinomial $ax^2 + bx + c$. |

**EXAMPLES**

**8.** Factor $x^2 - 2x - 15$ using the *ac*-method.

**Solution** $a = 1$, $b = -2$, and $c = -15$

**Step 1** Find the product $a \cdot c$: $1(-15) = -15$

**Step 2** Find two integers whose product is $-15$ and whose sum is $-2$:

$$(-5)(+3) = -15 \qquad \text{and} \qquad -5 + 3 = -2$$

**Step 3** Rewrite $-2x$ as $-5x + 3x$ giving
$x^2 - 2x - 15 = x^2 - 5x + 3x - 15$

**Step 4** Factor by grouping:

$$x^2 - 2x - 15 = x^2 - 5x + 3x - 15 = x(x - 5) + 3(x - 5)$$

**Step 5** Factor the common binomial factor $(x - 5)$:

$$x^2 - 2x - 15 = x(x - 5) + 3(x - 5) = (x - 5)(x + 3)$$

**9.** Factor $18x^3 - 39x^2 + 18x$ using the *ac*-method.

**Solution** First factor the greatest common factor $3x$.

$$18x^3 - 39x^2 + 18x = 3x(6x^2 - 13x + 6)$$

Now factor the trinomial $6x^2 - 13x + 6$ with $a = 6$, $b = -13$, and $c = 6$.

**Step 1** Find the product $a \cdot c$: $6(6) = 36$

**Step 2** Find two integers whose product is 36 and whose sum is $-13$. (**Note:** This may take some time and experimentation. We do know that both numbers must be negative since the product is positive and the sum is negative.)

$$(-9)(-4) = +36 \qquad \text{and} \qquad -9 + (-4) = -13$$

**Step 3** Rewrite $-13x$ as $-9x - 4x$ giving
$6x^2 - 13x + 6 = 6x^2 - 9x - 4x + 6$

**Step 4** Factor by grouping:

$$6x^2 - 13x + 6 = 6x^2 - 9x - 4x + 6 = 3x(2x - 3) - 2(2x - 3)$$

[**Note:** $-2$ is factored from the last two terms so that there will be a common binomial factor $(2x - 3)$.]

**Step 5** Factor the common binomial factor $(2x - 3)$:

$$\begin{aligned} 6x^2 - 13x + 6 &= 6x^2 - 9x - 4x + 6 \\ &= 3x(2x - 3) - 2(2x - 3) \\ &= (2x - 3)(3x - 2) \end{aligned}$$

Thus, for the original expression,

$$\begin{aligned} 18x^3 - 39x^2 + 18x &= 3x(6x^2 - 13x + 6) \\ &= 3x(2x - 3)(3x - 2) \end{aligned}$$

■

### *FOIL Method (Trial-and-Error)*

Now we will consider factoring a trinomial such as

$$x^2 + 17x + 30$$

by using the FOIL method of multiplication in reverse.

$$x^2 + 17x + 30 = (x \quad )(x \quad )$$

**O**, **I**, **L** $= 30$, **F** $= x^2$

Since the coefficient of $x^2$ is 1, we write F as $x \cdot x$. For O and I, we need the factors of 30 that add up to 17.

Since $15 \cdot 2 = 30$ and $15 + 2 = 17$, we have

$$x^2 + 17x + 30 = (x + 2)(x + 15)$$

**O** $= 15x$, **I** $= 2x$, **L** $= 30$, **F** $= x^2$

Similarly,

$$x^2 + 11x + 30 = (x + 6)(x + 5)$$

$5x$, $6x$, $30$, $x^2$

Here, we used 6 and 5 because $6 \cdot 5 = 30$ and $6 + 5 = 11$.

To factor a trinomial with leading coefficient not 1, we use the FOIL method also, but with more of a trial-and-error approach. For example, to factor $4x^2 - x - 5$, the product of the first two terms is to be $4x^2$, so we might have

$$(2x \quad )(2x \quad ) \quad \text{or} \quad (4x \quad )(x \quad )$$

**F** $= 4x^2$ **F** $= 4x^2$

The product of the last terms, $L$, is to be $-5$. The factors could be $-5(+1)$ or $+5(-1)$. We simply try each of the possible combinations until we find the correct product. If we exhaust all possibilities, then the trinomial is **not factorable using integer coefficients.**

1. $(2x + 1)(2x - 5) = 4x^2 - 10x + 2x - 5 \neq 4x^2 - x - 5$
2. $(2x - 1)(2x + 5) = 4x^2 + 10x - 2x - 5 \neq 4x^2 - x - 5$
3. $(4x + 1)(x - 5) = 4x^2 - 20x + x - 5 \neq 4x^2 - x - 5$
4. $(4x - 1)(x + 5) = 4x^2 + 20x - x - 5 \neq 4x^2 - x - 5$
5. $(4x + 5)(x - 1) = 4x^2 - 4x + 5x - 5 \neq 4x^2 - x - 5$
6. $(4x - 5)(x + 1) = 4x^2 + 4x - 5x - 5 = 4x^2 - x - 5$

We have found the factors on the last try. With practice, most of the steps can be done mentally and the final form can be found more quickly.

## EXAMPLES

**10.** Factor $x^2 - 10x + 16$.

**Solution** The coefficient of $x^2$ is 1 and $(-8)(-2) = 16$ and $(-8) + (-2) = -10$. We have

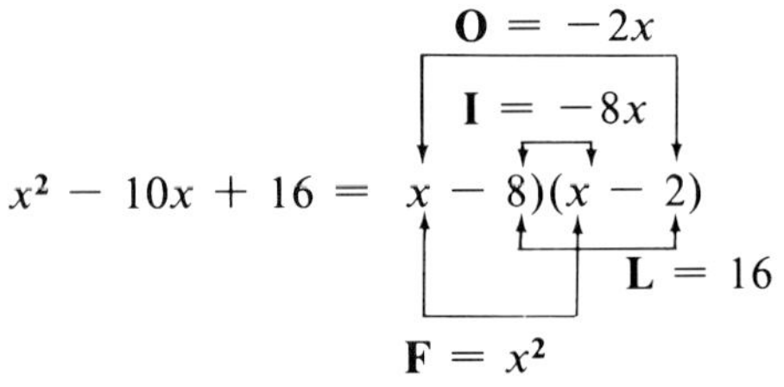

**O** $= -2x$

**I** $= -8x$

$$x^2 - 10x + 16 = (x - 8)(x - 2)$$

**L** $= 16$

**F** $= x^2$

**11.** Factor $3x^2 + 3x - 36$.

**Solution** $3x^2 + 3x - 36 = 3(x^2 + x - 12)$ — The greatest common factor is 3.

$= 3(x - 3)(x + 4)$ — $(-3)(4) = -12$ and $-3 + 4 = 1$

**12.** Factor $5x^2 + 23x - 10$.

**Solution**

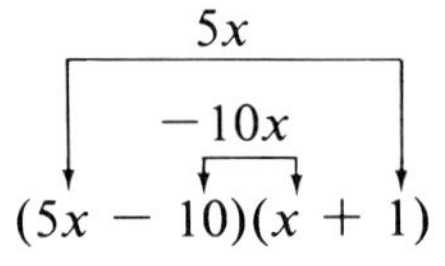

$5x$

$-10x$

$(5x - 10)(x + 1)$ — $5x - 10x = -5x \neq 23x$

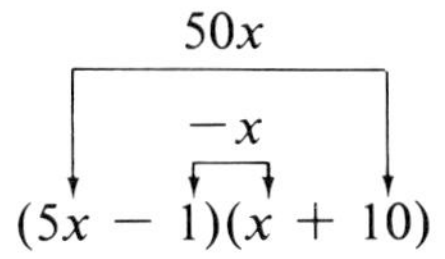

$50x$

$-x$

$(5x - 1)(x + 10)$ — $50x - x = 49x \neq 23x$

$$\overset{-25x}{\overset{2x}{(5x + 2)(x - 5)}}\qquad -25x + 2x = -23x \neq 23x$$

$$\overset{25x}{\overset{-2x}{(5x - 2)(x + 5)}}\qquad 25x - 2x = 23x$$

So, $5x^2 + 23x - 10 = (5x - 2)(x + 5)$. ■

**Practice Problems**

Factor completely.

1. $x^2 - 12x + 36$
2. $x^2 + 2x - 35$
3. $15 - 2x - x^2$
4. $50y^6 - 2y^4$
5. $8x^2 + 13x - 6$
6. $2x^2(x + 1) + 9x(x + 1) - 18(x + 1)$

## EXERCISES 2.4

Factor Exercises 1–60 completely.

1. $5x^2 + 15$
2. $7x^3 - 14x^2$
3. $x^2y - 2xy + xy^2$
4. $8x^2y - 4xy^2$
5. $5x^2y^2 + 20x^2y$
6. $6x^2y + 3xy + 9xy^2$
7. $3x^2y + 21x^3y^2 + 3x^2y^3$
8. $10x^3y^2 - 5x^2y - 15x^2$
9. $x^2 - 121$
10. $y^2 - 81$
11. $9x^2 - 25$
12. $4x^2 - 49$
13. $x^2 - 10x + 25$
14. $y^2 + 12y + 36$
15. $9y^2 + 12y + 4$
16. $49x^2 - 14x + 1$
17. $x^2 + 9x + 18$
18. $y^2 - 7y - 30$
19. $x^2 - 6x - 27$
20. $x^2 - 144$
21. $y^2 - 5y - 14$
22. $x^2 - 27x + 50$
23. $x^2 + 14x + 49$
24. $2x^2 + 15x - 8$
25. $6x^2 + 13x + 6$
26. $8x^2 + 10x - 25$
27. $2x^2 + 9x - 35$
28. $3x^2 + 2x - 8$
29. $16x^2 - 40x + 25$
30. $6x^2 - 35x - 5$
31. $35x^2 + 9x - 18$
32. $18x^2 - 7xy - y^2$
33. $4x^3 - 64x$
34. $2x^3y + 32x^2y + 128xy$
35. $21x^4 - 4x^3 - 32x^2$
36. $2x^4y^3 - 5x^3y^3 - 18x^2y^3$
37. $6x^5y^2 - 28x^4y^3 - 6x^3y^4$
38. $16x^4 + 8x^2y + y^2$
39. $12x^5 + 38x^3 + 20x$
40. $14x^5y^2 - 11x^3y^2$
41. $9x^5 - 16x$
42. $x^4 + 10x^2y + 25y^2$
43. $2x^4 + x^2y - 15y^2$
44. $4x^6 - 49$
45. $2x^4 - 7x^2y + 4y^2$
46. $2x^4 + 11x^2y - 21y^2$
47. $2x^6 + 9x^3y^2 + 4y^4$
48. $5x^4 + 17x^2y^2 + 6y^4$
49. $x^2(x - 5) + 4x(x - 5) - 21(x - 5)$
50. $x^2(x + 4) - 6x(x + 4) - 15(x + 4)$
51. $2x^2(2x + 1) - 9x(2x + 1) - 18(2x + 1)$
52. $3x^2(4x - 1) + 19x(4x - 1) + 28(4x - 1)$

**Answers to Practice Problems** **1.** $(x - 6)^2$ **2.** $(x + 7)(x - 5)$ **3.** $(5 + x)(3 - x)$ **4.** $2y^4(5y + 1)(5y - 1)$ **5.** $(8x - 3)(x + 2)$ **6.** $(x + 1)(2x - 3)(x + 6)$

**53.** $(x + y)^2 + 6(x + y) + 9$ **54.** $(x - y)^2 - 81$ **55.** $(x + 2y)^2 - 25$
**56.** $(4x + y)^2 + 4(4x + y) + 4$ **57.** $(x + 5y)^2 + 8(x + 5y) + 12$ **58.** $(6x - y)^2 + 8(6x - y) + 7$
**59.** $(3x - y)^2 + 3(3x - y) - 4$ **60.** $(2x + y)^2 + (2x + y) - 6$

## 2.5 Special Factoring

**OBJECTIVES**

In this section, you will be learning to:

1. Factor the sums or differences of cubes.
2. Factor by grouping.
3. Factor special expressions in which there are common factors with negative exponents.

In this section, we will develop factoring techniques for three different special cases: sums and differences of cubes, grouping with four terms, and negative exponents.

### *Sums and Differences of Cubes*

We know from forms IV and V in Section 2.3 that:

$$\textbf{IV.}\quad x^3 - a^3 = (x - a)(x^2 + ax + a^2)$$
$$\textbf{V.}\quad x^3 + a^3 = (x + a)(x^2 - ax + a^2)$$

By mentally substituting for $x$ and $a$ in these two forms, we can factor a variety of expressions, as shown in the following examples.

**EXAMPLES** Factor completely.

**1.** $x^3 - 27$

**Solution**

$$\begin{aligned} x^3 - 27 &= \overset{x^3}{x^3} - \overset{a^3}{(3)^3} \qquad \text{Difference of two cubes} \\ &= (x - 3)(x^2 + 3x + 9) \\ &\phantom{=}\ (x - a)(x^2 + ax + a^2) \end{aligned}$$

**2.** $8y^{12} - 125$

**Solution**

$$\begin{aligned} 8y^{12} - 125 &= \overset{x^3}{(2y^4)^3} - \overset{a^3}{(5)^3} \qquad \text{Difference of two cubes} \\ &= (2y^4 - 5)(4y^8 + 10y^4 + 25) \\ &\phantom{=}\ (x - a)(x^2 + ax + a^2) \end{aligned}$$

**3.** $x^6 + 64$

**Solution**

$$\begin{aligned} x^6 + 64 &= \overset{x^3}{(x^2)^3} + \overset{a^3}{(4)^3} \qquad \text{Sum of two cubes} \\ &= (x^2 + 4)(x^4 - 4x^2 + 16) \\ &\phantom{=}\ (x + a)(x^2 - ax + a^2) \end{aligned}$$

**4.** $x^{3k} + y^{6k}$

**Solution**

$$\begin{aligned} x^{3k} + y^{6k} &= (x^k)^3 + (y^{2k})^3 \\ &= (x^k + y^{2k})(x^{2k} - x^k y^{2k} + y^{4k}) \end{aligned}$$

(with $x^k$ playing the role of $x$ and $y^{2k}$ the role of $a$ in $(x + a)(x^2 - ax + a^2)$) ■

## Factoring by Grouping

Not all products are binomials or trinomials. For example,

$$\begin{aligned} (a + b)(c + d) &= (a + b)c + (a + b)d \\ &= ac + bc + ad + bd \end{aligned}$$

and

$$\begin{aligned} (x + 3)(y - 7) &= (x + 3)y + (x + 3)(-7) \\ &= xy + 3y - 7x - 21 \end{aligned}$$

In both examples, there are no like terms to be combined and the product has four terms. Now, given a product such as $xy + 3y - 7x - 21$, try to **group** terms in such a way that a common factor can be recognized.

**EXAMPLES** Factor each of the following by grouping terms.

**5.** $$\begin{aligned} ax - ay + bx - by &= a(x - y) + b(x - y) \\ &= (x - y)(a + b) \end{aligned}$$ $x - y$ is treated as a common factor.

**6.** $xy - 8x - 2y + 16 = x(y - 8) + 2(-y + 8)$

But $y - 8$ and $-y + 8$ are not the same factor. Factor $-2$ instead of $+2$ from the last two terms.

$$\begin{aligned} xy - 8x - 2y + 16 &= x(y - 8) - 2(y - 8) \\ &= (y - 8)(x - 2) \end{aligned}$$ $y - 8$ is a common factor.

**7.** $4xy - 28x + 3y - 15 = 4x(y - 7) + 3(y - 5)$

But $y - 7$ and $y - 5$ are not the same factor. Therefore, $4xy - 28x + 3y - 15$ is not factorable. That is, **some of the terms being factorable does not necessarily imply the entire expression is factorable.**

**8.** $$\begin{aligned} (x^2 + 6x + 9) - y^2 &= (x + 3)^2 - y^2 \\ &= [(x + 3) + y][(x + 3) - y] \end{aligned}$$ Difference of two squares

The special grouping provided here shows that the expression can be treated as the difference of two squares. ■

## Factoring with Negative Exponents

Students continuing their mathematics studies into trigonometry, algebra, or calculus will find expressions with negative exponents that can be factored. These expressions cannot be classified as polynomials; but many times, one of the factors is a polynomial. We will consider expressions in which a common term with negative exponents can be factored. For example,

$$x^2 + 4x^{-1} = x^{-1}(x^3 + 4)$$

(The powers of $x$ are $x^2$ and $x^{-1}$. The smallest power is $x^{-1}$.) The following format may help in understanding the technique. We multiply and divide by the **smallest power,** namely $x^{-1}$ in this case.

$$x^2 + 4x^{-1} = \frac{x^{-1}(x^2 + 4x^{-1})}{x^{-1}} = x^{-1}\left(\frac{x^2 + 4x^{-1}}{x^{-1}}\right)$$

$$= x^{-1}\left(\frac{x^2}{x^{-1}} + \frac{4x^{-1}}{x^{-1}}\right) = x^{-1}(x^3 + 4)$$

Similarly,

$$x^2 + 5 - x^{-2} = x^{-2}(x^4 + 5x^2 - 1)$$

(The powers of $x$ are $x^2$, $x^0$, and $x^{-2}$. The smallest power is $x^{-2}$.)

Again, multiplying and dividing may help clarify the technique.

$$x^2 + 5 - x^{-2} = \frac{x^{-2}(x^2 + 5 - x^{-2})}{x^{-2}} = x^{-2}\left(\frac{x^2 + 5 - x^{-2}}{x^{-2}}\right)$$

$$= x^{-2}\left(\frac{x^2}{x^{-2}} + \frac{5}{x^{-2}} - \frac{x^{-2}}{x^{-2}}\right) = x^{-2}(x^4 + 5x^2 - 1)$$

The rule is to **factor out the smallest power of any factor that is common to all terms.** Remember that negative numbers are smaller than positive numbers. You can check your factoring by multiplying to see if you get the original expression.

**EXAMPLES**

**9.** $2x^2 + x + 3x^{-1} = x^{-1}(2x^3 + x^2 + 3)$

Another idea in factoring $x^{-n}$ is to add $n$ to the exponents of each term. Here, adding 1 to each exponent gives the desired factor:

$$2x^{2+1} + x^{1+1} + 3^{-1+1} = 2x^3 + x^2 + 3$$

**10.** $x^2y^{-1} + 4xy^{-1} + y^{-2} = y^{-2}(x^2y + 4xy + 1)$

**11.** $9x^2y^{-2} - 36y^{-2} = 9y^{-2}(x^2 - 4)$ In this case, the polynomial
$= 9y^{-2}(x + 2)(x - 2)$ $x^2 - 4$ can be factored. ■

***Practice Problems***

Factor completely.

1. $x^6 - 125$
2. $4x^{3k} + 32$
3. $2xy - 6x + 4y - 12$
4. $49x^{-1} - x^{-3}$

**Answers to Practice Problems** **1.** $(x^2 - 5)(x^4 + 5x^2 + 25)$ **2.** $4(x^k + 2)(x^{2k} - 2x^k + 4)$ **3.** $2(y - 3)(x + 2)$ **4.** $x^{-3}(7x + 1)(7x - 1)$

## EXERCISES 2.5

Factor Exercises 1–68 completely.

1. $x^3 - 125$
2. $x^3 - 64$
3. $x^3 - 8y^3$
4. $x^3 - y^3$
5. $x^3 + 216$
6. $x^3 + 125$
7. $x^3 + y^3$
8. $x^3 + 27y^3$
9. $x^3 - 1$
10. $8x^3 + 1$
11. $27x^3 + 8$
12. $x^3 - 125$
13. $3x^3 + 81$
14. $4x^3 - 32$
15. $125x^3 - 64y^3$
16. $64x^3 + 27y^3$
17. $54x^3 - 2y^3$
18. $3x^4 + 375xy^3$
19. $x^3y + y^4$
20. $x^4y^3 - x$
21. $x^2y^2 - x^2y^5$
22. $2x^2 - 16x^2y^3$
23. $24x^4y + 81xy^4$
24. $x^6 - 64y^3$
25. $x^6 - y^9$
26. $x^3 + (x - y)^3$
27. $27x^3 + (y^2 - 1)^3$
28. $(x - 3y)^3 - 64z^3$
29. $(x + 2)^3 + (y - 3)^3$
30. $(x + y)^3 + (y - 4)^3$
31. $(x + 2y)^3 - (y + 4)^3$
32. $(3x + 2y)^3 - (x + y)^3$
33. $xy + 4x - 3y - 12$
34. $xy + 5y + 2x + 10$
35. $6x + 4y + xy + 24$
36. $7x - 3y + xy - 21$
37. $3xy + 6y - x - 1$
38. $2xy + 10x + 3y + 15$
39. $4xy - 28x + 3y - 21$
40. $6xy - 9x + 4y - 6$
41. $5x^2 + 10x + xy + y$
42. $(x^2 + 2x + 1) - y^2$
43. $(x^2 - 2xy + y^2) - 36$
44. $(x^2 + 4xy + 4y^2) - 25$
45. $(16x^2 + 8x + 1) - y^2$
46. $x^2 - y^2 - 6y - 9$
47. $x^3 - 5x^2 + 6x - 30$
48. $4x^3 - 6x^2 - 14x + 21$
49. $x^3 + 12x^2 - 4x - 48$
50. $x^3 + 3x^2 - 9x - 27$
51. $x^3 + 2x^2 - 4x - 8$
52. $x^3 + 7x^2 - 4x - 28$
53. $x^2y + 5x^2 - 9y - 45$
54. $x^{2k} - 4y^2$
55. $x^{3k} + 8$
56. $x^{3k} + 27y^{3k}$
57. $3x^{3k} - 24x^{3k}y^{3k}$
58. $108x^{3k} - 32y^{3k}$
59. $4x^{-1} + 8x^{-2}$
60. $3x^{-2} + 15x^{-4}$
61. $2x^{-2} + 3x^{-3}y - 6x^{-3}$
62. $x^{-1} + 3x^{-2} + x^{-3}$
63. $5x^{-2} - 20x^{-4}$
64. $8x^{-1} - 50x^{-3}$
65. $x^{-1} + 4x^{-2} + 3x^{-3}$
66. $2x^{-2} + 3x^{-3} + x^{-4}$
67. $2 - 8x^{-1} - 10x^{-2}$
68. $3 - 3x^{-1} - 18x^{-2}$

# 2.6 Applications (Variation)

**OBJECTIVE**

In this section, you will be learning to solve applied problems using the principles of variation: direct, inverse, and combined.

If a weight is hung on a spring, the spring will stretch. In the diagram in Figure 2.1, a weight of 5 grams stretches the spring 3 centimeters. If a heavier weight is hung on the spring, will the spring stretch farther? For example, if 10 grams is hung on the spring, would you expect to see the spring stretch

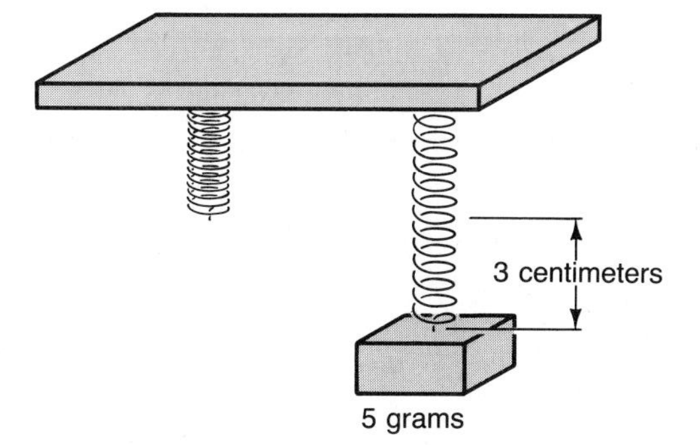


**Figure 2.1**

twice as far? If the spring is not stretched beyond its ability to spring back, a 10-gram weight will stretch the spring 6 centimeters.

The more weight added, the greater the distance the spring will stretch. We say that the distance of stretch **varies directly** with the weight. This is a property of springs studied in physics and is known as Hooke's Law. The relationship can be indicated with the formula

$$d = k \cdot w$$

where $k$ is called the **constant of proportionality** and will depend on the type and size of spring. For our spring, $d = 3$ centimeters for $w = 5$ grams.

$$d = k \cdot w$$

$$3 = k \cdot 5 \quad \text{Substitute the known values.}$$

$$\frac{3}{5} = k \quad \text{Find the value of } k \text{ to substitute into the formula.}$$

So, $d = \frac{3}{5}w$.

If $w = 10$ grams,

$$d = \frac{3}{5} \cdot 10 = 6 \text{ centimeters}$$

as we expected.

Whenever one quantity is equal to some constant times another quantity, the first quantity is said to **vary directly with** or to be **directly proportional to** or just **proportional to** the second quantity.

As examples:

| | |
|---|---|
| $d = \frac{3}{5}w$ | Hooke's Law for a spring where $k = \frac{3}{5}$ |
| $C = 2\pi r$ | The circumference of a circle varies directly as the radius. |
| $A = \pi r^2$ | The area of a circle is directly proportional to the **square** of the radius. |
| $P = 62.5d$ | Water pressure is proportional to the depth of the water. |

If one quantity is related to another so that the first is equal to some constant divided by the second quantity, the quantities **vary inversely** or are **inversely proportional.** For example, the gravitational force between an object and the earth is inversely proportional to the square of the distance from the object to the center of the earth.

$$F = \frac{k}{d^2}$$

If an astronaut weighs 200 pounds on the surface of the earth, what will he weigh 100 miles above the earth? Assume that the radius of the earth is 4000 miles.

**Solution** We know $F = 200 = 2 \times 10^2$ pounds when
$d = 4000 = 4 \times 10^3$ miles

$$2 \times 10^2 = \frac{k}{(4 \times 10^3)^2} \qquad \text{Find the value for } k.$$

$$k = 2 \times 10^2 \times 16 \times 10^6 = 32 \times 10^8 = 3.2 \times 10^9$$

So, $$F = \frac{3.2 \times 10^9}{d^2}$$

Let $d = 4100 = 4.1 \times 10^3$ miles. Then,

$$F = \frac{3.2 \times 10^9}{(4.1 \times 10^3)^2} = \frac{3.2 \times 10^9}{16.81 \times 10^6} \approx 0.190 \times 10^3 = 190 \text{ pounds}$$

If a variable varies either directly or inversely with more than one other variable, the variation is said to be a **combined variation.** If the combined variation is all direct variation (the variables are multiplied), then it is called **joint variation.** For example, the volume of a cylinder varies jointly as its height and the square of its radius.

$$V = k \cdot r^2 \cdot h$$

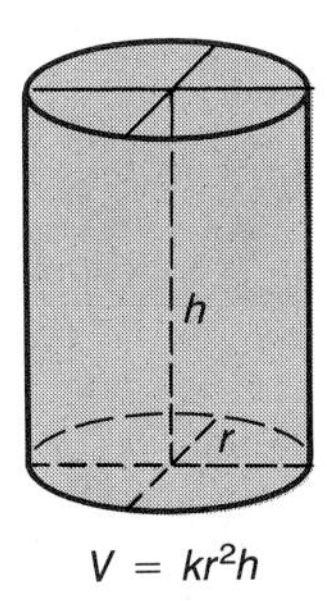


$V = kr^2h$

What is $k$, the constant of proportionality, if a cylinder has the measurements $V = 198$ cu ft, $r = 3$ ft, and $h = 7$ ft?

$$V = k \cdot r^2 \cdot h$$
$$198 = k \cdot 3^2 \cdot 7$$
$$\frac{198}{9 \cdot 7} = k$$
$$k = \frac{22}{7} \approx 3.14$$

We know from experience that $k = \pi$. The measurements can only be approximate.

The formula is $V = \pi r^2 h$.

**EXAMPLES**

**1.** If $y$ is **directly proportional** to $x^2$ and $y = 9$ when $x = 2$, what is $y$ when $x = 4$?

**Solution**

$$y = k \cdot x^2$$
$$9 = k \cdot 2^2$$
$$\frac{9}{4} = k$$

So, $$y = \frac{9}{4}x^2$$

If $x = 4$, then

$$y = \frac{9}{4} \cdot 4^2 = 36$$

**2.** The distance an object falls **varies directly** as the square of the time it falls (until it hits the ground and assuming little or no wind resistance). If an object fell 64 ft in 2 sec, how far will it have fallen by the end of 3 sec?

**Solution**

$$d = k \cdot t^2$$
$$64 = k \cdot 2^2$$
$$16 = k \qquad \text{Use this value for } k \text{ in the formula.}$$

So, $$d = 16t^2$$
$$d = 16 \cdot 3^2 = 144 \text{ ft}$$

The object fell 144 ft in 3 sec.

**3.** The volume of a gas in a container **varies inversely** as the pressure on the gas. If a gas has a volume of 200 cu in. under a pressure of 5 lb per sq in., what will be its volume if the pressure is increased to 8 lb per sq in.?

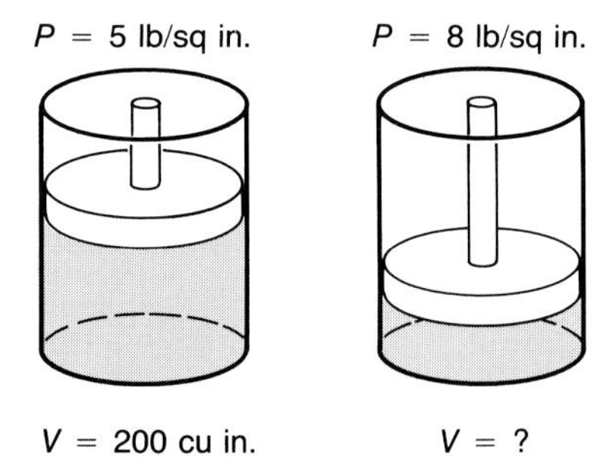

**Solution** $V = \dfrac{k}{P}$

$$200 = \frac{k}{5}$$

$$k = 1000 \qquad \text{Substitute 1000 for } k.$$

So, $V = \dfrac{1000}{P}$

$$V = \frac{1000}{8} = 125 \text{ cu in.}$$

The volume will be 125 cu in.

4. The illumination (in foot-candles, fc) of a light source **varies directly** as the intensity (in candlepower, cp) of the source and **inversely** as the square of the distance from the source. If a certain light source with intensity of 300 cp provides an illumination of 10 fc at a distance of 20 ft, what is the illumination at a distance of 40 ft?

**Solution** $I = \dfrac{k \cdot i}{d^2}$ where $I$ = illumination, $i$ = intensity, $d$ = distance

$$10 = \frac{k \cdot 300}{(20)^2}$$

$$k = \frac{400 \cdot 10}{300}$$

$$k = \frac{40}{3} \qquad \text{Value to be used in the formula}$$

So, $I = \dfrac{\frac{40}{3} \cdot i}{d^2}$

$$I = \frac{\frac{40}{3} \cdot 300}{(40)^2} = \frac{\cancel{40} \cdot 100}{\cancel{40} \cdot 40} = \frac{5}{2} = 2.5 \text{ fc}$$

The illumination at 40 ft is 2.5 fc. ■

## EXERCISES 2.6

Write an equation or formula that represents the general relationship indicated in Exercises 1–30. Then use the given information to find the unknown value.

1. If $y$ varies directly as $x$ and $y = 3$ when $x = 9$, find $y$ if $x = 7$.
2. If $y$ is directly proportional to $x^2$ and $y = 9$ when $x = 2$, what is $y$ when $x = 4$?
3. If $y$ varies inversely as $x$ and $y = 5$ when $x = 8$, find $y$ if $x = 20$.
4. If $y$ varies inversely as $x^2$ and $y = -8$ when $x = 2$, find $y$ if $x = 3$.
5. If $y$ is inversely proportional to $x$ and $y = 5$ when $x = 4$, what is $y$ when $x = 2$?
6. If $y$ is inversely proportional to $x^3$ and $y = 40$ when $x = \frac{1}{2}$, what is $y$ when $x = \frac{1}{3}$?
7. If $y$ is proportional to the square root of $x$ and $y = 6$ when $x = \frac{1}{4}$, what is $y$ when $x = 9$?
8. If $y$ is proportional to the square of $x$ and $y = 80$ when $x = 4$, find $y$ when $x = 6$.
9. If $y$ varies directly as $x^3$ and $y = 81$ when $x = 3$, find $y$ if $x = 2$.
10. $z$ varies jointly as $x$ and $y$, and $z = 60$ when $x = 2$ and $y = 3$. Find $z$ if $x = 3$ and $y = 4$.
11. $z$ varies jointly as $x$ and $y$. If $z = -6$ when $x = 5$ and $y = 8$, find $z$ if $x = 12$ and $y = 3$.
12. $z$ varies jointly as $x^2$ and $y$. If $z = 20$ when $x = 2$ and $y = 3$, find $z$ if $x = 4$ and $y = \frac{7}{10}$.
13. $z$ varies directly as $x$ and inversely as $y^2$. If $z = 5$ when $x = 1$ and $y = 2$, find $z$ if $x = 2$ and $y = 1$.
14. $z$ varies directly as $x^3$ and inversely as $y^2$. If $z = 24$ when $x = 2$ and $y = 2$, find $z$ if $x = 3$ and $y = 2$.
15. $z$ varies directly as $\sqrt{x}$ and inversely as $y$. If $z = 24$ when $x = 4$ and $y = 3$, find $z$ if $x = 9$ and $y = 2$.
16. $z$ is jointly proportional to $x^2$ and $y^3$. If $z = 192$ when $x = 4$ and $y = 2$, find $z$ when $x = 2$ and $y = 4$.
17. $z$ varies directly as $x^2$ and inversely as $\sqrt{y}$. If $z = 108$ when $x = 6$ and $y = 4$, find $z$ when $x = 4$ and $y = 9$.
18. $s$ varies directly as the sum of $r$ and $t$ and inversely as $w$. If $s = 24$ when $r = 7$, $t = 8$, and $w = 9$, find $s$ when $r = 9$, $t = 3$, and $w = 18$.
19. $L$ varies jointly as $m$ and $n$ and inversely as $p$. If $L = 6$ when $m = 7$, $n = 8$, and $p = 12$, find $L$ when $m = 15$, $n = 14$, and $p = 10$.
20. $W$ varies jointly as $x$ and $y$ and inversely as $\sqrt{z}$. If $W = 10$ when $x = 6$, $y = 5$, and $z = 4$, find $W$ when $x = 12$, $y = 6$, and $z = 9$.
21. The resistance, $R$, of a wire varies directly as its length and inversely as the square of the diameter. The resistance of a wire 500 ft long and diameter 0.01 in. is 20 ohms. What is the resistance of a wire 1500 ft long and diameter 0.02 in.?
22. The resistance of a wire 100 ft long and 0.01 in. in diameter is 8 ohms. What is the resistance of 150 ft of the same type of wire but with diameter 0.015 in.? (See Exercise 21.)
23. The lifting force, $F$, exerted on an airplane wing varies jointly as the area, $A$, of the surface and the square of the plane's velocity, $v$. The lift for a wing of area 120 sq ft is 1600 lb when the plane is going 80 mph. Find the lift if the speed is increased to 90 mph.
24. The lift for a wing of area 280 sq ft is 34,300 lb when the plane is going 210 mph. What is the lift if the speed is decreased to 180 mph? (See Exercise 23.)

**25.** The elongation, $E$, in a wire when a mass, $m$, is hung at its free end varies jointly as the mass and the length, $\ell$, of the wire and inversely as the cross-sectional area, $A$, of the wire. The elongation is 0.0055 cm when a mass of 120 gm is attached to a wire 330 cm long with a cross-sectional area of 0.4 sq cm. Find the elongation if a mass of 160 gm is attached to the same wire.

**26.** When a mass of 240 oz is suspended by a wire 49 in. long whose cross-sectional area is 0.035 sq in., the elongation of the wire is 0.016 in. Find the elongation if the same mass is suspended by a 28-in. wire of the same material with a cross-sectional area of 0.04 sq in. (See Exercise 25.)

**27.** The safe load, $L$, of a wooden beam supported at both ends varies jointly as the width, $w$, and the square of the depth, $d$, and inversely as the length, $\ell$. A wooden beam 2 in. wide, 8 in. deep, and 14 ft long holds up 2400 lb. What load would a beam 3 in. $\times$ 6 in. $\times$ 15 ft, of the same material, support?

**28.** A 4 in. $\times$ 6 in. beam 12 ft long supports a load of 4800 lb. What is the safe load of a beam of the same material that is 6 in. $\times$ 10 in. $\times$ 15 ft long? (See Exercise 27.)

**29.** The gravitational force of attraction, $F$, between two bodies varies directly as the product of their masses, $m_1$ and $m_2$, and inversely as the square of the distance, $d$, between them. The gravitational force between a 5-kg mass and a 2-kg mass 1 m apart is $1.5 \times 10^{-10}$ N. Find the force between a 24-kg mass and a 9-kg mass that are 6 m apart. (N represents a unit of force called a **newton.**)

**30.** In Exercise 29, what happens to the force if the distance between the bodies is cut in half?

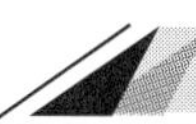

## CHAPTER 2 SUMMARY

### Key Terms and Formulas

In **scientific notation,** decimal numbers are written as the product of a number between 1 and 10 and an integer power of 10. [2.2]

The general form of a **monomial in** $x$ is $kx^n$ where $n$ is a whole number and $k$ is any number. $n$ is called the **degree** of the monomial, and $k$ is the **coefficient.** [2.3]

Any monomial or algebraic sum of monomials is a **polynomial.** The **degree of a polynomial** is the largest of the degrees of its terms after like terms have been combined. [2.3]

**Classification of Polynomials** [2.3]

Monomial: polynomial with one term

Binomial: polynomial with two terms

Trinomial: polynomial with three terms

**Special Products of Polynomials** [2.3], [2.4], [2.5]

**I.** $(x + a)(x - a) = x^2 - a^2$ — Difference of two squares

**II.** $(x + a)^2 = x^2 + 2ax + a^2$ — Perfect square trinomial

**III.** $(x - a)^2 = x^2 - 2ax + a^2$ — Perfect square trinomial

**IV.** $(x - a)(x^2 + ax + a^2) = x^3 - a^3$ — Difference of two cubes

**V.** $(x + a)(x^2 - ax + a^2) = x^3 + a^3$ — Sum of two cubes

Whenever one quantity is equal to some constant times another quantity, the first quantity is said to **vary directly with** or to be **directly proportional to** or just **proportional to** the second quantity. The constant is called the **constant of proportionality.** [2.6]

Whenever one quantity is equal to some constant divided by another quantity, the first quantity is said to **vary inversely with** or to be **inversely proportional to** the second quantity. [2.6]

If a variable varies either directly or inversely with more than one other variable, the variation is said to be a **combined variation.** If a combined variation is all direct variation, then it is called **joint variation.** [2.6]

## Properties and Rules

**Properties of Exponents** [2.1], [2.2]

If $a$ and $b$ are nonzero numbers and $m$ and $n$ are integers,

1. $a^m \cdot a^n = a^{m+n}$
2. $a^0 = 1$
3. $a^{-n} = \dfrac{1}{a^n}$
4. $\dfrac{a^m}{a^n} = a^{m-n}$
5. $(a^m)^n = a^{mn}$
6. $(ab)^n = a^n b^n$
7. $\left(\dfrac{a}{b}\right)^n = \dfrac{a^n}{b^n}$

## Procedures

In factoring polynomials, always **factor out the greatest common factor first.** [2.4]

The ***ac*-Method** (Grouping) of factoring trinomials of the form $ax^2 - bx + c$ involves the following five steps: [2.4]

**Step 1** Multiply $a \cdot c$.

**Step 2** Find two integers whose product is $ac$ and whose sum is $b$. If this is not possible, then the trinomial is **not factorable.**

**Step 3** Rewrite the middle term ($bx$) using the two numbers found in Step 2 as coefficients.

**Step 4** Factor by grouping the first two terms and the last two terms.

**Step 5** Factor the common binomial factor to find two binomial factors of the trinomial $ax^2 + bx + c$.

The **FOIL Method** of factoring trinomials is a reverse of the FOIL method of multiplication. It is a **trial-and-error** approach to factoring. [2.4]

## CHAPTER 2 REVIEW

Simplify each of the expressions in Exercises 1–27. [2.1], [2.2]

1. $(-3)^5(-3)^2$
2. $\dfrac{7^4}{7^5}$
3. $y^5 \cdot y^{-2}$
4. $x^3 \cdot x^{-2}$
5. $\dfrac{5^3 \cdot 5^0}{5^4}$
6. $\dfrac{(-6)^3 \cdot (-6)}{(-6)^2}$
7. $\dfrac{x^0 x^4}{x^2}$
8. $\dfrac{x^3}{x^2 x^4}$
9. $(4x^2y)^3$
10. $(7x^5y^2)^2$
11. $(-2x^3y^2)^{-3}$
12. $\left(\dfrac{6x^2}{y^5}\right)^2$
13. $\left(\dfrac{-3xy^4}{x^2}\right)^2$
14. $\left(\dfrac{2x^2y^0}{xy}\right)^{-3}$
15. $8\left(\dfrac{x^3y^2}{xy^4}\right)^5$
16. $(x^2y)(x^3y^{-2})$
17. $(x^{-1}y^{-1})(x^4y^{-3})$
18. $(x^2y^3)^{-1}(x^{-2}y)^2$
19. $(x^3y^2)^2(x^2y)^{-3}$
20. $\dfrac{x^2y^{-3}}{x^{-3}y^{-2}}$
21. $\dfrac{x^{-1}y^2}{x^3y^{-4}}$
22. $\dfrac{2^{-1}x^{-1}}{3x^2y}$
23. $\dfrac{5xy}{2^{-2}x^2y^{-3}}$
24. $\left(\dfrac{5x^2y}{3xy^2}\right)\left(\dfrac{10x^{-1}}{6y}\right)^{-1}$
25. $\left(\dfrac{x^2y^3}{x^{-1}y}\right)^{-1}\left(\dfrac{x^4y}{xy^2}\right)^2$
26. $\left(\dfrac{x^{-1}y^{-1}}{(xy)^{-1}}\right)\left(\dfrac{x^{-2}y^2}{x^3y^{-1}}\right)^2$
27. $\left(\dfrac{2}{x^{-1}y^{-1}}\right)^{-2}\left(\dfrac{x^5y}{x^{-1}y^2}\right)$

Write Exercises 28–31 in scientific notation and simplify if possible. [2.2]

28. 0.000000345
29. 6,820,000
30. $2100 \times 0.000005$
31. $\dfrac{270{,}000 \times 0.00014}{42{,}000}$

Simplify the polynomials in Exercises 32–41. [2.3]

**32.** $(7x^2 + 2xy - y^2) + (8xy - 4y^2)$

**33.** $(4x^2 + 2xy - 7y^2) + (5x^2 + xy - 2y^2)$

**34.** $(6x^2 + x - 10) - (x^3 + x^2 + x - 4)$

**35.** $(x^3 + 4x^2 - x) + (-2x^3 + 6x - 3)$

**36.** $(2x^2 - 6xy + y^2) - (4x^2 + 3xy - y^2 + 2y)$

**37.** $(5x^2y - 3xy^2 + xy) - (-2x^2y + 4xy^2 - 7y^2)$

**38.** $x^2 - (-2xz - 6z^2) + [(5x^2 - xz - 2z^2) - (8xz + 3z^2)]$

**39.** $[2x^2 - (5xz + 7z^2) + (x^2 + 9)] - (3x^2 - 4xz + z^2)$

**40.** $[(x^2 - 4x) - (2x^2 + 6x + 1)] + [(3x^2 - 2x + 7) - (x^2 - 3x + 1)]$

**41.** $[(xy + xz - yz) - (3xz + 2xy)] - [(yz + 4xy) - (3xy + 5yz)]$

Evaluate each polynomial in Exercises 42–45 for the specified values of the variable. [2.3]

**42.** $5x^2 - 3x - 4;\ x = -2$

**43.** $x^2 - 4x + 8;\ x = 4$

**44.** $x^3 - x - 3;\ x = -1$

**45.** $x^2 + 4xy - 3y^2;\ x = 3,\ y = -1$

Find the indicated product in Exercises 46–55. [2.3]

**46.** $(4x - 7y)(3x + 2y)$

**47.** $(3x + 8y)(5x + y)$

**48.** $x^2y(x + 3y)(6x - y)$

**49.** $xy^3(2x - 9y)(4x - 3y)$

**50.** $(2x + 5y)(4x^2 - 10xy + 25y^2)$

**51.** $(7x^2 - 6)(7x^2 + 6)$

**52.** $x^k(x + 5)(2x - 11)$

**53.** $x^{2k}(4x + 5)(2x - 9)$

**54.** $(3x^k - 1)(x^k + 6)$

**55.** $(2x^k - 5)(2x^k + 5)$

Factor Exercises 56–70 completely. [2.4], [2.5]

**56.** $x^2 - 9x - 36$

**57.** $x^2y - 5xy^2 - 14y^3$

**58.** $15x^2 + 29xy + 8y^2$

**59.** $6x^2y^2 - 96y^4$

**60.** $8x^3 + 1$

**61.** $40x^3 - 625$

**62.** $(x + 6)^2 + (x + 6) - 2$

**63.** $x^2 - 2x - xy + 2y$

**64.** $x^3 + x^2 - 4x - 4$

**65.** $(x + y)^2 - (x - y)^2$

**66.** $5x^{2k} - 3x^k - 14$

**67.** $8x^{k+2} + 2x^{k+1} - 45x^k$

**68.** $x^6 + y^6$

**69.** $9x^{-1} + 6x^{-2} + x^{-3}$

**70.** $5 - 7x^{-1} - 6x^{-2}$

[2.6]

**71.** $V$ varies directly as $s$ and inversely as $y$. $V = 27$ when $s = 10$ and $y = 2$. Find $V$ if $s = 11$ and $y = 6$.

**72.** $Z$ varies jointly as $x$ and the cube of $y$. If $Z = 154$ when $x = 7$ and $y = 2$, find $Z$ if $x = 4$ and $y = 3$.

**73.** $W$ varies jointly as $x$ and $y$ and inversely as $z$. $W = \frac{3}{16}$ when $x = 5$, $y = 3$, and $z = 14$. Find $W$ if $x = 6$, $y = 4$, and $z = 9$.

Write an equation for Exercises 74–76, then solve. [2.6]

**74.** The force of the wind, $F$, on a vertical surface, such as a sail, varies jointly as the area, $A$, of the surface and the square of the velocity, $v$, of the wind. The force on 21 sq ft is 120 lb when the wind is 40 mph. Find the force on 25 sq ft in a wind of 60 mph.

**75.** Using the relationship given in Exercise 74, find the force of the wind on 20 sq ft in a wind of 50 mph.

**76.** Torricelli's Theorem states that the rate, $R$, at which water flows through an orifice is jointly proportional to the area, $A$, of the orifice and the square root of the height, $h$, of the water above the orifice. Water leaks through a hole 1 sq cm in area at the bottom of a tank when the water level is 3 m, at a rate of $7.7 \times 10^{-4}$ cubic meters per second. Find the rate of flow if the height of the water is 4 m and the hole is 2 sq cm in area.

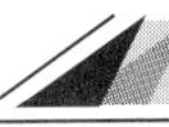

## CHAPTER 2 TEST

Simplify each of the expressions in Exercises 1–4.

1. $(4x^2y)(-5x^4y^{-3})$
2. $\dfrac{x^2y}{x^3y^{-1}}$
3. $\dfrac{(x^2y^{-2})^3}{(x^3y)^2}$
4. $\left(\dfrac{xy^{-1}}{x^2}\right)^{-3}\left(\dfrac{x^{-2}y^{-1}}{y^3}\right)^2$
5. Write the expression $\dfrac{0.27 \times 0.0016}{120}$ in scientific notation and simplify.

Simplify the polynomials in Exercises 6–8.

6. $(5x^2 - 3x + 7) - (2x^2 - 5x - 6)$
7. $[(3x - y) - (y - 2)] + [y + 3 - (x + 2)]$
8. $[(2x^2 - x) + (3x + 2)] + [(x - 3) - (x^2 + 1)]$

Find the indicated products in Exercises 9–14.

9. $(7x - 3)(4x + 5)$
10. $(2x - 7)^2$
11. $(8x + 3)(8x - 3)$
12. $(2x + 1)(3x^2 + x - 1)$
13. $(x^3 + 5)^2$
14. $[(x + 1) - y][(x + 1) + y]$

Factor Exercises 15–22 completely.

15. $30x^2y - 18xy^2 + 24xy$
16. $7x^2 - 26x - 8$
17. $3x^2y - 18xy^2 + 27y^3$
18. $9x^2 - 81y^2$
19. $(x + 2)^2 - (x - 2)^2$
20. $x^3 - 8y^3$
21. $3x^3 + 81y^3$
22. $x^2 - 4y^2 + x + 2y$

In Exercises 23 and 24, evaluate each polynomial for $x = -3$ and $y = -1$.

23. $2x^3 - x^2 - x + 5$
24. $2x^2 + xy - y^2$
25. $z$ varies directly as $x^2$ and inversely as $\sqrt{y}$. If $z = 24$ when $x = 3$ and $y = 4$, find $z$ if $x = 5$ and $y = 9$.
26. The pressure on one side of a metal plate submerged horizontally in water varies directly as the depth of the water. If a plate 10 feet below the surface has a total pressure of 625 ft lbs on one face, how much pressure will then be on that same plate 20 feet below the surface?

## CUMULATIVE REVIEW (2)

Perform the indicated operations in Exercises 1–4.

1. $\dfrac{1}{3} + \dfrac{11}{15} + \left(-\dfrac{7}{30}\right)$
2. $\dfrac{15}{16} - \dfrac{7}{12} - \left(-\dfrac{3}{8}\right)$
3. $\dfrac{13}{14} \cdot \dfrac{5}{26} \cdot \dfrac{7}{9}$
4. $\dfrac{2}{3} \cdot \dfrac{7}{5} \div \dfrac{7}{12}$

Simplify the expressions in Exercises 5–8, using the rules for order of operations.

5. $2 \cdot 3^2 \div 6 \cdot 3 - 3$
6. $6 + 3[4 - 2(3^3 - 1)]$
7. $\dfrac{3}{4} \div \dfrac{3}{16} - \dfrac{2}{3} \cdot \dfrac{3}{4}$
8. $\left(\dfrac{2}{3} - \dfrac{1}{5}\right) \div \left(\dfrac{5}{7} - \dfrac{1}{3}\right)$

Combine like terms in Exercises 9 and 10.

9. $8x - [2x + 4(x - 3) - 5]$
10. $9x + [8 - 5(3 - 2x) - 7x]$

Solve each of the equations in Exercises 11–14.

**11.** $4(2x - 3) + 2 = 5 - (2x + 6)$

**12.** $\frac{4x}{7} - 3 = 9$

**13.** $\frac{2x + 3}{6} - \frac{x + 1}{4} = 2$

**14.** $\left|\frac{2x}{5} - 1\right| = 3$

Perform the indicated operations in Exercises 15–20.

**15.** $(x^2 + 3x - 8) + (3x^2 - 2x - 6)$

**16.** $(5x^2 + 7x - 2) + (2x^3 - x^2 + 6x + 4)$

**17.** $(2x^2 + 9x - 3) - (5x^2 - 2x + 1)$

**18.** $(3x^2 + x - 9) - (-2x^2 + 5x - 3)$

**19.** $(5x + 3)(x - 8)$

**20.** $(4x + 1)(2x + 3)$

Factor Exercises 21–24 completely.

**21.** $4x^2 - 4x - 15$

**22.** $6x^2 - 7x + 2$

**23.** $6x^3 - 22x^2 - 8x$

**24.** $8x^3 + 125$

CHAPTER

# 3 RATIONAL EXPRESSIONS

## MATHEMATICAL CHALLENGES

A motorist on a trip covered one-third of the distance at 40 kph and two-thirds of the distance at 50 kph. What was the motorist's average speed?

Solve for all real values of $x$: $(x)(\sqrt{x^3}) = \dfrac{x^x}{x}$

The cost of a chartered bus was split evenly by twenty students going to Cleveland. Prior to leaving Oberlin, ten more students decided to go, thereby reducing the expense of each student in the original group by \$1.50. What was the charge for the use of the bus?

## CHAPTER OUTLINE



Rational expressions are fractions in which the numerator and denominator are polynomials. All the rules you learned in arithmetic about operating with fractions are going to be applied. For example, to add or subtract rational expressions, you need a common denominator, just as with fractions in arithmetic. To find common denominators and to multiply, divide, and simplify rational expressions, you are going to apply all the factoring skills you learned in Chapter 2. Thus, all those skills will be reinforced and, even if you had some difficulty with factoring, you probably will be very comfortable with factoring after Chapter 3.

The use of rational expressions opens the way to a variety of algebraic expressions, equations, and applications not available otherwise. In Section 3.6, you will learn to solve inequalities involving fractions, graph the solutions, and represent the solutions in interval notation in a manner similar to that discussed in Chapter 1. In Section 3.7, the applied problems involve fractions using some familiar formulas from a different point of view. You should find them particularly interesting.

## 3.1 Basic Properties

**OBJECTIVES**

In this section, you will be learning to:

1. Determine whether or not two rational expressions are equivalent.
2. Write rational expressions as equivalent rational expressions with specified denominators.
3. Reduce rational expressions to lowest terms.

Expressions such as

$$x - 5,\ x^2 + 1,\ x^5 - 3x^4 + 2, \text{ and } 2y^3 + 4y^2 - 8y + 6$$

are polynomials. Any fraction formed with a polynomial as numerator and a polynomial as denominator is called a **rational expression.** Thus,

$$\frac{2x}{x^2 + 1}, \quad \frac{x + 3}{x - 5}, \quad \text{and} \quad \frac{y^3 - 3y^2}{y^2 - 7y + 12}$$

are all rational expressions. **No denominator may be 0.** Otherwise the expression would be undefined. So, in the second example, $x \neq 5$, and in the third example, $y \neq 4$ or 3. There are no restrictions on $x$ in the first example since $x^2 + 1 \neq 0$ for any real number.

We will use the capital letters $P$, $Q$, $R$, $S$, and $D$ as general representations for polynomials in various formulas and properties. For example, we could write a general rational expression as $\frac{P}{Q}$ and a general product as $P \cdot R$. So, if

$$P = x^2 + 2x + 1, \quad Q = 3x - 6, \quad \text{and} \quad R = x^2 + 4$$

then

$$\frac{P}{Q} = \frac{x^2 + 2x + 1}{3x - 6} \quad \text{and} \quad P \cdot R = (x^2 + 2x + 1)(x^2 + 4)$$

Note that for $\frac{P}{Q}$, we assume that $Q = 3x - 6 \neq 0$. In other words, we assume that $x \neq 2$.

**Rational Expression**

A **rational expression** is an expression of the form $\frac{P}{Q}$ where $P$ and $Q$ are polynomials and $Q \neq 0$.

For the remainder of this chapter, the fractions $\frac{P}{Q}$ and $\frac{R}{S}$ represent rational expressions. **We will assume that $x$ will not equal any value that will cause a denominator to be 0.**

All the basic properties of real numbers are true for rational expressions since they represent real numbers. Several useful properties are listed here without proof, and examples are given using both real numbers and rational expressions. **Study the examples carefully.**

**Property 1**

$$\frac{P}{Q} = \frac{R}{S} \text{ if and only if } P \cdot S = Q \cdot R$$

**EXAMPLES**

**1.** $\frac{6}{8} = \frac{15}{20}$ Since $6 \cdot 20 = 8 \cdot 15$ or $120 = 120$

**2.** $\frac{x + 1}{x + 2} = \frac{x^2 + x}{x^2 + 2x}$ Since $(x + 1)(x^2 + 2x) = (x + 2)(x^2 + x)$ or $x^3 + 3x^2 + 2x = x^3 + 3x^2 + 2x$ (Assume $x \neq -2, 0$.) ■

Just as with rational numbers, a rational expression is said to be **reduced** if the numerator and denominator have no common factors other than 1 and −1. Property 2 is used both to "build up" rational expressions and to reduce rational expressions. Property 2 is known as the **Fundamental Principle of Fractions,** and the two fractions are said to be equivalent.

**Property 2**

$$\frac{P}{Q} = \frac{P \cdot R}{Q \cdot R} \quad \text{if } R \neq 0$$

**EXAMPLES** Use Property 2 to build up each expression as indicated.

**3.** $\frac{7}{8} = \frac{?}{24}$

**Solution** Since $24 = 8 \cdot 3$,

$$\frac{7}{8} = \frac{7 \cdot 3}{8 \cdot 3} = \frac{21}{24}$$ Building up the fraction using Property 2

**4.** $\frac{5x}{x + 3} = \frac{?}{x^2 - x - 12}$

**Solution** Since $x^2 - x - 12 = (x + 3)(\mathbf{x - 4})$,

$$\frac{5x}{x + 3} = \frac{5x(\mathbf{x - 4})}{(x + 3)(\mathbf{x - 4})}$$ Building up the expression using Property 2

$$= \frac{5x^2 - 20x}{x^2 - x - 12}$$ ($x \neq 3, 4$ because either of these values for $x$ would make the denominator 0.)

Use Property 2 to reduce each expression to lowest terms.

**5.** $\frac{2x - 10}{3x - 15} = \frac{2\cancel{(x - 5)}}{3\cancel{(x - 5)}} = \frac{2}{3} \quad (x \neq 5)$

[Note that $(x - 5)$ is a common **factor.** The key word here is **factor.** We can only reduce using **factors.**]

**6.** $\frac{x^2 - 16}{x^3 - 64} = \frac{(x + 4)\cancel{(x - 4)}}{\cancel{(x - 4)}(x^2 + 4x + 16)}$

$$= \frac{x + 4}{x^2 + 4x + 16} \quad (x \neq 4)$$

Reduce using the common **factor** $(x - 4)$. Note that $x^3 - 64$ is the difference of two cubes. ■

Property 3 and Property 4 are concerned with the placement of negative signs. These properties are particularly useful in addition and subtraction with rational expressions.

***Property 3***

$$-\frac{P}{Q} = \frac{P}{-Q} = \frac{-P}{Q}$$

In words, Property 3 states that a negative sign can be in front of an expression, or with the denominator, or with the numerator, and the expression will have the same value. This means that we can use the form that best suits our purposes when simplifying particular expressions.

**EXAMPLES**

**7.** $-\frac{35}{5} = \frac{35}{-5} = \frac{-35}{5} = -7$

**8.** $-\frac{y-3}{y^2} = \frac{y-3}{-y^2} = \frac{-(y-3)}{y^2}$ ■

***Property 4***

$$\frac{P}{Q} = \frac{-P}{-Q} = -\frac{-P}{Q} = -\frac{P}{-Q}$$

In words, Property 4 states that we can introduce negative signs into a rational expression as long as we place two in the positions as shown.

**EXAMPLES**

**9.** $\frac{24}{6} = \frac{-24}{-6} = -\frac{-24}{6} = -\frac{24}{-6} = 4$

**10.** $\frac{x-5}{2x+3} = \frac{-(x-5)}{-(2x+3)} = -\frac{-(x-5)}{2x+3} = -\frac{(x-5)}{-(2x+3)}$ ■

***Property 5***

In general,

$$\frac{-P}{P} = -1 \quad \text{if } P \neq 0.$$

In particular,

$$\frac{a-x}{x-a} = -1 \quad \text{if } x \neq a.$$

In words, Property 5 states that if a nonzero expression is divided by its opposite, then the result is $-1$. Note that $a - x$ and $x - a$ are opposites since $-(x - a) = -x + a = a - x$.

**EXAMPLES**

**11.** $\frac{x^2 - 2x - 3}{-x^2 + 2x + 3} = -1 \qquad [x^2 - 2x - 3 = -(-x^2 + 2x + 3)]$

**12.** $\frac{17 - x}{x - 17} = -1 \qquad [17 - x = -(x - 17)]$

In more detail,

$$\frac{17-x}{x-17} = \frac{-(-17+x)}{x-17} = \frac{-(x-17)}{x-17} = \frac{-1 \cdot \cancel{(x-17)}}{+1 \cdot \cancel{(x-17)}} = -1$$ ■

## Common Error

Reduce only **factors.** Do not reduce terms unless they are **factors.**

| | | |
|---|---|---|
| $\dfrac{x + \cancel{2}}{\cancel{2}}$ | WRONG | 2 is not a factor. |
| $\dfrac{\overset{x}{\cancel{x^2}} - \overset{3}{\cancel{9}}}{\cancel{x} - \cancel{3}}$ | WRONG | 3 is not a factor and $x$ is not a factor. |
| $\dfrac{x^2 - 9}{x - 3} = \dfrac{(x + 3)\cancel{(x - 3)}}{\cancel{x - 3}}$ | RIGHT | $(x - 3)$ is a factor. |
| $\dfrac{2x + 8}{2} = \dfrac{\cancel{2}(x + 4)}{\cancel{2}}$ | RIGHT | 2 is a factor. |

## Practice Problems

Reduce to lowest terms. State any restrictions on the variables.

1. $\dfrac{5x + 20}{7x + 28}$ 2. $\dfrac{4 - 2x}{2x - 4}$ 3. $\dfrac{x^2 + x - 2}{x^2 + 3x + 2}$

## EXERCISES 3.1

Which of the pairs of rational expressions in Exercises 1–10 are equal? Assume that no denominator is 0.

1. $\dfrac{-5}{16}, \dfrac{20}{-64}$

2. $\dfrac{35}{24}, \dfrac{-105}{-75}$

3. $\dfrac{8x^2y^3}{3z}, \dfrac{16x^3y^4z}{9z^2}$

4. $\dfrac{12x^2}{5y^2}, \dfrac{24x^3y^2}{10xy^4}$

5. $\dfrac{x^2 - 9}{x + 3}, \dfrac{3 - x}{-1}$

6. $\dfrac{6 - x - x^2}{2x^2 - 2x - 4}, \dfrac{2(x + 3)}{4(x + 1)}$

7. $\dfrac{3(x^2 - 2x)}{4 - 2x}, \dfrac{-6x}{4}$

8. $\dfrac{3x + y}{27x^3 + y^3}, \dfrac{1}{9x^2 - 3xy + y^2}$

9. $\dfrac{x^3 - 8y^3}{x - 2y}, x^2 + 4y^2$

10. $\dfrac{x^3 - 3x^2 - x + 3}{x - 1}, x^2 - 2x - 3$

Express each rational expression in Exercises 11–30 as an equivalent rational expression with the indicated denominator. Assume that no denominator is 0.

11. $\dfrac{3x^2}{-8y^2} = \dfrac{?}{32x^2y^3}$

12. $\dfrac{-6x^2}{19y^3} = \dfrac{?}{57x^5y^8}$

13. $\dfrac{5y}{12x} = \dfrac{?}{72x^2y^2}$

14. $\dfrac{3y}{16x^2} = \dfrac{?}{96x^3y^2}$

15. $\dfrac{4x}{x + 4} = \dfrac{?}{(x + 2)(x + 4)}$

16. $\dfrac{5x}{x - 3} = \dfrac{?}{(x - 3)(x + 5)}$

**Answers to Practice Problems** 1. $\dfrac{5}{7}, x \neq -4$ 2. $-1, x \neq 2$
3. $\dfrac{x - 1}{x + 1}, x \neq -1, -2$

**17.** $\dfrac{x+1}{x-2} = \dfrac{?}{(x-2)(x+1)}$

**18.** $\dfrac{x-3}{x+4} = \dfrac{?}{(x+3)(x+4)}$

**19.** $\dfrac{7x}{x-y} = \dfrac{?}{x^2-y^2}$

**20.** $\dfrac{x-1}{x+1} = \dfrac{?}{x^2+3x+2}$

**21.** $\dfrac{5}{9x^2-3x} = \dfrac{?}{9x^3+15x^2-6x}$

**22.** $\dfrac{-2}{5x^2+25x} = \dfrac{?}{5x^3+35x^2+50x}$

**23.** $\dfrac{8}{x-4} = \dfrac{?}{8+2x-x^2}$

**24.** $\dfrac{6}{3-y} = \dfrac{?}{y^2-11y+24}$

**25.** $\dfrac{7}{x+4} = \dfrac{?}{x^3+64}$

**26.** $\dfrac{12}{x+3} = \dfrac{?}{x^3+27}$

**27.** $\dfrac{-x}{2x-3} = \dfrac{?}{8x^3-27}$

**28.** $\dfrac{x}{2x+1} = \dfrac{?}{8x^3+1}$

**29.** $\dfrac{x}{5-x} = \dfrac{?}{x^3-125}$

**30.** $\dfrac{3}{x-2} = \dfrac{?}{4x^3-32}$

Reduce each rational expression in Exercises 31–70. State any restrictions on the variable using the fact that no denominator can be 0.

**31.** $\dfrac{9x^2y^3}{12xy^4}$

**32.** $\dfrac{18xy^4}{27x^2y}$

**33.** $\dfrac{20x^5}{30x^2y^3}$

**34.** $\dfrac{15y^4}{20x^3y^2}$

**35.** $\dfrac{34x^3y^2}{10x^3y^2}$

**36.** $\dfrac{33x^2y^4}{36x^2y^4}$

**37.** $\dfrac{2x+4}{5x+10}$

**38.** $\dfrac{5x+20}{7x+28}$

**39.** $\dfrac{6x-9}{4x-6}$

**40.** $\dfrac{14x-7}{10x-5}$

**41.** $\dfrac{x}{x^2-3x}$

**42.** $\dfrac{3x}{x^2+5x}$

**43.** $\dfrac{7x-14}{x-2}$

**44.** $\dfrac{4-2x}{2x-4}$

**45.** $\dfrac{9-3x}{4x-12}$

**46.** $\dfrac{6x^2-4x}{3xy+2y}$

**47.** $\dfrac{4xy+y^2}{3y^2+2y}$

**48.** $\dfrac{5xy-2y^2}{10y^2-4y^3}$

**49.** $\dfrac{x^2}{x^2-6x}$

**50.** $\dfrac{x^2y}{5x^2-3x}$

**51.** $\dfrac{4-x}{2x-8}$

**52.** $\dfrac{3x-9}{12-4x}$

**53.** $\dfrac{x+3y}{4x^2+12xy}$

**54.** $\dfrac{x^2+6x}{x^2+5x-6}$

**55.** $\dfrac{x^2-5x+6}{x^2-x-2}$

**56.** $\dfrac{x^2-y^2}{3x^2+3xy}$

**57.** $\dfrac{3-10x-8x^2}{2x-9x^2+4x^3}$

**58.** $\dfrac{3x^2+19x-14}{49-x^2}$

**59.** $\dfrac{9x^3-x}{3x^2-8x-3}$

**60.** $\dfrac{x^3y-x^2y}{x^2y(x-1)^2}$

**61.** $\dfrac{x^2-4}{x^3-8}$

**62.** $\dfrac{x^3+64}{2x^2+x-28}$

**63.** $\dfrac{3x^2+14x-24}{18-9x-2x^2}$

**64.** $\dfrac{x^3-2x^2+4x-8}{16-x^4}$

**65.** $\dfrac{xy-3y+2x-6}{y^2-4}$

**66.** $\dfrac{x^2-2xy+y^2}{x^2-xy-5x+5y}$

**67.** $\dfrac{x^3-2x^2+5x-10}{x^3-8}$

**68.** $\dfrac{x^2+4xy+3y^2}{ax+ay+bx+by}$

**69.** $\dfrac{9x^2-3x+1}{27x^3+1}$

**70.** $\dfrac{ax^2+bx^2-ay^2-by^2}{a^2x+a^2y-b^2x-b^2y}$

## 3.2 Multiplication and Division

**OBJECTIVES**

In this section, you will be learning to:

1. Multiply rational expressions.
2. Divide rational expressions.

How would you find the product $\frac{2}{3} \cdot \frac{5}{9}$ or the product $\frac{15}{7} \cdot \frac{49}{65}$?

$$\frac{2}{3} \cdot \frac{5}{9} = \frac{2 \cdot 5}{3 \cdot 9} = \frac{10}{27} \quad \text{and} \quad \frac{15}{7} \cdot \frac{49}{65} = \frac{3 \cdot \cancel{5} \cdot \cancel{7} \cdot 7}{\cancel{7} \cdot \cancel{5} \cdot 13} = \frac{21}{13}$$

The same techniques are used to multiply rational expressions.

$$\frac{2x}{x-6} \cdot \frac{x+5}{x-4} = \frac{2x(x+5)}{(x-6)(x-4)} = \frac{2x^2+10x}{x^2-10x+24}$$

$$\frac{y^2-4}{y^3} \cdot \frac{y^2-3y}{y^2-y-6} = \frac{\cancel{(y+2)}(y-2) \cdot \cancel{y(y-3)}}{\cancel{y^3}^{\,y^2}\cancel{(y-3)}\cancel{(y+2)}} = \frac{y-2}{y^2}$$

**Multiplication with Rational Expressions**

$$\frac{P}{Q} \cdot \frac{R}{S} = \frac{P \cdot R}{Q \cdot S} \quad \text{where } Q, S \neq 0$$

**EXAMPLES** Find the following products and reduce to lowest terms by factoring whenever possible. Assume that no denominator has a value of 0.

**1.** $\frac{5x^2y}{9xy^3} \cdot \frac{6x^3y^2}{15xy^4} = \frac{\overset{1}{\cancel{5}} \cdot \overset{2}{\cancel{6}}x^5y^3}{\underset{3}{\cancel{9}} \cdot \underset{3}{\cancel{15}}x^2y^7} = \frac{2x^{5-2}}{9y^{7-3}} = \frac{2x^3}{9y^4}$

**2.** $\frac{x}{x-2} \cdot \frac{x^2-4}{x^2} = \frac{\cancel{x}(x+2)\cancel{(x-2)}}{\cancel{(x-2)}\underset{x}{\cancel{x^2}}} = \frac{x+2}{x}$

**3.** $\frac{3x-3}{x^2+x} \cdot \frac{x^2+2x+1}{3x^2-6x+3} = \frac{\cancel{3(x-1)} \cdot \overset{(x+1)}{\cancel{(x+1)^2}}}{x\cancel{(x+1)} \cdot \cancel{3}\underset{(x-1)}{\cancel{(x-1)^2}}} = \frac{x+1}{x(x-1)}$

**4.** $\frac{4-2x}{x^2-25} \cdot \frac{x^2+7x+10}{x^2-4} = \frac{2\overset{-1}{\cancel{(2-x)}} \cdot \cancel{(x+5)}\cancel{(x+2)}}{\cancel{(x+5)}(x-5) \cdot \cancel{(x-2)}\cancel{(x+2)}} = \frac{-2}{x-5}$

$\left(\textbf{Note: } \frac{2-x}{x-2} = -1.\right)$

■

A fraction can be thought of as division or multiplication in the following manner:

$$\frac{a}{b} = a \div b \quad \text{and} \quad \frac{a}{b} = a \cdot \frac{1}{b}$$

And, if $b, c, d \neq 0$,

$$\frac{\frac{a}{b}}{\frac{c}{d}} = \frac{a}{b} \div \frac{c}{d} \quad \text{and} \quad \frac{\frac{a}{b}}{\frac{c}{d}} = \frac{a}{b} \cdot \frac{1}{\frac{c}{d}} = \frac{a}{b} \cdot \frac{d}{c}$$

$\frac{d}{c}$ is the **reciprocal** of $\frac{c}{d}$ and $\frac{1}{\frac{c}{d}} = \frac{d}{c}$. We have the result

$$\frac{a}{b} \div \frac{c}{d} = \frac{a}{b} \cdot \frac{d}{c}$$

**Dividing by a number is the same as multiplying by its reciprocal.** This same technique is used to divide rational expressions.

***Division with Rational Expressions***

$$\frac{P}{Q} \div \frac{R}{S} = \frac{P}{Q} \cdot \frac{S}{R} \quad \text{where } Q, R, S \neq 0$$

**EXAMPLES** Find the following quotients and reduce to lowest terms by factoring whenever possible. Assume that no denominator has a value of 0.

**5.** $\dfrac{12x^2y}{10xy^2} \div \dfrac{3x^4y}{xy^3} = \dfrac{12x^2y}{10xy^2} \cdot \dfrac{xy^3}{3x^4y}$

$$= \frac{\cancel{2} \cdot 2 \cdot \cancel{3}x^3y^4}{\cancel{2} \cdot 5 \cdot \cancel{3}x^5y^3} = \frac{2y^{4-3}}{5x^{5-3}} = \frac{2y}{5x^2}$$

**6.** $\dfrac{x^2}{x-1} \div \dfrac{x^2+x}{x^2-1} = \dfrac{x^2}{x-1} \cdot \dfrac{x^2-1}{x^2+x} = \dfrac{\overset{x}{\cancel{x^2}}\overset{1}{\cancel{(x+1)}}\overset{1}{\cancel{(x-1)}}}{\underset{1}{\cancel{(x-1)}}\cancel{x}\underset{1}{\cancel{(x+1)}}} = \dfrac{x}{1} = x$

**7.** $\dfrac{x^3-y^3}{x^3} \div \dfrac{y-x}{xy} = \dfrac{x^3-y^3}{x^3} \cdot \dfrac{xy}{y-x} = \dfrac{\overset{-1}{\cancel{(x-y)}}(x^2+xy+y^2)\cancel{x}y}{\underset{x^2}{\cancel{x^3}}\cancel{(y-x)}}$

$$= \frac{-y(x^2+xy+y^2)}{x^2}$$

**8.** $\dfrac{x^2 - 8x + 15}{2x^2 + 11x + 5} \div \dfrac{2x^2 - 5x - 3}{4x^2 - 1}$

$$= \frac{x^2 - 8x + 15}{2x^2 + 11x + 5} \cdot \frac{4x^2 - 1}{2x^2 - 5x - 3}$$

$$= \frac{\cancel{(x-3)}(x-5)(2x-1)\cancel{(2x+1)}}{(2x+1)(x+5)\cancel{(x-3)}\cancel{(2x+1)}}$$

$$= \frac{(x-5)(2x-1)}{(2x+1)(x+5)} = \frac{2x^2 - 11x + 5}{(2x+1)(x+5)}$$

(**Note:** Generally the denominator is left in factored form.) ■

***Practice Problems***

Perform the following operations and simplify the results. Assume that no denominator is 0.

1. $\dfrac{x - 7}{x^3} \cdot \dfrac{x^2}{49 - x^2}$
2. $\dfrac{y^2 - y - 6}{y^2 - 5y + 6} \cdot \dfrac{y^2 - 4}{y^2 + 4y + 4}$
3. $\dfrac{x^3 + 3x}{2x + 1} \div \dfrac{x^2 + 3}{x + 1}$
4. $\dfrac{x^2 + 2x - 3}{x^2 - 3x - 10} \cdot \dfrac{2x^2 - 9x - 5}{x^2 - 2x + 1} \div \dfrac{4x + 2}{x^2 - x}$

## *EXERCISES 3.2*

Perform the indicated operations in Exercises 1–60. Assume that no denominator is 0.

**1.** $\dfrac{ax^2}{b} \cdot \dfrac{b^2}{x^2y}$

**2.** $\dfrac{18x^3}{5y^2} \cdot \dfrac{30y^3}{9x^4}$

**3.** $\dfrac{24x^3}{25y^2} \cdot \dfrac{10y^5}{18x}$

**4.** $\dfrac{16x^8}{3y^{11}} \cdot \dfrac{-21y^9}{10x^7}$

**5.** $\dfrac{3}{x + 6} \cdot \dfrac{x + 6}{x}$

**6.** $\dfrac{4x}{x - 2} \cdot \dfrac{2x - 4}{3}$

**7.** $\dfrac{5x + 10}{4x} \cdot \dfrac{2x}{3x + 6}$

**8.** $\dfrac{x^2 - 4}{x + 3} \cdot \dfrac{6}{x - 2}$

**9.** $\dfrac{x^2 - 9}{x^2 + 2x} \cdot \dfrac{x + 2}{x - 3}$

**10.** $\dfrac{16x^2 - 9}{3x^2 - 15x} \cdot \dfrac{6}{4x + 3}$

**11.** $\dfrac{x^2 + 2x - 3}{x^2 + 3x} \cdot \dfrac{x}{x + 1}$

**12.** $\dfrac{4x + 16}{x^2 - 16} \cdot \dfrac{x - 4}{x}$

**13.** $\dfrac{x^2 + 6x - 16}{x^2 - 64} \cdot \dfrac{1}{2 - x}$

**14.** $\dfrac{4 - x^2}{x^2 - 4x + 4} \cdot \dfrac{3}{x + 2}$

**15.** $\dfrac{x^2 - 5x + 6}{x^2 - 4x} \cdot \dfrac{x - 4}{x - 3}$

**16.** $\dfrac{2x^2 + x - 3}{x^2 + 4x} \cdot \dfrac{2x + 8}{x - 1}$

**Answers to Practice Problems** 1. $\dfrac{-1}{x(x + 7)}$ 2. 1 3. $\dfrac{x^2 + x}{2x + 1}$ 4. $\dfrac{x^2 + 3x}{2(x + 2)}$

17. $\frac{2x^2 + 10x}{3x^2 + 5x + 2} \cdot \frac{6x + 4}{x^2}$

18. $\frac{x + 3}{x^2 - 16} \cdot \frac{x^2 - 3x - 4}{x^2 - 1}$

19. $\frac{x}{x^2 + 7x + 12} \cdot \frac{x^2 - 2x - 24}{x^2 - 7x + 6}$

20. $\frac{x^2 - 2x - 3}{x + 5} \cdot \frac{x^2 - 5x - 14}{x^2 - x - 6}$

21. $\frac{8 - 2x - x^2}{x^2 - 2x} \cdot \frac{x - 4}{x^2 - 3x - 4}$

22. $\frac{3x^2 + 21x}{x^2 - 49} \cdot \frac{x^2 - 5x + 4}{x^2 + 3x - 4}$

23. $\frac{(x - 2y)^2}{x^2 - 5xy + 6y^2} \cdot \frac{x + 2y}{x^2 - 4xy + 4y^2}$

24. $\frac{4x^2 + 6x}{x^2 + 3x - 10} \cdot \frac{x^2 + 4x - 12}{x^2 + 5x - 6}$

25. $\frac{2x^2 + 5x + 2}{3x^2 + 8x + 4} \cdot \frac{3x^2 - x - 2}{4x^3 - x}$

26. $\frac{x^2 + 5x}{4x^2 + 12x + 9} \cdot \frac{6x^2 + 7x - 3}{x^2 + 10x + 25}$

27. $\frac{2x^2 - 7x + 3}{x^2 - 9} \cdot \frac{3x^2 + 8x - 3}{6x^2 + x - 1}$

28. $\frac{12x^2y}{9xy^2} \div \frac{4x^4y}{x^2y^3}$

29. $\frac{35xy^3}{24x^3y} \div \frac{15x^4y^3}{84xy^4}$

30. $\frac{15x^3y^2}{19x^5y^2} \div \frac{40x^2y^3}{38xy^4}$

31. $\frac{45xy^4}{21x^2y^2} \div \frac{40x^4}{112xy^5}$

32. $\frac{x - 3}{15x} \div \frac{4x - 12}{5}$

33. $\frac{x - 1}{6x + 6} \div \frac{2x - 2}{x^2 + x}$

34. $\frac{7x - 14}{x^2} \div \frac{x^2 - 4}{x^3}$

35. $\frac{6x^2 - 54}{x^4} \div \frac{x - 3}{x^2}$

36. $\frac{x^2 - 25}{6x + 30} \div \frac{x - 5}{x}$

37. $\frac{2x - 1}{x^2 + 2x} \div \frac{10x^2 - 5x}{6x^2 + 12x}$

38. $\frac{x + 3}{x^2 + 3x - 4} \div \frac{x + 2}{x^2 + x - 2}$

39. $\frac{6x^2 - 7x - 3}{x^2 - 1} \div \frac{2x - 3}{x - 1}$

40. $\frac{x^2 - 9}{2x^2 + 7x + 3} \div \frac{x^2 - 3x}{2x^2 + 11x + 5}$

41. $\frac{x^2 - 8x + 15}{x^2 - 9x + 14} \div \frac{x^2 + 4x - 21}{7 - x}$

42. $\frac{2x + 1}{x^2 - 4x} \div \frac{4x^2 - 1}{x^2 - 16}$

43. $\frac{x^2 - 6x + 9}{x^2 - 4x + 3} \div \frac{2x^2 - 7x + 3}{x^2 - 3x + 2}$

44. $\frac{x^2 - 4x + 4}{x^2 + 5x + 6} \div \frac{x^2 + 2x - 8}{x^2 + 7x + 12}$

45. $\frac{x^2 - x - 6}{x^2 + 6x + 8} \div \frac{x^2 - 4x + 3}{x^2 + 5x + 4}$

46. $\frac{x^2 - x - 12}{6x^2 + x - 9} \div \frac{x^2 - 6x + 8}{3x^2 - x - 6}$

47. $\frac{6x^2 + 5x + 1}{4x^3 - x} \div \frac{3x^2 - 2x - 1}{2x^2 - 3x + 1}$

48. $\frac{8x^2 + 2x - 15}{3x^2 + 13x + 4} \div \frac{2x^2 + 5x + 3}{6x^2 - x - 1}$

49. $\frac{3x^2 + 13x + 14}{4x^3 - 3x^2} \div \frac{6x^2 - x - 35}{4x^2 + 5x - 6}$

50. $\frac{6 - 11x - 10x^2}{2x^2 + x - 3} \div \frac{5x^3 - 2x^2}{3x^2 - 5x + 2}$

51. $\frac{x - 6}{x^2 - 7x + 6} \cdot \frac{x^2 - 3x}{x + 3} \cdot \frac{x^2 - 9}{x^2 - 4x + 3}$

52. $\frac{3x^2 + 11x + 10}{2x^2 + x - 6} \cdot \frac{x^2 + 2x - 3}{2x - 1} \cdot \frac{2x - 3}{3x^2 + 2x - 5}$

53. $\frac{x^3 + 3x^2}{x^2 + 7x + 12} \cdot \frac{2x^2 + 7x - 4}{2x^2 - x} \div \frac{2x^2 - x - 1}{x^2 + 4x - 5}$

54. $\frac{x^2 + 2x - 3}{x^2 + 10x + 21} \div \frac{x^2 - 7x - 8}{x^2 + 6x + 5} \cdot \frac{x^2 - x - 56}{x^2 - 3x - 40}$

55. $\frac{2x^2 - 5x + 2}{4xy - 2y + 6x - 3} \div \frac{xy - 2y + 3x - 6}{2y^2 + 9y + 9}$

56. $\frac{xy + 4x - 3y - 12}{x^2 - 3x + 9} \div \frac{7x + xy - 3y - 21}{x^3 + 27}$

57. $\dfrac{x^3 + 3x^2 - 9x - 27}{2xy + x + 6y + 3} \div \dfrac{x^2 - x - 12}{2xy + x - 8y - 4}$

58. $\dfrac{x^3 - 8}{x^2 - 25} \div \dfrac{x^2 + 2x + 4}{x^3 - 5x^2 + 6x - 30}$

59. $\dfrac{x^3 - y^3}{x^2} \div \dfrac{x^2 + xy + y^2}{x^2 + 2xy + y^2} \cdot \dfrac{y^2}{x^2 - y^2}$

60. $\dfrac{x^3 + 27}{x^3 + 6x^2 + 9x} \div \dfrac{9 - x^2}{x^2 - 3x} \cdot \dfrac{x^2 + 4x + 3}{9x - 3x^2 + x^3}$

## 3.3 Addition and Subtraction

**OBJECTIVES**

In this section, you will be learning to:

1. Find the least common multiples of sets of algebraic expressions.
2. Add rational expressions.
3. Subtract rational expressions.

Just as with adding real numbers, a sum such as $\dfrac{x}{x+2} + \dfrac{3}{x+2}$ is found by adding the numerators and using the common denominator. Thus,

$$\frac{x}{x+2} + \frac{3}{x+2} = \frac{x+3}{x+2}$$

The denominator $x + 2$ cannot be 0. That is, $x \neq -2$. For any other value of $x$, the sum of the two fractions will equal the one fraction. For example, if $x = 5$,

$$\frac{x}{x+2} + \frac{3}{x+2} = \frac{5}{5+2} + \frac{3}{5+2} = \frac{5}{7} + \frac{3}{7} = \frac{5+3}{7} = \frac{8}{7}$$

For the following sum, the result can be reduced.

$$\frac{x}{x^2-1} + \frac{1}{x^2-1} = \frac{x+1}{x^2-1} \qquad \text{Factor and reduce.}$$

$$= \frac{\overset{1}{\cancel{(x+1)}}}{\cancel{(x+1)}(x-1)} = \frac{1}{x-1}$$

In this case, the denominator $x^2 - 1 \neq 0$ (or $x \neq \pm 1$).

A difference such as $\dfrac{x^2}{x^2+4x+4} - \dfrac{2x+8}{x^2+4x+4}$ is found by subtracting the numerators and using the common denominator.

$$\frac{x^2}{x^2+4x+4} - \frac{2x+8}{x^2+4x+4} = \frac{x^2-(2x+8)}{x^2+4x+4} = \frac{x^2-2x-8}{x^2+4x+4}$$

Again, the result can be reduced.

$$\frac{x^2-2x-8}{x^2+4x+4} = \frac{(x-4)\cancel{(x+2)}}{(x+2)\cancel{(x+2)}} = \frac{x-4}{x+2}$$

**Addition and Subtraction with Rational Expressions**

$$\frac{P}{Q} + \frac{R}{Q} = \frac{P+R}{Q} \quad \text{and} \quad \frac{P}{Q} - \frac{R}{Q} = \frac{P-R}{Q} \quad \text{where } Q \neq 0$$

Note that the expression $P - R$ indicates the difference of two polynomials and this will affect all the signs in $R$. **A good idea is to put $P$ in parentheses and $R$ in parentheses so that all changes in sign will be done correctly.**

**EXAMPLES** Find the following sums or differences. Reduce if possible. Assume that no denominator is 0.

**1.** $\frac{2x+1}{3x-3} + \frac{x+2}{3x-3} = \frac{(2x+1)+(x+2)}{3x-3}$

$= \frac{3x+3}{3x-3} = \frac{\not{3}(x+1)}{\not{3}(x-1)} = \frac{x+1}{x-1}$

**2.** $\frac{2x-5y}{x+y} - \frac{3x-7y}{x+y} = \frac{(2x-5y)-(3x-7y)}{x+y}$

$= \frac{2x-5y-3x+7y}{x+y} = \frac{-x+2y}{x+y}$ ■

In the following discussion, we will illustrate how to find the least common multiple (LCM) for sets of polynomials and how to add rational expressions that do not have the same denominator.

***To Find the LCM***

1. Find the complete factorization of each expression, including the prime factors of any constant factor.
2. Form the product of all the factors that appear in the complete factorizations using each factor the most number of times it appears in any one factorization.

Consider the three terms $\{18x^3, 24xy, 63\}$. To find the LCM:

**1.** Find the complete factorization of each term including prime factors.

$18x^3 = 2 \cdot 3 \cdot 3 \cdot x^3$ one 2, two 3s, $x^3$
$24xy = 2 \cdot 2 \cdot 2 \cdot 3 \cdot x \cdot y$ three 2s, one 3, $x$, $y$
$63 = 3 \cdot 3 \cdot 7$ two 3s, one 7

**2.** Form a product using each prime factor and each variable the most number of times it appears in **any one** of the factorizations.

$\text{LCM} = 2 \cdot 2 \cdot 2 \cdot 3 \cdot 3 \cdot 7 \cdot x^3 \cdot y$ three 2s, two 3s, one 7, $x^3$, $y$
$= 504x^3y$

This product, $504x^3y$, is the least common multiple. This is the very smallest number with the smallest positive exponents on the variables that is divisible by all three terms.

Now we employ the same technique to find the LCM for the polynomial expressions $x^2 + 6x + 9$, $x^2 - 9$, $2x + 6$.

1. Factor each expression completely.

$$x^2 + 6x + 9 = (x + 3)^2$$
$$x^2 - 9 = (x + 3)(x - 3)$$
$$2x + 6 = 2(x + 3)$$

2. For the LCM, we form the product of 2, $(x + 3)^2$, and $(x - 3)$. That is, we use each factor the most number of times it appears in **any one** factorization.

$$\text{LCM} = 2(x + 3)^2(x - 3)$$

This LCM is the least common denominator (LCD) for the rational expressions in the sum $\frac{1}{x^2 + 6x + 9} + \frac{1}{x^2 - 9} + \frac{1}{2x + 6}$.

By using the Fundamental Principle of Fractions (see Section 3.1), each rational expression is multiplied by 1 in a form that will give an equivalent expression with the desired denominator. Thus,

$$\frac{1}{x^2 + 6x + 9} = \frac{1 \cdot \mathbf{2(x - 3)}}{(x + 3)^2 \cdot \mathbf{2(x - 3)}}$$

$$\frac{1}{x^2 - 9} = \frac{1 \cdot \mathbf{2(x + 3)}}{(x + 3)(x - 3) \cdot \mathbf{2(x + 3)}}$$

$$\frac{1}{2x + 6} = \frac{1 \cdot \mathbf{(x + 3)(x - 3)}}{2(x + 3) \cdot \mathbf{(x + 3)(x - 3)}}$$

Each rational expression now has the same denominator, $2(x + 3)^2(x - 3)$.

$$\frac{1}{x^2 + 6x + 9} + \frac{1}{x^2 - 9} + \frac{1}{2x + 6}$$

$$= \frac{1}{(x + 3)^2} + \frac{1}{(x + 3)(x - 3)} + \frac{1}{2(x + 3)}$$

$$= \frac{1 \cdot \mathbf{2(x - 3)}}{(x + 3)^2 \cdot \mathbf{2(x - 3)}} + \frac{1 \cdot \mathbf{2(x + 3)}}{(x + 3)(x - 3) \cdot \mathbf{2(x + 3)}} + \frac{1 \cdot \mathbf{(x + 3)(x - 3)}}{2(x + 3) \cdot \mathbf{(x + 3)(x - 3)}}$$

$$= \frac{(2x - 6) + (2x + 6) + (x^2 - 9)}{2(x + 3)^2(x - 3)}$$

$$= \frac{x^2 + 4x - 9}{2(x + 3)^2(x - 3)}$$

**EXAMPLES** Find the following sums or differences. Reduce if possible. Assume that no denominator is 0.

**3.** $\dfrac{x}{x-3} + \dfrac{6}{x+4}$

**Solution**

$$\left.\begin{aligned} x-3 &= x-3 \\ x+4 &= x+4 \end{aligned}\right\} \quad \text{LCM} = (x-3)(x+4)$$

Here each factor appears only once so the LCM is the product of these factors.

$$\begin{aligned} \frac{x}{x-3} + \frac{6}{x+4} &= \frac{x\boldsymbol{(x+4)}}{(x-3)\boldsymbol{(x+4)}} + \frac{6\boldsymbol{(x-3)}}{(x+4)\boldsymbol{(x-3)}} \\ &= \frac{(x^2+4x)+(6x-18)}{(x-3)(x+4)} \\ &= \frac{x^2+4x+6x-18}{(x-3)(x+4)} \\ &= \frac{x^2+10x-18}{(x-3)(x+4)} \end{aligned}$$

(**Note:** The denominator is left in factored form as a convenience for possibly reducing or adding to some other expression later. The student may choose to multiply these factors. Either form is correct.)

**4.** $\dfrac{x+5}{x-5} - \dfrac{x^2+5x}{x^2-25}$

**Solution** $$\left.\begin{aligned} x-5 &= x-5 \\ x^2-25 &= (x+5)(x-5) \end{aligned}\right\} \quad \text{LCM} = (x+5)(x-5)$$

$$\begin{aligned} \frac{x+5}{x-5} - \frac{x^2+5x}{x^2-25} &= \frac{(x+5)\boldsymbol{(x+5)}}{(x-5)\boldsymbol{(x+5)}} - \frac{x^2+5x}{(x+5)(x-5)} \\ &= \frac{(x^2+10x+25)-(x^2+5x)}{(x+5)(x-5)} \\ &= \frac{x^2+10x+25-x^2-5x}{(x+5)(x-5)} \\ &= \frac{5x+25}{(x+5)(x-5)} = \frac{5\cancel{(x+5)}}{\cancel{(x+5)}(x-5)} = \frac{5}{x-5} \end{aligned}$$

**5.** $\dfrac{x+y}{(x-y)^2} + \dfrac{x}{2x^2-2y^2}$

**Solution**

$$\left.\begin{aligned}(x-y)^2 &= (x-y)^2\\ 2x^2-2y^2 &= 2(x+y)(x-y)\end{aligned}\right\}\text{LCM} = 2(x-y)^2(x+y)$$

$$\frac{x+y}{(x-y)^2} + \frac{x}{2x^2-2y^2}$$

$$= \frac{(x+y)\cdot \mathbf{2(x+y)}}{(x-y)^2\cdot \mathbf{2(x+y)}} + \frac{x\mathbf{(x-y)}}{2(x+y)(x-y)\mathbf{(x-y)}}$$

$$= \frac{2x^2+4xy+2y^2+x^2-xy}{2(x-y)^2(x+y)} = \frac{3x^2+3xy+2y^2}{2(x-y)^2(x+y)}$$

**6.** $\dfrac{3x-12}{x^2+x-20} - \dfrac{x^2+5x}{x^2+9x+20}$

**Hint:** In this problem, both expressions can be reduced before looking for the LCD.

**Solution**

$$\frac{3x-12}{x^2+x-20} - \frac{x^2+5x}{x^2+9x+20}$$

$$= \frac{3\cancel{(x-4)}}{(x+5)\cancel{(x-4)}} - \frac{x\cancel{(x+5)}}{\cancel{(x+5)}(x+4)}$$

$$= \frac{3}{x+5} - \frac{x}{x+4}$$

Now subtract these two expressions with LCD $(x+5)(x+4)$.

$$\frac{3}{x+5} - \frac{x}{x+4} = \frac{3\mathbf{(x+4)}}{(x+5)\mathbf{(x+4)}} - \frac{x\mathbf{(x+5)}}{(x+4)\mathbf{(x+5)}}$$

$$= \frac{(3x+12)-(x^2+5x)}{(x+5)(x+4)}$$

$$= \frac{3x+12-x^2-5x}{(x+5)(x+4)} = \frac{-x^2-2x+12}{(x+5)(x+4)}$$

**7.** $\dfrac{5}{x^2 - 1} + \dfrac{4}{1 - x}$

**Solution** $\left.\begin{aligned} x^2 - 1 &= (x + 1)(x - 1) \\ 1 - x &= -1(x - 1) \end{aligned}\right\}$ In this case, note that $1 - x = -1(x - 1)$.

Rewrite the problem by changing $1 - x$ to $-1(x - 1)$:

$$\frac{5}{x^2 - 1} + \frac{4}{1 - x}$$

$$= \frac{5}{(x + 1)(x - 1)} + \frac{4}{-1(x - 1)}$$

$$= \frac{5}{(x + 1)(x - 1)} - \frac{4}{x - 1}$$ For these expressions, LCM $= (x + 1)(x - 1)$.

$$= \frac{5}{(x + 1)(x - 1)} - \frac{4(\mathbf{x + 1})}{(x - 1)(\mathbf{x + 1})}$$

$$= \frac{5 - (4x + 4)}{(x + 1)(x - 1)} = \frac{5 - 4x - 4}{(x + 1)(x - 1)}$$

$$= \frac{-4x + 1}{(x + 1)(x - 1)}$$

**8.** $\dfrac{x + 1}{xy - 3y + 4x - 12} + \dfrac{x - 3}{xy + 6y + 4x + 24}$

**Solution**

$$\left.\begin{aligned} xy - 3y + 4x - 12 &= y(x - 3) + 4(x - 3) \\ &= (x - 3)(y + 4) \\ xy + 6y + 4x + 24 &= y(x + 6) + 4(x + 6) \\ &= (x + 6)(y + 4) \end{aligned}\right\} \text{LCM} = (y + 4)(x - 3)(x + 6)$$

$$\frac{(x + 1)(\mathbf{x + 6})}{(x - 3)(y + 4)(\mathbf{x + 6})} + \frac{(x - 3)(\mathbf{x - 3})}{(x + 6)(y + 4)(\mathbf{x - 3})}$$

$$= \frac{x^2 + 7x + 6 + x^2 - 6x + 9}{(y + 4)(x - 3)(x + 6)}$$

$$= \frac{2x^2 + x + 15}{(y + 4)(x - 3)(x + 6)}$$ ■

**Practice Problems**

Perform the indicated operations and reduce if possible. Assume that no denominator is 0.

1. $\dfrac{x}{x^2 - 1} + \dfrac{1}{x - 1}$
2. $\dfrac{x + 3}{x^2 + x - 6} + \dfrac{x - 2}{x^2 + 4x - 12}$
3. $\dfrac{1}{y + 2} - \dfrac{1}{y^3 + 8}$
4. $\dfrac{1}{1 - y} - \dfrac{2}{y^2 - 1}$

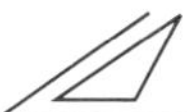

## EXERCISES 3.3

Perform the indicated operations and reduce if possible. Assume that no denominator is 0.

1. $\dfrac{3x}{x + 4} + \dfrac{12}{x + 4}$
2. $\dfrac{7x}{x + 5} + \dfrac{35}{x + 5}$
3. $\dfrac{x - 1}{x + 6} + \dfrac{x + 13}{x + 6}$
4. $\dfrac{3x - 1}{2x - 6} + \dfrac{x - 11}{2x - 6}$
5. $\dfrac{3x + 1}{5x + 2} + \dfrac{2x + 1}{5x + 2}$
6. $\dfrac{x^2 + 3}{x + 1} + \dfrac{4x}{x + 1}$
7. $\dfrac{x - 5}{x^2 - 2x + 1} + \dfrac{x + 3}{x^2 - 2x + 1}$
8. $\dfrac{2x^2 + 5}{x^2 - 4} + \dfrac{3x - 1}{x^2 - 4}$
9. $\dfrac{6x}{x - 5} - \dfrac{30}{x - 5}$
10. $\dfrac{3x}{2x + 4} - \dfrac{2x - 1}{2x + 4}$
11. $\dfrac{10}{x^2 - 5x + 6} - \dfrac{5x}{x^2 - 5x + 6}$
12. $\dfrac{x^2 + 2}{x^2 + x - 12} - \dfrac{x + 1}{x^2 + x - 12}$
13. $\dfrac{3x + 2}{2x + 8} - \dfrac{2x - 2}{2x + 8}$
14. $\dfrac{2x + 1}{x^2 - x - 6} - \dfrac{x - 1}{x^2 - x - 6}$
15. $\dfrac{x^2 + 2}{x^2 - 4} - \dfrac{4x - 2}{x^2 - 4}$
16. $\dfrac{2x + 5}{2x^2 - x - 1} - \dfrac{4x + 2}{2x^2 - x - 1}$
17. $\dfrac{x + 3}{7x - 2} + \dfrac{2x - 1}{14x - 4}$
18. $\dfrac{3x + 1}{4x + 10} + \dfrac{4 - x}{2x + 5}$
19. $\dfrac{5}{x - 3} + \dfrac{x}{x^2 - 9}$
20. $\dfrac{x + 1}{x^2 - 3x - 10} + \dfrac{x}{x - 5}$
21. $\dfrac{x}{x - 1} - \dfrac{4}{x + 2}$
22. $\dfrac{x - 1}{3x - 1} - \dfrac{8 + 4x}{x + 2}$
23. $\dfrac{x + 2}{x + 3} - \dfrac{4}{3 - x}$
24. $\dfrac{x - 1}{4 - x} + \dfrac{3x}{x + 5}$
25. $\dfrac{x + 2}{3x + 9} + \dfrac{2x - 1}{2x - 6}$
26. $\dfrac{x}{4x - 8} - \dfrac{3x + 2}{3x + 6}$
27. $\dfrac{3x}{6 + x} - \dfrac{2x}{x^2 - 36}$
28. $\dfrac{3x - 4}{x^2 - x - 20} - \dfrac{2}{5 - x}$
29. $\dfrac{4x + 1}{7 - x} + \dfrac{x - 1}{x^2 - 8x + 7}$
30. $\dfrac{4}{x + 5} - \dfrac{2x + 3}{x^2 + 4x - 5}$
31. $\dfrac{4x}{x^2 + 3x - 28} + \dfrac{3}{x^2 + 6x - 7}$
32. $\dfrac{3x}{x^2 + 2x + 1} - \dfrac{x}{x^2 + 5x + 4}$
33. $\dfrac{x + 1}{x^2 + 4x + 4} - \dfrac{x - 3}{x^2 - 4}$

**Answers to Practice Problems** 1. $\dfrac{2x + 1}{(x + 1)(x - 1)}$ 2. $\dfrac{2x + 4}{(x - 2)(x + 6)}$ 3. $\dfrac{y^2 - 2y + 3}{y^3 + 8}$ 4. $\dfrac{-y - 3}{(y + 1)(y - 1)}$

34. $\frac{x - 4}{x^2 - 5x + 6} + \frac{2x}{x^2 - 2x - 3}$

35. $\frac{3x}{9 - x^2} + \frac{5}{x^2 - 7x + 12}$

36. $\frac{4x}{3x^2 + 4x + 1} - \frac{x + 4}{x^2 + 7x + 6}$

37. $\frac{x - 6}{7x^2 - 3x - 4} + \frac{7 - x}{7x^2 + 18x + 8}$

38. $\frac{x + 5}{9x^2 - 26x - 3} - \frac{8x}{9x^2 + 37x + 4}$

39. $\frac{x - 3}{4x^2 - 5x - 6} - \frac{4x + 10}{2x^2 + x - 10}$

40. $\frac{2x + 1}{8x^2 - 37x - 15} + \frac{2 - x}{8x^2 + 11x + 3}$

41. $\frac{3x}{4 - x} + \frac{7x}{x + 4} - \frac{x + 3}{x^2 - 16}$

42. $\frac{x}{x + 3} + \frac{x + 1}{3 - x} + \frac{x^2 + 4}{x^2 - 9}$

43. $2 - \frac{4x + 1}{x - 4} + \frac{x - 3}{x^2 - 6x + 8}$

44. $-4 + \frac{1 - 2x}{x + 6} + \frac{x^2 + 1}{x^2 + 4x - 12}$

45. $\frac{2}{x^2 - 4} - \frac{3}{x^2 - 3x + 2} + \frac{x - 1}{x^2 + x - 2}$

46. $\frac{x}{x^2 + 4x - 21} + \frac{1 - x}{x^2 + 8x + 7} + \frac{3x}{x^2 - 2x - 3}$

47. $\frac{3(x + 3)}{x^2 - 5x + 4} + \frac{49}{12 + x - x^2} + \frac{3x + 21}{x^2 + 2x - 3}$

48. $\frac{4}{x^2 + 3x - 10} + \frac{3}{x^2 - 25} - \frac{5}{x^2 - 7x + 10}$

49. $\frac{5x + 22}{x^2 + 8x + 15} - \frac{4}{x^2 + 4x + 3} + \frac{6}{x^2 + 6x + 5}$

50. $\frac{x + 1}{2x^2 - x - 1} + \frac{2x}{2x^2 + 5x + 2} - \frac{2x}{3x^2 + 4x - 4}$

51. $\frac{x - 6}{3x^2 + 10x + 3} - \frac{2x}{5x^2 - 3x - 2} + \frac{2x}{3x^2 - 2x - 1}$

52. $\frac{x}{xy + x - 2y - 2} + \frac{x + 2}{xy + x + y + 1}$

53. $\frac{4x}{xy - 3x + y - 3} + \frac{x + 2}{xy + 2y - 3x - 6}$

54. $\frac{3y}{xy + 3y + 2x + 6} - \frac{x}{x^2 - 2x - 15}$

55. $\frac{2}{xy - 4x - 2y + 8} + \frac{5y}{y^2 - 3y - 4}$

56. $\frac{x + 6}{x^2 + x + 1} - \frac{3x^2 + x - 4}{x^3 - 1}$

57. $\frac{2x - 5}{8x^2 - 4x + 2} + \frac{x^2 - 2x + 5}{8x^3 + 1}$

58. $\frac{x + 1}{x^3 - 3x^2 + x - 3} + \frac{x^2 - 5x - 8}{x^4 - 8x^2 - 9}$

59. $\frac{x + 4}{x^3 - 5x^2 + 6x - 30} - \frac{x - 7}{x^3 - 2x^2 + 6x - 12}$

60. $\frac{x + 2}{9x^2 - 6x + 4} + \frac{10x - 5x^2}{27x^3 + 8} - \frac{2}{3x + 2}$

## 3.4 Long Division and Synthetic Division

**OBJECTIVES**

In this section, you will be learning to:

1. Divide polynomials by monomials.
2. Divide polynomials using the long division algorithm.
3. Divide polynomials using synthetic division.

We have divided with rational expressions by factoring and reducing if possible. There are times when we divide to determine whether or not the divisor is a factor of the dividend. (If it is, the remainder will be 0.) At other times, we simply want to change the form so that some other operation can be performed.

A rational expression (a form of fraction) indicates division of the numerator by the denominator. We begin with two examples in which the denominator (or divisor) is a monomial.

## EXAMPLES

**1.** Divide $x^3 - 6x^2 + 2x$ by $3x$.

**Solution** Write the quotient as a sum of fractions by dividing each term in the numerator by the denominator.

$$\frac{x^3 - 6x^2 + 2x}{3x} = \frac{x^3}{3x} - \frac{6x^2}{3x} + \frac{2x}{3x}$$ Divide each term in the numerator by $3x$.

$$= \frac{x^2}{3} - 2x + \frac{2}{3}$$ Simplify each fraction.

**2.** Divide $\dfrac{15y^3 + 20y^2 - 5y}{5y}$ by dividing each term in the numerator by the denominator.

**Solution** $$\frac{15y^3 + 20y^2 - 5y}{5y} = \frac{15y^3}{5y} + \frac{20y^2}{5y} - \frac{5y}{5y}$$

$$= 3y^2 + 4y - 1$$ ■

If the denominator is not a monomial and the degree of the numerator is equal to or larger than the degree of the denominator, the division can be performed in a manner similar to dividing with whole numbers. The method is called the **division algorithm** or **long division.** (An algorithm is a process or series of steps for solving a problem.) The procedure is illustrated in the following three examples.

## EXAMPLE

**3.** Simplify $\dfrac{6x^2 - 2x + 1}{2x - 4}$ by using long division.

**Solution**

**Step 1** $2x - 4 \overline{)\,6x^2 - 2x + 1}$

Write the expression in the long division format with both polynomials written in descending powers of the variables.

**Step 2**
$$\begin{array}{r} 3x \phantom{ - 2x + 1} \\ 2x - 4 \overline{)\,6x^2 - 2x + 1} \end{array}$$

Divide $6x^2$ by $2x$, $\left(\dfrac{6x^2}{2x} = 3x\right)$, and write $3x$ in the quotient.

**Step 3**
$$\begin{array}{r} 3x \phantom{ - 2x + 1} \\ 2x - 4 \overline{)\,6x^2 - 2x + 1} \\ \underline{6x^2 - 12x} \phantom{ + 1} \end{array}$$

Multiply $3x$ times $2x - 4$ and write the product $6x^2 - 12x$ below the polynomial $6x^2 - 2x + 1$.

**Step 4**

$$\begin{array}{r} 3x \phantom{\,-2x+1} \\ 2x-4\overline{)6x^2 - \phantom{1}2x + 1} \\ \underline{-6x^2 \pm 12x} \phantom{\,+1} \\ +10x + 1 \end{array}$$

Subtract $6x^2 - 12x$ by changing signs and adding. Bring down the next term, $+1$.

**Step 5**

$$\begin{array}{r} 3x + 5 \phantom{\,+1} \\ 2x-4\overline{)6x^2 - \phantom{1}2x + 1} \\ \underline{-6x^2 \pm 12x} \phantom{\,+1} \\ +10x + 1 \end{array}$$

Divide $10x$ by $2x$ $\left(\frac{10x}{2x} = 5\right)$, and write $+5$ in the quotient.

**Step 6**

$$\begin{array}{r} 3x + 5 \phantom{\,+1} \\ 2x-4\overline{)6x^2 - \phantom{1}2x + \phantom{2}1} \\ \underline{-6x^2 \pm 12x} \phantom{\,+\,21} \\ +10x + \phantom{2}1 \\ \underline{-10x \pm 20} \\ 21 \text{ Remainder} \end{array}$$

Multiply 5 times $2x - 4$. Write the product $10x - 20$ below $+10x + 1$ and subtract.

**Step 7**

$$\frac{6x^2 - 2x + 1}{2x - 4} = 3x + 5 + \frac{21}{2x - 4}$$

Write the remainder over the divisor and **add** this fraction to the quotient. ■

In Example 4 we show the same sequence of steps as in Example 3; but they are written in the normal compact form that should be used. Note that 0 is written as a place holder for any missing powers of the variable.

## EXAMPLES

**4.** Divide $\frac{25x^3 + 9x + 2}{5x + 1}$ by using long division.

**Solution**

$$\begin{array}{r} 5x^2 - 1x + \phantom{9x+}2 \\ 5x+1\overline{)25x^3 + 0x^2 + 9x + 2} \\ \underline{-25x^3 \mp 5x^2} \phantom{\,+9x+2} \\ -5x^2 + \phantom{1}9x \phantom{\,+2} \\ \underline{\pm 5x^2 \pm \phantom{1}1x} \phantom{\,+2} \\ +10x + 2 \\ \underline{\pm 10x \mp 2} \\ 0 \end{array}$$

Divide $\frac{25x^3}{5x} = 5x^2$. Write $5x^2$ in the quotient. Multiply $5x^2$ times $5x + 1$ and subtract. Continue the process **until the degree of the remainder is smaller than the degree of the divisor.**

$$\frac{25x^3 + 9x + 2}{5x + 1} = 5x^2 - x + 2$$

This means that $5x + 1$ and $5x^2 - x + 2$ are both factors of $25x^3 + 9x + 2$.

**5.** Divide $\dfrac{x^4 - 5x^3 + 2x^2 - 6x + 1}{x^2 + 1}$ by using the division algorithm.

**Solution**

$$\begin{array}{r}
x^2 - 5x + 1 \phantom{- 6x + 1} \\
x^2 + 1 \overline{\big)\, x^4 - 5x^3 + 2x^2 - 6x + 1} \\
\underline{-x^4 \phantom{- 5x^3} \mp x^2 \phantom{- 6x + 1}} \\
- 5x^3 + x^2 - 6x \phantom{+ 1} \\
\underline{\pm 5x^3 \phantom{+ x^2} \pm 5x} \phantom{+ 1} \\
x^2 - x + 1 \\
\underline{- x^2 \phantom{- x} \mp 1} \\
- x \phantom{+ 1}
\end{array}$$

Note that the divisor is second degree and the remainder is less than second degree.

So,

$$\frac{x^4 - 5x^3 + 2x^2 - 6x + 1}{x^2 + 1} = x^2 - 5x + 1 + \frac{-x}{x^2 + 1}$$ ■

In general, for polynomials $P$ and $D$, the division algorithm gives

$$\frac{P}{D} = Q + \frac{R}{D}, \; D \neq 0$$

where $Q$ and $R$ are polynomials and degree of $R <$ degree of $D$.

In the special case **when the divisor is first-degree with leading coefficient 1,** the division can be simplified by omitting the variables entirely and writing only certain coefficients. The procedure is called **synthetic division.** The following analysis describes how the procedure works.

**a.** With variables

$$\begin{array}{r}
5x^2 - 4x + 9 \phantom{+ 1} \\
x + 3 \overline{\big)\, 5x^3 + 11x^2 - 3x + 1} \\
\underline{5x^3 + 15x^2} \phantom{- 3x + 1} \\
- 4x^2 - 3x \phantom{+ 1} \\
\underline{- 4x^2 - 12x} \phantom{+ 1} \\
9x + 1 \\
\underline{9x + 27} \\
- 26
\end{array}$$

**b.** Without variables

$$\begin{array}{r}
5 - 4 + 9 \phantom{+ 1} \\
1 + 3 \overline{\big)\, 5 + 11 - 3 + 1} \\
\underline{\textcircled{5} + 15} \phantom{- 3 + 1} \\
- 4 \textcircled{- 3} \phantom{+ 1} \\
\underline{\textcircled{- 4} - 12} \phantom{+ 1} \\
9 \textcircled{+ 1} \\
\underline{\textcircled{9} + 27} \\
- 26
\end{array}$$

The circled numbers can be omitted since they are repetitions of the numbers directly above them.

**c.** Circled numbers omitted

$$\begin{array}{rrrrr} & 5 & -4 & +9 & \\ \hline 1+3) & 5 & +11 & -3 & +1 \\ & & \underline{+15} & & \\ & & -4 & & \\ & & & \underline{-12} & \\ & & & 9 & \\ & & & & \underline{+27} \\ & & & & -26 \end{array}$$

**d.** Numbers moved up to fill in spaces

$$\begin{array}{rrrrr} & 5 & -4 & +9 & \\ \hline 1+3) & 5 & +11 & -3 & +1 \\ & & +15 & -12 & +27 \\ \hline & & -4 & +9 & -26 \end{array}$$

Now we omit the 1 in the divisor; change $+3$ to $-3$; and write the opposites of the circled numbers so they can be added. The number 5 is written on the bottom line and the top line is omitted. The quotient and remainder can now be read from the bottom line.

**e.**

$$\begin{array}{rrrrr} & 5 & -4 & +9 & \\ \hline 1+3) & 5 & +11 & -3 & +1 \\ & & \textcircled{+15} & \textcircled{-12} & \textcircled{+27} \\ \hline & & -4 & +9 & -26 \end{array}$$

**f.**

$$\begin{array}{rrrrr} -3) & 5 & +11 & -3 & +1 \\ & \downarrow & -15 & +12 & -27 \\ \hline & 5 & -4 & +9 & -26 \end{array}$$

Represents

$$5x^2 - 4x + 9 + \frac{-26}{x+3}$$

The numbers on the bottom now represent the coefficients of a polynomial of one degree less than the dividend and the remainder. The last number to the right is the remainder.

In summary, synthetic division can be accomplished as follows:

**1.** Write only the coefficients of the dividend and the opposite of the constant in the divisor.

$$\begin{array}{r|rrrr} -3 & 5 & +11 & -3 & +1 \\ \hline & & & & \end{array}$$

**2.** Rewrite the first coefficient as the first coefficient in the quotient.

$$\begin{array}{r|rrrr} -3 & 5 & +11 & -3 & +1 \\ & \downarrow & & & \\ \hline & 5 & & & \end{array}$$

**3.** Multiply this coefficient by the constant divisor and **add** this product to the second coefficient.

$$\begin{array}{r|rrrr} -3 & 5 & +11 & -3 & +1 \\ & & -15 & & \\ \hline & 5 \nearrow & -4 & & \end{array}$$

4. Continue to multiply each new coefficient by the constant divisor and add this product to the next coefficient in the dividend.

$$\begin{array}{r|rrrr} -3 & 5 & +11 & -3 & +1 \\ & & -15 & +12 & -27 \\ \hline & 5 & -4 & 9 & -26 \end{array}$$

5. The constants on the bottom line are the coefficients of the quotient and the remainder.

$$\frac{5x^3 + 11x^2 - 3x + 1}{x + 3} = 5x^2 - 4x + 9 + \frac{-26}{x + 3}$$

**EXAMPLES** Use synthetic division to write each expression in the form $Q + \dfrac{R}{D}$.

6. $\dfrac{4x^3 + 10x^2 + 11}{x + 5}$

**Solution**

$$\begin{array}{r|rrrr} -5 & 4 & 10 & 0 & 11 \\ & \downarrow & -20 & 50 & -250 \\ \hline & 4 & -10 & 50 & -239 \end{array}$$

There is no $x$-term, so 0 is the coefficient. The coefficient is 0 for any missing term.

$$\frac{4x^3 + 10x^2 + 11}{x + 5} = 4x^2 - 10x + 50 + \frac{-239}{x + 5}$$

7. $\dfrac{2x^4 - x^3 - 5x^2 - 2x + 7}{x - 2}$

**Solution**

$$\begin{array}{r|rrrrr} +2 & 2 & -1 & -5 & -2 & 7 \\ & \downarrow & 4 & 6 & 2 & 0 \\ \hline & 2 & 3 & 1 & 0 & 7 \end{array}$$

$$\frac{2x^4 - x^3 - 5x^2 - 2x + 7}{x - 2} = 2x^3 + 3x^2 + x + \frac{7}{x - 2}$$

■

## EXERCISES 3.4

Perform the indicated division in Exercises 1–40 and write the answer in the form $Q + \dfrac{R}{D}$ where the degree of $R$ is less than the degree of $D$. Assume that none of the divisors is 0.

1. $\dfrac{8y^3 - 16y^2 + 24y}{8y}$
2. $\dfrac{18x^4 + 24x^3 + 36x^2}{6x^2}$
3. $\dfrac{34x^5 - 51x^4 + 17x^3}{17x^3}$
4. $\dfrac{14y^4 + 28y^3 + 12y^2}{2y^2}$
5. $\dfrac{110x^4 - 121x^3 + 11x^2}{11x}$
6. $\dfrac{15x^7 + 30x^6 - 25x^3}{15x^3}$
7. $\dfrac{-56x^4 + 98x^3 - 35x^2}{14x^2}$
8. $\dfrac{108x^6 - 72x^5 + 63x^4}{18x^4}$
9. $\dfrac{16y^6 - 56y^5 - 120y^4 + 64y^3}{16y^3}$

10. $\dfrac{20y^5 - 14y^4 + 21y^3 + 42y^2}{4y^2}$

11. $\dfrac{21x^2 + 25x - 3}{7x - 1}$

12. $\dfrac{15x^2 - 14x - 11}{3x - 4}$

13. $\dfrac{2x^3 + 7x^2 + 10x - 6}{2x + 3}$

14. $\dfrac{6x^3 - 7x^2 + 14x - 8}{3x - 2}$

15. $\dfrac{21x^3 + 41x^2 + 13x + 5}{3x + 5}$

16. $\dfrac{6x^3 - 4x^2 + 5x - 7}{x - 2}$

17. $\dfrac{x^3 - x^2 - 10x - 10}{x - 4}$

18. $\dfrac{2x^3 - 3x^2 + 7x + 4}{2x - 1}$

19. $\dfrac{10x^3 + 11x^2 - 12x + 9}{5x + 3}$

20. $\dfrac{6x^3 + 19x^2 - 3x - 7}{6x + 1}$

21. $\dfrac{2x^3 - 7x + 2}{x + 4}$

22. $\dfrac{2x^3 + 4x^2 - 9}{x + 3}$

23. $\dfrac{9x^3 - 19x + 9}{3x - 2}$

24. $\dfrac{16x^3 + 7x + 12}{4x + 3}$

25. $\dfrac{6x^3 + 11x^2 + 25}{2x + 5}$

26. $\dfrac{4x^3 - 8x^2 - 9x}{2x - 3}$

27. $\dfrac{3x^3 + 5x^2 + 7x + 9}{x^2 + 2}$

28. $\dfrac{2x^4 + 2x^3 + 3x^2 + 6x - 1}{2x^2 + 3}$

29. $\dfrac{x^4 + x^3 - 4x + 1}{x^2 + 4}$

30. $\dfrac{2x^4 + x^3 - 8x^2 + 3x - 2}{x^2 - 5}$

31. $\dfrac{6x^3 + 5x^2 - 8x + 3}{3x^2 - 2x - 1}$

32. $\dfrac{x^3 - 9x^2 + 20x - 38}{x^2 - 3x + 5}$

33. $\dfrac{3x^4 - 7x^3 + 5x^2 + x - 2}{x^2 + x + 1}$

34. $\dfrac{2x^4 + 9x^3 - x^2 + 6x + 9}{x^2 - 3x + 1}$

35. $\dfrac{x^4 + 3x - 7}{x^2 + 2x - 3}$

36. $\dfrac{3x^4 - 2x^3 + 4x^2 - x + 3}{3x^2 + x - 1}$

37. $\dfrac{x^3 - 27}{x - 3}$

38. $\dfrac{x^3 + 125}{x + 5}$

39. $\dfrac{x^5 - 1}{x^2 + 1}$

40. $\dfrac{x^6 - 1}{x^3 - 1}$

Divide using synthetic division in Exercises 41–60. Write the quotient in the form $Q + \dfrac{R}{D}$ where the degree of $R$ is less than the degree of $D$.

41. $\dfrac{x^2 - 12x + 27}{x - 3}$

42. $\dfrac{x^2 - 12x + 35}{x - 5}$

43. $\dfrac{x^3 + 4x^2 + x - 1}{x + 8}$

44. $\dfrac{x^3 - 6x^2 + 8x - 5}{x - 2}$

45. $\dfrac{4x^3 + 2x^2 - 3x + 1}{x + 2}$

46. $\dfrac{3x^3 + 6x^2 + 8x - 5}{x + 1}$

47. $\dfrac{x^3 + 6x + 3}{x - 7}$

48. $\dfrac{2x^3 - 7x + 2}{x + 4}$

49. $\dfrac{2x^3 + 4x^2 - 9}{x + 3}$

50. $\dfrac{4x^3 - x^2 + 13}{x - 1}$

51. $\dfrac{x^4 - 3x^3 + 2x^2 - x + 2}{x - 3}$

52. $\dfrac{x^4 + x^3 - 4x^2 + x - 3}{x + 6}$

53. $\dfrac{x^4 + 2x^2 - 3x + 5}{x - 2}$

54. $\dfrac{3x^4 + 2x^3 - x^2 + x - 1}{x + 1}$

55. $\dfrac{x^4 - x + 3}{x - 4}$

56. $\dfrac{x^5 + 2x^2 + 1}{x - 5}$

57. $\dfrac{x^5 - 1}{x - 1}$

58. $\dfrac{x^5 - x^3 + x}{x + 1}$

59. $\dfrac{x^6 - 2x^3 + 4}{x + 4}$

60. $\dfrac{x^6 + 1}{x + 1}$

## 3.5 Complex Fractions

**OBJECTIVE**

In the section, you will be learning to simplify complex fractions.

A **complex fraction** is a fraction in which the numerator and/or denominator are themselves fractions or the sums or differences of fractions. Examples of complex fractions are

$$\frac{\frac{6x}{5y^2}}{\frac{8x^2}{10y^3}}, \quad \frac{x+y}{x^{-1}+y^{-1}}, \quad \text{and} \quad \frac{\frac{1}{x+3}-\frac{1}{x}}{1+\frac{3}{x}}$$

In the first example, the numerator and denominator are both single fractions; no sum or difference is indicated. To simplify this expression, we simply divide.

**EXAMPLE**

**1.** $$\frac{\frac{6x}{5y^2}}{\frac{8x^2}{10y^3}} = \frac{6x}{5y^2} \cdot \frac{10y^3}{8x^2} = \frac{\cancel{2} \cdot 3 \cdot \cancel{2} \cdot \cancel{5}\cancel{x}\overset{y}{\cancel{y^3}}}{\cancel{5} \cdot \cancel{2} \cdot \cancel{2} \cdot 2\cancel{y^2}\underset{x}{\cancel{x^2}}} = \frac{3y}{2x}$$ ■

There are two methods for simplifying a complex fraction when sums or differences of fractions are indicated. The method you choose depends on which method you think is easier or seems to better fit the particular problem.

The **first method** is to

1. simplify the numerator and denominator separately so that the numerator and denominator are simple fractions;
2. divide the numerator by the denominator.

**EXAMPLES**

**2.** $$\frac{x+y}{x^{-1}+y^{-1}} = \frac{x+y}{\frac{1}{x}+\frac{1}{y}} = \frac{\frac{x+y}{1}}{\frac{1}{x}\cdot\frac{y}{y}+\frac{1}{y}\cdot\frac{x}{x}}$$

Add the two fractions in the denominator.

$$= \frac{\frac{x+y}{1}}{\frac{y}{xy}+\frac{x}{xy}} = \frac{\frac{x+y}{1}}{\frac{y+x}{xy}}$$

$$= \frac{\cancel{x+y}}{1} \cdot \frac{xy}{\cancel{y+x}} = \frac{xy}{1} = xy$$

**3.** $$\frac{\frac{1}{x+3}-\frac{1}{x}}{1+\frac{3}{x}} = \frac{\frac{1}{x+3}-\frac{1}{x}}{1+\frac{3}{x}}$$ Combine the fractions in the numerator and in the denominator separately.

$$= \frac{\frac{1\cdot x}{(x+3)\cdot x}-\frac{1(x+3)}{x(x+3)}}{\frac{x}{x}+\frac{3}{x}}$$ Note that $1 = \frac{x}{x}$.

$$= \frac{\frac{x-(x+3)}{x(x+3)}}{\frac{x+3}{x}} = \frac{\frac{x-x-3}{x(x+3)}}{\frac{x+3}{x}}$$

$$= \frac{-3}{\cancel{x}(x+3)}\cdot\frac{\cancel{x}}{x+3} = \frac{-3}{(x+3)^2}$$ ∎

The **second method** is to

1. find the LCM of all the denominators in the original numerator and denominator;
2. multiply both the numerator and the denominator by this LCM.

**EXAMPLES**

**4.** $$\frac{x+y}{x^{-1}+y^{-1}} = \frac{\frac{x+y}{1}}{\frac{1}{x}+\frac{1}{y}}$$ The LCM for $\{x, y, 1\}$ is $xy$.

$$= \frac{\left(\frac{x+y}{1}\right)xy}{\left(\frac{1}{x}+\frac{1}{y}\right)xy} = \frac{(x+y)xy}{\frac{1}{x}\cdot xy+\frac{1}{y}\cdot xy}$$

$$= \frac{\cancel{(x+y)}xy}{\cancel{y+x}} = xy$$

**5.** $$\frac{\frac{1}{x+3}-\frac{1}{x}}{1+\frac{3}{x}} = \frac{\left(\frac{1}{x+3}-\frac{1}{x}\right)\cdot x(x+3)}{\left(1+\frac{3}{x}\right)\cdot x(x+3)}$$

The LCM for $\{x, x+3\}$ is $x(x+3)$.

$$= \frac{\frac{1}{\cancel{x+3}}\cdot x\cancel{(x+3)} - \frac{1}{\cancel{x}}\cdot \cancel{x}(x+3)}{1\cdot x(x+3) + \frac{3}{\cancel{x}}\cdot \cancel{x}(x+3)}$$

$$= \frac{x-(x+3)}{x(x+3)+3(x+3)} = \frac{x-x-3}{(x+3)(x+3)}$$

$$= \frac{-3}{(x+3)^2}$$

■

## EXERCISES 3.5

Simplify each complex fraction in Exercises 1–34.

**1.** $\dfrac{\frac{2x}{3y^2}}{\frac{5x^2}{6y}}$ **2.** $\dfrac{\frac{6x^2}{5y}}{\frac{x}{10y^2}}$ **3.** $\dfrac{\frac{12x^3}{7y^2}}{\frac{3x^5}{2y}}$ **4.** $\dfrac{\frac{9x^2}{7y^3}}{\frac{3xy}{14}}$ **5.** $\dfrac{\frac{x+3}{2x}}{\frac{2x-1}{4x^2}}$

**6.** $\dfrac{\frac{x-2}{6x}}{\frac{x+3}{3x^2}}$ **7.** $\dfrac{\frac{3}{x}+\frac{1}{2x}}{1+\frac{2}{x}}$ **8.** $\dfrac{\frac{2x-1}{x}}{\frac{2}{x}+3}$ **9.** $\dfrac{1+\frac{1}{x}}{1-\frac{1}{x^2}}$ **10.** $\dfrac{\frac{2}{y}+1}{\frac{4}{y^2}-1}$

**11.** $\dfrac{\frac{1}{x}+\frac{1}{3x}}{\frac{x+6}{x^2}}$ **12.** $\dfrac{\frac{3}{x}-\frac{6}{x^2}}{\frac{x-2}{x^2}}$ **13.** $\dfrac{\frac{7}{x}-\frac{14}{x^2}}{\frac{1}{x}-\frac{4}{x^3}}$ **14.** $\dfrac{\frac{3}{x}-\frac{6}{x^2}}{\frac{1}{x}-\frac{2}{x^2}}$ **15.** $\dfrac{\frac{x}{y}-\frac{1}{3}}{\frac{6}{y}-\frac{2}{x}}$

**16.** $\dfrac{\frac{3}{x}+\frac{5}{2x}}{\frac{1}{x}+4}$ **17.** $\dfrac{\frac{2}{x}+\frac{3}{4y}}{\frac{3}{2x}-\frac{5}{3y}}$ **18.** $\dfrac{1+x^{-1}}{1-x^{-2}}$ **19.** $\dfrac{1}{x^{-1}+y^{-1}}$ **20.** $\dfrac{x^{-1}+y^{-1}}{x+y}$

**21.** $\dfrac{x^{-1}+y^{-1}}{x^{-1}-y^{-1}}$ **22.** $\dfrac{x^{-1}+y^{-1}}{x^{-2}-y^{-2}}$ **23.** $\dfrac{2-\frac{4}{x}}{\frac{x^2-4}{x^2+x}}$ **24.** $\dfrac{\frac{1}{x}}{1-\frac{1}{x-2}}$

**25.** $\dfrac{x + \dfrac{3}{x - 4}}{1 - \dfrac{1}{x}}$ **26.** $\dfrac{1 - \dfrac{4}{x + 3}}{1 - \dfrac{2}{x + 1}}$ **27.** $\dfrac{1 + \dfrac{4}{2x - 3}}{1 + \dfrac{x}{x + 1}}$ **28.** $\dfrac{\dfrac{1}{x + h} - \dfrac{1}{x}}{h}$

**29.** $\dfrac{\dfrac{1}{(x + h)^2} - \dfrac{1}{x^2}}{h}$ **30.** $\dfrac{\left(2 + \dfrac{1}{x + h}\right) - \left(2 + \dfrac{1}{x}\right)}{h}$ **31.** $\dfrac{x^2 - 4y^2}{1 - \dfrac{2x + y}{x - y}}$

**32.** $\dfrac{\dfrac{x + 1}{x - 1} - \dfrac{x - 1}{x + 1}}{\dfrac{x + 1}{x - 1} + \dfrac{x - 1}{x + 1}}$ **33.** $\dfrac{\dfrac{1}{x^2 - 1} - \dfrac{1}{x + 1}}{\dfrac{1}{x - 1} + \dfrac{1}{x^2 - 1}}$ **34.** $\dfrac{\dfrac{x}{x - 4} + \dfrac{1}{x - 1}}{\dfrac{x}{x - 1} + \dfrac{2}{x - 3}}$

## 3.6 Equations and Inequalities Involving Rational Expressions

**OBJECTIVES**

In this section, you will be learning to:

1. Solve equations containing rational expressions.
2. Solve inequalities containing rational expressions.
3. Graph the solutions for inequalities containing rational expressions.

To solve an equation that has fractions, such as $\frac{x}{5} - \frac{x}{2} = -6$, multiply both sides of the equation by the LCM of the denominators. Use the distributive property if there is more than one term on either side of the equation. This will eliminate all fractions from the equation and make the coefficients easier to work with. In this problem, the LCM is 10.

$$10\left(\frac{x}{5} - \frac{x}{2}\right) = 10(-6)$$

$$10 \cdot \frac{x}{5} - 10 \cdot \frac{x}{2} = 10(-6)$$

$$2x - 5x = -60$$

$$-3x = -60$$

$$x = 20$$

**If both sides of an equation are multiplied by the same nonzero number or expression, the solution set of the new equation will contain the solutions to the original equation.** If the multiplication is by a variable expression, then the new equation may have more solutions than the original equation. These extra solutions are called **extraneous solutions** or **extraneous roots.** Check all solutions in the original equation to be sure you have not picked up an extraneous solution or multiplied by 0.

**EXAMPLES** Find the solution set for each of the following equations. Multiply both sides of each equation by the LCM of the denominators.

**1.** $\dfrac{x-5}{2x} = \dfrac{6}{3x}$ (LCM $= 6x$)

**Solution** $6x \cdot \left(\dfrac{x-5}{2x}\right) = 6x \cdot \left(\dfrac{6}{3x}\right)$ $(x \neq 0)$

$$3(x-5) = 2(6)$$
$$3x - 15 = 12$$
$$3x = 27$$
$$x = 9$$

*Check* $\dfrac{9-5}{2 \cdot 9} = \dfrac{6}{3 \cdot 9}$

$$\frac{4}{18} = \frac{6}{27}$$

$$\frac{2}{9} = \frac{2}{9}$$

The solution is $x = 9$.

**2.** $\dfrac{3}{x-6} = \dfrac{5}{x}$ [LCM $= x(x-6)$]

**Solution** $\cancel{x(x-6)} \cdot \dfrac{3}{\cancel{x-6}} = \cancel{x}(x-6) \cdot \dfrac{5}{\cancel{x}}$ $(x \neq 0, 6)$

$$3x = 5x - 30$$
$$30 = 2x$$
$$15 = x$$

*Check* $\dfrac{3}{15-6} = \dfrac{5}{15}$

$$\frac{1}{3} = \frac{1}{3}$$

The solution is $x = 15$.

**3.** $\dfrac{2}{x^2 - 9} = \dfrac{1}{x^2} + \dfrac{1}{x^2 - 3x}$

$$\left.\begin{aligned} x^2 - 9 &= (x + 3)(x - 3) \\ x^2 &= x^2 \\ x^2 - 3x &= x(x - 3) \end{aligned}\right\} \text{LCM} = x^2(x + 3)(x - 3)$$

**Solution**

$$x^2\cancel{(x + 3)(x - 3)} \cdot \frac{2}{\cancel{(x + 3)(x - 3)}}$$

$$= \cancel{x^2}(x + 3)(x - 3) \cdot \frac{1}{\cancel{x^2}} + \overset{x}{\cancel{x^2}}(x + 3)\cancel{(x - 3)} \cdot \frac{1}{\cancel{x}\cancel{(x - 3)}}$$

$$\begin{aligned} 2x^2 &= (x + 3)(x - 3) + x(x + 3) \qquad (x \neq 0, 3, -3) \\ 2x^2 &= x^2 - 9 + x^2 + 3x \\ 2x^2 &= 2x^2 + 3x - 9 \\ 9 &= 3x \\ \cancel{3 = x} & \qquad \text{3 is not allowed since no denominator can be 0.} \end{aligned}$$

There is no solution. (Multiplying by the factor $x - 3$ was, in effect, multiplying by 0.)

**4.** $\dfrac{1}{x - 7} = \dfrac{2}{x^2 - 12x + 35} + \dfrac{x}{x^2 - 5x}$

$$\left.\begin{aligned} x - 7 &= x - 7 \\ x^2 - 12x + 35 &= (x - 5)(x - 7) \\ x^2 - 5x &= x(x - 5) \end{aligned}\right\} \text{LCM} = x(x - 5)(x - 7)$$

**Solution**

$$x(x - 5)\cancel{(x - 7)} \cdot \frac{1}{\cancel{x - 7}} = x\cancel{(x - 5)(x - 7)} \cdot \frac{2}{\cancel{(x - 5)(x - 7)}}$$

$$+ \cancel{x}\cancel{(x - 5)}(x - 7) \cdot \frac{x}{\cancel{x}\cancel{(x - 5)}}$$

$$\begin{aligned} x(x - 5) &= 2x + x(x - 7) \qquad (x \neq 0, 5, 7) \\ x^2 - 5x &= 2x + x^2 - 7x \\ x^2 - 5x &= x^2 - 5x \\ 0 &= 0 \end{aligned}$$

The equation $0 = 0$ is true for all real numbers. Therefore, $x$ can be any real number except 0, 5, or 7. We can write $x \neq 0, 5, 7$. These values are not included since any one of them would give a 0 denominator. The implication is that all other values are allowed. ■

In Chapter 1, we solved first-degree inequalities and introduced intervals and interval notation. For example,

if $\quad x - 4 \leq 0$

then $\quad x \leq 4 \quad$ or $\quad x$ is in $(-\infty, 4]$.

Graphically,

4

Notice that the solution consists of the point where $x - 4 = 0$ $(x = 4)$ and all the points to the left of this point. For first-degree inequalities, the solution always contains the points to one side or the other of the point where the inequality has value 0. We use this idea in the following discussion.

A rational inequality may involve the product and/or quotient of several first-degree expressions. For example, the inequality

$$\frac{x+3}{x-2} > 0$$

involves the two first-degree expressions $x + 3$ and $x - 2$.

The following procedure for solving such an inequality is based on the fact that an expression of the form $x - a$ changes sign when $x$ has values on either side of $a$. That is, if $x < a$, then $x - a$ is negative; if $x > a$, then $x - a$ is positive.

The steps are as follows:

1. Find the points where each first-degree expression has value 0.

$$x + 3 = 0 \qquad\qquad x - 2 = 0$$
$$x = -3 \qquad\qquad x = 2$$

2. Mark each of these points on a number line. (Consider these points as endpoints of intervals.)

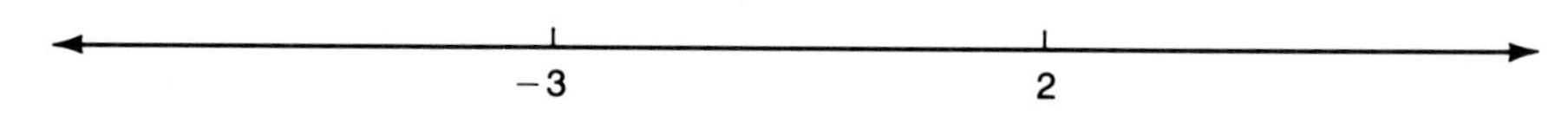


Three intervals, $(-\infty,-3)$, $(-3,2)$, and $(2,+\infty)$, are formed.

3. Use any one point from each interval as a **test point** to determine the sign of the expression for all points in that interval.

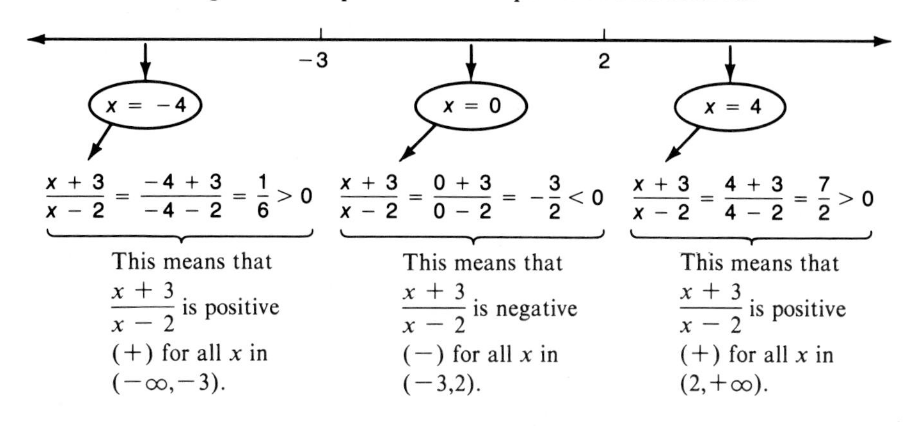

**4.** The solution to the inequality consists of all the intervals indicating the desired sign, $+$ (for $>0$) or $-$ (for $<0$). The solution for $\frac{x+3}{x-2} > 0$ is

$$\text{all } x \text{ in } (-\infty,-3) \quad \text{or} \quad (2,+\infty).$$

In algebraic notation: $x < -3$ or $x > 2$.
Graphically,

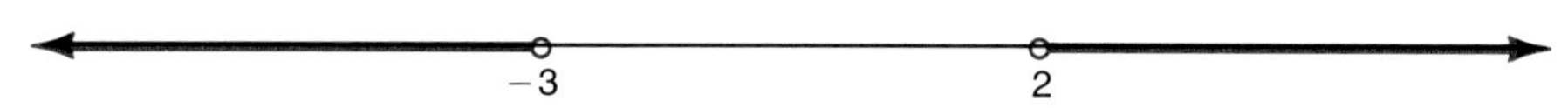


**EXAMPLES** Solve and graph the solution for each of the following inequalities.

**5.** $\frac{x+3}{x-2} < 0$

**Solution** From the previous discussion, we know that $\frac{x+3}{x-2}$ is negative whenever $x$ is in $(-3,2)$.

Or, $\frac{x+3}{x-2} < 0$ if $-3 < x < 2$.

Graphically,

**6.** $\frac{x+5}{x-4} \geq -1$

**Solution**

**a.** $\frac{x+5}{x-4} + 1 \geq 0$ — One side must be 0.

$\frac{x+5}{x-4} + \frac{x-4}{x-4} \geq 0$ — Simplify to get one fraction.

$$\frac{2x+1}{x-4} \geq 0$$

**b.** Now set each first-degree expression equal to 0 to find the interval endpoints.

$$2x + 1 = 0 \qquad \text{if } x = -\frac{1}{2}$$

$$x - 4 = 0 \qquad \text{if } x = 4$$

**c.** Now test a point from each of the intervals,
$\left(-\infty,-\frac{1}{2}\right), \left(-\frac{1}{2},4\right), (4,+\infty)$.

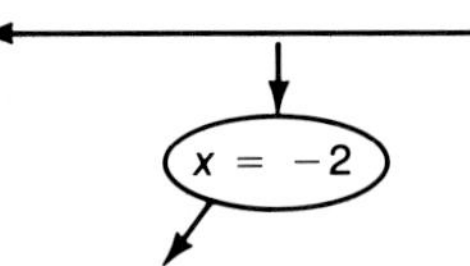
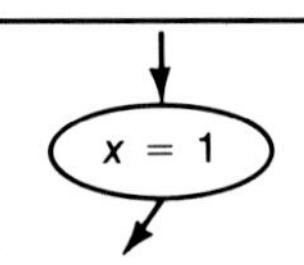
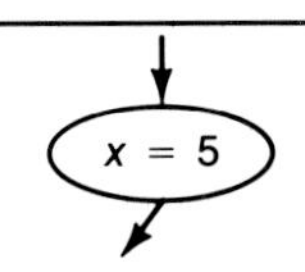

$\frac{2(-2)+1}{-2-4} = \frac{-3}{-6} > 0$

This means that $\frac{2x+1}{x-4} > 0$ for all $x$ in $\left(-\infty,-\frac{1}{2}\right)$.

$\frac{2(1)+1}{1-4} = \frac{3}{-3} < 0$

This means that $\frac{2x+1}{x-4} < 0$ for all $x$ in $\left(-\frac{1}{2},4\right)$.

$\frac{2(5)+1}{5-4} = \frac{11}{1} > 0$

This means that $\frac{2x+1}{x-4} > 0$ for all $x$ in $(4,+\infty)$.

**d.** $\frac{2x+1}{x-4} = 0$ if the numerator $2x+1=0$ or $x=-\frac{1}{2}$. So, the solution is all $x$ in $\left(-\infty,-\frac{1}{2}\right]$ or $(4,+\infty)$. In algebraic notation:

$x \leq -\frac{1}{2}$ or $x > 4$.

Graphically,

[**Special Note:** Notice that in step **a**, we do **not** multiply by the denominator $x-4$. The reason is that the variable expression is positive for some values and negative for other values.] ■

## EXERCISES 3.6

Solve the equations in Exercises 1–34.

**1.** $\frac{4x}{7} = \frac{x+5}{3}$

**2.** $\frac{3x+1}{4} = \frac{2x+1}{-3}$

**3.** $\frac{5x+2}{11x} = \frac{x-6}{4x}$

**4.** $\frac{x+3}{5x} = \frac{x-1}{6x}$

**5.** $\frac{5x}{4} - \frac{1}{2} = -\frac{3}{16}$

**6.** $\frac{x}{6} - \frac{1}{42} = \frac{1}{7}$

**7.** $\frac{4x}{3} - \frac{3}{4} = \frac{5x}{6}$

**8.** $\frac{x-2}{3} - \frac{x-3}{5} = \frac{13}{15}$

**9.** $\frac{2+x}{4} - \frac{5x-2}{12} = \frac{8-2x}{5}$

**10.** $\frac{8x+10}{5} = 2x+3-\frac{6x+1}{4}$

**11.** $\frac{2}{3x} = \frac{1}{4} - \frac{1}{6x}$

**12.** $\frac{x-4}{x} + \frac{3}{x} = 0$

**13.** $\frac{3}{8x} - \frac{7}{10} = \frac{1}{5x}$

**14.** $\frac{1}{x} - \frac{8}{21} = \frac{3}{7x}$

**15.** $\frac{3}{4x} - \frac{1}{2} = \frac{7}{8x} + \frac{1}{6}$

16. $\dfrac{7}{x-3} = \dfrac{6}{x-4}$

17. $\dfrac{2}{3x+2} = \dfrac{4}{5x+1}$

18. $\dfrac{-3}{2x+1} = \dfrac{4}{3x+1}$

19. $\dfrac{9}{5x-3} = \dfrac{5}{3x+7}$

20. $\dfrac{5x+2}{x-6} = \dfrac{11}{4}$

21. $\dfrac{x+9}{3x+2} = \dfrac{5}{8}$

22. $\dfrac{8}{2x+3} = \dfrac{9}{4x-5}$

23. $\dfrac{x}{x-4} - \dfrac{4}{2x-1} = 1$

24. $\dfrac{x}{x+3} + \dfrac{1}{x+2} = 1$

25. $\dfrac{x+2}{x+1} + \dfrac{x+2}{x+4} = 2$

26. $\dfrac{x-2}{x-3} + \dfrac{x-3}{x-2} = \dfrac{2x^2}{x^2-5x+6}$

27. $\dfrac{2}{4x-1} + \dfrac{1}{x+1} = \dfrac{3}{x+1}$

28. $\dfrac{3x-2}{15} - \dfrac{16-3x}{x+6} = \dfrac{x+3}{5}$

29. $\dfrac{x}{x-4} - \dfrac{12x}{x^2+x-20} = \dfrac{x-1}{x+5}$

30. $\dfrac{x-2}{x+4} - \dfrac{3}{2x+1} = \dfrac{x-7}{x+4}$

31. $\dfrac{3x+5}{3x+2} - \dfrac{4-2x}{3x^2+8x+4} = \dfrac{x+4}{x+2}$

32. $\dfrac{3}{3x-1} + \dfrac{1}{x+1} = \dfrac{4}{2x-1}$

33. $\dfrac{5}{2x+1} - \dfrac{1}{2x-1} = \dfrac{2}{x-2}$

34. $\dfrac{2}{x+1} + \dfrac{4}{2x-3} = \dfrac{4}{x-5}$

Solve and graph the solution set of each of the inequalities in Exercises 35–50.

35. $\dfrac{x+4}{2x} \geq 0$

36. $\dfrac{x}{x-4} \geq 0$

37. $\dfrac{x+6}{x^2} < 0$

38. $\dfrac{3-x}{x+1} < 0$

39. $\dfrac{x+3}{x+9} > 0$

40. $\dfrac{2x+3}{x-4} < 0$

41. $\dfrac{3x-6}{2x-5} < 0$

42. $\dfrac{4-3x}{2x+4} \leq 0$

43. $\dfrac{x+5}{x-7} \geq 1$

44. $\dfrac{x+4}{2x-1} > 2$

45. $\dfrac{2x+5}{x-4} \leq -3$

46. $\dfrac{3x+2}{4x-1} < 3$

47. $\dfrac{5-2x}{3x+4} < -1$

48. $\dfrac{8-x}{x+5} < -4$

49. $\dfrac{x(x+4)}{x-3} \leq 0$

50. $\dfrac{(x+3)(x-2)}{x+1} > 0$

## 3.7 Applications

**OBJECTIVE**

In this section, you will be learning to solve the following types of applied problems by using equations containing rational expressions:

a. fractions, b. jobs, and
c. distance-rate-time.

The Attack Plan for Word Problems given in Chapter 1 is valid for all word problems that involve algebraic equations (or inequalities). It is restated here for emphasis.

**Attack Plan for Word Problems**

1. Read the problem carefully. (Reread it several times if necessary.)
2. Decide what is asked for and assign a variable to the unknown quantity.
3. Draw a diagram or set up a chart whenever possible.
4. Form an equation (or inequality) that relates the information provided.
5. Solve the equation (or inequality).
6. Check your solution with the wording of the problem to be sure it makes sense.

We now introduce word problems involving rational expressions with problems relating the numerator and denominator of a fraction. Let one variable represent either the numerator or denominator; then write the equation to be solved using the information given in the problem.

**EXAMPLE**

1. The denominator of a fraction is 8 more than the numerator. If both the numerator and denominator are increased by 3, the resulting fraction is equal to $\frac{1}{2}$. Find the fraction.

**Solution** Reread the problem to be sure that you understand it. Assign variables to the unknown quantities.

Let $n$ = original numerator

$n + 8$ = original denominator

$\frac{n}{n+8}$ = original fraction

$$\frac{n+3}{(n+8)+3} = \frac{1}{2}$$

The numerator and denominator are each increased by 3, making a new fraction that is equal to $\frac{1}{2}$.

$$\frac{n+3}{n+11} = \frac{1}{2}$$

$$2\cancel{(n+11)} \cdot \left(\frac{n+3}{\cancel{n+11}}\right) = \cancel{2}(n+11) \cdot \frac{1}{\cancel{2}}$$

$$2n + 6 = n + 11$$

$$n = 5 \quad \longleftarrow \text{Original numerator}$$

$$n + 8 = 13 \quad \longleftarrow \text{Original denominator}$$

*Check* $\frac{5+3}{13+3} = \frac{8}{16} = \frac{1}{2}$

The original fraction is $\frac{5}{13}$. ■

Problems involving jobs (sometimes called **work problems**) usually translate into equations involving rational expressions. The basic idea is to **represent what part of a job is done in one unit of time.** For example, if a manuscript was typed in 35 hours, what part was typed in 1 hour? Assuming an even typing speed, $\frac{1}{35}$ of the manuscript was typed in 1 hour. If a boy can paint a fence in 2 days, he can do $\frac{1}{2}$ the job in 1 day.

## EXAMPLES

**2.** A man can wax his car three times as fast as his daughter can. Together they can do the job in 4 hours. How long does it take each of them working alone?

**Solution** Let $t$ = hr for man alone
$3t$ = hr for daughter alone

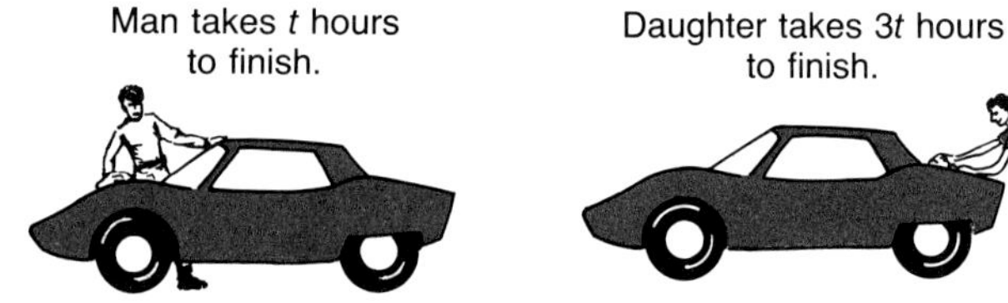


$$\underbrace{\text{part done by man alone in 1 hour}} + \underbrace{\text{part done by daughter alone in 1 hour}} = \underbrace{\text{part done working together in 1 hour}}$$

$$\frac{1}{t} + \frac{1}{3t} = \frac{1}{4}$$

$$\frac{1}{t}(12t) + \frac{1}{3t}(12t) = \frac{1}{4}(12t)$$

$$12 + 4 = 3t$$

$$16 = 3t$$

$$\frac{16}{3} = t$$

$$16 = 3t$$

*Check* Man's part in 1 hr $= \dfrac{1}{\frac{16}{3}} = \dfrac{3}{16}$

Man's part in 4 hr $= \dfrac{3}{16} \cdot 4 = \dfrac{3}{4}$

$$\text{Daughter's part in 1 hr} = \frac{1}{16}$$

$$\text{Daughter's part in 4 hr} = \frac{1}{16} \cdot 4 = \frac{1}{4}$$

$$\frac{3}{4} + \frac{1}{4} = 1 \text{ car waxed in 4 hours}$$

Working alone, the man takes $\frac{16}{3}$ hours, or $5\frac{1}{3}$ hours, and the daughter takes 16 hours.

**3.** An inlet pipe on a swimming pool can be used to fill the pool in 36 hours. The drain pipe can be used to empty the pool in 40 hours. If the pool is $\frac{2}{3}$ filled and then the drain pipe is accidentally opened, how long from that time will it take to fill the pool?

**Solution** Let $t$ = hours to fill pool with both pipes open

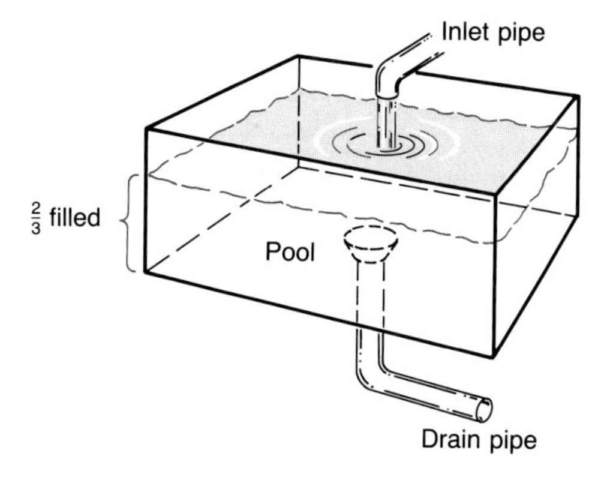


$$\underbrace{\begin{array}{c}\text{part filled}\\ \text{by inlet pipe}\\ \text{in 1 hour}\end{array}}_{\frac{1}{36}} - \underbrace{\begin{array}{c}\text{part emptied}\\ \text{by drain pipe}\\ \text{in 1 hour}\end{array}}_{\frac{1}{40}} = \underbrace{\begin{array}{c}\text{part filled in 1}\\ \text{hour when both}\\ \text{pipes are open}\end{array}}_{\frac{1}{t}}$$

$$\frac{1}{36}(360t) - \frac{1}{40}(360t) = \frac{1}{t}(360t)$$

$360t$ is the LCM of the denominators.

$$10t - 9t = 360$$
$$t = 360$$

However, 360 hours is the time it would take if the pool was empty at the beginning. Only $\frac{1}{3}$ this time will be used since the pool is $\frac{2}{3}$ filled.

$$\frac{1}{3} \cdot 360 \text{ hr} = 120 \text{ hr}$$

*Check* Part let in in 120 hr $= \frac{1}{36}(120) = \frac{10}{3} = 3\frac{1}{3}$

Part drained in 120 hr $= \frac{1}{40}(120) = 3$

$$\begin{array}{rl} 3\frac{1}{3} & \text{let in} \\ -3 & \text{drained} \\ \hline \frac{1}{3} & \text{filled in 120 hr} \end{array}$$

It will take 120 hours to fill the remaining third of the pool. ■

Problems involving distance, rate, and time were discussed in Section 1.5. The basic formula is $r \cdot t = d$. However, this relationship also can be stated in the forms $t = \frac{d}{r}$ and $r = \frac{d}{t}$.

If distance and rate are known or can be represented, then $t = \frac{d}{r}$ is the way to represent time. Similarly, if distance and time are known or can be represented, then $r = \frac{d}{t}$ is the way to represent rate.

**EXAMPLES**

**4.** A man can row his boat on a lake 5 miles per hour. On a river, it takes him the same time to row 5 miles downstream as it does to row 3 miles upstream. What is the speed of the river current in miles per hour?

**Solution** Let $c =$ rate of current
then $5 + c =$ rate going downstream (rower's rate plus rate of current)
and $5 - c =$ rate going upstream (rower's rate minus rate of current)

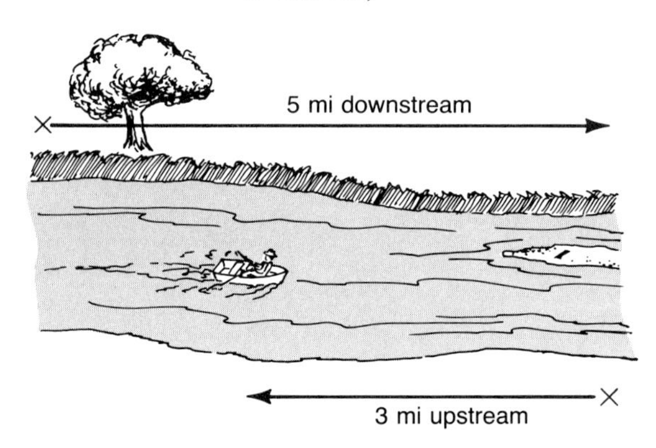

Rate and distance are represented.

| | Rate | | Distance |
|---|---|---|---|
| Downstream | $5 + c$ | | 5 |
| Upstream | $5 - c$ | | 3 |

Now represent the time.

| | Rate | $t = \frac{d}{r}$ | Distance |
|---|---|---|---|
| Downstream | $5 + c$ | $\frac{5}{5 + c}$ | 5 |
| Upstream | $5 - c$ | $\frac{3}{5 - c}$ | 3 |

$$\frac{5}{5 + c} = \frac{3}{5 - c} \quad \text{The times are equal.}$$

$$\cancel{(5 + c)}(5 - c) \cdot \frac{5}{\cancel{5 + c}} = (5 + c)\cancel{(5 - c)} \cdot \frac{3}{\cancel{5 - c}}$$

$$25 - 5c = 15 + 3c$$

$$10 = 8c$$

$$\frac{10}{8} = c$$

$$c = \frac{5}{4} \text{ miles per hour}$$

*Check*

$$\text{Time downstream} = \frac{5}{5 + \frac{5}{4}} = \frac{5}{6\frac{1}{4}} = \frac{5}{\frac{25}{4}} = 5 \cdot \frac{4}{25} = \frac{4}{5} \text{ hr}$$

$$\text{Time upstream} = \frac{3}{5 - \frac{5}{4}} = \frac{3}{3\frac{3}{4}} = \frac{3}{\frac{15}{4}} = 3 \cdot \frac{4}{15} = \frac{4}{5} \text{ hr}$$

The times are equal. The rate of the river current is $\frac{5}{4}$ miles per hour.

**5.** If a passenger train travels three times as fast as a freight train and the freight train takes 4 hours longer to travel 210 miles, what is the speed of each train?

**Solution** Let $r$ = rate of freight train in miles per hour
$3r$ = rate of passenger train in miles per hour

| | Rate | $t = \frac{d}{r}$ | Distance |
|---|---|---|---|
| Freight | $r$ | $\frac{210}{r}$ | 210 |
| Passenger | $3r$ | $\frac{210}{3r}$ | 210 |

$$\frac{210}{r} - \frac{210}{3r} = 4$$ The difference between their *times* is 4 hours.

$$\frac{210}{r} - \frac{70}{r} = 4$$

$$\frac{210}{r} \cdot r - \frac{70}{r} \cdot r = 4 \cdot r$$

$$210 - 70 = 4r$$

$$140 = 4r$$

$$35 = r$$

$$105 = 3r$$

*Check* Time for freight train $= \frac{210}{35} = 6$ hr

Time for passenger train $= \frac{210}{105} = 2$ hr

$6 - 2 = 4$ hours difference in time

The freight train travels 35 mph and the passenger train travels 105 mph. ∎

## EXERCISES 3.7

1. If 4 is subtracted from a certain number and the difference is divided by 2, the result is 1 more than $\frac{1}{5}$ of the original number. Find the original number.
2. What number must be added to both numerator and denominator of $\frac{16}{21}$ to make the resulting fraction equal to $\frac{5}{6}$?
3. Find the number that can be subtracted from both numerator and denominator of the fraction $\frac{69}{102}$ so that the result is $\frac{5}{8}$.
4. The denominator of a fraction exceeds the numerator by 7. If the numerator is increased by 3 and the denominator is increased by 5, the resulting fraction is equal to $\frac{1}{2}$. Find the original fraction.
5. The numerator of a fraction exceeds the denominator by 5. If the numerator is decreased by 4 and the denominator is increased by 3, the resulting fraction is equal to $\frac{4}{5}$. Find the original fraction.
6. One number is $\frac{3}{4}$ of another number. Their sum is 63. Find the numbers.
7. The sum of two numbers is 24. If $\frac{2}{5}$ the larger number is equal to $\frac{2}{3}$ the smaller number, find the two numbers.
8. One number exceeds another by 5. The sum of their reciprocals is equal to 19 divided by the product of the two numbers. Find the numbers.
9. One number is 3 less than another. The sum of their reciprocals is equal to 7 divided by the product of the two numbers. Find the numbers.
10. A manufacturer sold a group of shirts for \$1026. One-fifth of the shirts were priced at \$18 each, and the remainder at \$24 each. How many shirts were sold?
11. Luis spent $\frac{1}{5}$ of his monthly salary for rent and $\frac{1}{6}$ of his monthly salary for car payment. If \$950 was left, what was his monthly salary?
12. It takes Rosa, traveling at 50 mph, 45 minutes longer to go a certain distance than it takes Maria traveling at 60 mph. Find the distance traveled.
13. It takes a plane flying at 450 mph 25 minutes longer to go a certain distance than it does a second plane flying at 500 mph. Find the distance.
14. Toni needs 4 hours to complete the yard work. Her husband, Sonny, needs 6 hours to do the work. How long will the job take if they work together?
15. Ben's secretary can address the weekly newsletters in $4\frac{1}{2}$ hours. Charlie's secretary needs only 3 hours. How long will it take if they both work on the job?
16. Working together, Rick and Rod can clean the snow from the driveway in 20 minutes. It would have taken Rick, working alone, 36 minutes. How long would it have taken Rod alone?
17. A carpenter and his partner can put up a patio cover in $3\frac{3}{7}$ hours. If the partner needs 8 hours to complete the patio alone, how long will it take the carpenter working alone?
18. Beth can travel 208 miles in the same length of time it takes Anna to travel 192 miles. If Beth's speed is 4 mph greater than Anna's, find both rates.

19. A commercial airliner can travel 750 miles in the same length of time that it takes a private plane to travel 300 miles. The speed of the airliner is 60 mph more than twice the speed of the private plane. Find both speeds.

20. Luz travels 350 miles at a certain speed. If the average speed had been 9 mph less, she could have traveled only 300 miles in the same length of time. What was her average rate of speed?

21. A jet flies twice as fast as a propeller plane. On a trip of 1500 miles, the propeller plane took 3 hours longer than the jet. Find the speed of each plane.

22. A family travels 18 miles downriver and returns. It takes 8 hours to make the round trip. Their rate in still water is twice the rate of the current. Find the rate of the current. How long will the return trip take?

23. An airplane can fly 650 mph in calm air. If it can travel 2800 miles with the wind in the same time it can travel 2400 miles against the wind, find the speed of the wind.

24. Using a smaller inlet pipe, it takes 3 hours longer to fill a pool than if a larger pipe is used. If both are used, it takes $3\frac{3}{5}$ hours to fill the pool. Using each pipe alone, how long would it take to fill the pool?

25. A contractor hires two bulldozers to clear the trees from a 20-acre tract of land. One works twice as fast as the other. It takes them 3 days to clear the tract working together. How long would it take each of them alone?

26. John, Ralph, and Denny, working together, can clean a store in 6 hours. Working alone, Ralph takes twice as long to clean the store as does John. Denny needs three times as long as does John. How long would it take each man working alone?

27. Lam went 36 miles downstream and returned. The round trip took $5\frac{1}{4}$ hours. Find the speed of the boat in still water and the speed of the current if the speed of the current is $\frac{1}{7}$ the speed of the boat.

28. Town A is 12 miles upstream from Town B (on the same side of the river). A motorboat that can travel 8 mph in still water leaves A and travels downstream toward B. At the same time, another boat that can travel 10 mph in still water leaves B and travels upstream toward A. Each boat completes the trip at the same time. Find the rate of the current.

## CHAPTER 3 SUMMARY

### Key Terms and Formulas

A **rational expression** is an expression of the form $\frac{P}{Q}$ where $P$ and $Q$ are polynomials and $Q \neq 0$. [3.1]

$$\frac{P}{Q} \cdot \frac{R}{S} = \frac{P \cdot R}{Q \cdot S} \quad \text{where } Q, S \neq 0 \quad [3.2]$$

$$\frac{P}{Q} \div \frac{R}{S} = \frac{P}{Q} \cdot \frac{S}{R} \quad \text{where } Q, R, S \neq 0 \quad [3.2]$$

$$\frac{P}{Q} + \frac{R}{Q} = \frac{P + R}{Q} \quad \text{where } Q \neq 0 \quad [3.3]$$

$$\frac{P}{Q} - \frac{R}{Q} = \frac{P - R}{Q} \quad \text{where } Q \neq 0 \quad [3.3]$$

In general, for polynomials $P$ and $D$, the division algorithm gives

$$\frac{P}{D} = Q + \frac{R}{D}, D \neq 0$$

where $Q$ and $R$ are polynomials and degree of $R <$ degree of $D$. [3.4]

In the special case **when the divisor is first-degree with leading coefficient 1,** the division can be simplified by omitting the variables entirely and writing only certain coefficients. The procedure is called **synthetic division.** [3.4]

A **complex fraction** is a fraction in which the numerator and/or denominator are themselves fractions or the sums or differences of fractions. [3.5]

The basic idea when working with problems related to jobs is to represent **what part of a job is done in one unit of time.** [3.7]

## *Properties and Rules*

**Properties of Rational Expressions** [3.1]

Assume that no denominator has a value of 0.

1. $\dfrac{P}{Q} = \dfrac{R}{S}$ if and only if $P \cdot S = Q \cdot R$
2. $\dfrac{P}{Q} = \dfrac{P \cdot R}{Q \cdot R}$ if $R \neq 0$
3. $-\dfrac{P}{Q} = \dfrac{P}{-Q} = \dfrac{-P}{Q}$
4. $\dfrac{P}{Q} = \dfrac{-P}{-Q} = -\dfrac{-P}{Q} = -\dfrac{P}{-Q}$
5. $\dfrac{-P}{P} = -1$

   In particular, $\dfrac{a - x}{x - a} = -1$

## *Procedures*

**To Find the LCM** [3.3]

1. Find the complete factorization of each expression, including the prime factors of any constant factor.
2. Form the product of all the factors that appear in the complete factorizations using each factor the most number of times it appears in any one factorization.

There are two methods for simplifying complex fractions. The **first method** is to

1. simplify the numerator and denominator separately so that the numerator and denominator are simple fractions;
2. divide the numerator by the denominator.

The **second method** is to

1. find the LCM of all the denominators in the original numerator and denominator;
2. multiply both the numerator and the denominator by this LCM. [3.5]

If both sides of an equation are multiplied by the same nonzero number or expression, the solution set of the new equation will contain the solutions to the original equation. If the multiplication is by a variable expression, then the new equation may have more solutions than the original equation. These extra solutions are called **extraneous solutions** or **extraneous roots.** Check all solutions in the original equation to be sure you have not picked up an extraneous solution or multiplied by 0. [3.6]

Rational inequalities that involve the product and/or quotient of several first-degree expressions may be solved by testing points in intervals formed by the points where the various first-degree expressions are 0. [3.6]

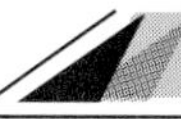

## CHAPTER 3 REVIEW

Express each rational expression in Exercises 1–4 as an equivalent rational expression with the indicated denominator. [3.1]

1. $\dfrac{7x^2y}{15z^2} = \dfrac{?}{60xyz^2}$
2. $\dfrac{5x}{3x + 1} = \dfrac{?}{9x^2 - 1}$
3. $\dfrac{11x}{x - 3} = \dfrac{?}{6 + x - x^2}$
4. $\dfrac{9y}{x + 4} = \dfrac{?}{x^3 + 64}$

Reduce each rational expression in Exercises 5–8 to lowest terms. [3.1]

5. $\dfrac{56x^3y^2}{80xy^5}$

6. $\dfrac{4x + y}{8x^2 + 2xy}$

7. $\dfrac{6 + x - 2x^2}{x^3 - 8}$

8. $\dfrac{4x^2 + 17x + 15}{4xy - 16x + 5y - 20}$

Perform the indicated operations in Exercises 9–20 and simplify if possible. [3.2–3.3]

9. $\dfrac{16x^3}{25y^2} \cdot \dfrac{15y^5}{64x^4}$

10. $\dfrac{17x^3y^2}{28xy^5} \div \dfrac{34xy^4}{7x^3y^2}$

11. $\dfrac{9}{5x^2} + \dfrac{3x}{35x^3}$

12. $\dfrac{3x + 1}{4x^2y} - \dfrac{2y + 5}{6xy^2}$

13. $\dfrac{5x^2 - 30x}{3x^4} \div \dfrac{x - 6}{x^5}$

14. $\dfrac{x + 5}{x^2 - 1} \cdot \dfrac{x + 1}{2x^2 + 11x + 5}$

15. $\dfrac{x}{3x - 6} - \dfrac{4x + 3}{2x + 4}$

16. $\dfrac{3x - 5}{1 - x} + \dfrac{2x + 7}{x^2 - 1}$

17. $\dfrac{-2}{x - 3} + \dfrac{3x^2 + 4x + 15}{x^3 - 27}$

18. $\dfrac{3x}{4 - x^2} + \dfrac{5}{x + 2} - \dfrac{x + 1}{2x^2 - x - 10}$

19. $\dfrac{3x - 2}{x^2 + 13x + 42} \cdot \dfrac{2x^2 - x - 1}{x + 1} \cdot \dfrac{x^2 + 8x + 7}{6x^2 - x - 2}$

20. $\dfrac{6x + 7}{4 + 11x - 3x^2} \cdot \dfrac{x^2 - 2x - 8}{5x^2 - 7x + 2} \div \dfrac{2x^2 - 3x - 14}{15x^2 - x - 2}$

Perform the indicated division in Exercises 21–24 and write the answer in the form $Q + \dfrac{R}{D}$ where the degree of $R$ is less than the degree of $D$. [3.4]

21. $\dfrac{2x^2 + 11x + 14}{2x + 3}$

22. $\dfrac{12x^2 - 4x - 1}{6x - 5}$

23. $\dfrac{x^3 - 8x^2 - 32}{x^2 + 4}$

24. $\dfrac{x^4 + x^3 + 5x^2 + 4x + 1}{x^2 + x + 2}$

Perform the indicated division in Exercises 25–27 using synthetic division. Write the quotient in the form $Q + \dfrac{R}{D}$ where the degree of $R$ is less than the degree of $D$. [3.4]

25. $\dfrac{x^2 + 10x - 7}{x - 5}$

26. $\dfrac{2x^3 + 5x^2 - x + 3}{x + 2}$

27. $\dfrac{x^3 - 4x^2 + 9}{x - 6}$

Simplify the complex fractions in Exercises 28–30. [3.5]

28. $\dfrac{1 + \dfrac{y}{x}}{\dfrac{1}{x^2} + \dfrac{y}{x^3}}$

29. $\dfrac{\dfrac{1}{x} - \dfrac{1}{x + 1}}{\dfrac{1}{x} + \dfrac{1}{x - 1}}$

30. $\dfrac{x + \dfrac{6}{x + 5}}{1 - \dfrac{3}{x}}$

Solve the equations in Exercises 31–38. [3.6]

31. $\dfrac{x + 3}{4} - \dfrac{x - 2}{3} = \dfrac{7}{12}$

32. $\dfrac{x}{7} - \dfrac{x + 3}{8} = \dfrac{5}{8}$

33. $\dfrac{2}{5x} + \dfrac{1}{20} = \dfrac{3}{4x}$

34. $\dfrac{3(x - 1)}{x + 7} = -\dfrac{3}{5}$

35. $\dfrac{1}{x} + \dfrac{2}{x + 4} = \dfrac{5}{3x}$

36. $\dfrac{x}{x - 1} - \dfrac{3}{x + 1} = 1$

37. $\dfrac{x}{x - 2} + \dfrac{x}{x + 1} = \dfrac{3}{x^2 - x - 2}$

38. $\dfrac{1}{x - 2} + \dfrac{4}{x + 2} = \dfrac{5}{x - 1}$

Solve and graph the solutions to the inequalities in Exercises 39–41. [3.6]

**39.** $\dfrac{4x + 5}{x - 2} < 0$

**40.** $\dfrac{x + 3}{x} \leq 4$

**41.** $\dfrac{(x + 5)(x - 1)}{x - 4} \geq 0$

Solve the word problems in Exercises 42–45. [3.7]

**42.** Judy waxes her car in 3 hours. Her brother, Jeff, can do it in 2 hours. If they work together, how long will it take?

**43.** It takes three times as long to drain a tank as it does to fill it. It takes 1 hour to fill the tank when the drain is accidentally left open. How long would it take to fill the tank with the drain closed?

**44.** A man drove 150 miles at a certain speed. He returned over the same route at twice his original rate. The return trip took two hours less time. What was his original rate of speed?

**45.** Brenda travels 4 miles downstream and returns. It takes 50 minutes to make the round trip. The rate of the boat in still water is 5 times the rate of the current. Find the rate of the boat.

## CHAPTER 3 TEST

Reduce each expression in Exercises 1 and 2 to lowest terms.

**1.** $\dfrac{x^2 + 3x}{x^2 + 7x + 12}$

**2.** $\dfrac{x^3 - 9x^2 + 20x}{16 - x^2}$

**3.** Determine the missing numerator that will make the following rational expressions equivalent.

$$\frac{x - 1}{3x + 1} = \frac{?}{3x^2 + 7x + 2}$$

Perform the indicated operations in Exercises 4–10. Reduce all answers to lowest terms.

**4.** $\dfrac{x + 3}{x^2 + 3x - 4} \cdot \dfrac{x^2 + x - 2}{x + 2}$

**5.** $\dfrac{6x^2 - x - 2}{12x^2 + 5x - 2} \div \dfrac{4x^2 - 1}{8x^2 - 6x + 1}$

**6.** $\dfrac{x}{x^2 + 3x - 10} + \dfrac{3x}{4 - x^2}$

**7.** $\dfrac{x - 4}{3x^2 + 5x + 2} - \dfrac{x - 1}{x^2 - 3x - 4}$

**8.** $\dfrac{x^2 - 16}{x^2 - 4x} \cdot \dfrac{x^2}{x + 4} \div \dfrac{x - 1}{2x^2 - 2x}$

**9.** $\dfrac{x}{x + 3} - \dfrac{x + 1}{x - 3} + \dfrac{x^2 + 4}{x^2 - 9}$

**10.** $(4x^3 - 6x^2 + x - 3) \div (2x^2 + 1)$

**11.** Divide using synthetic division:

$$\frac{2x^3 + 11x - 6}{x - 4}$$

**12.** Simplify the following complex fraction:

$$\frac{\dfrac{4}{3x} + \dfrac{1}{6x}}{\dfrac{1}{x^2} - \dfrac{1}{2x}}$$

Solve the equations in Exercises 13 and 14.

**13.** $\dfrac{4}{7} - \dfrac{1}{2x} = 1 + \dfrac{1}{x}$

**14.** $\dfrac{4}{x + 4} + \dfrac{3}{x - 1} = \dfrac{1}{x^2 + 3x - 4}$

Solve and graph each of the solution sets in Exercises 15 and 16.

**15.** $\dfrac{2x + 5}{x - 3} \geq 0$

**16.** $\dfrac{x - 3}{2x + 1} < 2$

**17.** The denominator of a fraction is three more than twice the numerator. If eight is added to both the numerator and denominator, the resulting fraction is equal to $\frac{2}{3}$. Find the original fraction.

**18.** Sonya can clean the apartment in 6 hours. It takes Lucy 12 hours to clean it. If they work together, how long will it take them?

**19.** Mario can travel 228 miles in the same time that Carlos travels 168 miles. If Mario's speed is 15 mph faster than Carlos', find their rates.

**20.** Bob travels 4 miles upstream. In the same length of time, he could have traveled 7 miles downstream. If the speed of the current is 3 mph, find the speed of the boat in still water.

## CUMULATIVE REVIEW (3)

Solve each of the equations in Exercises 1–3.

**1.** $4(3x - 1) = 2(2x - 5) - 3$

**2.** $\frac{4x - 1}{3} + \frac{x - 5}{2} = 2$

**3.** $\left|\frac{3x}{4} - 1\right| = 2$

Solve the inequalities in Exercises 4–6 and graph the solutions. Write the solutions in interval notation.

**4.** $x + 4 - 3x \geq 2x + 5$

**5.** $\frac{x}{5} - 3.4 > \frac{x}{2} + 1.6$

**6.** $\left|\frac{3x}{2} - 1\right| - 2 \leq 3$

Simplify each of the expressions in Exercises 7–9.

**7.** $(3x^2y^{-3})^{-2}$

**8.** $\left(\frac{x^4y}{x^2y^3}\right)^3$

**9.** $\left(\frac{5x^2y^{-1}}{2xy}\right)\left(\frac{3x^{-3}y}{5x^2y^0}\right)$

Write Exercises 10 and 11 in scientific notation.

**10.** $17000 \times 0.0004$

**11.** $\frac{6300}{0.006}$

Factor Exercises 12–15 completely.

**12.** $15x^2 + 22x + 8$

**13.** $2x^3 + 8x^2 + 3x + 12$

**14.** $16x^3 - 54$

**15.** $2x^{-1} - 5x^{-2} - 3x^{-3}$

**16.** Use synthetic division to find the quotient:
$(4x^3 - 5x^2 + 2x - 1) \div (x - 2)$

Perform the indicated operations in Exercises 17 and 18.

**17.** $\frac{x + 1}{x^2 + 3x - 4} \cdot \frac{2x^2 + 7x - 4}{x^2 - 1} \div \frac{x + 1}{x - 1}$

**18.** $\frac{x}{x^2 - 2x - 8} + \frac{x - 3}{2x^2 - 5x - 12} - \frac{2x - 5}{2x^2 + 7x + 6}$

**19.** Eric started walking to a town 10 miles away at a rate of 3 mph. After walking part of the way, he got a ride in a car. The car traveled at an average rate of 48 mph and Eric reached the town 50 minutes after he started. How long did he walk?

**20.** LeAnn is a carpet layer. She has agreed to carpet a house for \$18 per square yard. The carpet will cost her \$11 per square yard and she knows that there is approximately 12% waste in cutting and matching. If she plans to make a profit of \$580.80, how many yards of carpet will she buy?

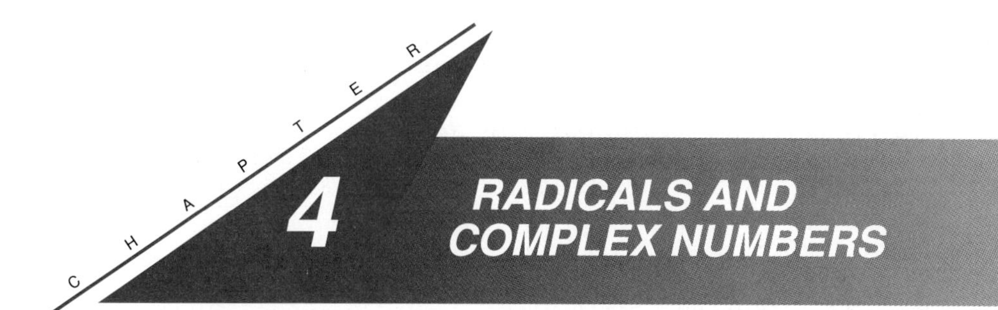

# Chapter 4 RADICALS AND COMPLEX NUMBERS

## MATHEMATICAL CHALLENGES

A child's age increased by three years gives a perfect square. The age decreased by three years gives the square root of that perfect square. What is the child's age?

If $4^x - 4^{x-1} = 24$, then find $(2x)^x$.

Find the unique two-digit number such that the square of the sum of its digits equals the number obtained by reversing its digits.

## CHAPTER OUTLINE



In this chapter, we will discuss expressions with exponents and their close relationship with radical expressions such as square roots ($\sqrt{x}$) and cube roots ($\sqrt[3]{x}$). This relationship allows translation from one type of expression to the other with relative ease and a choice for the form of an answer that best suits the purposes of the problem. For example, in higher level mathematics courses, particularly in calculus, we may be given an expression with a square root (such as $\sqrt{x^2 + 1}$); we will change it to an equivalent expression with a fractional exponent [such as $(x^2 + 1)^{1/2}$] so we can operate with it, then possibly choose to change the answer back to radical notation.

An important concern is under what conditions will an expression with fractional exponents (or radicals) be defined to be a real number. This concern leads to the definition of a new category of numbers called complex numbers. These numbers include the real numbers and a type of number called imaginary numbers. The term "imaginary" is somewhat unfortunate and misleading because these numbers have many practical applications. They are particularly useful in electrical engineering. As we will see in Chapter 5, complex numbers evolve quite naturally as solutions to second-degree equations.

## 4.1 Rational Exponents

**OBJECTIVE**

In this section, you will be learning to simplify expressions using the properties of rational exponents.

In this section, we will discuss the meaning of rational (or fractional) exponents in expressions such as

$$25^{1/2}, \quad 8^{1/3}, \quad (-8)^{2/3}, \quad \text{and} \quad x^{-3/4}$$

One of the properties of whole-number exponents is

$$(a^m)^n = a^{mn}$$

By letting $m = \dfrac{1}{n}$ and $a$ be positive, we get

$$(a^{1/n})^n = a^{1/n \cdot n} = a^1 = a$$

Thus, for positive $a$, $a^{1/n}$ is that number which raised to the $n$th power gives $a$. In other words, for positive $a$ and positive $b$,

$$b^n = a \text{ if and only if } b = a^{1/n}.$$

If $a$ is negative, special considerations must be made.

***Principal* n*th Root***

If $a$ is a positive real number and $n$ is a positive integer, $a^{1/n}$ is a positive real number called the **principal $n$th root of $a$,** and

$$(a^{1/n})^n = a$$

## EXAMPLES

**1.** $25^{1/2} = 5$ since $(5)^2 = 25$.

If the exponent is $\frac{1}{2}$, the root is called the **square root.** Thus, $25^{1/2}$ is the **principal square root** (or just square root) of 25. As a point of clarification, it is also true that $(-5)^2 = 25$. So, $-5$ is called the **negative square root** of 25. That is, 25 has two square roots, one positive and one negative. However, the notation $25^{1/2}$ represents only the positive square root, $+5$.

**2.** $8^{1/3} = 2$ since $2^3 = 8$.

If the exponent is $\frac{1}{3}$, the root is called the **cube root.** Thus, $8^{1/3}$ is the principal cube root (or just cube root) of 8.

**3.** $81^{1/4} = 3$ since $(81^{1/4})^4 = 81^1 = 81$ and $(3)^4 = 81$.

**4.** $125^{1/3} = 5$ since $(125^{1/3})^3 = 125^1 = 125$ and $(5)^3 = 125$. ■

A negative number raised to an odd power is negative. This means that if the base $a$ is negative and $n$ is an **odd** integer, $a^{1/n}$ will be negative.

## EXAMPLES

**5.** $(-8)^{1/3} = -2$ since $[(-8)^{1/3}]^3 = (-8)^1 = -8$ and $(-2)^3 = -8$.
The cube root of $-8$ is $-2$.

**6.** $(-243)^{1/5} = -3$ since $[(-243)^{1/5}]^5 = (-243)^1 = -243$ and $(-3)^5 = -243$.
The fifth root of $-243$ is $-3$. ■

If the base $a$ is **negative** and $n$ is an **even** integer, $a^{1/n}$ is **not a real number.** No real number can be raised to an even power and give a negative result. In Section 4.4, such numbers as $(-36)^{1/2}$ and $(-16)^{1/4}$ will be defined as complex numbers. For now, as the following examples illustrate, we simply say that they are not real numbers.

## EXAMPLES

**7.** Show that $(-36)^{1/2}$ is not a real number.

**Solution** The square of a real number is positive or 0. According to the rules of exponents,

$$[(-36)^{1/2}]^2 = (-36)^{1/2 \cdot 2} = (-36)^1 = -36$$

Since the square of $(-36)^{1/2}$ is a negative number, $(-36)^{1/2}$ is not a real number.

**8.** Show that $(-16)^{1/4}$ is not a real number.

**Solution** An even power of a real number is nonnegative. According to the rules of exponents,

$$[(-16)^{1/4}]^4 = (-16)^{1/4 \cdot 4} = (-16)^1 = -16$$

Since the 4th power of $(-16)^{1/4}$ is a negative number, $(-16)^{1/4}$ is not a real number. ■

***$a^{1/n}$ for Negative a***

If $a$ is a negative real number and $n$ is a positive integer:

**a.** If $n$ is even, $a^{1/n}$ is not a real number.
**b.** If $n$ is odd, $a^{1/n}$ is a negative real number and $(a^{1/n})^n = a$.

To make the discussion complete, we define $\mathbf{0^{1/n} = 0}$.
Summarizing, we have the following table:

If $a$ is a real number and $n$ is a positive integer:

| $a > 0$ | $a < 0$, $n$ even | $a < 0$, $n$ odd | $a = 0$ |
|---|---|---|---|
| $a^{1/n}$ is a positive real number and $(a^{1/n})^n = a$ | $a^{1/n}$ is not a real number | $a^{1/n}$ is a negative real number and $(a^{1/n})^n = a$ | $a^{1/n} = 0$ |

Note very carefully the difference between the two expressions $(-4)^{1/2}$ and $-4^{1/2}$:

$$(-4)^{1/2} \text{ is not a real number}$$
$$-4^{1/2} = -1 \cdot 4^{1/2} = -1 \cdot 2 = -2$$

In the expression $-4^{1/2}$, only 4 is the base of the exponent $\frac{1}{2}$, just as in $-x^3$, only $x$ is the base of 3, and $-x^3 = -1 \cdot x^3$.

Now consider the problem of evaluating the expression $8^{2/3}$ where the exponent is in the form $\frac{m}{n}$. We can write

$$8^{2/3} = (8^{1/3})^2 = (2)^2 = 4$$

or

$$8^{2/3} = (8^2)^{1/3} = (64)^{1/3} = 4$$

The result is the same with either approach. Generally, the first approach, taking a root first and then raising this root to a power, is easier because the numbers are smaller. For example,

$$81^{3/4} = (81^{1/4})^3 = (3)^3 = 27$$

is easier to calculate and work with than

$$(81)^{3/4} = (81^3)^{1/4} = (531{,}441)^{1/4} = 27$$

The fourth root of 81 is more commonly known than the fourth root of 531,441.

**$a^{m/n}$**

If $a$ is a real number, $n$ is a positive integer, $m$ is any nonzero integer, and $a^{1/n}$ is defined as a real number,

$$a^{m/n} = (a^{1/n})^m = (a^m)^{1/n}$$

Now we have $a^{m/n}$ defined for rational exponents $\frac{m}{n}$ and can consider simplifying expressions such as

$$x^{2/3} \cdot x^{1/6} \quad \text{and} \quad x^{-3/4} \cdot x^{1/2} \quad \text{and} \quad (2x^{1/4})^3$$

**All the previous properties of exponents apply to rational exponents.**

**EXAMPLES** Simplify each of the following expressions. Assume that all variables represent positive real numbers.

**9.** $x^{2/3} \cdot x^{1/6}$

**Solution** $x^{2/3} \cdot x^{1/6} = x^{2/3 + 1/6} = x^{4/6 + 1/6} = x^{5/6}$ Add the exponents.

**10.** $\dfrac{x^{3/4}}{x^{1/3}}$

**Solution** $\dfrac{x^{3/4}}{x^{1/3}} = x^{3/4 - 1/3} = x^{9/12 - 4/12} = x^{5/12}$ Subtract the exponents.

**11.** $(2x^{1/4})^3$

**Solution** $(2x^{1/4})^3 = 2^3 \cdot x^{3/4} = 8x^{3/4}$ Multiply the exponents.

**12.** $(27x^{-2/3})^{-1/3}$

**Solution** $(27x^{-2/3})^{-1/3} = 27^{-1/3} \cdot x^{2/9} = \dfrac{x^{2/9}}{27^{1/3}} = \dfrac{x^{2/9}}{3}$ or $\dfrac{1}{3}x^{2/9}$

**13.** $(-36)^{-1/2}$

**Solution** $(-36)^{-1/2} = \dfrac{1}{(-36)^{1/2}}$ Not a real number

This is not a real number because $(-36)^{1/2}$ is not a real number.

**14.** $-36^{-1/2}$

**Solution** $-36^{-1/2} = -1 \cdot 36^{-1/2} = \dfrac{-1}{36^{1/2}} = \dfrac{-1}{6} = -\dfrac{1}{6}$

**15.** $9^{2/4}$

**Solution** $9^{2/4} = 9^{1/2} = 3$ Reduce the exponent if the expression is defined.

**16.** $\left(\dfrac{49x^6y^{-2}}{z^{-4}}\right)^{1/2}$

**Solution** $\left(\dfrac{49x^6y^{-2}}{z^{-4}}\right)^{1/2} = \dfrac{49^{1/2}x^3y^{-1}}{z^{-2}} = \dfrac{7x^3z^2}{y}$ or $7x^3z^2y^{-1}$ ■

***Practice Problems*** Simplify each of the following expressions. Assume that any variable base is positive if its root is even.

**1.** $64^{2/3}$ **2.** $x^{3/4} \cdot x^{1/5} \cdot x^{1/2}$

**3.** $\dfrac{x^{1/6}y^{1/2}}{x^{1/3}y^{1/4}}$ **4.** $\left(\dfrac{16x^{-1/6}}{x^{-3/2}}\right)^{1/4}$

**5.** $(-81)^{1/4}$ **6.** $-81^{1/4}$

## EXERCISES 4.1

Simplify Exercises 1–31.

**1.** $9^{1/2}$ **2.** $36^{1/2}$ **3.** $100^{-1/2}$ **4.** $25^{-1/2}$ **5.** $-49^{1/2}$

**6.** $(-64)^{1/2}$ **7.** $\left(\dfrac{4}{25}\right)^{1/2}$ **8.** $\left(\dfrac{9}{49}\right)^{1/2}$ **9.** $16^{3/4}$ **10.** $(-8)^{2/3}$

**11.** $(-125)^{1/3}$ **12.** $64^{2/3}$ **13.** $(16)^{2/4}$ **14.** $25^{-3/2}$ **15.** $8^{-2/3}$

**16.** $\left(\dfrac{8}{125}\right)^{-1/3}$ **17.** $\left(\dfrac{16}{25}\right)^{-1/2}$ **18.** $-1000^{2/3}$ **19.** $(-4)^{5/2}$ **20.** $(81)^{2/4}$

**21.** $-100^{-3/2}$ **22.** $32^{-3/5}$ **23.** $81^{-3/4}$ **24.** $(-49)^{-5/2}$ **25.** $\left(\dfrac{27}{64}\right)^{2/3}$

**26.** $-\left(\dfrac{16}{81}\right)^{-3/4}$ **27.** $\left(-\dfrac{1}{32}\right)^{2/5}$ **28.** $3 \cdot 16^{-5/4}$ **29.** $-7 \cdot 10{,}000^{-3/4}$ **30.** $[(-27)^{2/3}]^{-2}$

**31.** $\left[\left(\dfrac{1}{32}\right)^{2/5}\right]^{-3}$

Simplify Exercises 32–70. Assume that all variables represent positive real numbers.

**32.** $(x^2)^{1/2}$ **33.** $(x^4)^{1/2}$ **34.** $(2x^{1/3})^3$ **35.** $(3x^{1/2})^4$ **36.** $(9x^4)^{1/2}$

**37.** $(16x^3)^{-1/4}$ **38.** $8x^2 \cdot x^{1/2}$ **39.** $3x^3 \cdot x^{2/3}$ **40.** $5x^2 \cdot x^{-1/3} \cdot x^{1/2}$ **41.** $x^{2/3} \cdot x^{-3/5} \cdot x^0$

**42.** $\dfrac{x^2}{x^{2/5}}$ **43.** $\dfrac{x^{3/4}}{x^{1/6}}$ **44.** $\dfrac{x^{2/5}}{x^{-1/10}}$ **45.** $\dfrac{x^{2/3}}{x^{1/9}}$ **46.** $\dfrac{x^{1/2}}{x^{-2/3}}$

**47.** $\dfrac{x^{3/4} \cdot x^{1/8}}{x^2}$ **48.** $\dfrac{x^{1/2} \cdot x^{-3/4}}{x^{-1/2}}$ **49.** $\dfrac{x^{2/3} \cdot x^{4/3}}{x^2}$ **50.** $\dfrac{x^{2/3}y}{x^2y^{1/2}}$ **51.** $\dfrac{x^{3/2}y^{4/5}}{x^{-1/2}y^2}$

**Answers to Practice Problems** 1. 16 2. $x^{29/20}$ 3. $\dfrac{y^{1/4}}{x^{1/6}}$ or $y^{1/4}x^{-1/6}$ 4. $2x^{1/3}$
5. Not a real number 6. $-3$

52. $\dfrac{x^{3/4}y^{-1/3}}{x^{3/2}y^{1/6}}$ 53. $(2x^{1/2}y^{1/3})^3$ 54. $(4x^{-3/4}y^{1/5})^{-2}$ 55. $(x^{1/2}x^{1/3})^6$ 56. $(-x^3y^6z^{-6})^{2/3}$

57. $\left(\dfrac{x^2y^{-3}}{z^4}\right)^{-1/2}$ 58. $\left(\dfrac{27x^3y^6}{z^9}\right)^{-1/3}$ 59. $(81x^{-8}y^2)^{-1/4}$ 60. $\left(\dfrac{16x^{-4}y^3}{z^4}\right)^{3/4}$ 61. $\left(\dfrac{-27x^2y^3}{z^{-3}}\right)^{1/3}$

62. $\dfrac{(x^{1/4}y^{1/2})^3}{x^{1/2}y^{1/4}}$ 63. $\dfrac{(x^{1/2}y)^{-1/3}}{x^{2/3}y^{-1}}$ 64. $\dfrac{(8x^2y)^{1/3}}{(5x^{1/3}y^{-1/2})^2}$ 65. $\dfrac{(25x^4y^{-1})^{1/2}}{(2x^{1/5}y^{3/5})^3}$

66. $\left(\dfrac{5x^{-3}}{21y^2}\right)^{-1} \cdot \left(\dfrac{49x^4}{100y^{-8}}\right)^{-1/2}$ 67. $\left(\dfrac{x^2y^{1/3}}{x^{1/2}y^{3/2}}\right)^{1/2} \cdot \left(\dfrac{x^{-1/2}y^{2/3}}{x^{-1}y^{3/4}}\right)^2$ 68. $\left(\dfrac{x^{-3}y^{1/3}}{x^{1/2}y}\right)^{1/2} \cdot \left(\dfrac{xy^{1/2}}{x^{-2/3}y^{-1}}\right)^{1/2}$

69. $\left(\dfrac{x^3y^{-2}}{xy^4}\right)^{1/6} \cdot \left(\dfrac{x^{1/5}y^{1/3}}{x^{-1/2}}\right)^3$ 70. $\dfrac{(27xy^{1/2})^{1/3}}{(25x^{-1/2}y)^{1/2}} \cdot \dfrac{(x^{1/2}y)^{1/6}}{(16x^{1/3}y)^{1/2}}$

## 4.2 Radicals

**OBJECTIVES**

In this section, you will be learning to:

1. Write exponential expressions as radicals.
2. Write radicals as exponential expressions.
3. Rationalize the denominators of radicals.
4. Simplify radical expressions.

As we discussed in Section 4.1, the $n$th root of $a$ can be written $a^{1/n}$. This notation is particularly useful in calculus and other areas of mathematics. Another common notation, called **radical notation,** is

$$\sqrt[n]{a}$$

The symbol $\sqrt{\phantom{a}}$ is called a **radical sign;**
$n$ is called the **index;** and
$a$ is called the **radicand.**

If no index is given, it is understood to be 2.

$$a^{1/2} = \sqrt{a}, \qquad a^{1/3} = \sqrt[3]{a}, \qquad a^{1/4} = \sqrt[4]{a}, \qquad \text{and so on}$$

**$\sqrt[n]{a}$**

If $a$ is a real number, $n$ is a positive integer, and $a^{1/n}$ is defined as a real number,

$$a^{1/n} = \sqrt[n]{a}$$

As a result of this definition and the definition of $a^{1/n}$,

$$(\sqrt[n]{a})^n = a.$$

We can say that, for positive $a$ and positive $b$,

$$b^n = a \text{ if and only if } b = \sqrt[n]{a}.$$

If $a$ is negative and $n$ is odd, $\sqrt[n]{a}$ will be negative.

**EXAMPLES**

1. $36^{1/2} = \sqrt{36} = 6$ since $6^2 = 36$
2. $64^{1/2} = \sqrt{64} = 8$ since $8^2 = 64$
3. $32^{1/5} = \sqrt[5]{32} = 2$ since $2^5 = 32$
4. $(-27)^{1/3} = \sqrt[3]{-27} = -3$ since $(-3)^3 = -27$ ■

Just as with $a^{1/n}$, $\sqrt[n]{a}$ is not a real number if $a$ is negative and $n$ is even. **An even root of a negative number is not a real number.** The relationship $a^{m/n} = (a^{1/n})^m = (a^m)^{1/n}$ can also be expressed in radical notation.

If $a$ is a real number, $n$ is a positive integer, $m$ is any integer, and $\sqrt[n]{a}$ is defined as a real number,

$$a^{m/n} = (\sqrt[n]{a})^m = \sqrt[n]{a^m}$$

**EXAMPLES** Change each expression to an equivalent expression in radical notation. Assume that each variable represents a positive real number.

**5.** $x^{2/3} = \sqrt[3]{x^2}$

**6.** $3x^{4/5} = 3\sqrt[5]{x^4}$

**7.** $-a^{3/2} = -\sqrt{a^3}$

Change each expression to an equivalent expression in exponential notation. Assume that each variable is positive (that is, each variable represents a positive real number).

**8.** $\sqrt[6]{x^5} = x^{5/6}$

**9.** $2\sqrt{x} = 2x^{1/2}$

**10.** $-\sqrt[3]{4} = -4^{1/3}$ ■

A positive real number has two square roots, one positive and one negative. For example,

$$\sqrt{9} = 3 \qquad \text{and} \qquad -\sqrt{9} = -3$$

since both $(3)^2 = 9$ and $(-3)^2 = 9$. Note that $\sqrt{9}$ indicates the positive or principal square root. The negative root must be indicated with a negative sign in front of the radical sign.

$$-\sqrt{4} = -2 \qquad \text{and} \qquad \sqrt{4} = 2 \qquad \text{but} \qquad \sqrt{4} \neq -2$$

The square root of a square deserves special mention also. In the case of $\sqrt{(-5)^2}$, we will agree to square $-5$ first and then take the square root. This will guarantee a positive result. Thus,

$$\sqrt{(-5)^2} = \sqrt{25} = 5 \qquad 5 = |-5|$$

and

$$\sqrt{(-6)^2} = \sqrt{36} = 6 \qquad 6 = |-6|$$

We can make a general statement by using absolute value.

For any real number $a$,

$$\sqrt{a^2} = |a|$$

**EXAMPLES**

**11.** $\sqrt{(-7)^2} = |-7| = 7$

**12.** $\sqrt{x^2} = |x|$

**13.** $\sqrt{32x^2y^2} = \sqrt{16}\sqrt{x^2}\sqrt{y^2}\sqrt{2} = 4|x||y|\sqrt{2}$ ■

To simplify a radical expression that involves a product or a quotient, either or both of the following two properties can be used.

If $a$ and $b$ are positive real numbers and $n$ is a positive integer:

1. $\sqrt[n]{ab} = \sqrt[n]{a}\sqrt[n]{b}$
2. $\sqrt[n]{\dfrac{a}{b}} = \dfrac{\sqrt[n]{a}}{\sqrt[n]{b}}$

**Radical expressions are considered to be simplified when there are no factors under the radical sign that can be expressed as a power with an exponent larger than or equal to the index.**

**EXAMPLES** Simplify each radical expression by finding the largest square factor.

**14.** $\sqrt{28} = \sqrt{4 \cdot 7} = \sqrt{4}\sqrt{7} = 2\sqrt{7}$

**15.** $\sqrt{72} = \sqrt{36 \cdot 2} = \sqrt{36}\sqrt{2} = 6\sqrt{2}$

We could have factored $\sqrt{72} = \sqrt{9 \cdot 8} = \sqrt{9 \cdot 4 \cdot 2} = \sqrt{9}\sqrt{4}\sqrt{2}$
$= 3 \cdot 2\sqrt{2} = 6\sqrt{2}$

The final result is the same.

Simplify each radical expression by finding the largest cube factor.

**16.** $\sqrt[3]{54x^5} = \sqrt[3]{27x^3 \cdot 2x^2} = \sqrt[3]{27x^3}\sqrt[3]{2x^2} = 3x\sqrt[3]{2x^2}$

**17.** $\sqrt[3]{\dfrac{16}{27y^6}} = \dfrac{\sqrt[3]{8 \cdot 2}}{\sqrt[3]{27y^6}} = \dfrac{\sqrt[3]{8} \cdot \sqrt[3]{2}}{\sqrt[3]{27y^6}} = \dfrac{2\sqrt[3]{2}}{3y^2}$

Simplify each radical so that no factor under the radical sign is a power larger than the index.

**18.** $\sqrt[4]{y^{11}} = \sqrt[4]{y^8 \cdot y^3} = \sqrt[4]{y^8} \cdot \sqrt[4]{y^3} = y^2\sqrt[4]{y^3}$

The exponent 3 under the radical sign is less than the index 4, so the expression is simplified. (The exponent 8 was used because it is a multiple of 4.)

**19.** $\sqrt[5]{y^{17}} = \sqrt[5]{y^{15} \cdot y^2} = \sqrt[5]{y^{15}} \cdot \sqrt[5]{y^2} = y^3\sqrt[5]{y^2}$

(The exponent 15 was used because it is a multiple of 5.) ■

An expression with a radical in the denominator may not be in the simplest form for further algebraic manipulations or operations. If this is the case, then we may want to **rationalize the denominator. The procedure consists of multiplying the numerator and denominator by another radical with index *n* that will make the denominator a perfect *n*th power.**

EXAMPLES Simplify the following radical expressions. Assume that each variable is positive.

**20.** $\sqrt{\dfrac{5}{4x}}$

**Solution** Multiply the numerator and denominator by $\sqrt{x}$ because $4x \cdot x = 4x^2$ and $4x^2$ is a perfect square.

$$\sqrt{\frac{5}{4x}} = \frac{\sqrt{5} \cdot \sqrt{x}}{\sqrt{4x} \cdot \sqrt{x}} = \frac{\sqrt{5x}}{\sqrt{4x^2}} = \frac{\sqrt{5x}}{2x}$$

Or, multiplying by $\dfrac{x}{x}$ under the radical,

$$\sqrt{\frac{5}{4x}} = \sqrt{\frac{5}{4x} \cdot \frac{x}{x}} = \frac{\sqrt{5x}}{\sqrt{4x^2}} = \frac{\sqrt{5x}}{2x}$$

**21.** $\dfrac{3}{\sqrt[3]{32x}}$

**Solution** Multiply the numerator and denominator by $\sqrt[3]{2x^2}$ because $32x \cdot 2x^2 = 64x^3$ and $64x^3$ is a perfect cube since $(4x)^3 = 64x^3$.

$$\frac{3}{\sqrt[3]{32x}} = \frac{3 \cdot \sqrt[3]{2x^2}}{\sqrt[3]{32x} \cdot \sqrt[3]{2x^2}} = \frac{3\sqrt[3]{2x^2}}{\sqrt[3]{64x^3}} = \frac{3\sqrt[3]{2x^2}}{4x}$$

**22.** $\sqrt[4]{\dfrac{16x}{9y}}$

**Solution** Multiply the numerator and denominator by $\sqrt[4]{9y^3}$ because $9y \cdot 9y^3 = 81y^4$ and $81y^4$ is a perfect 4th power since $(3y)^4 = 81y^4$.

$$\sqrt[4]{\frac{16x}{9y}} = \frac{\sqrt[4]{16x} \cdot \sqrt[4]{9y^3}}{\sqrt[4]{9y} \cdot \sqrt[4]{9y^3}} = \frac{\sqrt[4]{16} \cdot \sqrt[4]{9xy^3}}{\sqrt[4]{81y^4}} = \frac{2\sqrt[4]{9xy^3}}{3y}$$ ■

To simplify an expression such as $\sqrt[4]{\sqrt[3]{x}}$ where a root of a root is indicated, we can change the entire expression to one with fractional exponents and simplify using the rules of exponents.

EXAMPLE Simplify the following expression by using exponential notation.

**23.** $\sqrt[4]{\sqrt[3]{x}}$

**Solution** $\sqrt[4]{\sqrt[3]{x}} = (\sqrt[3]{x})^{1/4} = (x^{1/3})^{1/4} = x^{(1/3)(1/4)} = x^{1/12} = \sqrt[12]{x}$ ■

**Practice Problems**

Simplify the following expressions. Assume that the variables are positive.

1. $\sqrt[3]{125a^5}$
2. $\sqrt{-8x^3}$
3. $\sqrt[3]{-8x^3}$
4. $\sqrt{\dfrac{3}{2a^2}}$
5. $\sqrt{9y^2}$
6. $\sqrt[3]{-27y^6}$
7. $\sqrt[4]{x} \cdot \sqrt{x}$

## EXERCISES 4.2

Change each expression in Exercises 1–12 to an equivalent expression in radical notation. Assume that each variable is positive.

1. $x^{3/5}$ 2. $y^{3/2}$ 3. $y^{3/8}$ 4. $x^{3/4}$ 5. $-x^{4/5}$ 6. $-(xy)^{1/4}$
7. $4(xy)^{2/7}$ 8. $(13x)^{2/5}$ 9. $(8y)^{4/3}$ 10. $(4x^2)^{1/3}$ 11. $x^{2/3}y^{1/2}$ 12. $x^{1/3}y^{5/6}$

Change each expression in Exercises 13–24 to an equivalent expression in exponential notation. Assume that each variable is positive.

13. $\sqrt[3]{x}$ 14. $\sqrt[5]{x^3}$ 15. $\sqrt[5]{x}$ 16. $\sqrt[4]{y^6}$ 17. $4\sqrt[5]{y^2}$ 18. $-3\sqrt[3]{x^5}$
19. $\sqrt[3]{xy^2}$ 20. $\sqrt{x^3y^5}$ 21. $\sqrt[4]{7x^3}$ 22. $-\sqrt[3]{x^2y^5}$ 23. $\sqrt[6]{x^2y^3}$ 24. $\sqrt[4]{x^3y^2}$

Simplify the expressions in Exercises 25–47 and rationalize any denominators. Assume that the variables are positive.

25. $\sqrt{12}$ 26. $\sqrt{18}$ 27. $\sqrt{98}$ 28. $\sqrt{216}$ 29. $-\sqrt{162}$ 30. $-\sqrt{27}$
31. $\sqrt[3]{16}$ 32. $\sqrt[3]{40}$ 33. $\sqrt[3]{108}$ 34. $\sqrt[3]{-54}$ 35. $\sqrt[3]{8x^4}$ 36. $\sqrt{-4x^5}$
37. $\sqrt[4]{32x^5}$ 38. $-\sqrt[5]{x^{11}y}$ 39. $\dfrac{2}{\sqrt{3}}$ 40. $\sqrt{\dfrac{25}{x^2}}$ 41. $\sqrt{\dfrac{4}{x}}$ 42. $\sqrt{\dfrac{4x}{3y^2}}$
43. $-\sqrt{\dfrac{5y^2}{8x}}$ 44. $\sqrt[3]{\dfrac{y}{4x}}$ 45. $\sqrt[3]{\dfrac{7x}{y^4}}$ 46. $\sqrt[4]{\dfrac{3x^5}{8y^7}}$ 47. $\sqrt[4]{\dfrac{2y^4}{27x^6}}$

Simplify the expressions in Exercises 48–60. Assume that the variables may now be positive or negative.

48. $\sqrt{(7)^2}$ 49. $-\sqrt{(-3)^2}$ 50. $\sqrt{25y^2}$ 51. $-\sqrt{81x^2}$ 52. $\sqrt[3]{-64x^6}$
53. $-\sqrt[3]{81y^9}$ 54. $\sqrt{144x^2}$ 55. $\sqrt{(-9)^2y^4}$ 56. $\sqrt{18x^2y^2}$ 57. $\sqrt{32x^4y^4}$
58. $\sqrt[3]{-24x^5y^3}$ 59. $\sqrt[3]{108xy^6}$ 60. $\dfrac{-x}{\sqrt{2y^2}}$

**Answers to Practice Problems** 1. $5a\sqrt[3]{a^2}$ 2. Not a real number 3. $-2x$ 4. $\dfrac{\sqrt{6}}{2a}$
5. $3y$ 6. $-3y^2$ 7. $\sqrt[4]{x^3}$

Change each expression in Exercises 61–70 to an equivalent expression in exponential notation; then simplify. Assume that the variables are positive.

**61.** $\sqrt[3]{x} \cdot \sqrt{x}$ **62.** $\sqrt[3]{x^2} \cdot \sqrt[5]{x^3}$ **63.** $\dfrac{\sqrt[4]{x^3}}{\sqrt[6]{x}}$ **64.** $\dfrac{\sqrt[4]{x}}{\sqrt[3]{x^4}}$ **65.** $\dfrac{\sqrt[3]{x^2}\sqrt[5]{x^4}}{\sqrt{x^3}}$

**66.** $\dfrac{x\sqrt[4]{x}}{\sqrt[3]{x}\sqrt{x}}$ **67.** $\sqrt{\sqrt[3]{x}}$ **68.** $\sqrt[5]{\sqrt{x}}$ **69.** $\sqrt[3]{\sqrt[3]{x}}$ **70.** $\sqrt{\sqrt{x}}$

## 4.3 Arithmetic with Radicals

**OBJECTIVES**

In this section, you will be learning to:

1. Perform arithmetic operations with radical expressions.
2. Rationalize the denominators of radicals.

To find the sum $2x + 3x - 8x$, you can use the distributive property and write

$$\begin{aligned} 2x + 3x - 8x &= (2 + 3 - 8)x \\ &= -3x \end{aligned}$$

The terms $2x$, $3x$, and $-8x$ are called **like terms** because each term contains the same variable expression, $x$. Similarly,

$$\begin{aligned} 2\sqrt{5} + 3\sqrt{5} - 8\sqrt{5} &= (2 + 3 - 8)\sqrt{5} \\ &= -3\sqrt{5} \end{aligned}$$

and $2\sqrt{5}$, $3\sqrt{5}$, and $-8\sqrt{5}$ are called **like radicals** because each term contains the same radical expression, $\sqrt{5}$.

To find the sum of like radicals, you may need to simplify first:

$$\begin{aligned} 4\sqrt{12} + \sqrt{75} - \sqrt{108} &= 4\sqrt{4 \cdot 3} + \sqrt{25 \cdot 3} - \sqrt{36 \cdot 3} \\ &= 4 \cdot 2\sqrt{3} + 5\sqrt{3} - 6\sqrt{3} \\ &= (8 + 5 - 6)\sqrt{3} \\ &= 7\sqrt{3} \end{aligned}$$

**EXAMPLES** Simplify the following expressions. Assume that all variables are positive.

**1.** $\sqrt{32x} + \sqrt{18x}$

**Solution** 
$$\begin{aligned} \sqrt{32x} + \sqrt{18x} &= \sqrt{16 \cdot 2x} + \sqrt{9 \cdot 2x} \\ &= 4\sqrt{2x} + 3\sqrt{2x} \\ &= 7\sqrt{2x} \end{aligned}$$

**2.** $\sqrt{12} + \sqrt{18} + \sqrt{27}$

**Solution**

$$\begin{aligned} \sqrt{12} + \sqrt{18} + \sqrt{27} &= \sqrt{4 \cdot 3} + \sqrt{9 \cdot 2} + \sqrt{9 \cdot 3} \\ &= 2\sqrt{3} + 3\sqrt{2} + 3\sqrt{3} \\ &= 5\sqrt{3} + 3\sqrt{2} \end{aligned}$$

$\sqrt{3}$ and $\sqrt{2}$ are **not** like radicals.

**3.** $\sqrt[3]{5x} - \sqrt[3]{40x}$

**Solution** 
$$\begin{aligned} \sqrt[3]{5x} - \sqrt[3]{40x} &= \sqrt[3]{5x} - \sqrt[3]{8 \cdot 5x} \\ &= \sqrt[3]{5x} - 2\sqrt[3]{5x} \\ &= (1 - 2)\sqrt[3]{5x} \\ &= -\sqrt[3]{5x} \end{aligned}$$

■

To find a product such as $(\sqrt{3} + 5)(\sqrt{3} - 7)$, we can treat the two expressions as two binomials and multiply just as with polynomials. For example,

$$\begin{aligned} (\sqrt{3} + 5)(\sqrt{3} - 7) &= (\sqrt{3})^2 + 5\sqrt{3} - 7\sqrt{3} + 5(-7) \\ &= (\sqrt{3})^2 + 5\sqrt{3} - 7\sqrt{3} - 35 \\ &= 3 - 2\sqrt{3} - 35 \\ &= -32 - 2\sqrt{3} \end{aligned}$$

**EXAMPLES** Simplify the following expressions.

**4.** $(\sqrt{7} - 2)(\sqrt{7} + 3)$

**Solution** 
$$\begin{aligned} (\sqrt{7} - 2)(\sqrt{7} + 3) &= (\sqrt{7})^2 - 2\sqrt{7} + 3\sqrt{7} - 2(3) \\ &= 7 - 2\sqrt{7} + 3\sqrt{7} - 6 \\ &= 7 - 6 - 2\sqrt{7} + 3\sqrt{7} \\ &= 1 + \sqrt{7} \end{aligned}$$

**5.** $(\sqrt{6} + \sqrt{2})^2$

**Solution** 
$$\begin{aligned} (\sqrt{6} + \sqrt{2})^2 &= (\sqrt{6})^2 + 2\sqrt{6}\sqrt{2} + (\sqrt{2})^2 \\ &= 6 + 2\sqrt{12} + 2 \\ &= 8 + 2\sqrt{4} \cdot \sqrt{3} \\ &= 8 + 4\sqrt{3} \end{aligned}$$

**6.** $(\sqrt{2} + 5)(\sqrt{2} - 5)$

**Solution**

$$\begin{aligned} (\sqrt{2} + 5)(\sqrt{2} - 5) &= (\sqrt{2})^2 - (5)^2 \qquad \text{Difference of two squares} \\ &= 2 - 25 \\ &= -23 \end{aligned}$$

■

Now we want to **rationalize the denominator** of a fraction such as

$$\frac{2}{4 - \sqrt{2}}$$

in which the denominator is of the form $a - b = 4 - \sqrt{2}$ where square roots are involved.

Recall that the product $(a - b)(a + b)$ results in the difference of two squares.

$$(a - b)(a + b) = a^2 - b^2$$

We consider two cases:

1. If the denominator is of the form $a - b$, we multiply both the numerator and denominator by $a + b$.
2. If the denominator is of the form $a + b$, we multiply both the numerator and denominator by $a - b$.

In either case, the denominator becomes

$a^2 - b^2$, the difference of two squares

and we have a rational denominator. Thus,

$$\frac{2}{4 - \sqrt{2}} = \frac{2(\mathbf{4 + \sqrt{2}})}{(4 - \sqrt{2})(\mathbf{4 + \sqrt{2}})}$$

If $a - b = 4 - \sqrt{2}$, then $a + b = 4 + \sqrt{2}$.

$$= \frac{2(4 + \sqrt{2})}{4^2 - (\sqrt{2})^2}$$

The denominator is the difference of two squares.

$$= \frac{2(4 + \sqrt{2})}{16 - 2}$$

The denominator is a rational number.

$$= \frac{2(4 + \sqrt{2})}{14} = \frac{4 + \sqrt{2}}{7}$$

**EXAMPLES** Simplify the following expressions.

**7.** $\dfrac{31}{6 + \sqrt{5}}$

**Solution** Multiply the numerator and denominator by $6 - \sqrt{5}$.

$$\frac{31}{6 + \sqrt{5}} = \frac{31(\mathbf{6 - \sqrt{5}})}{(6 + \sqrt{5})(\mathbf{6 - \sqrt{5}})}$$

$$= \frac{31(6 - \sqrt{5})}{36 - 5}$$

$$= \frac{\cancel{31}(6 - \sqrt{5})}{\cancel{31}} = 6 - \sqrt{5}$$

**8.** $\dfrac{1}{\sqrt{7} - \sqrt{2}}$

**Solution** Multiply the numerator and denominator by $\sqrt{7} + \sqrt{2}$.

$$\frac{1}{\sqrt{7} - \sqrt{2}} = \frac{1(\mathbf{\sqrt{7} + \sqrt{2}})}{(\sqrt{7} - \sqrt{2})(\mathbf{\sqrt{7} + \sqrt{2}})}$$

$$= \frac{\sqrt{7} + \sqrt{2}}{7 - 2}$$

$$= \frac{\sqrt{7} + \sqrt{2}}{5}$$

**9.** $\dfrac{6}{1-\sqrt{x}}$

**Solution** $\dfrac{6}{1-\sqrt{x}} = \dfrac{6(1+\sqrt{x})}{(1-\sqrt{x})(1+\sqrt{x})}$

$= \dfrac{6(1+\sqrt{x})}{1-x}$ ■

***Practice Problems***

Simplify the following expressions. Assume that all variables are positive.

1. $2\sqrt{10} - 6\sqrt{10}$
2. $\sqrt{5} + \sqrt{45} - \sqrt{15}$
3. $\sqrt{8x} - 3\sqrt{2x} + \sqrt{18x}$
4. $\sqrt[3]{x^5} + x\sqrt[3]{27x^2}$
5. $\sqrt[3]{x^3y^6z} + 4xy^2\sqrt[3]{z}$
6. $(\sqrt{6} - 2\sqrt{5})(\sqrt{6} + \sqrt{5})$
7. $\dfrac{4}{\sqrt{2}+\sqrt{6}}$
8. $\dfrac{x-5}{\sqrt{x}-\sqrt{5}}$
9. $(\sqrt{3} + \sqrt{2})^2$
10. $(3 + \sqrt{2})^2$

## *EXERCISES 4.3*

Perform the indicated operations and simplify. Assume that all variables are positive.

1. $\sqrt{2} - 7\sqrt{2}$
2. $6\sqrt{11} + 4\sqrt{11} - 3\sqrt{11}$
3. $2\sqrt{x} + 4\sqrt{x} - \sqrt{x}$
4. $8\sqrt[3]{xy} - 3\sqrt[3]{xy} + 4\sqrt[3]{xy}$
5. $9\sqrt[3]{7x^2} - 4\sqrt[3]{7x^2} - 8\sqrt[3]{7x^2}$
6. $12\sqrt[3]{4x} - 10\sqrt[3]{4x} - 6\sqrt[3]{4x}$
7. $2\sqrt{3} + 4\sqrt{12}$
8. $2\sqrt{48} - 3\sqrt{75}$
9. $2\sqrt{18} + \sqrt{8} - 3\sqrt{50}$
10. $2\sqrt{12} + \sqrt{72} - \sqrt{75}$
11. $5\sqrt{48} + 2\sqrt{45} - 3\sqrt{20}$
12. $3\sqrt{28} - \sqrt{63} + 8\sqrt{10}$
13. $2\sqrt{96} + \sqrt{147} - \sqrt{150}$
14. $7\sqrt{12x} - 4\sqrt{27x} + \sqrt{108x}$
15. $6\sqrt{45x^3} + \sqrt{80x^3} - \sqrt{20x^3}$
16. $2\sqrt{18xy^2} + \sqrt{8xy^2} - 3y\sqrt{50x}$
17. $\sqrt{125} - \sqrt{63} + 3\sqrt{45}$
18. $5\sqrt{48} + 2\sqrt{24} - \sqrt{75}$
19. $\sqrt{32x} + 7\sqrt{12x} + \sqrt{98x}$
20. $\sqrt[3]{81x^2} - 5\sqrt[3]{48x^2} - 5\sqrt[3]{24x^2}$
21. $\sqrt[3]{16} - 5\sqrt[3]{54} + 2\sqrt[3]{40}$
22. $x\sqrt{y} + \sqrt{x^2y} - \sqrt{xy^3}$
23. $x\sqrt{2x^3} - 3\sqrt{8x^5} + x\sqrt{72x^3}$
24. $x\sqrt{y^3} - 2\sqrt{x^2y^3} - y\sqrt{x^2y}$
25. $x\sqrt{9x^3y^2} - 5x^2\sqrt{xy^2} + 6y\sqrt{x^5}$
26. $(3 + \sqrt{2})(5 - \sqrt{2})$
27. $(\sqrt{3x} - 8)(\sqrt{3x} - 1)$
28. $(2\sqrt{7} + 4)(\sqrt{7} - 3)$
29. $(6 + \sqrt{2x})(4 + \sqrt{2x})$
30. $(5\sqrt{3} - 2)(2\sqrt{3} - 7)$
31. $(\sqrt{6} + 2)(\sqrt{6} - 2)$
32. $(3\sqrt{2} + \sqrt{5})(\sqrt{2} + \sqrt{5})$
33. $(\sqrt{5} + 2\sqrt{2})^2$
34. $(2\sqrt{5} + 3\sqrt{2})^2$
35. $(3\sqrt{5} + 4\sqrt{3})(3\sqrt{5} - 4\sqrt{3})$
36. $(\sqrt{2} + \sqrt{3})(\sqrt{5} - \sqrt{3})$
37. $(\sqrt{6} + \sqrt{5})(\sqrt{6} - \sqrt{2})$
38. $(3\sqrt{7} + \sqrt{5})(3\sqrt{7} - \sqrt{5})$
39. $(\sqrt{11} + \sqrt{3})(\sqrt{11} - 2\sqrt{3})$
40. $(\sqrt{x} + \sqrt{6})(\sqrt{x} - 3\sqrt{6})$
41. $(7\sqrt{x} + \sqrt{2})(7\sqrt{x} - \sqrt{2})$
42. $(\sqrt{x} + 5\sqrt{y})^2$
43. $(3\sqrt{x} + \sqrt{y})^2$
44. $(4\sqrt{x} + 3\sqrt{y})(\sqrt{x} - 3\sqrt{y})$
45. $(2\sqrt{2x} + \sqrt{y})(3\sqrt{2x} + 2\sqrt{y})$
46. $\dfrac{1}{\sqrt{2}+1}$
47. $\dfrac{3}{\sqrt{3}-5}$
48. $\dfrac{\sqrt{3}}{\sqrt{5}-4}$
49. $\dfrac{\sqrt{6}}{\sqrt{7}+3}$

**Answers to Practice Problems** **1.** $-4\sqrt{10}$ **2.** $4\sqrt{5} - \sqrt{15}$ **3.** $2\sqrt{2x}$ **4.** $4x\sqrt[3]{x^2}$ **5.** $5xy^2\sqrt[3]{z}$ **6.** $-4 - \sqrt{30}$ **7.** $-(\sqrt{2} - \sqrt{6})$ or $-\sqrt{2} + \sqrt{6}$ **8.** $\sqrt{x} + \sqrt{5}$ **9.** $5 + 2\sqrt{6}$ **10.** $11 + 6\sqrt{2}$

**50.** $\dfrac{2}{\sqrt{2}+\sqrt{3}}$ **51.** $\dfrac{8}{\sqrt{5}-\sqrt{3}}$ **52.** $\dfrac{\sqrt{5}}{\sqrt{7}-\sqrt{3}}$ **53.** $\dfrac{\sqrt{10}}{\sqrt{5}-2\sqrt{2}}$

**54.** $\dfrac{2-\sqrt{6}}{\sqrt{6}-3}$ **55.** $\dfrac{7+2\sqrt{5}}{7-\sqrt{5}}$ **56.** $\dfrac{\sqrt{3}+\sqrt{7}}{\sqrt{3}-\sqrt{7}}$ **57.** $\dfrac{2\sqrt{3}+\sqrt{2}}{\sqrt{3}-\sqrt{2}}$

**58.** $\dfrac{2\sqrt{x}-y}{\sqrt{x}-y}$ **59.** $\dfrac{x+2\sqrt{y}}{x-2\sqrt{y}}$ **60.** $\dfrac{y}{\sqrt{x}+\sqrt{2y}}$

## 4.4 Complex Numbers

**OBJECTIVES**

In this section, you will be learning to:

1. Identify the real and imaginary parts of complex numbers.
2. Simplify the square roots of negative numbers.
3. Add complex numbers.
4. Subtract complex numbers.
5. Solve linear equations with complex numbers by setting the real and imaginary parts equal.

One of the properties of real numbers is that the square of any real number is nonnegative. That is, for any real number $x$, $x^2 \geq 0$. We have said that $\sqrt{-4}$ and $\sqrt{-5}$ are not real numbers. Now we will define these numbers by expanding the real number system into the system of **complex numbers.**

In Chapter 5, we will see how these numbers occur as solutions to equations with real coefficients. While at first they may seem to be somewhat impractical, complex numbers occur quite naturally in trigonometry and higher level mathematics and have practical applications in such fields as electrical engineering.

The first step is to define $\sqrt{-1}$.

***$\sqrt{-1}$*** $i = \sqrt{-1}$ and $i^2 = (\sqrt{-1})^2 = -1$

With this definition, we can make the following statement:

If $a$ is a positive real number, then

$$\sqrt{-a} = \sqrt{-1}\sqrt{a} = i\sqrt{a}$$

**EXAMPLES**

1. $\sqrt{-25} = \sqrt{-1}\sqrt{25} = i \cdot 5 = 5i$ $\quad (5i)^2 = 5^2i^2 = 25i^2 = -25$
2. $\sqrt{-36} = \sqrt{-1}\sqrt{36} = i \cdot 6 = 6i$
3. $\sqrt{-24} = \sqrt{-1}\sqrt{4 \cdot 6} = i \cdot 2 \cdot \sqrt{6} = 2i\sqrt{6}$
4. $\sqrt{-45} = \sqrt{-1}\sqrt{9 \cdot 5} = i \cdot 3 \cdot \sqrt{5} = 3i\sqrt{5}$ ■

***Complex Number*** If $a$ and $b$ are real numbers, $a + bi$ is a **complex number.** $a$ is called the **real part;** $b$ is called the **imaginary part.**

If $b = 0$, then $a + bi = a + 0i = a$ is a **real number.**

If $a = 0$, then $a + bi = 0 + bi = bi$ is called a **pure imaginary number** (or an **imaginary number**).

**EXAMPLES: Complex Numbers**

**5.** $4 - 2i$ — 4 is the real part; $-2$ is the imaginary part.

**6.** $\sqrt{5} + 3i\sqrt{2}$ — $\sqrt{5}$ is the real part; $3\sqrt{2}$ is the imaginary part.

**7.** $7 + 0i$ — 7 is the real part; 0 is the imaginary part. (If $b = 0$, the complex number is a real number.)

**8.** $0 - i\sqrt{3}$ — 0 is the real part; $-\sqrt{3}$ is the imaginary part. (If $a = 0$ and $b \neq 0$, the complex number is a **pure imaginary number.**) ■

In general, if $a$ is a real number, then we can write $a = a + 0i$. This means that $a$ is a complex number. Thus, **every real number is a complex number.**

If two complex numbers are equal, then their real parts are equal and their imaginary parts are equal. For example, if

$$x + yi = 7 + 2i$$

then

$$x = 7 \quad \text{and} \quad y = 2$$

This relationship can be used to solve equations involving complex numbers.

***Equality of Complex Numbers***

For complex numbers $a + bi$ and $c + di$,

$$a + bi = c + di \quad \text{if and only if} \quad a = c \text{ and } b = d$$

**EXAMPLES** Solve each equation for the unknown numbers.

**9.** $(x + 3) + 2yi = 7 - 6i$

**Solution** Equate the real parts and the imaginary parts.

$$\begin{aligned} x + 3 &= 7 \quad \text{and} \quad & 2y &= -6 \\ x &= 4 & y &= -3 \end{aligned}$$

**10.** $2y + 3 - 8i = 9 + 4xi$

**Solution**

$$\begin{aligned} 2y + 3 &= 9 \quad \text{and} \quad & -8 &= 4x \\ 2y &= 6 & -2 &= x \\ y &= 3 \end{aligned}$$

■

Adding complex numbers is similar to adding polynomials. Combine the like terms. For example,

$$\begin{aligned} (2 + 3i) + (9 - 8i) &= 2 + 9 + 3i - 8i \\ &= (2 + 9) + (3 - 8)i \\ &= 11 - 5i \end{aligned}$$

Similarly,

$$\begin{aligned}(5 - 2i) - (6 + 7i) &= 5 - 2i - 6 - 7i \\ &= (5 - 6) + (-2 - 7)i \\ &= -1 - 9i\end{aligned}$$

**Addition and Subtraction with Complex Numbers**

For complex numbers $a + bi$ and $c + di$,

$$(a + bi) + (c + di) = (a + c) + (b + d)i$$

and $$(a + bi) - (c + di) = (a - c) + (b - d)i$$

**EXAMPLES** Find the sums or differences as indicated.

**11.** $(6 - 2i) + (1 - 2i)$

**Solution** $$\begin{aligned}(6 - 2i) + (1 - 2i) &= (6 + 1) + (-2 - 2)i \\ &= 7 - 4i\end{aligned}$$

**12.** $(-8 - i\sqrt{2}) - (-8 + i\sqrt{2})$

**Solution** $$\begin{aligned}(-8 - i\sqrt{2}) - (-8 + i\sqrt{2}) &= -8 - i\sqrt{2} + 8 - i\sqrt{2} \\ &= -8 + 8 - i\sqrt{2} - i\sqrt{2} \\ &= 0 + (-\sqrt{2} - \sqrt{2})i \\ &= -2\sqrt{2}i \\ &= -2i\sqrt{2}\end{aligned}$$

**Note:** The form $-2i\sqrt{2}$ is written to avoid confusion about $i$ not being under the radical sign.

**13.** $(\sqrt{3} - 2i) + (1 + i\sqrt{5})$

**Solution** $$\begin{aligned}(\sqrt{3} - 2i) + (1 + i\sqrt{5}) &= \sqrt{3} + 1 - 2i + i\sqrt{5} \\ &= (\sqrt{3} + 1) + (\sqrt{5} - 2)i\end{aligned}$$

**Note:** Here, the real part is $\sqrt{3} + 1$, and the imaginary part is $\sqrt{5} - 2$. ■

***Practice Problems***

1. What is the imaginary part and what is the real part of $2 - i\sqrt{39}$?

Simplify as indicated.

2. $(-7 + i\sqrt{3}) + (5 - 2i)$
3. $(4 + i) - (5 + 2i)$

Solve for the unknown variables.

4. $x + yi = \sqrt{2} - 7i$
5. $3y + (x - 7)i = -9 + 2i$

**Answers to Practice Problems** **1.** Real part is 2; imaginary part is $-\sqrt{39}$. **2.** $-2 + (\sqrt{3} - 2)i$ **3.** $-1 - i$ **4.** $x = \sqrt{2}, y = -7$ **5.** $y = -3, x = 9$

## EXERCISES 4.4

What is the imaginary part and what is the real part of each of the complex numbers in Exercises 1–10?

**1.** $4 - 3i$
**2.** $6 + i\sqrt{3}$
**3.** $-11 + i\sqrt{2}$
**4.** $\frac{3}{4} + i$
**5.** $\frac{2}{3} + i\sqrt{17}$
**6.** $\frac{4}{7}i$
**7.** $\frac{4 + 7i}{5}$
**8.** $\frac{2 - i}{4}$
**9.** $\frac{3}{8}$
**10.** $-\sqrt{5} + \frac{\sqrt{2}}{2}i$

Simplify the radicals in Exercises 11–24.

**11.** $\sqrt{-49}$
**12.** $\sqrt{-121}$
**13.** $-\sqrt{-64}$
**14.** $\sqrt{-169}$
**15.** $3\sqrt{147}$
**16.** $\sqrt{128}$
**17.** $2\sqrt{-150}$
**18.** $4\sqrt{-99}$
**19.** $-2\sqrt{-108}$
**20.** $2\sqrt{175}$
**21.** $\sqrt{242}$
**22.** $\sqrt{-192}$
**23.** $\sqrt{-1000}$
**24.** $\sqrt{-243}$

Find the sums or differences as indicated in Exercises 25–46.

**25.** $(2 + 3i) + (4 - i)$
**26.** $(7 - i) + (3 + 6i)$
**27.** $(4 + 5i) - (3 - 2i)$
**28.** $(-3 + 2i) - (6 + 2i)$
**29.** $-3i + (2 - 3i)$
**30.** $(7 + 5i) + (6 - 2i)$
**31.** $(8 + 9i) - (8 - 5i)$
**32.** $(-6 + i) - (2 + 3i)$
**33.** $(\sqrt{5} - 2i) + (3 - 4i)$
**34.** $(4 + 3i) - (\sqrt{2} + 3i)$
**35.** $(7 + i\sqrt{6}) + (-2 + i)$
**36.** $(\sqrt{11} + 2i) + (5 - 7i)$
**37.** $(\sqrt{3} + i\sqrt{2}) - (5 + i\sqrt{2})$
**38.** $(\sqrt{5} + i\sqrt{3}) + (1 - i)$
**39.** $(5 + \sqrt{-25}) - (7 + \sqrt{-100})$
**40.** $(1 + \sqrt{-36}) + (-4 - \sqrt{-49})$
**41.** $(13 - 3\sqrt{-16}) + (-2 - 4\sqrt{-1})$
**42.** $(7 + \sqrt{-9}) - (3 - 2\sqrt{-25})$
**43.** $(4 + i) + (-3 - 2i) - (-1 - i)$
**44.** $(-2 - 3i) + (6 + i) - (2 + 5i)$
**45.** $(7 + 3i) + (2 - 4i) - (6 - 5i)$
**46.** $(-5 + 7i) + (4 - 2i) - (3 - 5i)$

Solve the equations in Exercises 47–60 for $x$ and $y$.

**47.** $x + 3i = 6 - yi$
**48.** $2x - 8yi = -2 + 4yi$
**49.** $\sqrt{5} - 2i = y + xi$
**50.** $\frac{2}{3} - 2yi = 2x + \frac{4}{5}$
**51.** $\sqrt{2} + i - 3 = x + yi$
**52.** $i\sqrt{5} - 3 + 4i = x + yi$
**53.** $2x + 3 + 6i = 7 - (y + 2)i$
**54.** $x + yi + 8 = 2i + 4 - 3yi$
**55.** $x + 2i = 5 - yi - 3 - 4i$
**56.** $3x + 2 - 7i = i - 2yi + 5$
**57.** $2 + 3i + x = 5 - 7i + yi$
**58.** $11i - 2x + 4 = 10 - 3i + 2yi$
**59.** $2x - 2yi + 6 = 6i - x + 2$
**60.** $x + 4 - 3x + i = 8 + yi$

# 4.5 Multiplication and Division with Complex Numbers

**OBJECTIVES**

In this section, you will be learning to:

1. Multiply complex numbers.
2. Divide complex numbers.
3. Simplify powers of *i*.

The product of two complex numbers can be found using the same procedure as in multiplying two binomials. **Remember that $i^2 = -1$.** For example,

$$\begin{aligned}(3 + 5i)(2 + i) &= (3 + 5i)2 + (3 + 5i)i \\ &= 6 + 10i + 3i + 5i^2 \\ &= 6 + 13i - 5 \qquad\qquad 5i^2 = -5 \\ &= 1 + 13i\end{aligned}$$

**Multiplication with Complex Numbers**

For complex numbers $a + bi$ and $c + di$,

$$(a + bi)(c + di) = (ac - bd) + (bc + ad)i$$

Rather than memorizing this definition, you might simply multiply two complex numbers by using the distributive property (or FOIL method) and do most of the steps mentally. **Remember that $i^2 = -1$.**

EXAMPLES Find the following products.

**1.** $(6 + 3i)(2 - 7i)$

**Solution** $$\begin{aligned}(6 + 3i)(2 - 7i) &= 12 + 6i - 42i - 21i^2 \\ &= 12 - 36i + 21 \\ &= 33 - 36i\end{aligned}$$

**2.** $(\sqrt{2} - i)(\sqrt{2} - i)$

**Solution** $$\begin{aligned}(\sqrt{2} - i)(\sqrt{2} - i) &= (\sqrt{2})^2 - 2\sqrt{2} \cdot i + i^2 \\ &= 2 - 2i\sqrt{2} - 1 \\ &= 1 - 2i\sqrt{2}\end{aligned}$$

**3.** $(-1 + i)(2 - i)$

**Solution** $$\begin{aligned}(-1 + i)(2 - i) &= -2 + 2i + i - i^2 \\ &= -2 + 3i + 1 \\ &= -1 + 3i\end{aligned}$$ ■

The powers of *i* form an interesting pattern. Regardless of the particular integer exponent, there are only four possible values for $i^n$: $i, -1, -i, +1$.

$$\begin{aligned}i^1 &= i \\ i^2 &= -1 \\ i^3 &= i^2 \cdot i = -1 \cdot i = -i \\ i^4 &= i^2 \cdot i^2 = (-1)(-1) = +1 \\ i^5 &= i^4 \cdot i^1 = +1 \cdot i = i \\ i^6 &= i^4 \cdot i^2 = (+1)(-1) = -1 \\ i^7 &= i^4 \cdot i^3 = (+1)(-i) = -i \\ i^8 &= i^4 \cdot i^4 = (+1)(+1) = +1\end{aligned}$$

We can simplify higher powers of $i$ using the following result:

$$i^{4m} = (i^4)^m = (+1)^m = 1$$

That is, when $i$ is raised to a power that is a multiple of 4, the result is 1.

**EXAMPLES**

**4.** $i^{45} = i^{44} \cdot i^1 = (i^4)^{11} \cdot i = 1^{11} \cdot i = i$

**5.** $i^{59} = i^{56} \cdot i^3 = (i^4)^{14} \cdot i^2 \cdot i = 1^{14} \cdot (-1)i = -i$

**6.** $i^{-6} = \dfrac{1}{i^6} = \dfrac{1}{i^4 \cdot i^2} = \dfrac{1}{1(-1)} = \dfrac{1}{-1} = -1$ ■

The two complex numbers $a + bi$ and $a - bi$ are called **complex conjugates** or simply **conjugates** of each other. **Their product will always be a nonnegative real number.**

$$\begin{aligned}(a + bi)(a - bi) &= (a + bi)a + (a + bi)(-bi)\\ &= a^2 + abi - abi - b^2i^2\\ &= a^2 - b^2i^2\\ &= a^2 + b^2\end{aligned}$$

$a^2 + b^2$ is a real number, and it is nonnegative since it is the sum of the squares of real numbers.

The form $a + bi$ is called the **standard form** of a complex number. The standard form allows easy identification of the real and imaginary parts. Thus,

$$\frac{1 + 3i}{5} = \frac{1}{5} + \frac{3}{5}i \qquad \text{in standard form}$$

The real part is $\dfrac{1}{5}$ and the imaginary part is $\dfrac{3}{5}$.

To write $\dfrac{1 + i}{2 - 3i}$ in standard form, multiply both the numerator and denominator by $2 + 3i$. This will give a positive real number in the denominator. ($2 + 3i$ is the conjugate of the denominator.)

$$\begin{aligned}\frac{1 + i}{2 - 3i} &= \frac{(1 + i)(2 + 3i)}{(2 - 3i)(2 + 3i)}\\ &= \frac{2 + 2i + 3i + 3i^2}{2^2 - 6i + 6i - 3^2i^2}\\ &= \frac{2 + 5i - 3}{2^2 + 3^2} = \frac{-1 + 5i}{13}\\ &= -\frac{1}{13} + \frac{5}{13}i\end{aligned}$$

You should memorize the fact that

$$\boldsymbol{(a + bi)(a - bi) = a^2 + b^2}$$

**EXAMPLES** Write the following quotients in standard form.

**7.** $$\frac{4}{-1-5i} = \frac{4(\mathbf{-1+5i})}{(-1-5i)(\mathbf{-1+5i})} = \frac{-4+20i}{(-1)^2+(5)^2} = \frac{-4+20i}{1+25} = \frac{-4+20i}{26} = -\frac{4}{26}+\frac{20}{26}i = -\frac{2}{13}+\frac{10}{13}i$$

**8.** $$\frac{\sqrt{3}+i}{\sqrt{3}-i} = \frac{(\sqrt{3}+i)(\mathbf{\sqrt{3}+i})}{(\sqrt{3}-i)(\mathbf{\sqrt{3}+i})} = \frac{3+2i\sqrt{3}+i^2}{(\sqrt{3})^2-i^2} = \frac{2+2i\sqrt{3}}{3+1} = \frac{2+2i\sqrt{3}}{4} = \frac{2}{4}+\frac{2i\sqrt{3}}{4} = \frac{1}{2}+\frac{\sqrt{3}}{2}i$$

**9.** $$\frac{6+i}{i} = \frac{(6+i)(\mathbf{-i})}{i(\mathbf{-i})} = \frac{-6i-i^2}{-i^2} = \frac{-6i+1}{1} = 1-6i$$

Since $i = 0 + i$ and $-i = 0 - i$, the number $-i$ is the conjugate of $i$.

**10.** $$\frac{\sqrt{2}+i}{-\sqrt{2}+i} = \frac{(\sqrt{2}+i)(\mathbf{-\sqrt{2}-i})}{(-\sqrt{2}+i)(\mathbf{-\sqrt{2}-i})} = \frac{-(\sqrt{2})^2-i\sqrt{2}-i\sqrt{2}-i^2}{(-\sqrt{2})^2+(1)^2} = \frac{-2-2i\sqrt{2}+1}{2+1} = \frac{-1-2i\sqrt{2}}{3} = -\frac{1}{3}-\frac{2\sqrt{2}}{3}i$$

■

**Practice Problems**

Write each of the following numbers in standard form.

1. $-2i(3 - i)$
2. $(2 + 4i)(1 + i)$
3. $i^{13}$
4. $i^{-2}$
5. $\dfrac{2}{1 + 5i}$
6. $\dfrac{7 + i}{2 - i}$

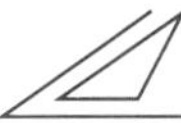

## EXERCISES 4.5

Write each of the following numbers in standard form. Assume that $n$ is an integer.

1. $8(2 + 3i)$
2. $-3(7 - 4i)$
3. $-7(\sqrt{2} - i)$
4. $5(\sqrt{3} + 2i)$
5. $3i(4 - i)$
6. $-4i(6 - 7i)$
7. $-i(\sqrt{3} + i)$
8. $2i(\sqrt{5} + 2i)$
9. $5i(2 - i\sqrt{2})$
10. $i\sqrt{3}(2 - i\sqrt{3})$
11. $(5 + 3i)(1 + i)$
12. $(2 + 7i)(6 + i)$
13. $(-3 + 5i)(-1 + 2i)$
14. $(6 + 2i)(3 - i)$
15. $(2 - 3i)(2 + 3i)$
16. $(4 + 3i)(7 - 2i)$
17. $(-2 + 5i)(i - 1)$
18. $(5 + 7i)^2$
19. $(3 + 2i)^2$
20. $(4 + 5i)(4 - 5i)$
21. $(\sqrt{3} + i)(\sqrt{3} - 2i)$
22. $(2\sqrt{5} + 3i)(\sqrt{5} - i)$
23. $(4 + i\sqrt{5})(4 - i\sqrt{5})$
24. $(\sqrt{7} + 3i)(\sqrt{7} + i)$
25. $(5 - i\sqrt{2})(5 - i\sqrt{2})$
26. $(7 + 2i\sqrt{3})(7 - 2i\sqrt{3})$
27. $(\sqrt{5} + 2i)(\sqrt{2} - i)$
28. $(2\sqrt{3} + i)(4 + 3i)$
29. $(3 + i\sqrt{5})(3 + i\sqrt{6})$
30. $(2 - i\sqrt{3})(3 - i\sqrt{2})$
31. $\dfrac{-3}{i}$
32. $\dfrac{7}{i}$
33. $\dfrac{5}{4i}$
34. $\dfrac{-3}{2i}$
35. $\dfrac{2 + i}{-4i}$
36. $\dfrac{3 - 4i}{3i}$
37. $\dfrac{-4}{1 + 2i}$
38. $\dfrac{7}{5 - 2i}$
39. $\dfrac{6}{4 - 3i}$
40. $\dfrac{-8}{6 + i}$
41. $\dfrac{2i}{5 - i}$
42. $\dfrac{-4i}{1 + 3i}$
43. $\dfrac{2 - i}{2 + 5i}$
44. $\dfrac{6 + i}{3 - 4i}$
45. $\dfrac{2 - 3i}{-1 + 5i}$
46. $\dfrac{-3 + i}{7 - 2i}$
47. $\dfrac{1 + 4i}{\sqrt{3} + i}$
48. $\dfrac{9 - 2i}{\sqrt{5} + i}$
49. $\dfrac{\sqrt{3} + 2i}{\sqrt{3} - 2i}$
50. $\dfrac{\sqrt{6} - 3i}{\sqrt{6} + 3i}$
51. $i^{13}$
52. $i^{20}$
53. $i^{30}$
54. $i^{15}$
55. $i^{-3}$
56. $i^{-5}$
57. $i^{4n}$
58. $i^{4n + 2}$
59. $i^{4n + 3}$
60. $i^{4n + 1}$

**Answers to Practice Problems** 1. $-2 - 6i$ 2. $-2 + 6i$ 3. $0 + i$ 4. $-1 + 0i$
5. $\dfrac{1}{13} - \dfrac{5}{13}i$ 6. $\dfrac{13}{5} + \dfrac{9}{5}i$

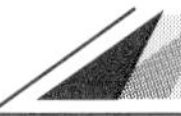

# CHAPTER 4 SUMMARY

## Key Terms and Formulas

For positive $a$, positive $b$, and positive integer $n$,

$$b^n = a \text{ if and only if } b = a^{1/n}. \quad [4.1]$$

The number $a^{1/n}$ is a positive real number called the **principal *n*th root of *a*,** and $(a^{1/n})^n = a$. [4.1]

If $a$ is a real number and $n$ is a positive integer:

| $a > 0$ | $a < 0$, $n$ even | $a < 0$, $n$ odd | $a = 0$ |
|---|---|---|---|
| $a^{1/n}$ is a positive real number and $(a^{1/n})^n = a$ | $a^{1/n}$ is not a real number | $a^{1/n}$ is a negative real number and $(a^{1/n})^n = a$ | $a^{1/n} = 0$ [4.1] |

If $a$ is a real number, $n$ is a positive integer, $m$ is any nonzero integer, and $a^{1/n}$ is defined as a real number,

$$a^{m/n} = (a^{1/n})^m = (a^m)^{1/n} \quad [4.1]$$

$\sqrt[n]{a}$ is called **radical notation.** [4.2]

The symbol $\sqrt{\phantom{a}}$ is called a **radical sign.** [4.2]

$n$ is called the **index,** and $a$ is called the **radicand.** [4.2]

If $a$ is a real number, $n$ is a positive integer, and $a^{1/n}$ is defined as a real number,

$$a^{1/n} = \sqrt[n]{a}$$

Thus, $b^n = a$ if and only if $b = \sqrt[n]{a}$. [4.2]

An even root of a negative number is not a real number. [4.2]

If $a$ is a real number, $n$ is a positive integer, $m$ is any integer, and $\sqrt[n]{a}$ is defined as a real number,

$$a^{m/n} = (\sqrt[n]{a})^m = \sqrt[n]{a^m} \quad [4.2]$$

For any real number $a$,

$$\sqrt{a^2} = |a| \quad [4.2]$$

$i = \sqrt{-1}$ and $i^2 = (\sqrt{-1})^2 = -1$ [4.4]

If $a$ and $b$ are real numbers, $a + bi$ is a **complex number.** $a$ is called the **real part;** $b$ is called the **imaginary part.**

If $b = 0$, then $a + bi = a + 0i = a$ is a **real number.**

If $a = 0$, then $a + bi = 0 + bi = bi$ is called a **pure imaginary number** (or an **imaginary number**). [4.4]

For complex numbers $a + bi$ and $c + di$,

$$a + bi = c + di \text{ if and only if } a = c \text{ and } b = d \quad [4.4]$$

For complex numbers $a + bi$ and $c + di$,

$$(a + bi) + (c + di) = (a + c) + (b + d)i$$

and $$(a + bi) - (c + di) = (a - c) + (b - d)i \quad [4.4]$$

For complex numbers $a + bi$ and $c + di$,

$$(a + bi)(c + di) = (ac - bd) + (bc + ad)i \quad [4.5]$$

The product of **complex conjugates** is always a nonnegative real number:

$$(a + bi)(a - bi) = a^2 + b^2 \quad [4.5]$$

## Properties and Rules

If $a$ and $b$ are positive real numbers and $n$ is a positive integer:

1. $\sqrt[n]{ab} = \sqrt[n]{a}\sqrt[n]{b}$
2. $\sqrt[n]{\dfrac{a}{b}} = \dfrac{\sqrt[n]{a}}{\sqrt[n]{b}}$ [4.2]

### Procedures

**To Rationalize a Denominator** [4.2]
Multiply the numerator and denominator by another radical with index $n$ to make the denominator a perfect $n$th power.

**Like radicals** are combined in the same manner as like terms. [4.3]

## CHAPTER 4 REVIEW

Simplify Exercises 1–17. Assume that all variables are positive. [4.1]

**1.** $121^{1/2}$ **2.** $16^{-1/2}$ **3.** $\left(\frac{64}{81}\right)^{1/2}$ **4.** $9^{3/2}$ **5.** $-27^{-2/3}$ **6.** $(-4)^{1/2}$

**7.** $16^{-3/4}$ **8.** $(4x^{1/3})^2$ **9.** $(5x^{1/2})^3$ **10.** $5x^2 \cdot x^{1/3}$ **11.** $3x^{1/2} \cdot x^{-1/3}$ **12.** $\frac{x^3}{x^{5/4}}$

**13.** $\frac{x^{1/2}}{x^{-2/3}}$ **14.** $\left(\frac{27x^3}{y^6}\right)^{1/3}$ **15.** $\left(\frac{16x^4y}{y^{-3}}\right)^{-3/4}$ **16.** $\frac{(x^{1/2}y)^{-2/3}}{x^{2/3}y^{-1}}$

**17.** $\left(\frac{x^2y^{1/3}}{x^{2/3}y^{1/2}}\right)^{1/2}\left(\frac{x^{-1/2}y^{2/3}}{x^{-1}y^2}\right)^{-1/2}$

Change each expression in Exercises 18–25 to an equivalent expression in radical notation or exponential notation. Assume that all variables are positive. [4.2]

**18.** $x^{8/3}$ **19.** $-x^{3/4}$ **20.** $4x^{1/3}y^{2/3}$ **21.** $7xy^{2/5}$ **22.** $6\sqrt{x}$ **23.** $3\sqrt[4]{x^3}$
**24.** $-2\sqrt[3]{x^2y}$ **25.** $11\sqrt[5]{x^3y^2}$

Change each expression in Exercises 26–28 to an equivalent expression in exponential notation; then simplify. Assume that all variables are positive. [4.2]

**26.** $\sqrt{x} \cdot \sqrt[4]{x}$ **27.** $\frac{\sqrt[3]{x^2}\sqrt[4]{x}}{x\sqrt{x}}$ **28.** $\frac{x\sqrt[4]{x}}{\sqrt{\sqrt[3]{x}}}$

Simplify the expressions in Exercises 29–42. Variables may be positive or negative. [4.2]

**29.** $\sqrt{125}$ **30.** $\sqrt{196}$ **31.** $\sqrt[3]{250}$ **32.** $\sqrt[3]{192}$ **33.** $4\sqrt{72}$ **34.** $3\sqrt{80}$

**35.** $\sqrt[3]{54x^5y^4}$ **36.** $\sqrt[3]{128x^7y^8}$ **37.** $\sqrt{64x^2y^4}$ **38.** $\sqrt{108x^4y^2}$ **39.** $\frac{3}{\sqrt{18}}$ **40.** $\frac{\sqrt{60}}{\sqrt{20}}$

**41.** $\frac{7}{\sqrt{2} + \sqrt{5}}$ **42.** $\frac{3}{\sqrt{2} + 5}$

Perform the indicated operations in Exercises 43–52 and simplify. Assume all variables are positive. [4.3]

**43.** $3\sqrt{50} + 5\sqrt{32}$ **44.** $4\sqrt{12} - 2\sqrt{75}$ **45.** $\sqrt[3]{54} + 5\sqrt[3]{16}$
**46.** $5\sqrt[3]{24} - 2\sqrt[3]{81}$ **47.** $\sqrt{8x^2y} - x\sqrt{2y} + 5\sqrt{18x^2y}$ **48.** $\sqrt{18x^3y} + \sqrt{3xy^3} - 4y\sqrt{3xy}$

**49.** $(\sqrt{10} + 3)(2\sqrt{10} - 4)$ **50.** $(\sqrt{3} + \sqrt{2})(\sqrt{5} + 6)$ **51.** $\frac{\sqrt{3} - \sqrt{5}}{\sqrt{3} + \sqrt{5}}$

**52.** $\frac{\sqrt{3} + \sqrt{2}}{\sqrt{6} - \sqrt{2}}$

Perform the indicated operations in Exercises 53–64 and simplify. Write the answers in standard form. [4.4, 4.5]

**53.** $(3 + 2i) + (-1 + 4i)$
**54.** $(-2 - i) - (4 - 2i)$
**55.** $(2 + \sqrt{-25}) - (3 + 2\sqrt{-25})$
**56.** $(6 - \sqrt{-4}) - (13 + 2\sqrt{-9})$
**57.** $(4 + i)(3 - 2i)$
**58.** $(3 - \sqrt{-16})(5 - \sqrt{-1})$
**59.** $\dfrac{1 + 2i}{2 + 3i}$
**60.** $\dfrac{2 - 5i}{-1 + i}$
**61.** $\dfrac{4 - \sqrt{-9}}{1 + \sqrt{-16}}$
**62.** $i^{27}$
**63.** $i^{-7}$
**64.** $i^{-24}$

Solve the equations in Exercises 65–70 for $x$ and $y$. [4.4]

**65.** $2x + 3i = yi - 2$
**66.** $5x - yi = 2x + 3 - 5i$
**67.** $3x + 2 + 7i = 4 - yi + 2i$
**68.** $(5 + i)(2 + 3i) = x + yi$
**69.** $2\sqrt{3} + (y + 1)i = (x - 3) + 5i$
**70.** $\dfrac{4 - i}{3 - 2i} = 2x + 3yi$

## CHAPTER 4 TEST

Simplify Exercises 1–5. Assume that all variables are positive.

**1.** $(-8)^{2/3}$
**2.** $9^{-3/2}$
**3.** $4x^{1/2} \cdot x^{2/3}$
**4.** $(49x^{1/2}y^{-2/3})^{1/2}$
**5.** $\left(\dfrac{16x^{-4}y}{y^{-1}}\right)^{3/4}$

**6.** Write $(2x)^{2/3}$ in radical notation.
**7.** Write $\sqrt[6]{8x^2y^4}$ in exponential notation.
**8.** Simplify $\sqrt[3]{x^2} \cdot \sqrt[4]{x}$.

Simplify Exercises 9–11. Assume that all variables are positive.

**9.** $\sqrt{112}$
**10.** $\sqrt[3]{48x^2y^5}$
**11.** $\sqrt{\dfrac{5y^2}{8x^3}}$

Perform the indicated operations and simplify in Exercises 12–15. Assume that all variables are positive.

**12.** $2\sqrt{75} + 3\sqrt{27} - \sqrt{12}$
**13.** $\sqrt{16x^3} + \sqrt{9x^3} - \sqrt{36x}$
**14.** $\sqrt[3]{24} + 2\sqrt[3]{81}$
**15.** $(\sqrt{3} - \sqrt{2})^2$
**16.** Rationalize the denominator and simplify: $\dfrac{2}{\sqrt{3} - \sqrt{5}}$

Perform the indicated operations and simplify Exercises 17–20.

**17.** $(5 + 8i) + (11 - 4i)$
**18.** $(2 + 3\sqrt{-4}) - (7 - 2\sqrt{-25})$
**19.** $(4 + 3i)(2 - 5i)$
**20.** $\dfrac{2 + i}{3 + 2i}$
**21.** Solve for $x$ and $y$:
$(2x + 3i) - (6 + 2yi) = 5 - 3i$
**22.** Simplify $i^{23}$. Write your answer in $a + bi$ form.

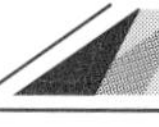

## CUMULATIVE REVIEW (4)

Perform the indicated operations in Exercises 1–3.

**1.** $(x^2 + 7x - 5) - (-2x^3 + 5x^2 - x - 1)$

**2.** $(2x + 7)(3x - 1)$

**3.** $(5x + 2)(4 - x)$

Solve for the indicated variable in Exercises 4 and 5.

**4.** $s = a + (n - 1)d$ for $n$

**5.** $A = p + prt$ for $p$

Factor Exercises 6–9 completely.

**6.** $12x^2 - 7x - 12$

**7.** $28 + x - 2x^2$

**8.** $5x^3 - 320$

**9.** $x^3 + 4x^2 - x - 4$

Perform the indicated operations in Exercises 10 and 11.

**10.** $\dfrac{x + 1}{x^2 + x - 6} + \dfrac{3x - 2}{x^2 - 2x - 15}$

**11.** $\dfrac{2x + 5}{4x^2 - 1} - \dfrac{2 - x}{2x^2 + 7x + 3}$

Simplify the complex fractions in Exercises 12 and 13.

**12.** $\dfrac{1 - \dfrac{1}{x^2}}{\dfrac{2}{x} - \dfrac{4}{x^2}}$

**13.** $\dfrac{x + 2 - \dfrac{12}{x + 3}}{x - 5 + \dfrac{16}{x + 3}}$

Solve the equations in Exercises 14 and 15.

**14.** $\dfrac{3}{x} + \dfrac{2}{x + 5} = \dfrac{8}{3x}$

**15.** $\dfrac{9}{x + 7} + \dfrac{3x}{x^2 + 4x - 21} = \dfrac{8}{x - 3}$

Solve the inequalities in Exercises 16 and 17 and graph the solutions on the number line. Write the solutions in interval notation.

**16.** $\dfrac{5x - 3}{2x + 4} \leq 0$

**17.** $\dfrac{x - 5}{3x - 1} \geq 1$

Simplify the expressions in Exercises 18–21. Assume all variables are positive.

**18.** $8^{-4/3}$

**19.** $(x^{1/2} \cdot x^{2/3})^2$

**20.** $\sqrt{288}$

**21.** $\sqrt[3]{16x^6y^{10}}$

Perform the indicated operations in Exercises 22 and 23 and simplify. Write the answers in standard form.

**22.** $(5 + 2i)(3 - 4i)$

**23.** $\dfrac{4 - 3i}{1 + 4i}$

**24.** Susan traveled 25 miles downstream. In the same length of time, she could have traveled 15 miles upstream. If the speed of the current is 2.5 mph, find the speed of Susan's boat in still water.

**25.** Robin can prepare a monthly sales report in 5 hours. If Mac helps her, together they can prepare the report in 3 hours. How long would it take Mac if he worked alone?

# CHAPTER 5 QUADRATIC EQUATIONS AND INEQUALITIES

## MATHEMATICAL CHALLENGES

Express, in simplest form, the quotient

$$\frac{x + x^2 + x^3 + x^4 + x^5 + x^6 + x^7}{x^3 + x^4 + x^5 + x^6 + x^7 + x^8 + x^9}.$$

Compute $n$ if $(10^{12} + 25)^2 - (10^{12} - 25)^2 = 10^n$.

The number $(2^{48} - 1)$ is exactly divisible by two numbers between 60 and 70. Find the two numbers.

## CHAPTER OUTLINE



Quadratic equations appear in one form or another in almost every course in mathematics and in many related courses in such fields as business, biology, and engineering. In this chapter, we will discuss three techniques for solving quadratic equations: factoring, completing the square, and the quadratic formula. Factoring, when possible, is generally considered the method of first choice because it is easier to apply, and factoring quadratic expressions is useful in other mathematical situations. However, the quadratic formula is very important and should be memorized. It works in all cases; in the discriminant, the formula gives easily understood information about the nature of the solutions; and it is easy to use in computer programs.

The wide variety of applications presented in Section 5.4 illustrates the practicality of knowing how to recognize and solve quadratic equations.

# 5.1 Solutions by Factoring

**OBJECTIVES**

In this section, you will be learning to:

1. Write quadratic equations given the roots.
2. Solve second-degree and third-degree equations by factoring.

An equation in which the largest degree is second degree, such as

$$2x^2 + 3x - 7 = 0, \quad 4x^2 + 9 = 0, \quad 6x^2 - 3x = 0, \quad 2x^2 = 0$$

is called a quadratic equation. In this chapter we will investigate three ways to **solve** or **find the roots of** quadratic equations, and then we will solve related word problems. The three ways are (1) factoring, (2) completing the square, and (3) using the quadratic formula. Generally, when factoring is possible, it is the easiest way. So, we discuss solving by factoring first.

***Quadratic Equation or Second-Degree Equation***

An equation of the form

$ax^2 + bx + c = 0$ where $a$, $b$, and $c$ are real numbers and $a \neq 0$

is called a **quadratic equation** or **second-degree equation.**

To solve a quadratic equation by factoring, we need the following important fact about the product of two numbers.

If $u$ and $v$ are complex numbers and $u \cdot v = 0$, then

$$u = 0 \text{ or } v = 0.$$

In words, **if the product of two numbers is 0, then one or both of the numbers must be 0.**

With this basic rule as a foundation, we can solve quadratic equations by factoring using the following steps.

***To Solve a Quadratic Equation by Factoring***

1. Add or subtract terms so that one side of the equation is 0.
2. Factor the quadratic expression.
3. Set each factor equal to 0 and solve for $x$.

**EXAMPLES** Solve the following quadratic equations by factoring.

**1.** $y^2 - 6y = 27$

**Solution**

$$y^2 - 6y = 27$$

$$y^2 - 6y - 27 = 0$$

Add $-27$ to both sides. **One side must be 0.**

$$(y - 9)(y + 3) = 0$$

Factor the left-hand side.

$$y - 9 = 0 \quad \text{or} \quad y + 3 = 0$$

Set each factor equal to 0.

$$y = 9 \qquad\qquad y = -3$$

Solve each first-degree equation.

*Check*

$$9^2 - 6 \cdot 9 \stackrel{?}{=} 27 \qquad (-3)^2 - 6(-3) \stackrel{?}{=} 27$$

$$81 - 54 \stackrel{?}{=} 27 \qquad 9 + 18 \stackrel{?}{=} 27$$

$$27 = 27 \qquad 27 = 27$$

**2.** $x^2 + 4 = 0$

**Solution**

$$x^2 + 4 = 0$$

$$(x + 2i)(x - 2i) = 0$$

Note that $x^2 + 4$ is the sum of two squares and can be factored as two conjugates of complex numbers. **$x^2 + 4$ cannot be factored with real coefficients.**

$$x + 2i = 0 \quad \text{or} \quad x - 2i = 0$$

$$x = -2i \qquad\qquad x = 2i$$

*Check*

$$(2i)^2 + 4 \stackrel{?}{=} 0 \qquad (-2i)^2 + 4 \stackrel{?}{=} 0$$

$$4i^2 + 4 \stackrel{?}{=} 0 \qquad 4i^2 + 4 \stackrel{?}{=} 0$$

$$-4 + 4 \stackrel{?}{=} 0 \qquad -4 + 4 \stackrel{?}{=} 0$$

$$0 = 0 \qquad 0 = 0$$

**3.** $(3z + 6)(4z + 12) = -3$

**Solution**

$$(3z + 6)(4z + 12) = -3$$

$$12z^2 + 60z + 72 = -3$$

Multiply factors on left-hand side.

$$12z^2 + 60z + 75 = 0$$

**One side must be 0.**

$$4z^2 + 20z + 25 = 0$$

Divide each term by 3.

$$(2z + 5)^2 = 0$$

Factor the left-hand side.

$$2z + 5 = 0$$

$$z = -\frac{5}{2}$$

There is only one root since the factor $2x + 5$ is repeated. In this case, the root is sometimes called a **double root.**

*Check* $12\left(-\frac{5}{2}\right)^2 + 75 \stackrel{?}{=} -60\left(-\frac{5}{2}\right)$

$$12\left(\frac{25}{4}\right) + 75 \stackrel{?}{=} -30(-5)$$

$$75 + 75 \stackrel{?}{=} 150$$

$$150 = 150$$

■

Sometimes higher-degree polynomial equations can be solved by factoring. In particular, if one of the factors is a monomial, factoring is very useful.

**EXAMPLE**

**4.** Solve the following **third-degree** (or **cubic**) equation by factoring:

$$4x^3 = 100x$$

**Solution**

$$4x^3 = 100x$$
$$4x^3 - 100x = 0$$
$$4x(x^2 - 25) = 0$$
$$4x(x + 5)(x - 5) = 0$$

$4x = 0$ or $x + 5 = 0$ or $x - 5 = 0$

$x = 0$ $\quad x = -5$ $\quad x = 5$

The student can check that each of these numbers is a solution. ■

Now, to aid in understanding the concepts of factors, factoring, and solutions to equations, we consider the problem of finding an equation that has certain given solutions (or roots). For example, to find an equation that has the roots

$$x = 4 \quad \text{and} \quad x = -7$$

we proceed as follows.

**1.** Write the equations.

$$x - 4 = 0 \quad \text{and} \quad x + 7 = 0$$

**2.** Form the product of the factors $(x - 4)$ and $(x + 7)$ and set this product equal to 0.

$$(x - 4)(x + 7) = 0$$

**3.** This quadratic equation (multiply the factors)

$$x^2 + 3x - 28 = 0$$

has the two roots 4 and $-7$.

The reasoning is based on the Factor Theorem, stated here without proof.

***Factor Theorem*** If $x = c$ is a root of a polynomial equation in which one side is 0, then $x - c$ is a factor of the polynomial.

**EXAMPLES** Find polynomial equations that have the given roots.

**5.** $x = 3$ and $x = -\frac{2}{3}$

**Solution** $x = 3$ $\qquad$ $x = -\frac{2}{3}$

$x - 3 = 0$ $\qquad$ $x + \frac{2}{3} = 0$ $\qquad$ Get 0 on one side of each equation.

$3x + 2 = 0$ $\qquad$ Multiply both sides by 3, the LCM of the denominators, to get integer coefficients.

Form an equation by setting the product of the two factors equal to 0.

$$(x - 3)(3x + 2) = 0$$
$$3x^2 - 7x - 6 = 0$$

This quadratic equation has $x = 3$ and $x = -\frac{2}{3}$ as roots.

**6.** $y = 3 + 2i$ and $y = 3 - 2i$

**Solution** $y = 3 + 2i$ $\qquad$ $y = 3 - 2i$

$y - 3 - 2i = 0$ $\qquad$ $y - 3 + 2i = 0$ $\qquad$ Get 0 on one side of each equation.

Set the product of the two factors equal to 0 and simplify.

$$[y - 3 - 2i][y - 3 + 2i] = 0$$
$$[(y - 3) - 2i][(y - 3) + 2i] = 0$$
$$(y - 3)^2 - 4i^2 = 0$$
$$y^2 - 6y + 9 + 4 = 0 \qquad i^2 = -1$$
$$y^2 - 6y + 13 = 0$$

This equation has the two solutions $y = 3 + 2i$ and $y = 3 - 2i$.

**7.** $x = 5 - \sqrt{2}$ and $x = 5 + \sqrt{2}$

**Solution**

$x = 5 - \sqrt{2}$ $\qquad$ $x = 5 + \sqrt{2}$

$x - 5 + \sqrt{2} = 0$ $\qquad$ $x - 5 - \sqrt{2} = 0$ $\qquad$ Get 0 on one side of each equation.

Set the product of the two factors equal to 0 and simplify.

$$[x - 5 + \sqrt{2}][x - 5 - \sqrt{2}] = 0$$
$$[(x - 5) + \sqrt{2}][(x - 5) - \sqrt{2}] = 0$$
$$(x - 5)^2 - (\sqrt{2})^2 = 0$$
$$x^2 - 10x + 25 - 2 = 0$$
$$x^2 - 10x + 23 = 0$$

This equation has the two solutions $x = 5 - \sqrt{2}$ and $x = 5 + \sqrt{2}$. ■

**Practice Problems**

Solve the following quadratic equations by factoring.

1. $y^2 - 4y = 21$
2. $3x^2 - 16x + 5 = 0$
3. $z^2 + 6z = -9$
4. $x^2 = -5$
5. Find a quadratic equation with integer coefficients that has the roots $\frac{2}{3}$ and $\frac{1}{2}$.

## EXERCISES 5.1

For Exercises 1–18, write a quadratic equation with integer coefficients that has the given roots.

1. $y = 3, y = -2$
2. $x = 5, x = 7$
3. $x = \frac{1}{2}, x = \frac{3}{4}$
4. $y = \frac{2}{3}, y = \frac{1}{6}$
5. $x = \sqrt{7}, x = -\sqrt{7}$
6. $x = \sqrt{5}, x = -\sqrt{5}$
7. $x = 1 + \sqrt{3}, x = 1 - \sqrt{3}$
8. $z = 2 + \sqrt{2}, z = 2 - \sqrt{2}$
9. $y = -2 + \sqrt{5}, y = -2 - \sqrt{5}$
10. $x = 1 + 2\sqrt{3}, x = 1 - 2\sqrt{3}$
11. $x = 4i, x = -4i$
12. $x = 7i, x = -7i$
13. $y = i\sqrt{6}, y = -i\sqrt{6}$
14. $x = i\sqrt{5}, x = -i\sqrt{5}$
15. $x = 2 + i, x = 2 - i$
16. $y = 3 + 2i, y = 3 - 2i$
17. $x = 1 + i\sqrt{2}, x = 1 - i\sqrt{2}$
18. $x = 2 + i\sqrt{3}, x = 2 - i\sqrt{3}$

Solve the equations in Exercises 19–60 by factoring.

19. $x^2 + 13x + 36 = 0$
20. $x^2 + 17x + 72 = 0$
21. $5x^2 - 70x + 240 = 0$
22. $2y^2 - 24y + 70 = 0$
23. $4x^2 = 20x + 200$
24. $7x^2 + 14x = 168$
25. $3x^2 = 147$
26. $64 - 49x^2 = 0$
27. $3x^2 + 10 = 17x$
28. $2x^2 = 3x - 1$
29. $6x^2 + 11x + 4 = 0$
30. $4y^2 = 14y - 6$
31. $x^2 + 25 = 0$
32. $x^2 + 9 = 0$
33. $4z^2 + 49 = 0$
34. $9x^2 + 16 = 0$
35. $2z^2 + 3 = 7z$
36. $32x - 5 = 12x^2$
37. $12x^2 = 6 - x$
38. $12x^2 + 5x = 3$
39. $6x^2 + x = 35$
40. $50y^2 - 98 = 0$
41. $150y^2 - 96 = 0$
42. $8y^2 + 6y = 35$

**Answers to Practice Problems** 1. $y = 7, y = -3$ 2. $x = \frac{1}{3}, x = 5$ 3. $z = -3$ 4. $x = i\sqrt{5}, x = -i\sqrt{5}$ 5. $6x^2 - 7x + 2 = 0$

**43.** $(x + 5)(x - 7) = 13$
**44.** $(2x + 3)(x - 1) = -2$
**45.** $x(x - 5) + 9 = 3(x - 1)$
**46.** $x(2x + 3) - 2 = 2(x + 4)$
**47.** $2x(x + 3) - 14 = x(x - 2) + 19$
**48.** $x(3x + 5) = x(x + 2) + 14$
**49.** $18y^2 - 15y + 2 = 0$
**50.** $14 + 11y = 15y^2$
**51.** $63x^2 = 40x + 12$
**52.** $12z^2 - 47z + 11 = 0$
**53.** $3x^3 + 15x^2 + 18x = 0$
**54.** $x^3 = 4x^2 + 12x$
**55.** $16x^3 + 100x = 0$
**56.** $112x - 2x^2 = 2x^3$
**57.** $12x^3 + 2x^2 = 70x$
**58.** $21x^3 = 13x^2 - 2x$
**59.** $63x = 3x^2 + 30x^3$
**60.** $144x^3 + 10x^2 = 50x$

## 5.2 Solutions by Completing the Square

**OBJECTIVES**

In this section, you will be learning to:

1. Determine the constant terms that will make incomplete trinomials perfect squares and factor these perfect squares.
2. Solve quadratic equations by using the definition of square root.
3. Solve quadratic equations by completing the square.

In Section 5.1, we solved quadratic equations by factoring. In the next two sections, we will develop two more techniques for solving quadratic equations. The particular technique used in this section, called **completing the square,** is used to develop the general formula in Section 5.3 called the **quadratic formula.** Both of these approaches are important; however, remember that factoring, whenever possible, is the easiest of the methods for solving quadratic equations.

We know that

$$(x + a)^2 = x^2 + 2ax + a^2$$

and

$$(x - a)^2 = x^2 - 2ax + a^2$$

In each case, the trinomial is called a **perfect square trinomial.** Note that the constant term $a^2$ is the square of one-half times $2a$ (the coefficient of $x$). That is,

$$a^2 = \left(\frac{1}{2} \cdot 2a\right)^2 = (a)^2$$

Thus,

$$(x + 8)^2 = x^2 + 16x + 64 \qquad \textbf{Note: } \frac{1}{2} \cdot 16 = 8 \text{ and } 8^2 = 64.$$

and

$$(x - 3)^2 = x^2 - 6x + 9 \qquad \textbf{Note: } \frac{1}{2}(-6) = -3 \text{ and } (-3)^2 = 9.$$

What constant should be added to $x^2 - 8x$ to get a perfect square trinomial? By following the notes just discussed, we find that $\frac{1}{2}(-8) = -4$ and $(-4)^2 = 16$. Therefore, 16 should be added to get a perfect square trinomial. In fact,

$$x^2 - 8x + 16 = (x - 4)^2$$

By adding 16, we have **completed the square** for $x^2 - 8x$.

**EXAMPLES** Add the constant that will complete the square for each expression, and write the new expression as the square of a binomial.

**1.** $x^2 + 10x$

**Solution**

$$x^2 + 10x + \underline{\qquad} = (\qquad)^2$$

$$\frac{1}{2}(10) = 5 \quad \text{and} \quad (5)^2 = 25$$

So, add 25:

$$x^2 + 10x + \underline{25} = (x + 5)^2$$

**2.** $x^2 - 7x$

**Solution**

$$x^2 - 7x + \underline{\qquad} = (\qquad)^2$$

$$\frac{1}{2}(-7) = -\frac{7}{2} \quad \text{and} \quad \left(-\frac{7}{2}\right)^2 = \frac{49}{4}$$

So, add $\frac{49}{4}$:

$$x^2 - 7x + \underline{\frac{49}{4}} = \left(x - \frac{7}{2}\right)^2$$

■

Now consider the equation

$$x^2 = 13$$

The definition of square root tells us that there are two solutions:

$$x = \sqrt{13} \quad \text{and} \quad x = -\sqrt{13}$$

Similarly, for the equation

$$(x - 3)^2 = 5$$

we get

$$x - 3 = \sqrt{5} \quad \text{or} \quad x - 3 = -\sqrt{5}$$

or, solving for $x$,

$$x = 3 + \sqrt{5} \quad \text{or} \quad x = 3 - \sqrt{5}$$

Thus, there are two real solutions: $3 + \sqrt{5}$ and $3 - \sqrt{5}$.

If a quadratic equation has a squared expression on one side and a constant on the other, this technique of using the definition of square root works very well. In fact, if there is no squared expression, we can complete the square so that there is a square expression. This is called **solving by completing the square.** Remember two basic steps:

1. Divide or multiply so that the coefficient of $x^2$ is 1 and then
2. add the constant that completes the square to both sides of the equation.

**EXAMPLES** Solve the following quadratic equations by completing the square.

**3.** $x^2 - 8x = 25$

**Solution** $x^2 - 8x = 25$ — The coefficient of $x^2$ is already 1.

$x^2 - 8x + \mathbf{16} = 25 + \mathbf{16}$ — $\frac{1}{2}(-8) = -4$ and $(-4)^2 = 16$. So, add 16 to both sides.

$(x - 4)^2 = 41$ — The left side is a perfect square.

$x - 4 = \sqrt{41}$ or $x - 4 = -\sqrt{41}$ — Use the definition of square root.

$x = 4 + \sqrt{41}$ $\qquad$ $x = 4 - \sqrt{41}$

There are two real solutions: $4 + \sqrt{41}$ and $4 - \sqrt{41}$.

**4.** $3x^2 + 6x - 15 = 0$

**Solution** $3x^2 + 6x - 15 = 0$

$\frac{3x^2}{3} + \frac{6x}{3} - \frac{15}{3} = \frac{0}{3}$ — Divide each term by 3. The leading coefficient must be 1.

$x^2 + 2x - 5 = 0$

$x^2 + 2x = 5$

$x^2 + 2x + \mathbf{1} = 5 + \mathbf{1}$ — Complete the square: $\frac{1}{2}(2) = 1$ and $1^2 = 1$. So, add 1 to both sides.

$(x + 1)^2 = 6$

$x + 1 = \pm\sqrt{6}$ — Use the definition of square root.

$x = -1 \pm \sqrt{6}$ — The **two equations** $x = -1 + \sqrt{6}$ and $x = -1 - \sqrt{6}$ can be represented in this form. Remember that this is shorthand for two equations.

**5.** $2x^2 + 2x - 7 = 0$

**Solution** $2x^2 + 2x - 7 = 0$

$x^2 + x - \frac{7}{2} = 0$ — Divide each term by 2 so that the leading coefficient will be 1.

$x^2 + x = \frac{7}{2}$

$$x^2 + x + \mathbf{\frac{1}{4}} = \frac{7}{2} + \mathbf{\frac{1}{4}}$$

Complete the square: $\frac{1}{2}(1) = \frac{1}{2}$ and $\left(\frac{1}{2}\right)^2 = \frac{1}{4}$.

$$\left(x + \frac{1}{2}\right)^2 = \frac{15}{4}$$

$$x + \frac{1}{2} = \pm\sqrt{\frac{15}{4}} = \pm\frac{\sqrt{15}}{2}$$

$$x = -\frac{1}{2} \pm \frac{\sqrt{15}}{2} = \frac{-1 \pm \sqrt{15}}{2}$$

6. $x^2 - 2x + 13 = 0$

**Solution**

$$x^2 - 2x + 13 = 0$$
$$x^2 - 2x = -13$$
$$x^2 - 2x + \mathbf{1} = -13 + \mathbf{1}$$
$$(x - 1)^2 = -12$$
$$x - 1 = \pm\sqrt{-12} = \pm i\sqrt{12} = \pm 2i\sqrt{3}$$
$$x = 1 \pm 2i\sqrt{3}$$

The solutions are nonreal complex numbers. ■

**Practice Problems**

Solve each of the following quadratic equations by completing the square.

1. $2x^2 + 5x - 3 = 0$
2. $x^2 + 2x + 2 = 0$
3. $x^2 - 24x + 72 = 0$
4. $x^2 - 3x + 1 = 0$
5. $3x^2 - 6x + 15 = 0$

## EXERCISES 5.2

Add the correct constant to complete the square in Exercises 1–10; then factor the trinomial as indicated.

1. $x^2 - 12x + \_\_\_\_ = (\quad)^2$
2. $y^2 + 14y + \_\_\_\_ = (\quad)^2$
3. $x^2 + 6x + \_\_\_\_ = (\quad)^2$
4. $x^2 + 8x + \_\_\_\_ = (\quad)^2$
5. $x^2 - 5x + \_\_\_\_ = (\quad)^2$
6. $x^2 + 7x + \_\_\_\_ = (\quad)^2$
7. $y^2 + y + \_\_\_\_ = (\quad)^2$
8. $x^2 + \frac{1}{2}x + \_\_\_\_ = (\quad)^2$
9. $x^2 + \frac{1}{3}x + \_\_\_\_ = (\quad)^2$
10. $y^2 + \frac{3}{4}y + \_\_\_\_ = (\quad)^2$

**Answers to Practice Problems** 1. $x = -3, x = \frac{1}{2}$ 2. $x = -1 \pm i$ 3. $x = 12 \pm 6\sqrt{2}$ 4. $x = \frac{3 \pm \sqrt{5}}{2}$ 5. $x = 1 \pm 2i$

Solve the equations in Exercises 11–20.

**11.** $(x - 2)^2 = 9$
**12.** $(x - 4)^2 = 25$
**13.** $(x + 1)^2 = 5$
**14.** $(x - 6)^2 = 8$
**15.** $2(x + 3)^2 = 6$
**16.** $3(x - 1)^2 = 15$
**17.** $(x - 3)^2 = -4$
**18.** $(x + 8)^2 = -9$
**19.** $(x + 2)^2 = -7$
**20.** $(x - 5)^2 = -10$

Solve the quadratic equations in Exercises 21–50 by completing the square.

**21.** $x^2 + 4x - 5 = 0$
**22.** $x^2 + 6x - 7 = 0$
**23.** $y^2 + 2y = 5$
**24.** $x^2 + 3 = 8x$
**25.** $x^2 - 10x + 3 = 0$
**26.** $z^2 + 4z = 2$
**27.** $x^2 - 6x + 10 = 0$
**28.** $x^2 + 2x + 6 = 0$
**29.** $x^2 + 11 = 12x$
**30.** $y^2 - 10y + 4 = 0$
**31.** $z^2 + 3z - 5 = 0$
**32.** $x^2 - 5x + 5 = 0$
**33.** $x^2 + 5x + 2 = 0$
**34.** $x^2 + x + 2 = 0$
**35.** $x^2 - 2x + 5 = 0$
**36.** $x^2 = 3 - 4x$
**37.** $3y^2 + 9y + 9 = 0$
**38.** $4x^2 + 8x + 16 = 0$
**39.** $x^2 = 6 - x$
**40.** $3y^2 = 4 - y$
**41.** $3x^2 - 10x + 5 = 0$
**42.** $7x + 2 = -4x^2$
**43.** $3y^2 + 5y - 3 = 0$
**44.** $4x^2 - 2x + 3 = 0$
**45.** $2x + 2 = -6x^2$
**46.** $5y^2 + 15y + 25 = 0$
**47.** $2x^2 + 9x + 4 = 0$
**48.** $2x^2 - 8x + 4 = 0$
**49.** $3 = 3x - 6x^2$
**50.** $4x^2 + 20x + 32 = 0$

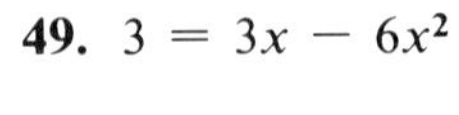

## 5.3 The Quadratic Formula

**OBJECTIVES**

In this section, you will be learning to:

1. Determine the nature of the solutions for quadratic equations by using the discriminant.
2. Solve quadratic equations by using the quadratic formula.

The general quadratic equation $ax^2 + bx + c = 0$ can be solved for $x$ in terms of $a$, $b$, and $c$ by completing the square. The result is called the **quadratic formula.**

$$ax^2 + bx + c = 0$$
The general quadratic formula

$$x^2 + \frac{b}{a}x + \frac{c}{a} = \frac{0}{a}$$
Divide each term on both sides of the equation by $a$. Since $a \neq 0$, this is permissible.

$$x^2 + \frac{b}{a}x = -\frac{c}{a}$$
Add $-\frac{c}{a}$ to both sides of the equation.

$$x^2 + \frac{b}{a}x + \frac{b^2}{4a^2} = \frac{b^2}{4a^2} - \frac{c}{a}$$
$\frac{1}{2}\left(\frac{b}{a}\right) = \frac{b}{2a}$ and $\left(\frac{b}{2a}\right)^2 = \frac{b^2}{4a^2}$. Add $\frac{b^2}{4a^2}$ to both sides of the equation.

$$\left(x + \frac{b}{2a}\right)^2 = \frac{b^2}{4a^2} - \frac{4ac}{4a^2}$$
Factor the left side. $4a^2$ is the common denominator on the right side.

$$\left(x + \frac{b}{2a}\right)^2 = \frac{b^2 - 4ac}{4a^2}$$
Simplify.

$$x + \frac{b}{2a} = \pm\sqrt{\frac{b^2 - 4ac}{4a^2}}$$
Use the definition of square root.

$$x + \frac{b}{2a} = \pm\frac{\sqrt{b^2 - 4ac}}{2a} \qquad \text{Simplify.}$$

$$x = -\frac{b}{2a} \pm \frac{\sqrt{b^2 - 4ac}}{2a} \qquad \text{Solve for } x.$$

$$x = \frac{-b \pm \sqrt{b^2 - 4ac}}{2a}$$

**The Quadratic Formula**

For the general quadratic equation

$$ax^2 + bx + c = 0 \qquad (a \neq 0)$$

the solutions are

$$x = \frac{-b \pm \sqrt{b^2 - 4ac}}{2a}$$

**The quadratic formula should be memorized.**

Applications of quadratic equations are found in such fields as economics, business, computer science, and chemistry, and in almost all branches of mathematics. Most instructors assume that their students know the quadratic formula and how to apply it.

**EXAMPLES**

**1.** $x^2 - 5x + 3 = 0$

**Solution** Substitute $a = 1$, $b = -5$, and $c = 3$ into the formula:

$$x = \frac{-b \pm \sqrt{b^2 - 4ac}}{2a} = \frac{-(-5) \pm \sqrt{(-5)^2 - 4 \cdot 1 \cdot 3}}{2 \cdot 1}$$

$$= \frac{5 \pm \sqrt{25 - 12}}{2}$$

$$= \frac{5 \pm \sqrt{13}}{2}$$

**2.** $2x^3 - 10x^2 + 6x = 0$

**Solution**

$$2x^3 - 10x^2 + 6x = 0$$

$$2x(x^2 - 5x + 3) = 0 \qquad \text{Factor out } 2x.$$

$$2x = 0 \quad \text{or} \quad x^2 - 5x + 3 = 0 \qquad \text{Set each factor equal to 0.}$$

$$x = 0 \quad \text{or} \quad x = \frac{5 \pm \sqrt{13}}{2} \qquad \text{Solve each equation. (We use the results from Example 1.)}$$

**3.** $7x^2 - 2x + 1 = 0$

**Solution** Substitute $a = 7$, $b = -2$, and $c = 1$ into the formula:

$$x = \frac{-b \pm \sqrt{b^2 - 4ac}}{2a} = \frac{-(-2) \pm \sqrt{(-2)^2 - 4 \cdot 7 \cdot 1}}{2 \cdot 7}$$

$$= \frac{2 \pm \sqrt{4 - 28}}{14} = \frac{2 \pm \sqrt{-24}}{14}$$

$$= \frac{2 \pm 2i\sqrt{6}}{14} = \frac{\not{2}(1 \pm i\sqrt{6})}{\not{2} \cdot 7} \quad \text{Factor and reduce.}$$

$$= \frac{1 \pm i\sqrt{6}}{7} \quad \text{The solutions are complex numbers.}$$

**4.** $\frac{3}{4}x^2 - \frac{1}{2}x = \frac{1}{3}$

**Solution** Multiply each term by the LCM, 12, so that the coefficients will be integers. The quadratic formula is easier to use with integer coefficients.

$$12 \cdot \frac{3}{4}x^2 - 12 \cdot \frac{1}{2}x = 12 \cdot \frac{1}{3}$$

$$9x^2 - 6x = 4$$

$$9x^2 - 6x - 4 = 0 \quad \text{To apply the formula, one side must be 0.}$$

$$x = \frac{-(-6) \pm \sqrt{(-6)^2 - 4(9)(-4)}}{2 \cdot 9} = \frac{6 \pm \sqrt{36 + 144}}{18}$$

$$= \frac{6 \pm \sqrt{180}}{18} = \frac{6 \pm 6\sqrt{5}}{18} = \frac{\not{6}(1 \pm \sqrt{5})}{\not{6} \cdot 3} = \frac{1 \pm \sqrt{5}}{3}$$

■

The expression $b^2 - 4ac$ in the quadratic formula is called the **discriminant. Since we have designated *a*, *b*, and *c* to be real numbers,** the discriminant tells what kind of numbers will be solutions to a quadratic equation. For example, in Example 1, the discriminant was positive, $b^2 - 4ac = (-5)^2 - 4(1)(3) = 13$, and there were two real solutions, $\frac{5 + \sqrt{13}}{2}$ and $\frac{5 - \sqrt{13}}{2}$. In Example 2, the discriminant was negative, $b^2 - 4ac = (-2)^2 - 4(7)(1) = -24$, and there were two nonreal solutions.

The discriminant gives the following information:

| *Discriminant* | *Nature of Solutions* |
|---|---|
| $b^2 - 4ac > 0$ | Two real solutions |
| $b^2 - 4ac = 0$ | One real solution, $x = -\frac{b}{2a}$ |
| $b^2 - 4ac < 0$ | Two nonreal, complex solutions |

**EXAMPLES** Find the discriminant and determine the nature of the solutions to each of the following quadratic equations.

**5.** $3x^2 + 11x - 4 = 0$

**Solution** $b^2 - 4ac = 11^2 - 4(3)(-4) = 121 + 48 = 169 > 0$

There are two real solutions.

**6.** $9x^2 + 18x + 9 = 0$

**Solution** $b^2 - 4ac = 18^2 - 4(9)(9) = 324 - 324 = 0$

There is one real solution.

**7.** $x^2 + 1 = 0$

**Solution** Here $b = 0$. We could write $x^2 + 0x + 1 = 0$.

$$b^2 - 4ac = 0^2 - 4 \cdot 1 \cdot 1 = -4$$

There are two nonreal, complex solutions.

**8.** Determine the values for $k$ so that $x^2 + 8x - k = 0$ will have one real solution. (**Hint:** Set the discriminant equal to 0.)

**Solution** $b^2 - 4ac = 8^2 - 4(1)(-k) = 0$

$$64 + 4k = 0$$
$$4k = -64$$
$$k = -16$$

*Check* $x^2 + 8x - (-16) = 0$

$$x^2 + 8x + 16 = 0$$
$$(x + 4)^2 = 0$$
$$x = -4$$

There is only one real solution. Using the quadratic formula was not necessary. ■

## Practice Problems

Solve the following quadratic equations by using the quadratic formula.

**1.** $x^2 + 2x - 4 = 0$

**2.** $2x^2 - 3x + 4 = 0$

**3.** $5x^2 - x - 4 = 0$

**4.** $\frac{1}{4}x^2 - \frac{1}{2}x = -\frac{1}{4}$

**5.** $3x^2 + 5 = 0$

**Answers to Practice Problems** **1.** $x = -1 \pm \sqrt{5}$ **2.** $x = \frac{3 \pm i\sqrt{23}}{4}$ **3.** $x = 1, x = -\frac{4}{5}$ **4.** $x = 1$ **5.** $x = \frac{\pm i\sqrt{15}}{3}$

## EXERCISES 5.3

Find the discriminant and determine the nature of the solutions to each of the quadratic equations in Exercises 1–12.

**1.** $x^2 + 6x - 8 = 0$
**2.** $x^2 + 3x + 1 = 0$
**3.** $x^2 - 8x + 16 = 0$
**4.** $x^2 + 3x + 5 = 0$
**5.** $2x^2 + 4x + 3 = 0$
**6.** $3x^2 - x + 2 = 0$
**7.** $5x^2 + 8x + 3 = 0$
**8.** $4x^2 + 12x + 9 = 0$
**9.** $100x^2 - 49 = 0$
**10.** $9x^2 + 121 = 0$
**11.** $3x^2 + x + 1 = 0$
**12.** $5x^2 - 3x - 2 = 0$

In Exercises 13–24, find the indicated values for $k$.

**13.** Determine the values for $k$ so that $x^2 - 8x + k = 0$ will have two real solutions.

**14.** Determine the values for $k$ so that $x^2 + 5x + k = 0$ will have two real solutions.

**15.** Determine the values for $k$ so that $x^2 + 9x + k = 0$ will have one real solution.

**16.** Determine the values for $k$ so that $x^2 - 7x + k = 0$ will have one real solution.

**17.** Determine the values for $k$ so that $kx^2 - 6x + 3 = 0$ will have two nonreal solutions.

**18.** Determine the values for $k$ so that $kx^2 + 4x - 2 = 0$ will have two nonreal solutions.

**19.** Determine the values for $k$ so that $kx^2 + x - 9 = 0$ will have two real solutions.

**20.** Determine the values for $k$ so that $kx^2 + 6x + 3 = 0$ will have two real solutions.

**21.** Determine the values for $k$ so that $kx^2 + 7x + 12 = 0$ will have one real solution.

**22.** Determine the value for $k$ so that $kx^2 - 2x + 8 = 0$ will have one real solution.

**23.** Determine the values for $k$ so that $3x^2 + 4x + k = 0$ will have two nonreal solutions.

**24.** Determine the values for $k$ so that $2x^2 + 3x + k = 0$ will have two nonreal solutions.

Solve the equations in Exercises 25–60.

**25.** $x^2 + 3x - 5 = 0$
**26.** $x^2 = 7x + 3$
**27.** $x^2 - 5x + 2 = 0$
**28.** $x^2 + 4x + 3 = 0$
**29.** $2x^2 + 7x + 2 = 0$
**30.** $3x^2 + 2x - 2 = 0$
**31.** $6x^2 = 5x + 1$
**32.** $4x^2 + x - 4 = 0$
**33.** $3x^2 - 4 = 4$
**34.** $7x^2 + 6x + 1 = 0$
**35.** $x^3 - 9x^2 + 4x = 0$
**36.** $x^3 - 8x^2 = 3x^2 + 3x$
**37.** $x^3 + 3x^2 + x = 0$
**38.** $4x^3 + 10x^2 - 3x = 0$
**39.** $x^2 - 3x - 4 = 0$
**40.** $9x^2 - 6x + 1 = 0$
**41.** $2x^2 + 8x + 9 = 0$
**42.** $3x^2 + 7x - 4 = 0$
**43.** $x^2 - 7 = 0$
**44.** $3x^2 - 6x + 4 = 0$
**45.** $x^2 + 4x = x - 2x^2$
**46.** $3x^2 + 4x = 0$
**47.** $5x^2 - 7x + 5 = 0$
**48.** $4x^2 - 5x + 3 = 0$
**49.** $6x^2 + 2x - 20 = 0$
**50.** $10x^2 + 35x + 30 = 0$
**51.** $4x^2 + 9 = 0$
**52.** $3x^2 - 8x + 6 = 0$

(**Hint:** In Exercises 53–60, multiply each side of the equation by the LCM of the denominator to get integer coefficients.)

**53.** $3x^2 - 4x + \frac{1}{3} = 0$
**54.** $\frac{3}{4}x^2 - 2x + \frac{1}{8} = 0$
**55.** $\frac{3}{7}x^2 - \frac{1}{2}x + 1 = 0$

**56.** $2x^2 + 3x + \frac{5}{4} = 0$

**57.** $\frac{1}{2}x^2 - x + \frac{1}{4} = 0$

**58.** $\frac{2}{3}x^2 - \frac{1}{3}x + \frac{1}{2} = 0$

**59.** $\frac{1}{4}x^2 + \frac{7}{8}x + \frac{1}{2} = 0$

**60.** $\frac{5}{12}x^2 - \frac{1}{2}x - \frac{1}{4} = 0$

## 5.4 Applications

**OBJECTIVE**

In this section, you will be learning to solve applied problems by using quadratic equations.

The following Attack Plan, given in Section 1.5, is still a valid approach to solving word problems.

***Attack Plan for Word Problems***

1. Read the problem carefully. (Reread it several times if necessary.)
2. Decide what is asked for and assign a variable to the unknown quantity.
3. Draw a diagram or set up a chart whenever possible.
4. Form an equation (or inequality) that relates the information provided.
5. Solve the equation (or inequality).
6. Check your solution with the wording of the problem to be sure that it makes sense. As an aid in checking, write your answer in a complete sentence.

The problems in this section can be solved by setting up quadratic equations and then solving these equations by factoring, or completing the square, or using the quadratic formula.

Three terms that appear in the exercises are **consecutive integers, consecutive even integers,** and **consecutive odd integers.** If $n$ is an integer:

**a.** Consecutive integers are $n$, $n + 1$, and $n + 2$.
(For example, if $n = 3$, then $n + 1 = 4$ and $n + 2 = 5$.)

**b.** Consecutive even integers are $n$, $n + 2$, and $n + 4$, if $n$ is even.
(For example, if $n = 8$, then $n + 2 = 10$ and $n + 4 = 12$.)

**c.** Consecutive odd integers are $n$, $n + 2$, and $n + 4$, if $n$ is odd.
(For example, if $n = 13$, then $n + 2 = 15$ and $n + 4 = 17$.)

Consecutive integers differ by 1, and consecutive even or odd integers differ by 2.

A geometric topic that generates quadratic equations is right triangles. In a **right triangle,** one of the angles is a right angle (measures 90°), and the side opposite this angle (the longest side) is called the **hypotenuse.** The other two sides are called **legs.** A famous Greek mathematician, Pythagoras, is given credit for proving the following very important and useful theorem. Now, there are entire books written that contain only proofs of the Pythagorean Theorem developed by mathematicians since the time of Pythagoras.

---

***Pythagorean Theorem***

In a right triangle, the square of the hypotenuse is equal to the sum of the squares of the other sides.

$$a^2 + b^2 = c^2$$

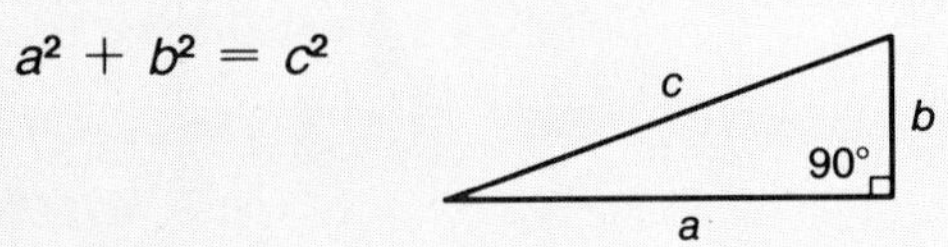


---

**EXAMPLE: The Pythagorean Theorem**

**1.** The length of a rectangle is 6 meters more than its width. If the diagonal is 30 meters, what are the dimensions of the rectangle?

**Solution** Let $w$ = width
$w + 6$ = length

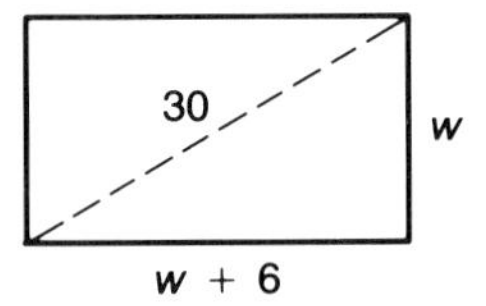


$$(w + 6)^2 + w^2 = 30^2$$ Use the Pythagorean Theorem.

$$w^2 + 12w + 36 + w^2 = 900$$
$$2w^2 + 12w - 864 = 0$$
$$w^2 + 6w - 432 = 0$$
$$(w + 24)(w - 18) = 0$$

~~$w = -24$~~ or $w = 18$ A negative number does not fit the condition of the problem.

$$w = 18 \text{ meters}$$
$$w + 6 = 24 \text{ meters}$$

The length is 24 meters and the width is 18 meters.

**EXAMPLE: Projectiles**

The formula $h = -16t^2 + v_0t + h_0$ is used in physics and relates to the height of a projectile such as a thrown ball, a bullet, or a rocket.

$h$ = height of object
$t$ = time
$v_0$ = beginning velocity
$h_0$ = beginning height    $h_0 = 0$ if the object is initially at ground level.

**2.** A bullet is fired straight up from ground level with a muzzle velocity of 320 ft per sec. (a) When will the bullet hit the ground? (b) When will the bullet be 1200 ft above the ground?

**Solution** In this problem, $v_0 = 320$ ft per sec
and $h_0 = 0$

**a.** The bullet hits the ground when $h = 0$.

$$0 = -16t^2 + 320t$$
$$0 = t^2 - 20t \qquad \text{Divide by } -16.$$
$$0 = t(t - 20)$$
$$t = 0 \quad \text{or} \quad t = 20$$

The bullet hits the ground in 20 seconds. The solution $t = 0$ confirms the fact that the bullet was fired from the ground.

**b.** Let $h = 1200$.

$$1200 = -16t^2 + 320t$$
$$0 = -16t^2 + 320t - 1200$$
$$0 = t^2 - 20t + 75$$
$$0 = (t - 5)(t - 15)$$
$$t = 5 \quad \text{or} \quad t = 15$$

Both solutions are meaningful. The bullet is at 1200 ft twice, once in 5 sec going up and once in 15 sec coming down.

**EXAMPLE: Geometry**

**3.** A rectangular sheet of metal is 6 in. longer than it is wide. An open box is to be made by cutting 3-in. squares at each corner and folding up the sides. If the box has a volume of 336 cu in., what were the dimensions of the sheet of metal? ($V = \ell wh$.)

**Solution** Let $x$ = width
$x + 6$ = length

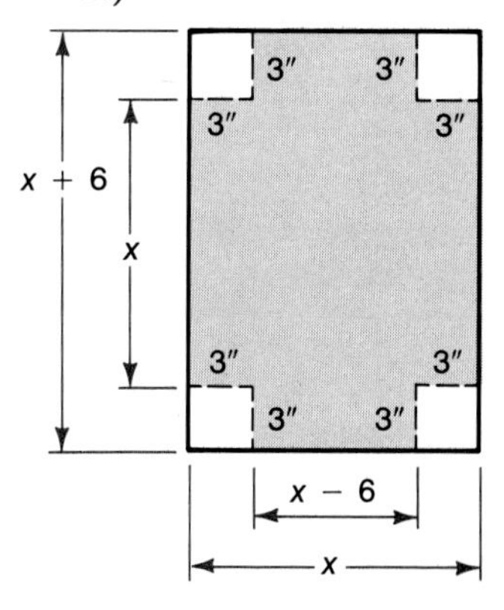


$$3(x - 6)x = 336$$
$$3x^2 - 18x = 336$$
$$3x^2 - 18x - 336 = 0$$
$$x^2 - 6x - 112 = 0$$
$$(x + 8)(x - 14) = 0$$
$$\cancel{x = -8} \quad \text{or} \quad x = 14$$
$$x + 6 = 20$$

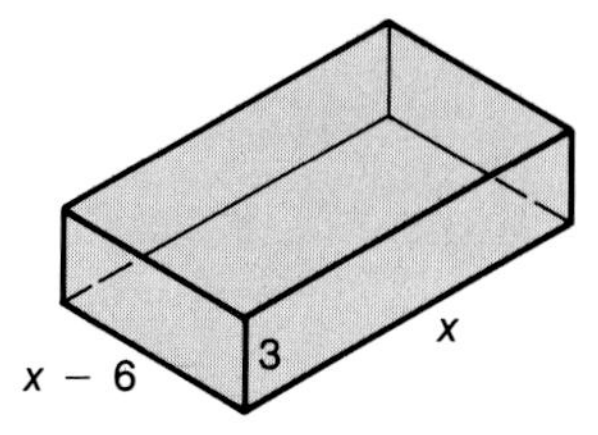


The length of the sheet is 20 in. and the width is 14 in. ■

In the following example, note that the **cost per person** is found by dividing the total cost by the number of people going to the tournament. The **cost per person** changes because the total cost remains fixed but the number of people changes.

**EXAMPLE: Cost per Person**

**4.** The members of a bowling club were going to fly commercially to a tournament at a total cost of \$2420 which was to be divided equally among the members. At the last minute, two of the members decided to fly their own private planes. The cost to the remaining members increased \$11 each. How many members flew commercially?

**Solution** Let $x$ = number of club members

$x - 2$ = number of club members that flew commercially

$$\underbrace{\text{final cost per member}}_{\frac{2420}{x-2}} - \underbrace{\text{initial cost per member}}_{\frac{2420}{x}} = \underbrace{\text{difference in cost per member}}_{11}$$

$$\cancel{x(x-2)} \cdot \frac{2420}{\cancel{x-2}} - \cancel{x}(x-2) \cdot \frac{2420}{\cancel{x}} = x(x-2) \cdot 11$$

$$\begin{aligned} 2420x - 2420(x-2) &= 11x(x-2) \\ 2420x - 2420x + 4840 &= 11x^2 - 22x \\ 0 &= 11x^2 - 22x - 4840 \\ 0 &= 11(x^2 - 2x - 440) \\ 0 &= 11(x-22)(x+20) \end{aligned}$$

$$x = 22 \qquad \cancel{x = -20}$$
$$x - 2 = 20$$

($-20$ does not fit the conditions. That is, the number of people in a club is a positive number.)

*Check* Final cost per member $= \dfrac{2420}{20} = \$121$

Initial cost per member $= \dfrac{2420}{22} = \$110$

\$121 − \$110 = \$11 difference in cost per member

Twenty members flew commercially. ■

## EXERCISES 5.4

1. A positive integer is one more than twice another. Their product is 78. Find the two integers.
2. One number is equal to the square of another. Find the numbers if their sum is 72.
3. Find two positive numbers whose difference is 10 and whose product is 56.
4. One number is three more than twice a second number. Their product is 119. Find the numbers.
5. The sum of two numbers is $-17$. Their product is 66. Find the numbers.
6. Find two consecutive positive integers the sum of whose squares is 113.
7. The product of two consecutive even integers is 168. Find the integers.
8. The product of two consecutive odd integers is 420 more than three times the smaller integer. Find the integers.
9. Find three consecutive positive integers such that twice the product of the two smaller integers is 88 more than the product of the larger integers.
10. The sum of three positive integers is 68. The second is one more than twice the first. The third is three less than the square of the first. Find the integers.
11. A right triangle has two equal sides. The hypotenuse is 14 centimeters. Find the length of the sides.
12. The length of one leg of a right triangle is twice the length of the second leg. The hypotenuse is 15 meters. Find the lengths of the two legs.
13. Mel and John leave Desert Point at the same time. Mel drives north and John drives east. Mel averages 10 mph slower than John. At the end of one hour they are 50 miles apart. Find the rates of each driver.
14. A flagpole was bent over at a point $\frac{4}{9}$ of the distance from its base to the top. The top of the pole reached a point on the ground 9 meters from the base of the pole. What was the original height of the pole?
15. The length of a rectangle is 2 feet less than three times the width. If the area of the rectangle is 40 square feet, find the dimensions.
16. A rectangle is 3 meters longer than it is wide. If the width is doubled and the length decreased by 4 meters, the area is unchanged. Find the original dimensions.
17. A rectangular piece of cardboard twice as long as it is wide has a small square 4 cm by 4 cm cut from each corner. The edges are then folded up to form an open box with a volume of 1536 cu cm. What are the dimensions of the box? (See Example 3.)
18. An orchard has 2030 trees. The number of trees in each row exceeds twice the number of rows by 12. How many trees are in each row?
19. A rectangular auditorium seats 960 people. The number of seats in each row exceeds the number of rows by 16. Find the number of seats in each row.
20. The perimeter of a rectangle is 24 meters; its area is 27 square meters. Find the dimensions of the rectangle.
21. The area of a rectangle is 102 square inches. The perimeter of the rectangle is 46 inches. Find the length and width.
22. The length of a rectangle is 2 cm greater than the width. If the length and width are each increased by 3 cm, the area is increased by 57 sq cm. Find the dimensions of the original rectangle.

23. A picture 9 in. wide and 12 in. long is surrounded by a frame of uniform width. The area of the frame only is 162 sq in. Find the width of the frame.

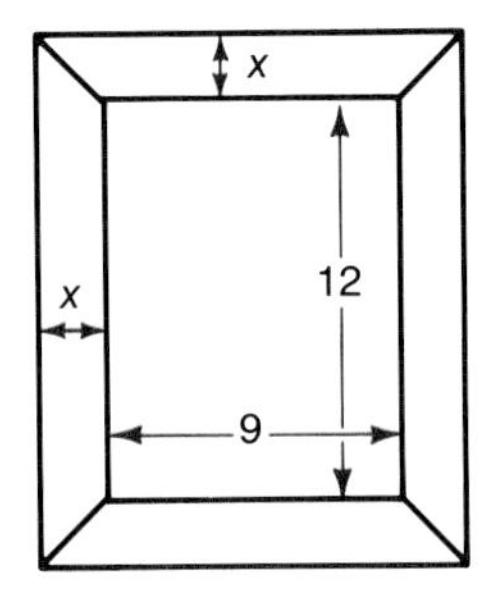


24. A 40-volt generator with a resistance of 4 ohms delivers power externally of $40I - 4I^2$ watts, where $I$ is the current measured in amperes. Find the current needed for the generator to deliver 100 watts of power.

25. Find the current needed for the 40-volt generator in Exercise 24 to deliver 64 watts of power.

26. Vince operates a small sign-making business. He finds that if he charges $x$ dollars for each sign, he can sell $40 - x$ signs per week. What is the least number of signs he can sell to have an income of $336 in one week?

27. Sam operates a small hot dog stand. He estimates that he can sell 600 hot dogs per day if he charges 50¢ apiece. He determines that he can sell 20 more hot dogs for each 1-cent reduction in price. What should he charge in order to have receipts of $315?

28. It costs Ms. Snow $3 to build a picture frame. She estimates that if she charges $x$ dollars each, she can sell $60 - x$ frames per week. What is the lowest price necessary to make a profit of $432 each week?

29. J. B. bought some shirts and trousers. He bought two more pairs of trousers than shirts. The trousers cost $154 and the shirts cost $65. Find the price of each if the price of a pair of trousers exceeds the price of a shirt by $9.

30. The Piton Rock Climbing Club planned a climbing expedition. The total cost was $900, which was to be divided equally among the members going. While practicing, three members fell and were hurt so were unable to go. If the cost per person increased by $15, how many people went on the expedition?

31. Mack traveled 240 miles to a convention. Later his wife, Ann, drove up to meet him. Ann's average speed exceeded Mack's by 4 mph, and the trip took her 15 minutes less time. Find Ann's speed.

32. In two hours, a motorboat can travel 8 miles down a river and return 4 miles back. If the river flows at a rate of 2 miles per hour, how fast can the boat travel in still water?

33. It takes Bob 5 hours longer to assemble a machine than it does Sam. After Bob works for as many hours as it would take Sam to do the entire job, Sam can finish the job in 3 hours. How long would it take each man working alone to assemble the machine? (**Hint:** Represent the total job by the number 1.)

34. Two employees together can prepare a large order in 2 hr. Working alone, one employee takes three hours longer than the other. How long does it take each person working alone?

35. Fern can fly her plane 240 miles against the wind in the same time it takes her to fly 360 miles with the wind. The speed of the plane in still air is 30 mph more than four times the speed of the wind. Find the speed of the plane in still air.

36. A grocer mixes $9.00 worth of Grade A coffee with $12.00 worth of Grade B coffee to obtain 30 pounds of a blend. If Grade A costs 30¢ a pound more than Grade B, how many pounds of each were used?

37. The Andersonville Little Theater Group sold 340 tickets to their spring production. Receipts from the sale of reserved tickets were $855. Receipts from general admission tickets were $375. How many of each type ticket were sold if the cost of a reserved ticket is $2 more than a general admission ticket?

**38.** It takes a young man 2 hours longer to build a wall than it does his father. After the son has worked for 1 hour, his father joins him and they finish the job together in 6 more hours. How long would it take the father working alone?

For Exercises 39–42, use the formula $h = -16t^2 + v_0t + h_0$.

**39.** A ball is thrown vertically upward from ground level with an initial speed of 108 ft per sec. (a) When will the ball hit the ground? (b) When will the ball be 180 ft above the ground?

**40.** A ball is thrown vertically upward from the ground with an initial speed of 160 ft per sec. (a) When will the ball strike the ground? (b) When will it be 400 ft above the ground?

**41.** An arrow is shot vertically upward from a platform 40 ft high at a rate of 224 ft per sec. (a) When will the arrow be 824 ft above the ground? (b) When will it be 424 ft above the ground?

**42.** A stone is dropped from a platform 196 ft high. (a) When will it hit the ground? (b) How far will it fall during the third second of time? (**Hint:** Since the stone is dropped, $v_0 = 0$.)

## 5.5 Equations with Radicals

**OBJECTIVE**

In this section, you will be learning to solve equations that contain one or more radical expressions.

Each of the following equations involves at least one radical expression:

$$x + 3 = \sqrt{x + 5}, \qquad \sqrt{x} - \sqrt{2x - 14} = 1, \qquad \sqrt[3]{x + 1} = 5$$

If the radicals are square roots, we solve by squaring both sides of the equation. If the radical is some other root, we solve by raising both sides of the equation to the integer power corresponding to the root or by using one of the techniques discussed in the next section.

Squaring both sides of an equation may introduce new solutions. For example, the first-degree equation $x = -3$ has only one solution, namely, $-3$. However, squaring both sides gives

$$x^2 = (-3)^2 \qquad \text{or} \qquad x^2 = 9$$

The quadratic equation $x^2 = 9$ has two solutions, $+3$ and $-3$. Thus, a new solution, not a solution to the original equation, has been introduced. Such a solution is called an **extraneous solution.**

**When both sides of an equation are squared, an extraneous solution may be introduced. Be sure to check all solutions in the original equation.**

The following examples illustrate a variety of situations involving radicals. The steps used are related to the following general method.

***Method for Solving Equations with Radicals***

**Step 1** Isolate one of the radicals on one side of the equation. (An equation may have more than one radical.)
**Step 2** Raise both sides of the equation to the power corresponding to the index of the radical.
**Step 3** If the equation still contains a radical, repeat Steps 1 and 2.
**Step 4** Solve the equation. (No radicals are left.)
**Step 5** Be sure to check all possible solutions in the original equation and eliminate any extraneous solutions.

**EXAMPLES** Solve the following equations.

**1.** $x + 3 = \sqrt{x + 5}$

**Solution** The one radical is by itself on one side of the equation; square both sides.

$$(x + 3) = \sqrt{x + 5}$$
$$(x + 3)^2 = (\sqrt{x + 5})^2 \quad \text{Square both sides.}$$
$$x^2 + 6x + 9 = x + 5 \quad \text{You now have an equation with no radical.}$$
$$x^2 + 5x + 4 = 0$$
$$(x + 4)(x + 1) = 0 \quad \text{Solve by factoring.}$$
$$x = -4 \quad \text{or} \quad x = -1$$

**Check both answers** in the original equation:

$$-4 + 3 \stackrel{?}{=} \sqrt{-4 + 5} \qquad -1 + 3 \stackrel{?}{=} \sqrt{-1 + 5}$$
$$-1 \stackrel{?}{=} \sqrt{1} \qquad 2 \stackrel{?}{=} \sqrt{4}$$
$$-1 \neq 1 \qquad 2 = 2$$

$-4$ is **not** a solution. The only solution is $-1$.

**2.** $\sqrt{y^2 - 10y - 11} = 1 + y$

**Solution** Since there is only one radical and it is by itself on one side of the equation, square both sides.

$$\sqrt{y^2 - 10y - 11} = 1 + y$$
$$(\sqrt{y^2 - 10y - 11})^2 = (1 + y)^2$$
$$y^2 - 10y - 11 = 1 + 2y + y^2$$
$$-12y - 12 = 0 \quad \text{Simplifying gives a first-degree equation.}$$
$$-12y = 12$$
$$y = -1$$

**Check** in the original equation:

$$\sqrt{(-1)^2 - 10(-1) - 11} \stackrel{?}{=} 1 + (-1)$$
$$\sqrt{1 + 10 - 11} \stackrel{?}{=} 0$$
$$\sqrt{0} \stackrel{?}{=} 0$$
$$0 = 0$$

There is one solution, $-1$.

**3.** $\sqrt{x + 4} = \sqrt{3x - 2}$

**Solution** There are two radicals and they are on opposite sides of the equation. Squaring both sides will give a new equation with no radicals.

$$\sqrt{x + 4} = \sqrt{3x - 2}$$
$$(\sqrt{x + 4})^2 = (\sqrt{3x - 2})^2$$
$$x + 4 = 3x - 2 \quad \text{Simplifying gives a first-degree equation.}$$
$$6 = 2x \quad \text{Simplify.}$$
$$3 = x$$

**Check** in the original equation:

$$\sqrt{3 + 4} \stackrel{?}{=} \sqrt{3 \cdot 3 - 2}$$
$$\sqrt{7} = \sqrt{7}$$

There is one solution, 3.

**4.** $\sqrt{3x + 13} + 3 = 2x$

**Solution** $\sqrt{3x + 13} + 3 = 2x$

$$\sqrt{3x + 13} = 2x - 3 \quad \text{Isolate the radical.}$$
$$(\sqrt{3x + 13})^2 = (2x - 3)^2 \quad \text{Square both sides.}$$
$$3x + 13 = 4x^2 - 12x + 9$$
$$0 = 4x^2 - 15x - 4$$
$$0 = (4x + 1)(x - 4)$$

$$x = -\frac{1}{4} \quad \text{or} \quad x = 4$$

**Check both answers** in the original equation:

$$\sqrt{3\left(-\frac{1}{4}\right)+13}+3 \stackrel{?}{=} 2\left(-\frac{1}{4}\right) \qquad \sqrt{3(4)+13}+3 \stackrel{?}{=} 2(4)$$

$$\sqrt{\frac{49}{4}}+3 \stackrel{?}{=} -\frac{1}{2} \qquad \sqrt{25}+3 \stackrel{?}{=} 8$$

$$\frac{7}{2}+3 \stackrel{?}{=} -\frac{1}{2} \qquad 5+3 \stackrel{?}{=} 8$$

$$\frac{13}{2} \neq -\frac{1}{2} \qquad 8 = 8$$

$-\frac{1}{4}$ is **not** a solution. The only solution is 4.

**5.** $\sqrt[3]{2x+1}+1=3$

**Solution** First, get the radical by itself on one side of the equation. Then, since this radical is a cube root, cube both sides of the equation.

$$\sqrt[3]{2x+1}+1=3$$

$$\sqrt[3]{2x+1}=2 \qquad \text{Add } -1 \text{ to both sides.}$$

$$(\sqrt[3]{2x+1})^3=(2)^3 \qquad \text{Cube both sides.}$$

$$2x+1=8 \qquad \text{Simplify.}$$

$$x=\frac{7}{2}$$

**Check** in the original equation:

$$\sqrt[3]{2\left(\frac{7}{2}\right)+1}+1 \stackrel{?}{=} 3$$

$$\sqrt[3]{7+1}+1 \stackrel{?}{=} 3$$

$$\sqrt[3]{8}+1 \stackrel{?}{=} 3$$

$$2+1 \stackrel{?}{=} 3$$

$$3=3$$

There is one solution, $\frac{7}{2}$.

**6.** $\sqrt{x} - \sqrt{2x - 14} = 1$

**Solution** Where there is a sum or difference of radicals, squaring is easier if the radicals are on different sides of the equation. Also, squaring both sides of the equation is easier if one of the radicals is by itself on one side of the equation.

| | |
|---|---|
| $\sqrt{x} - \sqrt{2x - 14} = 1$ | |
| $\sqrt{x} = 1 + \sqrt{2x - 14}$ | Arrange a radical on each side. |
| $(\sqrt{x})^2 = (1 + \sqrt{2x - 14})^2$ | Square both sides. |
| $x = 1 + 2\sqrt{2x - 14} + (2x - 14)$ | Remember, the right-hand side is the square of a binomial. |
| $-x + 13 = 2\sqrt{2x - 14}$ | Simplify so that the radical is on one side by itself. |
| $(-x + 13)^2 = 2\sqrt{2x - 14}$ | Square both sides **again.** |
| $x^2 - 26x + 169 = 4(2x - 14)$ | |
| $x^2 - 26x + 169 = 8x - 56$ | |
| $x^2 - 34x + 225 = 0$ | |
| $(x - 9)(x - 25) = 0$ | Solve by factoring. |
| $x = 9 \quad \text{or} \quad x = 25$ | The quadratic formula will always give the same results if you do not see the factors. |

**Check both answers** in the original equation:

$$\sqrt{9} - \sqrt{2 \cdot 9 - 14} \stackrel{?}{=} 1 \qquad \sqrt{25} - \sqrt{2 \cdot 25 - 14} \stackrel{?}{=} 1$$

$$3 - 2 \stackrel{?}{=} 1 \qquad 5 - 6 \stackrel{?}{=} -1$$

$$1 = 1 \qquad -1 \neq 1$$

25 is **not** a solution. 9 is the only solution. ■

***Practice Problems***

Solve the following equations.

1. $2\sqrt{x + 4} = x + 1$
2. $\sqrt{3x + 1} + 1 = \sqrt{x}$
3. $\sqrt[3]{2x - 9} + 4 = 3$

**Answers to Practice Problems** **1.** $x = 5$ **2.** No solution **3.** $x = 4$

## EXERCISES 5.5

Solve the following equations.

1. $\sqrt{8x + 1} = 5$
2. $\sqrt{7x + 1} = 6$
3. $\sqrt{4x - 3} = 7$
4. $\sqrt{5x - 6} = 8$
5. $\sqrt{2x + 5} = \sqrt{4x - 1}$
6. $\sqrt{5x - 1} = \sqrt{x + 7}$
7. $\sqrt{3x + 2} = \sqrt{9x - 10}$
8. $\sqrt{2 - x} = \sqrt{2x - 7}$
9. $\sqrt{x(x + 3)} = 2$
10. $\sqrt{x(x - 5)} = 6$
11. $\sqrt{x(2x + 5)} = 5$
12. $\sqrt{x(3x - 14)} = 7$
13. $\sqrt{x + 6} = x + 4$
14. $\sqrt{x + 7} = 2x - 1$
15. $\sqrt{x^2 - 16} = 3$
16. $\sqrt{x^2 - 25} = 12$
17. $5x + \sqrt{x + 7} - 13 = 0$
18. $x - 6 - \sqrt{x + 4} = 2$
19. $2x = \sqrt{7x - 3} + 3$
20. $x - \sqrt{3x - 8} = 4$
21. $\sqrt{3x + 1} = 1 - \sqrt{x}$
22. $\sqrt{x} = \sqrt{x + 16} - 2$
23. $\sqrt{x + 4} = \sqrt{x + 11} - 1$
24. $\sqrt{1 - x} + 2 = \sqrt{13 - x}$
25. $\sqrt{x + 1} = \sqrt{x + 6} + 1$
26. $\sqrt{x + 4} = \sqrt{x + 20} - 2$
27. $\sqrt{x + 5} + \sqrt{x} = 5$
28. $\sqrt{5x - 18} - 4 = \sqrt{5x + 6}$
29. $\sqrt{2x + 3} = 1 + \sqrt{x + 1}$
30. $\sqrt{x} + \sqrt{x - 3} = 3$
31. $\sqrt{3x + 1} - \sqrt{x + 4} = 1$
32. $\sqrt{3x + 4} - \sqrt{x + 5} = 1$
33. $\sqrt{5x - 1} = 4 - \sqrt{x - 1}$
34. $\sqrt{2x - 5} - 2 = \sqrt{x - 2}$
35. $\sqrt{2x - 3} + \sqrt{x + 3} = 6$
36. $\sqrt{2x + 3} - \sqrt{x + 5} = 1$
37. $\sqrt[3]{4 + 3x} = -2$
38. $\sqrt[3]{2x + 9} = 3$
39. $\sqrt[3]{5x + 4} = 4$
40. $\sqrt[3]{7x + 1} = -5$

## 5.6 Equations in Quadratic Form

**OBJECTIVES**

In this section, you will be learning to:

1. Solve equations that can be written in quadratic form by appropriate substitutions.
2. Solve equations that contain rational expressions.

The general quadratic equation is

$$ax^2 + bx + c = 0 \qquad \text{where } a \neq 0$$

The equations

$$x^4 - 7x^2 + 12 = 0$$

and

$$x^{2/3} - 4x^{1/3} - 21 = 0$$

are not quadratic equations, but they are in **quadratic form** because the degree of the middle term is one-half the degree of the first term. Specifically,

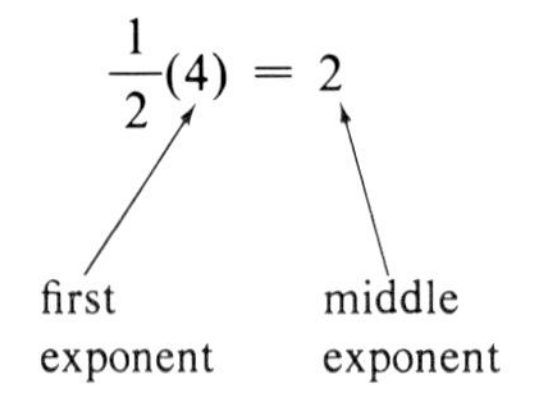


and

$$\frac{1}{2}\left(\frac{2}{3}\right) = \frac{1}{3}$$

first exponent — middle exponent

Equations in quadratic form can be solved by using the quadratic formula or by factoring just as if they were quadratic equations. In each case, a substitution can be made to clarify the problem. Try to follow these suggestions:

1. Look at the middle term.
2. Substitute a first-degree variable—say, $u$—for the variable expression in the middle term.
3. Substitute the square of this variable—say, $u^2$—for the variable expression in the first term.

The following examples illustrate how such a substitution may help. Study each example carefully and note the variety of algebraic manipulations used.

**EXAMPLES** Solve the following equations. Each one is in quadratic form, and a substitution will help.

**1.** $x^4 - 7x^2 + 12 = 0$

**Solution** $x^4 - 7x^2 + 12 = 0$ Substitute $u = x^2$ and $u^2 = x^4$.

$u^2 - 7u + 12 = 0$

$(u - 3)(u - 4) = 0$ Solve for $u$ by factoring.

$u = 3$ or $u = 4$ Now substitute $x^2$ for $u$ and solve for $x$.

$x^2 = 3$ or $x^2 = 4$

$x = \pm\sqrt{3}$ or $x = \pm 2$

There are four solutions: $+\sqrt{3}$, $-\sqrt{3}$, $+2$, $-2$.

**2.** $x^{2/3} - 4x^{1/3} - 21 = 0$

**Solution** $x^{2/3} - 4x^{1/3} - 21 = 0$ Let $u = x^{1/3}$ and $u^2 = x^{2/3}$.

$u^2 - 4u - 21 = 0$

$(u - 7)(u + 3) = 0$ Solve for $u$ by factoring.

$u = 7$ or $u = -3$ Now substitute $x^{1/3}$ for $u$.

$x^{1/3} = 7$ or $x^{1/3} = -3$

$(x^{1/3})^3 = 7^3$ $(x^{1/3})^3 = (-3)^3$ Cube both sides.

$x = 343$ $x = -27$

There are two solutions: 343 and $-27$.

**3.** $x^{-4} - 7x^{-2} + 10 = 0$

**Solution** $x^{-4} - 7x^{-2} + 10 = 0$ Let $u = x^{-2}$ and $u^2 = x^{-4}$.

$$u^2 - 7u + 10 = 0$$

$$(u - 2)(u - 5) = 0 \quad \text{Solve for } u \text{ by factoring.}$$

$$u = 2 \quad \text{or} \quad u = 5 \quad \text{Now substitute } x^{-2} \text{ for } u.$$

$$x^{-2} = 2 \qquad x^{-2} = 5$$

$$\frac{1}{x^2} = 2 \qquad \frac{1}{x^2} = 5 \quad \text{Remember, } x^{-2} = \frac{1}{x^2}.$$

$$x^2 = \frac{1}{2} \qquad x^2 = \frac{1}{5} \quad \text{Reciprocals.}$$

$$x = \pm\sqrt{\frac{1}{2}} \qquad x = \pm\sqrt{\frac{1}{5}}$$

$$x = \pm\frac{1}{\sqrt{2}} \qquad x = \pm\frac{1}{\sqrt{5}}$$

or

$$x = \pm\frac{\sqrt{2}}{2} \qquad x = \pm\frac{\sqrt{5}}{5} \quad \text{Rationalizing denominators.}$$

There are four solutions: $\frac{\sqrt{2}}{2}, \frac{-\sqrt{2}}{2}, \frac{\sqrt{5}}{5}, \frac{-\sqrt{5}}{5}$.

**4.** $x^5 - 16x = 0$

**Solution** No substitution is needed here. The factoring is straightforward.

$$x^5 - 16x = 0$$

$$x(x^4 - 16) = 0 \quad \text{Factor the common term } x.$$

$$x(x^2 + 4)(x^2 - 4) = 0 \quad \text{Factor the difference of two squares.}$$

$$x = 0 \quad \text{or} \quad x^2 = -4 \quad \text{or} \quad x^2 = 4$$

$$x = \pm 2i \qquad x = \pm 2$$

There are five solutions: 0, $2i$, $-2i$, 2, $-2$.

**5.** $(x + 2)^2 - (x + 2) - 12 = 0$

**Solution** $(x + 2)^2 - (x + 2) - 12 = 0$ Let $u = x + 2$.

$$u^2 - u - 12 = 0$$

$$(u - 4)(u + 3) = 0 \quad \text{Solve for } u \text{ by factoring.}$$

$$u = 4 \quad \text{or} \quad u = -3$$

$$x + 2 = 4 \qquad x + 2 = -3 \quad \text{Now substitute } x + 2 \text{ for } u.$$

$$x = 2 \qquad x = -5$$

There are two solutions: 2 and $-5$.

6. $\frac{2}{3x-1}+\frac{1}{x+1}=\frac{x}{x+1}$

**Solution** This equation is not in quadratic form; however, multiplying both sides of the equation by the LCM of the denominators gives a quadratic equation.

$$(x+1)\cancel{(3x-1)}\cdot\frac{2}{\cancel{3x-1}}+\cancel{(x+1)}(3x-1)\cdot\frac{1}{\cancel{x+1}}=\cancel{(x+1)}(3x-1)\cdot\frac{x}{\cancel{x+1}}$$

$$\begin{aligned}2(x+1)+3x-1&=(3x-1)x\\2x+2+3x-1&=3x^2-x\\0&=3x^2-6x-1\end{aligned}$$

$$\begin{aligned}x&=\frac{6\pm\sqrt{36-4\cdot3(-1)}}{6}\\&=\frac{6\pm\sqrt{48}}{6}\\&=\frac{6\pm4\sqrt{3}}{6}\\&=\frac{3\pm2\sqrt{3}}{3}\end{aligned}$$

■

**Practice Problems**

Solve the following equations.

1. $x-x^{1/2}-2=0$
   (Let $u=x^{1/2}$ and $u^2=x$.)
2. $x^4+16x^2=-48$
   (Let $u=x^2$ and $u^2=x^4$.)
3. $\frac{3(x-2)}{x-1}=\frac{2(x+1)}{x-2}+2$

**Answers to Practice Problems** 1. $x=4$ 2. $x=\pm2i, x=\pm2i\sqrt{3}$
3. $x=-3\pm\sqrt{19}$

## EXERCISES 5.6

Solve the following equations.

1. $x^4 - 13x^2 + 36 = 0$
2. $x^4 - 29x^2 + 100 = 0$
3. $x^4 - 9x^2 + 20 = 0$
4. $y^4 - 11y^2 + 18 = 0$
5. $y^4 - 3y^2 - 28 = 0$
6. $y^4 + y^2 - 12 = 0$
7. $y^4 - 25 = 0$
8. $x^{-2} - 12x^{-1} + 35 = 0$
9. $z^{-2} - 2z^{-1} - 24 = 0$
10. $16x^3 + 100x = 0$
11. $2x + 9x^{1/2} + 10 = 0$
12. $2x - 3x^{1/2} + 1 = 0$
13. $x^3 - 9x^{3/2} + 8 = 0$
14. $y^3 - 28y^{3/2} + 27 = 0$
15. $2x^{2/3} + 3x^{1/3} - 2 = 0$
16. $2x^{-2/3} + x^{-1/3} - 6 = 0$
17. $x^{-1} + 5x^{-1/2} - 50 = 0$
18. $2x^{-2} - 7x^{-1} + 6 = 0$
19. $3x^{-2} + x^{-1} - 24 = 0$
20. $3y^{-1} + 7y^{-1/2} + 2 = 0$
21. $3x^{5/3} + 15x^{4/3} + 18x = 0$
22. $2x^2 - 2x^{3/2} - 112x = 0$
23. $(3x - 5)^2 + (3x - 5) - 2 = 0$
24. $(x - 1)^2 + (x - 1) - 6 = 0$
25. $(2x + 3)^2 + 7(2x + 3) + 12 = 0$
26. $(5x - 4)^2 + 2(5x - 4) - 8 = 0$
27. $(x - 3)^2 - 2(x - 3) - 15 = 0$
28. $(x + 4)^2 - 2(x + 4) = 3$
29. $(2x + 1)^2 + (2x + 1) = 20$
30. $(x + 7)^2 + 5(x + 7) = 50$
31. $x^4 - 2x^2 + 2 = 0$
32. $x^4 - 4x^2 + 5 = 0$
33. $x^4 - 2x^2 + 10 = 0$
34. $x^4 + 16 = 0$
35. $x^4 - 4x^2 + 7 = 0$
36. $x^4 - 6x^2 + 11 = 0$
37. $x^{-4} - 6x^{-2} + 5 = 0$
38. $3x^{-4} - 5x^{-2} + 2 = 0$
39. $3x^{-4} + 25x^{-2} - 18 = 0$
40. $2x^{-4} + 3x^{-2} - 20 = 0$
41. $\dfrac{2}{4x - 1} + \dfrac{1}{x + 1} = \dfrac{-x}{x + 1}$
42. $\dfrac{3x - 2}{15} - \dfrac{16 - 3x}{x + 6} = \dfrac{x + 3}{5}$
43. $\dfrac{2x}{x - 4} - \dfrac{12x}{x^2 + x - 20} = \dfrac{x - 1}{x + 5}$
44. $\dfrac{x + 1}{x + 3} + \dfrac{2x - 1}{x - 2} = \dfrac{12x - 2}{x^2 + x - 6}$
45. $\dfrac{x + 5}{3x + 2} - \dfrac{4 - 2x}{3x^2 + 8x + 4} = \dfrac{x + 4}{x + 2}$
46. $\dfrac{x + 5}{3x + 4} + \dfrac{16x^2 + 5x + 6}{3x^2 - 2x - 8} = \dfrac{4x}{x - 2}$
47. $\dfrac{4x + 1}{x - 6} - \dfrac{3x^2 - 8x + 20}{2x^2 - 13x + 6} = \dfrac{3x + 7}{2x - 1}$
48. $\dfrac{3x + 2}{x + 3} + \dfrac{22x - 31}{x^2 - x - 12} = \dfrac{3(x + 4)}{x + 3}$
49. $\dfrac{5(x - 10)}{x - 7} = \dfrac{2(x + 1)}{x - 4} + 3$
50. $2 + \dfrac{2 - x}{x + 2} = \dfrac{x - 3}{x + 5}$

## 5.7 Inequalities

**OBJECTIVES**

In this section, you will be learning to:

1. Solve second-, third-, and fourth-degree inequalities.
2. Graph the solutions for second-, third-, and fourth-degree inequalities.
3. Solve inequalities that contain rational expressions.
4. Graph the solutions for inequalities that contain rational expressions.

In section 1.6, we solved first-degree inequalities and in Section 3.6, we solved inequalities involving rational expressions. In this section, we will solve inequalities such as

$$x^2 + 3x + 2 \geq 0, \qquad x^2 - 2x > 8, \qquad \text{and} \qquad x^3 + 4x^2 - 5x < 0$$

that involve quadratic expressions or other expressions that can be factored.

The technique used is based on the same concept that was used in Section 3.6. That is, the sign of the binomial expression $x - a$ changes for $x$ on either side of $a$. If $x > a$, then $x - a$ is positive. If $x < a$, then $x - a$ is negative. The procedure involves the following steps:

1. Arrange the terms so that one side of the inequality is 0.
2. Factor the algebraic expression, if possible, and find the points where each factor is 0. Otherwise, use the quadratic formula to find the points (if any) where the expression is 0.
3. Mark these points on a number line. (Consider these points as endpoints of intervals.)
4. Test one point from each interval to determine the sign of the expression for that interval.
5. The solution consists of those intervals where the test points satisfy the original inequality.

The following examples illustrate the technique.

**EXAMPLES** Solve the following inequalities using a number line and graph each solution set on a number line.

**1.** $x^2 - 2x > 8$

**Solution**

$$x^2 - 2x > 8$$

$$x^2 - 2x - 8 > 0 \quad \text{Add } -8 \text{ to both sides so that one side is 0.}$$

$$(x + 2)(x - 4) > 0 \quad \text{Factor.}$$

Set each factor equal to 0 to find endpoints of intervals.

$$x + 2 = 0 \qquad x - 4 = 0$$

$$x = -2 \qquad x = 4$$

Mark these points on a number line and test one point from the intervals formed.

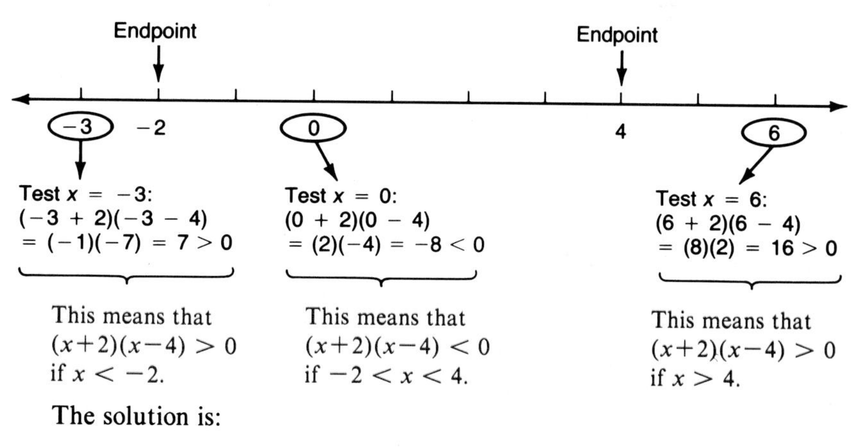


The solution is:

(algebraic notation) $x < -2$ or $x > 4$ or (interval notation) $x$ is in $(-\infty, -2)$ or $(4, +\infty)$

-2 0 4

**2.** $2x^2 + 15 \leq 13x$

**Solution**

$$2x^2 + 15 \leq 13x$$

$$2x^2 - 13x + 15 \leq 0 \quad \text{Add } -13x \text{ to both sides so that one side is 0.}$$

$$(2x - 3)(x - 5) \leq 0 \quad \text{Factor.}$$

Set each factor equal to 0 to locate interval endpoints.

$$2x - 3 = 0 \qquad x - 5 = 0$$

$$x = \frac{3}{2} \qquad x = 5$$

Test one point from each interval formed.

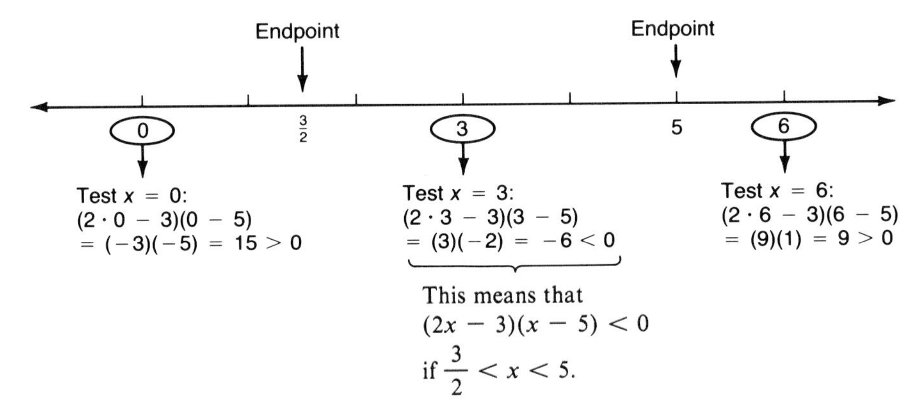


The solution includes both endpoints since the inequality ($\leq$) includes 0:

(algebraic notation) or (interval notation)

$$\frac{3}{2} \leq x \leq 5 \qquad x \text{ is in } \left[\frac{3}{2}, 5\right]$$

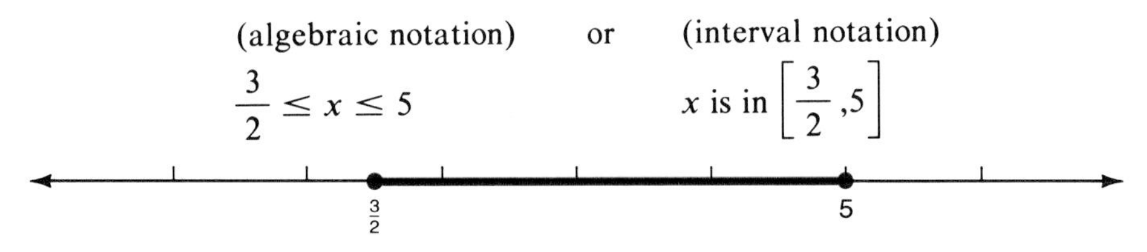


**3.** $x^3 + 4x^2 - 5x < 0$

**Solution**

$$x^3 + 4x^2 - 5x < 0$$

$$x(x^2 + 4x - 5) < 0 \quad \text{Factor.}$$

$$x(x + 5)(x - 1) < 0 \quad \text{There are three factors.}$$

Set each factor equal to 0 to locate endpoints of intervals.

$$x = 0 \qquad x + 5 = 0 \qquad x - 1 = 0$$

$$x = -5 \qquad x = 1$$

Test one point from each interval.

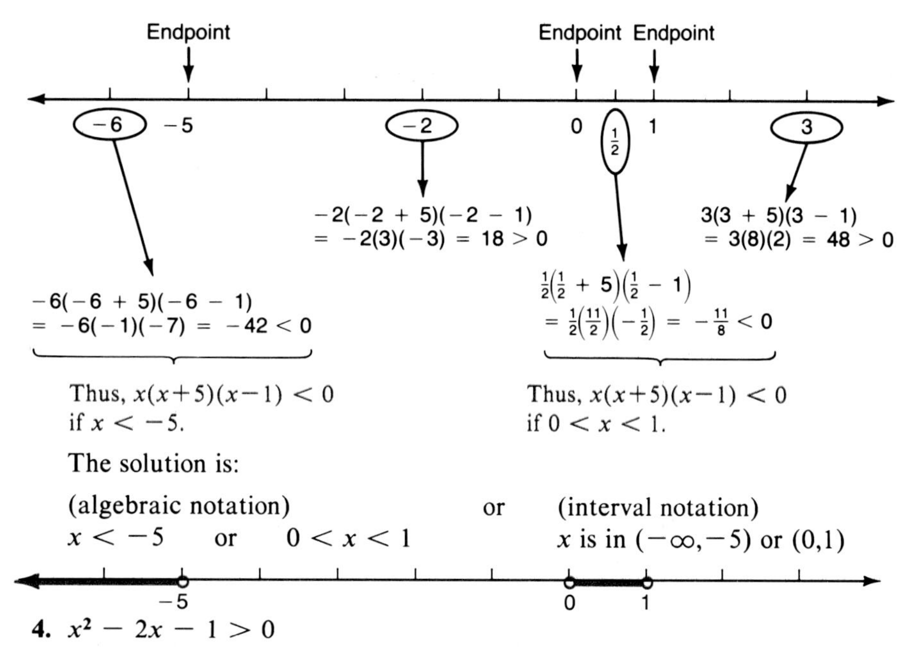


The solution is:

(algebraic notation) or (interval notation)

$x < -5$ or $0 < x < 1$ $\qquad$ $x$ is in $(-\infty, -5)$ or $(0,1)$

**4.** $x^2 - 2x - 1 > 0$

**Solution** The quadratic expression $x^2 - 2x - 1$ will not factor with integer coefficients. Use the quadratic formula and find the roots of the equation $x^2 - 2x - 1 = 0$. Then, use these roots as endpoints for the intervals.

$$x^2 - 2x - 1 = 0$$

$$x = \frac{2 \pm \sqrt{(-2)^2 - 4(1)(-1)}}{2} \qquad \text{Use the quadratic formula.}$$

$$x = \frac{2 \pm \sqrt{4 + 4}}{2}$$

$$x = \frac{2 \pm \sqrt{2}}{2} = 1 \pm \sqrt{2}$$

The endpoints are $x = 1 - \sqrt{2}$ and $x = 1 + \sqrt{2}$.

Test one point from each interval in the expression $x^2 - 2x - 1$. (**Note:** $1 - \sqrt{2} \approx 1 - 1.414 = -0.414$ and $1 + \sqrt{2} \approx 1 + 1.414 = 2.414$.)

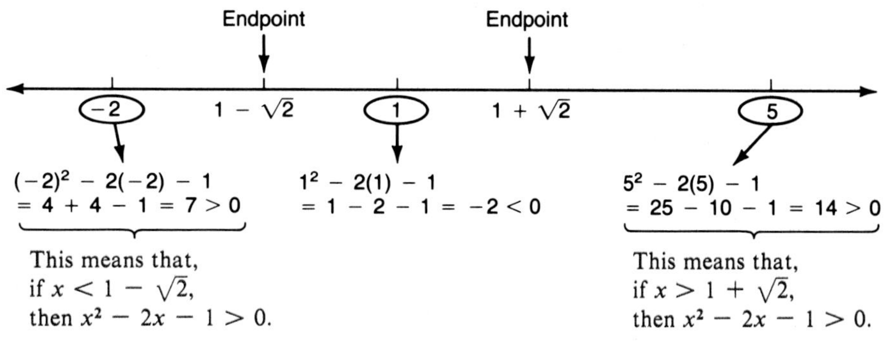

The solution is:

| (algebraic notation) | or | (interval notation) |
|---|---|---|
| $x < 1 - \sqrt{2}$ or $x > 1 + \sqrt{2}$ | | $x$ is in $(-\infty, 1 - \sqrt{2})$ or $(1 + \sqrt{2}, +\infty)$ |

**5.** $x^2 - 2x + 13 > 0$

**Solution** To find where $x^2 - 2x + 13 = 0$, we use the quadratic formula:

$$x = \frac{2 \pm \sqrt{(-2)^2 - 4(1)(13)}}{2} = \frac{2 \pm \sqrt{-48}}{2}$$

$$= \frac{2 \pm 4i\sqrt{3}}{2} = 1 \pm 2i\sqrt{3}$$

Since these values are nonreal, the polynomial is either always positive or always negative for real values of $x$. So, we need to test only one point. If that point satisfies the inequality, then the solution is all real numbers. Otherwise, there is no solution. Test $x = 5$:

$$5^2 - 2(5) + 13 = 28 > 0$$

The solution is all real numbers: $(-\infty, +\infty)$.

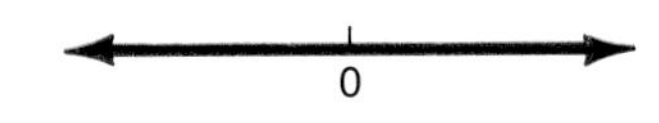


■

## EXERCISES 5.7

Solve the inequalities in Exercises 1–16 and graph each solution set.

**1.** $(x - 6)(x + 2) < 0$
**2.** $(x + 4)(x - 2) > 0$
**3.** $(3x - 2)(x - 5) > 0$
**4.** $(4x + 1)(x + 1) \leq 0$
**5.** $(x + 7)(2x - 5) \geq 0$
**6.** $(x - 3)(5x - 3) \leq 0$
**7.** $(3x + 1)(x + 2) \leq 0$
**8.** $(x - 4)(3x - 8) > 0$
**9.** $x(3x + 4)(x - 5) < 0$
**10.** $(x - 1)(x + 4)(2x + 5) < 0$
**11.** $x^2 + 4x + 4 \leq 0$
**12.** $5x^2 + 4x - 12 > 0$
**13.** $2x^2 > x + 15$
**14.** $6x^2 + x > 2$
**15.** $8x^2 < 10x + 3$
**16.** $2x^2 < x + 10$

Solve the inequalities in Exercises 17–34.

**17.** $2x^2 - 5x + 2 \geq 0$
**18.** $15y^2 - 21y - 18 < 0$
**19.** $6y^2 + 7y < -2$
**20.** $3x^2 + 3 \geq 10x$
**21.** $4z^2 - 20z + 25 > 0$
**22.** $14 + 11x \geq 15x^2$
**23.** $8x^2 + 6x \leq 35$
**24.** $7x < 6x^2 + x^3$
**25.** $x^3 > 2x^2 + 3x$
**26.** $x^3 < 6x^2 - 9x$
**27.** $x^3 > 5x^2 - 4x$
**28.** $4x^2 \leq x^3 + 3x$
**29.** $(x + 2)(x - 2) > 3x$
**30.** $(x + 4)(x - 1) < 2x + 2$
**31.** $x^4 - 5x^2 + 4 > 0$
**32.** $x^4 - 25x^2 + 144 < 0$
**33.** $y^4 - 13y^2 + 36 \leq 0$
**34.** $y^4 - 13y^2 - 48 \geq 0$

Solve the inequalities in Exercises 35–54. You may need to use the quadratic formula to find factors.

**35.** $(x + 1)^2 - 9 \geq 0$
**36.** $(3x - 1)^2 - 16 < 0$
**37.** $(2x - 3)(3x + 2) - (3x + 2) < 0$
**38.** $2(x - 1)(x - 3) > (x - 1)(x - 6)$
**39.** $x^2 + 2x - 4 > 0$
**40.** $x^2 - 8x + 14 < 0$
**41.** $x^2 + 6x + 7 \geq 0$
**42.** $2x^2 + 4x - 3 < 0$
**43.** $3x^2 + 5x + 1 < 0$
**44.** $3x^2 + 8x + 5 \geq 0$
**45.** $2x^3 \leq 7x^2 + 4x$
**46.** $2x^2 > 9x - 8$
**47.** $\dfrac{x^2 - 9}{x} < 0$
**48.** $\dfrac{x^2 + 2x - 3}{x} \leq 0$
**49.** $\dfrac{x^2 - 4x + 3}{x - 2} \geq 0$
**50.** $\dfrac{x^2 + x - 12}{x + 1} > 0$
**51.** $x^2 - 2x + 2 > 0$
**52.** $x^2 + 3x + 3 < 0$
**53.** $2x - 1 > 3x^2$
**54.** $6x - 10 < x^2$

## CHAPTER 5 SUMMARY

### Key Terms and Formulas

An equation of the form

$ax^2 + bx + c = 0$ where $a$, $b$, and $c$ are real numbers and $a \neq 0$

is called a **quadratic equation** or **second-degree equation.** [5.1]

**Factor Theorem** [5.1]
If $x = c$ is a root of a polynomial equation in which one side is 0, then $x - c$ is a factor of the polynomial.

**The Quadratic Formula** [5.3]
For the general quadratic equation

$$ax^2 + bx + c = 0 \qquad (a \neq 0)$$

the solutions are

$$x = \frac{-b \pm \sqrt{b^2 - 4ac}}{2a}$$

| **Discriminant** | **Nature of Solutions** [5.3] |
|---|---|
| $b^2 - 4ac > 0$ | Two real solutions |
| $b^2 - 4ac = 0$ | One real solution, $x = -\dfrac{b}{2a}$ |
| $b^2 - 4ac < 0$ | Two nonreal, complex solutions |

**Pythagorean Theorem** [5.4]
In a right triangle, the square of the hypotenuse is equal to the sum of the squares of the other two sides.

$$a^2 + b^2 = c^2$$

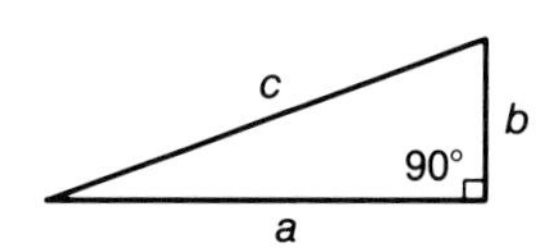


### Procedures

**To Solve a Quadratic Equation by Factoring** [5.1]

1. Add or subtract terms so that one side of the equation is 0.
2. Factor the quadratic expression.
3. Set each factor equal to 0 and solve for $x$.

**To Solve a Quadratic Equation by Completing the Square** [5.2]

1. Divide or multiply so that the coefficient of $x^2$ is 1.
2. Add the constant that completes the square to both sides of the equation.

**Method for Solving Equations with Radicals** [5.5]

**Step 1** Isolate one of the radicals on one side of the equation. (An equation may have more than one radical.)

**Step 2** Raise both sides of the equation to the power corresponding to the index of the radical.

**Step 3** If the equation still contains a radical, repeat Steps 1 and 2.

**Step 4** Solve the equation. (No radicals are left.)

**Step 5** Be sure to check all possible solutions in the original equation and eliminate any extraneous solutions.

Inequalities that involve quadratic expressions or other expressions that can be factored may be solved by testing points in intervals formed by points where the various factors are 0. [5.7]

## CHAPTER 5 REVIEW

Solve Exercises 1–5 by factoring. [5.1]

**1.** $4x^2 - 49 = 0$ **2.** $8x^2 + 14x + 3 = 0$ **3.** $3x^2 + 7x = 0$

**4.** $2x(x + 2) = 3(x + 5)$ **5.** $5x^3 = 7x^2 + 6x$

Write a quadratic equation with the given roots for Exercises 6–9. [5.1]

**6.** $x = 2, x = -\frac{3}{4}$ **7.** $x = 1 \pm \sqrt{5}$ **8.** $x = -3 \pm i\sqrt{2}$ **9.** $y = -2 \pm 4i$

Add the correct constant to complete the square and then factor the trinomial as indicated in Exercises 10–12. [5.2]

**10.** $x^2 + 10x + \_\_\_\_ = (\quad)^2$ **11.** $y^2 - 7y + \_\_\_\_ = (\quad)^2$

**12.** $3y^2 + 4y + \_\_\_\_ = 3(\quad)^2$

Find the discriminant and determine the nature of the solution for each of the quadratic equations in Exercises 13–16. [5.3]

**13.** $2x^2 + 7x - 3 = 0$ **14.** $3x^2 + 12x + 12 = 0$

**15.** $x^2 - 4x + 5 = 0$ **16.** $4x^2 + 9 = 0$

**17.** Determine the values for $k$ so that $kx^2 + x + 4 = 0$ has two real solutions.

**18.** Determine the values for $k$ so that $2x^2 + kx + 9 = 0$ has one real solution.

**19.** Determine the values for $k$ so that $2x^2 + 5x + k = 0$ has two nonreal complex solutions.

Solve Exercises 20–25 by completing the square. [5.2]

**20.** $x^2 + 3x - 2 = 0$ **21.** $x^2 + 3x - 5 = 0$ **22.** $z^2 - 5z + 7 = 0$

**23.** $2x^2 + 4x - 1 = 0$ **24.** $5x^2 - 6x - 3 = 0$ **25.** $2x^2 + 7x + 2 = 0$

Solve the equations in Exercises 26–37. [5.3, 5.5, 5.6]

**26.** $28y^2 + 3y - 18 = 0$ **27.** $4y^2 + 6y - 3 = 0$ **28.** $2x^2 + 5x + 6 = 0$

**29.** $3y^2 + y + 1 = 0$ **30.** $4y^{-2} + 8y^{-1} - 21 = 0$ **31.** $3x^2 - 4x - 8x = 0$

**32.** $(4x + 1) - 2\sqrt{4x + 1} = 15$ **33.** $x^4 = 4x^2 + 12$ **34.** $\sqrt{x + 4} = \sqrt{x} + 1$

**35.** $\sqrt{6x + 1} + \sqrt{x} = \sqrt{10x + 9}$ **36.** $\frac{5x}{2x + 3} = \frac{x}{x - 6}$ **37.** $\frac{2}{3x + 1} - \frac{1}{x + 2} + \frac{1}{x - 3} = 0$

Solve the inequalities in Exercises 38–44. [5.7]

**38.** $4x^2 - 19x - 30 \geq 0$

**39.** $6z^2 - 7z - 5 < 0$

**40.** $x^2 - 4x - 7 < 0$

**41.** $3x^2 - 2x - 2 \geq 0$

**42.** $6x^3 + 13x^2 < 15x$

**43.** $2x^3 + x^2 - 4x < 0$

**44.** $\dfrac{x^2 + 3}{x} \leq 4$

**45.** Find three consecutive even integers such that the square of the first added to the product of the second and third gives a result of 368. [5.4]

**46.** The base of a triangle is 3 centimeters more than twice the altitude. The area is 76 sq cm. Find the base and altitude of the triangle. [5.4]

**47.** Jim leaves Bryan's corner traveling north on his bicycle. One hour later, Beth leaves the same point traveling west. If Jim averages 16 mph and Beth averages 20 mph, when will they be 50 miles apart? [5.4]

**48.** Mr. Williams owns a 16-unit apartment complex. The rent is currently \$200 per month and all units are rented. Each time the rent is increased by \$20, Mr. Williams will lose one tenant. What is the rental rate if he expects to receive \$3360 monthly in rent? [5.4]

**49.** A discount store sells 85 radios per month at \$20 each. They estimate that for each \$1 increase in price, they will sell 3 fewer radios per month. Find the selling price if the sales for last month were \$1738. [5.4]

**50.** An airplane travels 200 mph in calm air. The plane travels 750 miles with the wind and returns against the wind. It takes 2 hours more to make the return trip. Find the speed of the wind. [5.4]

## CHAPTER 5 TEST

Solve Exercises 1–3 by factoring.

**1.** $7x^2 - 6x - 1 = 0$

**2.** $36x^3 = 9x$

**3.** $(x + 3)(2x - 1) = 22$

**4.** Write a quadratic equation having roots $x = 1 + \sqrt{2}$ and $x = 1 - \sqrt{2}$.

**5.** Solve by completing the square: $x^2 + 4x + 1 = 0$

**6.** Use the discriminant to determine the nature of the solutions to $4x^2 + 5x - 3 = 0$.

**7.** Determine the values for $k$ so that $2x^2 - kx + 3 = 0$ has exactly one real solution.

Solve Exercises 8–15 using any method.

**8.** $2x^2 + x + 1 = 0$

**9.** $2x^2 - 3x - 4 = 0$

**10.** $2x^2 + 3 = 4x$

**11.** $\sqrt{x + 8} - 2 = x$

**12.** $\sqrt{2x + 9} = \sqrt{x} + 3$

**13.** $x^4 = 10x^2 - 9$

**14.** $3x^{-2} + x^{-1} - 2 = 0$

**15.** $\dfrac{2x}{x - 3} - \dfrac{2}{x - 2} = 1$

Solve the inequalities in Exercises 16 and 17.

**16.** $\dfrac{x^2 - 4}{x + 1} \geq 0$

**17.** $\dfrac{x^2 + 3x - 10}{x - 4} < 0$

**18.** The difference between two positive numbers is 11. If their product is 126, find the numbers.

**19.** The length of a rectangle is 4 inches more than the width. If the diagonal is 20 inches, find the dimensions.

**20.** Sandy made a business trip to a city 200 miles away and then returned home. Her average speed on the return trip was 10 mph less than her speed going. If the total travel time was 9 hours, find her rate each way.

## CUMULATIVE REVIEW (5)

Solve the equations in Exercises 1 and 2.

**1.** $7(2x - 5) = 5(x + 3) + 4$

**2.** $(2x + 1)(x - 4) = (2x - 3)(x + 6)$

Solve the inequalities in Exercises 3 and 4 and graph the solutions. Write the solutions in interval notation.

**3.** $4(x + 3) - 1 \geq 2(x - 4)$

**4.** $\frac{7}{2}x + 3 \leq x + \frac{13}{2}$

Solve for the indicated variable in Exercises 5 and 6.

**5.** $3x + 2y = 6$ for $y$

**6.** $\frac{3}{4}x + \frac{1}{2}y = 5$ for $y$

Simplify Exercises 7–14. Assume that all variables are positive.

**7.** $(4x^{-3})(2x)^{-2}$

**8.** $\frac{x^{-2}y^4}{x^{-5}y^{-2}}$

**9.** $(27x^{-3}y^{3/4})^{2/3}$

**10.** $\left(\frac{9x^{4/3}}{4y^{2/3}}\right)^{3/2}$

**11.** $\sqrt[3]{-27x^6y^8}$

**12.** $\sqrt[4]{32x^9y^{15}}$

**13.** $\frac{\sqrt{72}}{3} + 5\sqrt{\frac{1}{2}}$

**14.** $\frac{1}{2}\sqrt{\frac{4}{3}} + 3\sqrt{\frac{1}{3}}$

Solve the equations in Exercises 15–18.

**15.** $10x^2 + 11x - 6 = 0$

**16.** $4x^2 + 7x + 1 = 0$

**17.** $\sqrt{x + 5} - 2 = x + 1$

**18.** $8x^{-2} - 2x^{-1} - 3 = 0$

**19.** A car rental agency rents 200 cars per day at a rate of \$30 per day. For each \$1 increase in rate, five fewer cars are rented. If the receipts for one day were \$6125, find the daily rental rate. (**Hint:** Let $x =$ the number of \$1 increases.)

**20.** The area of a rectangle is 520 square meters. The perimeter of the rectangle is 92 meters. Find the dimensions of the rectangle.

# CHAPTER 6 LINEAR EQUATIONS AND INEQUALITIES

## MATHEMATICAL CHALLENGES

How many segments have endpoints that are vertices of a given cube?

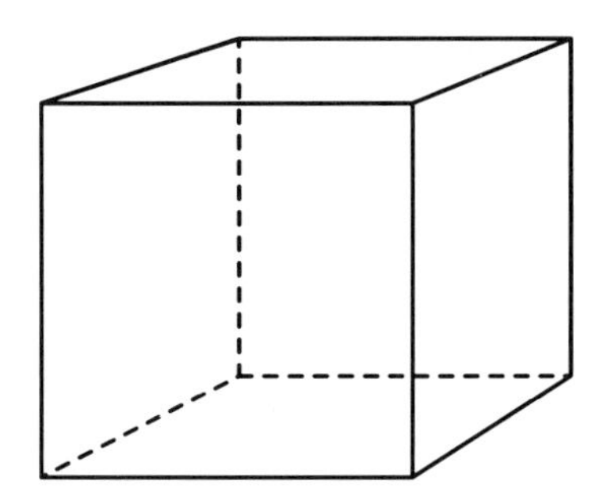

*These twenty-five dots form a square.* Connect twelve of these dots with line segments. Form a perfect cross having five dots inside it and eight dots outside it.

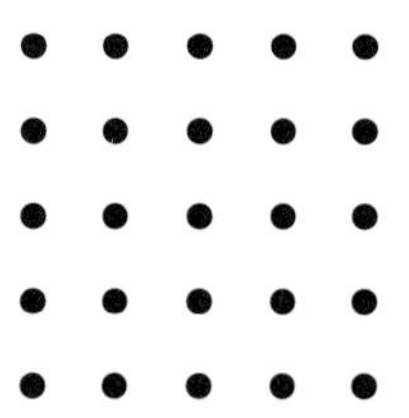

Plant nineteen trees in an arrangement that has 8 rows with five trees along each row.

## CHAPTER OUTLINE

**6.1** Graphing Straight Lines: $Ax + By = C$

**6.2** Slope-intercept Form: $y = mx + b$

**6.3** Point-slope Form: $y - y_1 = m(x - x_1)$

**6.4** Linear Inequalities

This chapter is designed to develop graphing skills with linear equations and linear inequalities in two variables. The graphs of the linear equations are straight lines. The graphs of the linear inequalities are half-planes separated by straight lines. The related skills will be particularly helpful in dealing with applications involving linear equations and linear inequalities that are discussed in Chapter 7.

Three forms of linear equations are the standard form, the slope-intercept form, and the point-slope form. Each form is equally important and useful depending on the information given and the use to be made of the equation. Your algebraic skills should allow you to change from one form to another and to recognize when the same equation is in a different form.

The concept of slope underlies all our work with straight lines. Even in the cases of horizontal lines (with slope 0) and vertical lines (with slope undefined), we can relate to the slopes. Relationships between two or more lines can be discussed in terms of their slopes; parallel lines have the same slope, while perpendicular lines have slopes that are negative reciprocals of each other.

Slope also is a part of our daily lives as in the slope of the roof of a building, the approaching slope of a landing airplane, or the slope of a mountain road.

## 6.1 Graphing Straight Lines: Ax + By = C

### OBJECTIVES

In this section, you will be learning to:

1. Find ordered pairs that satisfy given linear equations.
2. Graph lines in a Cartesian coordinate system by locating two points that satisfy given linear equations.

The equations and inequalities we have discussed so far in this text have involved only one variable. In this chapter, we will discuss equations and inequalities that contain two variables. We begin with the concept of **ordered pairs** and show how these are related to the solutions and graphs of equations.

The pair of numbers (3,9), enclosed in parentheses, is called an **ordered pair.** The first number, 3, is called the **first-coordinate, first component,** or **abscissa.** The second number, 9, is called the **second coordinate, second component,** or **ordinate.** The ordered pair (3,9) is not the same as the ordered pair (9,3) because the order of the numbers is not the same.

Now, consider the equation in two variables

$$2x + 3y = 6$$

The solution to this equation consists of an infinite set of ordered pairs in the form $(x,y)$ where the first component represents the variable $x$ and the second component represents the variable $y$. To find some of these solutions, we form a table by choosing arbitrary values for $x$ and finding the corresponding values for $y$ using the equation. We say that these ordered pairs **satisfy the equation.**

| $x = 0$: | $x = -3$: | $x = 3$: | $x = \frac{1}{2}$: |
|---|---|---|---|
| $2(0) + 3y = 6$ | $2(-3) + 3y = 6$ | $2(3) + 3y = 6$ | $2\left(\frac{1}{2}\right) + 3y = 6$ |
| $0 + 3y = 6$ | $-6 + 3y = 6$ | $6 + 3y = 6$ | $1 + 3y = 6$ |
| $3y = 6$ | $3y = 12$ | $3y = 0$ | $3y = 5$ |
| $y = 2$ | $y = 4$ | $y = 0$ | $y = \frac{5}{3}$ |

| $x$ | 0 | $-3$ | 3 | $\frac{1}{2}$ |
|---|---|---|---|---|
| $y$ | 2 | 4 | 0 | $\frac{5}{3}$ |

Thus, (0,2), $(-3,4)$, (3,0), and $\left(\frac{1}{2}, \frac{5}{3}\right)$ are four ordered pairs that satisfy the equation $2x + 3y = 6$.

***Solution of an Equation in Two Variables***

The **solution** (or **solution set**) of an equation in two variables, $x$ and $y$, consists of all those ordered pairs of real numbers $(x,y)$ that satisfy the equation.

**EXAMPLES**

**1.** Given the equation $3x + y = 9$, find the missing component of each ordered pair so that the ordered pair belongs to the solution set of the equation.

**a.** (2, ) **b.** (0, ) **c.** (6, ) **d.** ( ,0)

**Solution**

**a.** For (2, ), $x = 2$:

$$3(2) + y = 9$$
$$6 + y = 9$$
$$y = 3$$

The ordered pair is (2,3).

**b.** For (0, ), $x = 0$:

$$3(0) + y = 9$$
$$0 + y = 9$$
$$y = 9$$

The ordered pair is (0,9).

**c.** For (6, ), $x = 6$:

$$3(6) + y = 9$$
$$18 + y = 9$$
$$y = -9$$

The ordered pair is $(6,-9)$.

**d.** For ( ,0), $y = 0$:

$$3x + 0 = 9$$
$$3x = 9$$
$$x = 3$$

The ordered pair is (3,0).

**2.** Suppose that $x$ belongs to the set $\left\{0, \frac{2}{3}, 1, 1.6\right\}$. Find the corresponding ordered pairs that satisfy the equation $x + y = 2$.

**Solution** We solve for $y$ to make evaluations easier: $y = 2 - x$. In table form:

| $x$ | $y = 2 - x$ | $(x,y)$ |
|---|---|---|
| 0 | $y = 2 - 0 = 2$ | (0,2) |
| $\frac{2}{3}$ | $y = 2 - \frac{2}{3} = \frac{4}{3}$ | $\left(\frac{2}{3}, \frac{4}{3}\right)$ |
| 1 | $y = 2 - 1 = 1$ | (1,1) |
| 1.6 | $y = 2 - 1.6 = 0.4$ | (1.6, 0.4) |

■

Ordered pairs are graphed as points in a plane using the **Cartesian coordinate system.** This system uses two perpendicular number lines to separate a plane into four **quadrants.** By convention, the horizontal number line is the ***x*-axis** and the vertical number line is the ***y*-axis.** The point (0,0) is called the **origin.** The graphs of several points are shown in Figure 6.1.

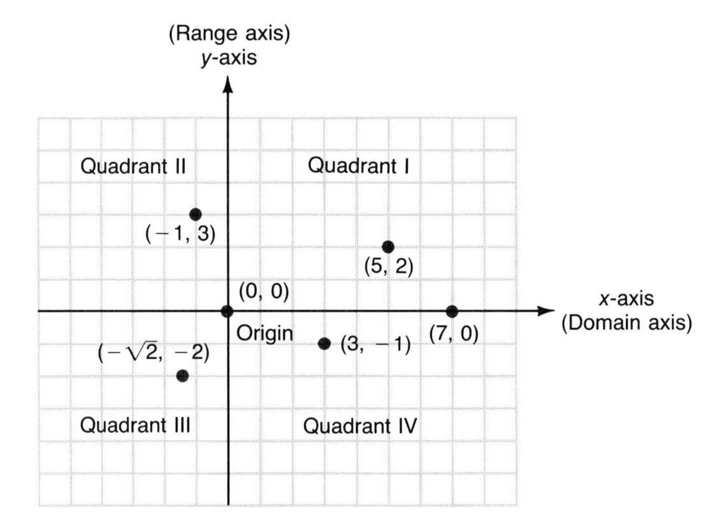


**Figure 6.1**

One important fact that we use when working with the Cartesian coordinate system is that

> **there is a one-to-one correspondence between the points in a plane and ordered pairs of real numbers.**

This fact assures us that there are no holes in the plane unaccounted for by some ordered pair of real numbers.

Just as we use the terms **ordered pair** and **point** (graph of an ordered pair) interchangeably, we use the terms **equation** and **graph of an equation** interchangeably. The equations

$$2x + 3y = 4, \quad y = 7, \quad x = -1, \quad \text{and} \quad y = 3x + 5$$

are called **linear equations** (or first-degree equations in two variables) and their graphs will be straight lines.

***Standard Form of a Linear Equation***

Any equation of the form

$$\boldsymbol{Ax + By = C} \qquad \text{where not both } A \text{ and } B \text{ are equal to 0}$$

is called the **standard form** of a **linear equation** in two variables.

Every straight line corresponds to some linear equation, and the graph of every linear equation is a straight line. Thus, since we know from geometry that two points determine a line, we can find the graph of an equation by locating any two points that satisfy the equation. (We will learn later how to find an equation given two points.)

***To Graph a Linear Equation in Two Variables***

1. Locate any two points that satisfy the equation. Choose values for $x$ or $y$ that lead to simple solutions.
2. Plot these two points on a Cartesian coordinate system.
3. Draw a straight line through these two points. (**Note:** Every point on that line will satisfy the equation.)
4. To check: Locate a third point that satisfies the equation and check to see that it does lie on the line you have drawn.

**EXAMPLES** Graph each of the following linear equations.

**3.** $2x + 3y = 6$

**Solution** Make a table with headings $x$ and $y$ and, whenever possible, choose values for $x$ and $y$ that lead to simple solutions. (In our previous discussion, we found several ordered pairs that satisfy this equation.)

| $x$ | $y$ |
|---|---|
| 0 | 2 |
| −3 | 4 |
| 3 | 0 |

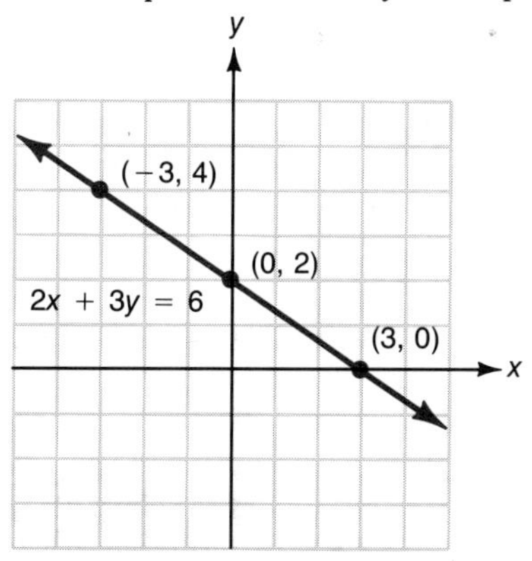

**4.** $x - 2y = 1$

**Solution** In this example, with the coefficient of $x$ as 1, it is easier to choose $y$-values first.

| $x$ | $y$ |
|---|---|
| 1 | 0 |
| 3 | 1 |
| 5 | 2 |

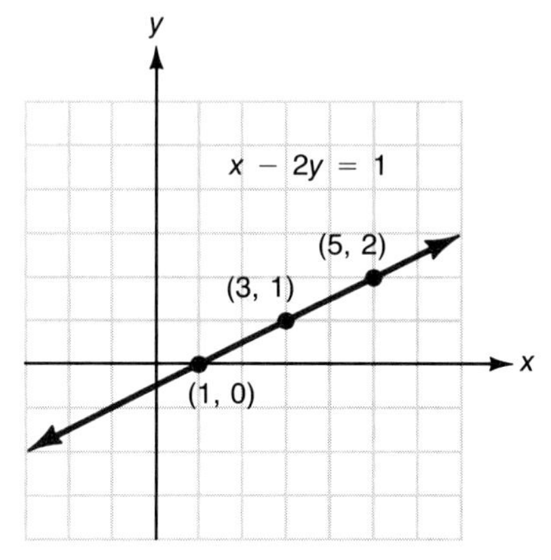


**5.** $y = 2x$

**Solution**

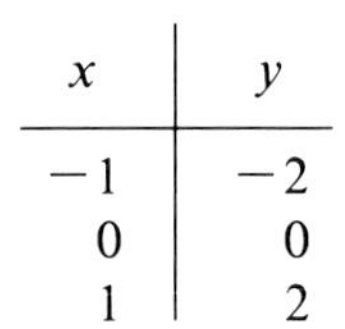

| $x$ | $y$ |
|---|---|
| −1 | −2 |
| 0 | 0 |
| 1 | 2 |

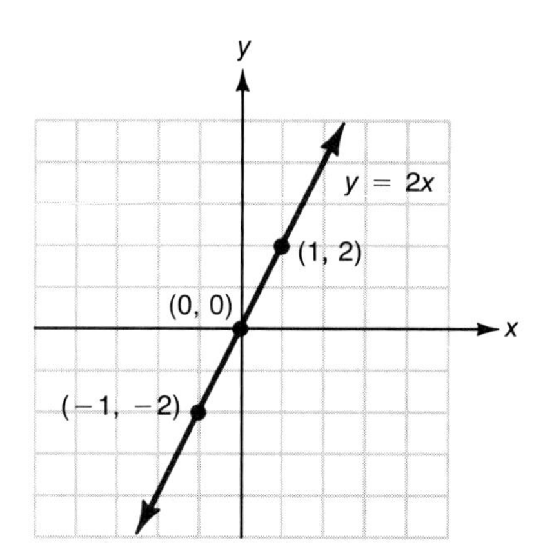


■

**Practice Problems**

1. If $x$ belongs to the set $\{-1, 2, 3\}$, find the corresponding ordered pairs that satisfy the equation $x - 2y = 3$.
2. Find the missing component of each ordered pair so that it belongs to the solution set of the equation $2x + y = 4$: (0, ), ( ,0), ( ,8), (−1, ).
3. Does the ordered pair $\left(1, \frac{3}{2}\right)$ satisfy the equation $3x + 2y = 6$?

**Answers to Practice Problems** **1.** $(-1,-2)$, $\left(2,-\frac{1}{2}\right)$, $(3,0)$ **2.** $(0,4)$, $(2,0)$, $(-2,8)$, $(-1,6)$ **3.** Yes

## EXERCISES 6.1

Find the missing component of each ordered pair so that the ordered pair belongs to the solution set of the equation in Exercises 1–8.

**1.** $2x + y = 5$
**a.** (0, ) **b.** ( ,0) **c.** (−2, ) **d.** ( ,3)

**2.** $x + 2y = 6$
**a.** (0, ) **b.** ( ,0) **c.** (4, ) **d.** ( ,−2)

**3.** $3x - y = 4$
**a.** (0, ) **b.** ( ,0) **c.** (2, ) **d.** ( ,5)

**4.** $x - 3y = 9$
**a.** (0, ) **b.** ( ,0) **c.** (−3, ) **d.** ( ,−1)

**5.** $y = 5 - 2x$
**a.** (0, ) **b.** ( ,0) **c.** (2, ) **d.** ( ,7)

**6.** $y = 5x - 3$
**a.** (0, ) **b.** ( ,0) **c.** (−1, ) **d.** ( ,7)

**7.** $3x - 2y = 6$
**a.** (0, ) **b.** ( ,0) **c.** (−2, ) **d.** ( ,3)

**8.** $5x + 2y = 10$
**a.** (0, ) **b.** ( ,0) **c.** (4, ) **d.** ( ,10)

Graph each of the linear equations in Exercises 9–34.

**9.** $x + y = 3$
**10.** $x + y = 4$
**11.** $y = 3x$
**12.** $2y = x$
**13.** $2x + y = 0$
**14.** $3x + 2y = 0$
**15.** $2x + 3y = 7$
**16.** $4x + 3y = 11$
**17.** $3x - 4y = 12$
**18.** $2x - 5y = 10$
**19.** $-4x + y = 4$
**20.** $-3x + 2y = 6$
**21.** $3y = 2x - 4$
**22.** $4x = 3y + 8$
**23.** $3x + 5y = 6$
**24.** $2x + 7y = -4$
**25.** $2x + 3y = 1$
**26.** $5x - 3y = -1$
**27.** $5x - 2y = 7$
**28.** $3x + 4y = 7$
**29.** $\frac{2}{3}x - y = 4$
**30.** $x + \frac{3}{4}y = 6$
**31.** $2x + \frac{1}{2}y = 3$
**32.** $\frac{2}{5}x - 3y = 5$
**33.** $5x = y + 2$
**34.** $4x = 3y - 5$

# 6.2 Slope-intercept Form: $y = mx + b$

In this section, you will be learning to:

1. Find the slope of a line containing two given points.
2. Graph lines using the slope-intercept method.
3. Graph horizontal lines.
4. Graph vertical lines.

In Section 6.1, we discussed the **standard form** of a linear equation in two variables

$$Ax + By = C$$

In this section, we will analyze linear equations in another form using the concept of **slope.**

The term **slope** is common in phrases such as the slope of a roof, the slope of a road, or the slope of a putting green in golf. For example, a roof that is to be a 7:12 roof is constructed so that for every 7 inches of rise (vertical distance), there are 12 inches of run (horizontal distance). That is, the ratio of the rise to the run is

$$\frac{\text{rise}}{\text{run}} = \frac{7}{12}$$

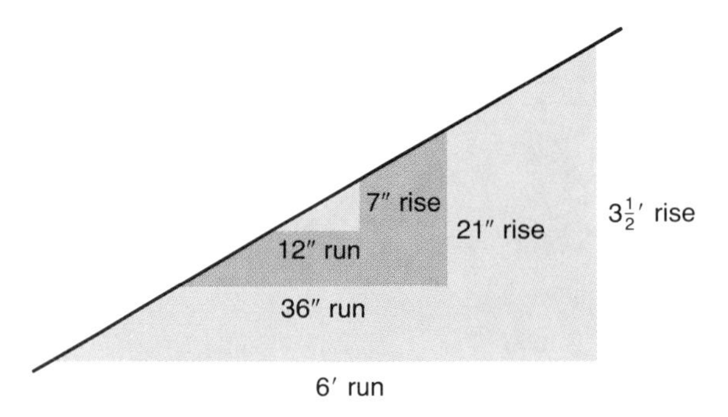


**Figure 6.2**

If we know any two points on a line, $(x_1,y_1)$ and $(x_2,y_2)$, we can calculate the **slope** using the following formula. (**Note:** The letter $m$ is standard notation for representing the slope of a line.)

$$\text{slope} = m = \frac{\text{rise}}{\text{run}} = \frac{y_2 - y_1}{x_2 - x_1} = \frac{y_1 - y_2}{x_1 - x_2} \qquad \text{where } x_1 \neq x_2$$

(**Note:** Be sure to subtract the coordinates in the same order in both the numerator and denominator.)

The slope of a line is illustrated in Figure 6.3.

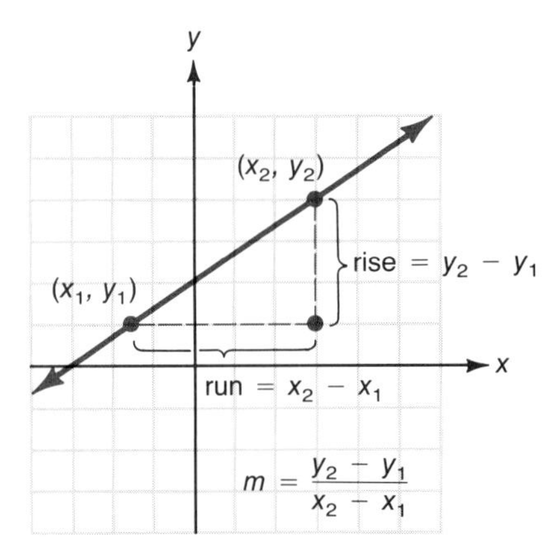


**Figure 6.3**

**EXAMPLES**

1. Find the slope of the line that contains the two points $(-1,2)$ and $(3,5)$ and graph the line.

   **Solution** Using $(x_1,y_1) = (-1,2)$ and $(x_2,y_2) = (3,5)$,

$$\text{slope} = m = \frac{5 - 2}{3 - (-1)} = \frac{3}{4}$$

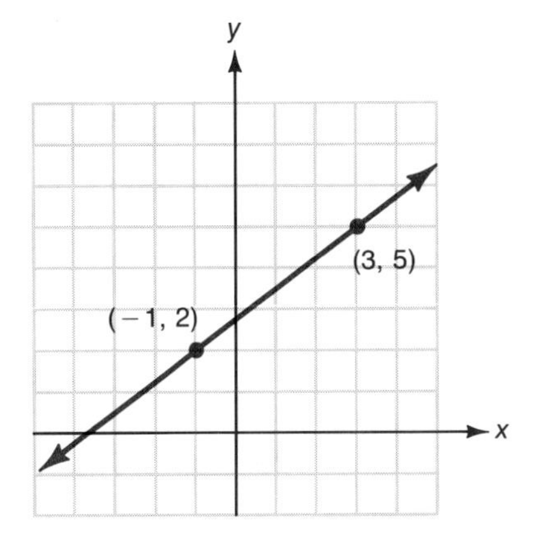


2. Find the slope of the line that contains the two points $(0,1)$ and $(2,6)$ and graph the line.

   **Solution** Using $(x_1,y_1) = (0,1)$ and $(x_2,y_2) = (2,6)$,

$$\text{slope} = m = \frac{1 - 6}{0 - 2} = \frac{-5}{-2} = \frac{5}{2}$$

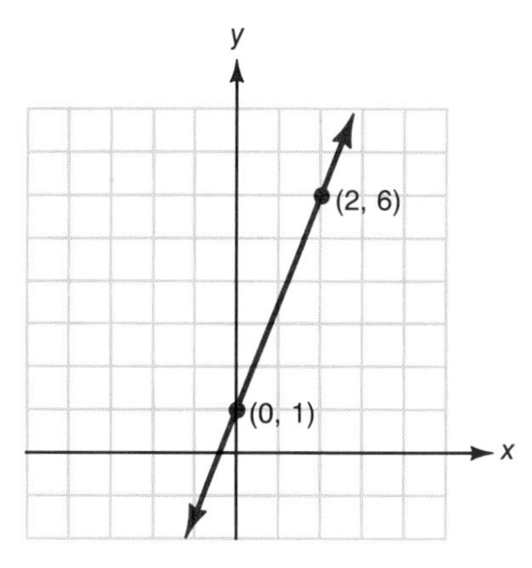

**3.** Find the slope of the line through the points (5,1) and (1,3) and graph the line.

**Solution** Using $(x_1,y_1) = (5,1)$ and $(x_2,y_2) = (1,3)$,

$$\text{slope} = m = \frac{1-3}{5-1} = \frac{-2}{4} = -\frac{1}{2}$$

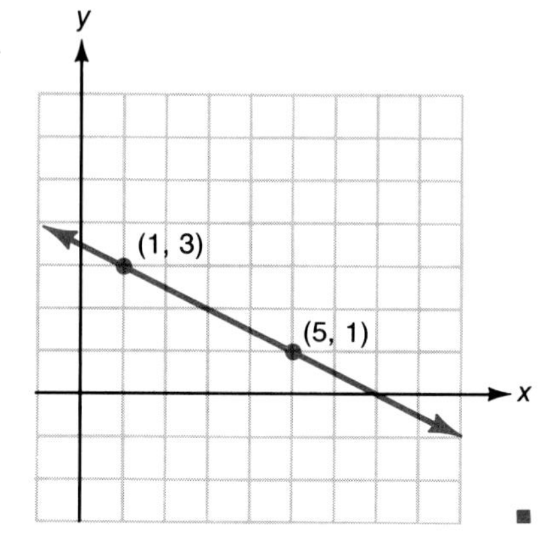


Now we want to find out if there is any connection between the equation of a line and the slope of that line. Consider the equation

$$y = 5x - 7$$

Find two points on the line and calculate the slope:

$(0,-7)$ and $(2,3)$ both satisfy the equation.

$$\text{slope} = m = \frac{-7-3}{0-2} = \frac{-10}{-2} = 5$$

or

$$m = \frac{3-(-7)}{2-0} = \frac{10}{2} = 5$$

In this case, the slope $m$ is the same as the coefficient of $x$ in the equation $y = 5x - 7$. This is not just a coincidence. The following explanation proves a general relationship between the slope and a linear equation in the form

$$y = mx + b$$

Let $(x_1,y_1)$ and $(x_2,y_2)$ be two points on the line. Then,

$$y_1 = mx_1 + b \qquad \text{and} \qquad y_2 = mx_2 + b$$

$$\begin{aligned} \text{slope} &= \frac{y_2 - y_1}{x_2 - x_1} \\ &= \frac{(mx_2 + b) - (mx_1 + b)}{x_2 - x_1} \\ &= \frac{mx_2 + b - mx_1 - b}{x_2 - x_1} = \frac{m\cancel{(x_2 - x_1)}}{\cancel{x_2 - x_1}} = m \end{aligned}$$

If $x = 0$, then $y = m \cdot 0 + b = b$. The point $(0,b)$ is the point where the graph crosses the $y$-axis and is called the ***y*-intercept.** The number $b$ is also called the $y$-intercept with the understanding that $x = 0$ when $y = b$. We are led to the following statement.

***Slope-intercept Form*** **$y = mx + b$** is called the **slope-intercept** form for the equation of a line with slope $m$ and $y$-intercept $b$.

An equation in the standard form

$$Ax + By = C \qquad \text{with } B \neq 0$$

can be written in the slope-intercept form by solving for $y$.

$$\begin{aligned} Ax + By &= C \\ By &= -Ax + C \\ y &= -\frac{A}{B}x + \frac{C}{B} \end{aligned}$$

**EXAMPLES**

**4.** Find the slope, $m$, and $y$-intercept, $b$, of the line $-2x + 3y = 6$ and graph the line.

**Solution** Solving for $y$,

$$\begin{aligned} -2x + 3y &= 6 \\ 3y &= 2x + 6 \\ y &= \frac{2}{3}x + 2 \end{aligned}$$

Thus,

$$m = \frac{2}{3} \qquad \text{and} \qquad b = 2$$

To draw the graph, we could find two points that satisfy the equation, as we did in Section 6.1. However, another technique is to locate the $y$-intercept, then use the slope as $\frac{\text{rise}}{\text{run}}$ to locate a second point on the line.

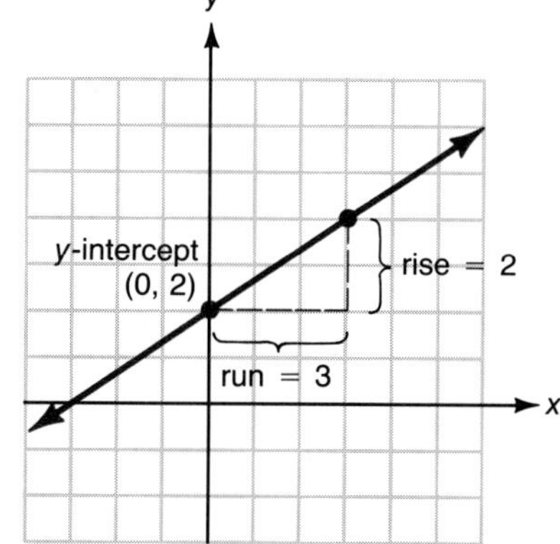


From the $y$-intercept, count 3 units right (run) and 2 units up (rise) to locate a second point on the line.

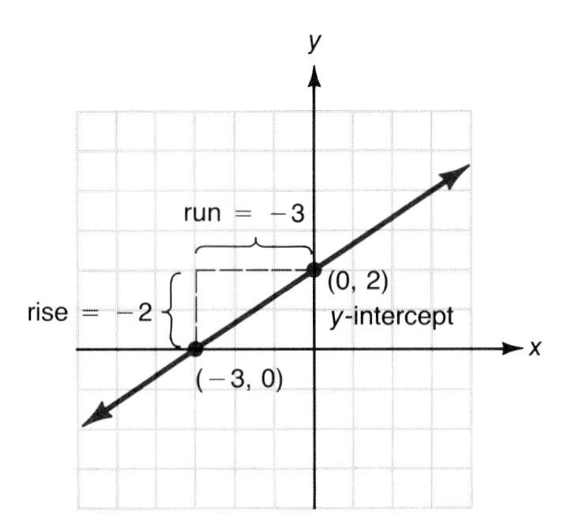


Or we can locate a second point by counting 3 units left (run) and 2 units down (rise). That is, we can interpret the slope $m$ as $m = \frac{-2}{-3}$.

**5.** Find the slope, $m$, and $y$-intercept, $b$, of the line $y = -3x + 2$ and graph the line.

**Solution** The equation is already in the slope-intercept form. Thus,

$$m = -3 \qquad \text{and} \qquad b = 2$$

To draw the graph, we note that the slope

$$m = -3 = \frac{-3}{1} = \frac{3}{-1} = \frac{\text{rise}}{\text{run}}$$

We can treat the rise as $-3$ and the run as 1 or treat the rise as 3 and the run as $-1$.

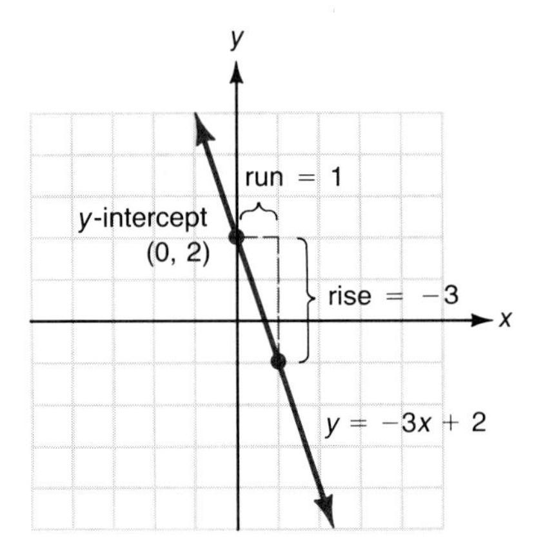


From the $y$-intercept, count 1 unit to the right and 3 units down to locate a second point.

■

Now, suppose that two points on a line have the same $y$-coordinate, such as $(-2,3)$ and $(5,3)$ as in Figure 6.4. Then the slope is

$$m = \frac{3 - 3}{5 - (-2)} = \frac{0}{7} = 0$$

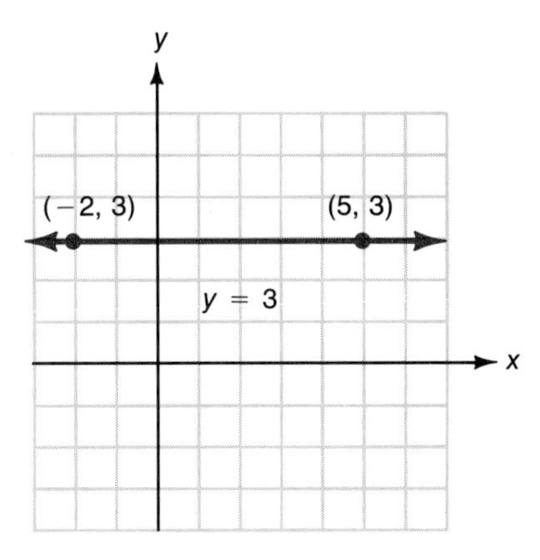


**Figure 6.4**

The line is horizontal and the equation of the line is in the form $y = 0x + b$ or $y = b$. In this case, the $y$-coordinates are all 3, and the equation of the line is $y = 3$.

Next, consider a line with points that have the same $x$-coordinates, such as $(1,3)$ and $(1,-2)$ as in Figure 6.5. Then the slope is

$$m = \frac{-2 - 3}{1 - 1} = \frac{-5}{0} \quad \text{which is undefined.}$$

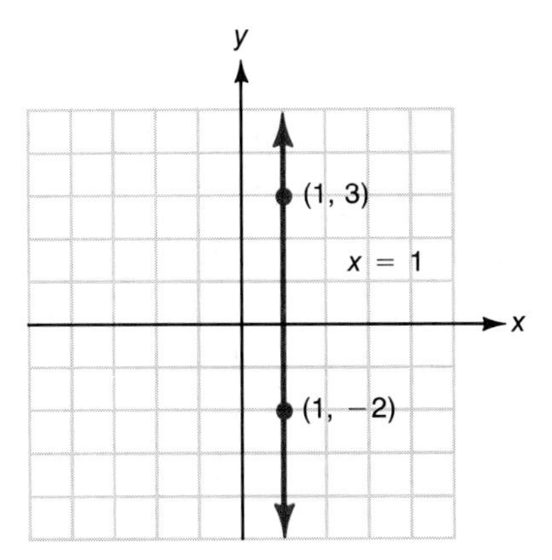


**Figure 6.5**

The line is vertical and the equation of the line is $x = 1$ (in standard form, $x + 0y = 1$).

In general, the equation $y = b$ restricts $y$-values to $b$ with no restrictions on $x$, giving a horizontal line. Similarly, the equation $x = a$ restricts $x$-values to $a$ with no restrictions on $y$, giving a vertical line.

**Horizontal lines** are of the form

$$\boldsymbol{y = b} \quad \text{with slope 0.}$$

**Vertical lines** are of the form

$$\boldsymbol{x = a} \quad \text{with slope undefined.}$$

**EXAMPLES**

**6.** Find the slope and $y$-intercept and **graph the line** $3y + 6 = 0$.

**Solution** $3y + 6 = 0$

$3y = -6$

$y = -2 = 0x - 2$

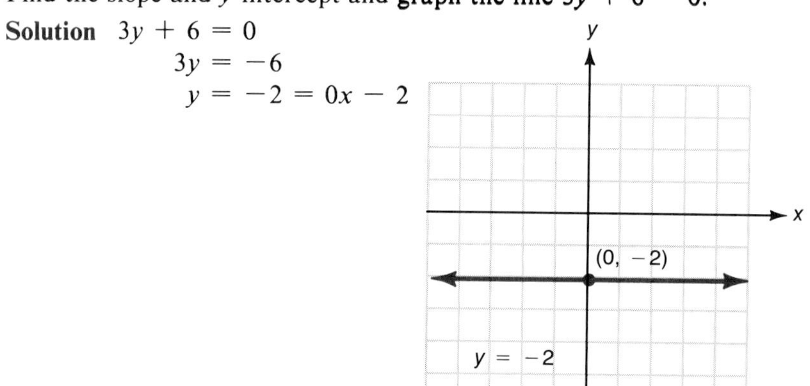


The slope is 0. The $y$-intercept is $-2$ [or $(0,-2)$].

**7.** Graph the line $x = -2$.

**Solution** The line is a vertical line with slope undefined.

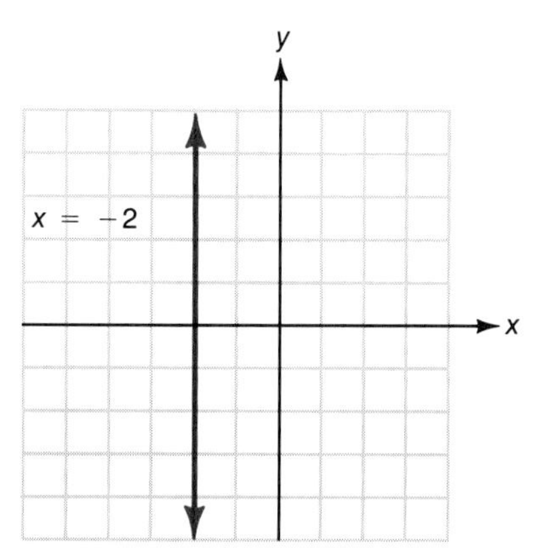


■

**General Comments about Slopes and Graphs of Lines**

1. Lines that have positive slope slant "upward" to the right.
2. Lines that have negative slope slant "downward" to the right.
3. Horizontal lines have slope 0.
4. Vertical lines have undefined slopes.

**Practice Problems**

1. Find the slope of the line through the two points (1,3) and 4,6).
2. What is the slope and the $y$-intercept of the line with the equation $2x + y = 5$?
3. Find the slope and $y$-intercept of the line with equation $5x - 4y = 20$.

## EXERCISES 6.2

Graph the line determined by each pair of points in Exercises 1–12. Then find the slope of the line.

**1.** $(2,4), (1,-1)$ **2.** $(5,1), (3,0)$ **3.** $(-3,7), (4,-1)$ **4.** $(-6,3), (1,2)$

**5.** $(-5,8), (3,8)$ **6.** $(0,0), (-2,-3)$ **7.** $\left(4,\frac{1}{2}\right), (-1,2)$ **8.** $\left(\frac{3}{4},\frac{3}{2}\right), (1,2)$

**9.** $(-2,3), (-2,-1)$ **10.** $(1,-2), (1,4)$ **11.** $\left(\frac{3}{2},\frac{4}{5}\right), \left(-2,\frac{1}{10}\right)$ **12.** $\left(\frac{7}{2},\frac{3}{4}\right), \left(\frac{1}{2},-3\right)$

Write each equation in Exercises 13–40 in slope-intercept form, if possible. Find the slope and the $y$-intercept, then draw the graph.

**13.** $y = 2x - 1$ **14.** $y = 3x - 4$ **15.** $y = 5 - 4x$ **16.** $y = 4 - x$

**17.** $y = \frac{2}{3}x + 2$ **18.** $y = \frac{2}{5}x - 3$ **19.** $x + y = 5$ **20.** $x - 2y = 6$

**21.** $x + 5y = 10$ **22.** $4x + y + 3 = 0$ **23.** $2y - 8 = 0$ **24.** $2x + 7y + 7 = 0$

**25.** $4x + y = 0$ **26.** $3y - 9 = 0$ **27.** $2x = 3y + 6$ **28.** $4x = y + 2$

**29.** $3x + 9 = 0$ **30.** $3x + 6 = 4y$ **31.** $5x - 6y = 10$ **32.** $4x + 7 = 0$

**33.** $5 - 3x = 4y$ **34.** $5x = 11 - 2y$ **35.** $6x + 4y = -7$ **36.** $7x + 2y = 4$

**37.** $6y = 4 + 3x$ **38.** $6x + 5y = -15$ **39.** $5x - 2y + 5 = 0$ **40.** $4x = 3y - 7$

**Answers to Practice Problems** 1. $m = 1$ 2. $m = -2, b = 5$ 3. $m = \frac{5}{4}, b = -5$

## 6.3 Point-slope Form: $y - y_1 = m(x - x_1)$

### OBJECTIVES

In this section, you will be learning to:

1. Write the equations of lines in standard form given any of the following:
   a. the slope and one point,
   b. two points, or
   c. the equation of either a parallel or perpendicular line and one point.
2. Graph lines by using a point and the slope.
3. Graph lines by finding the *x*- and *y*-intercepts.

We have discussed and analyzed linear equations in the **standard form** ($Ax + By = C$) and in the **slope-intercept form** ($y = mx + b$). In this section, we are going to develop a third form for linear equations and discuss parallel lines and perpendicular lines.

Suppose that a point (8,3) and a slope $m = -\frac{3}{4}$ are given. We want to find the equation of the line through the given point with the given slope. If $(x,y)$ represents a point on the line other than (8,3), then using the formula for slope, we have

$$\frac{y-3}{x-8} = -\frac{3}{4} \qquad \frac{y-y_1}{x-x_1} = m$$

or $$y - 3 = -\frac{3}{4}(x-8) \qquad y - y_1 = m(x - x_1)$$

or $$y - 3 = -\frac{3}{4}x + 6$$

or $$y = -\frac{3}{4}x + 9 \qquad \text{Slope-intercept form}$$

or $$3x + 4y = 36 \qquad \text{Standard form}$$

**Point-slope Form**

$y - y_1 = m(x - x_1)$ is called the **point-slope** form for the equation of a line that contains point $(x_1,y_1)$ and has slope $m$.

### EXAMPLES

1. Find the equation of the line containing the two points $(-1,2)$ and $(4,-2)$.

   **Solution** First find the slope.

$$m = \frac{2-(-2)}{-1-4}$$
$$= \frac{4}{-5}$$
$$= -\frac{4}{5}$$

Now, use one of the line's points and the point-slope form for the equation of a line [(−1,2) is used here].

$$y - 2 = -\frac{4}{5}[x - (-1)] \quad \text{Point-slope form}$$

$$y - 2 = -\frac{4}{5}x - \frac{4}{5}$$

$$y = -\frac{4}{5}x - \frac{4}{5} + 2$$

$$y = -\frac{4}{5}x + \frac{6}{5} \quad \text{Slope-intercept form}$$

or $\quad 4x + 5y = 6 \quad$ Standard form

**2.** Find the equation of the line with $m = -\frac{1}{2}$ and passing through the point (2,3). Graph the line using the point and the slope.

**Solution** Substitute into the point-slope form:

$$y - y_1 = m(x - x_1)$$

$$y - 3 = -\frac{1}{2}(x - 2) \quad \text{Point-slope form}$$

$$y = -\frac{1}{2}x + 4 \quad \text{Slope-intercept form}$$

or $\quad x + 2y = 8 \quad$ Standard form

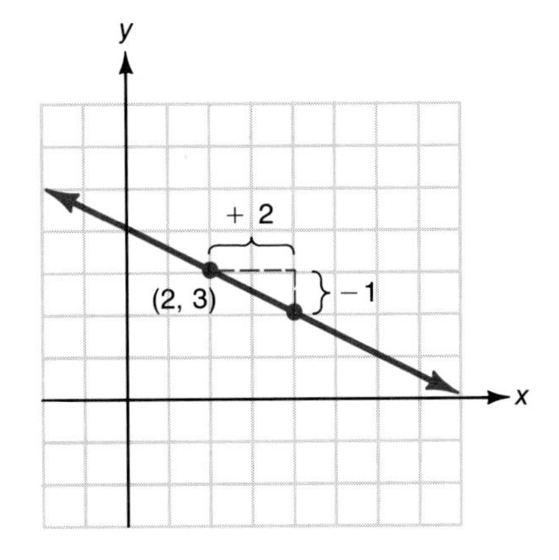


The point two units right and one unit down from (2,3) will be on the line because the slope is $-\frac{1}{2}$.

With a negative slope, either the run is positive and the rise is negative as was just shown, or the run is negative and the rise is positive. In either case, the line is the same.

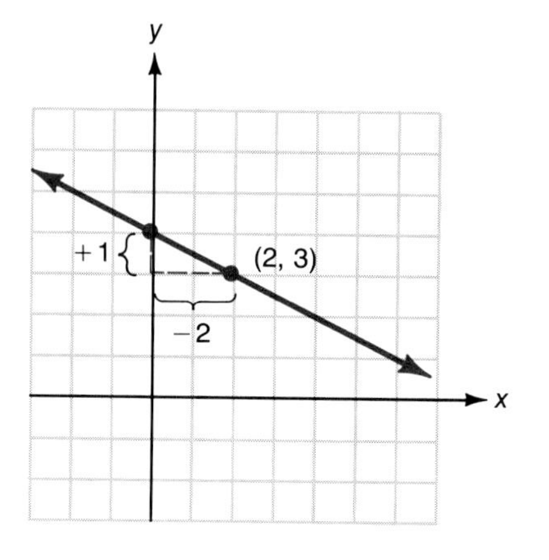


The point two units left and one unit up from (2,3) is on the line because the slope is $-\frac{1}{2}$.

**3.** Graph the line $3x - 2y = 12$ by finding the $x$-intercept and the $y$-intercept and then drawing the line through these two points.

**Solution** $3x - 2y = 12$

| | |
|---|---|
| $0 - 2y = 12$<br>$y = -6$ | The $y$-intercept occurs where $x = 0$.<br>$(0,-6)$ is the $y$-intercept. |
| $3x - 0 = 12$<br>$x = 4$ | The $x$-intercept occurs where $y = 0$.<br>$(4,0)$ is the $x$-intercept. |

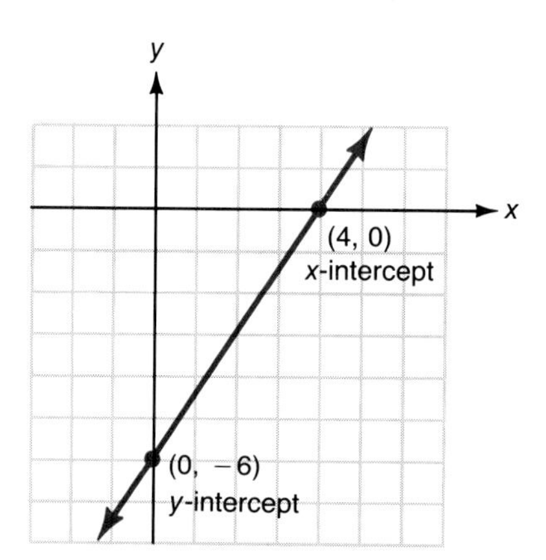


■

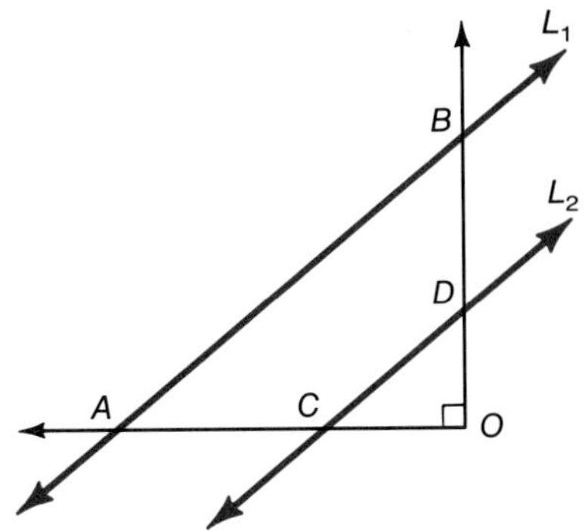
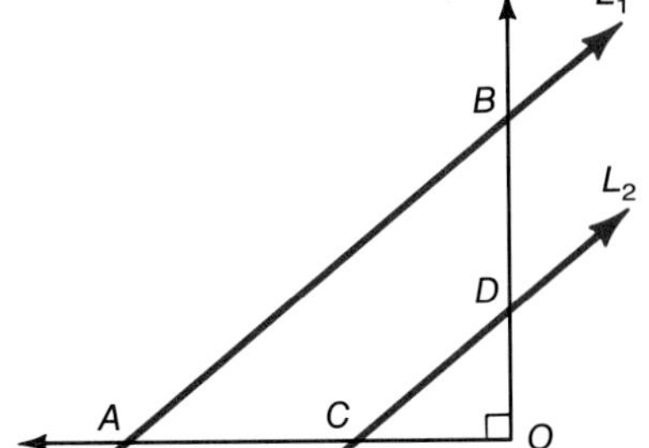


Figure 6.6

Now, consider two nonvertical parallel lines, $L_1$ and $L_2$, with slopes $m_1$ and $m_2$, respectively, as shown in Figure 6.6. Since the lines are parallel, the two triangles $AOB$ and $COD$ are similar triangles (from plane geometry). Therefore, the corresponding sides are proportional; that is,

$$\frac{BO}{DO} = \frac{AO}{CO} \quad \text{or equivalently} \quad \frac{BO}{AO} = \frac{DO}{CO}$$

But,

$$\frac{BO}{AO} = m_1 \quad \text{and} \quad \frac{DO}{CO} = m_2$$

Thus, $m_1 = m_2$, and **parallel lines have the same slope.**

It can also be shown (using plane geometry) that lines with the same slope are parallel.

In Figure 6.7, the two lines $L_1$ and $L_2$ with slopes $m_1$ and $m_2$, respectively, are **perpendicular.** Using the Pythagorean Theorem three times, we find the following relationships.

$$(AO)^2 = (m_1)^2 + 1^2$$
$$(BO)^2 = (m_2)^2 + 1^2$$

and
$$(AB)^2 = (AO)^2 + (BO)^2$$

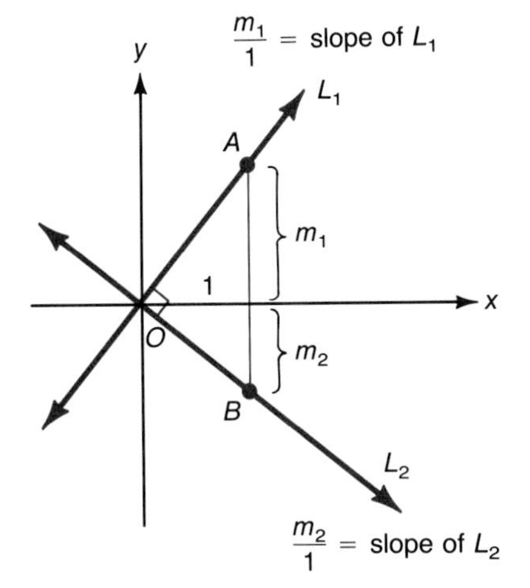


But, $AB = m_1 - m_2$ (since $m_2$ is negative). So,

$$(m_1 - m_2)^2 = (AO)^2 + (BO)^2 = (m_1)^2 + 1^2 + (m_2)^2 + 1^2$$
$$m_1^2 - 2m_1m_2 + (m_2)^2 = (m_1)^2 + (m_2)^2 + 2$$
$$-2m_1m_2 = 2$$
$$m_1m_2 = -1 \quad \text{or} \quad m_2 = \frac{-1}{m_1}$$

Thus, except for horizontal and vertical lines, **perpendicular lines have slopes that are negative reciprocals of each other.**

The lines $y = 2x + 1$ and $y = 2x - 3$ are **parallel** because they have the same slope. The lines $y = \frac{2}{3}x + 1$ and $y = -\frac{3}{2}x - 2$ are **perpendicular** because their slopes are negative reciprocals of each other. Or, the product of their slopes is $-1$: $\left[\frac{2}{3}\left(-\frac{3}{2}\right) = -1\right]$ (Figure 6.8).

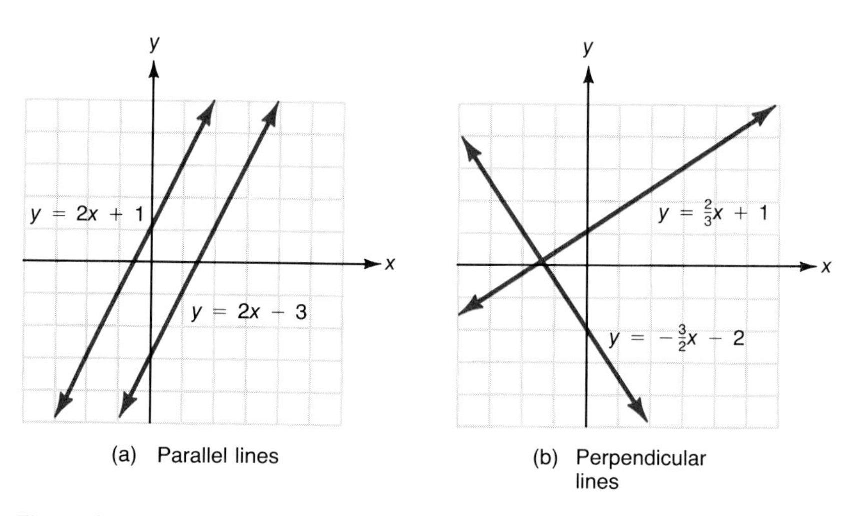


(a) Parallel lines

(b) Perpendicular lines

**Figure 6.8**

**EXAMPLES** Graph the following pairs of lines and state the slope of each line.

**4.** $y = -1$
$x = \sqrt{2}$

**Solution**

**5.** $2y = x + 4$
$2y - x = 10$

**Solution**

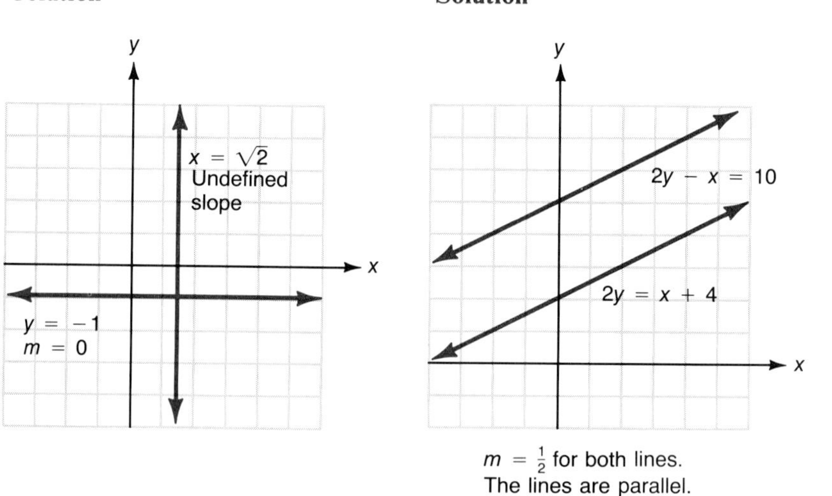


$m = \frac{1}{2}$ for both lines.
The lines are parallel.

**6.** $3x - 4y = 8$
$3y = -4x + 3$

**Solution**

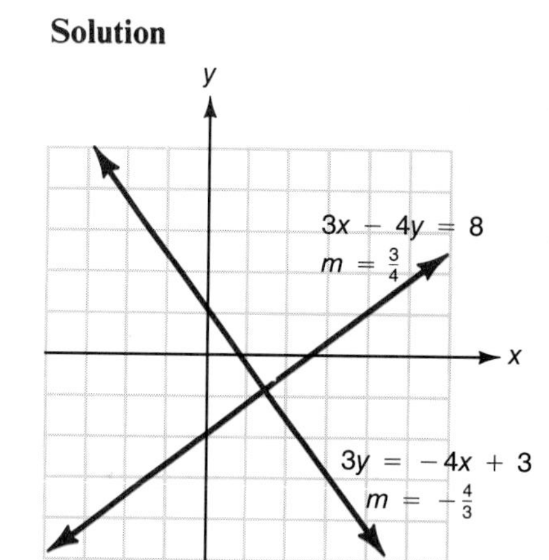


The lines are perpendicular.
$-\frac{4}{3} \cdot \frac{3}{4} = -1$

■

The discussion of straight lines can be summarized as follows:

1. $Ax + By = C$ — Standard form
2. $m = \dfrac{y_2 - y_1}{x_2 - x_1} = \dfrac{y_1 - y_2}{x_1 - x_2}$ — Slope of a line
3. $y = mx + b$ — Slope-intercept form
4. $y - y_1 = m(x - x_1)$ — Point-slope form
5. $y = b$ — Horizontal line, slope 0
6. $x = a$ — Vertical line, undefined slope
7. Parallel lines have the same slope ($m_1 = m_2$).
8. Perpendicular lines have slopes that are negative reciprocals of each other $\left(m_2 = \dfrac{-1}{m_1} \text{ or } m_1m_2 = -1\right)$.

**EXAMPLE**

**7.** Find the equation in standard form for the line parallel to the line $5x + 3y = 1$ and passing through the point (2,3).

**Solution** First solve for $y$ to find the slope.

$$5x + 3y = 1$$
$$3y = -5x + 1$$
$$y = -\frac{5}{3}x + \frac{1}{3}$$

Now use the point-slope form with $m = -\frac{5}{3}$ and $(x_1,y_1) = (2,3)$.

$$y - y_1 = m(x - x_1)$$

$$y - 3 = -\frac{5}{3}(x - 2) \quad \text{Point-slope form}$$

$$3(y - 3) = -5(x - 2) \quad \text{Multiply both sides by 3.}$$

$$3y - 9 = -5x + 10$$

$$5x + 3y = 19 \quad \text{Standard form}$$

■

**Practice Problems**

Find a linear equation in standard form that satisfies the given conditions.

1. Passing through $(4,-1)$ with $m = 2$
2. Parallel to $y = -3x + 4$ and passing through $(-1,5)$
3. Perpendicular to $2x + y = 1$ and passing through $(0,0)$
4. Passing through the two points $(6,-2)$ and $(2,0)$

## EXERCISES 6.3

Given the slope and a point on the line or two points on the line, find an equation in standard form for the lines in Exercises 1–15. Graph each line.

**1.** $m = -2$; $(-2,1)$ **2.** $m = 3$; $(3,4)$ **3.** $(-5,2)$; $(3,6)$ **4.** $(-3,4)$; $(2,1)$

**5.** $m = -\frac{1}{3}$; $(5,-1)$ **6.** $m = \frac{3}{4}$; $\left(0,\frac{1}{2}\right)$ **7.** $(4,2)$; $(4,-3)$ **8.** $(5,2)$; $(1,-3)$

**9.** $m = 0$; $(2,3)$ **10.** $m = -\frac{5}{7}$; $\left(-2,\frac{1}{2}\right)$ **11.** $(-2,7)$; $(3,1)$ **12.** $(2,-5)$; $(4,-5)$

**13.** $\left(\frac{5}{2},0\right)$; $\left(-2,\frac{1}{3}\right)$ **14.** $m = 0$; $(-3,-1)$ **15.** $m = -\frac{4}{3}$; $\left(\frac{2}{3},1\right)$

Graph each line in Exercises 16–23 by finding the $x$-intercept and the $y$-intercept.

**16.** $2x + y = 4$ **17.** $2x + y = 6$ **18.** $3x - 2y = 6$ **19.** $2x - 3y = 6$

**20.** $-2x + 5y = 10$ **21.** $-3x + 5y = 15$ **22.** $3x + 4y = 9$ **23.** $3x - 4y = 9$

Find an equation in standard form for each line satisfying the conditions in Exercises 24–38.

**24.** Parallel to $3x + y = 5$ and passing through $(2,1)$

**25.** Parallel to $2x + 4y = 9$ and passing through $(1,6)$

**26.** Parallel to $7x - 3y = 1$ and passing through $(1,0)$

**27.** Parallel to $5x = 7 + y$ and passing through $(-1,-3)$

**Answers to Practice Problems** 1. $2x - y = 9$ 2. $3x + y = 2$ 3. $x - 2y = 0$ 4. $x + 2y = 2$

**28.** Parallel to the $x$-axis and passing through $(-1,3)$

**29.** Parallel to the $y$-axis and passing through $(2,-4)$

**30.** Perpendicular to $4x + 3y = 1$ and passing through $(2,2)$

**31.** Perpendicular to $x - 3y + 4 = 0$ and passing through $(4,-1)$

**32.** Perpendicular to $5x - 2y - 4 = 0$ and passing through $(-3,5)$

**33.** Perpendicular to $8 - 3x - 2y = 0$ and passing through $(-4,-2)$

**34.** Perpendicular to $3x - y = 4$ and passing through the origin

**35.** Perpendicular to $2x - y = 7$ and having the same $y$-intercept as $x - 3y = 6$

**36.** Perpendicular to $3x - 2y = 4$ and having the same $y$-intercept as $5x + 4y = 12$

**37.** Show that the points $(-2,4)$, $(0,0)$, $(4,7)$, and $(6,3)$ are the vertices of a rectangle. (A rectangle is a quadrilateral with opposite sides parallel and adjacent sides perpendicular.)

**38.** Show that the points $(0,-1)$, $(3,-4)$, $(6,3)$, and $(9,0)$ are the vertices of a parallelogram. (A parallelogram is a quadrilateral with opposite sides parallel.)

## 6.4 Linear Inequalities in Two Variables

**OBJECTIVE**

In this section, you will be learning to graph linear inequalities.

A straight line separates a plane into two **half-planes.** The points on one side of the line are in one of the half-planes, and the points on the other side of the line are in the other half-plane. The line itself is called the **boundary line.** If the boundary line is included with a half-plane, then the half-plane is said to be **closed.** If the boundary line is not included, then the half-plane is said to be **open.** (Note the similarity between the terminology for open and closed intervals.)

For example, all the points that satisfy the linear inequality

$$5x - 3y < 15$$

lie in an open half-plane on one side of the line

$$5x - 3y = 15$$

To decide which side of the line is the graph of the solution set, choose any **one test-point** that is obviously on one side of the line. If this point satisfies the inequality, graph the solution set by shading all points on that side of the line. Otherwise, shade the points on the other side of the line (Figure 6.9).

***To Graph a Linear Inequality in Two Variables***

1. Graph the boundary line (dashed if the inequality is $<$ or $>$, solid if the inequality is $\leq$ or $\geq$).
2. Test any point obviously on one side of the line.
3. If the test-point satisfies the inequality, shade the half-plane on that side of the line. Otherwise, shade the other half-plane.

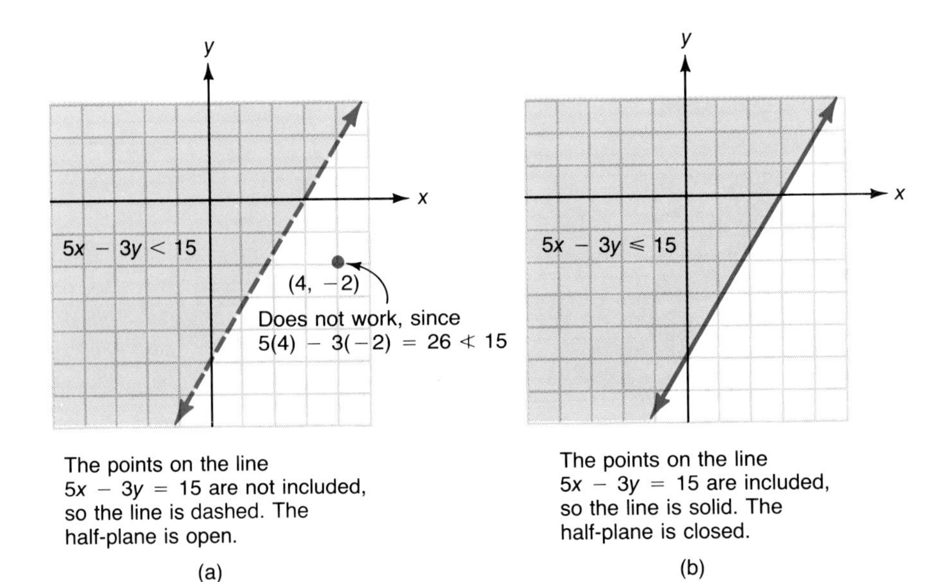


The points on the line $5x - 3y = 15$ are not included, so the line is dashed. The half-plane is open.

(a)

The points on the line $5x - 3y = 15$ are included, so the line is solid. The half-plane is closed.

(b)

**Figure 6.9**

**EXAMPLES** Graph the following linear inequalities.

**1.** Graph the half-plane that satisfies the inequality $2x + y \leq 6$.

**Solution**

**Step 1** Graph the line $2x + y = 6$ as a solid line.

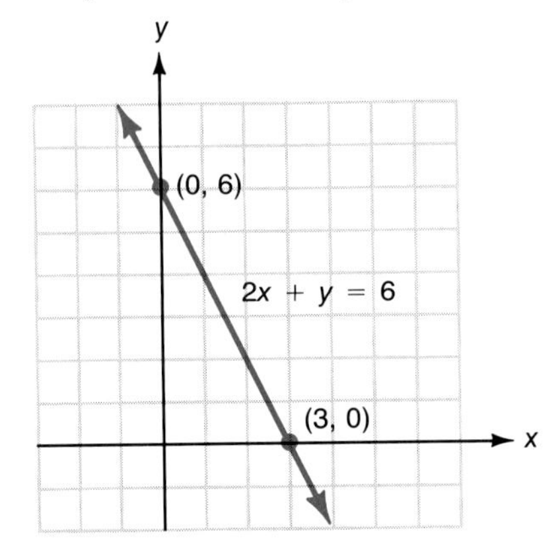

**Step 2** Test (0,0):
$2 \cdot 0 + 0 \leq 6$
$0 \leq 6$
This is a true statement.

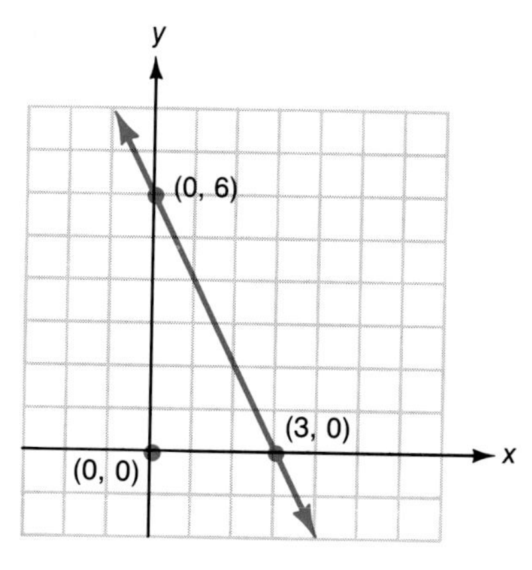


**Step 3** Shade the points on the same side as (0,0).

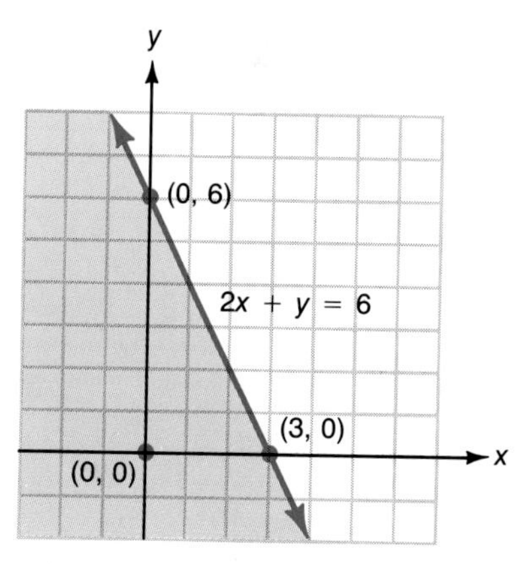


**2.** Graph the solution set to the inequality $y > 2x$.

**Solution**

**Step 1** Graph the line $y = 2x$ as a dashed line.

**Step 2** Test (3,0):
$0 > 2 \cdot 3$
$0 > 6$
This is a false statement.

**Step 3** Shade the points on the opposite side as (3,0).

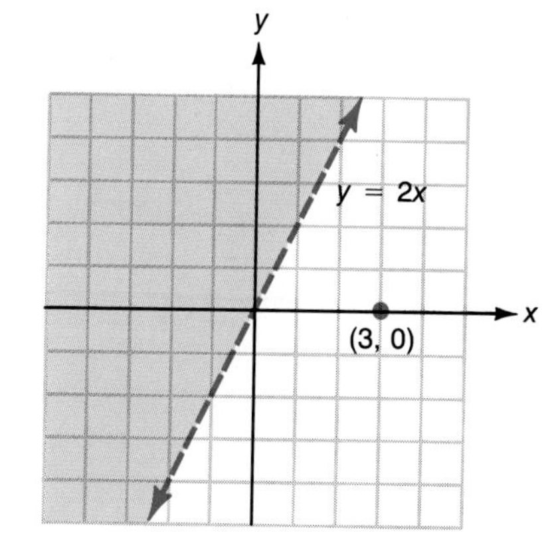

3. Graph the half-plane that satisfies the inequality $y > 1$.

**Solution**

**Step 1** Graph the horizontal line $y = 1$ as a dashed line.

**Step 2** Test (0,0):
$0 \geq 1$
This is false.

**Step 3** Shade the half-plane above the line.

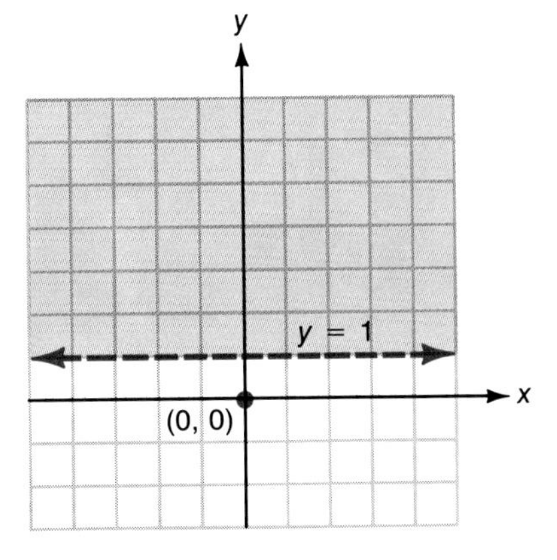


4. Graph the solution set to the inequality $x \leq 0$.

**Solution**

**Step 1** Graph the line $x = 0$ as a solid line. This is the $y$-axis.

**Step 2** Test $(-2,1)$:
$-2 \leq 0$
This is true.

**Step 3** Shade the half-plane on the same side as $(-2,1)$. These are all the points when the $x$-coordinate is 0 or negative.

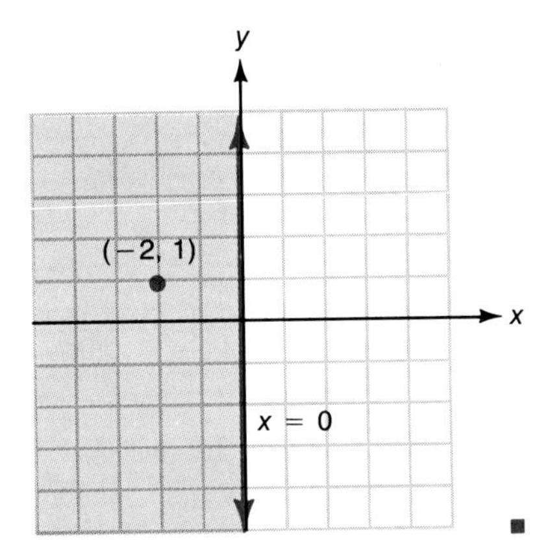


■

***Practice Problems***

1. Which of the following points satisfy the inequality $x + y < 3$: (2,1), $\left(\frac{1}{2},3\right)$, (0,5), $(-5,2)$?
2. Which of the following points satisfy the inequality $x - 2y \geq 0$: (2,1), (4,2), (1,3), (3,1)?
3. Which of the following points satisfy the inequality $x - 3 < 0$: (1,0), (0,1), $(4,-1)$, (2,3)?

**Answers to Practice Problems** **1.** $(-5,2)$ **2.** (2,1), (4,2), (3,1) **3.** (1,0), (0,1), (2,3)

## EXERCISES 6.4

Graph the linear inequalities in Exercises 1–30.

1. $x + y \leq 7$
2. $x - y > -2$
3. $x - y > 4$
4. $x + y \leq 6$
5. $y < 4x$
6. $y < -2x$
7. $y \geq -3x$
8. $y > x$
9. $x - 2y > 5$
10. $x + 3y \leq 7$
11. $4x + y \geq 3$
12. $5x - y < 4$
13. $y \leq 5 - 3x$
14. $y \geq 8 - 2x$
15. $2y - x \leq 0$
16. $x + y > 0$
17. $x + 4 \geq 0$
18. $x - 5 \leq 0$
19. $y \geq -2$
20. $y + 3 < 0$
21. $4x + 3y < 8$
22. $3x < 2y - 4$
23. $3y > 4x + 6$
24. $5x < 2y - 5$
25. $x + 3y < 7$
26. $3x + 4y > 11$
27. $\frac{1}{2}x - y > 1$
28. $\frac{1}{3}x + y \geq 3$
29. $\frac{2}{3}x + y \geq 4$
30. $2x - \frac{4}{3}y > 8$

## CHAPTER 6 SUMMARY

### Key Terms and Formulas

In an **ordered pair,** such as $(x,y)$, the first number is called the **first coordinate, first component,** or **abscissa.** The second number is called the **second coordinate, second component,** or **ordinate.** [6.1]

The **solution** (or **solution set**) of an equation in two variables, $x$ and $y$, consists of all those ordered pairs of real numbers $(x,y)$ that satisfy the equation. [6.1]

The **Cartesian coordinate system** uses two perpendicular lines, called **axes,** to separate a plane into four **quadrants.** The system is based on the fact that **there is a one-to-one correspondence between the points in a plane and ordered pairs of real numbers.** [6.1]

The discussion of straight lines can be summarized as follows: [6.1, 6.2, 6.3]

1. $Ax + By = C$ — Standard form
2. $m = \dfrac{y_2 - y_1}{x_2 - x_1} = \dfrac{y_1 - y_2}{x_1 - x_2}$ — Slope of a line $(x_1 \neq x_2)$
3. $y = mx + b$ — Slope-intercept form
4. $y - y_1 = m(x - x_1)$ — Point-slope form
5. $y = b$ — Horizontal line, slope 0
6. $x = a$ — Vertical line, undefined slope
7. Parallel lines have the same slope $(m_1 = m_2)$.
8. Perpendicular lines have slopes that are negative reciprocals of each other $\left(m_2 = \dfrac{-1}{m_1} \text{ or } m_1m_2 = -1\right)$.

A straight line separates a plane into two **half-planes.** The line is called the **boundary line.** A **closed half-plane** includes its boundary line. An **open half-plane** does not include its boundary line. [6.4]

**To Graph a Linear Inequality in Two Variables** [6.4]

1. Graph the boundary line (dashed if the inequality is $<$ or $>$, solid if the inequality is $\leq$ or $\geq$).
2. Test any point obviously on one side of the line.
3. If the test-point satisfies the inequality, shade the half-plane on that side of the line. Otherwise, shade the other half-plane.

## CHAPTER 6 REVIEW

Find the missing component of each ordered pair so that the ordered pair belongs to the solution set of the equation in Exercises 1–4. [6.1]

**1.** $2x - y = 4$
**a.** (0, )
**b.** ( ,0)
**c.** (1, )
**d.** ( ,−2)

**2.** $x + 3y = 6$
**a.** (0, )
**b.** ( ,0)
**c.** (2, )
**d.** ( ,−1)

**3.** $3x + y = 9$
**a.** (0, )
**b.** ( ,0)
**c.** (−2, )
**d.** ( ,−2)

**4.** $4x - y = 6$
**a.** (0, )
**b.** ( ,0)
**c.** (−2, )
**d.** ( ,4)

Graph each of the linear equations in Exercises 5–14. [6.1]

**5.** $x - y = 2$
**6.** $x + y = 7$
**7.** $x - 3y = 3$
**8.** $5x + y = 3$
**9.** $3x - 7y = 7$
**10.** $3x + 2y = 9$
**11.** $3y = 2x$
**12.** $y = -\frac{3}{4}x$
**13.** $\frac{4}{5}x + y = 3$
**14.** $2x - \frac{3}{2}y = -1$

Graph the line determined by each pair of points in Exercises 15–20. Then find the slope of the line. [6.2]

**15.** $(1,3), (-2,5)$
**16.** $(-4,1), (3,6)$
**17.** $(0,3), (4,-3)$
**18.** $(4,5), (-3,7)$
**19.** $\left(\frac{3}{2},5\right), (1,2)$
**20.** $\left(\frac{1}{4},-1\right), \left(2,-\frac{1}{2}\right)$

Write each equation in Exercises 21–32 in slope-intercept form, if possible. Find the slope and the $y$-intercept; then draw the graph. [6.2]

**21.** $x + 5y = 10$
**22.** $3x + y = 1$
**23.** $3x - 7y = 7$
**24.** $5x + 3y = 6$
**25.** $4y - 9 = 0$
**26.** $2y - 3 = 1$
**27.** $3x + 2y = 5$
**28.** $4x - 2y = 9$
**29.** $4x - 3 = 0$
**30.** $-\frac{x}{4} = 1$
**31.** $\frac{2}{3}x + 2y = 9$
**32.** $\frac{x}{2} - 2y = 0$

Find an equation in standard form for the line determined by the given point and slope or two points in Exercises 33–46: [6.3]

**33.** $(2,-3), m = -\frac{1}{4}$
**34.** $(3,1), m = 0$
**35.** $(4,0), m = 0$
**36.** $(6,-1), m = \frac{2}{5}$
**37.** $(-1,2), m = \frac{4}{3}$
**38.** $(0,0), m = 2$
**39.** $(5,2)$, undefined slope
**40.** $(-2,4), m = -3$
**41.** $(0,3), (5,-1)$
**42.** $(5,-2), (1,6)$
**43.** $(4,-1), (2,-7)$
**44.** $(-2,1), (0,-3)$
**45.** $(5,5), (-1,3)$
**46.** $(-4,0), (7,3)$

Find an equation in standard form for each line satisfying the conditions in Exercises 47–54. [6.3]

**47.** Parallel to $3x + 2y - 6 = 0$, passing through $(2,3)$
**48.** Parallel to the $y$-axis, passing through $(1,-7)$
**49.** Parallel to $2x - 5y = 1$, passing through the origin
**50.** Parallel to the $x$-axis, passing through $(2,-5)$
**51.** Perpendicular to $4x + 3y = 5$, passing through $(4,0)$
**52.** Perpendicular to $3x - 5y = 1$, passing through $(6,-2)$
**53.** Perpendicular to the $y$-axis, passing through $(-3,1)$
**54.** Perpendicular to $3x - y = 1$, having the same $y$-intercept as $7x - 2y = 4$

Graph the linear inequalities in Exercises 55–60. [6.4]

**55.** $y \geq 4x$

**56.** $3x + y < 2$

**57.** $2x + 5y < 10$

**58.** $4y - 12 > 0$

**59.** $x + \frac{2}{3}y \leq 2$

**60.** $\frac{1}{2}x - \frac{1}{4}y \geq 2$

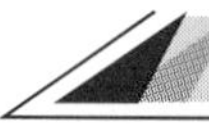

## CHAPTER 6 TEST

Find the missing component of each ordered pair so that the ordered pair belongs to the solution set of the equation in Exercises 1 and 2.

**1.** $3x + y = 2$
- **a.** (0, )
- **b.** ( ,0)
- **c.** (−2, )
- **d.** ( ,−7)

**2.** $x - 5y = 6$
- **a.** (0, )
- **b.** ( ,0)
- **c.** (11, )
- **d.** ( ,−2)

Graph each of the linear equations in Exercises 3–5.

**3.** $x + 4y = 5$

**4.** $2x - 5y = 1$

**5.** $3x - y = 8$

Graph the line determined by each pair of points in Exercises 6–8. Then find the slope of the line.

**6.** $(1,-2)$, $(9,7)$

**7.** $(-2,5)$, $(8,3)$

**8.** $(6,-1)$, $(6,3)$

Write each equation in Exercises 9–11 in slope-intercept form, if possible. Find the slope and the $y$-intercept; then draw the graph.

**9.** $x - 3y = 4$

**10.** $4x - 3y = 3$

**11.** $5x + 8 = 0$

Find an equation in standard form for the line determined by the given point and slope or two points in Exercises 12–14.

**12.** $(3,7)$, $m = -\frac{5}{3}$

**13.** $(0,0)$, $m = \frac{2}{3}$

**14.** $(-4,6)$, $(3,-2)$

Find an equation in standard form for the line satisfying the conditions in Exercises 15–18.

**15.** Passing through $(-1,6)$ and $(4,-2)$

**16.** Passing through $(0,5)$ with slope $-\frac{4}{3}$

**17.** Parallel to $3x + 2y = -1$, passing through $(2,4)$

**18.** Perpendicular to the $y$-axis, passing through $(3,-2)$

Graph the linear inequalities in Exercises 19 and 20.

**19.** $3x - 5y \leq 10$

**20.** $3x + 4y \geq 7$

## CUMULATIVE REVIEW (6)

Solve the equations in Exercises 1 and 2.

**1.** $3(2x + 7) - 1 = x + 3(1 - x)$

**2.** $4(2x + 1) = 2x - 8(2 + x)$

Solve the inequalities in Exercises 3 and 4 and graph the solution set.

**3.** $2(5x - 3) \leq 14 + (3x - 4)$

**4.** $-3(x + 4) + 9 \geq 7x - 16$

Perform the indicated operations in Exercises 5–8.

**5.** $(3x - 7y - 2) + (x + 4y - 9)$

**6.** $(5x + y - 11) - (2x + 4y - 2)$

**7.** $2(5x + y - 14) + (6x - 2y + 11)$

**8.** $4(2x + 3y + 6) - 3(3x - 4y + 9)$

Solve for the indicated variable in Exercises 9 and 10.

**9.** $6x + y = 9$ for $y$

**10.** $2x + 5y = 8$ for $x$

**11.** The sum of two numbers is 14. Twice the larger number added to three times the smaller is equal to 31. Find both numbers.

**12.** Melanie traveled upstream in her boat for $3\frac{1}{2}$ hours and returned. The return trip took 2 hours. If the speed of the current is 3 mph, find the speed of her boat in still water.

**13.** At the sports arena, two sizes of hot dogs are sold. The regular hot dog sells for \$1.50 and the "foot-long" hot dog sells for \$2.50. At a recent basketball game, one concession stand sold a total of 1200 hot dogs. If the total income from the sale of hot dogs was \$2250, how many hot dogs of each size were sold?

Graph each of the linear equations in Exercises 14–16.

**14.** $3x - 4y = 24$

**15.** $4x + 5y = 15$

**16.** $6y + 12 = 0$

**17.** Graph the line with slope $m = -\frac{4}{5}$, passing through $(-2,2)$. Then find an equation in standard form.

**18.** Graph the line passing through $(5,-1)$ and $(2,-6)$. Then find an equation in standard form.

**19.** Find an equation in standard form for the line parallel to $-3x + 5y = 1$ and passing through $(2,7)$.

**20.** Find an equation in standard form for the line perpendicular to $5x + 6y + 10 = 0$ and passing through $(-3,2)$.

# 7 SYSTEMS OF EQUATIONS AND INEQUALITIES

## MATHEMATICAL CHALLENGES

The area of a rectangular field is 42 000 $m^2$. Julie finds that by taking a diagonal path instead of walking halfway around the field, she walks 120 m less. How long is the diagonal?

What is the first two-digit prime number for which the product of its digits is equal to the sum of its digits plus 7?

*ABCD* is a square. *A* is folded onto the midpoint of *AB*. The perimeter of the smaller figure formed is 25. Find the area of *ABCD*.

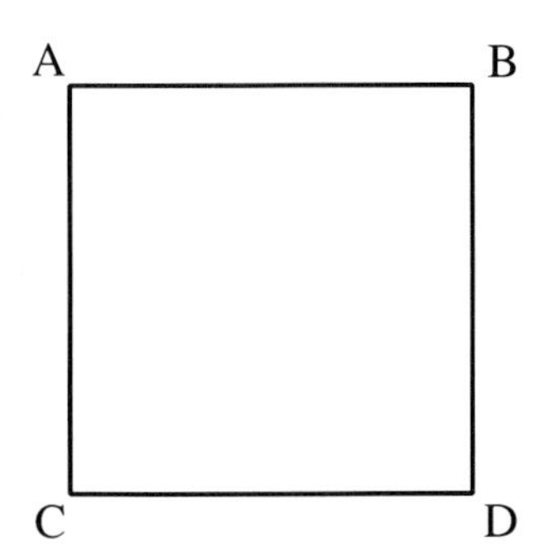

## Chapter Outline



Many applications involve two (or more) quantities and, by using two (or more) variables, we can form linear equations using the information given. Such a set of equations is called a system, and in this chapter we will develop techniques for solving systems of linear equations.

Graphing systems of equations in two variables is helpful in visualizing the relationships between two equations. However, this approach is somewhat limited since numbers might be quite large or solutions might involve fractions that must be estimated on the graph. Therefore, other algebraic techniques are necessary. We will discuss two of these in this chapter, two others in Chapter 11, and one other in Chapter 12. These algebraic techniques are also necessary when we try to solve three (or more) equations with three (or more) variables. Graphing in three dimensions will be left to later courses.

Systems of linear inequalities represent relationships between variables that have implied or stated restrictions. For example, a menu in a dormitory cafeteria may require a minimum amount of certain vitamins, minerals, and nutrients yet be restricted by an upper dollar figure. We will investigate an interesting concept called linear programming that allows us to solve problems such as the best menu for the price at the cafeteria.

## 7.1 Systems of Linear Equations (Two Variables)

**OBJECTIVE**

In this section, you will be learning to solve systems of linear equations in two variables using three methods:

a. graphing,
b. substitution,
c. addition.

In Chapter 6, we discussed linear equations and inequalities and their graphs. In this chapter, we will discuss equations and inequalities in systems. That is, we want to consider several equations (or inequalities) at one time and determine any solutions that satisfy all the conditions.

Consider the two linear equations

$2x + y = 5$ [Some solutions are (1,3), (4,−3), and (0,5).]

and $x - y = 1$ [Some solutions are (1,0), (3,2), and (−1,−2).]

Are there any ordered pairs that satisfy both equations? The answer is yes, because the ordered pair (2,1) satisfies both equations:

$$2(2) + 1 = 5 \qquad \text{and} \qquad 2 - 1 = 1$$

We want to develop techniques for finding such solutions.

Two (or more) linear equations considered at one time are said to form a **system of linear equations** or a **set of simultaneous equations.** If the system has a unique solution (that is, only one pair of values for $x$ and $y$ satisfy both equations), the system is **consistent.** If there are no solutions, the system is **inconsistent.** If there are an infinite number of solutions, the system is **dependent.**

Summary Table 7.1 summarizes the basic ideas and terminology.

***Summary Table 7.1***

| System | Graph | Intersection | Terms |
|---|---|---|---|
| $\begin{cases} 2x + y = 5 \\ x - y = 1 \end{cases}$ | | $(2,1)$ or $\begin{cases} x = 2 \\ y = 1 \end{cases}$ | Consistent |
| $\begin{cases} 3x - 2y = 2 \\ 6x - 4y = -4 \end{cases}$ | | No solution. (The lines are parallel.) | Inconsistent |
| $\begin{cases} 2x - 4y = 6 \\ x - 2y = 3 \end{cases}$ | | Any ordered pair that satisfies $x - 2y = 3$ (The lines are the same.) | Dependent |

We will discuss the following three basic techniques for solving systems of linear equations in two variables:

1. **graphing** (as illustrated in Summary Table 7.1)
2. **substitution** (algebraically substituting from one equation into the other)
3. **addition** (combining like terms from both equations)

## Graphing

To solve a system of two linear equations by graphing, graph both equations on the same set of axes and observe the point of intersection of the two lines (if there is one). As illustrated in Summary Table 7.1, if the lines are parallel, then the system is inconsistent; if the lines are the same, then the system is dependent.

Solving by graphing can involve estimating the solutions whenever the intersection of the two lines is at a point not represented by a pair of integers. (See Example 2.)

**EXAMPLES** Solve each of the following systems by graphing.

**1.** $\begin{cases} x + y = 6 \\ y = x + 4 \end{cases}$

**Solution** The two lines intersect at the point (1,5).

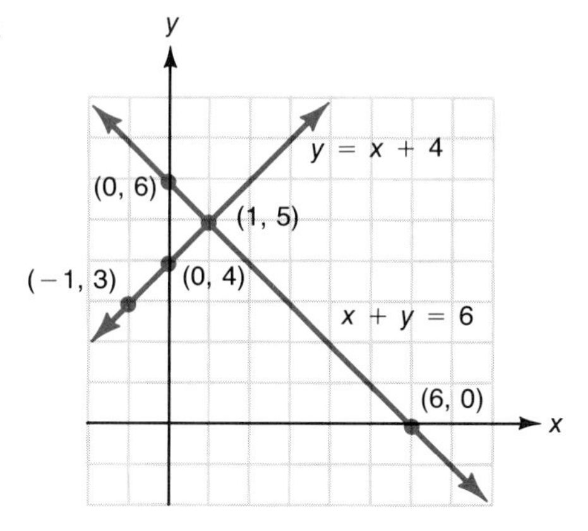


The system is consistent and the solution is $x = 1$ and $y = 5$.

**Note:** Substitution shows that (1,5) satisfies both of the equations in the system.

$$\begin{cases} 1 + 5 = 6 \\ 5 = 1 + 4 \end{cases}$$

**2.** $\begin{cases} x - 3y = 4 \\ 2x + y = 3 \end{cases}$

**Solution** The two lines intersect, but we can only estimate the point of intersection at $\left(2, -\frac{1}{2}\right)$.

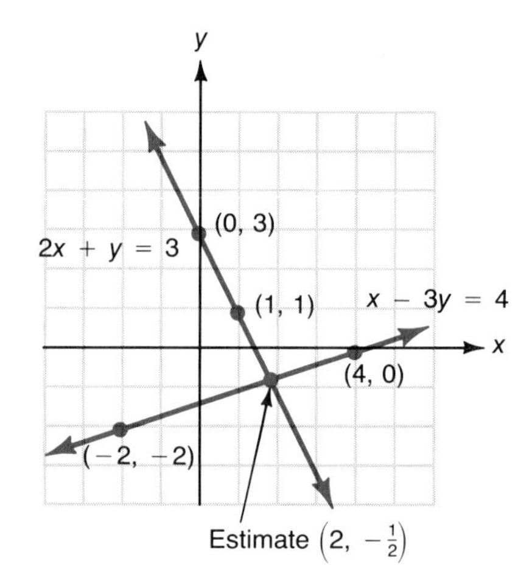


In this situation, we must be aware that, while graphing gives a good "picture," we are not likely to find exact solutions to the system.

Substituting $x = 2$ and $y = -\frac{1}{2}$ gives:

$$2 - 3\left(-\frac{1}{2}\right) = 4 \quad \text{or} \quad \frac{7}{2} = 4 \quad \text{which is not true.}$$

$$2(2) + \left(-\frac{1}{2}\right) = 3 \quad \text{or} \quad \frac{7}{2} = 3 \quad \text{which is not true.}$$

We will find in the next discussion using an algebraic technique that the correct solution is

$$\left(\frac{13}{7}, -\frac{5}{7}\right) \quad \text{or} \quad x = \frac{13}{7} \quad \text{and} \quad y = -\frac{5}{7}$$

■

## Substitution

**To solve a system of linear equations by substitution,** solve one of the equations for one of the variables; then substitute the resulting expression into the other equation. This will give a single equation in one variable.

**EXAMPLES** Solve the following systems by substitution.

**3.** $\begin{cases} x - 3y = 4 \\ 2x + y = 3 \end{cases}$

**Solution**

$$x - 3y = 4$$
$$x = 4 + 3y$$

Solve the first equation for $x$.

$$2(4 + 3y) + y = 3$$

Substitute $4 + 3y$ for $x$ in the second equation.

$$8 + 6y + y = 3$$
$$7y = -5$$
$$y = -\frac{5}{7}$$

Solve this equation for $y$.

$$x - 3\left(-\frac{5}{7}\right) = 4$$

To find $x$, substitute $-\frac{5}{7}$ for $y$ in one of the original equations.

$$x = 4 - \frac{15}{7} = \frac{13}{7}$$

The solution is $x = \frac{13}{7}$ and $y = -\frac{5}{7}$ or $\left(\frac{13}{7}, -\frac{5}{7}\right)$.

We could have solved the second equation for y and substituted into the first equation. The solution would have been the same. **In fact, we can solve for either variable in either equation, but for simplicity, we generally solve for the variable that has a coefficient of 1 if there is such a variable.**

**4.** $\begin{cases} 6x + 3y = 14 \\ 2x + y = -3 \end{cases}$

**Solution**

$$2x + y = -3$$
$$y = -3 - 2x$$

Solve the second equation for $y$.

$$6x + 3(-3 - 2x) = 14$$

Substitute $-3 - 2x$ for $y$ in the first equation.

$$6x - 9 - 6x = 14$$
$$-9 = 14$$

Solve this equation for $x$. This last equation is never true. Therefore, the system is inconsistent.

There is no solution. ■

## Addition

In arithmetic, we can show that if $a = b$ and $c = d$, then $a + c = b + d$. Similarly, in algebra, we can show that any solution to **both**

$$a_1x + b_1y = c_1$$

and

$$a_2x + b_2y = c_2$$

will also be a solution to the equation

$$k_1(a_1x + b_1y) + k_2(a_2x + b_2y) = k_1c_1 + k_2c_2$$

where $k_1$ and $k_2$ are not both 0.

Thus, to find a common solution to the original two equations, we form a new equation by combining like terms of the two equations. The procedure can be accomplished in the following way:

**To solve a system of linear equations by addition,** set the equations one under the other so that like terms are aligned. Multiply all terms of either equation (or both equations) by some constant so that two like terms have opposite coefficients. Then add like terms and solve the resulting equation.

**EXAMPLES** Solve the following systems by addition.

**5.** $\begin{cases} 4x + 3y = -6 \\ 5x + y = -13 \end{cases}$

**Solution** Multiply each term in the second equation by $-3$ so that the $y$-coefficients will be opposites.

$$\begin{cases} \phantom{[-3]}\ 4x + 3y = -6 \\ [-3]\ 5x + y = -13 \end{cases} \qquad \begin{array}{r} 4x + 3y = -6 \\ \underline{-15x - 3y = 39} \\ -11x \phantom{- 3y} = 33 \\ x = -3 \end{array}$$

Add like terms.

To find $y$, substitute $x = -3$ in one equation.

$$5(-3) + y = -13$$
$$y = 2$$

The solution is $x = -3$ and $y = 2$.

(By adding like terms, we reduced the system to solving one equation in one variable.)

**6.** $\begin{cases} 2x = -17 - 3y \\ 3x - \dfrac{51}{2} = 4y \end{cases}$

**Solution** Rearrange the equations in standard form.

$$2x + 3y = -17$$

$$3x - 4y = \frac{51}{2}$$

Multiply each term in the first equation by $-3$ and each term in the second equation by 2 so that the $x$ coefficients will be opposites.

$$\begin{cases} [-3] & 2x + 3y = -17 \\ [2] & 3x - 4y = \dfrac{51}{2} \end{cases} \qquad \begin{aligned} -6x - 9y &= 51 \\ 6x - 8y &= 51 \\ \hline -17y &= 102 \\ y &= -6 \end{aligned}$$

To find $x$, substitute $y = -6$ in one equation.

$$\begin{aligned} 2x + 3(-6) &= -17 \\ 2x &= -17 + 18 \\ 2x &= 1 \\ x &= \frac{1}{2} \end{aligned}$$

The solution is $x = \frac{1}{2}$ and $y = -6$.

(We could have multiplied the first equation by 4 and the second equation by 3 and eliminated the $y$-terms.)

**7.** $\begin{cases} 3x - \frac{1}{2}y = 6 \\ 6x - y = 12 \end{cases}$

**Solution** Multiply the first equation by $-2$.

$$\begin{cases} [-2] & 3x - \dfrac{1}{2}y = 6 \\ & 6x - y = 12 \end{cases} \qquad \begin{aligned} -6x + y &= -12 \\ 6x - y &= 12 \\ \hline 0 &= 0 \end{aligned}$$

The last equation $0 = 0$ is always true, so the system is dependent. There are an infinite number of solutions. Any ordered pair that satisfies the equation $6x - y = 12$ is a solution of the system. ■

***Practice Problems***

Solve each of the following systems algebraically.

**1.** $\begin{cases} y = 3x + 4 \\ 2x + y = -1 \end{cases}$

**2.** $\begin{cases} 2x - 3y = 0 \\ 6x + 3y = 4 \end{cases}$

**3.** $\begin{cases} 4x + y = 3 \\ 4x + y = 2 \end{cases}$

**Answers to Practice Problems** **1.** $x = -1, y = 1$ **2.** $x = \frac{1}{2}, y = \frac{1}{3}$ **3.** Inconsistent

## EXERCISES 7.1

Solve each of the systems in Exercises 1–10 by graphing.

1. $\begin{cases} x + y = 5 \\ x - 4y = 5 \end{cases}$
2. $\begin{cases} 3x - y = 6 \\ 2x + y = -1 \end{cases}$
3. $\begin{cases} 2x - y = 8 \\ y = 2x \end{cases}$
4. $\begin{cases} 5x + 2y = 21 \\ x = y \end{cases}$
5. $\begin{cases} y = \frac{5}{6}x + 1 \\ x - 2y = 2 \end{cases}$
6. $\begin{cases} 4x - 2y = 10 \\ -6x + 3y = -15 \end{cases}$
7. $\begin{cases} 2x + y + 1 = 0 \\ 3x + 4y - 1 = 0 \end{cases}$
8. $\begin{cases} 2x + 3y = 4 \\ 4x - y = 1 \end{cases}$
9. $\begin{cases} x - 2y = 11 \\ 2x - 3y = 18 \end{cases}$
10. $\begin{cases} 4x + 3y + 7 = 0 \\ 5x = 2y - 3 \end{cases}$

Solve the systems of equations in Exercises 11–36. State which systems are dependent or inconsistent.

11. $\begin{cases} x + 4y = 6 \\ 2x + y = 5 \end{cases}$
12. $\begin{cases} 2x + y = 0 \\ x - 2y = -10 \end{cases}$
13. $\begin{cases} 5x - y = -2 \\ x + 2y = -7 \end{cases}$
14. $\begin{cases} 7x - y = 18 \\ x + 2y = 9 \end{cases}$
15. $\begin{cases} x + 2y = 3 \\ 4x + 8y = 8 \end{cases}$
16. $\begin{cases} 2x + 3y = 3 \\ x + 4y = 4 \end{cases}$
17. $\begin{cases} 6x + 2y = 16 \\ 3x + y = 8 \end{cases}$
18. $\begin{cases} 4x - y = 18 \\ 3x + 5y = 2 \end{cases}$
19. $\begin{cases} y = 3x + 3 \\ y = -2x + 8 \end{cases}$
20. $\begin{cases} x = 5 - 4y \\ x = 2y - 7 \end{cases}$
21. $\begin{cases} 2x + y = 4 \\ 4x + 5y = 11 \end{cases}$
22. $\begin{cases} x + 6y = 4 \\ 2x + 3y = 5 \end{cases}$
23. $\begin{cases} 3x + 4y = 6 \\ x - 8y = 9 \end{cases}$
24. $\begin{cases} 3x + 5y = 3 \\ 9x - y = -7 \end{cases}$
25. $\begin{cases} 2x = 5y - 1 \\ 4x - 10y = 0 \end{cases}$
26. $\begin{cases} 6x + 2y = 5 \\ 2x + y = 1 \end{cases}$
27. $\begin{cases} 4x + 12y = 5 \\ 5x - 6y = 1 \end{cases}$
28. $\begin{cases} 2x - 3y = 18 \\ 5x + 4y = -1 \end{cases}$
29. $\begin{cases} x + y = 7 \\ 2x + 3y = 16 \end{cases}$
30. $\begin{cases} 5x - 7y = 8 \\ 3x + 11y = -12 \end{cases}$
31. $\begin{cases} 6x - y = 15 \\ 0.2x + 0.5y = 2.1 \end{cases}$
32. $\begin{cases} 3x + y = 14 \\ 0.1x - 0.2y = 1.4 \end{cases}$
33. $\begin{cases} x + y = 12 \\ 0.05x + 0.25y = 1.6 \end{cases}$
34. $\begin{cases} x + y = 20 \\ 0.1x + 0.25y = 3.8 \end{cases}$
35. $\begin{cases} 0.6x + 0.5y = 5.9 \\ 0.8x + 0.4y = 6 \end{cases}$
36. $\begin{cases} 0.5x - 0.3y = 7 \\ 0.3x + 0.4y = 2 \end{cases}$

## 7.2 Applications

**OBJECTIVE**

In this section, you will be learning to solve applied problems using systems of linear equations in two variables.

Throughout this text we have solved a variety of word problems by using one variable and one equation. Now we will solve word problems by using two variables and two equations. That is, we will represent the information given in the problem with a system of linear equations, then solve the system.

**EXAMPLE: Mixture**

1. A manufacturer receives an order for 30 tons of a 40% copper alloy. He has only 20% alloy and 50% alloy in stock. How much of each will he need to fill the order?

**Solution** Let $x =$ amount of 20% alloy
$y =$ amount of 50% alloy

Form two equations based on the information given.

$$\begin{cases} x + y = 30 \\ 0.20x + 0.50y = 0.40(30) \end{cases}$$

The total order is 30 tons.
The total amount of copper is 40% of 30 or 0.40(30). This amount is equal to 20% of $x$ plus 50% of $y$.

Now solve the system. (Use either the method of addition or substitution.)

$$\begin{cases} [-2] & x + y = 30 \\ [10] & 0.20x + 0.50y = 0.40(30) \end{cases} \qquad \begin{aligned} -2x - 2y &= -60 \\ 2x + 5y &= 120 \\ \hline 3y &= 60 \\ y &= 20 \end{aligned}$$

Substituting $y = 20$:

$$x + 20 = 30$$
$$x = 10$$

He will use 10 tons of the 20% alloy and 20 tons of the 50% alloy.

**EXAMPLE: Interest**

**2.** A savings and loan company pays 7% interest on long-term savings, and a high-risk stock indicates that it should yield 12% interest. If a woman has $40,000 to invest and wants an annual income of $3550 from her investments, how much should she put in the savings and loan and how much in the stock?

**Solution** Let $x =$ amount invested at 7%.
$y =$ amount invested at 12%.

$$\begin{cases} x + y = 40{,}000 \\ 0.07x + 0.12y = 3550 \end{cases}$$

$40,000 is the total invested.
$3550 is the total interest.

$$\begin{cases} [-7] & x + y = 40{,}000 \\ [100] & 0.07x + 0.12y = 3550 \end{cases} \qquad \begin{aligned} -7x - 7y &= -280{,}000 \\ 7x + 12y &= 355{,}000 \\ \hline 5y &= 75{,}000 \\ y &= 15{,}000 \end{aligned}$$

Substituting $y = 15{,}000$:

$$x + 15{,}000 = 40{,}000$$
$$x = 25{,}000$$

She should put $25,000 in the savings and loan at 7% and $15,000 in the stock at 12% to earn an income of $3550.

**EXAMPLE: Work**

**3.** Working his way through school, Richard works two part-time jobs for a total of 25 hours a week. Job A pays \$4.50 per hour and job B pays \$6.30 per hour. How many hours did he work at each job the week he made \$137.70?

**Solution** Let $x$ = number of hours at job A, \$4.50 per hour
$y$ = number of hours at job B, \$6.30 per hour

$$\begin{cases} x + y = 25 \\ 4.50x + 6.30y = 137.70 \end{cases}$$ Total hours worked is 25. Total earnings is \$137.70.

$$\begin{cases} [-45] \quad x + y = 25 \\ [10] \quad 4.50x + 6.30y = 137.70 \end{cases}$$

$$\begin{aligned} -45x - 45y &= 1125 \\ 45x + 63y &= 1377 \\ \hline 18y &= 252 \\ y &= 14 \end{aligned}$$

Substituting $y = 14$:

$$\begin{aligned} x + 14 &= 25 \\ x &= 11 \end{aligned}$$

He worked 11 hours at job A and 14 hours at job B.

**EXAMPLE: Algebra**

**4.** Determine the values of $a$ and $b$ such that the straight line $ax + by = 11$ passes through the point $(3, -1)$ and has slope $-\frac{5}{4}$.

**Solution**

**a.** First substitute $x = 3$ and $y = -1$ into the equation $ax + by = 11$.

$$3a - b = 11$$

**b.** To find the slope in terms of $a$ and $b$, solve the equation $ax + by = 11$ for $y$.

$$\begin{aligned} ax + by &= 11 \\ by &= -ax + 11 \\ y &= -\frac{a}{b}x + \frac{11}{6} \end{aligned}$$

Thus, the slope is $-\frac{a}{b}$ and

$$-\frac{a}{b} = -\frac{5}{4}$$

or $$4a = 5b$$

or $$4a - 5b = 0$$

**c.** Now solve the system

$$\begin{cases} [-5] & 3a - b = 11 \\ & 4a - 5b = 0 \end{cases} \qquad \begin{aligned} -15a + 5b &= -55 \\ 4a - 5b &= 0 \\ \hline -11a &= -55 \\ a &= 5 \end{aligned}$$

Substituting $a = 5$:

$$\begin{aligned} 3 \cdot 5 - b &= 11 \\ -b &= -4 \\ b &= 4 \end{aligned}$$

Thus, $a = 5$ and $b = 4$, so the line $5x + 4y = 11$ passes through the point $(3, -1)$ and has slope $-\frac{5}{4}$. ■

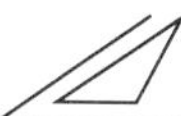

## EXERCISES 7.2

1. How many liters each of a 12% iodine solution and a 30% iodine solution must be used to produce 90 liters of a 22% iodine solution?
2. A meat market has ground beef that is 40% fat and extra lean ground beef that is only 15% fat. How many pounds of each will be needed to obtain 50 pounds of "lean" ground beef that is 25% fat?
3. A dairy needs 360 pounds of milk containing 4% butterfat. How many pounds each of milk containing 5% butterfat and milk containing 2% butterfat must be used to obtain the desired 360 pounds?
4. A druggist has two solutions of alcohol. One is 25% alcohol and the other is 45% alcohol. How many ounces of each must be used to obtain 36 ounces that will be 30% alcohol?
5. Marilyn inherited $12,400 from her Uncle Jake. She invested a portion in bonds and the remainder in a long-term certificate account. The amount invested in bonds was $2600 less than 3 times the amount invested in certificates. How much was invested in bonds and how much in certificates?
6. Ann has $4800 to invest, part at 6% and the remainder at 7%. How much should be invested at each rate to yield $306 annual interest?
7. An investor bought 500 shares of stock, some at $3.50 per share and some at $6.00 per share. If the total cost was $2187.50, how many shares of each stock did the investor buy?
8. Ruth has 10 coins consisting of dimes and quarters. How many of each type does she have if all together she has $2.05?
9. A confectioner mixes candy worth $1.96 per pound with candy worth $1.16 per pound to obtain 30 pounds of candy worth $1.48 per pound. How many pounds of each kind does the confectioner use?
10. A secretary purchased 48 stamps consisting of 25-cent and 45-cent stamps. If the total cost was $15.60, how many of each type stamp did she buy?
11. A grocer mixes two kinds of nuts; one costs 35 cents per pound and the other costs 65 cents per pound. If the mixture weighs 20 pounds and costs 41 cents per pound, how many pounds of each kind does the grocer use?

**12.** A manufacturing plant uses two stamping machines. One produces 100 units per hour, while the other produces 75 units per hour. How long will it take to complete an order of 975 units if the faster machine was shut down for $2\frac{1}{2}$ hours for repairs? (**Hint:** The two variables should represent time.)

**13.** The bookstore can buy a popular book with either paperback or hardback cover. A hardback book costs \$3.50 more than the paperback book. What is the cost of each if 90 paperback books cost the same as 55 hardback books?

**14.** In an election, the winner received 430 votes more than twice as many votes as the loser. If there was a total of 2290 votes cast, how many did each candidate receive?

**15.** A bill was defeated by the legislature by 50 votes. If $\frac{1}{5}$ of those voting against the bill had voted for it, the bill would have passed by 30 votes. How many legislators voted for the bill?

**16.** Barbara made a trip of 440 kilometers. She averaged 64 kilometers per hour for the first part of the trip and 80 kilometers per hour for the second part. If the total trip took 6 hours, how long was she traveling at 80 kilometers per hour?

**17.** A boat left Dana Point Marina at 11:00 A.M. traveling at 10 knots per hour. Two hours later, a Coast Guard boat left the same marina traveling at 14 knots per hour. If they both traveled the same course, when did the Coast Guard boat overtake the other boat?

**18.** Determine $a$ and $b$ such that the straight line $ax + by = 7$ passes through the points $(2,1)$ and $(-1,10)$.

**19.** Determine $a$ and $b$ such that the straight line $ax + by = 6$ passes through the points $(-6,-2)$ and $(3,4)$.

**20.** Determine $a$ and $b$ such that the straight line $ax + by = 4$ passes through the point $(5,2)$ and has slope $\frac{2}{3}$.

**21.** Determine $a$ and $b$ such that the straight line $ax + by = 11$ passes through the point $(3,-1)$ and has slope $-\frac{5}{4}$.

**22.** A manufacturer produces two kinds of radios, Model A and Model B. Model A takes 4 hours to produce and costs \$8 each. Model B takes 3 hours to produce and costs \$7 each. If the manufacturer allots a total of 58 hours and \$126 each week, how many of each model will be produced?

**23.** A company manufactures two products. One requires 2.5 hours of labor, 3 pounds of raw materials, and costs \$28.40 each. The second product requires 4 hours of labor, 4 pounds of raw materials, and costs \$41.60 each. Find the cost of labor per hour and the cost of raw materials per pound.

**24.** A furniture shop refinishes chairs. Employees use two methods to refinish a chair. Method I takes 1 hour and the material costs \$3. Method II takes $1\frac{1}{2}$ hours and the material costs \$1.50. Last week, they took 36 hours and spent \$60 refinishing chairs. How many did they refinish with each method?

**25.** A large feed lot uses two feed supplements, Ration I and Ration II. Each pound of Ration I contains 4 units of protein and 2 units of carbohydrate. Each pound of Ration II contains 3 units of protein and 6 units of carbohydrate. If the dietary requirement calls for 42 units of protein and 30 units of carbohydrate, how many pounds of each ration should be used?

## 7.3 Systems of Linear Equations (Three Variables)

**OBJECTIVES**

In this section, you will be learning to:

1. Solve systems of linear equations in three variables.
2. Solve applied problems by using systems of linear equations in three variables.

The equation $2x + 3y - z = 16$ is called a **linear equation in three variables.** The solutions to such equations are called **ordered triples** and are of the form $(x_0,y_0,z_0)$ or $x = x_0$, $y = y_0$, and $z = z_0$. One ordered triple that satisfies the equation $2x + 3y - z = 16$ is $(1,4,-2)$. To check this, substitute $x = 1$, $y = 4$, and $z = -2$.

$$\begin{aligned} 2(1) + 3(4) - (-2) &= 2 + 12 + 2 \\ &= 16 \end{aligned}$$

There are an infinite number of ordered triples that satisfy any linear equation in three variables in which at least two of the coefficients are nonzero. You may substitute any two values you like for two of the variables and then solve the resulting equation for the third variable. For example, by letting $x = -1$ and $y = 5$, we find

$$\begin{aligned} 2(-1) + 3(5) - z &= 16 \\ -2 + 15 - z &= 16 \\ -z &= 3 \\ z &= -3 \end{aligned}$$

So, the ordered triple $(-1,5,-3)$ satisfies the equation $2x + 3y - z = 16$.

Graphs can be drawn in three dimensions using a coordinate system involving three mutually perpendicular number lines labeled as the $x$-axis, $y$-axis, and $z$-axis. These three axes separate space into eight regions called **octants.** The point represented by the ordered triple $(2,3,1)$ is shown in Figure 7.1.

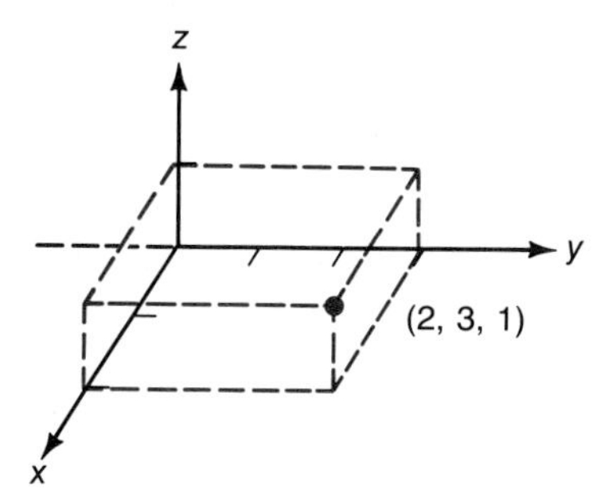


**Figure 7.1**

The graphs of linear equations in three variables are planes in three dimensions. The graph of $2x + 3y - z = 16$ appears in Figure 7.2.

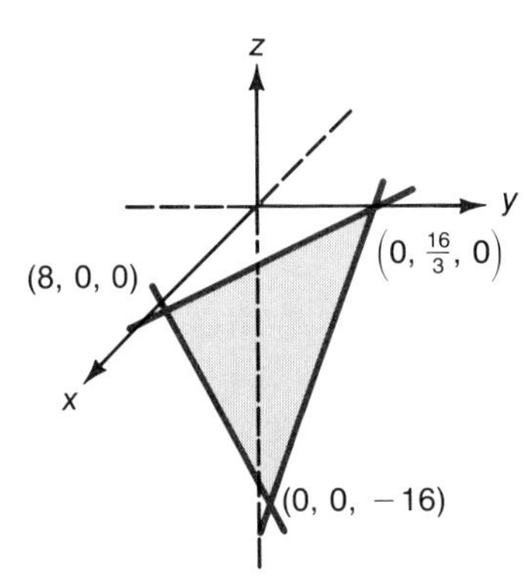


**Figure 7.2**

Two distinct planes will either be parallel or intersect. If they intersect, their intersection will be a straight line. If three distinct planes intersect, they will intersect in a straight line, or they will intersect in a single point represented by an ordered triple.

We are interested here in systems of three linear equations in three variables. Although the graphs are interesting, they can be difficult to sketch and points of intersection difficult to estimate. Only the algebraic techniques for solving these systems will be discussed.

***To Solve Three Linear Equations in Three Variables***

1. Select two equations and eliminate one variable using the addition method.
2. Select two other equations and eliminate **the same** variable.
3. From steps 1 and 2, you have **two** linear equations in **two** variables. Solve these two equations by either addition or substitution.
4. Substitute your values from step 3 into any one of the original equations and find the value of the third variable.

The solution possibilities are as follows:

1. There will be exactly one ordered triple solution. (Graphically, the three planes intersect in one point.)
2. There will be an infinite number of solutions. (Graphically, the three planes intersect in a line or are the same plane.)
3. There will be no solutions. (Graphically, the three planes have no points in common.)

Now to illustrate the technique, we will solve the system

$$\begin{cases} 2x + 3y - z = 16 & \textbf{(a)} \\ x - y + 3z = -9 & \textbf{(b)} \\ 5x + 2y - z = 15 & \textbf{(c)} \end{cases}$$

**Step 1** Using equations (a) and (b), eliminate $y$.

$$\begin{cases} & 2x + 3y - z = 16 \\ [3] & x - y + 3z = -9 \end{cases} \qquad \begin{array}{r} 2x + 3y - z = 16 \\ \underline{3x - 3y + 9z = -27} \\ 5x \qquad + 8z = -11 \end{array}$$

**Step 2** Using equations (b) and (c), eliminate $y$.

$$\begin{cases} [2] & x - y + 3z = -9 \\ & 5x + 2y - z = 15 \end{cases} \qquad \begin{array}{r} 2x - 2y + 6z = -18 \\ \underline{5x + 2y - z = 15} \\ 7x \qquad + 5z = -3 \end{array}$$

**Step 3** Using the results of steps 1 and 2, solve for $x$ and $z$.

$$\begin{cases} [-7] & 5x + 8z = -11 \\ [5] & 7x + 5z = -3 \end{cases} \qquad \begin{array}{r} -35x - 56z = 77 \\ \underline{35x + 25z = -15} \\ -31z = 62 \\ z = -2 \end{array} \qquad \begin{array}{r} 5x + 8(-2) = -11 \\ 5x = 5 \\ x = 1 \end{array}$$

**Step 4** Using $x = 1$ and $z = -2$, find $y$.

$$\begin{array}{r} 1 - y + 3(-2) = -9 \\ -y = -4 \\ y = 4 \end{array} \qquad \text{Using equation (b)}$$

The solution is $(1,4,-2)$ or $x = 1$, $y = 4$, and $z = -2$. The solution can be checked by substituting the results into **all three** of the original equations.

$$\begin{cases} 2(1) + 3(4) - (-2) = 16 \\ 1 - (4) + 3(-2) = -9 \\ 5(1) + 2(4) - (-2) = 15 \end{cases}$$

**EXAMPLES** Solve the following systems of linear equations.

**1.** $\begin{cases} x - y + 2z = -4 & \textbf{(a)} \\ 2x + 3y + z = \frac{1}{2} & \textbf{(b)} \\ x + 4y - 2z = 4 & \textbf{(c)} \end{cases}$

**Solution** Using **(a)** and **(c)**, eliminate $z$.

$$\begin{cases} x - y + 2z = -4 \\ x + 4y - 2z = 4 \end{cases}$$
$$2x + 3y = 0$$

Using **(a)** and **(b)**, eliminate $z$.

$$\begin{cases} & x - y + 2z = -4 \\ [-2] & 2x + 3y + z = \frac{1}{2} \end{cases} \qquad \begin{array}{r} x - y + 2z = -4 \\ -4x - 6y - 2z = -1 \\ \hline -3x - 7y = -5 \end{array}$$

$$\begin{cases} [3] & 2x + 3y = 0 \\ [2] & -3x - 7y = -5 \end{cases} \qquad \begin{array}{r} 6x + 9y = 0 \\ -6x - 14y = -10 \\ \hline -5y = -10 \\ y = 2 \end{array} \qquad \begin{array}{r} 2x + 3(2) = 0 \\ 2x = -6 \\ x = -3 \end{array}$$

Using **(a)** and substituting $x = -3$ and $y = 2$,

$$-3 - 2 + 2z = -4$$
$$2z = 1$$
$$z = \frac{1}{2}$$

The solution is $\left(-3,2,\frac{1}{2}\right)$.

**2.** $\begin{cases} 3x - 5y + z = 6 & \textbf{(a)} \\ x - y + 3z = -1 & \textbf{(b)} \\ 2x - 2y + 6z = 5 & \textbf{(c)} \end{cases}$

**Solution** Using **(a)** and **(b)**, eliminate $z$.

$$\begin{cases} [-3] & 3x - 5y + z = 6 \\ & x - y + 3z = -1 \end{cases} \qquad \begin{array}{r} -9x + 15y - 3z = -18 \\ x - y + 3z = -1 \\ \hline -8x + 14y = -19 \end{array}$$

Using **(b)** and **(c)**, eliminate $z$.

$$\begin{cases} [-2] & x - y + 3z = -1 \\ & 2x - 2y + 6z = 5 \end{cases} \qquad \begin{array}{r} -2x + 2y - 6z = 2 \\ 2x - 2y + 6z = 5 \\ \hline 0 = 7 \end{array}$$

The system does not have a solution.

**3.** A cash register contains \$341 in \$20, \$5, and \$2 bills. There are twenty-eight bills in all, and three more twos than fives. How many bills of each kind are there?

**Solution** Let $x$ = number of \$20 bills
$y$ = number of \$5 bills
$z$ = number of \$2 bills

$$\begin{cases} x + y + z = 28 & \textbf{(a)} \quad \text{There are twenty-eight bills.} \\ 20x + 5y + 2z = 341 & \textbf{(b)} \quad \text{The total value is \$341.} \\ y = z - 3 & \textbf{(c)} \quad \text{There are three more twos than fives.} \end{cases}$$

Rewrite the system:

$$\begin{cases} x + y + z = 28 & \textbf{(a)} \\ 20x + 5y + 2z = 341 & \textbf{(b)} \\ y - z = -3 & \textbf{(c)} \end{cases}$$

Using **(a)** and **(b)**, eliminate $x$.

$$\begin{cases} [-20] \quad x + y + z = 28 \\ 20x + 5y + 2z = 341 \end{cases} \qquad \begin{aligned} -20x - 20y - 20z &= -560 \\ 20x + 5y + 2z &= 341 \\ \hline -15y - 18z &= -219 \end{aligned}$$

Using **(c)** along with the results just found,

$$\begin{cases} [15] \quad y - z = -3 \\ -15y - 18z = -219 \end{cases} \qquad \begin{aligned} 15y - 15z &= -45 \\ -15y - 18z &= -219 \\ \hline -33z &= -264 \\ z &= 8 \end{aligned} \qquad \begin{aligned} y - 8 &= -3 \\ y &= 5 \end{aligned}$$

Using **(a)** and substituting $z = 8$ and $y = 5$,

$$\begin{aligned} x + 5 + 8 &= 28 \\ x &= 15 \end{aligned}$$

There are 15 \$20 bills, 5 \$5 bills, and 8 \$2 bills. ■

***Practice Problem***

Solve the following system of equations.

$$\begin{cases} 2x + y + z = 4 \\ x + 2y + z = 1 \\ 3x + y - z = -3 \end{cases}$$

**Answer to Practice Problem** $x = 1$, $y = -2$, $z = 4$

## EXERCISES 7.3

Solve each of the systems of equations in Exercises 1–20. State which systems have no solution or an infinite number of solutions.

1. $\begin{cases} x + y - z = 0 \\ 3x + 2y + z = 4 \\ x - 3y + 4z = 5 \end{cases}$

2. $\begin{cases} x - y + 2z = 3 \\ -6x + y + 3z = 7 \\ x + 2y - 5z = -4 \end{cases}$

3. $\begin{cases} 2x - y - z = 1 \\ 2x - 3y - 4z = 0 \\ x + y - z = 4 \end{cases}$

4. $\begin{cases} y + z = 6 \\ x + 5y - 4z = 4 \\ x - 3y + 5z = 7 \end{cases}$

5. $\begin{cases} x + y - 2z = 4 \\ 2x + y = 1 \\ 5x + 3y - 2z = 6 \end{cases}$

6. $\begin{cases} 2y + z = -4 \\ 3x + 4z = 11 \\ x + y = -2 \end{cases}$

7. $\begin{cases} x - y + 5z = -6 \\ x + 2z = 0 \\ 6x + y + 3z = 0 \end{cases}$

8. $\begin{cases} x - y + 2z = -3 \\ 2x + y - z = 5 \\ 3x - 2y + 2z = -3 \end{cases}$

9. $\begin{cases} y + z = 2 \\ x + z = 5 \\ x + y = 5 \end{cases}$

10. $\begin{cases} x - y - 2z = 3 \\ x + 2y + z = 1 \\ 3y + 3z = -2 \end{cases}$

11. $\begin{cases} 2x - y + 5z = -2 \\ x + 3y - z = 6 \\ 4x + y + 3z = -2 \end{cases}$

12. $\begin{cases} 2x - y + 5z = 5 \\ x - 2y + 3z = 0 \\ x + y + 4z = 7 \end{cases}$

13. $\begin{cases} 3x + y + 4z = -6 \\ 2x + 3y - z = 2 \\ 5x + 4y + 3z = 2 \end{cases}$

14. $\begin{cases} 2x + y - z = -3 \\ -x + 2y + z = 5 \\ 2x + 3y - 2z = -3 \end{cases}$

15. $\begin{cases} x - 2y + z = 4 \\ x - y - 4z = 1 \\ 2x - 4y + 2z = 8 \end{cases}$

16. $\begin{cases} 2x - 2y + 3z = 4 \\ x - 3y + 2z = 2 \\ x + y + z = 1 \end{cases}$

17. $\begin{cases} 2x - 3y + z = -1 \\ 6x - 9y - 4z = 4 \\ 4x + 6y - z = 5 \end{cases}$

18. $\begin{cases} x + y + z = 3 \\ 2x - y - 2z = -3 \\ 3x + 2y + z = 4 \end{cases}$

19. $\begin{cases} 2x + 3y + z = 4 \\ 3x - 5y + 2z = -5 \\ 4x - 6y + 3z = -7 \end{cases}$

20. $\begin{cases} x + 6y + z = 6 \\ 2x + 3y - 2z = 8 \\ 2x + 4z = 3 \end{cases}$

21. The sum of three numbers is 67. The sum of the first and second numbers exceeds the third number by 13. The third number is 7 less than the first. Find the three numbers.

22. The sum of three numbers is 189. The first number is 28 less than the second number. The second number is 21 less than the sum of the first and third numbers. Find the three numbers.

23. Bud has 23 coins in his pocket, including nickels, dimes, and quarters. He has two more dimes than quarters. The total value of the coins is \$2.50. How many of each kind of coin does he have?

24. A wallet contains \$218 in \$10, \$5, and \$1 bills. There are forty-six bills in all and four more fives than tens. How many bills of each kind are there?

25. Find values for $a$, $b$, and $c$ so that the points $(-1,-4)$, $(2,8)$, and $(-2,-4)$ lie on the graph of $y = ax^2 + bx + c$.

26. Find the values for $a$, $b$, and $c$ so that the points $(1,1)$, $(-3,13)$, and $(0,-2)$ lie on the graph of $y = ax^2 + bx + c$.

27. The perimeter of a triangle is 73 cm. The longest side is 13 cm less than the sum of the other two sides. The shortest side is 11 cm less than the longest side. Find the lengths of the three sides.

28. At Tony's Fruit Stand, 4 pounds of bananas, 2 pounds of apples, and 3 pounds of grapes cost \$8.20. Five pounds of bananas, 4 pounds of apples, and 2 pounds of grapes cost \$8.30. Two pounds of bananas, 3 pounds of apples, and 1 pound of grapes cost \$4.80. Find the price per pound of each kind of fruit.

**29.** The Marshalls are having a house built. The cost of building the house is \$12,000 more than three times the cost of the lot. The cost of landscaping, sidewalks, and so on, is one-half the cost of the lot. If the total cost is \$61,500, what was the cost of each part (the home, the lot, and the improvements)?

**30.** At the Happy Burger Drive-In, you can buy 2 hamburgers, 1 chocolate shake, and 2 orders of fries, or 3 hamburgers and 1 order of fries, for \$4.75. One hamburger, 2 chocolate shakes, and 1 order of fries cost \$3.65. How much does a hamburger cost?

## 7.4 Systems of Linear Inequalities and Linear Programming

**OBJECTIVES**

In this section, you will be learning to:

1. Solve systems of linear inequalities graphically.
2. Solve applications of linear programming that involve systems of linear inequalities.

From Section 6.4, we know that the graph of the set of all points satisfying a linear inequality such as

$$4x - 3y \leq 12$$

is a closed half-plane. The boundary line $4x - 3y = 12$ is included in the graph (Figure 7.3).

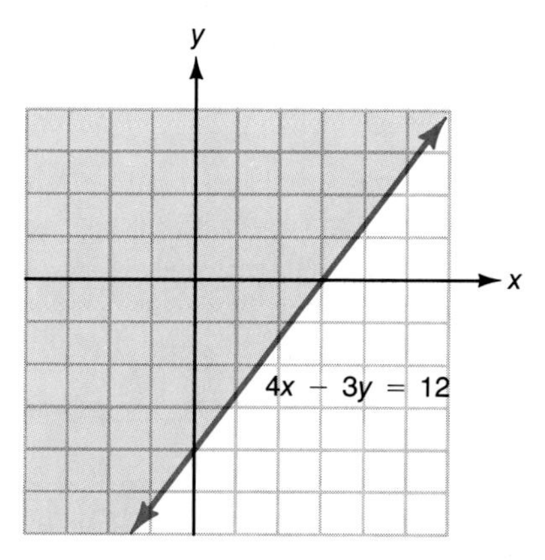


Closed half-plane $4x - 3y \leq 12$

**Figure 7.3**

Now we want to graph a **system of inequalities.** That is, given a system of two or more inequalities, we want to find the points that satisfy **all** the inequalities. We want the intersection of all the half-planes represented. For example, the solution to the system

$$\begin{cases} x \geq 0 \\ y \leq 0 \\ 4x - 3y \leq 12 \end{cases}$$

is graphed in Figure 7.4.

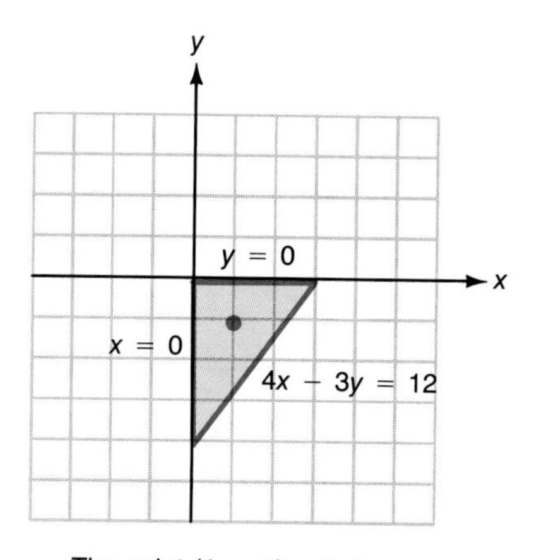


The point $(1, -1)$ satisfies all three inequalities.

**Figure 7.4**

The shaded region represents the points that are to the right of the line $x = 0$ ($x \geq 0$), below the line $y = 0$ ($y \leq 0$), and above the line $4x - 3y = 12$ ($4x - 3y \leq 12$). Each border line segment is also included.

**EXAMPLE**

**1.** Solve the system $\begin{cases} 2x + y \leq 6 \\ x + y \leq 4 \\ x \geq 0 \\ y \geq 0 \end{cases}$ graphically.

**Solution**

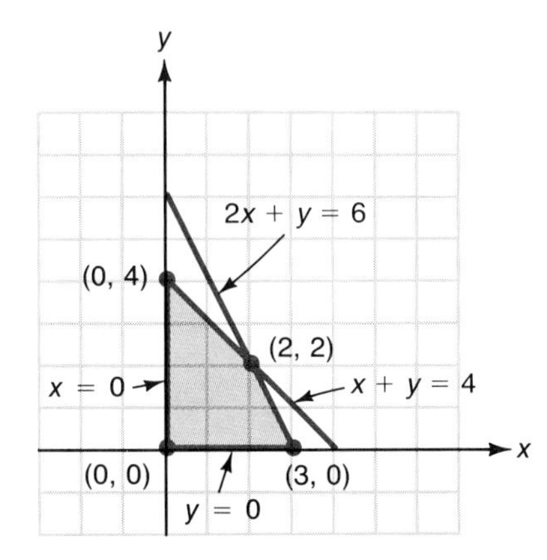


The region shaded in Example 1 is bounded by straight line segments. These segments form a polygon, and the points where the segments meet, namely, (0,0), (3,0), (2,2), and (0,4), are called **vertices.** These can be found by solving simultaneously in pairs the linear equations associated with each inequality.

$$\begin{cases} 2x + y = 6 \\ x + y = 4 \end{cases} \qquad \begin{cases} 2x + y = 6 \\ y = 0 \end{cases} \qquad \begin{cases} x + y = 4 \\ x = 0 \end{cases} \qquad \begin{cases} x = 0 \\ y = 0 \end{cases}$$

A look at the graph tells you which pairs to solve simultaneously to find each vertex of the polygon. For example, solving

$$\begin{cases} 2x + y = 6 \\ x = 0 \end{cases}$$

gives (0,6), but this is not a vertex and therefore not under consideration. ■

The system of inequalities in Example 1 has an infinite number of solutions. But in **linear programming,** which has a great many applications in business, dietary problems, and science, we are concerned mainly with the vertices. In such applications, we are generally dealing with an **objective function** that relates some quantity such as cost or profit to the variables $x$ and $y$. The following theorem is basic to linear programming and we will not attempt any proof here. We are using it only to illustrate the usefulness of systems of inequalities. (**Note:** We will discuss functions in detail in Chapter 8.)

***Theorem*** In a linear program (which involves an objective function and a set of conditions given as a system of inequalities), the maximum and minimum values of the **objective function,** if they exist, occur at the vertices or along one edge of the solution polygon.

**EXAMPLE**

**2.** A farmer can grow two crops, A and B. He has 20 acres to plant. The cost of fertilizing for crop A is $16 per acre and $4 for crop B, and he has $160 for fertilizer. If the profit from crop A is $100 per acre and the profit from crop B is $75 per acre, how many acres of each crop should the farmer plant to **maximize his profit?**

**Solution** Let $x$ = number of acres of crop A
$y$ = number of acres of crop B

The objective function is $P = 100x + 75y$ where P represents profit. The conditions are

| | |
|---|---|
| $x + y \leq 20$ | He has a total of 20 acres. |
| $16x + 4y \leq 160$ | He has $160 to spend. |
| $x \geq 0$ | He can plant all or none in crop A. |
| $y \geq 0$ | He can plant all or none in crop B. |

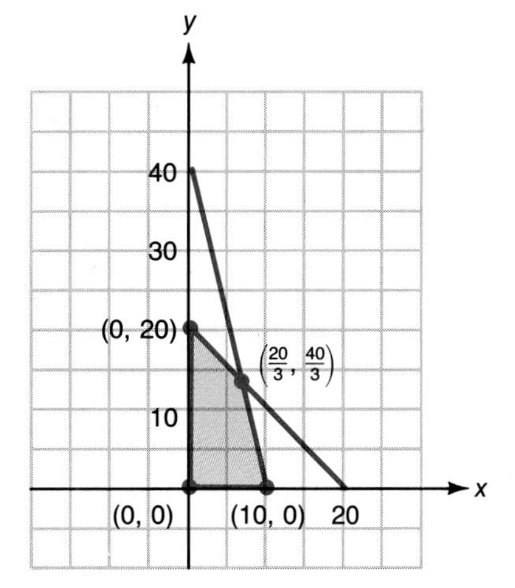

Solving

$$\begin{cases} x + y = 20 \\ 16x + 4y = 160 \end{cases} \quad \text{Gives } x = \frac{20}{3}, y = \frac{40}{3}.$$

$$\begin{cases} x = 0 \\ x + y = 20 \end{cases} \quad \text{Gives } x = 0, y = 20.$$

$$\begin{cases} y = 0 \\ 16x + 4y = 160 \end{cases} \quad \text{Gives } x = 10, y = 0.$$

$$\begin{cases} x = 0 \\ y = 0 \end{cases} \quad \text{Gives } (0,0).$$

Evaluating $P$ at each vertex gives

$$P = 100\left(\frac{20}{3}\right) + 75\left(\frac{40}{3}\right) = \frac{5000}{3} = \$1666\frac{2}{3} \quad \text{Maximum profit}$$
$$P = 100(0) + 75(20) = \$1500$$
$$P = 100(10) + 75(0) = \$1000$$
$$P = 100(0) + 75(0) = \$0 \quad \text{Minimum profit}$$

His maximum profit lies in planting $\frac{20}{3}$ acres of crop A and $\frac{40}{3}$ acres of crop B. This profit is $\$1666\frac{2}{3}$. ■

## EXERCISES 7.4

Solve the systems of inequalities in Exercises 1–15 graphically.

1. $\begin{cases} 3x + y < 10 \\ 5x - y \geq 6 \end{cases}$

2. $\begin{cases} y \geq 2x - 5 \\ 3x + 2y > -3 \end{cases}$

3. $\begin{cases} 3x + 4y \geq -7 \\ y < 2x + 1 \end{cases}$

4. $\begin{cases} 2x - 3y \geq 0 \\ 8x - 3y < 36 \end{cases}$

5. $\begin{cases} x + y < 4 \\ 2x - 3y < 3 \end{cases}$

6. $\begin{cases} 2x + 3y < 12 \\ 3x + 2y \leq 13 \end{cases}$

7. $\begin{cases} x \geq 0 \\ 3x - 5y \leq 10 \\ 4x + 3y \leq 23 \end{cases}$

8. $\begin{cases} x \geq 0 \\ 3x + 2y \leq 15 \\ 2x + 5y \geq 10 \end{cases}$

9. $\begin{cases} y \geq 0 \\ 4x - 3y \geq -6 \\ 3x - y \leq 3 \end{cases}$

10. $\begin{cases} y \geq 0 \\ 3x + 4y < 24 \\ 2x + y \leq 11 \end{cases}$

11. $\begin{cases} x \geq 0 \\ y \geq 0 \\ 3x - 4y \geq -6 \\ 3x + 2y \leq 12 \end{cases}$

12. $\begin{cases} x \geq 0 \\ y \geq 0 \\ y \leq 2x + 2 \\ x + 2y \leq 11 \end{cases}$

13. $\begin{cases} x \geq 0 \\ y \geq 0 \\ x + y \leq 8 \\ 3x - 2y \geq -6 \end{cases}$

14. $\begin{cases} x \geq 0 \\ y \geq 0 \\ y \leq 2x + 1 \\ x + y \leq 7 \\ 2x - y \leq 8 \end{cases}$

15. $\begin{cases} x \geq 0 \\ y \geq 0 \\ y \leq x \\ 2x + 5y \leq 21 \\ 2x + y \leq 17 \end{cases}$

Maximize the function $F$, subject to the indicated system of inequalities in Exercises 16–21.

**16.** $\begin{cases} x \geq 0 \\ y \geq 0 \\ 2x - y \geq -3 \\ 3x + y \leq 24 \end{cases}$ $F = 5x - 2y$

**17.** $\begin{cases} x \geq 0 \\ y \geq 0 \\ x - y \geq -2 \\ 4x - y \leq 16 \end{cases}$ $F = -3x - y$

**18.** $\begin{cases} x \geq 0 \\ y \geq 0 \\ x + 3y \leq 18 \\ 4x + y \leq 28 \end{cases}$ $F = 4x + 3y$

**19.** $\begin{cases} x \geq 0 \\ y \geq 0 \\ x \leq 6 \\ x - 2y \geq -6 \\ 2x + 3y \leq 18 \end{cases}$ $F = 7x + 14y$

**20.** $\begin{cases} x \geq 0 \\ y \geq 0 \\ y \leq x + 2 \\ x + 2y \leq 17 \\ 4x - y \leq 32 \end{cases}$ $F = -3x + 8y$

**21.** $\begin{cases} x \geq 0 \\ y \geq 0 \\ 3y \leq x + 12 \\ x + 2y \leq 13 \\ 3x + y \leq 24 \end{cases}$ $F = 6x - 2y$

In each of the word problems in Exercises 22–25, find a system of inequalities and the objective function; then solve.

**22.** A small retail butcher shop sells two grades of hamburger. Grade A costs $0.90 per pound and Grade B costs $0.70 per pound. The daily order can be no more than 20 pounds and can cost no more than $15.20. If the profit is $0.15 per pound on Grade B and $0.30 per pound on Grade A, what is the maximum daily profit on hamburger?

**23.** A manufacturer produces two kinds of radios. Model X takes 4 hours to produce, costs $8, and sells for $15. Model Y takes 3 hours to produce, costs $7, and sells for $12. If the manufacturer allots a total of 58 hours and $126 per week to produce radios, find his maximum income.

**24.** Larry sells sound systems. Brand X takes 2 hours to install and $\frac{1}{2}$ hour to adjust. Brand Y takes $1\frac{1}{2}$ hours to install and $\frac{3}{4}$ hour to adjust. Larry spends no more than 30 hours per week for installation and 12 hours per week for adjusting. He sells a total of at least 8 systems per week. Brand X sells for $160 and Brand Y sells for $240. Find the maximum weekly income.

**25.** A small shop specializes in refinishing antique chairs. They use two types of finishes. Type 1 takes 2 hours, and material costs $6.00. Type 2 takes 3 hours, and material costs $3.00. The shop charges $30 per chair for Type 1 finish and $36 per chair for Type 2 finish. If they allow no more than 40 hours per week for labor and no more than $84 for material, what is the shop's maximum income?

## CHAPTER 7 SUMMARY

### Key Terms and Formulas

Two (or more) linear equations considered at one time are said to form a **system of linear equations** or a **set of simultaneous equations.** If the system has a unique solution, the system is **consistent.** If there are no solutions, the system is **inconsistent.** If there are an infinite number of solutions, the system is **dependent.** [7.1]

In a linear program (which involves an objective function and a set of conditions given as a system of inequalities), the maximum and minimum values of the **objective function,** if they exist, occur at the vertices or along one edge of the solution polygon. [7.4]

### Procedures

Three Techniques for Solving a System of Two Linear Equations [7.1]

1. **Graphing**
   Graph both equations on the same set of axes and observe the point of intersection (if there is one). Graphical solutions may be only estimates.
2. **Substitution**
   Solve one of the equations for one of the variables; then substitute the resulting expression into the other equation.
3. **Addition**
   Multiply the terms of each equation so that two like terms have opposite coefficients; then add like terms.

**To Solve Three Linear Equations in Three Variables** [7.3]

1. Select two equations and eliminate one variable using the addition method.
2. Select two other equations and eliminate **the same** variable.
3. From steps 1 and 2, you have **two** linear equations in **two** variables. Solve these two equations by either addition or substitution.
4. Substitute your values from step 3 into any one of the original equations and find the value of the third variable.

## CHAPTER 7 REVIEW

Solve each of the systems in Exercises 1–4 by graphing. [7.1]

**1.** $\begin{cases} x - y = 4 \\ x + 2y = -2 \end{cases}$ **2.** $\begin{cases} 3x - y = 5 \\ 2x + y = 10 \end{cases}$ **3.** $\begin{cases} 3y = 5 - 2x \\ x - 4y = -14 \end{cases}$ **4.** $\begin{cases} 5x + 3y = 7 \\ 3x + 4y = 2 \end{cases}$

Solve the systems of equations in Exercises 5–10. State which systems are dependent or inconsistent. [7.1]

**5.** $\begin{cases} x + y = 2 \\ x - y = -1 \end{cases}$ **6.** $\begin{cases} x + 2y = -8 \\ 3x - 4y = 6 \end{cases}$ **7.** $\begin{cases} 6x - 2y = -3 \\ 10x + 3y = 14 \end{cases}$

**8.** $\begin{cases} 5x + 8y = 9 \\ \frac{5}{2}x + 4y = 3 \end{cases}$ **9.** $\begin{cases} 2x + y = 12 \\ 0.25x + 0.1y = 1.3 \end{cases}$ **10.** $\begin{cases} 9x + 6y = 12 \\ 6x + 4y = 8 \end{cases}$

Solve the systems of equations in Exercises 11–16. State which systems have no solution or an infinite number of solutions. [7.3]

**11.** $\begin{cases} 3x - 5y + 2z = 3 \\ 2x + 2z = 3 \\ -x + 5y - 4z = 2 \end{cases}$ **12.** $\begin{cases} x - 3y + 2z = -1 \\ -2x + y + 3z = 1 \\ x - y + 4z = 9 \end{cases}$ **13.** $\begin{cases} 2x - 3y - 2z = 2 \\ x + 2y + 4z = 3 \\ 8x - 12y - 8z = 9 \end{cases}$

**14.** $\begin{cases} 4x + 3y - 2z = -14 \\ 2x + 2y - 3z = -17 \\ x + 5y + 6z = 29 \end{cases}$ **15.** $\begin{cases} x + y + z = -1 \\ x + y + 3z = 1 \\ 2x + 2y + 4z = 0 \end{cases}$ **16.** $\begin{cases} 2x + y - z = 5 \\ x + 2y - 2z = 4 \\ 4x + 5y - 5z = 13 \end{cases}$

Solve each system of inequalities in Exercises 17–19 graphically. [7.4]

**17.** $\begin{cases} 3x - y \leq 15 \\ 2x + y \leq 5 \end{cases}$ **18.** $\begin{cases} x \geq 0 \\ y \geq 0 \\ x - 4y \leq -2 \\ 2x + 3y \geq 18 \end{cases}$ **19.** $\begin{cases} x \geq 0 \\ y \geq 0 \\ 3x + y \geq 7 \\ x + y \leq 7 \\ 3x - y \geq 5 \end{cases}$

In Exercises 20–22, maximize the function $F$, subject to the indicated system of inequalities. [7.4]

**20.** $\begin{cases} x \geq 0 \\ y \geq 0 \\ 2x + 3y \leq 18 \\ x - 2y \geq -5 \end{cases}$
$F = 13x + 15y$

**21.** $\begin{cases} x \geq 0 \\ y \geq 0 \\ x + 2y \leq 10 \\ x - 4y \leq -2 \end{cases}$
$F = 8x + 20y$

**22.** $\begin{cases} x \geq 0 \\ y \geq 0 \\ x + 2y \leq 12 \\ 3x + 2y \leq 16 \\ x + 6y \geq 16 \end{cases}$
$F = 18x + 24x$

**23.** Determine $a$ and $b$ such that the straight line $ax + by = 11$ passes through the points $(1,-3)$ and $(2,5)$. [7.2]

**24.** MeiLing bought 2 blouses and 1 skirt for a total of \$86. If she had bought 1 blouse and 2 skirts, she would have paid \$91. What was the price of each blouse and each skirt? [7.2]

**25.** Jeanne has 11 coins in her purse, consisting of quarters and nickels. If the total value of the coins is \$1.95, how many of each type of coin does she have? [7.2]

**26.** A farmer can buy two grades of fertilizer. Grade 1 contains 25% nitrogen and 10% phosphate. Grade 2 contains 15% nitrogen and 22% phosphate. How many kilograms of each should be used to obtain 1000 kilograms of a fertilizer that is 21% nitrogen and 14.8% phosphate? [7.2]

**27.** A watch and ring together cost \$860. The cost of the ring is \$10 more than $1\frac{1}{2}$ times the cost of the watch. Find the cost of each. [7.2]

**28.** In her nutrition class, Susan is to plan a diet supplement using three foods, labeled A, B, and C. Each ounce of food A contains 3 units of protein and 5 units of vitamin C. Each ounce of food B contains 5 units of protein and 4 units of vitamin C. Each ounce of food C contains 2 units of protein and 7 units of vitamin C. If the total intake is to be 21 ounces per day containing 72 units of protein and 108 units of vitamin C, how many ounces of each food should be used? [7.3]

**29.** The points $(0,4)$, $(-2,6)$, and $(1,9)$ lie on the curve described by $y = ax^2 + bx + c$. Find $a$, $b$, and $c$. [7.3]

**30.** A company manufactures two products, A and B. Product A takes 4 hours to produce, costs \$18, and sells for \$25. Product B takes 2 hours to produce, costs \$12, and sells for \$17. If there are 44 labor-hours and \$228 available, how many of each product should be produced to maximize revenue? [7.4]

**31.** A farmer grows two crops. Crop I costs \$72 per acre to plant and makes a profit of \$104 per acre. Crop II costs \$120 per acre to plant and makes a profit of \$130 per acre. Government regulations require that the farmer plant no more than a total of 110 acres. If the farmer has only \$10,800 to spend on planting, how many acres of each crop should he plant to maximize profit? [7.4]

## CHAPTER 7 TEST

In Exercises 1 and 2, solve each system of equations by graphing.

**1.** $\begin{cases} 4x - y = 13 \\ 2x - 3y = 9 \end{cases}$

**2.** $\begin{cases} x - y = 7 \\ 5x + 7y = -1 \end{cases}$

Solve the systems of equations in Exercises 3–7. State which systems are dependent or inconsistent.

**3.** $\begin{cases} x + y = 9 \\ x - y = 5 \end{cases}$

**4.** $\begin{cases} 6x + 3y = 5 \\ 4x + 2y = -3 \end{cases}$

**5.** $\begin{cases} 7x - 6y = 2 \\ 5x + 2y = 3 \end{cases}$

**6.** $\begin{cases} 2x + 3y = 1 \\ 5x + 7y = 6 \end{cases}$

**7.** $\begin{cases} 3x - 4y = 12 \\ 2x + 6y = 21 \end{cases}$

**8.** Determine $a$ and $b$ such that the straight line $ax + by = 17$ passes through the points $(-3,2)$ and $(1,5)$.

**9.** The length of a rectangle is 7 ft more than twice the width. The perimeter is 62 ft. Find the width and length of the rectangle.

**10.** One pair of trousers and 2 shirts cost \$79. Three pairs of trousers and 3 shirts cost \$168. Find the price of each.

Solve the systems of equations in Exercises 11–14. State which systems have no solution or an infinite number of solutions.

**11.** $\begin{cases} 2x + y = 3 \\ x + 2z = -6 \\ 3y - z = 1 \end{cases}$

**12.** $\begin{cases} x - 2y - 3z = 3 \\ x + y - z = 2 \\ 2x - 3y - 5z = 5 \end{cases}$

**13.** $\begin{cases} x + 2y - 2z = 0 \\ x - y + z = 4 \\ -x + 4y - 4z = -8 \end{cases}$

**14.** $\begin{cases} x + 2y - 3z = -11 \\ x - y - z = 2 \\ x + 3y + 2z = -4 \end{cases}$

**15.** Alma bought 100 stamps in denominations of 18¢, 20¢, and 22¢. She bought three times as many 18¢ stamps as 20¢ stamps. The total cost of the stamps was \$19.90. How many stamps of each denomination did she buy?

In Exercises 16 and 17, solve each system of inequalities graphically.

**16.** $\begin{cases} 2x + y \leq 8 \\ x - y \geq -2 \end{cases}$

**17.** $\begin{cases} 4x + 3y \leq 18 \\ 2x + 5y \geq 16 \end{cases}$

In Exercises 18 and 19, maximize the function $F$, subject to the indicated system of inequalities.

**18.** $\begin{cases} x \geq 0 \\ y \geq 0 \\ 3x + y \leq 24 \\ 2x - y \geq -4 \end{cases}$

$F = 12x + 9y$

**19.** $\begin{cases} x \geq 0 \\ y \geq 0 \\ y \geq x \\ x + 3y \leq 24 \\ 2x - 3y \geq -15 \end{cases}$

$F = 11x + 14y$

**20.** A furniture shop refinishes chairs. Two methods are used in refinishing. Method I takes 1 hour and materials cost \$3. Method II takes $1\frac{1}{2}$ hours and materials cost \$1.50. For a typical week, 36 hours can be devoted to labor and \$60 for materials. If they charge \$32 to refinish a chair using Method I and \$42.50 using Method II what is their maximum revenue?

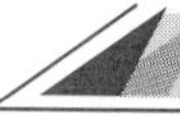

## CUMULATIVE REVIEW (7)

Evaluate each expression in Exercises 1 and 2 for the indicated values of $x$.

**1.** $4x^2 - 3x + 6, x = -2$

**2.** $2x^2 + 5x - 4, x = \frac{3}{2}$

Factor completely each expression in Exercises 3–6.

**3.** $6x^2 - 7x - 20$

**4.** $2x^3 - 54$

**5.** $8x^3 + 125$

**6.** $x^3 + 3x^2 - 4x - 12$

Add the correct constant in Exercises 7 and 8; then factor as indicated.

**7.** $x^2 + 7x +$ _____ $= (\quad)^2$

**8.** $3x^2 - 24x +$ _____ $= 3(\quad)^2$

**9.** Determine the value of $K$ so that $x^2 + 9x + K$ will have one real solution.

**10.** Determine the values of $K$ so that $4x^2 + 3x + K$ will have two real solutions.

Solve the equations in Exercises 11 and 12.

**11.** $2x^2 - 7x - 15 = 0$

**12.** $3x^2 - 5x - 4 = 0$

Graph the linear equations in Exercises 13 and 14 by locating the $x$-intercept and the $y$-intercept.

**13.** $3x - 4y = 12$

**14.** $5x + 2y = 15$

Graph the linear equations in Exercises 15 and 16.

**15.** $2x + 3y - 7 = 0$

**16.** $6y - 18 = 4x$

Solve each of the systems in Exercises 17 and 18 by graphing.

**17.** $\begin{cases} 2x + y = 6 \\ 3x - 2y = -5 \end{cases}$

**18.** $\begin{cases} x + 3y = 10 \\ 5x - y = 2 \end{cases}$

**19.** Bill makes two kinds of candy. One recipe requires 4 oz of peanuts for each 10 oz of chocolate chips. The other recipe requires 12 oz of peanuts for each 8 oz of chocolate chips. How many "batches" of each can he make if he has 46 oz of chocolate chips and 36 oz of peanuts?

**20.** Olivia has \$7000 invested, some at 7% and the remainder at 8%. The interest from the 7% investment exceeds the interest from the 8% investment by \$70. How much is invested at each rate?

CHAPTER

# 8 FUNCTIONS

## MATHEMATICAL CHALLENGES

My watch is 1 second fast each hour, and my friend's is 1.5 seconds slow each hour. Right now they show the same time. When will they show the same time again.

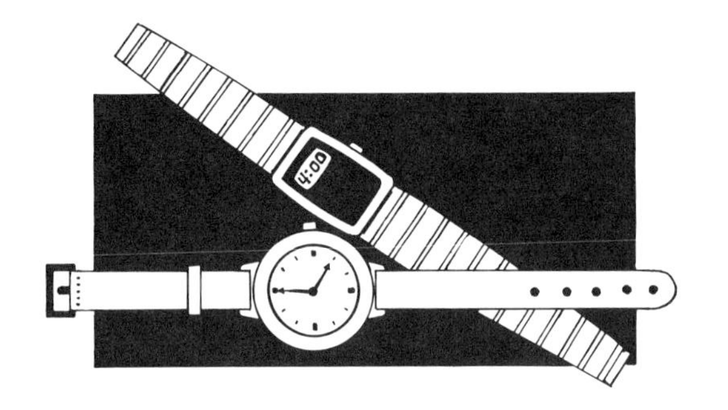


Henry threw six darts and all six hit the target shown. Which of the following could be his score?

4, 17, 56,

28, 29, 31

Given two odd integers $a$ and $b$, prove that $a^3 - b^3$ is divisible by $2^n$ if and only if $a - b$ is divisible by $2^n$.

## CHAPTER OUTLINE



The speed at which you drive your car is a function of how much gasoline you give it; your energy level is a function of the amount and type of food you eat; your grade in this class is a function of the quality time you spend studying. Obviously, the concept of a function is present in many aspects of our daily lives. In this chapter, we will quantify these ideas by dealing only with functions involving pairs of real numbers. This restriction allows us to analyze functions in terms of their graphs in the Cartesian coordinate system. We will be particularly interested in a special category called quadratic functions whose graphs are parabolas. However, we will also build a list of general properties that form the basis for all our work with functions and lead to the topics of exponential functions and logarithmic functions in Chapter 9.

Many mathematical models for applications depend on the concept of functions and employ function notation. For example, we might represent the depreciated value of a piece of equipment as $f(t) = 10{,}000 + 4000.2^{-0.3t}$ where $t$ is time in years. [$f(t)$ is read "$f$ of $t$."] Or, the profit in selling $x$ items might be $P(x) = 500 + 3x - 2x^2$. In any case, we will find that functions and function notation provide the basic tools for dealing with graphs and understanding applied problems.

## 8.1 Introduction to Functions

**OBJECTIVES**

In this section, you will be learning to:

1. State the domains and ranges of relations and functions.
2. Use the vertical line test to determine whether or not a graph represents a function.
3. Write functions as sets of ordered pairs.

In Chapter 6, we discussed linear equations and linear inequalities and their graphs using the Cartesian coordinate system. In this system, the graphs are represented by ordered pairs of real numbers. In this chapter, we will introduce new terminology related to ordered pairs and discuss certain types of second-degree equations and their graphs. We will also develop the important concept of a **function** and the related **function notation.**

***Relation, Domain, and Range***

A **relation** is a set of ordered pairs.
The **domain, D,** of a relation is the set of all first components in the relation.
The **range, R,** of a relation is the set of all second components in the relation.

Thus, in graphing relations, the $x$-axis is called the **domain axis** and the $y$-axis is called the **range axis.**

**EXAMPLES** Find the domain and range for each of the following relations.

**1.** $r = \{(5,7), (\sqrt{6},2), (\sqrt{6},3), (-1,2)\}$

**Solution** $D = \{5,\sqrt{6},-1\}$ All the first components in $r$

$R = \{7,2,3\}$ All the second components in $r$

Note that we write $\sqrt{6}$ only once in the domain and 2 only once in the range, even though each appears more than once in the relation.

**2.** $f = \{(-1,1), (1,5), (0,3)\}$

**Solution** $D = \{-1,1,0\}$

$R = \{1,5,3\}$ ■

The relation $f = \{(-1,1), (1,5), (0,3)\}$ used in Example 2 meets a particular condition in that each first component has only one corresponding second component. Such a relation is called a **function.** Notice that $r$ in Example 1 is **not** a function because the first component $\sqrt{6}$ has more than one corresponding second component.

The following three definitions of a function are all equivalent.

***Function***

A **function** is a relation in which each domain element has only one corresponding range element.

OR

A **function** is a relation in which each first component appears only once.

OR

A **function** is a relation in which no two ordered pairs have the same first component.

**EXAMPLES** Determine whether or not each of the following relations is a function.

**3.** $r = \{(2,3), (1,6), (2,\sqrt{5}), (0,-1)\}$

**Solution** $r$ is not a function. The first component, 2, appears more than once.

**4.** $s = \{(0,0), (1,1), (2,4), (3,9)\}$

**Solution** $s$ is a function. Each first component has only one corresponding second component.

**5.** $t = \{(1,5), (3,5), (\sqrt{2},5), (-1,5), (-4,5)\}$

**Solution** $t$ is a function. Each first component appears only once. The fact that the second components are all the same has no effect on the definition of a function. ■

If one point in the graph of a relation is directly above or below another point in the graph, then these points have the same first coordinate (or $x$-coordinate). Such a relation would **not** be a function. Therefore, we can tell whether or not a graph represents a function by using the **vertical line test.** If **any** vertical line intersects a graph of a relation in more than one point, then the relation graphed is **not** a function (Figure 8.1).

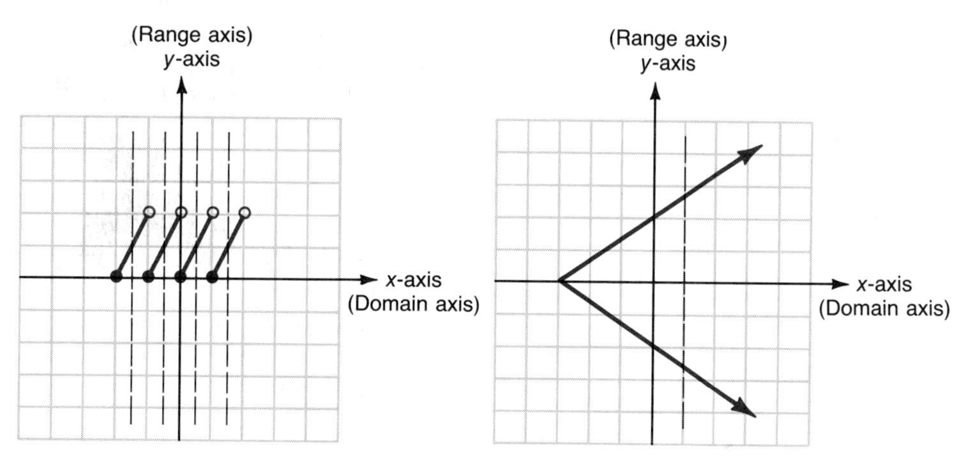


This graph represents a function. Several vertical lines are drawn to illustrate that vertical lines do not intersect the graph in more than one point. Each $x$-value has only one corresponding $y$-value.

(a)

This graph is not a function because the vertical line drawn intersects the graph in more than one point. Thus for some $x$-value, there is more than one corresponding $y$-value.

(b)

**Figure 8.1**

**EXAMPLES** Use the vertical line test to determine whether or not each of the following graphs represents a function.

**6.**

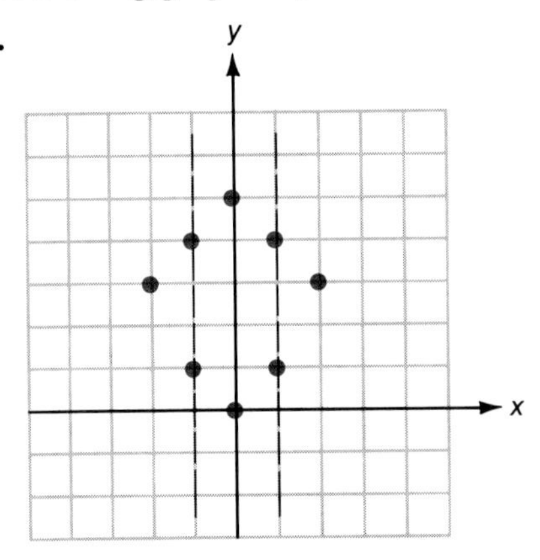


**Solution** Not a function. Some vertical line intersects the graph in more than one point. Listing the ordered pairs shows that several $x$-coordinates are repeated:

$$\{(-2,3), (-1,1), (-1,4), (0,0), (0,5), (1,1), (1,4), (2,3)\}$$

**7.**

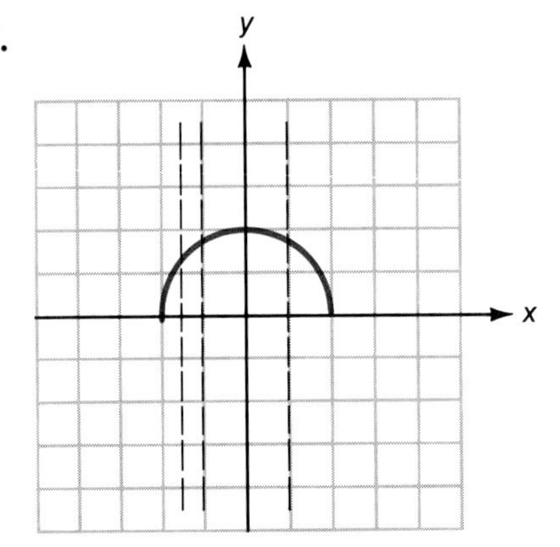


**Solution** A function. No vertical line will intersect the graph in more than one point. Several vertical lines are drawn to illustrate this.

**8.**

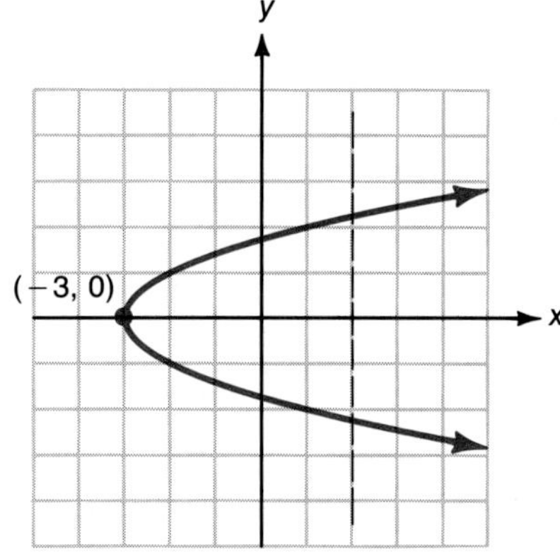


**Solution** Not a function. Some vertical line intersects the graph in more than one point.

**9.**

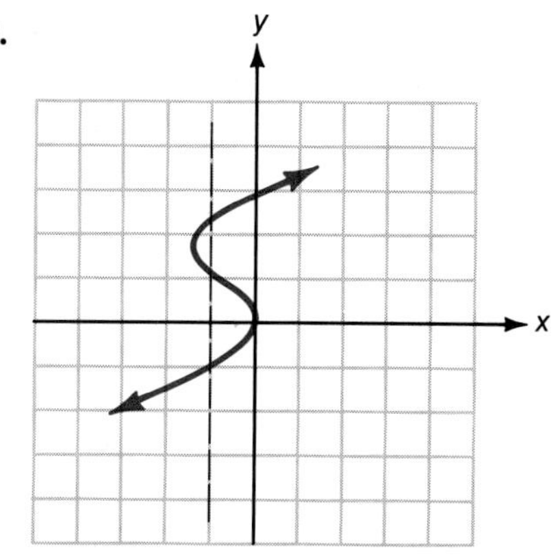


**Solution** Not a function. Some vertical line will intersect the graph in more than one point.

**10.**

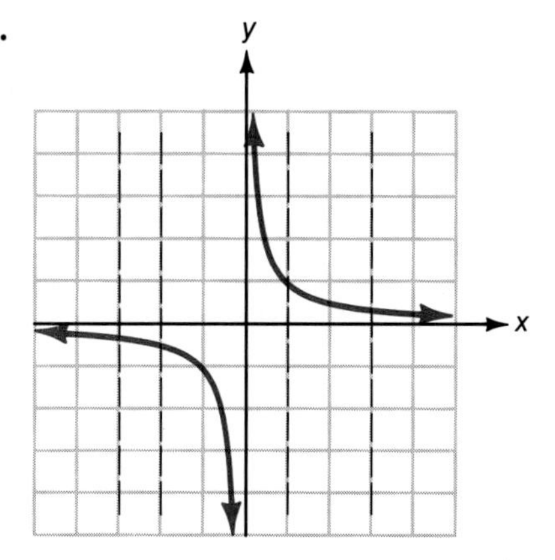


**Solution** A function. No vertical line will intersect the graph in more than one point. Several vertical lines are drawn to illustrate this. ■

If we are given the domain and range of a function, there is no way to determine the function because we do not know how to pair the components. However, given a rule or equation relating two variables, we can then determine the pairing of $x$ and $y$. For example,

given the equation $y = \frac{2}{x - 1}$

and the domain (or $x$-values) $\{2,3,5\}$,

we can determine each corresponding $y$-value by substituting 2, 3, and 5 for $x$.

$x = 2$ $\quad y = \frac{2}{2 - 1} = 2$

$x = 3$ $\quad y = \frac{2}{3 - 1} = 1$

$x = 5$ $\quad y = \frac{2}{5 - 1} = \frac{1}{2}$

The function is $\left\{(2,2), (3,1), \left(5,\frac{1}{2}\right)\right\}$.

In general, we are not interested in relations or functions involving only a few points. We adopt the following rule concerning equations and domains:

**Unless a specific domain is explicitly stated, the domain will be considered to be the set of all real $x$-values for which the given equation is defined; that is, the domain consists of all values for $x$ that give real values for $y$.**

## EXAMPLES

**11.** Given the equation (or rule relating $x$ and $y$)

$$y = x^2 + 1 \qquad \text{and} \qquad D = \{-1,0,1,2,3\},$$

find the function as a set of ordered pairs.

**Solution** We set up a table.

| $x$ | $y = x^2 + 1$ |
|---|---|
| $-1$ | $(-1)^2 + 1 = 2$ |
| 0 | $(0)^2 + 1 = 1$ |
| 1 | $(1)^2 + 1 = 2$ |
| 2 | $(2)^2 + 1 = 5$ |
| 3 | $(3)^2 + 1 = 10$ |

The function is $\{(-1,2), (0,1), (1,2), (2,5), (3,10)\}$.

Find the domain of the function represented by each equation.

**12.** $y = \dfrac{2x + 1}{x - 5}$

**Solution** The domain is all real numbers for which the expression $\dfrac{2x + 1}{x - 5}$ is defined.

Thus, $D = \{x \mid x \neq 5\}$.

**13.** $y = \sqrt{x - 2}$

**Solution** For $\sqrt{x - 2}$ to be a real number, we must have

$$x - 2 \geq 0 \qquad \text{or} \qquad x \geq 2.$$

Thus, $D = \{x \mid x \geq 2\}$. ■

***Practice Problems***

1. State the domain and range of the relation $\{(5,6), (7,8), (9,10), (10,11)\}$. Is the relation a function?
2. Write the function as a set of ordered pairs given $y = 2x - 5$ and $D = \left\{-4, 0, \dfrac{1}{2}, 3\right\}$.
3. State the domain of the function represented by the equation $y = \sqrt{x + 3}$.

## *EXERCISES 8.1*

List the sets of ordered pairs corresponding to the points in Exercises 1–10. Give the domain and range, and indicate which of the relations are also functions.

**1.**

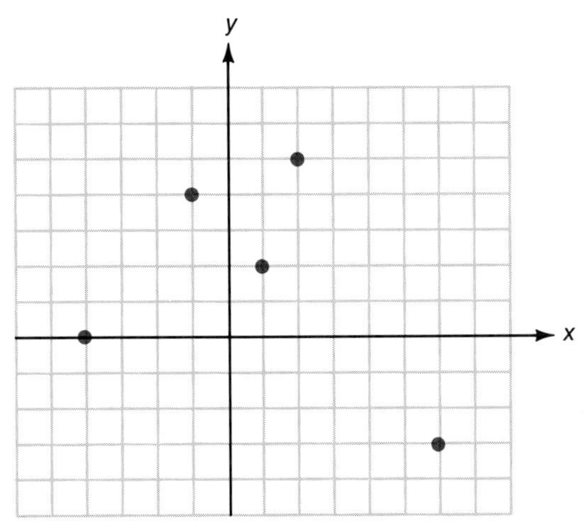


**2.**

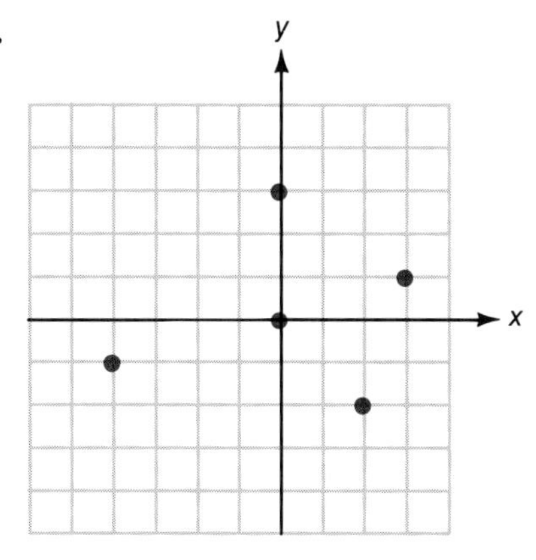


**Answers to Practice Problems** 1. $D = \{5, 7, 9, 10\}$; $R = \{6, 8, 10, 11\}$; Yes, the relation is a function. 2. $\{(-4, -13), (0, -5), \left(\frac{1}{2}, -4\right), (3, 1)\}$ 3. $D = \{x \mid x \geq -3\}$

**3.**

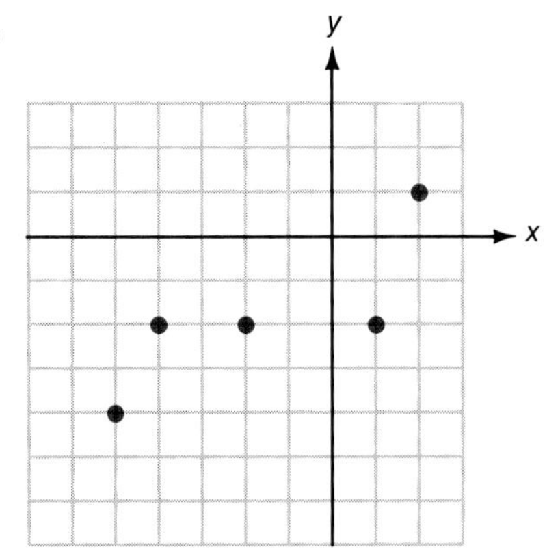


**4.**

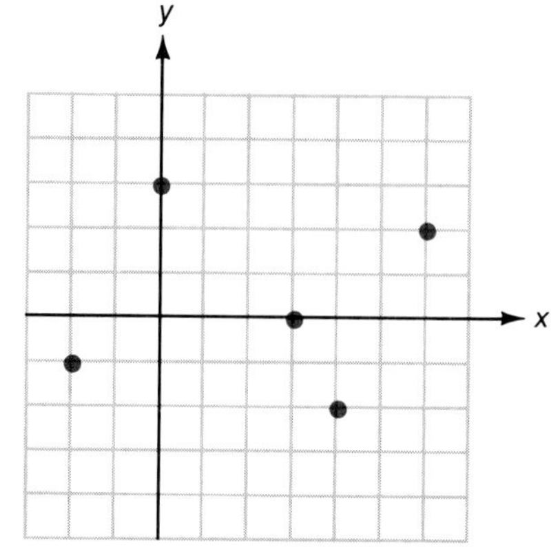


**5.**

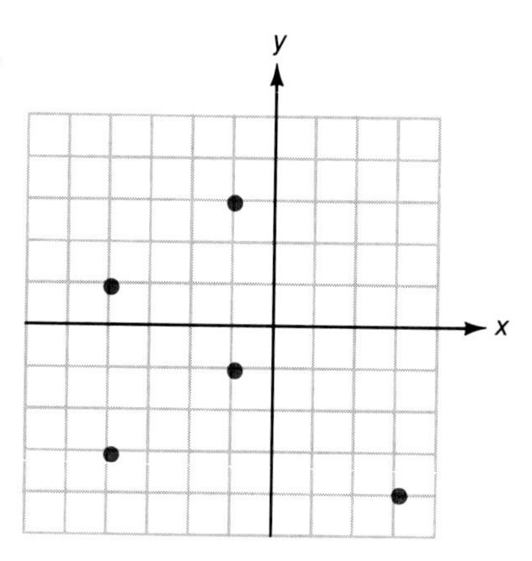


**6.**

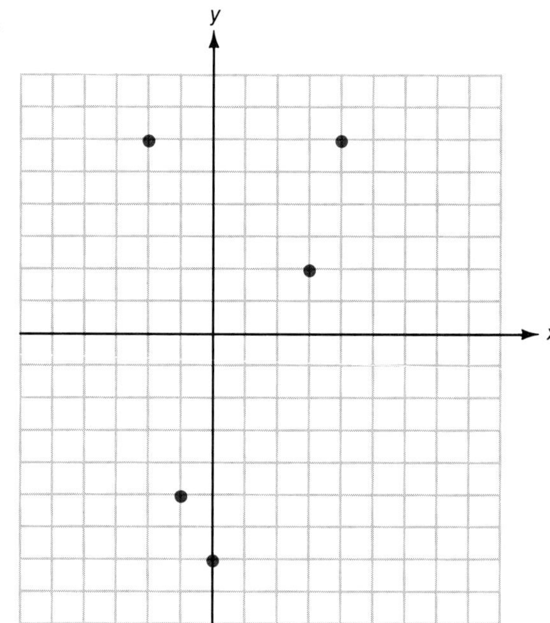


**7.**

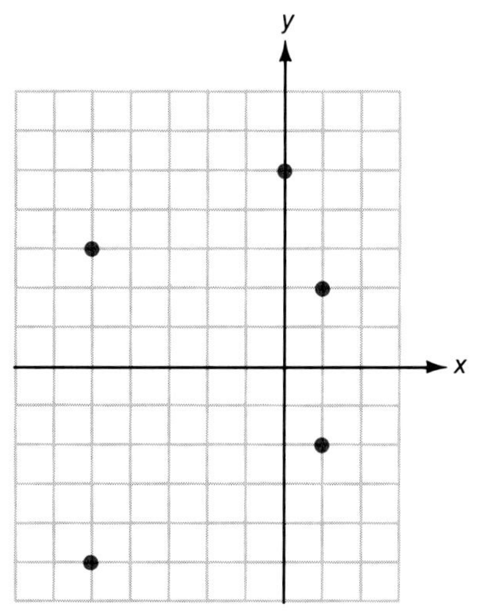


**8.**

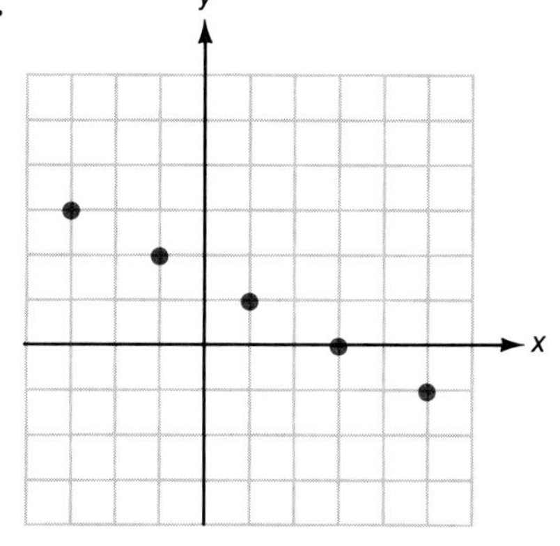

**9.**

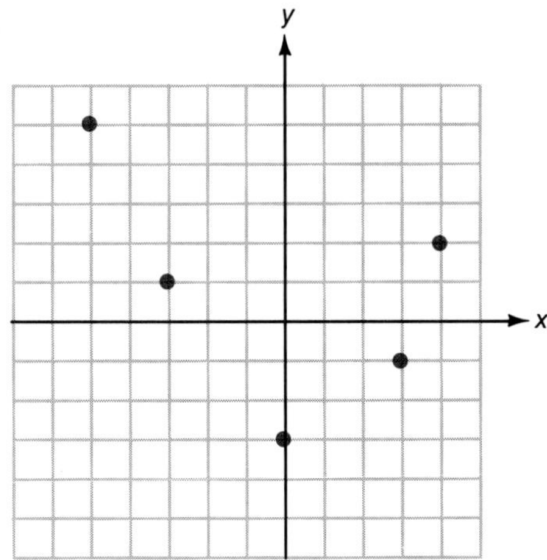


**10.**

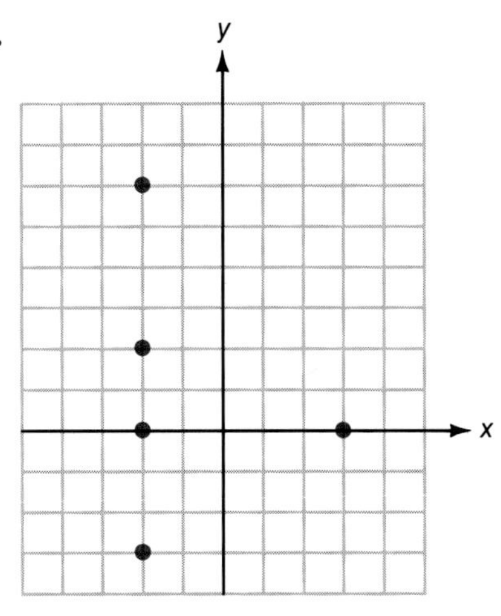


Graph the relations in Exercises 11–22. State the domain and range, and indicate which of the relations are also functions.

**11.** $\{(0,0), (1,6), (4,-2), (-3,5), (2,-1)\}$

**12.** $\{(1,-5), (2,-3), (-1,-3), (0,2), (4,3)\}$

**13.** $\{(-4,4), (3,-2), (1,0), (2,-3), (3,1)\}$

**14.** $\{(-3,-3), (0,1), (-2,1), (3,1), (5,1)\}$

**15.** $\{(0,2), (-1,1), (2,4), (3,5), (-3,5)\}$

**16.** $\{(-1,-4), (0,-3), (2,-1), (4,1), (1,1)\}$

**17.** $\{(-1,-1), (-3,-2), (1,3), (0,0), (2,5)\}$

**18.** $\{(0,3), (-1,2), (3,4), (2,-5), (-2,-2)\}$

**19.** $\{(2,-4), (2,-2), (2,5), (2,0), (2,3)\}$

**20.** $\{(0,5), (-2,6), (1,-5), (3,1), (-4,3)\}$

**21.** $\{(-1,-4), (-1,2), (-1,0), (-1,6), (-1,-2)\}$

**22.** $\{(0,0), (-2,-5), (2,0), (4,-6), (5,2)\}$

Use the vertical line test to determine whether or not each of the graphs in Exercises 23–34 represents a function.

**23.**

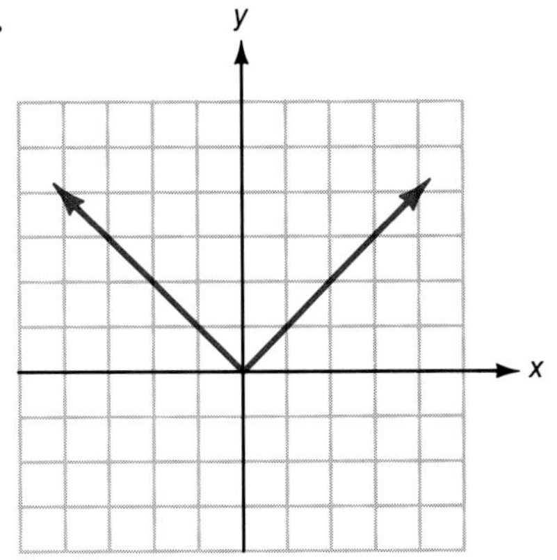


**24.**

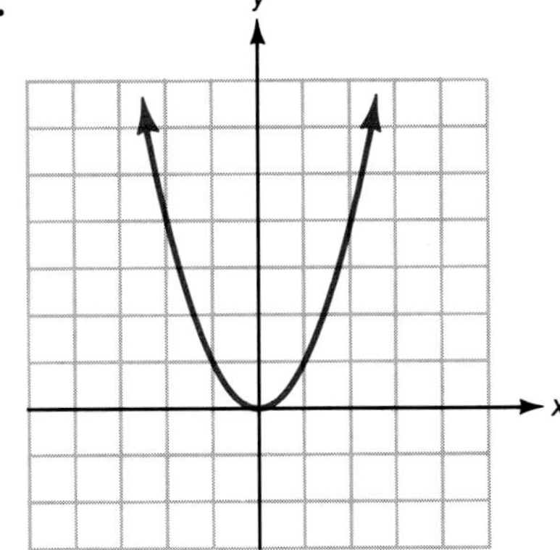

**25.**

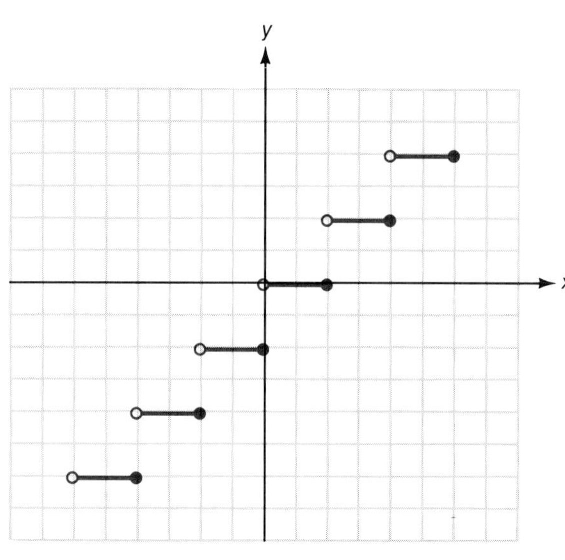


**26.**

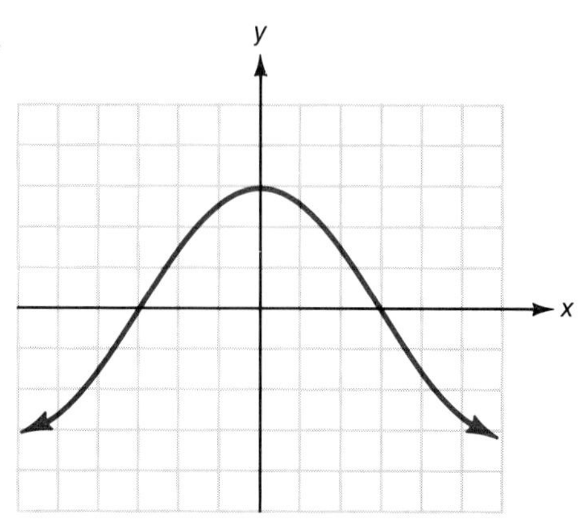


**27.**

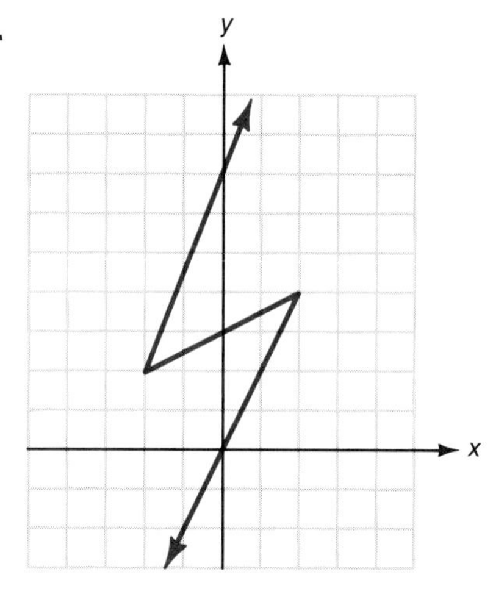


**28.**

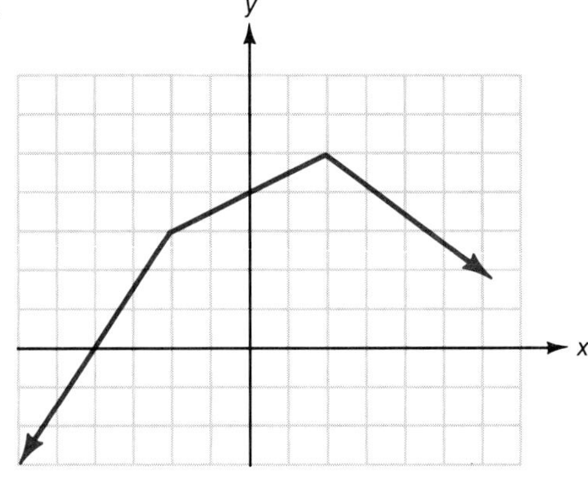


**29.**

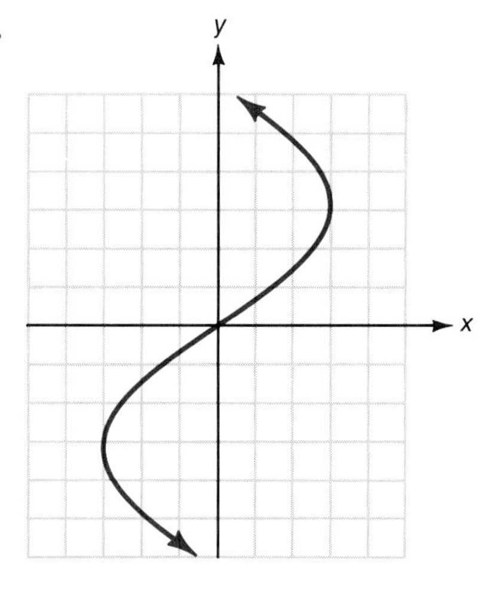


**30.**

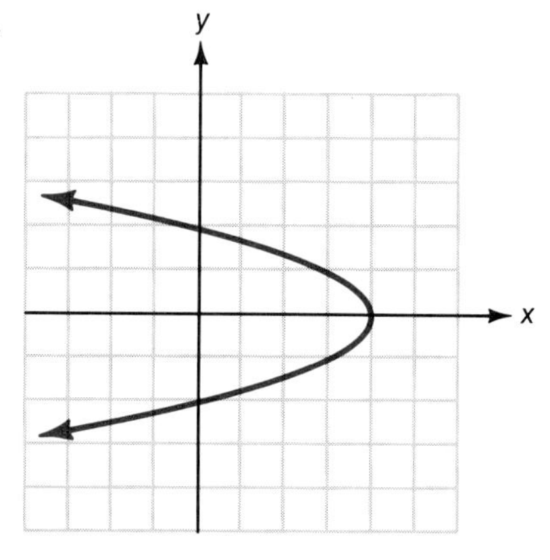

**31.**

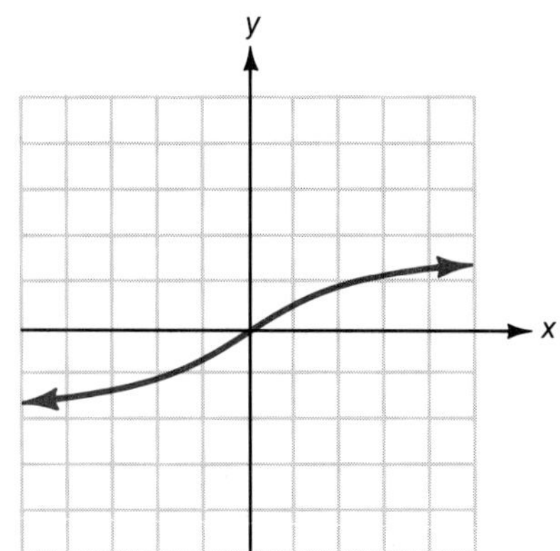


**32.**

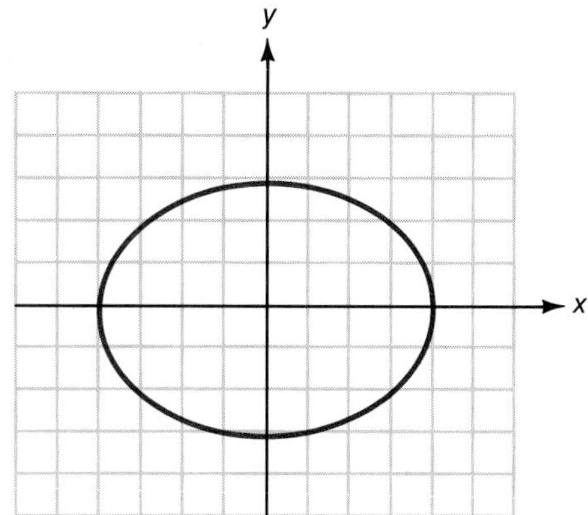


**33.**

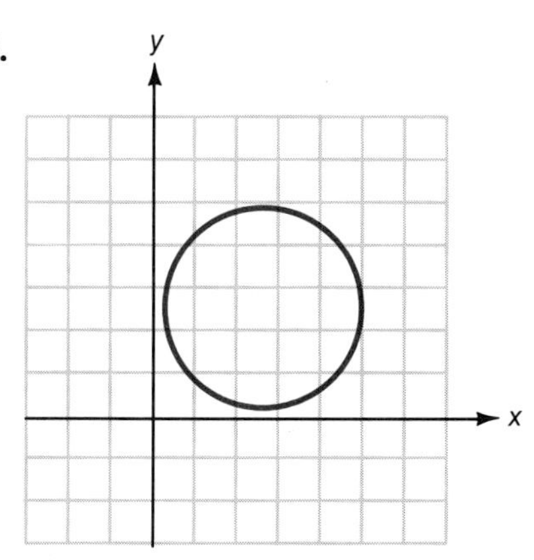


**34.**

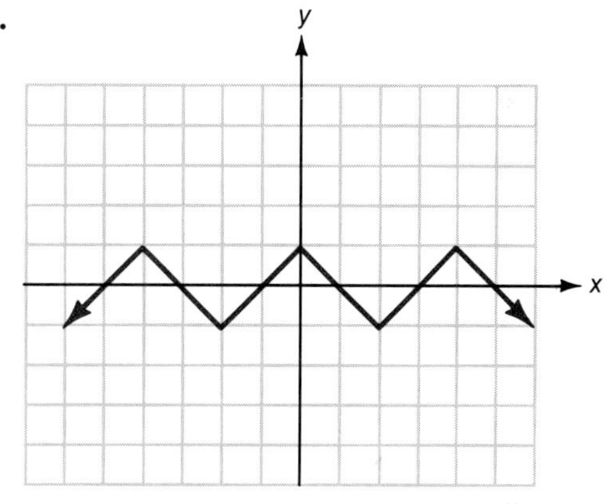


In Exercises 35–42, find the function as a set of ordered pairs for the given equation and domain.

**35.** $y = 3x + 1$

$D = \left\{-9, -\frac{1}{3}, 0, \frac{4}{3}, 2\right\}$

**36.** $y = -\frac{3}{4}x + 2$

$D = \{-4, -2, 0, 3, 4\}$

**37.** $y = 1 - 3x^2$

$D = \{-2, -1, 0, 1, 2\}$

**38.** $y = x^3 - 4x$

$D = \left\{-1, 0, \frac{1}{2}, 1, 2\right\}$

**39.** $y = \frac{x^2 + 3}{x + 2}$

$D = \left\{-1, 0, \frac{1}{2}, 2, 4\right\}$

**40.** $y = \frac{3x + 2}{x - 2}$

$D = \left\{-3, -2, 0, \frac{2}{3}, 1\right\}$

**41.** $y = \sqrt{2x + 4}$
$D = \{-2, -1, 0, 2, 3\}$

**42.** $y = \sqrt{5 - x^2}$
$D = \{-2, -1, 0, 1, 2\}$

Find the domain of each function represented by the equation in Exercises 43–50.

**43.** $y = \frac{x + 5}{x + 3}$

**44.** $y = \frac{x - 6}{x^2 + 1}$

**45.** $y = \frac{7 - 2x}{x^2 - 4}$

**46.** $y = \frac{3x - 1}{x^2 - 2x - 8}$

**47.** $y = \sqrt{2x + 5}$

**48.** $y = \sqrt{4 - 3x}$

**49.** $y = \sqrt{9 - x^2}$

**50.** $y = \sqrt{x^2 - 1}$

## 8.2 Quadratic Functions

**OBJECTIVES**

In this section, you will be learning to:

1. Graph parabolas by finding the zeros (if any) and completing the square, if necessary, to determine the vertices, ranges, and lines of symmetry.
2. Solve applied problems by using quadratic functions.

Linear equations in two variables were discussed in Chapter 6 and, except for vertical lines in the form $x = a$, we can now conclude that linear equations in two variables represent functions. We now refer to an equation in the form

$$y = mx + b$$

as a **linear function.**

We are now interested in functions in which $y$ is first-degree and $x$ is second-degree. For example, consider the function

$$y = x^2 - 4x + 3$$

What is the graph of this function? Since the equation is not linear, the graph will not be a straight line. The nature of the graph can be seen by plotting several points (Figure 8.2).

| $x$ | $y = x^2 - 4x + 3$ |
|---|---|
| $-1$ | $(-1)^2 - 4(-1) + 3 = 8$ |
| $0$ | $0^2 - 4(0) + 3 = 3$ |
| $\frac{1}{2}$ | $\left(\frac{1}{2}\right)^2 - 4\left(\frac{1}{2}\right) + 3 = \frac{5}{4}$ |
| $1$ | $1^2 - 4(1) + 3 = 0$ |
| $2$ | $2^2 - 4(2) + 3 = -1$ |
| $3$ | $3^2 - 4(3) + 3 = 0$ |
| $\frac{7}{2}$ | $\left(\frac{7}{2}\right)^2 - 4\left(\frac{7}{2}\right) + 3 = \frac{5}{4}$ |
| $4$ | $4^2 - 4(4) + 3 = 3$ |
| $5$ | $5^2 - 4(5) + 3 = 5$ |

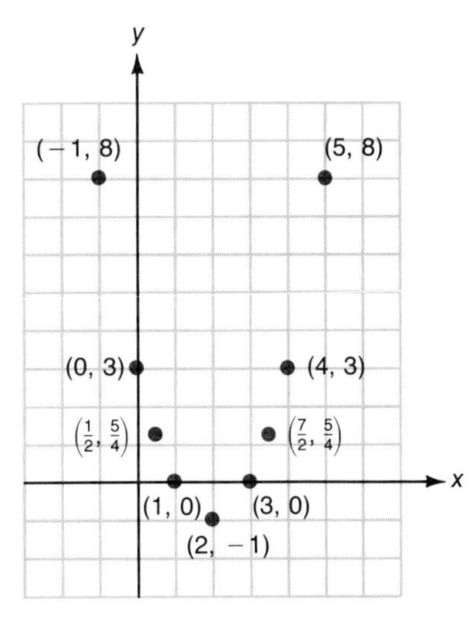


Figure 8.2

The complete graph of $y = x^2 - 4x + 3$ is shown in Figure 8.3. The curve is called a **parabola.** The point $(2, -1)$ is the "turning point" of the parabola and is called the **vertex** of the parabola. The line $x = 2$ is the **line of symmetry** or **axis of symmetry** for the parabola. That is, the curve is a "mirror image" of itself on either side of the line $x = 2$.

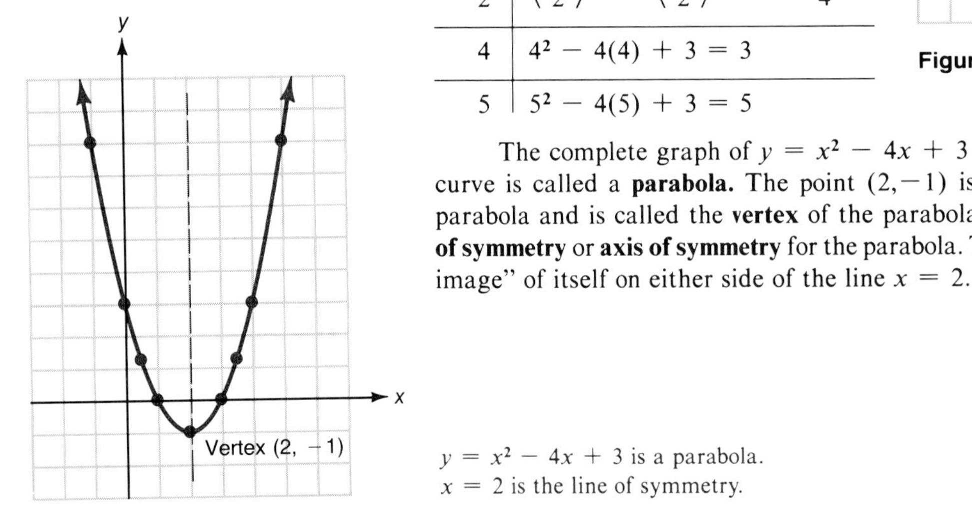


Figure 8.3

**Quadratic Function**

Any function of the form

$$y = ax^2 + bx + c \qquad (x \text{ is any real number})$$

where $a$, $b$, and $c$ are real constants and $a \neq 0$ is a **quadratic function.**

The graph of every quadratic function is a parabola. The position of the parabola, its shape, and whether it "opens up" or "opens down" can be determined by investigating the function itself. We will discuss quadratic functions of the form $y = ax^2$, $y = ax^2 + k$, $y = a(x - h)^2$, $y = a(x - h)^2 + k$, and $y = ax^2 + bx + c$.

## $y = ax^2$

For any real number $x$, $x^2 \geq 0$. So, $ax^2 \geq 0$ if $a > 0$ and $ax^2 \leq 0$ if $a < 0$. This means that the graph of $y = ax^2$ is "above" the $x$-axis if $a > 0$ and "below" the $x$-axis if $a < 0$. The vertex is at (0,0) in either case and is the one point where the graph touches the $x$-axis.

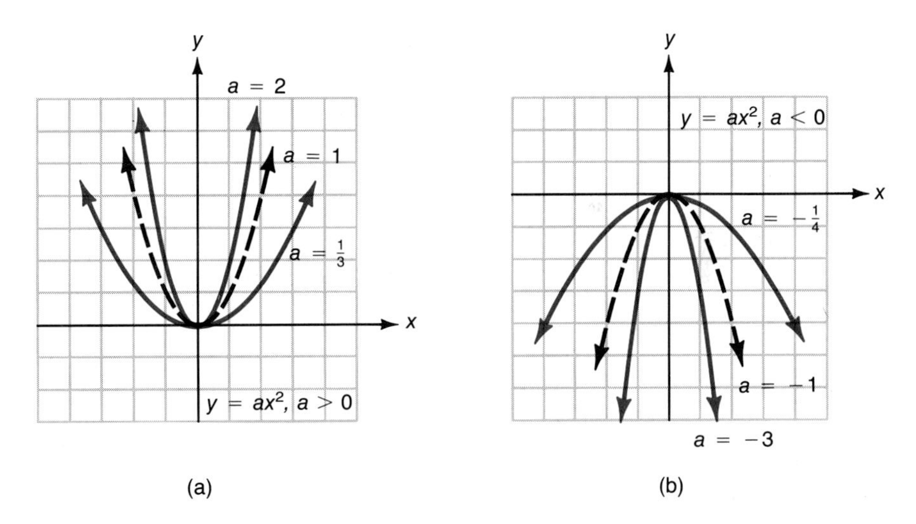


**Figure 8.4**

Figure 8.4 illustrates several properties of quadratic functions of the form $y = ax^2$. If $a > 0$, the parabola "opens upward." If $a < 0$, the parabola "opens downward." The bigger $|a|$ is, the narrower the opening. The smaller $|a|$ is, the wider the opening. The line $x = 0$ (the $y$-axis) is the line of symmetry.

## $y = ax^2 + k$

Adding $k$ to $ax^2$ simply increases (or decreases) each $y$-value of $y = ax^2$ by $k$ units. That is, the graph of $y = ax^2 + k$ can be found by "sliding" the graph of $y = ax^2$ up $|k|$ units if $k > 0$ or down $|k|$ units if $k < 0$ (Figure 8.5).

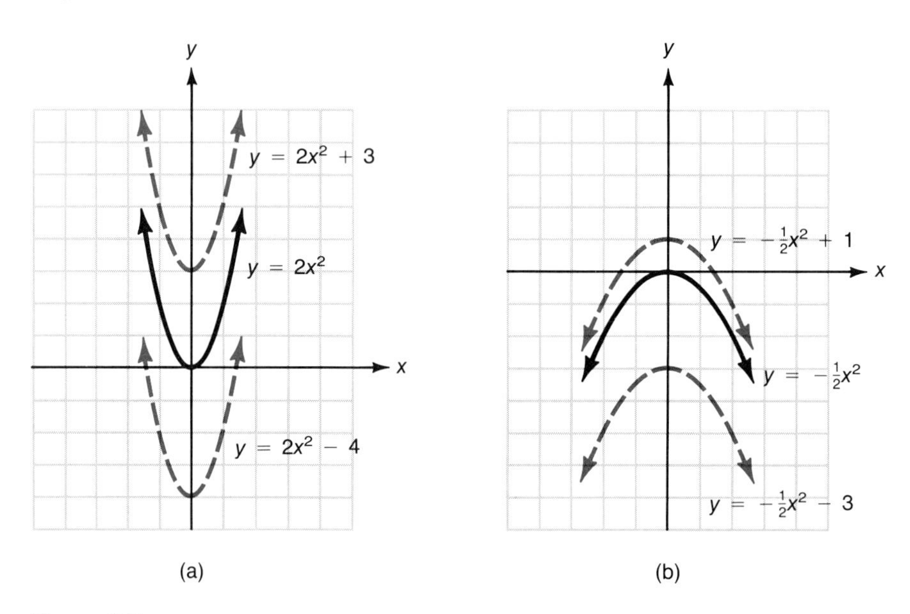


**Figure 8.5**

The vertex of $y = ax^2 + k$ is at the point $(0,k)$. The graph of $y = ax^2 + k$ is a **vertical shift** of the graph of $y = ax^2$. The line $x = 0$ is still the line of symmetry.

## $y = a(x - h)^2$

We know that $(x - h)^2 \geq 0$. So, if $a > 0$, then $y = a(x - h)^2 \geq 0$. If $a < 0$, then $y = a(x - h)^2 \leq 0$. Thus, for $y = a(x - h)^2$:

> The vertex is at $(h,0)$, and the parabola "opens upward" if $a > 0$ and "opens downward" if $a < 0$ (Figure 8.6).

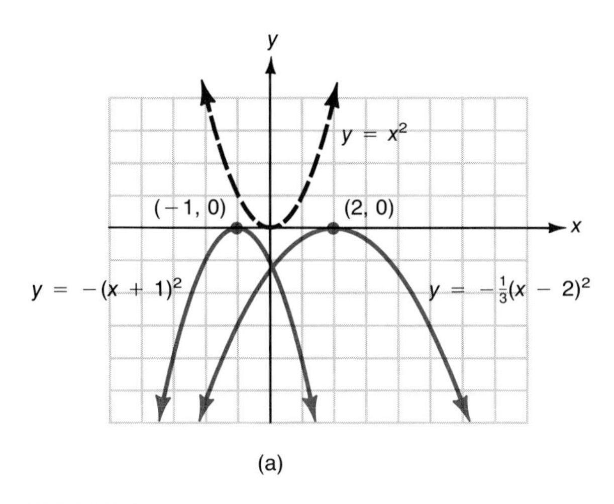


(a)

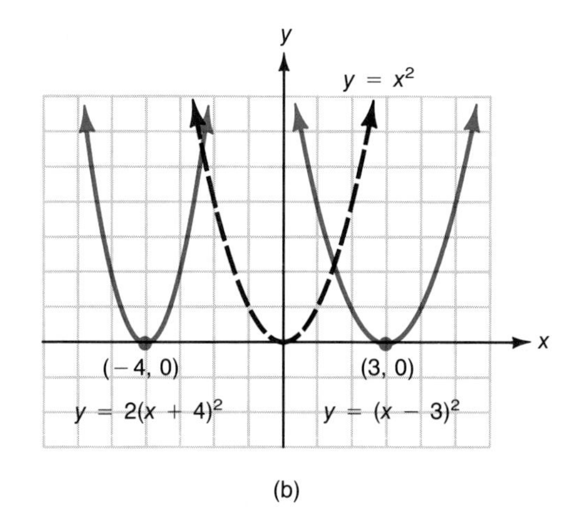


(b)

**Figure 8.6**

The graph of $y = a(x - h)^2$ is a **horizontal shift** of the graph of $y = ax^2$. The shift is to the right if $h > 0$ and to the left if $h < 0$. As a special comment, note that if $h = -3$, then $y = a(x - h)^2$ gives $y = a[x - (-3)]^2$ or $y = a(x + 3)^2$. Thus, if $h$ is negative, the expression $(x - h)^2$ appears with a positive sign. The line $x = h$ is the **line of symmetry.**

**EXAMPLES** Graph the following quadratic functions. Set up a table of values for $x$ and $y$ as an aid, and choose values of $x$ on each side of the line of symmetry.

**1.** $y = 2x^2 - 3$

**Solution** Line of symmetry is $x = 0$.

| $x$ | $y$ |
|---|---|
| 0 | −3 |
| $\frac{1}{2}$ | $-\frac{5}{2}$ |
| $-\frac{1}{2}$ | $-\frac{5}{2}$ |
| 1 | −1 |
| −1 | −1 |
| 2 | 5 |
| −2 | 5 |

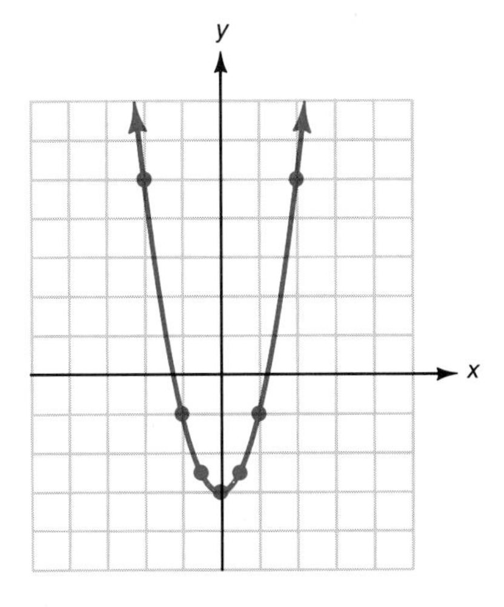


Note that the parabola opens upward for $a = 2 > 0$.

**2.** $y = -\left(x - \frac{5}{2}\right)^2$

**Solution** Line of symmetry is $x = \frac{5}{2}$.

| $x$ | $y$ |
|---|---|
| $\frac{5}{2}$ | 0 |
| 2 | $-\frac{1}{4}$ |
| 3 | $-\frac{1}{4}$ |
| 1 | $-\frac{9}{4}$ |
| 4 | $-\frac{9}{4}$ |
| 0 | $-\frac{25}{4}$ |
| 5 | $-\frac{25}{4}$ |

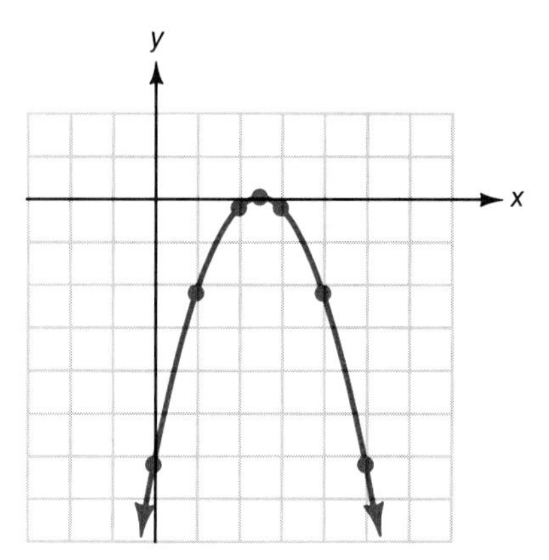


Note that the parabola opens downward for $a = -1 < 0$. ■

## $y = a(x - h)^2 + k$ *and* $y = ax^2 + bx + c$

The graph of $y = a(x - h)^2 + k$ combines both the vertical shift of $k$ units and the horizontal shift of $h$ units. The vertex is at $(h,k)$. For example, the graph of the function $y = -2(x - 3)^2 + 5$ is a shift of the graph of $y = -2x^2$ up 5 units and to the right 3 units and has its vertex at (3,5). The graph of $y = \left(x + \frac{1}{2}\right)^2 - 2$ is the same as the graph of $y = x^2$ shifted left $\frac{1}{2}$ unit and down 2 units. The vertex is at $\left(-\frac{1}{2}, -2\right)$ (Figure 8.7).

The vertex is either the lowest point or the highest point on a parabola. If the function is written in the form

$$y = a(x - h)^2 + k$$

then the vertex occurs at the point $(h,k)$. Thus,

if $a > 0$, then the range is all $y \geq k$;
if $a < 0$, then the range is all $y \leq k$.

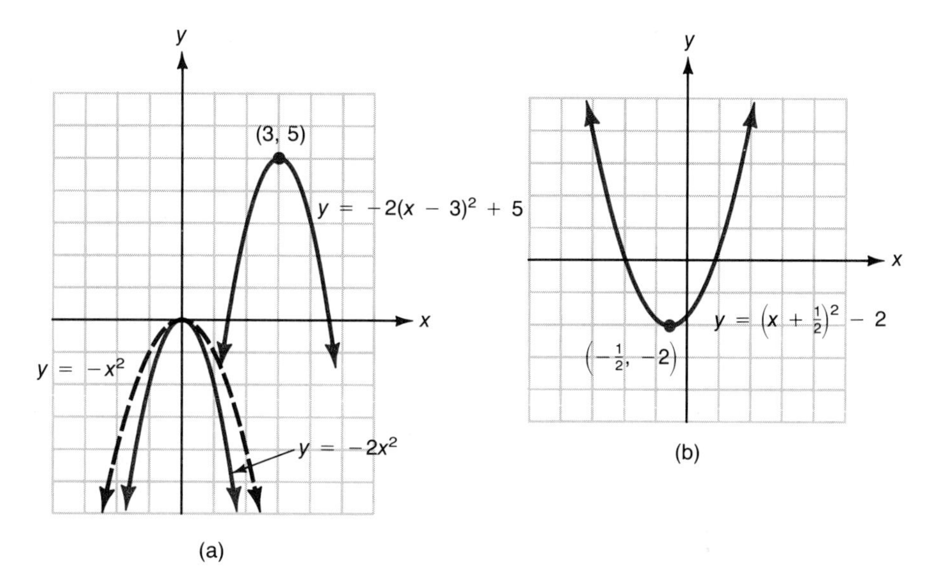


**Figure 8.7**

In either case, the domain of a quadratic function is all real numbers. That is, $x$ can be any real number.

To easily find the vertex, line of symmetry, and range, and to graph the parabola, change the general form $y = ax^2 + bx + c$ into the form $y = a(x - h)^2 + k$. This can be accomplished by completing the square using the following technique. This technique is also useful in other courses in mathematics and should be studied carefully. (Be aware that you are not solving an equation. You do not "do something" to both sides. You are **changing the form** of a function.)

$y = ax^2 + bx + c$ — Write the function.

$= a\left(x^2 + \frac{b}{a}x\right) + c$ — Factor $a$ from the first two terms.

$= a\left(x^2 + \frac{b}{a}x + \frac{b^2}{4a^2} - \frac{b^2}{4a^2}\right) + c$ — Complete the square of $x^2 + \frac{b}{a}x$. $\frac{1}{2} \cdot \frac{b}{a} = \frac{b}{2a}$ and $\left(\frac{b}{2a}\right)^2 = \frac{b^2}{4a^2}$. Add and subtract $\frac{b^2}{4a^2}$.

$= a\left(x^2 + \frac{b}{a}x + \frac{b^2}{4a^2}\right) - \frac{b^2}{4a} + c$ — Multiply $a\left(\frac{-b^2}{4a^2}\right)$ and write this term outside the parentheses.

$= a\left(x + \frac{b}{2a}\right)^2 + \frac{4ac - b^2}{4a}$ — Factor and simplify to get the form $y = a(x - h)^2 + k$.

In terms of the coefficients $a$, $b$, and $c$,

$$x = -\frac{b}{2a} \quad \text{is the line of symmetry,}$$

and

$$(h,k) = \left(-\frac{b}{2a}, \frac{4ac - b^2}{4a}\right) \quad \text{is the vertex.}$$

The points where a parabola crosses the $x$-axis, if it crosses the $x$-axis, are the $x$-intercepts where $y = 0$. These points, if they exist, are called the **zeros of the function.** We find these points by setting the function equal to 0 and solving the resulting quadratic equation:

$$y = ax^2 + bx + c = 0$$

The equation can be solved by either factoring or using the quadratic formula.

$$x = \frac{-b \pm \sqrt{b^2 - 4ac}}{2a}$$

If the solutions are nonreal complex numbers, then the graph does not cross the $x$-axis. It is either entirely above the $x$-axis or entirely below the $x$-axis.

The following examples illustrate how to apply all our knowledge about quadratic functions.

## EXAMPLES

**3.** $y = x^2 - 6x + 1$

**Solution** $x^2 - 6x + 1 = 0$

$$x = \frac{6 \pm \sqrt{36 - 4}}{2} = \frac{6 \pm \sqrt{32}}{2}$$

$$= \frac{6 \pm 4\sqrt{2}}{2} = 3 \pm 2\sqrt{2}$$

The zeros are $3 \pm 2\sqrt{2}$.

$$\begin{aligned} y &= x^2 - 6x + 1 \\ &= (x^2 - 6x + 9 - 9) + 1 \\ &= (x - 3)^2 - 8 \end{aligned}$$

Vertex: $(3, -8)$
Domain: all real numbers
Range: $y \geq -8$
Line of symmetry is $x = 3$.

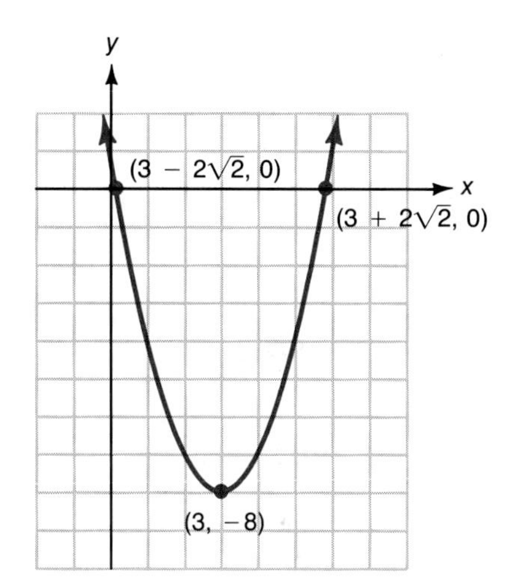


**4.** $y = 2x^2 - 6x + 5$

**Solution** $2x^2 - 6x + 5 = 0$

$$x = \frac{6 \pm \sqrt{36 - 40}}{4}$$

No real zeros since $36 - 40 < 0$.

$$\begin{aligned} y &= 2x^2 - 6x + 5 \\ &= 2(x^2 - 3x) + 5 && \text{Factor 2 from only the first two terms.} \\ &= 2\left(x^2 - 3x + \frac{9}{4} - \frac{9}{4}\right) + 5 && \text{Add } 0 = +\frac{9}{4} - \frac{9}{4} \text{ inside the parentheses.} \\ &= 2\left(x^2 - 3x + \frac{9}{4}\right) + 2\left(-\frac{9}{4}\right) + 5 && \text{Multiply } 2\left(-\frac{9}{4}\right). \\ &= 2\left(x - \frac{3}{2}\right)^2 + \frac{1}{2} && 2\left(-\frac{9}{4}\right) + 5 = -\frac{9}{2} + \frac{10}{2} = \frac{1}{2} \end{aligned}$$

Vertex: $\left(\frac{3}{2}, \frac{1}{2}\right)$

Domain: all real numbers

Range: $y \geq \frac{1}{2}$

Line of symmetry is $x = \frac{3}{2}$.

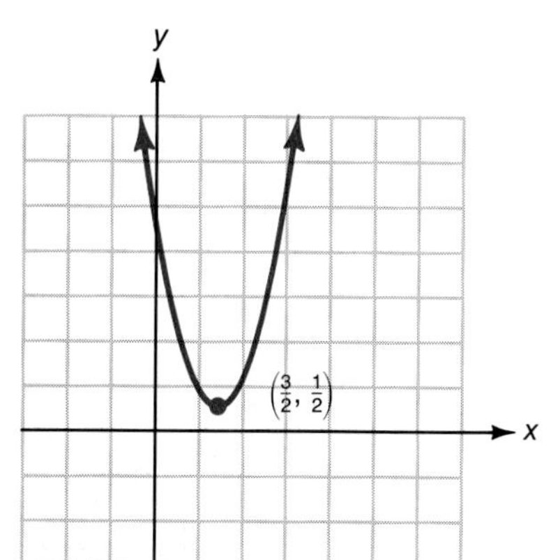

**5.** $y = -x^2 - 4x + 2$

**Solution** $-x^2 - 4x + 2 = 0$

$$x = \frac{4 \pm \sqrt{16 + 8}}{-2}$$

$$= \frac{4 \pm 2\sqrt{6}}{-2}$$

$$= -2 \pm \sqrt{6}$$

The zeros are $-2 \pm \sqrt{6}$.

$$\begin{aligned} y &= -x^2 - 4x + 2 \\ &= -(x^2 + 4x) + 2 \\ &= -(x^2 + 4x + 4 - 4) + 2 \\ &= -(x^2 + 4x + 4) + 4 + 2 \\ &= -(x + 2)^2 + 6 \end{aligned}$$

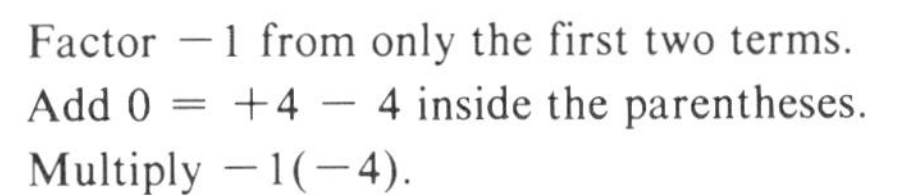

Factor $-1$ from only the first two terms.
Add $0 = +4 - 4$ inside the parentheses.
Multiply $-1(-4)$.

Vertex: $(-2,6)$
Domain: all real numbers
Range: $y \leq 6$
Line of symmetry is $x = -2$.

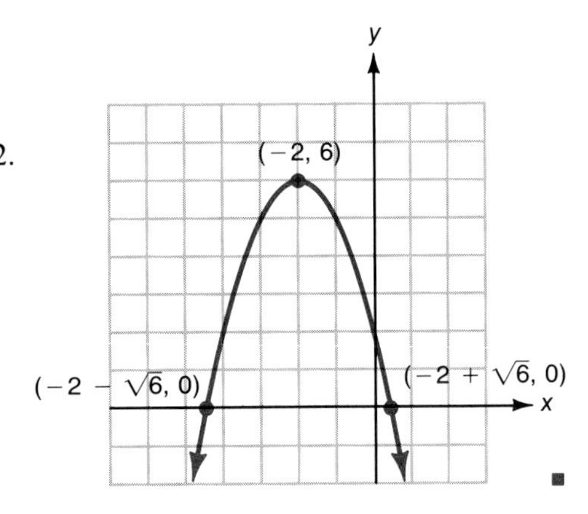


If a parabola opens upward, the vertex is the lowest point on the graph and the corresponding $y$-value is called the **minimum value** of the function. Likewise, for a parabola that opens downward, the $y$-value at the vertex is called the **maximum value** of the function.

## EXAMPLE

**6.** A sandwich company sells hot dogs at the local baseball stadium for \$3.00 each and sells 2000 hot dogs per game. The company estimates that each time the price is raised by 25¢, they will sell 100 fewer hot dogs. (**a**) What price should they charge to maximize their revenue (income) per game? (**b**) What will be the maximum revenue?

**Solution** Let $x$ = number of 25¢-increases in price

Then,

$$3.00 + 0.25x = \text{price per hot dog}$$
$$2000 - 100x = \text{number of hot dogs sold}$$

Now,

$$\text{Revenue} = (\text{price per unit}) \cdot (\text{number of units sold})$$

So,

$$\begin{aligned} R &= (3.00 + 0.25x)(2000 - 100x) \\ &= 6000 + 500x - 300x - 25x^2 \\ &= 6000 + 200x - 25x^2 \end{aligned}$$

Thus, the revenue is represented by a quadratic function and the maximum revenue occurs at the vertex where

$$x = -\frac{b}{2a} = -\frac{200}{-50} = 4$$

For $x = 4$,

the price per hot dog is $3.00 + 0.25(4) = \$4.00$

and the revenue is $R = (4)(2000 - 400) = \$6400$. ■

**Practice Problems**

1. Write the function $y = 2x^2 - 4x + 3$ in the form $y = a(x - h)^2 + k$.
2. Find the zeros of the function $y = x^2 - 7x + 10$.
3. Find the vertex and the range of the function $y = -x^2 + 4x - 5$.

## EXERCISES 8.2

For each of the quadratic functions in Exercises 1–20, determine the line of symmetry, the vertex, and the range.

**1.** $y = 3x^2 - 4$

**2.** $y = \frac{2}{3}x^2 + 7$

**3.** $y = 7x^2 + 9$

**4.** $y = 5x^2 - 1$

**5.** $y = -4x^2 + 1$

**6.** $y = -2x^2 - 6$

**7.** $y = -\frac{3}{4}x^2 + \frac{1}{5}$

**8.** $y = \frac{5}{3}x^2 + \frac{7}{8}$

**9.** $y = (x + 1)^2$

**10.** $y = (x - 1)^2$

**11.** $y = -\frac{2}{3}(x - 4)^2$

**12.** $y = -5(x + 2)^2$

**13.** $y = 2(x + 3)^2 - 2$

**14.** $y = 4(x - 5)^2 + 1$

**15.** $y = \frac{3}{4}(x + 2)^2 - 6$

**16.** $y = -2(x + 1)^2 - 4$

**17.** $y = -\frac{1}{2}\left(x - \frac{3}{2}\right)^2 + \frac{7}{2}$

**18.** $y = -\frac{5}{3}\left(x - \frac{9}{2}\right)^2 + \frac{3}{4}$

**19.** $y = \frac{1}{4}\left(x - \frac{4}{5}\right)^2 - \frac{11}{5}$

**20.** $y = \frac{10}{3}\left(x + \frac{7}{8}\right)^2 - \frac{9}{16}$

**Answers to Practice Problems** **1.** $y = 2(x - 1)^2 + 1$ **2.** $x = 5, x = 2$
**3.** Vertex: $(2, -1)$, Range: $y \leq -1$

**21.** Graph the function $y = x^2$. Without additional computation, graph:
**a.** $y = x^2 - 2$
**b.** $y = (x - 3)^2$
**c.** $y = -(x - 1)^2$
**d.** $y = 5 - (x + 1)^2$

**22.** Graph the function $y = 2x^2$. Without additional computation, graph:
**a.** $y = 2x^2 - 3$
**b.** $y = 2(x - 4)^2$
**c.** $y = -2(x + 1)^2$
**d.** $y = -2(x + 2)^2 - 4$

**23.** Graph the function $y = \frac{1}{2}x^2$. Without additional computation, graph:
**a.** $y = \frac{1}{2}x^2 + 3$
**b.** $y = \frac{1}{2}(x + 2)^2$
**c.** $y = -\frac{1}{2}x^2$
**d.** $y = \frac{1}{2}(x - 1)^2 - 4$

**24.** Graph the function $y = \frac{1}{4}x^2$. Without additional computation, graph:
**a.** $y = -\frac{1}{4}x^2$
**b.** $y = \frac{1}{4}x^2 - 5$
**c.** $y = \frac{1}{4}(x + 4)^2$
**d.** $y = 2 - \frac{1}{4}(x + 2)^2$

Rewrite each of the quadratic functions in Exercises 25–32 in the form $y = a(x - h)^2 + k$. Find the vertex, range, and zeros of each function. Graph the function.

**25.** $y = 2x^2 - 4x + 2$
**26.** $y = -3x^2 + 12x - 12$
**27.** $y = x^2 - 2x - 3$
**28.** $y = x^2 - 4x + 5$
**29.** $y = x^2 + 6x + 5$
**30.** $y = x^2 - 8x + 12$
**31.** $y = 2x^2 - 8x + 5$
**32.** $y = 2x^2 - 6x + 5$

Find the vertex, range, and zeros of each function in Exercises 33–42. Graph the function.

**33.** $y = -3x^2 - 12x - 9$
**34.** $y = 3x^2 - 6x - 1$
**35.** $y = 5x^2 - 10x + 8$
**36.** $y = -4x^2 + 16x - 11$
**37.** $y = -x^2 - 5x - 2$
**38.** $y = x^2 + 3x - 1$
**39.** $y = 2x^2 + 7x + 5$
**40.** $y = 2x^2 + x - 3$
**41.** $y = 3x^2 + 7x + 2$
**42.** $y = 4x^2 - 12x + 9$

In Exercises 43–46, use the formula $h = -16t^2 + v_0t + h_0$ where $h$ is the height of the object after time, $t$; $v_0$ is the initial velocity; and $h_0$ is the initial height.

**43.** A ball is thrown vertically upward from the ground with an initial velocity of 112 ft per sec. (**a**) When will the ball reach its maximum height? (**b**) What will be the maximum height?

**44.** A ball is thrown vertically upward from the ground with an initial velocity of 104 ft per sec. (**a**) When will the ball reach its maximum height? (**b**) What will be the maximum height?

**45.** A stone is projected vertically upward from a platform that is 32 ft high, at a rate of 128 ft per sec. (**a**) When will the stone reach its maximum height? (**b**) What will be the maximum height?

**46.** A stone is projected vertically upward from a platform that is 20 ft high at a rate of 160 ft per sec. (**a**) When will the stone reach its maximum height? (**b**) What will be the maximum height?

**47.** A store owner estimates that by charging $x$ dollars each for a certain lamp, he can sell $40 - x$ lamps each week. What price will give him maximum receipts?

**48.** A manufacturer produces radios. He estimates that by selling them for $x$ dollars each, he will be able to sell $100 - x$ radios each month. (**a**) What price will yield a maximum revenue? (**b**) What will be the maximum revenue?

**49.** Ms. Richey can sell 72 picture frames each month if she charges \$24 each. She estimates that for each \$1 increase in price, she will sell 2 fewer frames. (**a**) Find the price that will yield a maximum revenue. (**b**) What will be the maximum revenue?

**50.** The perimeter of a rectangle is 40 feet. What are the dimensions that will yield the maximum area?

## 8.3 f(x) Notation and Translations

**OBJECTIVES**

In this section, you will be learning to:

1. Evaluate functions for given values of the independent variables.
2. Graph functions of the form $y - k = \pm f(x - h)$ given the related graphs of $y = f(x)$.

We have used the ordered pair notation $(x,y)$ to represent points on the graphs of relations and functions. Equations of the form

$y = mx + b$ represent linear functions

and $y = ax^2 + bx + c$ represent quadratic functions

The use of $x$ and $y$ in this manner is standard in mathematics and particularly useful when graphing.

Another notation, called **functional notation,** is more convenient for indicating calculations of values of a function and for indicating operations performed with functions. In functional notation,

instead of writing $y$, write $f(x)$, read "$f$ of $x$."

(The letter $f$ is the name of the function. Any letter will do, and $f, g, h, F, G,$ and $H$ are commonly used.)

Suppose that $y = 2x - 7$ is a given linear function. Then you can write

$$f(x) = 2x - 7$$

By $f(3)$ we mean to replace $x$ by 3 in the function.

$$f(3) = 2 \cdot 3 - 7 = 6 - 7 = -1$$

Thus, $(3,f(3))$ is the same as $(3,-1)$.

**EXAMPLES**

**1.** Let $f(x) = 3x + 5$. Find the following:

**a.** $f(2)$ **b.** $f(-1)$
**c.** $f(a)$ **d.** $f(a + 1)$

**Solution** In each case, replace $x$ with whatever number or expression is in the parentheses.

$$f(x) = 3x + 5$$

**a.** $f(2) = 3 \cdot 2 + 5 = 6 + 5 = 11$
**b.** $f(-1) = 3(-1) + 5 = -3 + 5 = 2$
**c.** $f(a) = 3a + 5$
**d.** $f(a + 1) = 3(a + 1) + 5 = 3a + 3 + 5 = 3a + 8$

**2.** Suppose that $g(x) = 2x^2 - 4$. Find the following:

**a.** $g(2)$ **b.** $g(-1)$
**c.** $g(a)$ **d.** $g(a + 1)$

**Solution** Substitute whatever is in parentheses for $x$.

$$g(x) = 2x^2 - 4$$

**a.** $g(2) = 2 \cdot 2^2 - 4 = 8 - 4 = 4$
**b.** $g(-1) = 2(-1)^2 - 4 = 2 \cdot 1 - 4 = 2 - 4 = -2$
**c.** $g(a) = 2a^2 - 4$
**d.** $g(a + 1) = 2(a + 1)^2 - 4 = 2(a^2 + 2a + 1) - 4$
$= 2a^2 + 4a + 2 - 4 = 2a^2 + 4a - 2$

**3.** Let $h(x) = \sqrt{2 - x}$. Find the following:

**a.** $h(3)$ **b.** $h(-3)$ **c.** $h(x + 5)$

**Solution**

$$h(x) = \sqrt{2 - x}$$

**a.** $h(3) = \sqrt{2 - 3} = \sqrt{-1}$, a nonreal number. At this time, we say that 3 is not in the domain of $h$.
**b.** $h(-3) = \sqrt{2 - (-3)} = \sqrt{2 + 3} = \sqrt{5}$
**c.** $h(x + 5) = \sqrt{2 - (x + 5)} = \sqrt{2 - x - 5} = \sqrt{-x - 3}$
This substitution is allowed as long as $-x - 3 \geq 0$ or $-3 \geq x$. ■

The formula

$$\frac{f(x + h) - f(x)}{h}$$

is particularly useful in calculus. Simplifying the resulting expression is a good algebraic exercise that is a necessary step for finding a new function in calculus called a **derivative.** Be careful to supply parentheses and brackets so that all the algebra is done correctly.

## EXAMPLES

**4.** Apply $f(x) = 2x^2 - 5x$ to the formula $\dfrac{f(x + h) - f(x)}{h}$ and simplify.

**Solution** $f(x + h) = 2(x + h)^2 - 5(x + h)$

and
$$\frac{f(x + h) - f(x)}{h} = \frac{[2(x + h)^2 - 5(x + h)] - [2x^2 - 5x]}{h}$$
$$= \frac{2x^2 + 4xh + 2h^2 - 5x - 5h - 2x^2 + 5x}{h}$$
$$= \frac{4xh + 2h^2 - 5h}{h} = \frac{h(4x + 2h - 5)}{h}$$
$$= 4x + 2h - 5$$

**5.** Use the function $f(x) = 2 - 6x$ in the formula $\dfrac{f(x + h) - f(x)}{h}$ and simplify.

**Solution** $f(x + h) = 2 - 6(x + h)$

$$\frac{f(x + h) - f(x)}{h} = \frac{[2 - 6(x + h)] - [2 - 6x]}{h}$$

$$= \frac{2 - 6x - 6h - 2 + 6x}{h} = \frac{-6h}{h} = -6$$

■

Now we will use function notation to help in understanding a technique for graphing functions called **translation.** There are two kinds of translations: **horizontal translations** (or horizontal shifts) and **vertical translations** (or vertical shifts) (Figure 8.8). We have already discussed the fact that the graph of $y = a(x - h)^2$ is a horizontal shift (or translation) and the graph of $y = ax^2 + k$ is a vertical shift (or translation) of the graph of $y = ax^2$. (See Section 8.2.) In function notation,

if $\quad f(x) = ax^2$,
then $\quad f(x - h) = a(x - h)^2$ is a horizontal translation of $f(x)$;
and $\quad f(x) + k = ax^2 + k$ is a vertical translation of $f(x)$.

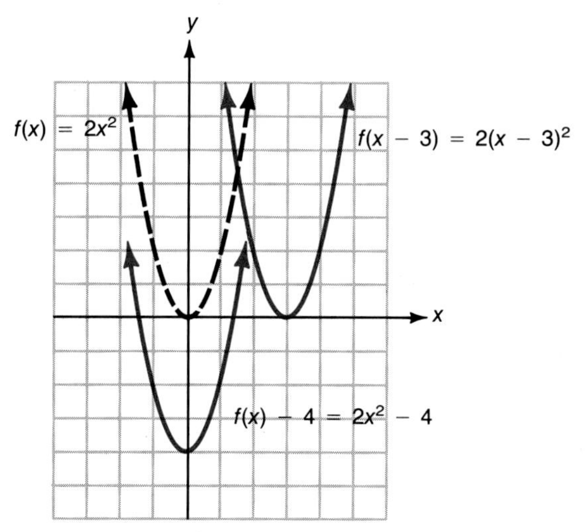


$f(x - 3)$ is a horizontal translation 3 units right
$f(x) - 4$ is a vertical translation 4 units down

**Figure 8.8**

We will illustrate in some detail the following general approach.

**Horizontal and Vertical Translations**

If the graph of $y = f(x)$ is known, then the graph of $y - k = f(x - h)$ is

1. a horizontal translation of $h$ units, and
2. a vertical translation of $k$ units of the graph of $y = f(x)$.

Think of the origin (0,0) being moved to the point $(h,k)$. Then draw the graph of $y = f(x)$ in the same relation to $(h,k)$ as it was to (0,0). This new graph will be the graph of $y - k = f(x - h)$.

A good example to use is the function $f(x) = |x|$. First we need to know what the graph of $f(x) = |x|$ or $y = |x|$ looks like. The definition of $|x|$ gives

$$f(x) = |x| = \begin{cases} x & \text{if } x \geq 0 \\ -x & \text{if } x < 0 \end{cases}$$

The graph of $f(x) = x$ or of $y = x$ is a straight line, as shown in Figure 8.9(a), but we want only the part where $x \geq 0$, as shown in Figure 8.9(b).

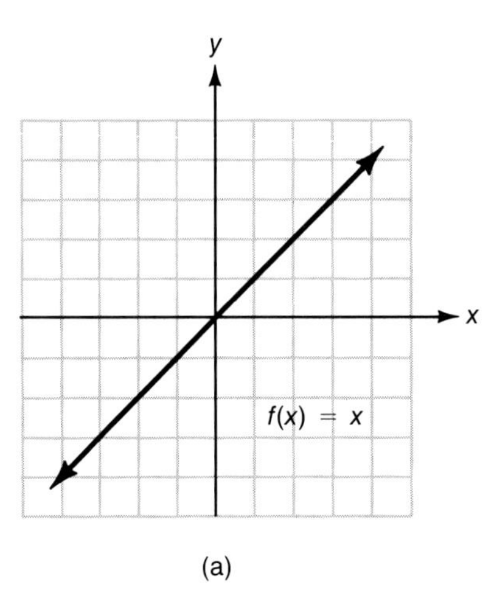


(a)

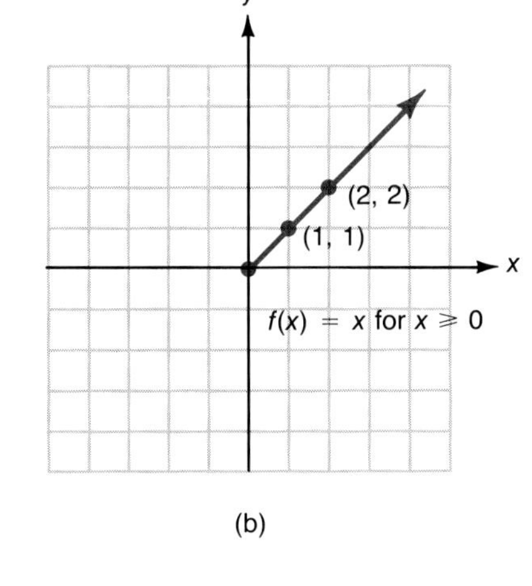


(b)

**Figure 8.9**

The graph of $f(x) = -x$ or of $y = -x$ is also a straight line, as shown in Figure 8.10(a), but we want only the part where $x < 0$, as shown in Figure 8.10(b).

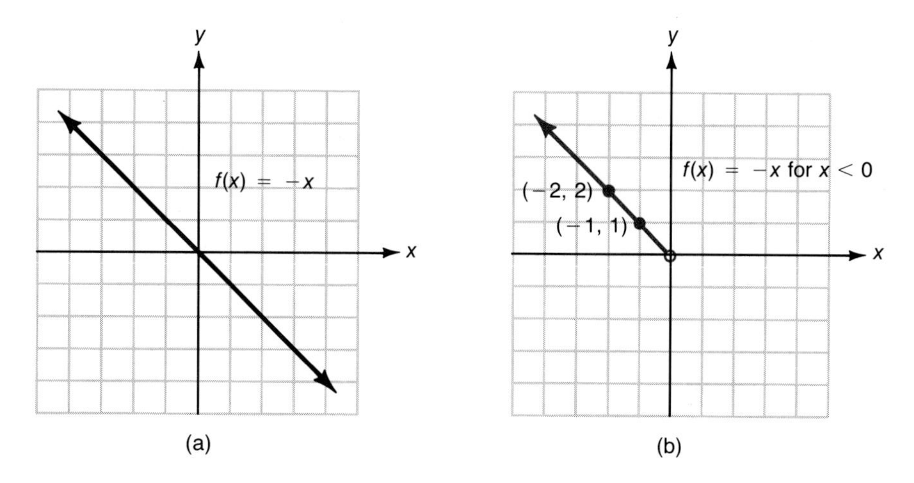


**Figure 8.10**

The two graphs in Figures 8.9(b) and 8.10(b) together give the graph of $f(x) = |x|$, as shown in Figure 8.11.

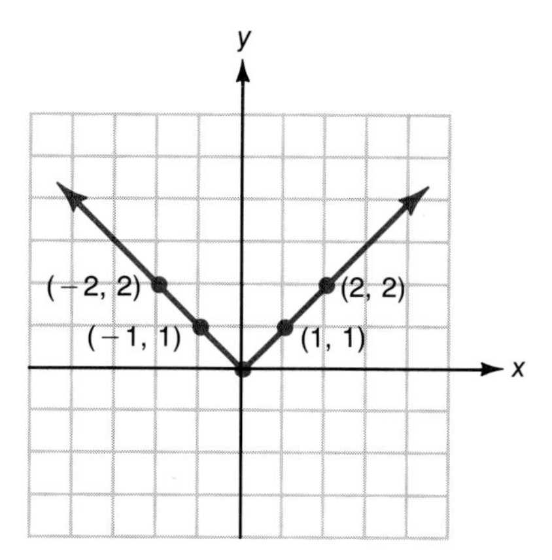


**Figure 8.11**

The following examples illustrate how to graph horizontal and vertical translations of the function $f(x) = |x|$.

**EXAMPLES** Graph each of the following functions.

**6.** $y - 2 = |x - 3|$

**Solution** Here $(h,k) = (3,2)$, so there is a horizontal translation of 3 units and a vertical translation of 2 units. In effect, (3,2) is the "vertex" of the graph just as (0,0) is for $y = |x|$. You should check that the points (2,3) and (4,3) both satisfy the function.

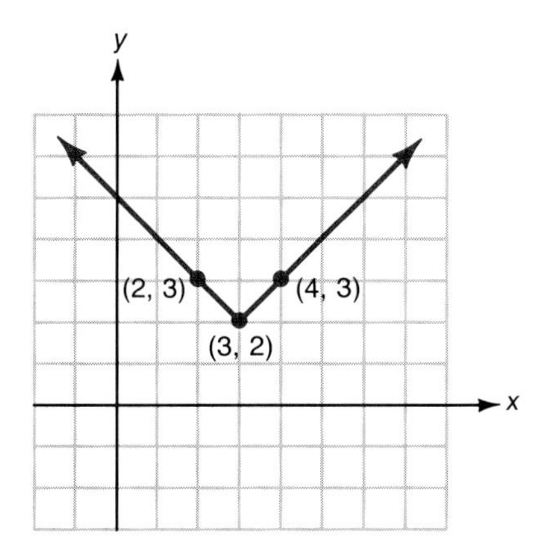


**7.** $y + 1 = |x + 4|$

**Solution** Here $(h,k) = (-4,-1)$, so there is a horizontal translation of $-4$ units and a vertical translation of $-1$ unit. The effect is that the "vertex" is now at $(-4,-1)$.

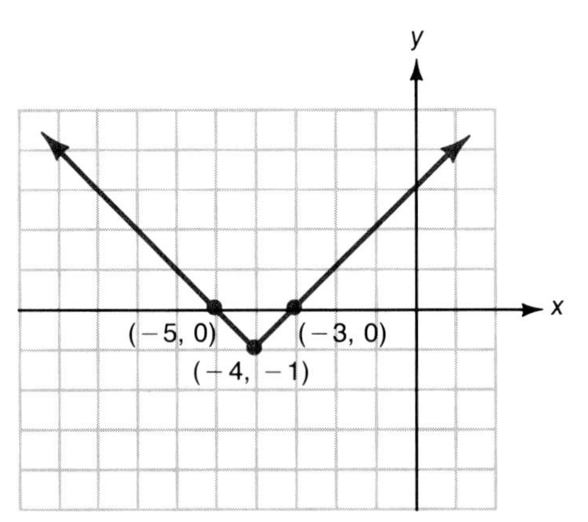

**8.** $y = |x + 2|$

**Solution** Here $(h,k) = (-2,0)$, so there is a horizontal translation of $-2$ units.

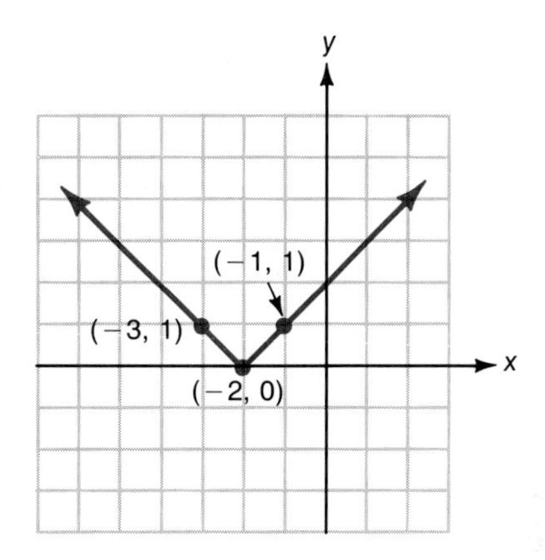


■

Example 9 illustrates the fact that many times, the original function is not given in the form $y - k = f(x - h)$. In many situations, not related particularly to translations, the functions are simply solved for $y$, as in $y = |x + 4| - 7$. By adding an appropriate constant to both sides of the equation, we can change the form to indicate the correct translation.

**EXAMPLE**

**9.** Graph the function $y = |x + 4| - 7$.

**Solution** By adding 7 to both sides of the equation, we get the form $y - k = f(x - h)$.

$$y = |x + 4| - 7$$
$$y + 7 = |x + 4|$$

Now we see that $(h,k) = (-4,-7)$, so the graph of $y = |x|$ is translated $-4$ units horizontally and $-7$ units vertically.

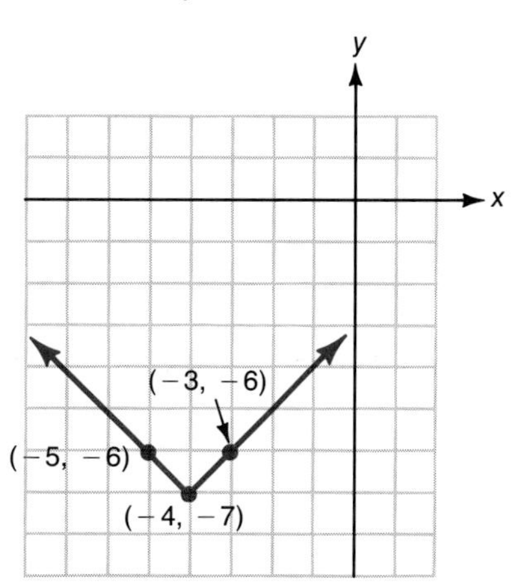


■

The graph of $y = -f(x)$ is a reflection in the $x$-axis of the graph of $y = f(x)$. In the case of $y = |x|$, the graph of $y = -|x|$ is the mirror image in the $x$-axis of $y = |x|$ and "opens" downward instead of upward. Example 10 shows such a reflection along with a translation.

**EXAMPLE**

**10.** Graph the function $y = -|x + 2| + 5$.

**Solution** By adding $-5$ to both sides of the equation, we get

$$y - 5 = -|x + 2|$$

which is in the form

$$y - k = -f(x - h)$$

Here $(h,k) = (-2,5)$, and the graph is reflected in the $x$-axis and "opens" downward.

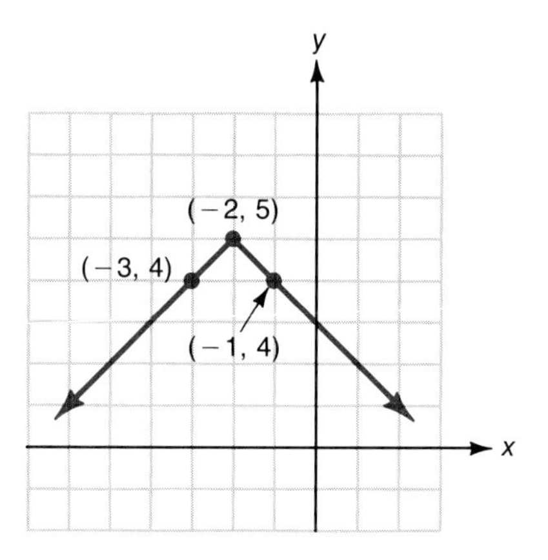


■

The following two examples show how the concept of translation can be applied to the graph of any function if the graph of $y = f(x)$ is known.

**EXAMPLES**

**11.** Graph the function $y - 1 = \sqrt{x - 2}$. The graph of $y = \sqrt{x}$ is given.

**Solution** If $y = \sqrt{x}$ is written $y = f(x)$, then $y - 1 = \sqrt{x - 2}$ is the same as $y - 1 = f(x - 2)$. So, $(h,k) = (2,1)$, and there is a horizontal translation of 2 units and a vertical translation of 1 unit.

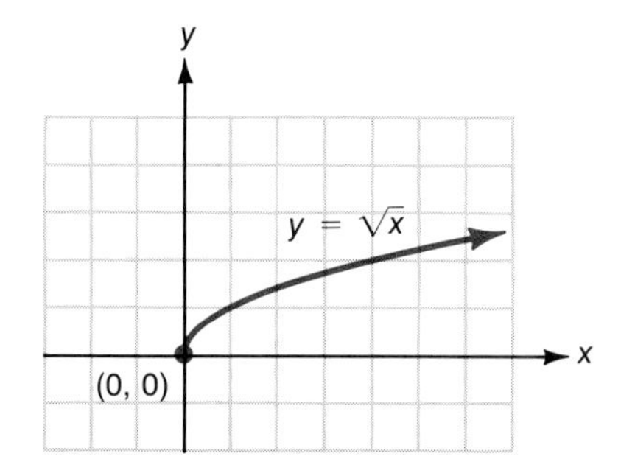


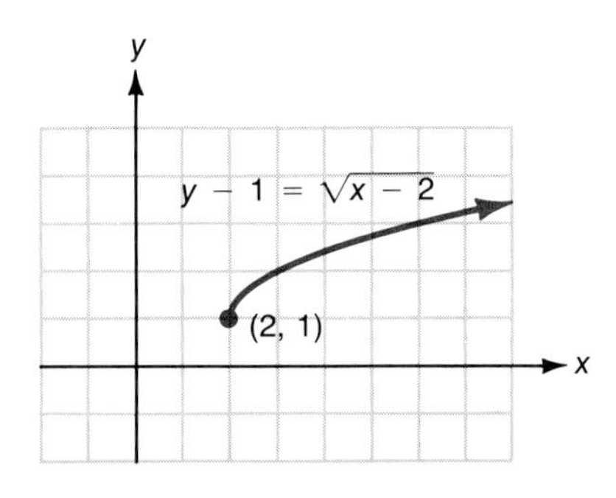

**12.** The graph of $y = f(x)$ is given. Draw the graph of $y + 2 = f(x - 3)$.

**Solution** $(h,k) = (3,-2)$, so translate horizontally 3 units and vertically $-2$ units. (Add 3 to each $x$-value and $-2$ to each $y$-value.)

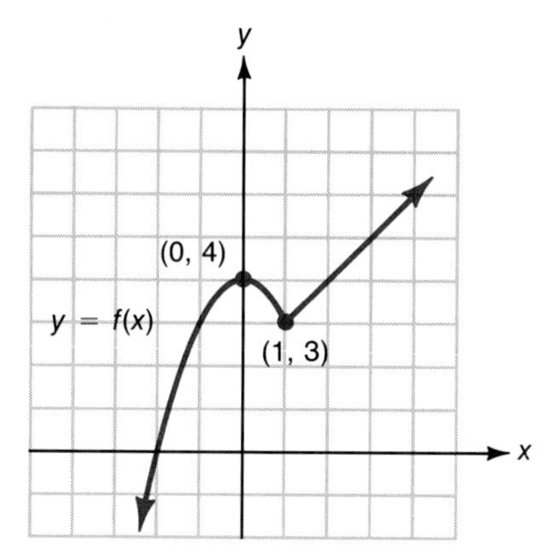


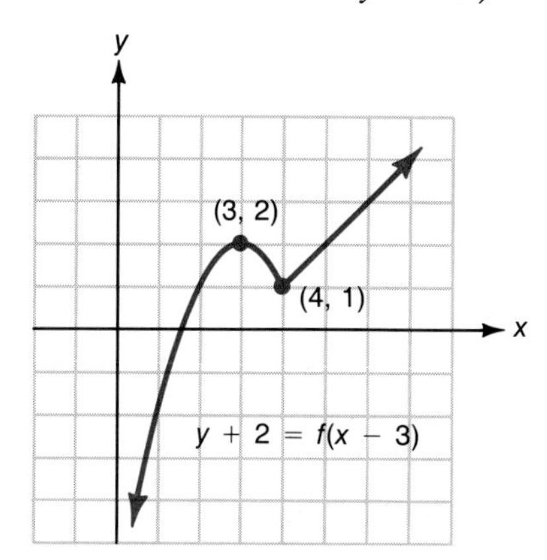


## Practice Problems

1. For $f(x) = x^2 - 5$, find:
   a. $f(0)$
   b. $f(a)$
   c. $f(a + 2)$

2. If $g(x) = 3x + 7$, find:
   a. $g(0)$
   b. $g(x + h)$
   c. $\dfrac{g(x + h) - g(x)}{h}$

## EXERCISES 8.3

**1.** Let $f(x) = x + 7$. Find:
a. $f(5)$
b. $f(-3)$
c. $f(a + 1)$

**2.** Let $f(x) = 2x - 3$. Find:
a. $f(-4)$
b. $f(7)$
c. $f(x + 4)$

**3.** Let $f(x) = x^2 - 5$. Find:
a. $f(3)$
b. $f(-6)$
c. $f(x - 2)$

**4.** Let $g(x) = x^2 + 1$. Find:
a. $g(-4)$
b. $g(5)$
c. $g(x - 3)$

**5.** Let $g(x) = 4x - 3$. Find:
a. $g(1)$
b. $g(a + 2)$
c. $g(x + h)$
d. $\dfrac{g(x + h) - g(x)}{h}$

**6.** Let $f(x) = 5 - 2x$. Find:
a. $f(-1)$
b. $f(a + 4)$
c. $f(x + h)$
d. $\dfrac{f(x + h) - f(x)}{h}$

**7.** Let $F(x) = x^2 - 4$. Find:
a. $F(-2)$
b. $F(a - 3)$
c. $F(x + h)$
d. $\dfrac{F(x + h) - F(x)}{h}$

**8.** Let $G(x) = 2 - x^2$. Find:
a. $G(\sqrt{2})$
b. $G(a - 1)$
c. $G(x + h)$
d. $\dfrac{G(x + h) - G(x)}{h}$

**9.** Let $f(x) = 2x^2 - 3$. Find:
a. $f(0)$
b. $f(a - 2)$
c. $f(x + h)$
d. $\dfrac{f(x + h) - f(x)}{h}$

**Answers to Practice Problems** 1. a. $f(0) = -5$ b. $f(a) = a^2 - 5$ c. $f(a + 2) = (a + 2)^2 - 5$ 2. a. $g(0) = 7$ b. $g(x + h) = 3(x + h) + 7$ c. $\dfrac{g(x + h) - g(x)}{h} = 3$

**10.** Let $f(x) = 3x^2 - x$. Find:
 **a.** $f(4)$
 **b.** $f(a + 2)$
 **c.** $f(x + h)$
 **d.** $\dfrac{f(x + h) - f(x)}{h}$

Using the graph of $y = |x|$, graph the functions in Exercises 11–20 without additional computations (see Example 6).

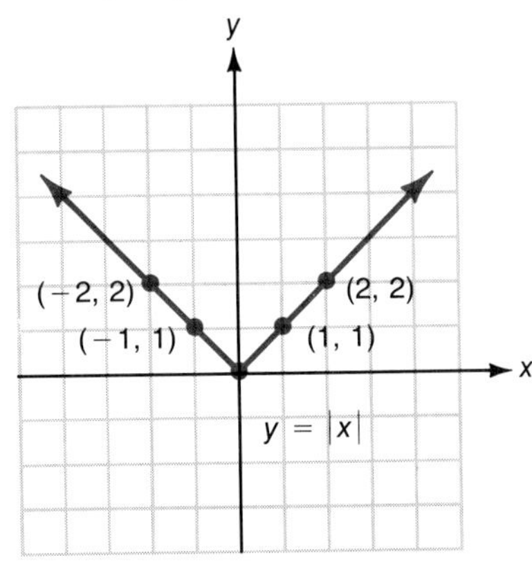


**11.** $y + 2 = |x - 1|$

**12.** $y - 6 = |x - 2|$

**13.** $y = -|x + 3|$

**14.** $y = -|x - 4|$

**15.** $y - 4 = -|x + 5|$

**16.** $y + 3 = \left|x + \dfrac{3}{4}\right|$

**17.** $y = |x - 3| + 5$

**18.** $y = |x + 2| - 3$

**19.** $y = \left|x + \dfrac{1}{2}\right| - \dfrac{3}{2}$

**20.** $y = \dfrac{5}{2} + \left|x - \dfrac{2}{3}\right|$

Using the graph of $y = \sqrt{x}$, graph the functions in Exercises 21–30 without additional computations (see Example 11).

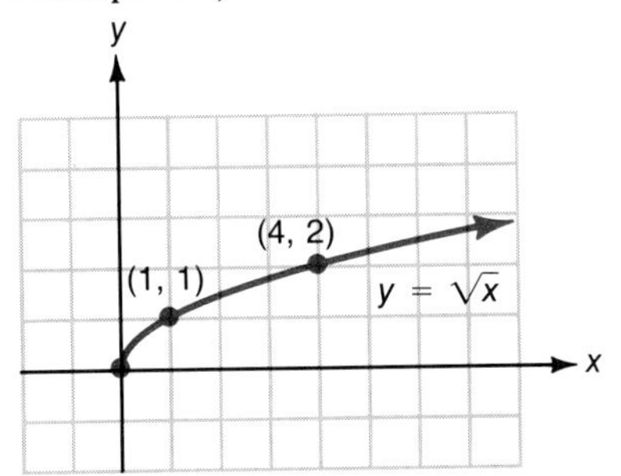


**21.** $y + 2 = \sqrt{x}$

**22.** $y - 1 = \sqrt{x}$

**23.** $y = \sqrt{x + 1}$

**24.** $y = \sqrt{x} - 6$

**25.** $y + 3 = \sqrt{x - 4}$

**26.** $y + 4 = \sqrt{x - 2}$

**27.** $y - \dfrac{1}{2} = \sqrt{x - 3}$

**28.** $y - 2 = \sqrt{x + \dfrac{3}{2}}$

**29.** $y = 5 + \sqrt{x + 2}$

**30.** $y = \sqrt{x + 4} - 3$

Using the graph of $y = \frac{1}{x}$, graph the functions in Exercises 31–40 without additional computations.

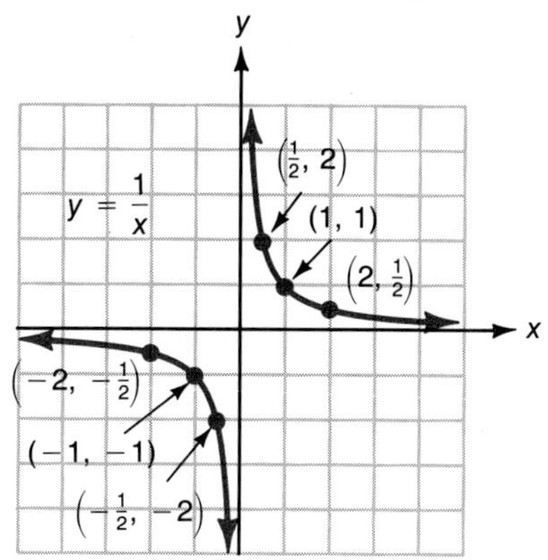


**31.** $y + 3 = \frac{1}{x}$ **32.** $y - 5 = \frac{1}{x}$ **33.** $y = \frac{1}{x - 1}$ **34.** $y = \frac{1}{x + 2}$

**35.** $y - 1 = \frac{1}{x - 3}$ **36.** $y + 2 = \frac{1}{x + 5}$ **37.** $y + 4 = \frac{1}{x + 1}$ **38.** $y - 3 = \frac{1}{x - 2}$

**39.** $y = \frac{1}{x + 4} - 5$ **40.** $y = \frac{1}{x - 5} + 2$

Using the graph of $y = f(x)$, graph the functions in Exercises 41–50.

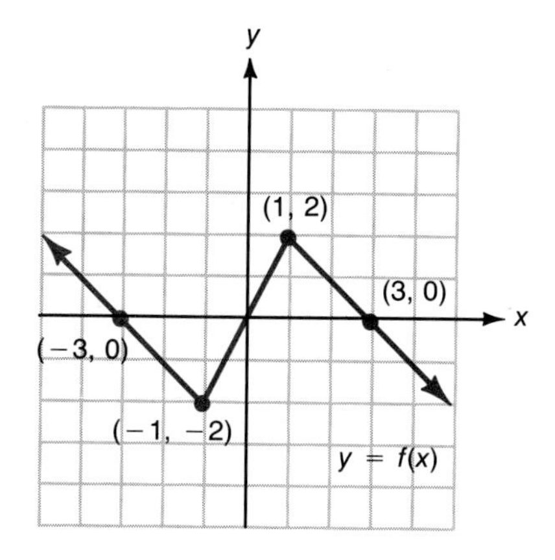


**41.** $y + 1 = f(x)$ **42.** $y - 2 = f(x)$ **43.** $y = f(x - 3)$ **44.** $y = f(x + 1)$

**45.** $y = -f(x)$ **46.** $y = -f(x - 4)$ **47.** $y - 3 = f(x + 5)$ **48.** $y - 5 = f(x - 1)$

**49.** $y = f(x + 2) - 4$ **50.** $y = f(x + 3) + 2$

## 8.4 Composition of Functions and Inverse Functions

**OBJECTIVES**

In this section, you will be learning to:

1. Form the compositions of two functions.
2. Determine if functions are one-to-one by using the horizontal line test.
3. Show that two functions, $f$ and $g$, are inverse functions by verifying that $f[g(x)] = g[f(x)] = x$.
4. Find the inverses of one-to-one functions.
5. State the domains and ranges of one-to-one functions and their inverses.
6. Graph the inverses of one-to-one functions by reflecting the graphs of the functions in the line $y = x$.

In Section 8.3, we used the $f(x)$ notation in evaluating functions for specified values of $x$. Now we want to use this same notation to indicate a new function created from two (or more) given functions. For example,

if $f(x) = x^2 - 4x + 1$ and $g(x) = 2x - 4$

then $f(2) = 2^2 - 4(2) + 1 = -3$ (substituting 2 for $x$) and $g(3) = 2(3) - 4 = 2$ (substituting 3 for $x$)

Using this same idea, we can find

$f[g(3)]$ read "$f$ of $g$ of 3"

We know that $g(3) = 2$ and $f(2) = -3$.
So,

$$f[g(3)] = f[2] = -3$$

More generally, given $f(x)$ and $g(x)$, a new function, $f[g(x)]$, called the **composition of $f$ and $g$,** is found by substituting the entire function $g(x)$ in place of $x$ in the function $f(x)$. Thus, for $f(x) = x^2 - 4x + 1$ and $g(x) = 2x - 4$,

substituting $g(x)$ for $x$

$$\begin{aligned} f[g(x)] &= [g(x)]^2 - 4[g(x)] + 1 \\ &= [2x - 4]^2 - 4[2x - 4] + 1 \\ &= 4x^2 - 16x + 16 - 8x + 16 + 1 \\ &= 4x^2 - 24x + 33 \end{aligned}$$

The **composition of $g$ and $f$** is indicated by

$g[f(x)]$ read "$g$ of $f$ of $x$"

and is found by substituting the entire function $f(x)$ in place of $x$ in the function $g(x)$.
Thus,

substituting $f(x)$ for $x$

$$\begin{aligned} g[f(x)] &= 2[f(x)] - 4 \\ &= 2[x^2 - 4x + 1] - 4 \\ &= 2x^2 - 8x + 2 - 4 \\ &= 2x^2 - 8x - 2 \end{aligned}$$

As illustrated in the previous discussion, the two compositions $f[g(x)]$ and $g[f(x)]$ generally give entirely different results. This means that substitutions must be done carefully and accurately.

We also write

$$f[g(x)] = (f \circ g)(x) \qquad \text{and} \qquad g[f(x)] = (g \circ f)(x)$$

**EXAMPLES**

**1.** Form the compositions $(f \circ g)(x)$ and $(g \circ f)(x)$ if $f(x) = 5x + 2$ and $g(x) = 3x - 7$.

**Solution**

$$\begin{aligned}(f \circ g)(x) &= f[g(x)] = 5 \cdot g(x) + 2 = 5(3x - 7) + 2 \\ &= 15x - 33 \\ (g \circ f)(x) &= g[f(x)] = 3 \cdot f(x) - 7 = 3(5x + 2) - 7 \\ &= 15x - 1\end{aligned}$$

**2.** Form the compositions $f[g(x)]$ and $g[f(x)]$ if $f(x) = \sqrt{x - 3}$ and $g(x) = x^2 + 4$.

**Solution**

$$\begin{aligned}f[g(x)] &= \sqrt{g(x) - 3} \\ &= \sqrt{x^2 + 4 - 3} = \sqrt{x^2 + 1} \\ g[f(x)] &= [f(x)]^2 + 4 \\ &= (\sqrt{x - 3})^2 + 4 = x - 3 + 4 = x + 1\end{aligned}$$

**3.** Find $(f \circ g)(x)$ and $(g \circ f)(x)$ if $f(x) = \sqrt{x + 3}$ and $g(x) = 2x - 5$.

**Solution**

$$\begin{aligned}(f \circ g)(x) &= \sqrt{g(x) + 3} \\ &= \sqrt{2x - 5 + 3} = \sqrt{2x - 2} \\ (g \circ f)(x) &= 2 \cdot \sqrt{x + 3} - 5\end{aligned}$$

■

In any function, for each $x$ there is only one $y$. That is, for each domain element, there is only one corresponding range element. Graphically, we have used the **vertical line test** to help determine quickly whether or not the graph represents a function. (See Section 8.1.)

Some functions have an additional property of having only one $x$ for each $y$. That is, for each range element, there is only one corresponding domain element. These functions are called **one-to-one functions** (or **1–1 functions**).

In terms of ordered pairs, in a 1–1 function no two ordered pairs have the same second component. For example,

$$f = \{(1,4), (2,8), (3,12)\} \qquad \text{is a 1–1 function}$$

but

$$g = \{(1,6), (2,6), (3,5)\} \qquad \text{is not 1–1}$$

because 6 appears more than once as a second component, or the second component 6 has more than one corresponding first component.

Graphically, we can use the **horizontal line test** to help determine whether or not a function is 1–1. If **any** horizontal line intersects the graph of a function in more than one point, it is not 1–1. If no horizontal line intersects the graph in more than one point, it is 1–1 (Figure 8.12).

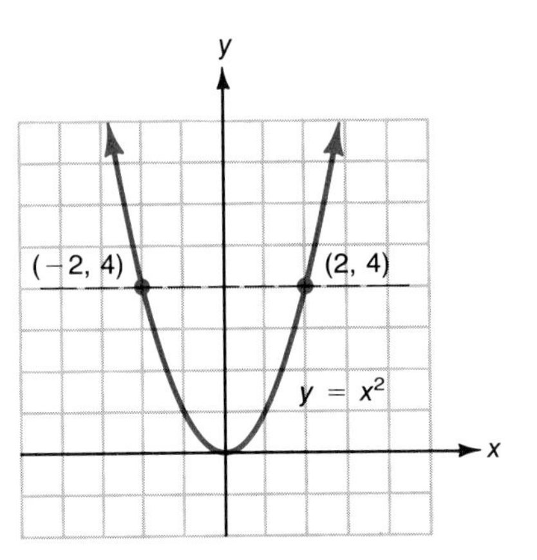


$f(x) = x^2$ is not 1–1.
$f(2) = 4$ and $f(-2) = 4$, so for the $y$-value 4, there are two corresponding $x$-values, $-2$ and 2.

(a)

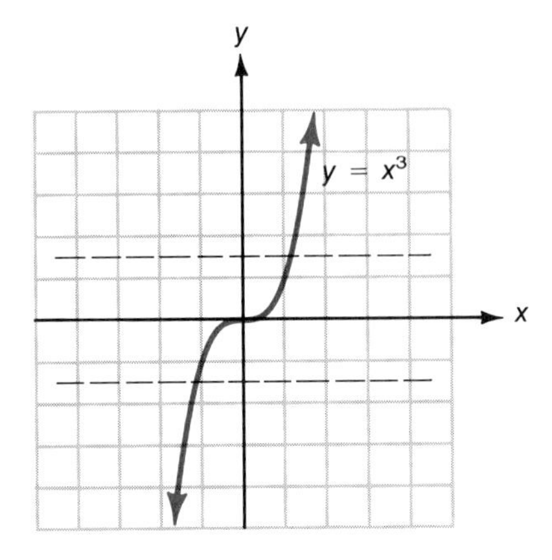


$f(x) = x^3$ is 1–1.
No horizontal line will intersect the graph in more than one point.

(b)

**Figure 8.12**

If we interchange the components of a 1–1 function, then a new function is formed. For example, if

$$f = \{(-1,1), (0,2), (1,4)\}$$

then interchanging the components gives the function

$$g = \{(1,-1), (2,0), (4,1)\}$$

The functions $f$ and $g$ are called **inverses** of each other. The notation for $g$ is

$$f^{-1} \qquad \text{read "}f\text{ inverse"}$$

Graphically, the points of $f^{-1}$ are reflections of the points of $f$ across the line $y = x$, as shown in Figure 8.13.

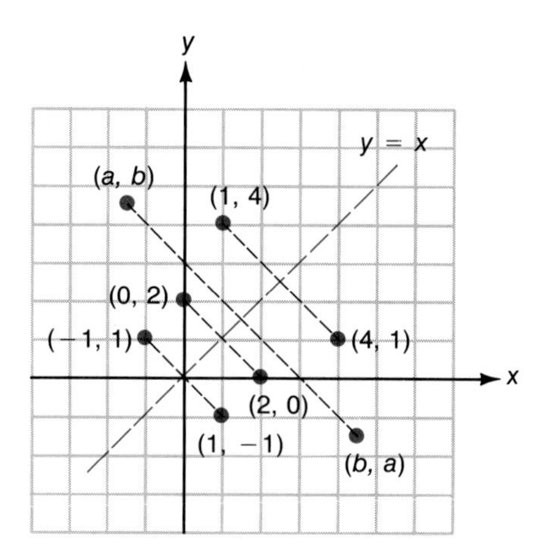


The points of $f$ and $f^{-1}$ are reflections across the line $y = x$. In general, the points $(a, b)$ and $(b, a)$ are reflections across the line $y = x$.

**Figure 8.13**

Algebraic proofs to determine whether or not a function is 1–1 are beyond the scope of this text.

You can find the graph of the inverse, $f^{-1}$, of a 1–1 function, $f$, by reflecting the graph of $f$ across the line $y = x$, as shown in Figure 8.14.

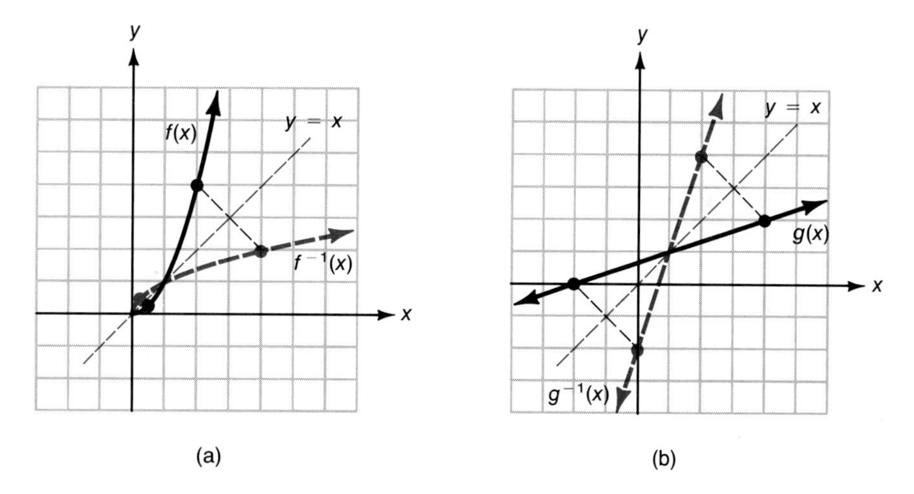


**Figure 8.14**

For a more algebraic approach to inverse functions, we need the following definition. In this definition, we introduce the notation

$D_f$ for the domain of function $f$ and
$R_f$ for the range of function $f$.

This notation is particularly useful when more than one function is being considered.

**Inverse Functions**

If $f$ and $g$ are functions and

$$f[g(x)] = x \quad \text{for } x \text{ in } D_g$$
$$g[f(x)] = x \quad \text{for } x \text{ in } D_f$$
$$R_f = D_g$$
$$D_f = R_g$$

then $g = f^{-1}$ and $f = g^{-1}$ (or $f$ and $g$ are **inverse functions**).

Note that if a number belongs to the range of $f$, then it is in the domain of $f^{-1}$. If a number belongs to the domain of $f$, then it is in the range of $f^{-1}$.

$$R_f = D_{f^{-1}} \qquad \text{and} \qquad D_f = R_{f^{-1}}$$

This relationship is apparent from the discussion of inverses as the interchange of $x$ and $y$ components:

$$f = \{(-1,1), (0,2), (1,4)\}, \quad D_f = \{-1,0,1\}, \quad R_f = \{1,2,4\}$$
$$f^{-1} = \{(1,-1), (2,0), (4,1)\}, \quad D_{f^{-1}} = \{1,2,4\}, \quad R_{f^{-1}} = \{-1,0,1\}$$

**EXAMPLES** Using the definition, show that in each case $f$ and $g$ are inverses of each other.

**4.** $f(x) = 2x + 6$ and $g(x) = \dfrac{x-6}{2}$

**Solution** The domain and range of both functions are all real numbers.

$$f[g(x)] = 2\left(\frac{x-6}{2}\right) + 6 = (x - 6) + 6 = x$$

$$g[f(x)] = \frac{(2x+6)-6}{2} = \frac{2x}{2} = x$$

So, $g = f^{-1}$ and $f = g^{-1}$.

**5.** $f(x) = \sqrt{x-3}$ and $g(x) = x^2 + 3$ for $x \geq 0$.

**Solution** $D_f$: $x \geq 3$ $\quad R_f$: $y \geq 0$ $\quad$ Since $\sqrt{x-3}$ is nonnegative.
$D_g$: $x \geq 0$ $\quad R_g$: $y \geq 3$ $\quad$ Since $x^2 + 3$ will never be less than 3 for $x \geq 0$.

$$f[g(x)] = \sqrt{(x^2+3)-3} = \sqrt{x^2} = x \quad \text{for } x \geq 0$$
$$g[f(x)] = (\sqrt{x-3})^2 + 3 = x - 3 + 3 = x \quad \text{for } x \geq 3$$

So, $g = f^{-1}$ and $f = g^{-1}$. ■

Figure 8.15 shows the graphs of the functions and their inverses from Examples 4 and 5. In each case, the line $y = x$ is a line of symmetry for the two graphs.

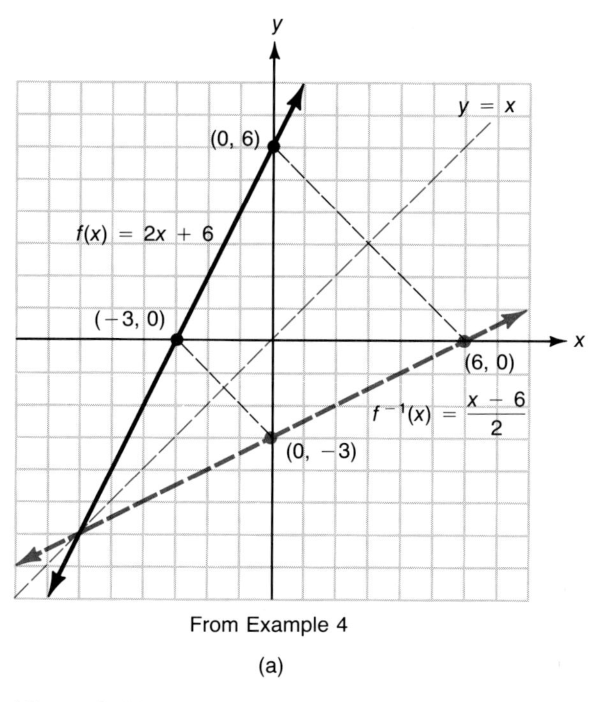


From Example 4

(a)

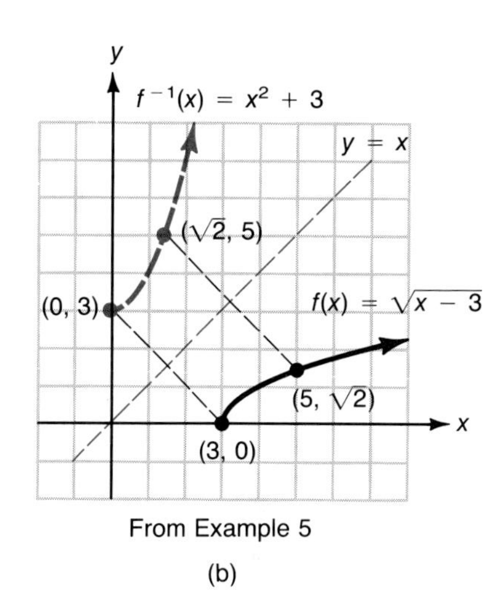


From Example 5

(b)

**Figure 8.15**

***To Find the Inverse of a 1–1 Function***

1. Let $y = f(x)$. [Substitute $y$ for $f(x)$.]
2. Interchange $x$ and $y$.
3. In the new equation, solve for $y$ in terms of $x$.
4. Substitute $f^{-1}(x)$ for $y$. (This new expression for $y$ is the inverse.)

**EXAMPLES**

**6.** Find $f^{-1}$ if $f(x) = 5x - 7$.

**Solution**

$$f(x) = 5x - 7$$

$$y = 5x - 7 \qquad \text{Substitute } y \text{ for } f(x).$$

$$x = 5y - 7 \qquad \text{Interchange } x \text{ and } y.$$

$$x + 7 = 5y \qquad \text{Solve for } y \text{ in terms of } x.$$

$$\frac{x + 7}{5} = y$$

$$f^{-1}(x) = \frac{x + 7}{5} \qquad \text{Substitute } f^{-1}(x) \text{ for } y.$$

**7.** Find $g^{-1}$ if $g(x) = x^2 - 2$ for $x \geq 0$.

**Solution**

| | |
|---|---|
| $g(x) = x^2 - 2$ | For $x \geq 0$. |
| $y = x^2 - 2$ | Substitute $y$ for $g(x)$. |
| $x = y^2 - 2$ | Interchange $x$ and $y$. |
| $x + 2 = y^2$ | Solve for $y$ in terms of $x$. |
| $\sqrt{x + 2} = y$ | Take the positive square root since we must have $y \geq 0$. (The domain of $g$ is $x \geq 0$.) |
| $g^{-1}(x) = \sqrt{x + 2}$ | |

■

***Practice Problems***

1. For $f(x) = 2x - 1$ and $g(x) = x^2$, find:
   **a.** $f[g(x)]$ **b.** $g[f(x)]$
2. Find the inverse of the function $f(x) = 3x - 1$.
3. Find $f^{-1}(x)$ if $f(x) = x^2 - 4$, $x \geq 0$.

## *EXERCISES 8.4*

Form the compositions $f[g(x)]$ and $g[f(x)]$ for Exercises 1–20.

**1.** $f(x) = 3x + 5$, $g(x) = \frac{x + 4}{2}$

**2.** $f(x) = \frac{1}{4}x + 1$, $g(x) = 6x - 7$

**3.** $f(x) = x^2$, $g(x) = 2x + 3$

**4.** $f(x) = x^2 + 1$, $g(x) = x - 6$

**5.** $f(x) = \frac{1}{x}$, $g(x) = 5x - 8$

**6.** $f(x) = \frac{1}{x + 1}$, $g(x) = x^2 + x - 3$

**7.** $f(x) = x - 1$, $g(x) = \frac{1}{x^2}$

**8.** $f(x) = \frac{1}{x^2}$, $g(x) = x^2 + 1$

**9.** $f(x) = x^3 + x + 1$, $g(x) = x + 1$

**10.** $f(x) = x^3$, $g(x) = 2x - 1$

**11.** $f(x) = \sqrt{x}$, $g(x) = x - 2$

**12.** $f(x) = \sqrt{x}$, $g(x) = x^2 - 9$

**13.** $f(x) = \sqrt{x}$, $g(x) = x^2$

**14.** $f(x) = \frac{1}{\sqrt{x}}$, $g(x) = x^2$

**15.** $f(x) = \frac{1}{\sqrt{x}}$, $g(x) = x^2 - 4$

**16.** $f(x) = x^{3/2}$, $g(x) = 2x - 6$

**17.** $f(x) = \frac{1}{x}$, $g(x) = \frac{1}{x}$

**18.** $f(x) = x^{1/3}$, $g(x) = 4x + 7$

**19.** $f(x) = x^3$, $g(x) = \sqrt{x - 8}$

**20.** $f(x) = x^3 + 1$, $g(x) = \frac{1}{x}$

Determine which of the functions in Exercises 21–30 are 1–1 by inspecting the graph.

**21.** $f(x) = 2x + 3$

**22.** $f(x) = 7 - 4x$

**23.** $g(x) = x^2 - 2$

**24.** $g(x) = 9 - x^2$

**25.** $f(x) = x^3 + 2$

**26.** $g(x) = \frac{1}{x}$

**27.** $g(x) = \sqrt{x - 3}$

**28.** $f(x) = \sqrt{x + 5}$

**Answers to Practice Problems** **1. a.** $f[g(x)] = 2x^2 - 1$ **b.** $g[f(x)] = (2x - 1)^2$
**2.** $f^{-1}(x) = \frac{x + 1}{3}$ **3.** $f^{-1}(x) = \sqrt{x + 4}$, $x \geq -4$

**29.** $f(x) = |x + 1|$ **30.** $f(x) = |x - 5|$

Determine the inverse of each of the functions in Exercises 31–44. Graph $f(x)$ and $f^{-1}(x)$ on the same set of axes.

**31.** $f(x) = 2x - 3$ **32.** $f(x) = 2x - 5$ **33.** $g(x) = x$ **34.** $g(x) = 1 - 3x$

**35.** $f(x) = 5x + 1$ **36.** $g(x) = \frac{2}{3}x + 2$ **37.** $f(x) = x^3$ **38.** $f(x) = (x + 1)^3$

**39.** $f(x) = \frac{1}{x}$ **40.** $f(x) = \frac{1}{x - 3}$ **41.** $f(x) = x^2, x \geq 0$ **42.** $f(x) = x^2 + 2, x \geq 0$

**43.** $g(x) = \sqrt{x + 7}$ **44.** $f(x) = \sqrt{x + 5}$

Find $f^{-1}(x)$ and state $D_f$, $R_f$, $D_{f^{-1}}$ and $R_{f^{-1}}$ for Exercises 45–50. Then show that $f[f^{-1}(x)] = x$ for $x$ in $D_{f^{-1}}$ and $f^{-1}[f(x)] = x$ for $x$ in $D_f$.

**45.** $f(x) = -\sqrt{x}$ **46.** $f(x) = -2x + 4$ **47.** $f(x) = x^2 + 1, x \geq 0$

**48.** $f(x) = x^2 - 1, x \geq 0$ **49.** $f(x) = -x - 2$ **50.** $f(x) = -\sqrt{x - 2}$

Using the horizontal line test, determine which of the graphs in Exercises 51–60 are graphs of 1–1 functions. If the graph represents a 1–1 function, graph the inverse function by reflecting the graph of the function in the line $y = x$.

**51.**

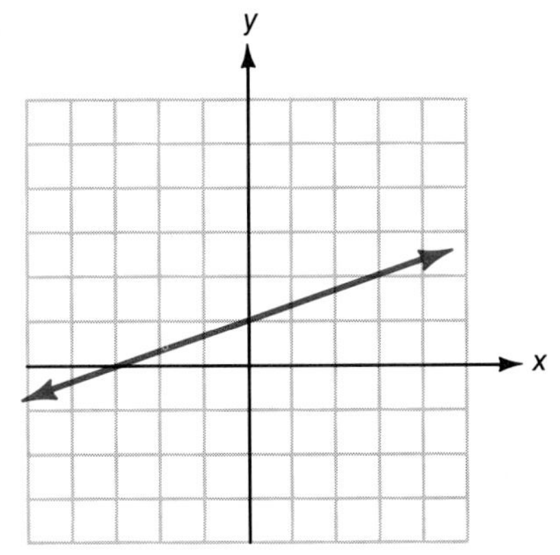


**52.**

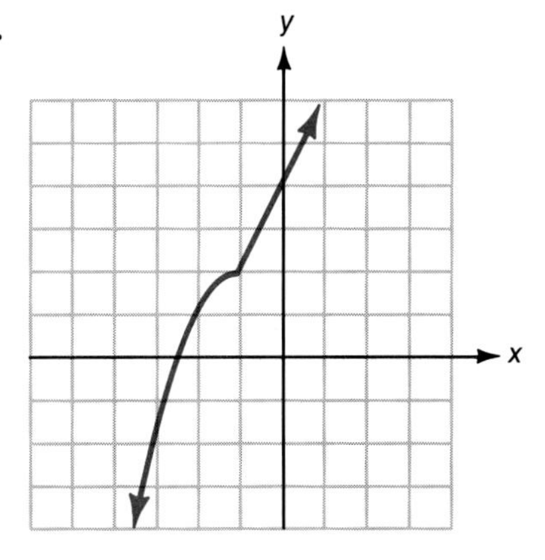


**53.**

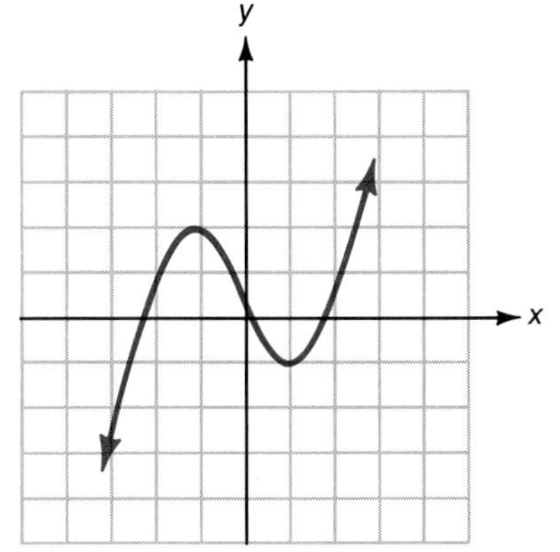


**54.**

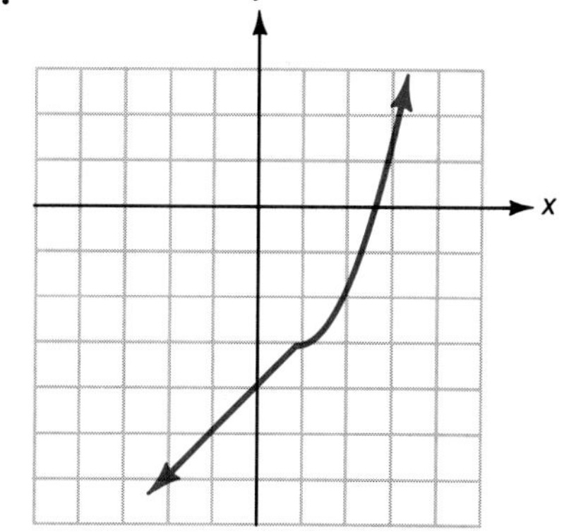

**55.**

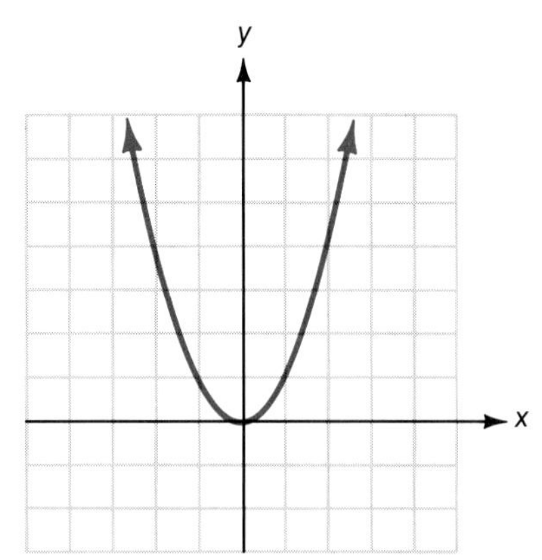


**56.**

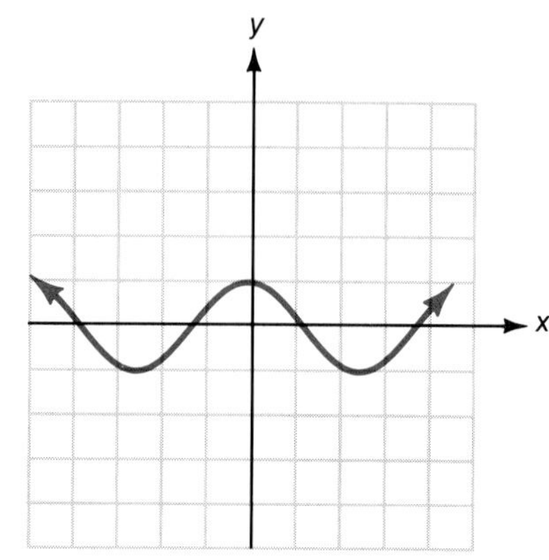


**57.**

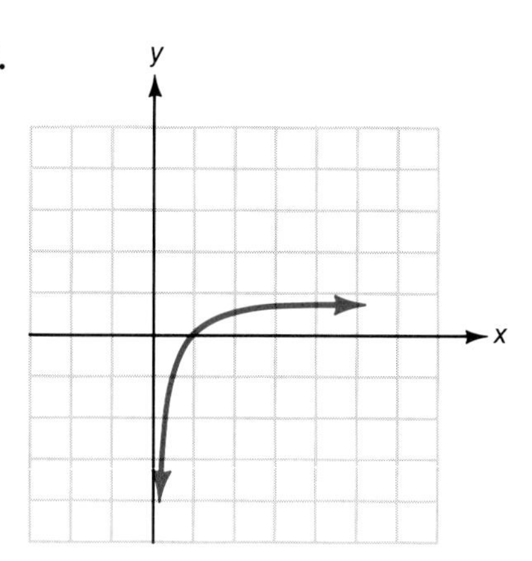


**58.**

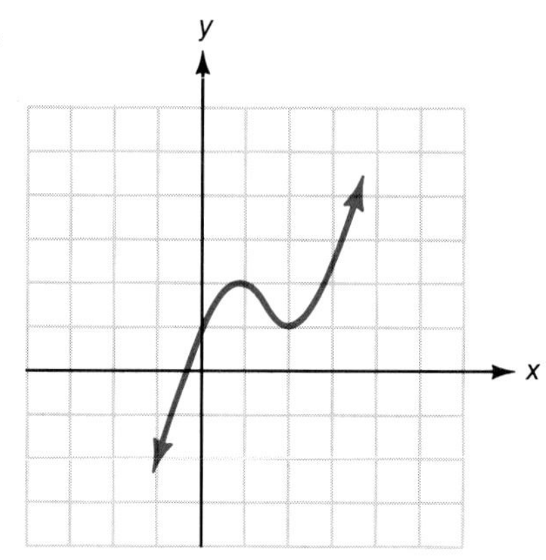


**59.**

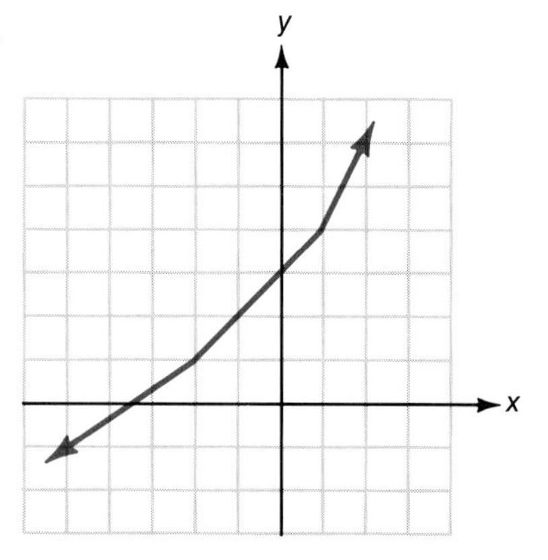


**60.**

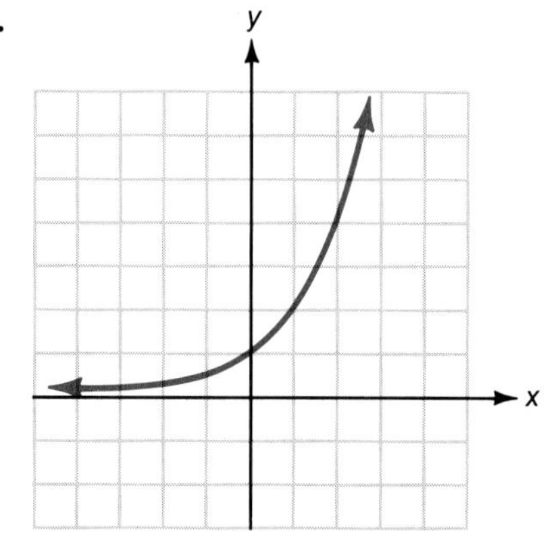

# CHAPTER 8 SUMMARY

## Key Terms and Formulas

A **relation** is a set of ordered pairs.

The **domain, D,** of a relation is the set of all first components in the relation.

The **range, R,** of a relation is the set of all second components in the relation. [8.1]

A **function** is a relation in which each domain element has only one corresponding range element.

OR

A **function** is a relation in which each first component appears only once.

OR

A **function** is a relation in which no two ordered pairs have the same first component. [8.1]

**Vertical line test** If any vertical line intersects the graph of a relation in more than one point, then the graph does **not** represent a function. [8.1]

Equations in the form $y = mx + b$ represent **linear functions.** [8.2]

Equations in the form $y = ax^2 + bx + c$ where $a \neq 0$ represent **quadratic functions.** [8.2]

The graph of every quadratic function is a parabola. If the function is written in the form $y = a(x - h)^2 + k$, then the **vertex** is at $(h,k)$, and the parabola "opens upward" if $a > 0$ and "opens downward" if $a < 0$. The vertical line $x = h$ is the **line of symmetry.** The **zeros of the function** are the values of $x$ that correspond to $y = 0$. If these values of $x$ are real, they are the points where the graph crosses the $x$-axis. If they are not real, then the graph does not cross the $x$-axis. [8.2]

In terms of the coefficients $a$, $b$, and $c$,

$$x = -\frac{b}{2a} \quad \text{is the line of symmetry,}$$

and

$$(h,k) = \left(-\frac{b}{2a}, \frac{4ac - b^2}{4a}\right) \quad \text{is the vertex.} \quad [8.2]$$

The function $f[g(x)]$ is the **composition** or the **composite** of the two functions $f$ and $g$.

We also write

$$f[g(x)] = (f \circ g)(x) \qquad \text{and} \qquad g[f(x)] = (g \circ f)(x)$$

If a function has only one $x$ for each $y$, then it is a **one-to-one function** (or **1–1 function**). [8.4]

If $f$ and $g$ are functions and

$$\begin{aligned} f[g(x)] &= x \quad \text{for } x \text{ in } D_g \\ g[f(x)] &= x \quad \text{for } x \text{ in } D_f \\ R_f &= D_g \\ D_f &= R_g \end{aligned}$$

then $g = f^{-1}$ and $f = g^{-1}$ (or $f$ and $g$ are **inverse functions**). [8.4]

## Procedures

Suppose that the graph of $y = f(x)$ is known. Then the graph of $y - k = f(x - h)$ is

1. a horizontal translation of $h$ units, and
2. a vertical translation of $k$ units of the graph of $y = f(x)$.

Think of the origin (0,0) being moved to the point $(h,k)$. Then draw the graph of $y = f(x)$ in the same relation to $(h,k)$ as it was to (0,0). This new graph will be the graph of $y - k = f(x - h)$. [8.3]

**To Find the Inverse of a 1–1 Function** [8.4]

1. Let $y = f(x)$. [Substitute $y$ for $f(x)$.]
2. Interchange $x$ and $y$.
3. In the new equation, solve for $y$ in terms of $x$.
4. Substitute $f^{-1}(x)$ for $y$. (This new expression for $y$ is the inverse.)

## CHAPTER 8 REVIEW

List the sets of ordered pairs corresponding to the points in Exercises 1 and 2. Give the domain and range, and indicate which of the relations are also functions. [8.1]

**1.**

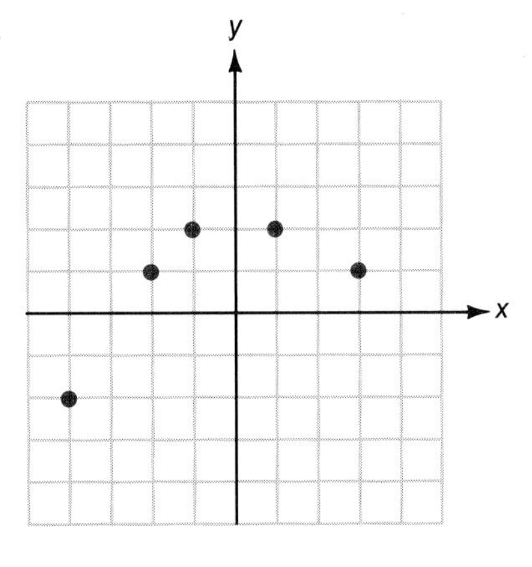


**2.**

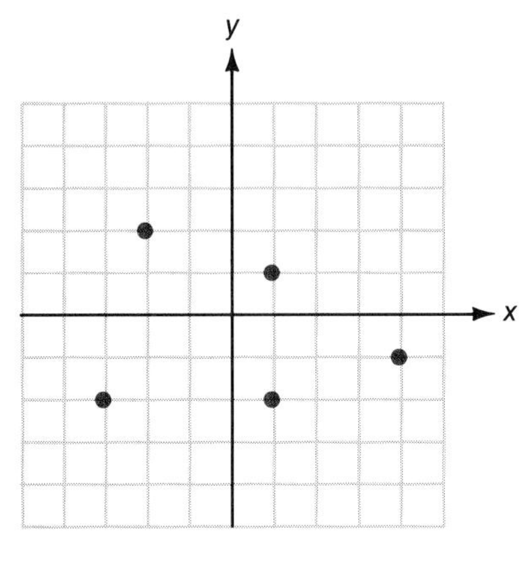


Graph the relations in Exercises 3 and 4. State the domain and range, and indicate which of the relations are also functions. [8.1]

**3.** $\{(0,3), (2,1), (4,-1), (2,2), (1,6)\}$

**4.** $\{(1,-3), (2,5), (-1,-1), (-2,5), (0,3)\}$

Use the vertical line test to determine whether or not each of the graphs in Exercises 5–8 represents a function. [8.1]

**5.**

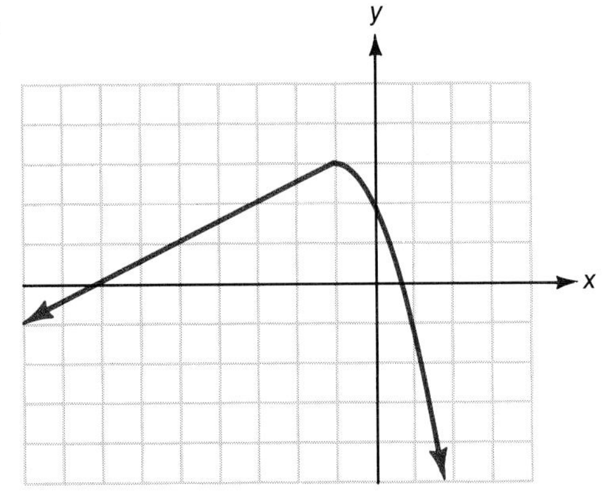


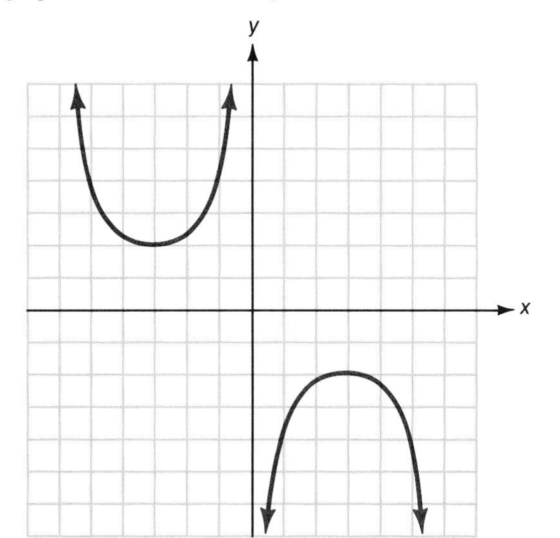

**7.**

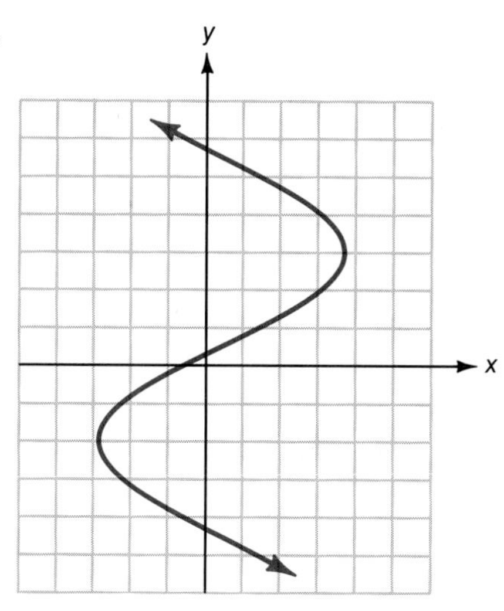


**8.**

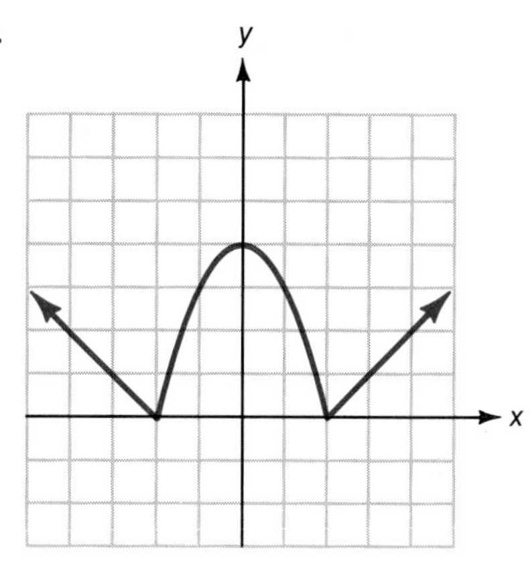


In Exercises 9–12, find the function as a set of ordered pairs for the given equation and domain. [8.1]

**9.** $y = x^2 - 2x + 5$
$D = \{-2,-1,0,1,2\}$

**10.** $y = x^3 - 5x^2$
$D = \{-2,-1,0,1,2\}$

**11.** $y = \dfrac{x}{x-6}$
$D = \left\{-2,-1,0,\dfrac{1}{2},4\right\}$

**12.** $y = \sqrt{2x+5}$
$D = \left\{-2,-\dfrac{1}{2},0,1,2\right\}$

Find the domain of each function represented by the equation in Exercises 13–16. [8.1]

**13.** $y = \dfrac{x^2-3}{2x+7}$ **14.** $y = \dfrac{3x-2}{x^2-2x-3}$ **15.** $y = \sqrt{3x+4}$ **16.** $y = \sqrt{x^2-4x-5}$

Write each of the quadratic functions in Exercises 17–19 in the form $y = a(x-h)^2 + k$. Find the vertex, axis of symmetry, range, and zeros for each function. Graph the function. [8.2]

**17.** $y = x^2 + 4x + 2$ **18.** $y = -2x^2 + 4x - 5$ **19.** $y = 2x^2 - 6x - 1$

Graph each of the quadratic functions in Exercises 20–23. [8.2]

**20.** $y = x^2 - 6x + 1$ **21.** $y = 2x^2 + 8x + 5$ **22.** $y = x^2 + 5x + 3$ **23.** $y = -3x^2 + 4x + 2$

**24.** The height of a ball projected vertically is given by the function $h = -16t^2 + 80t + 48$ where $h$ is the height in feet and $t$ is the time in seconds. **(a)** When will the ball reach its maximum height? **(b)** What will be the maximum height? [8.2]

**25.** A manufacturer estimates that by charging $x$ dollars for a certain shirt, he can sell $60 - 2x$ shirts each week. What price will give him maximum receipts? [8.2]

In Exercises 26 and 27, let $f(x) = 4x - 7$ and $g(x) = 2x^2 + 3x - 1$. [8.3, 8.4]

**26.** Find:
**a.** $f(3)$
**b.** $f(-1)$
**c.** $f[g(x)]$
**d.** $f(x+h)$
**e.** $\dfrac{f(x+h) - f(x)}{h}$

**27.** Find:
**a.** $g(2)$
**b.** $g(-2)$
**c.** $g[f(x)]$
**d.** $g(x+h)$
**e.** $\dfrac{g(x+h) - g(x)}{h}$

Using the graph of $y = f(x)$, graph the functions in Exercises 28–31. [8.3]

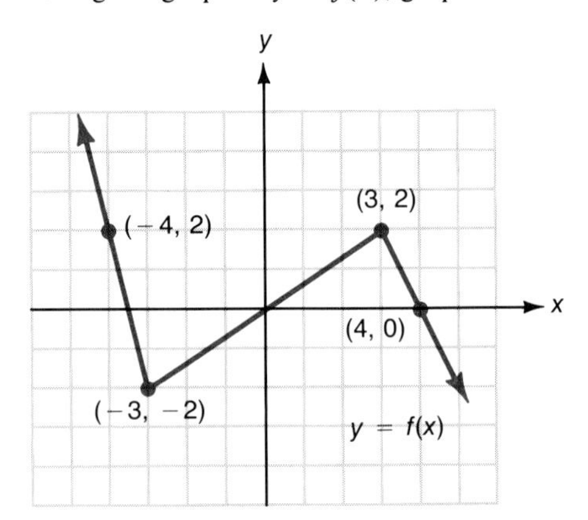


**28.** $y - 2 = f(x)$

**29.** $y = f(x - 1)$

**30.** $y = -f(x)$

**31.** $y + 3 = f(x + 2)$

Which of the functions in Exercises 32–35 are 1–1 functions? If a function is 1–1, draw the graph of its inverse. [8.4]

**32.**

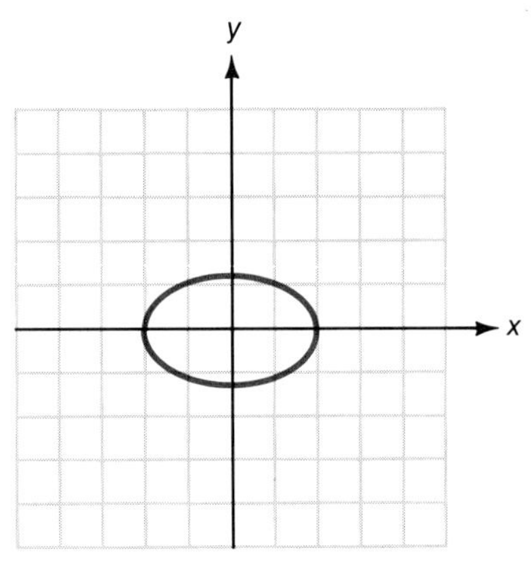


**33.**

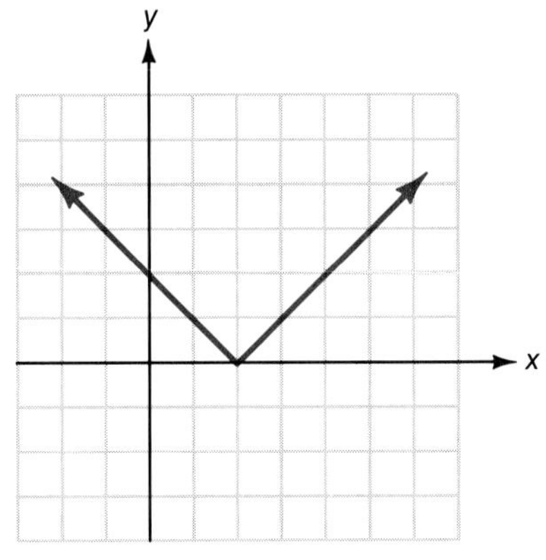


**34.**

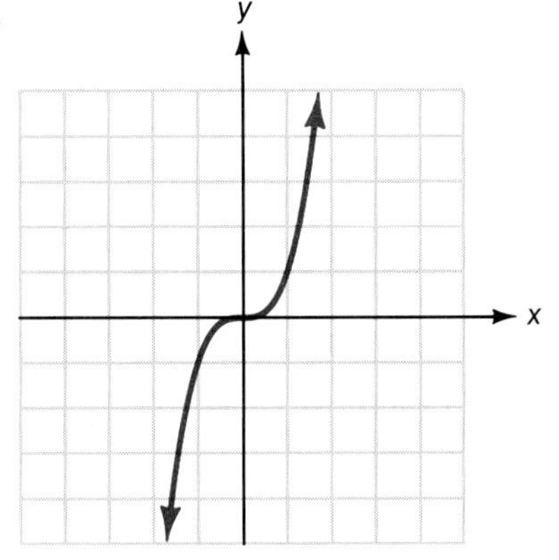


**35.**

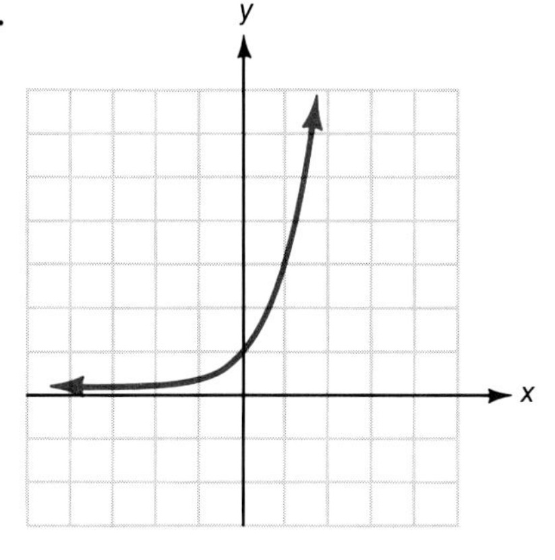

Form the composition $f[g(x)]$ in Exercises 36–39. [8.4]

**36.** $f(x) = x + 6,\ g(x) = \dfrac{x - 3}{2}$

**37.** $f(x) = x^2 + 3x,\ g(x) = x - 4$

**38.** $f(x) = \dfrac{1}{x},\ g(x) = x^2 - 25$

**39.** $f(x) = x^{3/2},\ g(x) = 2x - 5$

Determine the inverse of each of the functions in Exercises 40–44. Graph $f(x)$ and $f^{-1}(x)$ on the same set of axes. Also, show that $f^{-1}[f(x)] = x$ for $x$ in $D_f$. [8.4]

**40.** $f(x) = 3x + 2$

**41.** $g(x) = (x - 2)^3$

**42.** $f(x) = \sqrt{x - 4}$

**43.** $f(x) = x^2 - 4,\ x \geq 0$

**44.** $f(x) = \dfrac{1}{x + 3}$

## CHAPTER 8 TEST

**1.** List the set of ordered pairs corresponding to the points on the graph. Give the domain and range, and indicate if the relation is a function.

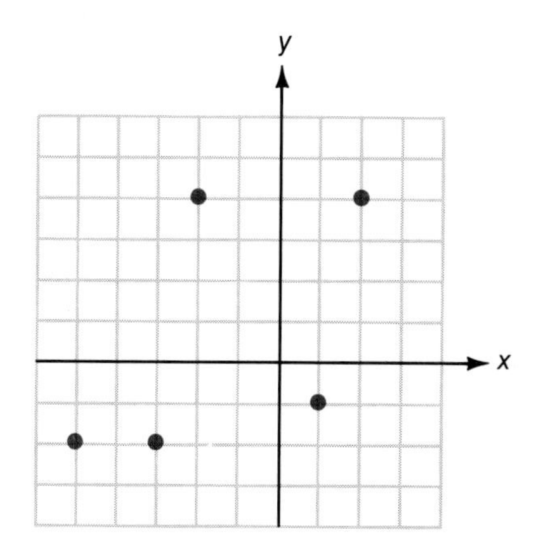


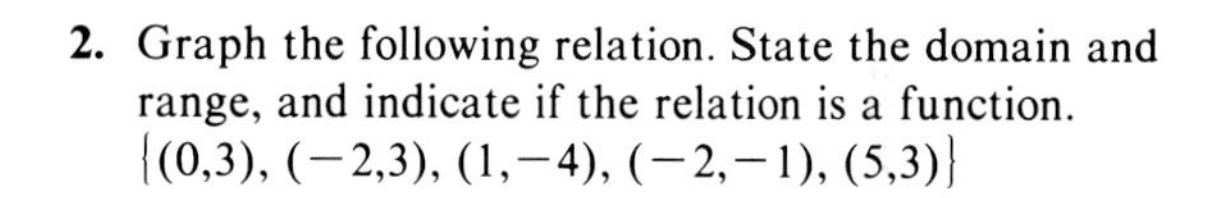

**2.** Graph the following relation. State the domain and range, and indicate if the relation is a function. $\{(0,3), (-2,3), (1,-4), (-2,-1), (5,3)\}$

Use the vertical line test to determine whether or not each of the graphs in Exercises 3 and 4 represents a function.

**3.**

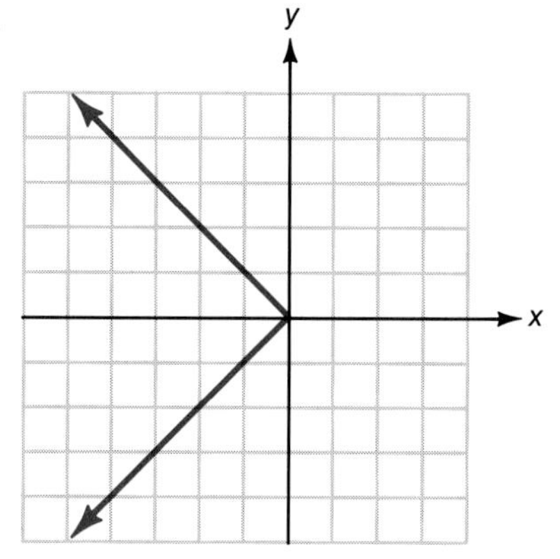


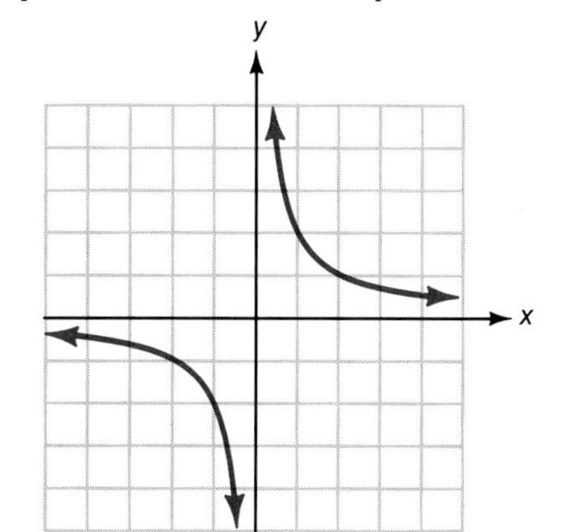

In Exercises 5 and 6, find the function as a set of ordered pairs for the given equation and domain.

**5.** $y = x^3 + 2x^2 - x - 4$
$D = \{-1,0,1,2,3\}$

**6.** $y = \dfrac{x - 7}{x^2}$

$D = \left\{-2,-1,-\dfrac{1}{2},\dfrac{1}{2},2\right\}$

**7.** Find the domain of the function represented by $y = \dfrac{x}{\sqrt{5x - 4}}$.

Graph each of the quadratic functions in Exercises 8 and 9.

**8.** $y = x^2 - 6x + 8$

**9.** $y = -2x^2 + 6x + 3$

**10.** Write the quadratic function $y = 2x^2 - 12x + 19$ in the form $y = a(x - h)^2 + k$. Find the vertex, axis of symmetry, range, and zeros. Graph the function.

**11.** One number exceeds another by 10. Find the minimum product of the two numbers.

**12.** The perimeter of a rectangle is 22 inches. Find the dimensions that will maximize the area.

**13.** If $f(x) = 2x^2 + 5$ and $g(x) = 3x + 1$, find:
**a.** $f(-2)$
**b.** $g(4)$
**c.** $f[g(x)]$
**d.** $f(x + h)$
**e.** $\dfrac{f(x + h) - f(x)}{h}$

Using the graph of $y = f(x)$, graph the functions in Exercises 14–16.

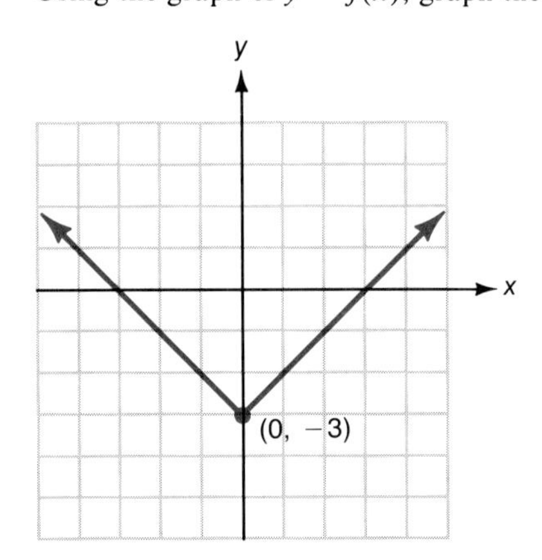


**14.** $y = -f(x)$

**15.** $y + 3 = -f(x)$

**16.** $y - 2 = f(x + 1)$

**17.** If $f(x) = 3 - 2x^2$ and $g(x) = \dfrac{1}{x + 2}$, find:
**a.** $f[g(x)]$
**b.** $g[f(x)]$

**18.** Which of the following pairs of functions are inverses?
**a.** $f(x) = x^2,\ g(x) = -x^2$
**b.** $f(x) = 5x - 3,\ g(x) = \dfrac{x + 3}{5}$
**c.** $f(x) = (x + 1)^3,\ g(x) = \sqrt[3]{x} - 1$
**d.** $f(x) = \dfrac{1}{x},\ g(x) = \dfrac{1}{x}$

**19.** Find $f^{-1}(x)$ if $f(x) = \dfrac{1}{x - 2}$.

**20.** The following is the graph for $y = f(x)$. Draw the graph of $y = f^{-1}(x)$.

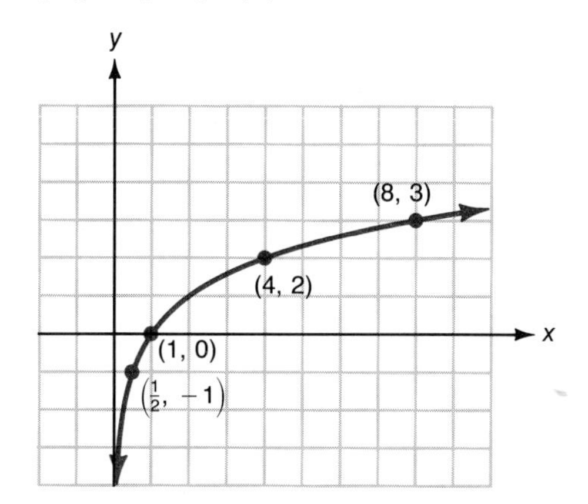


## CUMULATIVE REVIEW (8)

Simplify each of the expressions in Exercises 1–4. Assume all variables are positive.

**1.** $\dfrac{x^{-3} \cdot x}{x^2 \cdot x^{-4}}$ **2.** $\left(\dfrac{2x^{-1}y^2}{3x^3y^{-2}}\right)^{-2}$ **3.** $5x^{1/2} \cdot x^{1/4}$ **4.** $(4x^{-2/3}y^{2/5})^{3/2}$

**5.** Write $(7x^3y)^{2/3}$ in radical notation.

**6.** Write $\sqrt[3]{32x^6y}$ in exponential notation.

Perform the indicated operations in Exercises 7 and 8.

**7.** $\dfrac{x}{2x^2 - 5x - 12} - \dfrac{x + 1}{6x^2 + 5x - 6}$

**8.** $\dfrac{x^2 + 2x - 3}{x^2 + x - 2} \div \dfrac{9 - x^2}{x^2 - x - 6}$

Solve each of the equations in Exercises 9–11.

**9.** $3x^2 + 2x - 2 = 0$ **10.** $\dfrac{5}{x - 3} - \dfrac{3}{x + 2} = \dfrac{1}{x^2 - x - 6}$ **11.** $\sqrt{x + 14} - 2 = x$

**12.** Solve the system of equations: $\begin{cases} 2x - 3y = 5 \\ -5x + y = 7 \end{cases}$

Graph each of the quadratic equations in Exercises 13 and 14.

**13.** $y = \dfrac{1}{2}x^2 - 3$

**14.** $y = -x^2 + 4x - 4$

**15.** If $f(x) = \sqrt{x}$ and $g(x) = 10 - 3x^2$, find:
**a.** $f[g(x)]$
**b.** $g[f(x)]$

**16.** Given the graph of $y = f(x)$, draw the graph of $y = f^{-1}(x)$.

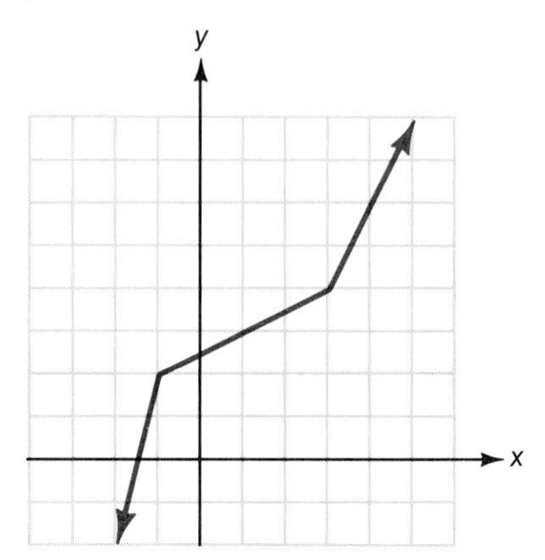


Determine the inverse of each of the functions in Exercises 17 and 18. Graph $y = f(x)$ and $y = f^{-1}(x)$ on the same set of axes.

**17.** $f(x) = \frac{5}{2}x - 3$

**18.** $f(x) = x^3 + 2$

**19.** Lisha invested $9000 in two different accounts. One account pays interest at the rate of 7%; the other pays at the rate of 8%. If her annual interest is $684, how much is invested at each rate?

**20.** If the average of a number and its square root is 21, find the number.

C H A P T E R 9

# EXPONENTIAL AND LOGARITHMIC FUNCTIONS

## MATHEMATICAL CHALLENGES

How many digits are in $5^{5^5}$?

Solve for $x$: $(\log x)^2 = \log (x^2)$

What is the product of $(x - a)(x - b)(x - c)$ $(x - d) \ldots (x - z)$?

## CHAPTER OUTLINE



Logarithms are exponents. Traditionally, logarithmic and exponential values were calculated with the extensive use of printed tables and techniques for estimating values not found in the tables. Now, hand-held calculators have these tables, with even greater accuracy, stored in their electronic memories, and complicated expressions can be evaluated by pushing a few keys. In this chapter, you will develop the skills and understanding for manipulating with exponents and logarithms so you will know you are pushing the right keys and that the resulting answer is sensible. The irrational number $e = 2.7181828459...$ will be discussed in detail. This number appears quite naturally in many applications and is the base for natural logarithms, indicated on calculators by the key marked [ln] or [ln $x$].

Of all the topics we discuss in algebra, logarithmic functions and exponential functions probably have the most value in terms of applied problems. For example, the bell-shaped curve (or normal curve) studied in statistics is the graph of an exponential function. Exponential growth and decay are basic concepts in biology and medicine. (Cancer cells grow exponentially and radium decays exponentially.) In business, depreciation and continuously compounded interest can be calculated using exponential functions. Learning curves, important in business and education, can be described with logarithmic and exponential functions. Computers use logarithmic and exponential concepts in their design and implementation. Obviously, you are likely to encounter these concepts at some time in almost any field you choose to study.

## 9.1 Exponential Functions and the Number $e$

**Authors' Note**

This chapter is written with the assumption that every student has access to a hand-held calculator with the keys [$y^x$], [log $x$], and [ln $x$]. Not all calculators have keys labeled in the manner assumed here, and some operations vary with different calculators. The student should be aware that, depending on the calculator used, some adjustment to various procedures discussed in this chapter may be necessary.

**OBJECTIVES**

In this section, you will be learning to:

1. Graph exponential functions.
2. Solve exponential equations.
3. Solve applied problems involving the number $e$ and involving exponential equations.

We begin this chapter with a discussion of a type of function in which exponents are variables and bases are constants. Such functions are called **exponential functions.** The inverses of these functions are called **logarithmic functions.** Both types of functions have important applications in a broad spectrum of studies in fields such as biology, medicine, finance, archaeology, physics, chemistry, and statistics. We will investigate some of these applications.

Until recently, most of the calculations related to exponential and logarithmic functions have been done with the aid of tables. Now, with the use of hand-held calculators, the need for such printed tables has diminished dramatically. The calculators contain the tables in electronic storage.

Calculators aside temporarily, to understand these new types of functions, we need to develop the meaning of expressions with exponents that are real numbers, including irrational numbers such as $\sqrt{2}$ and $\pi$. That is, we need to know the meaning of expressions such as $2^{\sqrt{2}}$ and $2^{\pi}$. Toward this objective, we state, without proof, that all the properties of exponents that have been discussed earlier are also valid for real exponents with a positive base.

***Properties of Real Exponents***

If $a$ and $b$ are positive real numbers and $x$ and $y$ are any real numbers, then:

1. $b^0 = 1$
2. $b^{-x} = \dfrac{1}{b^x}$
3. $b^x \cdot b^y = b^{x+y}$
4. $\dfrac{b^x}{b^y} = b^{x-y}$
5. $(b^x)^y = b^{xy}$
6. $(ab)^x = a^x b^x$
7. $\left(\dfrac{a}{b}\right)^x = \dfrac{a^x}{b^x}$

The following additional property of exponents is useful in solving equations with variable exponents.

For $b > 0$ and $b \neq 1$ and real numbers $x$ and $y$:

1. If $b^x = b^y$, then $x = y$.
2. If $x = y$, then $b^x = b^y$.

**EXAMPLES** Solve each of the following equations for $x$.

**1.** $3^{x-2} = 3^{1/2}$

**Solution**

$$3^{x-2} = 3^{1/2}$$ Both bases are 3.

$$x - 2 = \frac{1}{2}$$ Since the bases are the same, the exponents must be equal.

$$x = \frac{5}{2}$$ Solve for $x$.

**2.** $2^{x^2-7} = 2^{6x}$

**Solution**

$$2^{x^2-7} = 2^{6x}$$ Both bases are 2.

$$x^2 - 7 = 6x$$ The exponents are equal because the bases are the same.

$$x^2 - 6x - 7 = 0$$ Solve for $x$.

$$(x-7)(x+1) = 0$$

$$x = 7 \quad \text{or} \quad x = -1$$

**3.** $8^{4-2x} = 4^{x+2}$

**Solution**

| | |
|---|---|
| $8^{4-2x} = 4^{x+2}$ | Here, the bases are different. |
| $(2^3)^{4-2x} = (2^2)^{x+2}$ | Rewrite both sides so that the bases are the same. $8 = 2^3$ and $4 = 2^2$ |
| $2^{12-6x} = 2^{2x+4}$ | Use the property $(b^x)^y = b^{xy}$. |
| $12 - 6x = 2x + 4$ | The exponents must be equal. |
| $8 = 8x$ | Solve for $x$. |
| $1 = x$ | ■ |

To define a number with an irrational exponent, such as $2^{\sqrt{2}}$, we need the concept of a limit as developed in calculus. A formal discussion is beyond the scope of this text. Informally, the idea is to approximate $\sqrt{2}$ with rational numbers that get closer and closer to $\sqrt{2}$, such as 1.4, 1.41, 1.414, and so on. Then, $2^{1.4}$, $2^{1.41}$, $2^{1.414}$, and so on, get closer and closer to $2^{\sqrt{2}}$. In radical notation,

$$2^{1.4} = 2^{14/10} = \sqrt[10]{2^{14}}, \qquad \sqrt[10]{2^{14}} \approx 2.6390$$

and $$2^{1.41} = 2^{141/100} = \sqrt[100]{2^{141}}, \qquad \sqrt[100]{2^{141}} \approx 2.6574, \qquad \text{and so on}$$

To six-decimal-place accuracy, $2^{\sqrt{2}} = 2.665144$.

These concepts allow us to define $2^x$ as a real number for any real number $x$.

The quadratic function $f(x) = x^2$ has a variable base and a constant exponent. If the base is constant and the exponent is a variable, then we have an entirely different kind of function, such as $f(x) = 2^x$, called an **exponential function.**

***Exponential Function***

A function of the form

$$f(x) = b^x \qquad \text{where } b > 0,\ b \neq 1, \text{ and } x \text{ is any real number}$$

is called an **exponential function.**

Exponential functions are 1–1 and have inverses which were discussed in Section 8.4. The function $f(x) = 1^x$ is the constant function $f(x) = 1$, which is not 1–1 and is of no interest in this chapter.

We are now interested in graphing exponential functions and then discussing some very interesting applications. First a few points are plotted for the function $f(x) = 2^x$ so that the resulting graph will seem reasonable.

| $x$ | $y = 2^x$ |
|---|---|
| 3 | $2^3 = 8$ |
| 2 | $2^2 = 4$ |
| 1 | $2^1 = 2$ |
| $\frac{1}{2}$ | $2^{1/2} = \sqrt{2} \approx 1.41$ |
| 0 | $2^0 = 1$ |
| $-\frac{1}{2}$ | $2^{-1/2} = \frac{1}{\sqrt{2}} \approx 0.707$ |
| $-1$ | $2^{-1} = \frac{1}{2}$ |
| $-2$ | $2^{-2} = \frac{1}{2^2} = \frac{1}{4}$ |
| $-3$ | $2^{-3} = \frac{1}{2^3} = \frac{1}{8}$ |
| $-4$ | $2^{-4} = \frac{1}{2^4} = \frac{1}{16}$ |

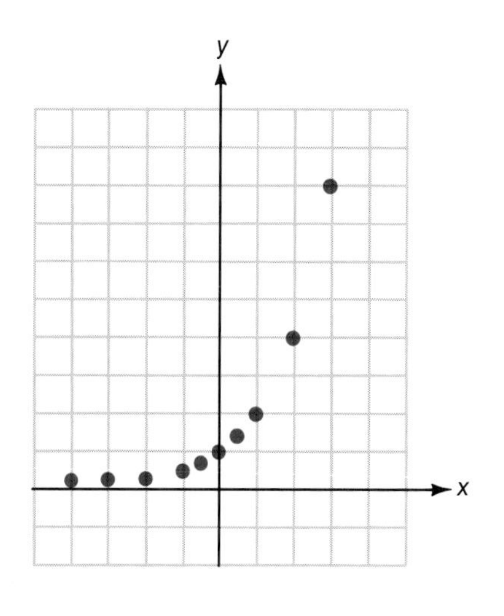


Since we know that $2^x$ is defined for all real exponents, points such as $(\sqrt{2}, 2^{\sqrt{2}})$, $(\pi, 2^{\pi})$, and $(\sqrt{5}, 2^{\sqrt{5}})$ are on the graph, and the graph for $y = 2^x$ is a smooth curve, as shown in Figure 9.1(a).

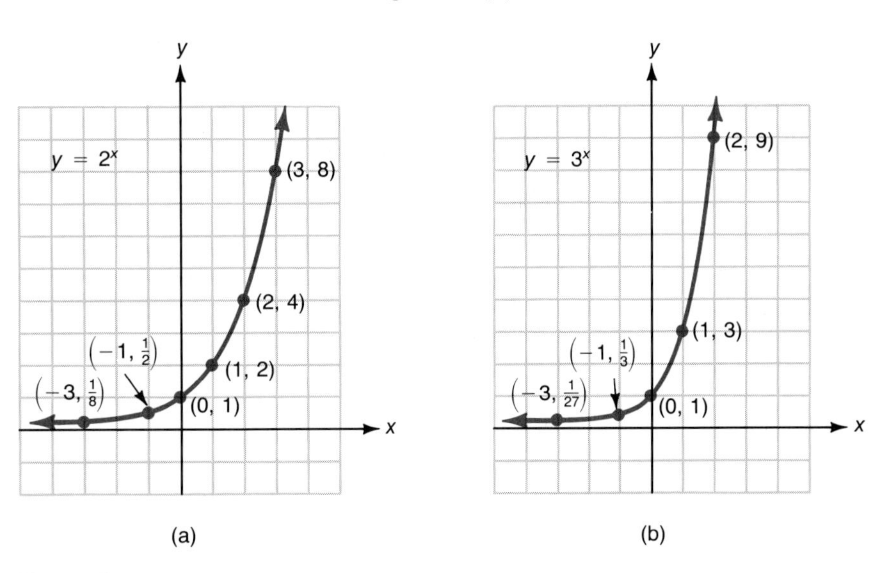


**Figure 9.1**

Now consider the exponential function $f(x) = \left(\frac{1}{2}\right)^x$ where the base $b = \frac{1}{2}$ is between 0 and 1. We can write

$$y = \left(\frac{1}{2}\right)^x = (2^{-1})^x = 2^{-x}$$

The following table and the plotted points indicate the nature of the graph.

| $x$ | $y = \left(\frac{1}{2}\right)^x = 2^{-x}$ |
|---|---|
| $-3$ | $2^{-(-3)} = 2^3 = 8$ |
| $-2$ | $2^{-(-2)} = 2^2 = 4$ |
| $-1$ | $2^{-(-1)} = 2^1 = 2$ |
| $-\frac{1}{2}$ | $2^{-(-1/2)} = 2^{1/2} = \sqrt{2} \approx 1.41$ |
| $0$ | $2^{-0} = 2^0 = 1$ |
| $\frac{1}{2}$ | $2^{-1/2} = \frac{1}{\sqrt{2}} \approx 0.707$ |
| $1$ | $2^{-1} = \frac{1}{2}$ |
| $2$ | $2^{-2} = \frac{1}{2^2} = \frac{1}{4}$ |
| $3$ | $2^{-3} = \frac{1}{2^3} = \frac{1}{8}$ |

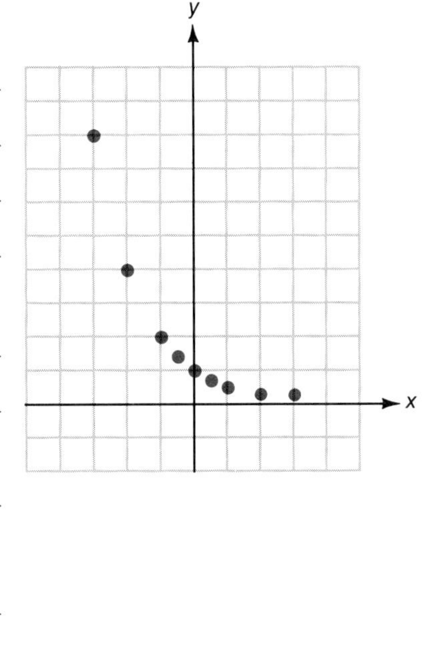


Figure 9.2 illustrates the graphs of the two functions $y = \left(\frac{1}{2}\right)^x = 2^{-x}$ and $y = \left(\frac{1}{3}\right)^x = 3^{-x}$. Notice that these graphs are the reflections (mirror images) of the graphs of $y = 2^x$ and $y = 3^x$, respectively, shown in Figure 9.1.

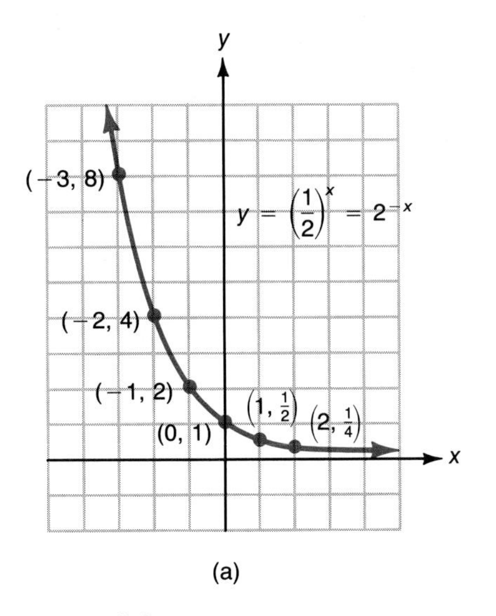

(a)

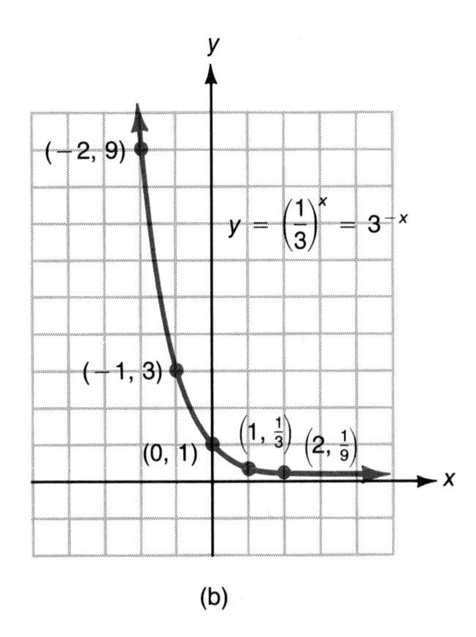

(b)

**Figure 9.2**

The following general concepts are helpful in understanding the graphs of exponential functions.

For $b > 1$:

1. $b^x > 0$.
2. $b^x$ increases to the right and is called an **increasing function.**
3. $b^0 = 1$, so (0,1) is on the graph.
4. $b^x$ approaches the $x$-axis for negative values of $x$. See Figure 9.1.

For $0 < b < 1$:

1. $b^x > 0$.
2. $b^x$ decreases to the right and is called a **decreasing function.**
3. $b^0 = 1$, so (0,1) is on the graph.
4. $b^x$ approaches the $x$-axis for positive values of $x$. See Figure 9.2.

Exponential functions have many practical applications, among which are bacterial growth, radioactive decay, compound interest, and light absorption. For example, a bacteria culture kept at a certain temperature may grow according to the function

$$y = y_0 \cdot 2^{0.3t} \qquad \text{where} \quad t = \text{time in hours}$$
$$y_0 = \text{amount of bacteria present when } t = 0$$

If the temperature is raised 10° C, then the bacteria may grow according to

$$y = y_0 \cdot 2^{0.5t}$$

The following examples indicate how such functions can be used. More applications of exponential functions will be discussed in Section 9.5.

**EXAMPLES**

**4.** A scientist has 10,000 bacteria present when $t = 0$, and she knows the bacteria grow according to the function $y = y_0 \cdot 2^{0.5t}$ where $t$ is measured in hours. How many bacteria will be present at the end of one day?

**Solution** Using $t = 24$ hours and $y_0 = 10{,}000$,

$$\begin{aligned} y &= 10{,}000 \cdot 2^{0.5(24)} = 10{,}000 \cdot 2^{12} \\ &= 10{,}000(4096) \\ &= 40{,}960{,}000 \\ &= 4.096 \times 10^7 \text{ bacteria} \end{aligned}$$

[To use your calculator to evaluate $2^{12}$: (a) enter 2, (b) press the key $\boxed{y^x}$, (c) enter 12, (d) press the key $\boxed{=}$.]

**5.** Determine the exponential function that fits the following information: $y_0 = 5000$ bacteria, and there are 135,000 bacteria present after 3 days.

**Solution** Use $y = y_0 b^t$ where $t$ is measured in days.

$$\begin{aligned} 135{,}000 &= 5000b^3 \\ 27 &= b^3 \\ \sqrt[3]{27} &= b \\ 3 &= b \end{aligned}$$

The function is $y = 5000 \cdot 3^t$.

**6.** If $P$ dollars are invested at a rate of interest $r$ (as a decimal) compounded annually for $t$ years, the amount $A$ becomes $A = P(1 + r)^t$. Find the value of \$1000 invested at 6% for 3 years.

**Solution**

$$\begin{aligned} A &= 1000(1 + 0.06)^3 && r = 0.06 \text{ as a decimal.} \\ &= 1000(1.06)^3 \\ &= 1000(1.191016) \approx \$1191.02 \end{aligned}$$

[To use your calculator to evaluate $(1.06)^3$: (a) enter 1.06, (b) press the key $\boxed{y^x}$, (c) enter 3, (d) press the key $\boxed{=}$.] ■

## The Number e

The topic of compound interest leads to a particularly interesting (and useful) result. In Example 6, we use the formula $A = P(1 + r)^t$ for finding the value of a principal when interest is compounded annually. A more practical formula is

$$A = P\left(1 + \frac{r}{n}\right)^{nt}$$

where

$P$ = original principal (original investment)
$r$ = annual rate of interest (in decimal form)
$n$ = number of times per year interest is compounded
$t$ = number of years
$A$ = amount accumulated

**EXAMPLES**

**7.** What will be the value of a principal investment of \$1000 invested at 6% for 3 years if interest is compounded monthly (12 months per year)?

**Solution** We have

$$P = 1000,\ r = 0.06,\ n = 12, \text{ and } t = 3$$

Using the formula for compound interest,

$$\begin{aligned} A &= 1000\left(1 + \frac{0.06}{12}\right)^{12(3)} \\ &= 1000(1 + 0.005)^{36} \\ &= 1000(1.005)^{36} \\ &= 1000(1.19668) \qquad \text{Using a calculator} \\ &= \$1196.68 \end{aligned}$$

**8.** Find the value of $A$ if \$1000 is invested at 6% for 3 years and interest is compounded daily. (Banks and savings institutions use 360 days per year.)

**Solution** Using the formula for compound interest with $n = 360$,

$$\begin{aligned} A &= 1000\left(1 + \frac{0.06}{360}\right)^{360(3)} \\ &= 1000(1.000166666)^{1080} \\ &= 1000(1.197198545) \qquad \text{Using a calculator} \\ &\approx \$1197.20 \end{aligned}$$

■

Examples 6, 7, and 8 illustrate the effects of compounding interest over smaller and smaller periods of time over 3 years. The formula gives

$$\begin{aligned} A &= \$1191.02 \qquad \text{for } n = 1 \\ A &= \$1196.68 \qquad \text{for } n = 12 \\ A &= \$1197.20 \qquad \text{for } n = 360 \end{aligned}$$

The effects are not as dramatic as might be expected by someone not familiar with the concept of a limit, as illustrated in Table 9.1.

Table 9.1

| $n$ | $\left(1 + \frac{1}{n}\right)$ | $\left(1 + \frac{1}{n}\right)^n$ |
|---|---|---|
| 1 | $\left(1 + \frac{1}{1}\right) = 2$ | $(2)^1 = 2$ |
| 2 | $\left(1 + \frac{1}{2}\right) = 1.5$ | $(1.5)^2 = 2.25$ |
| 5 | $\left(1 + \frac{1}{5}\right) = 1.2$ | $(1.2)^5 = 2.48832$ |
| 10 | $\left(1 + \frac{1}{10}\right) = 1.1$ | $(1.1)^{10} = 2.59374246$ |
| 100 | $\left(1 + \frac{1}{100}\right) = 1.01$ | $(1.01)^{100} = 2.704813829$ |
| 1000 | $\left(1 + \frac{1}{1000}\right) = 1.001$ | $(1.001)^{1000} = 2.716923932$ |
| 10,000 | $\left(1 + \frac{1}{10{,}000}\right) = 1.0001$ | $(1.0001)^{10{,}000} = 2.718145927$ |
| 100,000 | $\left(1 + \frac{1}{100{,}000}\right) = 1.00001$ | $(1.00001)^{100{,}000} = 2.718268237$ |
| $\downarrow$ $\infty$ | | $\downarrow$ $e = 2.718281828459...$ |

Table 9.1 shows that the **limit,** or **limiting value,** of the expression $\left(1 + \frac{1}{n}\right)^n$ is the irrational number $e = 2.718281828459...$ as $n$ approaches infinity. The student should understand that the symbol for infinity, $\infty$, does not represent a number. This symbol is used to indicate that numbers under discussion are to continue indefinitely, without end.

***The Number e***

The number $e$ is defined to be

$$e = \lim_{n \to \infty} \left(1 + \frac{1}{n}\right)^n = 2.718281828459...$$

We can show that $e$ is the base when interest is compounded **continuously** by rewriting the formula for $A$ as follows:

$$A = P\left(1 + \frac{r}{n}\right)^{nt} = P\left(1 + \frac{r}{n}\right)^{n/r \cdot rt} = P\left[\left(1 + \frac{1}{\frac{n}{r}}\right)^{n/r}\right]^{rt}$$

Now, substituting $m = \frac{n}{r}$, we can write

$$A = P\left[\left(1 + \frac{1}{m}\right)^{m}\right]^{rt}$$

The value of $m = \frac{n}{r} \longrightarrow \infty$ as $n \longrightarrow \infty$. This means that the expression in brackets approaches $e$ as $n \longrightarrow \infty$, and the formula for **continuously compounded interest** is

$$A = Pe^{rt}$$

**EXAMPLE**

**9.** Find the value of \$1000 invested at 6% for 3 years if interest is compounded continuously. (Use $e = 2.718281828$.)

**Solution**

$$\begin{aligned} A &= Pe^{rt} \\ &= 1000e^{0.06(3)} \\ &= 1000(2.718281828)^{0.18} \\ &= 1000(1.197217363) \\ &\approx \$1197.22 \end{aligned}$$

(Comparing Examples 8 and 9, we see that there is only 2¢ difference in $A$ when compounding interest daily or continuously over 3 years at 6%.) ■

***Practice Problems***

1. Solve the equation $5^x \cdot 5^{-2} = \frac{1}{125}$.
2. Solve the equation $10^{x^2 + x} = 10^6$.
3. Find the value of \$5000 invested at 8% for 10 years if interest is compounded continuously.

**Answers to Practice Problems** **1.** $x = -1$ **2.** $x = 2, x = -3$ **3.** \$11,128

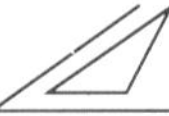

## EXERCISES 9.1

Graph each of the exponential functions in Exercises 1–10.

**1.** $y = 4^x$ **2.** $y = \left(\frac{1}{3}\right)^x$ **3.** $y = \left(\frac{1}{5}\right)^x$ **4.** $y = 5^x$ **5.** $y = 10^x$

**6.** $y = \left(\frac{2}{3}\right)^x$ **7.** $y = \left(\frac{5}{2}\right)^x$ **8.** $y = \left(\frac{1}{2}\right)^{-x}$ **9.** $y = 2^{x-1}$ **10.** $y = 3^{x+1}$

Solve each of the equations in Exercises 11–30 for $x$.

**11.** $2^4 \cdot 2^7 = 2^x$ **12.** $3^7 \cdot 3^{-2} = 3^x$ **13.** $(3^5)^2 = 3^{x+1}$

**14.** $(5^x)^2 = 5^6$ **15.** $(2^x)^3 = \sqrt{2}$ **16.** $\dfrac{10^4 \cdot 10^{1/2}}{10^x} = 10$

**17.** $(10^2)^x = \dfrac{10 \cdot 10^{2/3}}{10^{1/2}}$ **18.** $2^{5x} = 4^3$ **19.** $(25)^x = 5^3 \cdot 5^4$

**20.** $7^{3x} = 49^4$ **21.** $10^x \cdot 10^8 = 100^3$ **22.** $8^{x+3} = 2^{x-1}$

**23.** $27^x = 3 \cdot 9^{x-2}$ **24.** $100^{2x+1} = 1000^{x-2}$ **25.** $2^{3x+5} = 2^{x^2+1}$

**26.** $10^{x^2+x} = 10^{x+9}$ **27.** $10^{2x^2+3} = 10^{x+6}$ **28.** $3^{x^2+5x} = 3^{2x-2}$

**29.** $(3^{x+1})^x = (3^{x+3})^2$ **30.** $(10^x)^{x+3} = (10^{x+2})^{-2}$

**31.** A biologist knows that in the laboratory, bacteria in a culture grow according to the function $y = y_0 \cdot 5^{0.2t}$, where $y_0$ is the initial number of bacteria present and $t$ is measured in hours. How many bacteria will be present in a culture at the end of 5 hours if there were 5000 present initially?

**32.** In Exercise 31, how many bacteria were present initially if at the end of 15 hours, there were 2,500,000 bacteria present?

**33.** Four thousand dollars is deposited into a savings account at the rate of 8% per year. Find the total amount, $A$, on deposit at the end of 5 years if the interest is compounded
**a.** annually
**b.** semiannually
**c.** quarterly
**d.** daily
**e.** continuously

**34.** Find the amount, $A$, in a savings account if \$2000 is invested at 7% for 4 years and the interest is compounded
**a.** annually
**b.** semiannually
**c.** quarterly
**d.** daily
**e.** continuously

**35.** Find the value of \$1800 invested at 7% for 3 years if the interest is compounded continuously.

**36.** Find the value of \$2500 invested at 8% for 5 years if the interest is compounded continuously.

**37.** The revenue function is given by $R(x) = x \cdot p(x)$ dollars, where $x$ is the number of units sold and $p(x)$ is the unit price. If $p(x) = 25(2)^{-x/5}$, find the revenue if 15 units are sold.

**38.** In Exercise 37, if $p(x) = 40(3)^{-x/6}$, find the revenue if 12 units are sold.

**39.** A radio station knows that during an intense advertising campaign, the number of people, $N$, who will hear a commercial is given by $N = A(1 - 2^{-0.05t})$, where $A$ is the number of people in the broadcasting area and $t$ is the number of hours the commercial has been run. If there are 500,000 people in the area, how many will hear a commercial during the first 20 hours?

**40.** Statistics show that the fractional part of flashlight batteries, $P$, that is still good after $t$ hours of use is given by $P = 4^{-0.02t}$. What fractional part of the batteries is still operating after 150 hours of use?

**41.** If an amount of money, $P$, is invested at a rate, $r$ (expressed as a decimal), compounded continuously, the amount of money, $A$, after $t$ years is given approximately by $A = P(2.7)^{rt}$. How much will an investment of $1000 invested at 10% be worth after 20 years?

**42.** If $P$ dollars are invested at a rate, $r$ (expressed as a decimal), and compounded $k$ times a year, the amount, $A$, due at the end of $t$ years is given by $A = P\left(1 + \frac{r}{k}\right)^{kt}$. Find $A$ if $100 is invested at 8% compounded quarterly for 2 years.

**43.** The value of a machine, $V$, at the end of $t$ years is given by $V = C(1 - r)^t$, where $C$ is the original cost and $r$ is the rate of depreciation. Find the value at the end of 4 years of a machine that cost $1200 originally if $r = 0.20$.

**44.** In Exercise 43, find the value at the end of 3 years of a machine that cost $2000 originally if $r = 0.15$.

## 9.2 Logarithmic Functions and Evaluating Logarithms

**OBJECTIVES**

In this section, you will be learning to:

1. Write exponential expressions in logarithmic form.
2. Write logarithmic expressions in exponential form.
3. Graph exponential functions and logarithmic functions on the same set of axes.
4. Use a calculator to find the values of common logarithms.
5. Use a calculator to find the values of natural logarithms.
6. Use a calculator to find antilogarithms for common logarithms and for natural logarithms.

In this section, we will use the concept of inverse function (see Section 8.4) to develop **logarithmic functions.** Then we will use the resulting properties and a calculator to evaluate expressions involving logarithms.

The exponential function $y = b^x$ is a 1–1 function and, therefore, has an inverse. To find the inverse function, we interchange the $x$ and $y$ and solve for $y$. Thus, for

$$y = b^x$$

the inverse is

$$x = b^y$$

Now we try to solve for $y$, but we have no algebraic technique for doing this. So, mathematicians have simply created a name for this new function and called it a **logarithm,** abbreviated as **log.**

**Logarithm** If $b > 0$, $b \neq 1$, and $x = b^y$, then $y$ is the logarithm (base-$b$) of $x$:

$$y = \log_b x \qquad \text{where } x > 0$$

Thus, a **logarithm is the name of an exponent,** and the equations

$$x = b^y \qquad \text{and} \qquad y = \log_b x$$

are equivalent.

**REMEMBER, a logarithm is an exponent.** For example,

$$100 = 10^2 \qquad \text{and} \qquad 2 = \log_{10} 100$$

are equivalent. In words,

2 is the **exponent** of the base-10 to get $100(10^2 = 100)$; and
2 is the **logarithm** to the base-10 of $100(2 = \log_{10} 100)$.

EXAMPLES

| | Exponential Form | Logarithmic Form |
|---|---|---|
| **1.** | $2^3 = 8$ | $\log_2 8 = 3$ |
| **2.** | $2^4 = 16$ | $\log_2 16 = 4$ |
| **3.** | $10^3 = 1000$ | $\log_{10} 1000 = 3$ |
| **4.** | $10^4 = 10{,}000$ | $\log_{10} 10{,}000 = 4$ |
| **5.** | $2^0 = 1$ | $\log_2 1 = 0$ |
| **6.** | $3^0 = 1$ | $\log_3 1 = 0$ |
| **7.** | $10^1 = 10$ | $\log_{10} 10 = 1$ |
| **8.** | $5^1 = 5$ | $\log_5 5 = 1$ |
| **9.** | $2^{-2} = \frac{1}{4}$ | $\log_2 \frac{1}{4} = -2$ |
| **10.** | $10^{-1} = \frac{1}{10}$ | $\log_{10} \frac{1}{10} = -1$ |

■

**REMEMBER, a logarithm is an exponent.**

We can evaluate logarithmic expressions by changing them to the equivalent exponential form, as the following examples illustrate.

EXAMPLES

**11.** Evaluate $\log_2 32$.

**Solution** Let $\log_2 32 = x$

Then

$$2^x = 32$$
$$2^x = 2^5$$
$$x = 5$$

Thus, $\log_2 32 = 5$.

**12.** Evaluate $\log_{10}(0.01)$.

**Solution** Let $\log_{10}(0.01) = x$

Then
$$10^x = 0.01$$
$$10^x = \frac{1}{100}$$
$$10^x = 10^{-2}$$
$$x = -2$$

Thus, $\log_{10}(0.01) = -2$.

**13.** Find the value of $x$ if $\log_{16}x = \frac{3}{4}$.

**Solution** $\log_{16}x = \frac{3}{4}$ — $\frac{3}{4}$ is the **logarithm** of $x$.

$x = 16^{3/4}$ — $\frac{3}{4}$ is the **exponent** of the base-16.

$$x = (2^4)^{3/4}$$
$$x = 2^3 = 8$$

Thus, $\log_{16}8 = \frac{3}{4}$. ■

The following discussion leads us to three basic properties of logarithms. We know that

$$b^0 = 1 \qquad \text{so } \log_b 1 = 0$$

and

$$b^1 = b \qquad \text{so } \log_b b = 1$$

Also, the definition tells us that if $x = b^y$, then

$$y = \log_b x$$

By substituting for $y$,

$$x = b^y \quad \leftarrow \quad (y = \log_b x)$$

we get

$$x = b^{\log_b x}$$

which states that $\log_b x$ is the exponent of $b$ that will give $x$ as a result.

The following summarizes the three properties.

For $b > 0$, $b \neq 1$, and $x > 0$:

1. $\log_b 1 = 0$
2. $\log_b b = 1$
3. $x = b^{\log_b x}$

**A logarithm is an exponent.**

## EXAMPLES

**14.** $\log_3 1 = 0$ Property 1, $\log_b 1 = 0$

**15.** $\log_8 8 = 1$ Property 2, $\log_b b = 1$

**16.** $10^{\log_{10} 20} = 20$ Property 3, $b^{\log_b x} = x$

Later in this section, we will discuss how to use a calculator to find $\log_{10} 20 \approx 1.301029996$.
Thus,

$$10^{1.301029996} \approx 20$$

**17.** $10^{\log_{10} 300} = 300$ Property 3, $b^{\log_b x} = x$

A calculator will show that $\log_{10} 300 \approx 2.477121255$.
So,

$$10^{2.477121255} \approx 300$$

■

In Section 9.1, we discussed exponential functions of the form $y = b^x$ and their graphs. Since logarithms are the inverses of exponential functions, their graphs can be found by reflections across the line $y = x$. Figure 9.3 shows the graphs for $y = 2^x$ and its inverse, $x = 2^y$ (or $y = \log_2 x$).

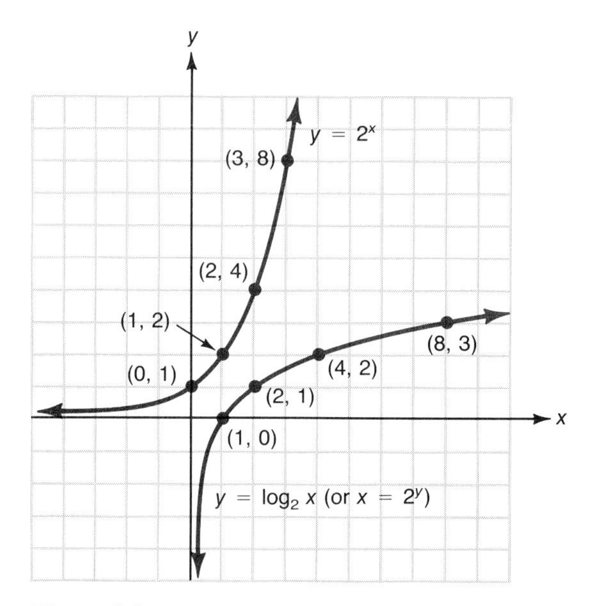


**Figure 9.3**

For the exponential function $y = b^x$,

the domain is all real $x$, and
the range is all $y > 0$.

For the logarithmic function $y = \log_b x$ (or $x = b^y$),

the domain is all $x > 0$, and
the range is all real $y$.

The domain of any log function is the set of positive real numbers. Negative numbers and 0 are not in the domain. **The logarithm of a negative number or zero is undefined.** For example, $\log_{10}(-2)$ is undefined. There is no way to take 10 to some power and get $-2$. $10^x = -2$ is impossible with real exponents.

However, logarithms are exponents and they may be negative or 0. Thus,

$$10^{-2} = \frac{1}{100} \quad \text{and} \quad -2 = \log_{10}\frac{1}{100}$$

$$10^0 = 1 \quad \text{and} \quad 0 = \log_{10}1$$

The graphs for the exponential and logarithmic functions, base-10, are shown in Figure 9.4. Note that $\log_{10}x$ is negative for $0 < x < 1$.

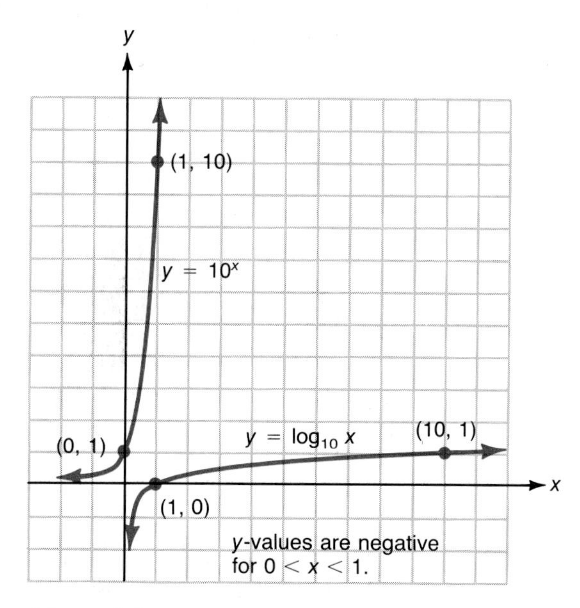


**Figure 9.4**

## Common Logarithms (Base-10)

Base-10 logarithms are called **common logarithms.** The notation log $x$ is used to indicate common logarithms. That is,

$$\log x = \log_{10} x$$

Finding values of common logarithms is a two-step process with most calculators:

1. Enter the number.
2. Press the [log x] (or [log]) key.

**EXAMPLE**

**18.** Use a calculator to find the values of the following common logarithms. To check your understanding, write your estimate of each value on your paper before you use the calculator.

**a.** log 200
**b.** log 50,000
**c.** log 0.0006
**d.** log (−25)

(**Note:** The accuracy of the calculator may be impractical for many situations. You may be instructed in class to report rounded-off answers.)

**Solution**

**a.** $\log 200 = 2.301029996$ (Note that this means $10^{2.301029996} = 200$.)

**b.** $\log 50{,}000 = 4.698970004$

**c.** $\log 0.0006 = -3.22184876$ (Note that a logarithm can be negative.)

**d.** $\log(-25)$ Your calculator will indicate an error. There are no logarithms of negative numbers. ■

If $\log x = N$, then we know that $x = 10^N$. The number $x$ is called the **antilog** of $N$ (or inverse log of $N$). Finding the antilog (or inverse log) of $N$ using most calculators is a three-step process:

1. Enter $N$.
2. Press the [INV] key (or the [2nd] key on some calculators).
3. Press the [log x] key.

**EXAMPLE**

**19.** Use a calculator to find the value of $x$ in each of the following expressions. To check your understanding, write your estimate of $x$ on your paper before you use the calculator.

**a.** $\log x = 5$
**b.** $\log x = -2$
**c.** $\log x = 2.4142$
**d.** $\log x = 16.5$

**Solution**

**a.** $x = \text{antilog }(5) = 100{,}000 \qquad (x = 10^5)$
**b.** $x = \text{antilog }(-2) = 0.01 \qquad (x = 10^{-2})$
**c.** $x = \text{antilog }(2.4142) = 259.5374301$
**d.** $x = \text{antilog }(16.5) = 3.16227766\ 16$
This is the calculator version of scientific notation. It means that $x = 3.16227766 \times 10^{16}$. ■

## Natural Logarithms (Base-e)

Base-$e$ logarithms are called **natural logarithms.** The notation $\ln x$ is used to indicate natural logarithms. That is,

$$\ln x = \log_e x.$$

Figure 9.5 shows the two functions $y = e^x$ and $y = \ln x$.

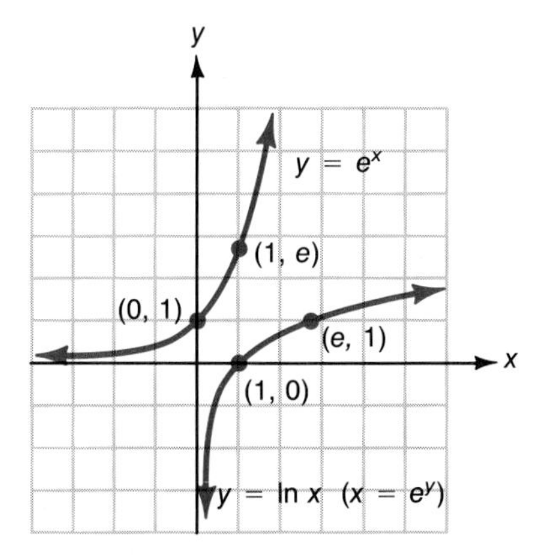


**Figure 9.5**

To find natural logarithms, use the [ln x] (or [ln]) key. The two steps are:

1. Enter the number.
2. Press the [ln x] key.

**EXAMPLE**

**20.** Use a calculator to find the following natural logarithms. To check your understanding, write your estimate of each value on your paper before you use the calculator.

**a.** $\ln 1$
**b.** $\ln 3$
**c.** $\ln (-2)$
**d.** $\ln 0.02$

**Solution**

**a.** $\ln 1 = 0$ (This means that $e^0 = 1$.)

**b.** $\ln 3 = 1.098612289$

**c.** $\ln (-2) =$ error. There are no logarithms of negative numbers.

**d.** $\ln 0.02 = -3.912023005$ (This means that $e^{-3.912023005} = 0.02$.) ■

If $\ln x = N$, then we know that $x = e^N$. The number $x$ is called the **antiln** of $N$ (or **inverse ln** of $N$). Finding the antiln (or inverse ln) of $N$ using a calculator is a three-step process similar to that for finding the antilog:

1. Enter $N$.
2. Press the [INV] key (or the [2nd] key on some calculators).
3. Press the [ln $x$] key.

**EXAMPLE**

**21.** Use a calculator to find the value of $x$ in each of the following expressions. To check your understanding, write your estimate of $x$ on your paper before you use the calculator.

**a.** $\ln x = 3$
**b.** $\ln x = -1$
**c.** $\ln x = -0.1$
**d.** $\ln x = 10$

**Solution**

**a.** $x = \text{antiln } (3) = 20.08553692$ $(x = e^3)$

**b.** $x = \text{antiln } (-1) = 0.367879441$ $(x = e^{-1})$

**c.** $x = \text{antiln } (-0.1) = 0.904837418$ $(x = e^{-0.1})$

**d.** $x = \text{antiln } (10) = 22{,}026.46579$ $(x = e^{10})$ ■

***Practice Problems***

1. Express the equation $4^2 = 16$ in logarithmic form.
2. Express the equation $\log_5 x = -1$ in exponential form.
3. Find the value of $x$ if $x = \ln 0.11$.
4. Find $x$ if $\log x = 2.46$.

**Answers to Practice Problems** **1.** $\log_4 16 = 2$ **2.** $x = 5^{-1}$ **3.** $x = -2.207274913$ **4.** $x = 288.4031503$

## EXERCISES 9.2

Express Exercises 1–12 in logarithmic form.

**1.** $7^2 = 49$ **2.** $3^3 = 27$ **3.** $5^{-2} = \frac{1}{25}$ **4.** $10^2 = 100$

**5.** $2^{-5} = \frac{1}{32}$ **6.** $1 = \pi^0$ **7.** $\left(\frac{2}{3}\right)^2 = \frac{4}{9}$ **8.** $10^k = 23$

**9.** $17 = e^x$ **10.** $e^k = 11.6$ **11.** $10^{\log_{10}6} = 6$ **12.** $12 = 10^{\log_{10}12}$

Express Exercises 13–24 in exponential form.

**13.** $\log_3 9 = 2$ **14.** $\log_5 125 = 3$ **15.** $\log_9 3 = \frac{1}{2}$ **16.** $\log_8 4 = \frac{2}{3}$

**17.** $\log_7 \frac{1}{7} = -1$ **18.** $\log_{1/2} 8 = -3$ **19.** $\ln N = 1.74$ **20.** $\ln 42.3 = x$

**21.** $\log_b 18 = 4$ **22.** $\log_b 39 = 10$ **23.** $\log_y y^x = x$ **24.** $\log_b a = x$

Solve Exercises 25–40 by first changing each equation to exponential form.

**25.** $\log_4 x = 2$ **26.** $\log_3 x = 4$ **27.** $\log_{14} 196 = x$ **28.** $\log_5 \frac{1}{125} = x$

**29.** $\log_{36} x = -\frac{1}{2}$ **30.** $\log_x 32 = 5$ **31.** $\log_x 121 = 2$ **32.** $\log_{81} x = -\frac{3}{4}$

**33.** $\log_8 x = \frac{5}{3}$ **34.** $\log_{25} 125 = x$ **35.** $\log_3 \frac{1}{9} = x$ **36.** $\log_8 8^{3.7} = x$

**37.** $\log_{10} 10^{1.52} = x$ **38.** $\log_5 5^{\log_5 25} = x$ **39.** $\log_4 4^{\log_2 8} = x$ **40.** $\log_p p^{\log_3 81} = x$

In Exercises 41–50, graph each function and its inverse on the same set of axes. (The inverse of a 1–1 function is the reflection across the line $y = x$. See Figure 9.3.)

**41.** $y = 6^x$ **42.** $y = 2^x$ **43.** $y = \left(\frac{2}{3}\right)^x$ **44.** $y = \left(\frac{1}{4}\right)^x$ **45.** $y = \log_4 x$

**46.** $y = \log_5 x$ **47.** $y = \log_{1/2} x$ **48.** $y = \log_{1/3} x$ **49.** $y = \log_8 x$ **50.** $y = \log_7 x$

Use a calculator to find the values of the logarithms in Exercises 51–62.

**51.** $\log 173$ **52.** $\log 396$ **53.** $\log 88.4$ **54.** $\log 0.0061$

**55.** $\log 0.0573$ **56.** $\log(-8.47)$ **57.** $\ln 37.5$ **58.** $\ln 96$

**59.** $\ln(-14.9)$ **60.** $\ln 157.6$ **61.** $\ln 0.00461$ **62.** $\ln 0.0139$

Use a calculator to find the value of $x$ in each of the Exercises 63–74.

**63.** $\log x = 2.31$ **64.** $\log x = -3$ **65.** $\log x = -1.7$ **66.** $\log x = 4.1$

**67.** $2 \log x = -0.038$ **68.** $5 \log x = 9.4$ **69.** $\ln x = 5.17$ **70.** $\ln x = 4.9$

**71.** $\ln x = -8.3$ **72.** $\ln x = 6.74$ **73.** $0.2 \ln x = 0.0079$ **74.** $3 \ln x = -0.066$

## 9.3 Properties of Logarithms

**OBJECTIVES**

In this section, you will be learning to:

1. Evaluate logarithms using the log properties and given values.
2. Write logarithmic expressions as the sums and/or differences of logarithms.
3. Write the sums and/or differences of logarithms as single logarithmic expressions.

In Section 9.2, we defined logarithms as exponents and showed how calculators can be used to find common logarithms and natural logarithms and antilogs. While calculators are certainly effective in giving numerical evaluations, they generally do not simplify or solve equations. In this section, we will discuss several properties of logarithms that are helpful in solving equations and simplifying expressions involving logarithms and/or exponential functions.

The following three basic properties of logarithms were discussed in Section 9.2.

For $b > 0$, $b \neq 1$, and $x > 0$:

1. $\log_b 1 = 0$
2. $\log_b b = 1$
3. $x = b^{\log_b x}$

Because **logarithms are exponents,** their properties are similar to those of exponents. In fact, to prove these properties, we will use the exponential form of numbers with the knowledge that the exponents involved are, indeed, logarithms. For example, when multiplying two exponential expressions with the same base, we add the exponents (the logarithms):

$$10^{0.3010} \times 10^{0.4771} = 10^{0.3010 + 0.4771} = 10^{0.7781}$$

But these exponents are logarithms. In fact,

$$0.3010 = \log 2 \qquad \text{and} \qquad 0.4771 = \log 3$$

[**Note:** In the remainder of this chapter, we will round off logarithms to 4 decimal place accuracy.]

Now, we will show that

$$\log 6 = \log (2 \cdot 3) = \log 2 + \log 3$$

Using Property 3,

$$6 = 10^{\log 6}, \qquad 2 = 10^{\log 2}, \qquad \text{and} \qquad 3 = 10^{\log 3}$$

Thus, we can write

$$\begin{array}{ccccc} 6 & = & 2 & \times & 3 \\ \downarrow & & \downarrow & & \downarrow \\ 10^{\log 6} & = & 10^{\log 2} & \times & 10^{\log 3} \end{array} = 10^{\log 2 + \log 3} \qquad \text{Adding exponents}$$

and, equating exponents gives the result

$$\begin{array}{ccccc} \log 6 & = & \log 2 & + & \log 3 \\ \downarrow & & \downarrow & & \downarrow \\ (\text{or} \quad 0.7781 & = & 0.3010 & + & 0.4771) \end{array}$$

Using this technique, we can prove Property 4:

**The logarithm of a product is equal to the sum of the logarithms of the numbers being multiplied.**

**Property 4 of Logarithms** For $b > 0$, $b \neq 1$, and $x,y > 0$,

$$\log_b xy = \log_b x + \log_b y$$

**Proof of Property 4**

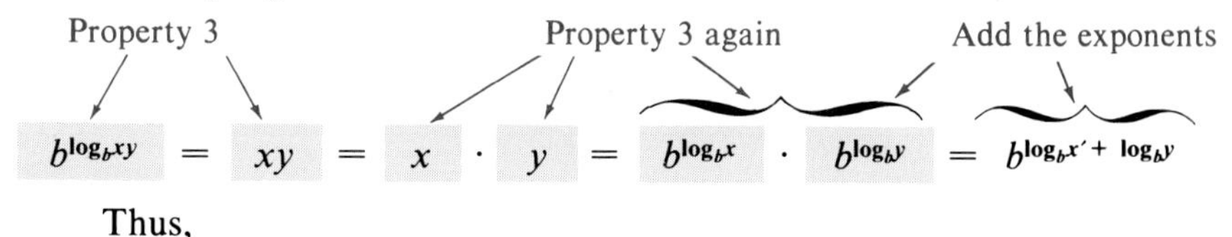


$$b^{\log_b xy} = xy = x \cdot y = b^{\log_b x} \cdot b^{\log_b y} = b^{\log_b x + \log_b y}$$

Thus,

$$b^{\log_b xy} = b^{\log_b x + \log_b y}$$

and we can equate the exponents

$$\log_b xy = \log_b x + \log_b y$$

To find $\log \frac{3}{2}$ (without a calculator), we can proceed in the following manner:

$$10^{\log 3/2} = \frac{3}{2} = \frac{10^{\log 3}}{10^{\log 2}} = 10^{\log 3 - \log 2}$$

Equating exponents gives

$$\log \frac{3}{2} = \log 3 - \log 2$$
$$= 0.4771 - 0.3010 = 0.1761$$

This leads to Property 5 of logarithms. The proof is left as an exercise for the student.

**Property 5 of Logarithms** For $b > 0$, $b \neq 1$, and $x,y > 0$,

$$\log_b \frac{x}{y} = \log_b x - \log_b y$$

Property 6 of logarithms involves a number raised to a power, such as $2^3$, and **multiplying exponents.** For example,

Property 3 — Property 3 again — Multiply the exponents

$$10^{\log(2^3)} = 2^3 = [10^{\log 2}]^3 = 10^{3(\log 2)}$$

Equating exponents gives

$$\log(2^3) = 3(\log 2)$$

**Property 6 of Logarithms** For $b > 0$, $b \neq 1$, $x > 0$, and any real number $r$,

$$\log_b x^r = r \cdot \log_b x$$

**Proof of Property 6**

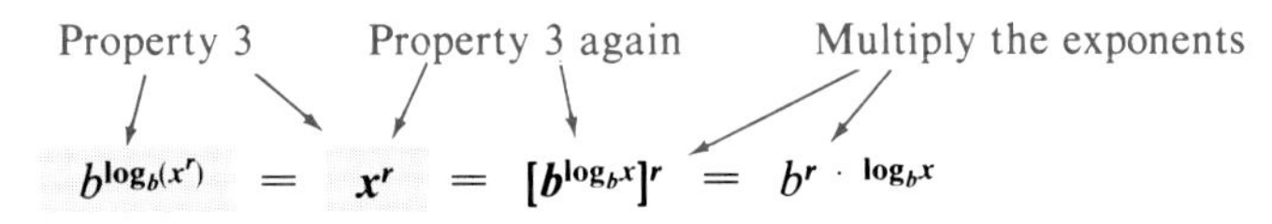


$$b^{\log_b(x^r)} = x^r = [b^{\log_b x}]^r = b^{r \cdot \log_b x}$$

Equating the exponents gives the result

$$\log_b(x^r) = r \cdot \log_b x$$

Table 9.2 summarizes the properties of logarithms for base $b$. **To emphasize that $b$ could be $e$, we list the properties separately for natural logarithms.**

Table 9.2 Properties of Logarithms

| For $b > 0$, $b \neq 1$, $x,y > 0$ and any real number $r$: | For natural logarithms: |
|---|---|
| **1.** $\log_b 1 = 0$ | **1.** $\ln 1 = 0$ |
| **2.** $\log_b b = 1$ | **2.** $\ln e = 1$ |
| **3.** $x = b^{\log_b x}$ | **3.** $x = e^{\ln x}$ |
| **4.** $\log_b xy = \log_b x + \log_b y$ | **4.** $\ln xy = \ln x + \ln y$ |
| **5.** $\log_b \frac{x}{y} = \log_b x - \log_b y$ | **5.** $\ln \frac{x}{y} = \ln x - \ln y$ |
| **6.** $\log_b x^r = r \cdot \log_b x$ | **6.** $\ln x^r = r \cdot \ln x$ |

EXAMPLES Use the properties of logarithms to write each expression as a sum and/or difference of logarithmic expressions.

**1.** $\log 2x^3$

**Solution** $\log 2x^3 = \log 2 + \log x^3$ Property 4

$= \log 2 + 3 \log x$ Property 6

**2.** $\log\dfrac{ab^2}{c}$

**Solution** 
$$\begin{aligned}\log\frac{ab^2}{c} &= \log(ab^2) - \log c && \text{Property 5}\\ &= \log a + \log b^2 - \log c && \text{Property 4}\\ &= \log a + 2\log b - \log c && \text{Property 6}\end{aligned}$$

**3.** $\ln(xy)^{-3}$

**Solution** 
$$\begin{aligned}\ln(xy)^{-3} &= -3\ln(xy) && \text{Property 6}\\ &= -3(\ln x + \ln y) && \text{Property 5}\end{aligned}$$

Use the properties of logarithms to write each expression as a single logarithm.

**4.** $2\log_b x - 3\log_b y$

**Solution** 
$$\begin{aligned}2\log_b x - 3\log_b y &= \log_b x^2 - \log_b y^3 && \text{Property 6}\\ &= \log_b\frac{x^2}{y^3} && \text{Property 5}\end{aligned}$$

**5.** $\dfrac{1}{2}\ln 4 + \ln 5 - \ln y$

**Solution**

$$\begin{aligned}\frac{1}{2}\ln 4 + \ln 5 - \ln y &= \ln 4^{1/2} + \ln 5 - \ln y && \text{Property 6}\\ &= \ln(4^{1/2}\cdot 5) - \ln y && \text{Property 4}\\ &= \ln\frac{10}{y} && \text{Property 5 } (4^{1/2} = 2)\end{aligned}$$

**6.** $\log(x + 1) + \log(x - 1)$

**Solution**

$$\begin{aligned}\log(x + 1) + \log(x - 1) &= \log(x + 1)(x - 1) && \text{Property 4}\\ &= \log(x^2 - 1)\end{aligned}$$

■

***Practice Problems***

1. Write $\log(x^3y)$ as a sum and/or difference of logarithmic expressions.
2. Write the expression $2\ln 5 + \ln x - \ln 3$ as a single logarithm.
3. Write $2\log x - \log(x + 1)$ as a single logarithm.

**Answers to Practice Problems** **1.** $3\log x + \log y$ **2.** $\ln\dfrac{25x}{3}$ **3.** $\log\dfrac{x^2}{x + 1}$

## EXERCISES 9.3

Use the properties of logarithms to write each expression in Exercises 1–20 as a sum and/or difference of logarithmic expressions. Assume that all variables are positive.

**1.** $\log 5x^4$ **2.** $\log 3x^2y$ **3.** $\ln 2x^{-3}y$ **4.** $\ln xy^2z^{-1}$ **5.** $\log \dfrac{2x}{y^3}$

**6.** $\log \dfrac{xy}{4z}$ **7.** $\ln \dfrac{x^2}{yz}$ **8.** $\ln \dfrac{xy^2}{z^2}$ **9.** $\log(xy)^{-2}$ **10.** $\log(x^2y)^4$

**11.** $\log \sqrt[3]{xy^2}$ **12.** $\log \sqrt{2x^3y}$ **13.** $\ln \sqrt{\dfrac{xy}{z}}$ **14.** $\ln \sqrt[3]{\dfrac{x^2}{y}}$ **15.** $\log 21x^2y^{2/3}$

**16.** $\log 15x^{-1/2}y^{1/3}$ **17.** $\log \dfrac{x}{\sqrt{x^3y^5}}$ **18.** $\log \dfrac{1}{\sqrt{x^4y}}$ **19.** $\ln\left(\dfrac{x^3y^2}{z}\right)^{-3}$ **20.** $\ln\left(\dfrac{x^{1/2}y}{z^2}\right)^{-2}$

Use the properties of logarithms to write each expression in Exercises 21–42 as a single logarithm.

**21.** $2 \ln 3 + \ln x - \ln 5$ **22.** $\dfrac{1}{2} \ln 25 + \ln 3 - \ln x$ **23.** $\log 7 - \log 8 + 2 \log x$

**24.** $\log 4 + \log 6 + \log y$ **25.** $2 \log x + \log y$ **26.** $\log x + 3 \log y$

**27.** $3 \ln x - 2 \ln y$ **28.** $3 \ln y - \dfrac{1}{2} \ln x$ **29.** $\dfrac{1}{2}(\ln x - \ln y)$

**30.** $\dfrac{1}{3}(\ln x - 2 \ln y)$ **31.** $\log x - \log y + \log z$ **32.** $\log x + 2 \log y - \dfrac{1}{2} \log z$

**33.** $\log x - 2 \log y - 2 \log z$ **34.** $-\dfrac{2}{3} \log x - \dfrac{1}{3} \log y + \dfrac{2}{3} \log z$

**35.** $\log x + \log(2x + 1)$ **36.** $\log(x + 3) + \log(x - 3)$

**37.** $\ln(x - 1) + \ln(x + 3)$ **38.** $\ln(3x + 1) + 2 \ln x$

**39.** $\log(x^2 - 2x - 3) - \log(x - 3)$ **40.** $\log(x - 4) - \log(x^2 - 2x - 8)$

**41.** $\log(x + 6) - \log(2x^2 + 9x - 18)$ **42.** $\log(3x^2 + 5x - 2) - \log(3x - 1)$

**43.** Prove Property 5: For $b > 0$, $b \neq 1$, and $x,y > 0$.

$$\log_b \frac{x}{y} = \log_b x - \log_b y$$

## 9.4 Logarithmic and Exponential Equations and Change-of-Base

**OBJECTIVES**

In this section, you will be learning to:

1. Solve exponential equations in which the bases are not necessarily the same.
2. Solve logarithmic equations without necessarily changing to exponential form.
3. Use the change-of-base formula and a calculator to evaluate logarithmic expressions.

In Section 9.1, we solved exponential equations in which both sides of the equation contained the same base. If different bases are involved, then logarithms and the properties of logarithms we discussed in Section 9.3 can be used to solve the equations.

We need the following property of logarithms to help in solving equations.

**Property 7 of Logarithms**

For $b > 0$, $b \neq 1$, and $x, y > 0$:

1. If $\log_b x = \log_b y$, then $x = y$.
2. If $x = y$, then $\log_b x = \log_b y$.

The following examples illustrate how to solve exponential equations with different bases by using Property 7 and **taking the log (or ln) of both sides.** If the bases are the same, there is no need to deal with logarithms.

**EXAMPLES** Solve each of the following exponential equations.

**1.** $10^{3x} = 2.1$

**Solution** Since the base of $3x$ is 10, we can solve by taking log of both sides.

$$
\begin{aligned}
10^{3x} &= 2.1 \\
\log 10^{3x} &= \log 2.1 && \text{Take log of both sides.} \\
3x \log 10 &= \log 2.1 && \log x^r = r \log x \\
3x &= \log 2.1 && \log 10 = 1 \\
x &= \frac{\log 2.1}{3}
\end{aligned}
$$

Using a calculator,

$$x = \frac{\log 2.1}{3} \approx \frac{0.3222}{3} = 0.1074$$

We could also have simply used the definition of a logarithm as an exponent and stated

$$
\begin{aligned}
10^{3x} &= 2.1 \\
3x &= \log 2.1 && \text{By the definition} \\
x &= \frac{\log 2.1}{3}
\end{aligned}
$$

**2.** $e^{0.2x} = 50$

**Solution** Using the definition of a logarithm as an exponent, we have

$$\begin{aligned} e^{0.2x} &= 50 \\ 0.2x &= \ln 50 && \text{By the definition} \\ x &= \frac{\ln 50}{0.2} \end{aligned}$$

Using a calculator,

$$x = \frac{\ln 50}{0.2} \approx \frac{3.9120}{0.2} = 19.560$$

**3.** $6^x = 18$

**Solution** Neither base is 10 or $e$, so we can solve by taking log of both sides or by taking ln of both sides. The result is the same.

**a.** Taking log of both sides:

$$\begin{aligned} 6^x &= 18 \\ \log 6^x &= \log 18 \\ x \cdot \log 6 &= \log 18 \\ x &= \frac{\log 18}{\log 6} \end{aligned}$$

Using a calculator,

$$\begin{aligned} x = \frac{\log 18}{\log 6} &\approx \frac{1.2553}{0.7782} \\ &\approx 1.6131 \end{aligned}$$

**b.** Taking ln of both sides:

$$\begin{aligned} 6^x &= 18 \\ \ln 6^x &= \ln 18 \\ x \cdot \ln 6 &= \ln 18 \\ x &= \frac{\ln 18}{\ln 6} \end{aligned}$$

Using a calculator,

$$\begin{aligned} x = \frac{\ln 18}{\ln 6} &\approx \frac{2.8904}{1.7918} \\ &\approx 1.6131 \end{aligned}$$

**4.** $5^{2x-1} = 10^x$

**Solution**

$$\begin{aligned} 5^{2x-1} &= 10^x \\ \log 5^{2x-1} &= \log 10^x && \text{By Property 7} \\ (2x - 1)\log 5 &= x \log 10 && \log x^r = r \log x \\ 2x \cdot \log 5 - 1 \cdot \log 5 &= x && \log 10 = 1 \\ 2x \log 5 - x &= \log 5 && \text{Arrange } x\text{-terms on one side.} \\ x(2 \log 5 - 1) &= \log 5 && \text{Factor } x. \\ x &= \frac{\log 5}{2 \log 5 - 1} \end{aligned}$$

As a decimal approximation,

$$x = \frac{\log 5}{2 \log 5 - 1} \approx \frac{0.6990}{2(0.6990) - 1} \approx 1.73$$

**5.** $5^{x+2} = \frac{1}{25}$

**Solution** $5^{x+2} = \frac{1}{25}$

$5^{x+2} = 5^{-2}$ — Here the bases are the same, so there is no need to use Property 7.

$x + 2 = -2$

$x = -4$ ■

All the various properties of logarithms can be used to solve equations that involve logarithms. Remember that logarithms are defined only for positive real numbers, so each solution should be checked in the original equation.

**EXAMPLES** Solve the following logarithmic equations.

**6.** $\log(x - 1) + \log(x - 4) = 1$

**Solution** $\log(x - 1) + \log(x - 4) = 1$

$\log(x - 1)(x - 4) = 1$ — $\log x + \log y = \log xy$

$(x - 1)(x - 4) = 10^1$ — Changing to exponential form using base 10

$x^2 - 5x + 4 = 10$ — Solve by factoring.

$x^2 - 5x - 6 = 0$

$(x - 6)(x + 1) = 0$

$x = 6$ or ~~$x = -1$~~ — Checking $x = -1$ shows $\log(-1 - 1) = \log(-2)$ which is undefined.

**7.** $\log x - \log(x - 1) = \log 3$

**Solution** $\log x - \log(x - 1) = \log 3$

$\log\left(\frac{x}{x-1}\right) = \log 3$ — $\log x - \log y = \log\frac{x}{y}$

$\frac{x}{x-1} = 3$ — By Property 7

$x = 3(x - 1)$ — Solve for $x$.

$x = 3x - 3$

$3 = 2x$

$\frac{3}{2} = x$ — Checking will show that $\frac{3}{2}$ is the solution.

**8.** $\ln(x^2 - x - 6) - \ln(x + 2) = 2$

**Solution**

$$\ln(x^2 - x - 6) - \ln(x + 2) = 2$$

$$\ln\left(\frac{x^2 - x - 6}{x + 2}\right) = 2 \qquad \ln x - \ln y = \ln\left(\frac{x}{y}\right)$$

$$\ln\left(\frac{\cancel{(x+2)}(x - 3)}{\cancel{x + 2}}\right) = 2 \qquad \text{Factor } x^2 - x - 6.$$

$$\ln(x - 3) = 2 \qquad \text{Simplify.}$$

$$x - 3 = e^2 \qquad \text{Change to exponential form by using base } e.$$

$$x = 3 + e^2$$

(Or, using a calculator,

$$x = 3 + e^2 \approx 3 + 7.3891 = 10.3891)$$

■

Because a calculator can be used to evaluate common logarithms and natural logarithms, we have restricted most of the examples to base 10 or base $e$ expressions. If an equation involves logarithms to other bases, the following discussion shows how to rewrite each logarithm using any base you choose.

***Change-of-Base Formula***

$$\log_b x = \frac{\log_a x}{\log_a b}$$

The change-of-base formula can be derived using properties of logarithms as follows:

$$b^{\log_b x} = x \qquad \text{By Property 3}$$

$$\log_a(b^{\log_b x}) = \log_a x \qquad \text{By Property 7 using } a \text{ as the base}$$

$$\log_b x(\log_a b) = \log_a x \qquad \text{By Property 6 where } \log_b x \text{ is treated as the exponent } r$$

$$\log_b x = \frac{\log_a x}{\log_a b} \qquad \text{Divide both sides by } \log_a b \text{ to arrive at the change-of-base formula.}$$

**EXAMPLES** Use the change-of-base formula to evaluate each of the following expressions.

**9.** $\log_2 3.42$

**Solution** We can evaluate this expression using either base 10 or base $e$ since both are easily available on a calculator.

$$\log_2 3.42 = \frac{\ln 3.42}{\ln 2} = \frac{1.2296}{0.6931} = 1.7741 \qquad \text{Using rounded-off values}$$

(The student can show that $\frac{\log 3.42}{\log 2}$ gives the same result.) In exponential form:

$$2^{1.7741} = 3.42$$

**10.** $\log_3 0.3333$

**Solution** $\log_3 0.3333 = \frac{\log 0.3333}{\log 3} = \frac{-0.4772}{0.4771} = -1.0002$ ■

**Practice Problems**

Solve each of the following equations.

1. $4^x = 64$
2. $10^x = 64$
3. $2^{3x-1} = 0.1$
4. $15 \log x = 45.15$
5. $\ln(x^2 - x - 6) - \ln(x - 3) = 1$

## EXERCISES 9.4

Solve each of the equations in Exercises 1–64.

**1.** $3^x = 9$ **2.** $2^{5x-8} = 4$ **3.** $4^{x^2} = \left(\frac{1}{2}\right)^{3x}$ **4.** $25^{x^2+2x} = 5^{-x}$

**5.** $5^{2x-x^2} = \frac{1}{125}$ **6.** $10^{x^2-2x} = 1000$ **7.** $10^{3x} = 140$ **8.** $10^{2x} = 97$

**9.** $10^{0.32x} = 253$ **10.** $10^{-0.48x} = 88.6$ **11.** $4 \cdot 10^{-0.94x} = 126.2$ **12.** $3 \cdot 10^{-2.1x} = 83.5$

**13.** $e^{0.03x} = 2.1$ **14.** $e^{-0.5x} = 47$ **15.** $e^{-0.06t} = 50.3$ **16.** $e^{4t} = 184$

**17.** $3e^{-0.12t} = 3.6$ **18.** $5e^{2.4t} = 44$ **19.** $2^x = 10$ **20.** $3^{x-2} = 100$

**21.** $5^{2x} = \frac{1}{100}$ **22.** $7^{2x-3} = 10$ **23.** $5^{1-x} = 1$ **24.** $4^{2x+5} = 0.01$

**25.** $4^{2-3x} = 0.1$ **26.** $14^{3x-1} = 10^3$ **27.** $12^{2x+7} = 10^4$ **28.** $2^{5x+2} = 1$

**29.** $7^x = 9$ **30.** $2^x = 20$ **31.** $3^{x-2} = 35$ **32.** $5^{2x} = 23$

**33.** $6^{2x-1} = 14.8$ **34.** $4^{7-3x} = 26.3$ **35.** $5 \log x = 7$ **36.** $3 \log x = 13.2$

**37.** $4 \log x - 6 = 0$ **38.** $2 \log x - 15 = 0$ **39.** $4 \log x^{1/2} + 8 = 0$

**40.** $\frac{2}{3} \log x^{2/3} + 9 = 0$ **41.** $5 \ln x - 8 = 0$ **42.** $2 \ln x + 3 = 0$

**43.** $\ln x^2 + 2.2 = 0$ **44.** $\ln x^2 - 4.16 = 0$ **45.** $\log x + \log 2x = \log 18$

**46.** $\log(x + 4) + \log(x - 4) = \log 9$ **47.** $\log x^2 - \log x = 2$

**48.** $\log x + \log x^2 = 3$ **49.** $\ln(x - 3) + \ln x = \ln 18$

**50.** $\ln(x^2 - 3x + 2) - \ln(x - 1) = \ln 4$ **51.** $\log(x - 15) + \log x = 2$

**52.** $\log(3x - 5) - \log(x - 1) = 1$ **53.** $\log(2x - 17) = 2 - \log x$

**Answers to Practice Problems** 1. $x = 3$ 2. $x = 1.8062$ 3. $x = -0.7740$ 4. $x = 10^{3.01} = 1023.2930$ 5. $x = e - 2 = 0.7183$

**54.** $\log(x - 3) - 1 = \log(x + 1)$

**55.** $\log(x^2 + 2x - 3) = 3 + \log(x + 3)$

**56.** $\log(x^2 - 9) - \log(x - 3) = -2$

**57.** $\log(x^2 - x - 12) + 2 = \log(x - 4)$

**58.** $\log(x^2 - 4x - 5) - \log(x + 1) = 2$

**59.** $\ln(x^2 + 4x - 5) - \ln(x + 5) = -2$

**60.** $\ln(x + 1) + \ln(x - 1) = 0$

**61.** $\ln(x^2 - 4) - \ln(x + 2) = 3$

**62.** $\ln(x^2 + 2x - 3) = 1 + \ln(x - 1)$

**63.** $\log \sqrt[3]{x^2 + 2x + 20} = \dfrac{2}{3}$

**64.** $\log \sqrt{x^2 - 24} = \dfrac{3}{2}$

Use the change-of-base formula to evaluate each of the expressions in Exercises 65–74.

**65.** $\log_3 12$ **66.** $\log_4 36$ **67.** $\log_5 1.68$ **68.** $\log_{11} 39.6$ **69.** $\log_8 0.271$

**70.** $\log_7 0.849$ **71.** $\log_{15} 739$ **72.** $\log_2 14.6$ **73.** $\log_{20} 0.0257$ **74.** $\log_9 2.384$

## 9.5 Applications

**OBJECTIVE**

In this section, you will be learning to: solve applied problems using base *e* exponential equations.

In Section 9.1, we found that the number $e$ appears in a surprisingly natural way in the formula for continuously compounding interest

$$A = Pe^{rt}$$

There are many formulas that involve exponential functions. A few are shown here and in the exercises.

$A = A_0e^{-0.04t}$ This is a law for the decomposition of radium where $t$ is measured in centuries.

$A = A_0e^{-0.1t}$ This is one law for skin healing where $t$ is measured in days.

$A = A_0 2^{-t/5600}$ This law is used for carbon-14 dating to determine the age of fossils where $t$ is measured in years.

$T = Ae^{-kt} + C$ This is Newton's law of cooling where $C$ is the constant temperature of the surrounding medium. The values of $A$ and $k$ depend on the particular object that is cooling.

**EXAMPLES**

1. Suppose that the formula $y = y_0e^{0.4t}$ represents the number of bacteria present after $t$ days, where $y_0$ is the initial number of bacteria. In how many days will the bacteria double in number?

**Solution**

$$y = y_0e^{0.4t}$$

$$2y_0 = y_0e^{0.4t} \qquad 2y_0 \text{ is double the initial number present.}$$

$$2 = e^{0.4t}$$

$$\ln 2 = 0.4t$$

$$t = \frac{\ln 2}{0.4} = \frac{0.69315}{0.4}$$

$$t = 1.73$$

The number of bacteria will double in approximately 1.73 days.

**2.** Suppose that the room temperature is 70° and the temperature of a cup of tea is 150° when it is placed on the table. In 5 minutes, the tea cools to 120°. How long will it take for the tea to cool to 100°?

**Solution** Using the formula $T = Ae^{-kt} + C$, we first find $A$ and $k$.

**a.** We know that $C = 70°$ and that $T = 150°$ when $t = 0$. We find $A$ as follows:

$$\begin{aligned} 150 &= Ae^{-k(0)} + 70 \\ 150 &= A \cdot 1 + 70 \\ 80 &= A \end{aligned}$$

**b.** We also know that $T = 120°$ when $t = 5$. This allows us to find $k$:

$$\begin{aligned} 120 &= 80e^{-k(5)} + 70 \\ 50 &= 80e^{-5k} \\ \frac{50}{80} &= e^{-5k} \\ \ln e^{-5k} &= \ln \frac{5}{8} && \text{Take ln of both sides.} \\ -5k &= \ln 0.625 \\ k &= \frac{\ln 0.625}{-5} \\ &= \frac{-0.4700}{-5} = 0.0940 \end{aligned}$$

**c.** We now have the formula

$$T = 80e^{-0.0940t} + 70$$

This allows us to find $t$ when $T = 100°$:

$$\begin{aligned} 100 &= 80e^{-0.0940t} + 70 \\ 30 &= 80e^{-0.0940t} \\ \frac{30}{80} &= e^{-0.0940t} \\ \ln e^{-0.0940t} &= \ln \frac{3}{8} && \text{Take ln of both sides.} \\ -0.0940t &= \ln 0.375 \\ t &= \frac{\ln 0.375}{-0.0940} \\ &= \frac{-0.9808}{-0.0940} = 10.43 \text{ minutes} \end{aligned}$$

The tea will cool to 100° in about 10.43 minutes.

**3.** If \$1000 is invested at 6% compounded continuously, in how many years will it grow to \$5000?

**Solution** Using $A = Pe^{rt}$,

$$\begin{aligned} 5000 &= 1000e^{0.06t} \\ 5 &= e^{0.06t} \\ \ln 5 &= 0.06t \\ t &= \frac{\ln 5}{0.06} = \frac{1.60944}{0.06} \\ t &= 26.82 \end{aligned}$$

\$1000 will grow to \$5000 in approximately 26.82 years. ■

## EXERCISES 9.5

1. If \$2000 is invested at a rate of 7% compounded continuously, what will be the balance after 10 years?
2. Find the amount, $A$, in a savings account if \$3200 is invested at 6.5% for 6 years and the interest is compounded continuously.
3. How long does it take \$1000 to double if it is invested at 5% compounded continuously?
4. Four thousand dollars is invested at 6% compounded continuously. How long will it take for the balance to be \$8000?
5. The reliability of a certain type of flashlight battery is given by $f = e^{-0.03x}$, where $f$ is the fractional part of the batteries produced that last $x$ hours. What fraction of the batteries produced is good after 40 hours of use?
6. From Exercise 5, how long will at least one-half of the batteries last?
7. The concentration of a drug in the body fluids is given by $C = C_0e^{-0.8t}$, where $C_0$ is the initial dosage and $t$ is the time in hours elapsed after administering the dose. If 20 mg of a drug is given, how much time elapses until 5 mg of the drug remains?
8. Using the formula in Exercise 7, determine the amount of insulin present after 3 hours if 0.60 mℓ are given.
9. A swarm of bees grows according to the formula $P = P_0e^{0.35t}$, where $P_0$ is the number present initially and $t$ is the time in days. How many bees will be present in 6 days if there were 1000 present initially?
10. If inversion of raw sugar is given by $A = A_0e^{-0.03t}$, where $A_0$ is the initial amount and $t$ is the time in hours, how long will it take for 1000 lb of raw sugar to be reduced to 800 lb?
11. Atmospheric pressure $P$ is related to the altitude $h$ by the formula $P = P_0e^{-0.00004h}$, where $P_0$, the pressure at sea level, is approximately 15 lb per sq in. Determine the pressure at 5000 ft.
12. One law for skin healing is $A = A_0e^{-0.1t}$, where $A$ = number of sq cm of unhealed area after $t$ days and $A_0$ is the number of sq cm of the original wound. Find the number of days needed to reduce the wound to one-third the original size.
13. A radioactive substance decays according to $A = A_0e^{-0.0002t}$, where $A_0$ is the initial amount and $t$ is the time in years. If $A_0 = 640$ grams and $A = 400$ grams, find $t$.
14. A substance decays according to $A = A_0e^{-0.045t}$, where $t$ is in hours and $A_0$ is the initial amount. Determine the time required for one-half of the amount to decay.

**15.** An employee learning to assemble remote-control units after $t$ days of training is given by $N = 80(1 - e^{-0.3t})$. How many days of training will be needed before the employee is able to assemble 40 units per day?

**16.** The temperature of a casserole is 350° when it is removed from the oven. The temperature in the room is 72°. In 10 minutes, the casserole cools to 280°. How long will it take for the casserole to cool to 160°?

**17.** How long does it take \$10,000 to double if it is invested at 8% compounded quarterly?

**18.** If \$1000 is deposited at 6% compounded monthly, how long before the balance is \$1520?

**19.** The value of a machine, $V$, at the end of $t$ years is given by $V = C(1 - r)^t$, where $C$ is the original cost of the machine and $r$ is the rate of depreciation. A machine that originally cost \$12,000 is now valued at \$3800. How old is the machine if $r = 0.12$?

**20.** Using the formula in Exercises 19, determine the age of a machine valued at \$5800 if its original value was \$18,000 and $r = 0.09$.

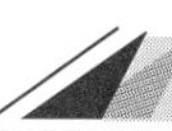

## CHAPTER 9 SUMMARY

### Key Terms and Formulas

**Exponential functions** are of the form

$f(x) = b^x$ where $b > 0$, $b \neq 0$, and $x$ is any real number. [9.1]

The formula for **compound interest** is

$$A = P\left(1 + \frac{r}{n}\right)^{nt} \quad [9.1]$$

The **number $e$** is defined to be

$$e = \lim_{n \to \infty}\left(1 + \frac{1}{n}\right)^n = 2.718281828459\ldots \quad [9.1]$$

The formula for **continuously compounded interest** is

$$A = Pe^{rt} \quad [9.1]$$

If $b > 0$, $b \neq 1$, and $x = b^y$, then $y$ is the logarithm with base $b$ of $x$:

$$y = \log_b x \qquad \text{where } x > 0 \quad [9.2]$$

A **logarithm is the name of an exponent** and the equations

$$x = b^y \qquad \text{and} \qquad y = \log_b x$$

are equivalent. [9.2]

Base-10 logarithms are called **common logarithms.** [9.2]

Base-$e$ logarithms are called **natural logarithms.** [9.2]

**Change-of-Base Formula** [9.4]

$$\log_b x = \frac{\log_a x}{\log_a b}$$

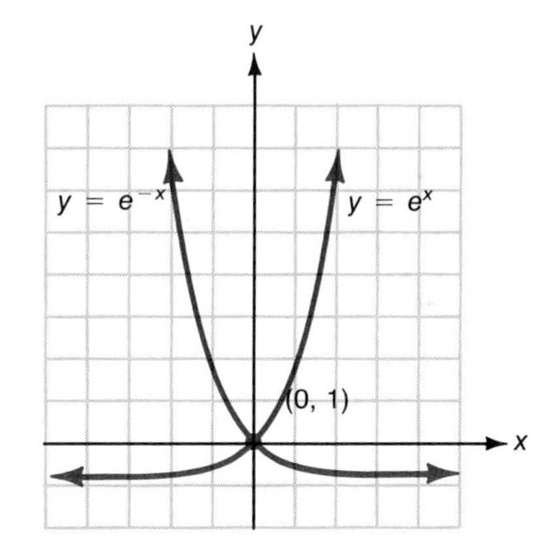


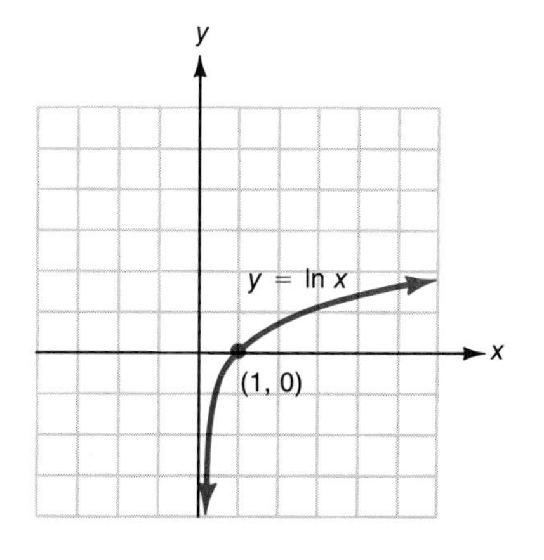

## Properties and Rules

**Properties of Real Exponents** [9.1]

If $a$ and $b$ are positive real numbers and $x$ and $y$ are any real numbers, then:

1. $b^0 = 1$
2. $b^{-x} = \dfrac{1}{b^x}$
3. $b^x \cdot b^y = b^{x+y}$
4. $\dfrac{b^x}{b^y} = b^{x-y}$
5. $(b^x)^y = b^{xy}$
6. $(ab)^x = a^x b^x$
7. $\left(\dfrac{a}{b}\right)^x = \dfrac{a^x}{b^x}$

For $b > 0$ and $b \neq 1$ and real numbers $x$ and $y$:

1. If $b^x = b^y$, then $x = y$.
2. If $x = y$, then $b^x = b^y$. [9.1]

The following general concepts are helpful in understanding the graphs of exponential functions. [9.1]

**For $b > 1$:**

1. $b^x > 0$.
2. $b^x$ increases to the right and is called an **increasing function.**
3. $b^0 = 1$, so (0,1) is on the graph.
4. $b^x$ approaches the $x$-axis for negative values of $x$.

**For $0 < b < 1$:**

1. $b^x > 0$.
2. $b^x$ decreases to the right and is called a **decreasing function.**
3. $b^0 = 1$, so (0,1) is on the graph.
4. $b^x$ approaches the $x$-axis for positive values of $x$.

**The Properties of Logarithms** [9.3], [9.4]

For $b > 0$, $b \neq 1$, $x, y > 0$ and any real number $r$:

1. $\log_b 1 = 0$
2. $\log_b b = 1$
3. $x = b^{\log_b x}$
4. $\log_b xy = \log_b x + \log_b y$
5. $\log_b \dfrac{x}{y} = \log_b x - \log_b y$
6. $\log_b x^r = r \cdot \log_b x$
7. a. If $\log_b x = \log_b y$, then $x = y$.
   b. If $x = y$, then $\log_b x = \log_b y$.

For natural logarithms:

1. $\ln 1 = 0$
2. $\ln e = 1$
3. $x = e^{\ln x}$
4. $\ln xy = \ln x + \ln y$
5. $\ln \dfrac{x}{y} = \ln x - \ln y$
6. $\ln x^r = r \cdot \ln x$

## CHAPTER 9 REVIEW

Sketch the graph of the functions in Exercises 1 and 2. [9.1]

1. $y = 3^x$
2. $y = 5^x$

Solve the equations in Exercises 3–8. [9.1]

3. $3^5 \cdot 3^2 = 3^x$
4. $5^3 \cdot 5^{-4} = 5^x$
5. $10^4 \cdot 10^x = 10^{-6}$
6. $2^{3x} = 4^6$
7. $10^{2x+1} = 1000^{x-2}$
8. $4^{x^2+x} = 4^{x+9}$

9. Studies show that the fractional part of light bulbs, $P$, that have burned out after $t$ hours of use is given by $P = 1 - 2^{-0.03t}$. What fractional part of the light bulbs have burned out after 100 hours of use? [9.1]

10. If $P$ dollars are invested at a rate, $r$ (expressed as a decimal), and compounded $k$ times a year, the amount, $A$, due at the end of $t$ years is given by $A = P\left(1 + \dfrac{r}{k}\right)^{kt}$ dollars. Find $A$ if \$500 is invested at 10% compounded quarterly for 2 years. [9.1]

Write Exercises 11–14 in logarithmic form. [9.2]

**11.** $8^2 = 64$ **12.** $5^{-3} = \frac{1}{125}$ **13.** $10^a = 95$ **14.** $e^3 = 20.086$

Write Exercises 15–18 in exponential form. [9.2]

**15.** $\log_4 64 = 3$ **16.** $\log_8 32 = \frac{5}{3}$ **17.** $\log_{10} 271 = 2.433$ **18.** $\ln 1 = 0$

Solve Exercises 19–21 by first changing them to exponential form. [9.2]

**19.** $\log_4 x = 3$ **20.** $\log_x 81 = 4$ **21.** $\log_{32} 64 = x$

In Exercises 22 and 23, graph each function and its inverse on the same set of axes. [9.2]

**22.** $y = 3^x$ **23.** $y = \log_5 x$

Use a calculator to find the values of the logarithms in Exercises 24–26. [9.2]

**24.** $\log 368.4$ **25.** $\ln 0.0652$ **26.** $\ln 77.9$

Use a calculator to find the value of $x$ in each of the Exercises 27–29. [9.2]

**27.** $\log x = 2.53$ **28.** $\ln x = 4.91$ **29.** $6 \ln x = 8.22$

Write Exercises 30–33 as single logarithms. [9.3]

**30.** $3 \log x + \log 2x$

**31.** $\frac{1}{2} \ln x - 2 \ln y$

**32.** $\log(x + 2) + \log(x - 2) - 2 \log x$

**33.** $\ln(x + 3) + \ln(2x - 1)$

Write Exercises 34–37 as a sum and/or difference of logarithmic expressions. [9.3]

**34.** $\log x\sqrt{y}$ **35.** $\log \sqrt[4]{x^3 y}$ **36.** $\ln \frac{6x^2}{y}$ **37.** $\ln \frac{x^2}{\sqrt{2yz}}$

Solve the equations in Exercises 38–47. [9.4]

**38.** $2^{x+6} = 8^3$

**39.** $25^x = 125^{(2x+1)}$

**40.** $10^{-3x} = 67.3$

**41.** $15^{3x+1} = 364$

**42.** $7^{1-4x} = 112$

**43.** $3e^{0.04x} = 8.4$

**44.** $5 \ln(2x) - 7 = 1$

**45.** $\log(x^2 + x - 2) - \log(x - 1) = 1$

**46.** $\ln(x - 2) + \ln(x + 2) = 0$

**47.** $\ln(2x^2 - 5x - 12) - \ln(2x + 3) = 2$

**48.** Radium decomposes according to $A = A_0 e^{-0.04t}$, where $t$ is measured in centuries and $A_0$ is the initial amount. Determine the time required for one-half of the initial amount to decompose.

**49.** When friction is used to stop the motion of a wheel, the velocity may be given by $V = V_0 e^{-0.35t}$, where $V_0$ is the initial velocity and $t$ is the number of seconds the friction has been applied. How long will it take to slow a wheel from 75 ft per sec to 15 ft per sec?

**50.** A manufacturer knows that the additional sales of a product resulting from a radio advertising campaign is given by $S = 600e^{-0.24t}$, where $t$ is the number of days after the campaign has ended. How many additional sales will there be after 5 days?

## CHAPTER 9 TEST

1. Sketch the graph of $y = 4^x$

Solve the equations in Exercises 2 and 3.

2. $7^3 \cdot 7^x = 7^{-1}$

3. $6^{x-5} = 36^{x+1}$

4. A scientist knows that a certain strain of bacteria grows according to the function $y = y_0 \cdot 3^{0.25t}$, where $t$ is a measurement in hours. If she starts a culture with 5000 bacteria, how many will be present after 6 hours?

5. Write in logarithmic form: $\left(\frac{1}{2}\right)^{-3} = 8$.

6. Write in exponential form: $\log_3 \frac{1}{9} = -2$.

Solve Exercises 7 and 8 by first changing them to exponential form.

7. $\log_7 x = 3$

8. $\log_9 27 = x$

9. Graph the function $y = \left(\frac{1}{2}\right)^x$ and its inverse on the same set of axes.

10. Use a calculator to find the value of log 579.

11. Use a calculator to find the value of $x$: $5 \ln x = 9.35$

12. Write $\log \sqrt[3]{\frac{x^2}{y}}$ as the sum and/or difference of logarithms.

13. Write $\ln(x + 5) + \ln(x - 4)$ as a single logarithm.

Solve the equations in Exercises 14–18.

14. $10^{x+2} = 283$

15. $2e^{0.24x} = 26$

16. $4^x = 12$

17. $\log(2x + 3) - \log(x + 1) = 0$

18. $\ln(x^2 + 3x - 4) - \ln(x + 4) = 3$

19. If \$1000 is invested at 12% compounded continuously, when will the amount be \$1800?

20. A substance decomposes according to $A = A_0 e^{-0.0035t}$, where $t$ is measured in years and $A_0$ is the initial amount. How long will it take for 800 grams to decompose to 500 grams?

## CUMULATIVE REVIEW (9)

Perform the indicated operations in Exercises 1–4.

1. $\dfrac{2x^2 + 7x + 3}{x^2 - 3x - 18} \cdot \dfrac{x^2 - x - 30}{2x^2 + 11x + 5}$

2. $\dfrac{9 - x^2}{x^2 + 7x + 6} \div \dfrac{x - 3}{x + 6}$

3. $\dfrac{1}{x - 1} + \dfrac{x - 6}{x^2 + 3x - 4}$

4. $\dfrac{2x}{2x + 3} - \dfrac{7x + 12}{2x^2 + 5x + 3}$

Simplify Exercises 5 and 6. Assume all variables are positive.

5. $\dfrac{x^{2/3}y^{1/3}}{x^{1/2}y^{2/3}}$

6. $(x^2y^{-1})^{1/2}(4xy^3)^{-1/2}$

Change each expression in Exercises 7 and 8 to an equivalent expression in exponential notation. Assume that each variable is positive.

**7.** $\sqrt{x^4y^3}$

**8.** $\sqrt[3]{x^2y^3}$

Find the vertex, range, and zeros for the functions in Exercises 9 and 10. Then draw the graph.

**9.** $y = x^2 - 6x - 2$

**10.** $y = 2x^2 + 8x + 3$

Solve each of the equations in Exercises 11–16.

**11.** $x^2 + 5x - 2 = 0$

**12.** $x^4 - 13x^2 + 36 = 0$

**13.** $x - 2 = \sqrt{x + 10}$

**14.** $\dfrac{2}{x - 2} + \dfrac{3}{x - 1} = 1$

**15.** $6^{3x + 5} = 55$

**16.** $\ln(x^2 - 7x + 10) - \ln(x - 2) = 2.5$

Solve each system of equations in Exercises 17 and 18.

**17.** $\begin{cases} 2x - y = -1 \\ x + 2y = 12 \end{cases}$

**18.** $\begin{cases} x + y + z = 2 \\ 2x + y - z = -1 \\ x + 3y + 2z = 2 \end{cases}$

**19.** A book is available in both clothbound and paperback. A bookstore sold a total of 43 books during the week. The total receipts were $297.50. If clothbound books sell for $12.50 and paperbacks sell for $4.50, how many of each were sold?

**20.** Nadine traveled 8 miles upstream and then returned. Her average speed on the return trip was 6 mph faster than her speed upstream. If her total travel time was 2.8 hours, find her rate each way.

CHAPTER

# 10 CONIC SECTIONS

##  MATHEMATICAL CHALLENGES

What is the radius of the circumscribed circle in a 3-4-5 triangle?

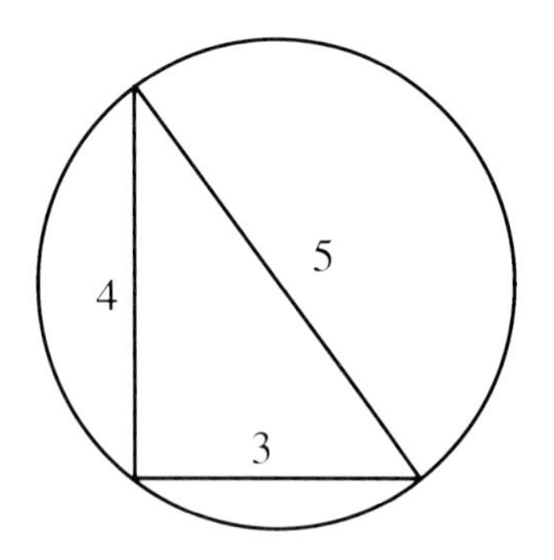


Two concentric circles are such that a chord of the larger, 36 cm long, is trisected by the smaller. The sum of the radii of the two circles is also 36 cm. Find the length of the radius of the larger circle.

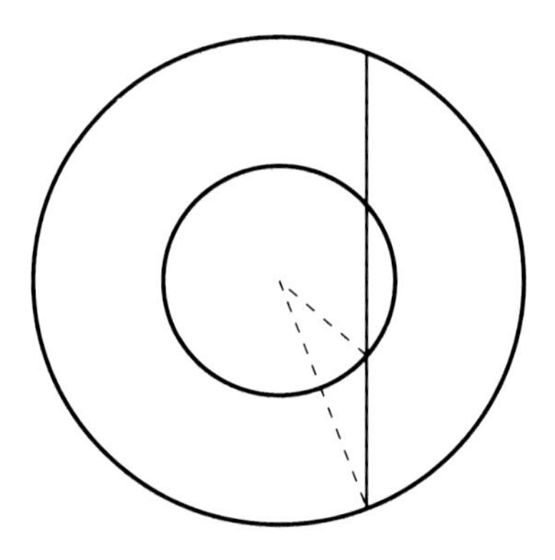

A very important and practical reason explains why utility-hole covers are round and not some polygonal shape. What is it? (No, it is *not* because the holes are round.)

## CHAPTER OUTLINE



Parabolas, circles, ellipses, and hyperbolas are all curves that can be found by intersecting a plane with a cone. Thus, they are all termed conic sections. The corresponding equations, called quadratic equations, for these conic sections are second-degree in $x$ and/or $y$. One of the goals of this chapter is for you to be able to look at one of these equations and tell immediately what type of curve it represents and where the curve is located in relation to a Cartesian coordinate system. These equations are in certain forms, or formulas, and the technique is similar to what we used in Chapter 6 in discussing straight lines. By looking at a linear equation, possibly with some algebraic manipulations, you know the slope of the line, its $y$-intercept, and where it is located. For circles, we will identify the center and radius; for parabolas, the vertex and line of symmetry; for ellipses, the center and intercepts; for hyperbolas, the vertices and asymptotes.

Applied problems using conic sections occur in science and engineering and conic sections appear frequently in higher-level courses in mathematics.

## 10.1 Parabolas as Conic Sections

### OBJECTIVES

In this section, you will be learning to:

1. Graph parabolas with lines of symmetry parallel to the $x$-axis.
2. Find the vertices, $y$-intercepts, and lines of symmetry of parabolas that open left or right.

We have discussed quadratic functions and the corresponding graphs of parabolas in some detail in Section 8.2. In this section, we will discuss parabolas from the general view of a **conic section.**

Conic sections are curves on a plane that are found when the plane intersects a cone. Four such sections are the circle, ellipse, parabola, and hyperbola, as shown in Figure 10.1.

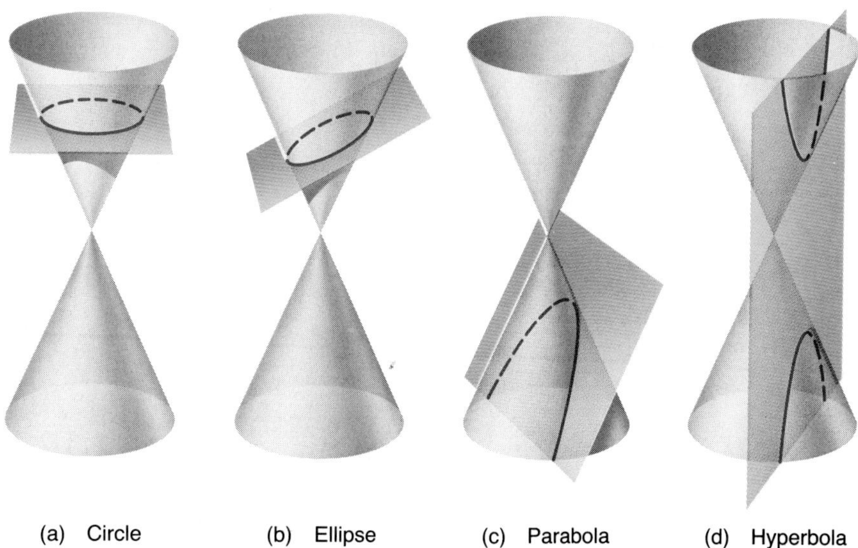


**Figure 10.1**

In a manner similar to what we did with linear functions and quadratic functions, we can represent each of these conic sections with equations involving $x$ and $y$. Such equations may or may not represent functions.

In Section 8.2, we discussed the equations and graphs of parabolas that opened upward or downward. These equations were in the form $y = ax^2 + bx + c$ or $y = a(x - h)^2 + k$. Now we will find that, by exchanging the roles of $x$ and $y$, we can write the equations of parabolas in the form

$$x = a(y - k)^2 + h$$

or $$x = ay^2 + by + c \qquad \text{(where } a \neq 0\text{)}$$

and the parabola opens left (if $a < 0$) or right (if $a > 0$) with vertex at $(h,k)$. The line $y = k$ is the line of symmetry.

In Figure 10.2, the graph of $x = 2(y - 3)^2 - 1$ is shown with a table of $y$- and $x$-values. The $y$-values are chosen on each side of the line of symmetry, $y = 3$.

$$x = 2(y - 3)^2 - 1$$

or $$x + 1 = 2(y - 3)^2$$

and the vertex is at $(h,k) = (-1,3)$.

| $y$ | $x$ |
|---|---|
| 3 | $2(3 - 3)^2 - 1 = -1$ |
| 4 | $2(4 - 3)^2 - 1 = 1$ |
| 2 | $2(2 - 3)^2 - 1 = 1$ |
| 5 | $2(5 - 3)^2 - 1 = 7$ |
| 1 | $2(1 - 3)^2 - 1 = 7$ |

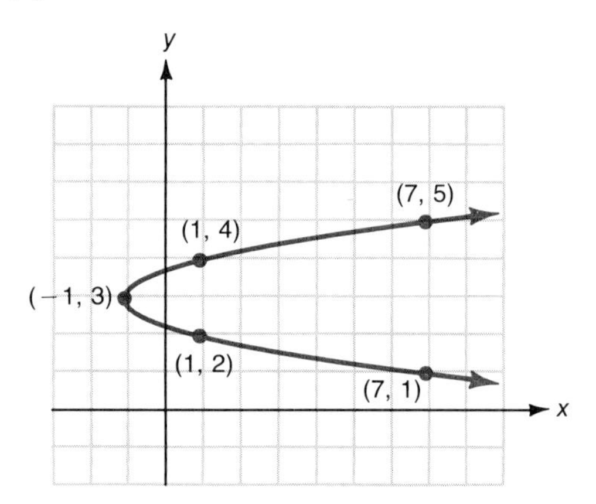


**Figure 10.2**

Just as we discussed in Section 8.3, replacing $x$ with $x - h$ and replacing $y$ with $y - k$ gives an equation whose graph is a horizontal translation of $h$ units and a vertical translation of $k$ units of the original graph. In this section, the original equation is of the form $x = ay^2$. The graphs of several of these equations are shown in Figure 10.3. These graphs do not represent functions, as the vertical line test will confirm.

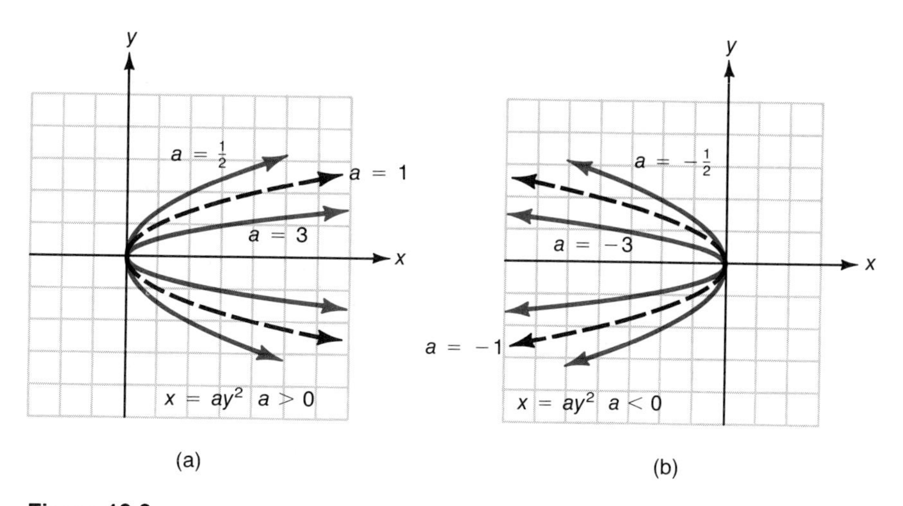


**Figure 10.3**

The graph of an equation of the form

$$x = ay^2 + by + c$$

can be found by completing the square (as in Section 8.2) and writing the equation in the form

$$x = a(y - k)^2 + h$$

or

$$x - h = a(y - k)^2$$

Also, by solving the quadratic equation

$$ay^2 + by + c = 0$$

we can determine at what points, if any, the graph intersects the ***y*-axis.** These points are called ***y*-intercepts.**

## EXAMPLES

**1.** For $x = y^2 - 6y + 6$, find the vertex, the points where the graph intersects the $y$-axis, and the line of symmetry. Then sketch the graph.

**Solution**

To find the vertex:

$x = y^2 - 6y + 6$
$x = (y^2 - 6y + 9) - 9 + 6$
$x = (y - 3)^2 - 3$

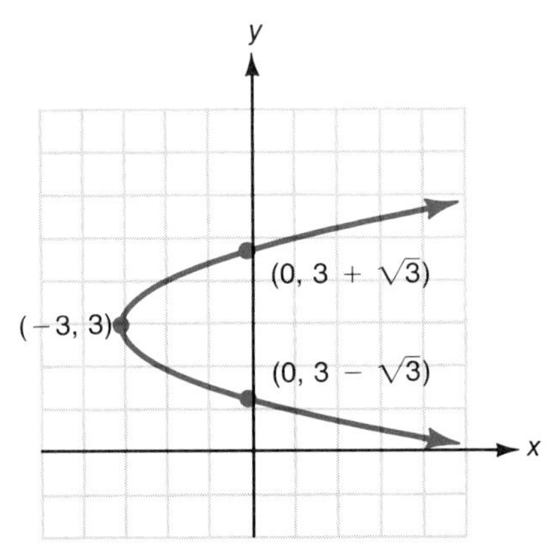


To find the $y$-intercepts, let $x = 0$:

$y^2 - 6y + 6 = 0$

$$y = \frac{6 \pm \sqrt{(-6)^2 - 4 \cdot 1 \cdot 6}}{2}$$

$$y = \frac{6 \pm \sqrt{12}}{2}$$

$y = 3 \pm \sqrt{3}$

Vertex $= (-3,3)$.
Intersects $y$-axis at $(0,3 + \sqrt{3})$ and $(0,3 - \sqrt{3})$.
Line of symmetry is $y = 3$.

**2.** For $x = -2y^2 - 4y + 6$, locate the vertex, the points where the graph intersects the $y$-axis, and the line of symmetry. Then sketch the graph.

**Solution**

To find the vertex:

$x = -2y^2 - 4y + 6$
$x = -2(y^2 + 2y) + 6$
$x = -2(y^2 + 2y + 1 - 1) + 6$
$x = -2(y^2 + 2y + 1) + 8$
$x = -2(y + 1)^2 + 8$

To find the $y$-intercepts, let $x = 0$:

$-2y^2 - 4y + 6 = 0$
$-2(y^2 + 2y - 3) = 0$
$-2(y + 3)(y - 1) = 0$
$y = -3 \quad \text{or} \quad y = 1$

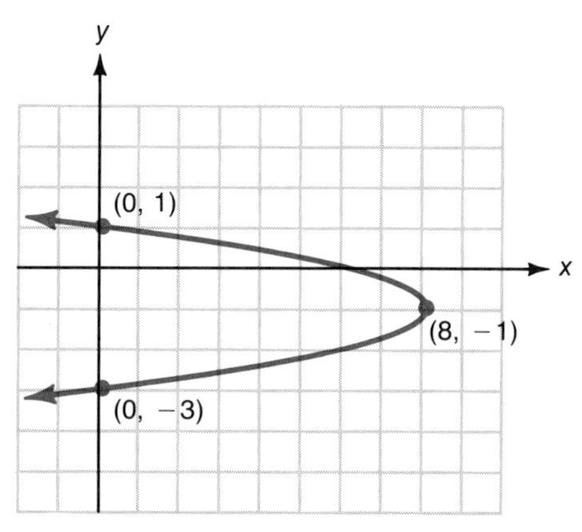


Vertex $= (8, -1)$.
The $y$-intercepts are $(0, -3)$ and $(0, 1)$.
Line of symmetry is $y = -1$.

**Practice Problems**

1. Write the equation $x = -y^2 - 10y - 24$ in the form $x = a(y - k)^2 + h$.
2. Find the vertex, $y$-intercepts, and line of symmetry for the curve $x = y^2 - 4$.
3. Find the $y$-intercepts for the curve $x = y^2 + 2y + 2$.

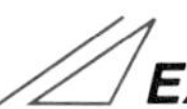

## EXERCISES 10.1

In each of the Exercises 1–30, find the vertex, $y$-intercepts, and the line of symmetry; then draw the graph.

**1.** $x = y^2 + 4$
**2.** $x = y^2 - 5$
**3.** $x + 3 = y^2$
**4.** $x - 2 = y^2$
**5.** $x = 2y^2 + 3$
**6.** $x = 3y^2 + 1$
**7.** $x = (y - 3)^2$
**8.** $x = (y - 2)^2$
**9.** $x - 4 = (y + 2)^2$
**10.** $x + 3 = (y - 5)^2$
**11.** $x + 1 = (y - 1)^2$
**12.** $x - 5 = (y + 4)^2$
**13.** $x = y^2 + 4y + 4$
**14.** $x = -y^2 + 10y - 25$
**15.** $x = y^2 - 8y + 16$
**16.** $x = y^2 + 6y + 1$
**17.** $y = -x^2 - 4x + 5$
**18.** $y = x^2 + 5x + 6$
**19.** $y = x^2 + 6x + 5$
**20.** $y = x^2 - 2x - 5$
**21.** $x = -y^2 + 4y - 3$
**22.** $x = y^2 + 8y + 10$
**23.** $y = 2x^2 + x - 1$
**24.** $y = -2x^2 + x + 3$
**25.** $x = 3y^2 + 6y - 5$
**26.** $x = 3y^2 + 5y + 2$
**27.** $x = -2y^2 + 5y - 2$
**28.** $x = 4y^2 - 4y - 15$
**29.** $y = 4x^2 - 12x + 9$
**30.** $y = -5x^2 + 10x + 2$

**Answers to Practice Problems** **1.** $x = -(y + 5)^2 + 1$ **2.** Vertex: $(-4, 0)$, $y$-intercepts: $(0, 2)$ and $(0, -2)$, Line of symmetry: $y = 0$ **3.** There are no $y$-intercepts. The solutions to the equation $0 = y^2 + 2y + 2$ are nonreal.

## 10.2 Distance Formula and Circles

**OBJECTIVES**

In this section, you will be learning to:

1. Find the distance between any two points.
2. Write the equations of circles given their centers and radii.
3. Graph circles centered at $(h,k)$.

Before discussing equations for circles, we need the formula for the distance between two points on a plane. The **Pythagorean Theorem,** previously discussed in Section 5.4, is the basis for the formula. The theorem is repeated here for easy reference.

***Pythagorean Theorem*** In a right triangle, the square of the hypotenuse is equal to the sum of the squares of the two legs.

$$c^2 = a^2 + b^2$$

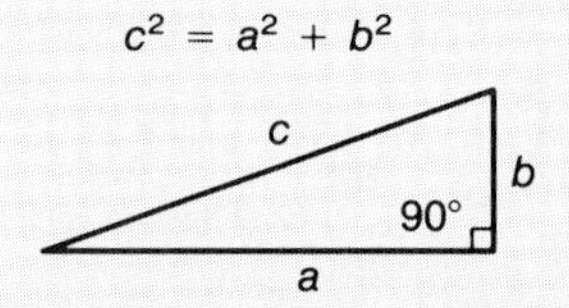


Suppose that we want to find the distance between the two points $P_1(-1,2)$ and $P_2(5,6)$, as shown in Figure 10.4(a). First form a right triangle, as shown in Figure 10.4(b), and find the lengths of the sides $a$ and $b$. Then, using $a$ and $b$ and the Pythagorean Theorem, we can find the length of the hypotenuse, which is the desired distance.

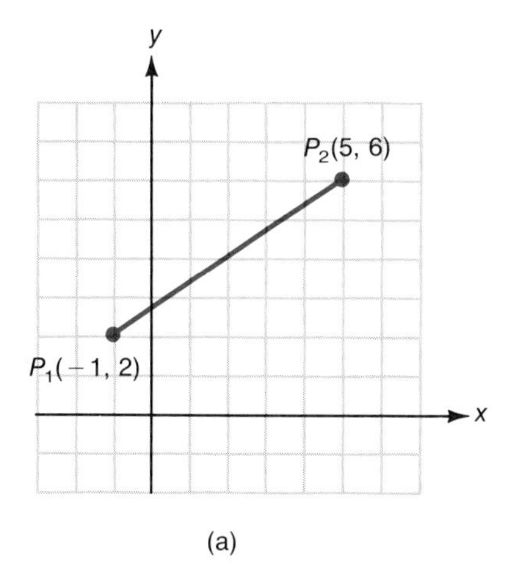


(a)

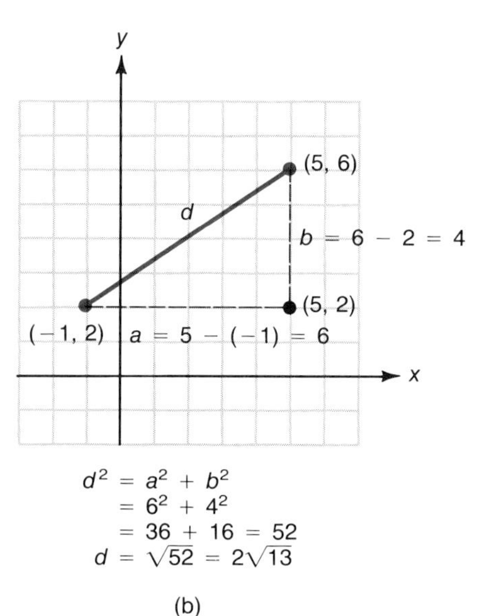


$$\begin{aligned} d^2 &= a^2 + b^2 \\ &= 6^2 + 4^2 \\ &= 36 + 16 = 52 \\ d &= \sqrt{52} = 2\sqrt{13} \end{aligned}$$

(b)

**Figure 10.4**

We could go directly to the formula

$$d = \sqrt{a^2 + b^2}$$

Carrying this idea one step further, we can write the formula for $d$ involving the coordinates of two general points $P_1(x_1,y_1)$ and $P_2(x_2,y_2)$. As illustrated in Figure 10.5, the distance formula is

$$d = \sqrt{(x_2 - x_1)^2 + (y_2 - y_1)^2}$$

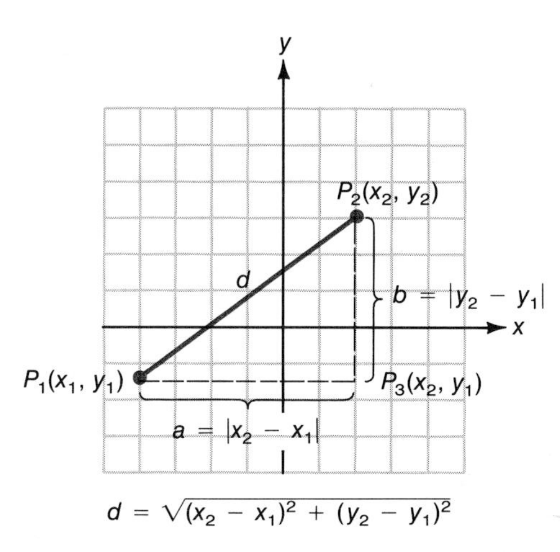


**Figure 10.5**

Note that in Figure 10.5, the calculations for $a$ and $b$ involve absolute value. These absolute values guarantee nonnegative values for $a$ and $b$ to represent the lengths of the legs. In the formula, the absolute values are disregarded because $x_2 - x_1$ and $y_2 - y_1$ are squared. **In the actual calculations of *d*, be sure to add the squares before taking the square root.**

EXAMPLES Each of the following examples illustrates how to use the distance formula $d = \sqrt{(x_2 - x_1)^2 + (y_2 - y_1)^2}$.

1. Find the distance between the two points (3,4) and (−2,7).

   **Solution** $d = \sqrt{[3 - (-2)]^2 + (4 - 7)^2}$
   $= \sqrt{5^2 + (-3)^2} = \sqrt{25 + 9} = \sqrt{34}$

**2.** Determine whether or not the triangle determined by the three points $A(-5,-1)$, $B(2,1)$, and $C(0,7)$ is a right triangle.

**Solution** Find the lengths of the three line segments $AB$, $AC$, and $BC$, and decide whether or not the Pythagorean Theorem is satisfied.

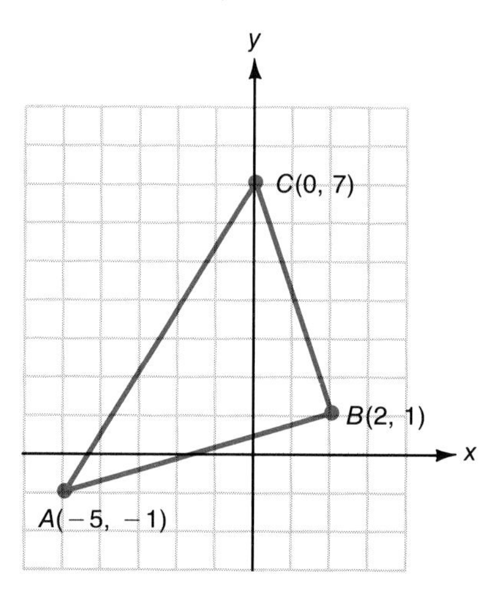


$$\begin{aligned} AB &= \sqrt{(-5-2)^2 + (-1-1)^2} = \sqrt{(-7)^2 + (-2)^2} \\ &= \sqrt{49 + 4} = \sqrt{53} \\ AC &= \sqrt{(-5-0)^2 + (-1-7)^2} = \sqrt{(-5)^2 + (-8)^2} \\ &= \sqrt{25 + 64} = \sqrt{89} \\ BC &= \sqrt{(2-0)^2 + (1-7)^2} = \sqrt{2^2 + (-6)^2} \\ &= \sqrt{4 + 36} = \sqrt{40} \end{aligned}$$

The longest side is $AC = \sqrt{89}$.

The triangle is **not** a right triangle since

$$(\sqrt{89})^2 \neq (\sqrt{53})^2 + (\sqrt{40})^2$$

or

$$89 \neq 53 + 40$$

■

We now give a formal definition of a circle, one of the conic sections, and use the distance formula to develop the equation of a circle.

***Circle, Center, and Radius***

A **circle** is the set of all points in a plane that are a fixed distance from a fixed point.
The fixed point is called the **center** of the circle.
The fixed distance is called the **radius** of the circle.

Suppose that we want to find the equation of a circle with its center at the origin (0,0) and radius 5. If $(x,y)$ is any point on the circle, the distance from $(x,y)$ to (0,0) must be 5. So, using the distance formula,

$$\sqrt{(x - x_2)^2 + (y - y_2)^2} = d$$
$$\sqrt{(x - 0)^2 + (y - 0)^2} = 5$$
$$\sqrt{x^2 + y^2} = 5$$
$$x^2 + y^2 = 25 \qquad \text{Squaring both sides}$$

Thus, as shown in Figure 10.6, all points on the circle satisfy the equation $x^2 + y^2 = 25$.

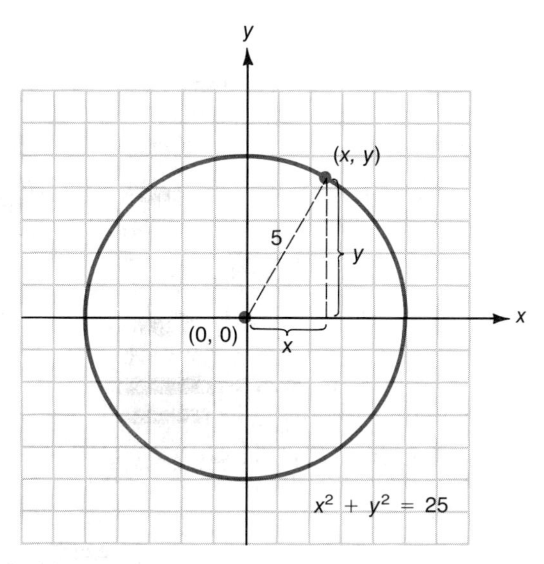


**Figure 10.6**

In general, if a circle has its center at $(h,k)$ and radius $r > 0$, then for any point $(x,y)$ on the circle,

$$\sqrt{(x - h)^2 + (y - k)^2} = r$$

or, squaring both sides, we get the **standard form**

$$(x - h)^2 + (y - k)^2 = r^2$$

(See Figure 10.7.)

The equation for a circle of radius $r$ with **center at the origin** is

$$x^2 + y^2 = r^2$$

We can think of this circle translated to have **center at $(h,k)$** by substituting $(x - h)$ for $x$ and $(y - k)$ for $y$ to give the **standard form**

$$(x - h)^2 + (y - k)^2 = r^2$$

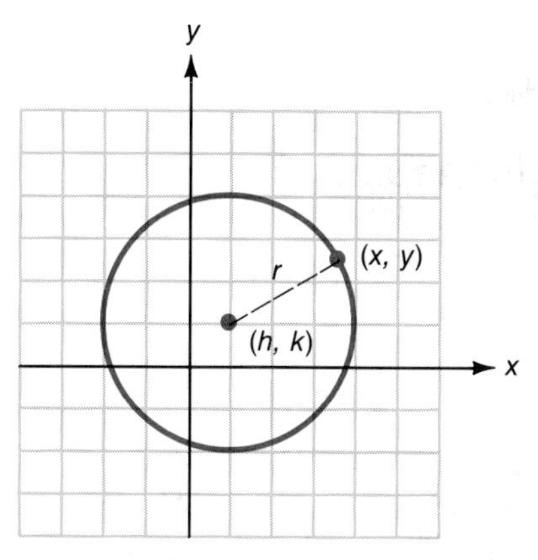


$(x - h)^2 + (y - k)^2 = r^2$
Circle with center at $(h, k)$ and radius $r$

**Figure 10.7**

## EXAMPLES

**3.** Find the equation of the circle with center at the origin and radius $\sqrt{3}$. Are the points $(\sqrt{2},1)$ and $(1,2)$ on the circle?

**Solution** The equation is $x^2 + y^2 = 3$.

$$\begin{aligned}(\sqrt{2})^2 + (1)^2 &= 2 + 1 \\ &= 3 \\ (1)^2 + (2)^2 &= 1 + 4 \\ &= 5 \neq 3\end{aligned}$$

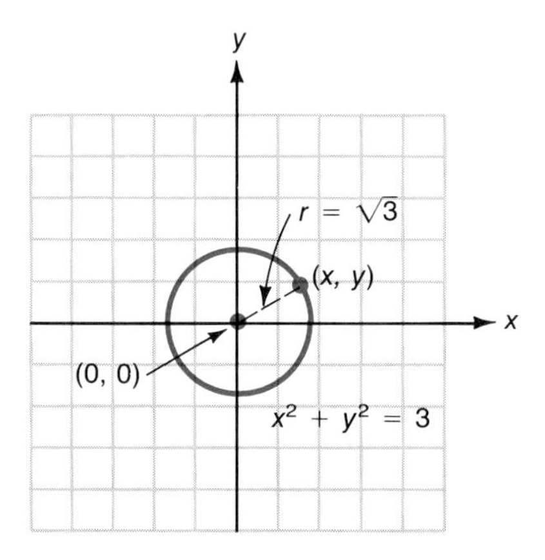


So, $(\sqrt{2},1)$ is on the circle and $(1,2)$ is not on the circle.

**4.** Find the equation of the circle with center at $(5,2)$ and radius 3. Is the point $(5,5)$ on the circle?

**Solution** The equation is $(x - 5)^2 + (y - 2)^2 = 9$

$$\begin{aligned}(5 - 5)^2 + (5 - 2)^2 &= 0^2 + 3^2 \\ &= 9\end{aligned}$$

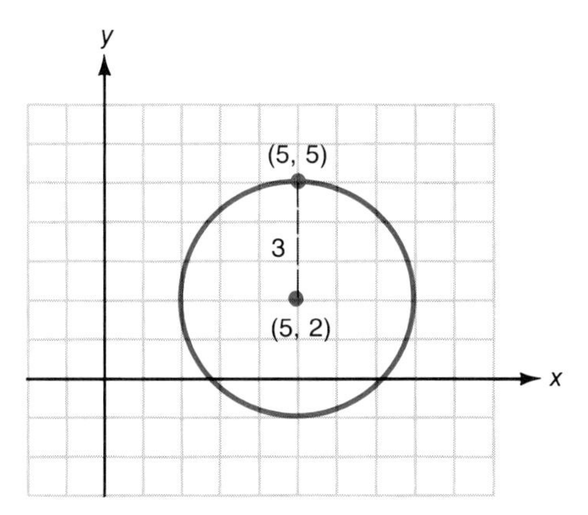


Yes, $(5,5)$ is on the circle.

**5.** Show that $x^2 + y^2 - 8x + 2y = 0$ represents a circle. Find the center of the circle and the radius. Graph the circle.

**Solution** Rearrange the terms and complete the square for $x^2 - 8x$ and $y^2 + 2y$.

$$x^2 + y^2 - 8x + 2y = 0$$
$$x^2 - 8x + y^2 + 2y = 0$$
$$x^2 - 8x + 16 + y^2 + 2y + 1 = 16 + 1$$

Completes the square. Completes the square.

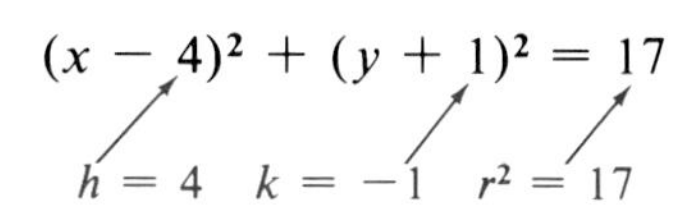


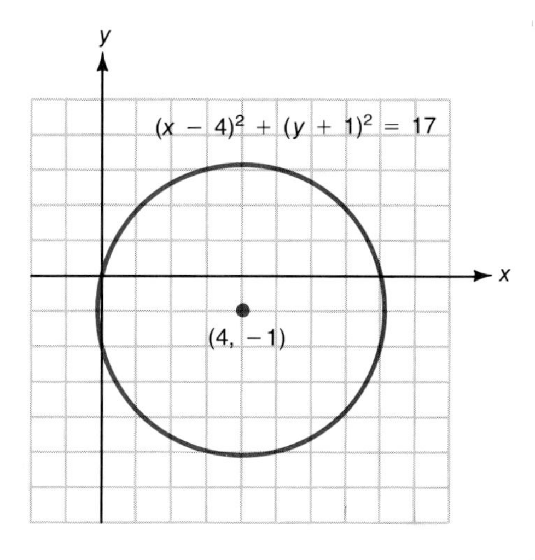


Center is at $(4, -1)$.
Radius $= \sqrt{17}$ ■

**Practice Problems**

1. Find the equation of the circle with center at $(-2,3)$ and radius 6.
2. Write the equation in standard form: $x^2 + y^2 + 6y = 7$.
3. Find the distance between the two points $(5,3)$ and $(-1,-3)$.

**Answers to Practice Problems** **1.** $(x + 2)^2 + (y - 3)^2 = 36$ **2.** $x^2 + (y + 3)^2 = 16$; center at $(0, -3)$ and radius 4 **3.** $\sqrt{72} = 6\sqrt{2}$

## EXERCISES 10.2

In Exercises 1–12, find the distance between the two points.

**1.** $(2,4)$, $(6,7)$
**2.** $(1,0)$, $(6,12)$
**3.** $(-3,2)$, $(9,7)$
**4.** $(-6,3)$, $(-2,0)$
**5.** $(1,7)$, $(3,2)$
**6.** $(-2,1)$, $(3,-4)$
**7.** $(4,-3)$, $(7,-3)$
**8.** $(-2,6)$, $(5,6)$
**9.** $(5,-2)$, $(7,-5)$
**10.** $(6,4)$, $(8,-5)$
**11.** $(-7,3)$, $(1,-12)$
**12.** $(3,8)$, $(-2,-4)$

Find equations for each of the circles in Exercises 13–32.

**13.** Center $(0,0)$; $r = 4$
**14.** Center $(0,0)$; $r = 6$
**15.** Center $(0,0)$; $r = \sqrt{3}$
**16.** Center $(0,0)$; $r = \sqrt{7}$
**17.** Center $(0,0)$; $r = \sqrt{11}$
**18.** Center $(0,0)$; $r = \sqrt{13}$
**19.** Center $(0,0)$; $r = \frac{2}{3}$
**20.** Center $(0,0)$; $r = \frac{7}{4}$
**21.** Center $(0,2)$; $r = 2$
**22.** Center $(0,5)$; $r = 5$
**23.** Center $(4,0)$; $r = 1$
**24.** Center $(-3,0)$; $r = 4$
**25.** Center $(-2,0)$; $r = \sqrt{8}$
**26.** Center $(5,0)$; $r = \sqrt{2}$
**27.** Center $(3,1)$; $r = 6$
**28.** Center $(-1,2)$; $r = 5$
**29.** Center $(3,5)$; $r = \sqrt{12}$
**30.** Center $(4,-2)$; $r = \sqrt{14}$
**31.** Center $(7,4)$; $r = \sqrt{10}$
**32.** Center $(-3,2)$; $r = \sqrt{7}$

Write each of the equations in Exercises 33–48 in standard form. Find the center and radius of the circle; then sketch the graph.

**33.** $x^2 + y^2 = 9$
**34.** $x^2 + y^2 = 16$
**35.** $x^2 = 49 - y^2$
**36.** $y^2 = 25 - x^2$
**37.** $x^2 + y^2 = 18$
**38.** $x^2 + y^2 = 12$
**39.** $x^2 + y^2 + 2x = 8$
**40.** $x^2 + y^2 - 4x = 12$
**41.** $x^2 + y^2 - 4y = 0$
**42.** $x^2 + y^2 + 6x = 0$
**43.** $x^2 + y^2 + 2x + 4y = 11$
**44.** $x^2 + y^2 - 4x + 10y + 20 = 0$
**45.** $x^2 + y^2 + 4x + 4y - 8 = 0$
**46.** $x^2 + y^2 - 6x - 8y + 9 = 0$
**47.** $x^2 + y^2 - 4x - 6y + 5 = 0$
**48.** $x^2 + y^2 + 10x - 2y + 14 = 0$

In Exercises 49 and 50, use the Pythagorean Theorem to decide if the triangle determined by the given points is a right triangle.

**49.** $A(1,-2)$, $B(7,1)$, $C(5,5)$
**50.** $A(-5,-1)$, $B(2,1)$, $C(-1,6)$

In Exercises 51 and 52, show that the triangle determined by the given points is an isosceles triangle (has two equal sides).

**51.** $A(1,1)$, $B(5,9)$, $C(9,5)$
**52.** $A(1,-4)$, $B(3,2)$, $C(9,4)$

In Exercises 53 and 54, show that the triangle determined by the given points is an equilateral triangle (all sides equal).

**53.** $A(1,0)$, $B(3,\sqrt{12})$, $C(5,0)$
**54.** $A(0,5)$, $B(0,-3)$, $C(\sqrt{48},1)$

In Exercises 55 and 56, show that the diagonals ($AC$ and $BD$) of the rectangle $ABCD$ are equal.

**55.** $A(2,-2)$, $B(2,3)$, $C(8,3)$, $D(8,-2)$
**56.** $A(-1,1)$, $B(-1,4)$, $C(4,4)$, $D(4,1)$

In Exercises 57–60, find the perimeter of the triangle determined by the given points.

**57.** $A(-5,0)$, $B(3,4)$, $C(0,0)$

**58.** $A(-6,-1)$, $B(-3,3)$, $C(6,4)$

**59.** $A(-2,5)$, $B(3,1)$, $C(2,-2)$

**60.** $A(1,4)$, $B(-3,3)$, $C(-1,7)$

## 10.3 Ellipses and Hyperbolas

**OBJECTIVES**

In this section, you will be learning to:

1. Graph ellipses centered at the origin.
2. Graph hyperbolas centered at the origin.
3. Find the equations for the asymptotes of hyperbolas.

In this section, two more conic sections, ellipses and hyperbolas, are discussed. We will restrict the discussion to basic forms with the "center" of each curve at the origin. Any translations, such as those we discussed with parabolas and circles, would introduce a degree of difficulty that we leave to later courses.

Now, consider the equation

$$\frac{x^2}{25} + \frac{y^2}{9} = 1$$

Several points that satisfy this equation are given below in tabular form and are graphed in Figure 10.8.

| $x$ | $y$ |
|---|---|
| 5 | 0 |
| $-5$ | 0 |
| 0 | 3 |
| 0 | $-3$ |
| 3 | $\frac{12}{5}$ |
| 3 | $-\frac{12}{5}$ |
| $-3$ | $\frac{12}{5}$ |
| $-3$ | $-\frac{12}{5}$ |

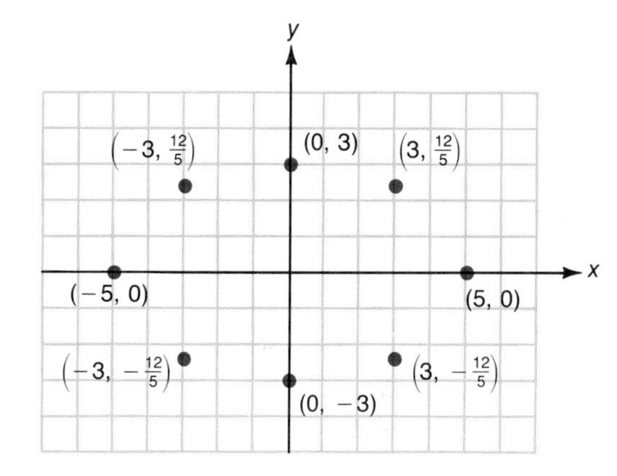


**Figure 10.8**

By joining the points in Figure 10.8 with a smooth curve, we get the graph called an **ellipse,** as shown in Figure 10.9. The points $(5,0)$ and $(-5,0)$ are $x$-intercepts, and the points $(0,3)$ and $(0,-3)$ are $y$-intercepts.

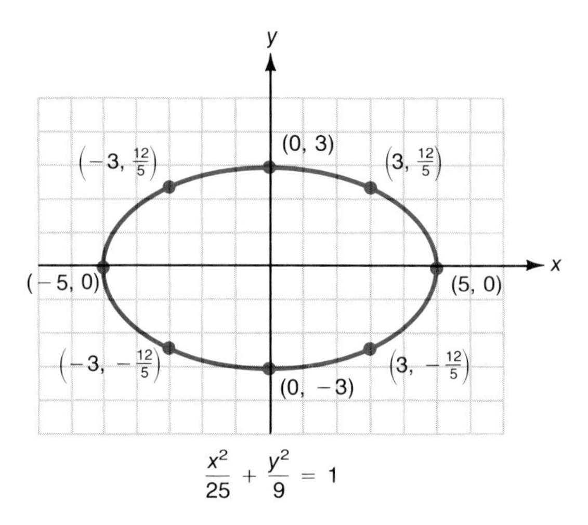


**Figure 10.9**

In general, the **standard form** for the equations of an ellipse with **center at the origin** is

$$\frac{x^2}{a^2} + \frac{y^2}{b^2} = 1 \qquad \text{where } a^2 \geq b^2$$

The points $(a,0)$ and $(-a,0)$ are the $x$-intercepts.
The points $(0,b)$ and $(0,-b)$ are the $y$-intercepts.
The segment of length $2a$ joining the $x$-intercepts is called the **major axis.**
The segment of length $2b$ joining the $y$-intercepts is called the **minor axis.**

[**Note:** Example 2 illustrates a second form $\frac{x^2}{b^2} + \frac{y^2}{a^2} = 1$ and corresponding adjustments in the related terminology.]

## EXAMPLE

1. Graph the equation $4x^2 + 16y^2 = 64$.

   **Solution** First divide both sides of the given equation by 64 to find the standard form.

$$4x^2 + 16y^2 = 64$$

$$\frac{4x^2}{64} + \frac{16y^2}{64} = \frac{64}{64}$$

$$\frac{x^2}{16} + \frac{y^2}{4} = 1$$

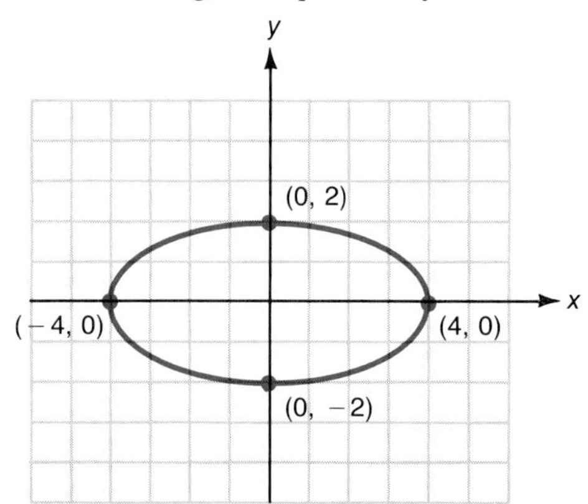

The curve is an ellipse. The end points of the major axis are $(-4,0)$ and $(4,0)$. The endpoints of the minor axis are $(0,-2)$ and $(0,2)$. ■

In the discussion and in Example 1, the major axis was horizontal and the minor axis vertical. For ellipses in standard form, the larger denominator is treated as $a^2$ and the smaller denominator as $b^2$. Thus, if the larger denominator is below $y^2$, then the major axis is vertical and the minor axis is horizontal. This situation is illustrated in Example 2.

**EXAMPLE**

**2.** Graph the equation $\frac{x^2}{1} + \frac{y^2}{9} = 1$.

**Solution** The equation is in standard form; however, since the larger denominator, 9, is below $y^2$, the major axis is vertical. The ellipse is elongated along the $y$-axis. The points $(0,-3)$ and $(0,3)$ are the endpoints of the major axis while $(-1,0)$ and $(0,0)$ are the endpoints of the minor axis.

$$\frac{x^2}{b^2} + \frac{y^2}{a^2} = 1$$

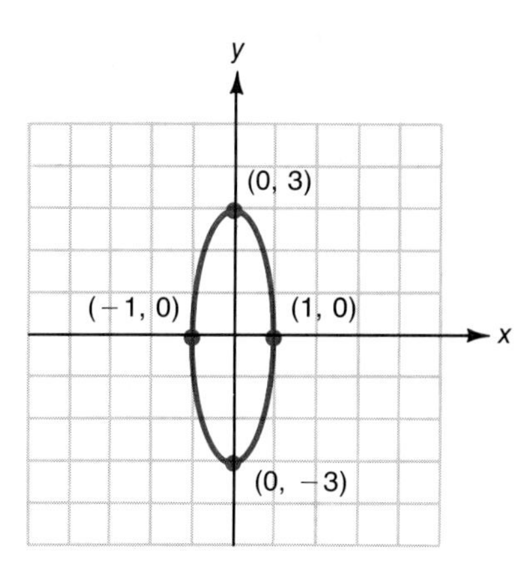


■

In the equation of an ellipse

$$\frac{x^2}{a^2} + \frac{y^2}{b^2} = 1 \qquad \left(\text{or } \frac{x^2}{b^2} + \frac{y^2}{a^2} = 1\right)$$

the coefficients for $x^2$ and $y^2$ are both positive. If one of these coefficients is negative, then the equation represents a **hyperbola.**

Several points that satisfy the equation

$$\frac{x^2}{25} - \frac{y^2}{9} = 1$$

and the curve joining three points (a hyperbola) are shown in Figure 10.10. The origin $(0,0)$ is called the **center** of the hyperbola.

| $x$ | $y$ |
|---|---|
| 5 | 0 |
| $-5$ | 0 |
| 7 | $\frac{6\sqrt{6}}{5}$ |
| 7 | $\frac{-6\sqrt{6}}{5}$ |
| $-7$ | $\frac{6\sqrt{6}}{5}$ |
| $-7$ | $\frac{-6\sqrt{6}}{5}$ |

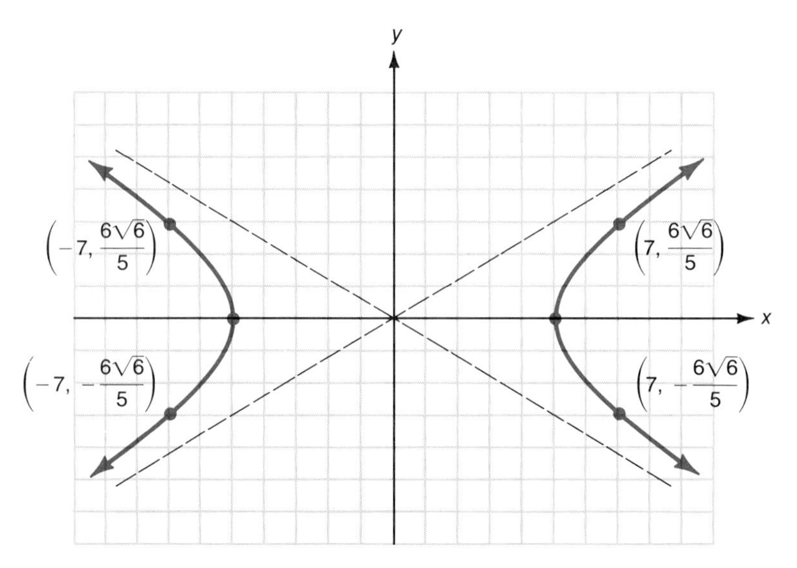


**Figure 10.10**

The two dotted lines shown in Figure 10.10 are called **asymptotes.** These lines are not part of the hyperbola, but they serve as guidelines for the graph because the curve gets closer and closer to these lines without ever touching them. The equations of these lines are $y = \frac{3}{5}x$ and $y = -\frac{3}{5}x$.

In general, there are **two standard forms** for equations of hyperbolas with their **centers at the origin:**

**1.** $\frac{x^2}{a^2} - \frac{y^2}{b^2} = 1$

$x$-intercepts (vertices) at $(a,0)$ and $(-a,0)$; no $y$-intercepts;

asymptotes $y = \frac{b}{a}x$ and $y = -\frac{b}{a}x$;

the curve "opens" left and right.

**2.** $\frac{y^2}{b^2} - \frac{x^2}{a^2} = 1$

$y$-intercepts (vertices) at $(0,b)$ and $(0,-b)$; no $x$-intercepts;

asymptotes $y = \frac{b}{a}x$ and $y = -\frac{b}{a}x$;

the curve "opens" up and down.

## Geometrical Aid for Sketching Asymptotes

The asymptotes $y = \frac{b}{a}x$ and $y = -\frac{b}{a}x$ pass through the diagonals of the rectangle formed with vertical and horizontal lines through the points $(a,0)$, $(-a,0)$, $(0,b)$, and $(0,-b)$. Such rectangles are shown in Examples 3 and 4.

**EXAMPLES**

**3.** Graph the curve $x^2 - 4y^2 = 4$.

**Solution** First we write the equation in standard form by dividing by 4.

$$\frac{x^2}{4} - \frac{y^2}{1} = 1$$

Here, $a^2 = 4$ and $b^2 = 1$. So, using $a = 2$ and $b = 1$, the asymptotes are $y = \frac{1}{2}x$ and $y = -\frac{1}{2}x$. Vertices are (2,0) and (−2,0). The curve "opens" left and right.

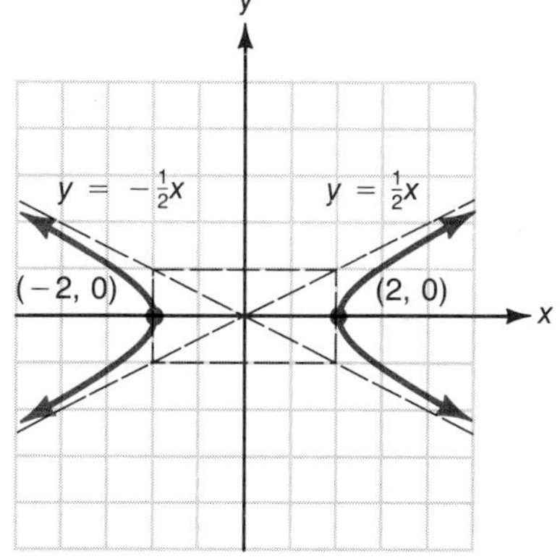


**4.** Graph the curve $\frac{y^2}{1} - \frac{x^2}{4} = 1$.

**Solution** First locate the asymptotes and the vertices; then sketch the curve. Again, $a = 2$ and $b = 1$. The asymptotes are $y = \frac{1}{2}x$ and $y = -\frac{1}{2}x$. Vertices are (0,1) and (0,−1). The curve "opens" up and down.

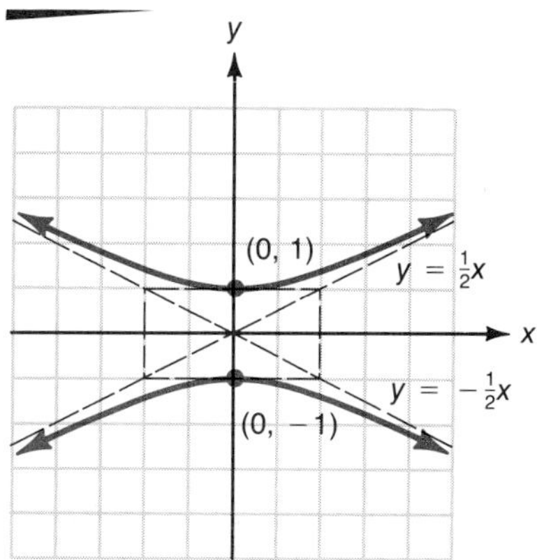


■

**Practice Problems**

1. Write the equation $2x^2 + 9y^2 = 18$ in standard form. State the length of the major axis and the length of the minor axis.
2. Write the equation $x^2 - 9y^2 = 9$ in standard form. Write the equations of the asymptotes.

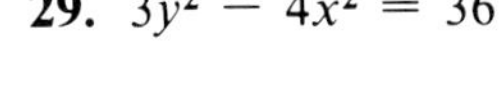

## EXERCISES 10.3

Write each of the equations in Exercises 1–30 in standard form; then sketch the graph. Write the equations and graph the asymptotes for each hyperbola.

**1.** $x^2 + 9y^2 = 36$ **2.** $x^2 + 4y^2 = 16$ **3.** $4x^2 + 25y^2 = 100$ **4.** $4x^2 + 9y^2 = 36$
**5.** $16x^2 + y^2 = 16$ **6.** $25x^2 + 9y^2 = 36$ **7.** $x^2 - y^2 = 1$ **8.** $x^2 - y^2 = 4$
**9.** $9x^2 - y^2 = 9$ **10.** $4x^2 - y^2 = 4$ **11.** $4x^2 - 9y^2 = 36$ **12.** $9x^2 - 16y^2 = 144$
**13.** $2x^2 + y^2 = 8$ **14.** $3x^2 + y^2 = 12$ **15.** $x^2 + 5y^2 = 20$ **16.** $x^2 + 7y^2 = 28$
**17.** $y^2 - x^2 = 9$ **18.** $y^2 - x^2 = 16$ **19.** $y^2 - 2x^2 = 8$ **20.** $y^2 - 3x^2 = 12$
**21.** $y^2 - 2x^2 = 18$ **22.** $y^2 - 5x^2 = 20$ **23.** $3x^2 + 2y^2 = 18$ **24.** $4x^2 + 3y^2 = 12$
**25.** $4x^2 + 5y^2 = 20$ **26.** $3x^2 + 8y^2 = 48$ **27.** $3x^2 - 5y^2 = 75$ **28.** $4x^2 - 7y^2 = 28$
**29.** $3y^2 - 4x^2 = 36$ **30.** $9y^2 - 8x^2 = 72$

## 10.4 Nonlinear Systems of Equations

**OBJECTIVE**

In this section, you will be learning to: solve systems of either two quadratic equations or one quadratic and one linear equation in two variables.

The equations for the conic sections that we have discussed all have at least one term that is second-degree. These equations are called **quadratic equations.** (Only the equations for parabolas of the form $y = ax^2 + bx + c$ are **quadratic functions.**) A summary of the equations with their related graphs is shown in Figure 10.11.

A system of two equations involving one or more quadratic equations can be solved either by substitution or by addition. (At least those systems given in this text can be so solved.) If one of the equations is linear, then the substitution method should be used. If both are quadratic equations, then the method used depends on the form of the equations. The graphs of the curves are particularly useful in indicating approximate solutions and the exact number of solutions.

**Answers to Practice Problems** **1.** $\frac{x^2}{9} + \frac{y^2}{2} = 1$, Major axis: 6, Minor axis: $2\sqrt{2}$
**2.** $\frac{x^2}{9} - \frac{y^2}{1} = 1$, The asymptotes are $y = \frac{1}{3}x$ and $y = -\frac{1}{3}x$.

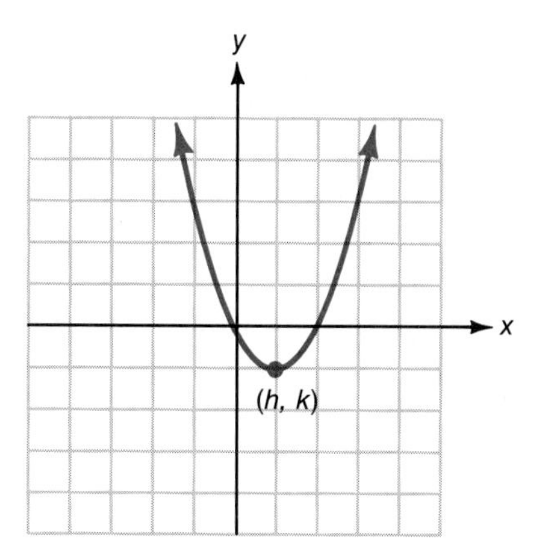


$y = a(x - h)^2 + k$
Parabola
$a > 0$, opens upward
$a < 0$, opens downward

(a)

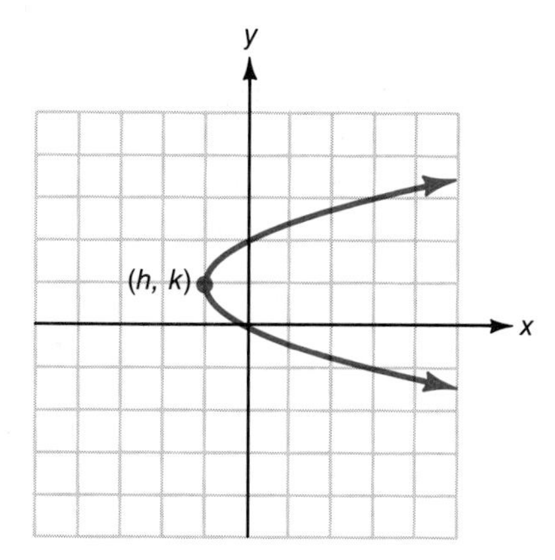


$x = a(y - k)^2 + h$
Parabola
$a > 0$, opens right
$a < 0$, opens left

(b)

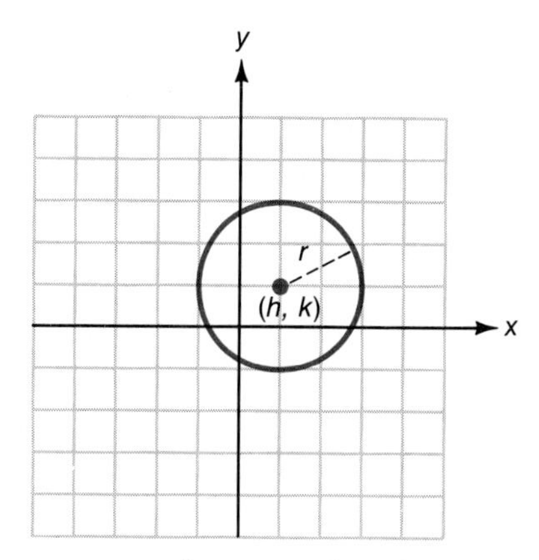


$(x - h)^2 + (y - k)^2 = r^2$
Circle

(c)

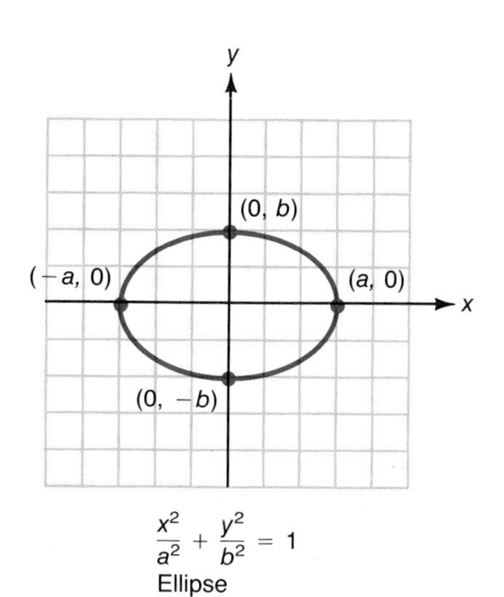


$$\frac{x^2}{a^2} + \frac{y^2}{b^2} = 1$$
Ellipse

(d)

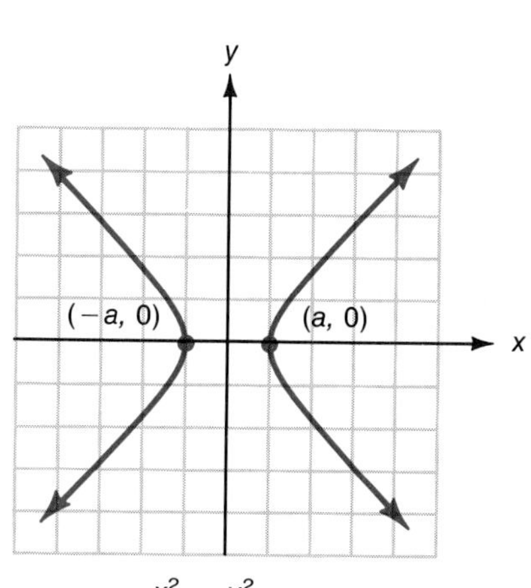


$$\frac{x^2}{a^2} - \frac{y^2}{b^2} = 1$$
Hyperbola

(e)

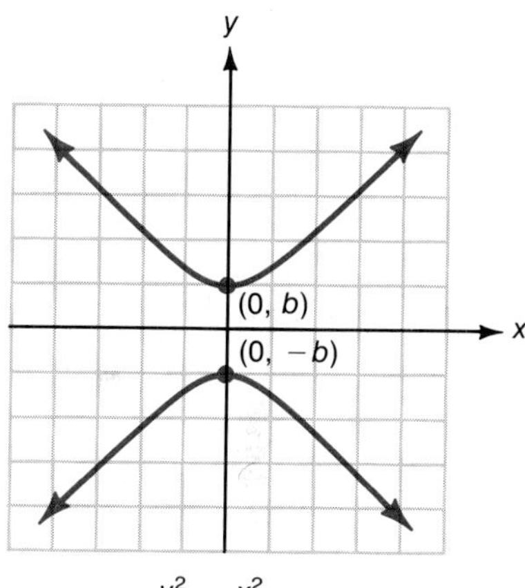


$$\frac{y^2}{b^2} - \frac{x^2}{a^2} = 1$$
Hyperbola

(f)

**Figure 10.11**

**EXAMPLES** Solve the following systems and graph both curves in each system.

**1.** $\begin{cases} x^2 + y^2 = 25 \\ x + y = 5 \end{cases}$

**Solution** Solve $x + y = 5$ for $y$ (or $x$). Then substitute into the other equation.

$$\begin{aligned} y &= 5 - x \\ x^2 + (5 - x)^2 &= 25 \qquad \text{Now solve for } x. \\ x^2 + 25 - 10x + x^2 &= 25 \\ 2x^2 - 10x &= 0 \\ 2x(x - 5) &= 0 \end{aligned}$$

$$\begin{cases} x = 0 \\ y = 5 - 0 = 5 \end{cases} \quad \text{or} \quad \begin{cases} x = 5 \\ y = 5 - 5 = 0 \end{cases}$$

The solutions (points of intersection) are (0,5) and (5,0).

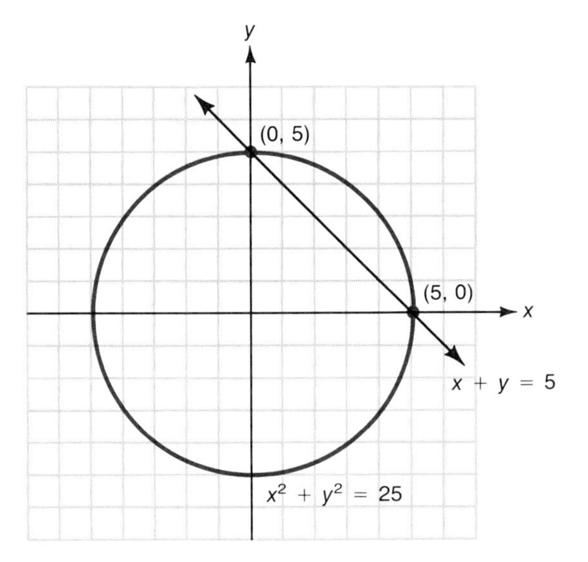

**2.** $\begin{cases} x + y = -7 \\ y = x^2 - 4x - 5 \end{cases}$

**Solution** Solve the linear equation for $y$; then substitute. (In this case, the quadratic equation is already solved for $y$, and the substitution could be made the other way.)

$$\begin{aligned} y &= -x - 7 \\ -x - 7 &= x^2 - 4x - 5 \\ 0 &= x^2 - 3x + 2 \\ 0 &= (x - 2)(x - 1) \end{aligned}$$

$$\begin{cases} x = 2 \\ y = -2 - 7 = -9 \end{cases} \quad \text{or} \quad \begin{cases} x = 1 \\ y = -1 - 7 = -8 \end{cases}$$

The solutions are $(2,-9)$ and $(1,-8)$.

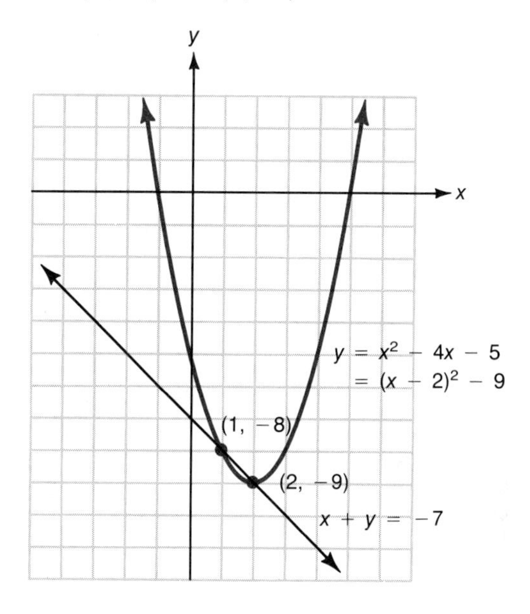

**3.** $\begin{cases} x^2 - y^2 = 4 \\ x^2 + y^2 = 36 \end{cases}$

**Solution** Here, addition will eliminate $y^2$.

$$\begin{array}{r} x^2 - y^2 = \phantom{0}4 \\ \underline{x^2 + y^2 = 36} \\ 2x^2 \qquad = 40 \end{array}$$

$$x^2 = 20$$
$$x^2 = \pm\sqrt{20} = \pm 2\sqrt{5}$$

$x = 2\sqrt{5}$, $\quad 20 + y^2 = 36$ $\quad$ or $\quad$ $x = -2\sqrt{5}$, $\quad 20 + y^2 = 36$

$\qquad y^2 = 16 \qquad\qquad\qquad y^2 = 16$

$\qquad y = \pm 4 \qquad\qquad\qquad y = \pm 4$

There are four points of intersection:

$$(2\sqrt{5},4), \qquad (2\sqrt{5},-4), \qquad (-2\sqrt{5},4), \qquad (-2\sqrt{5},-4)$$

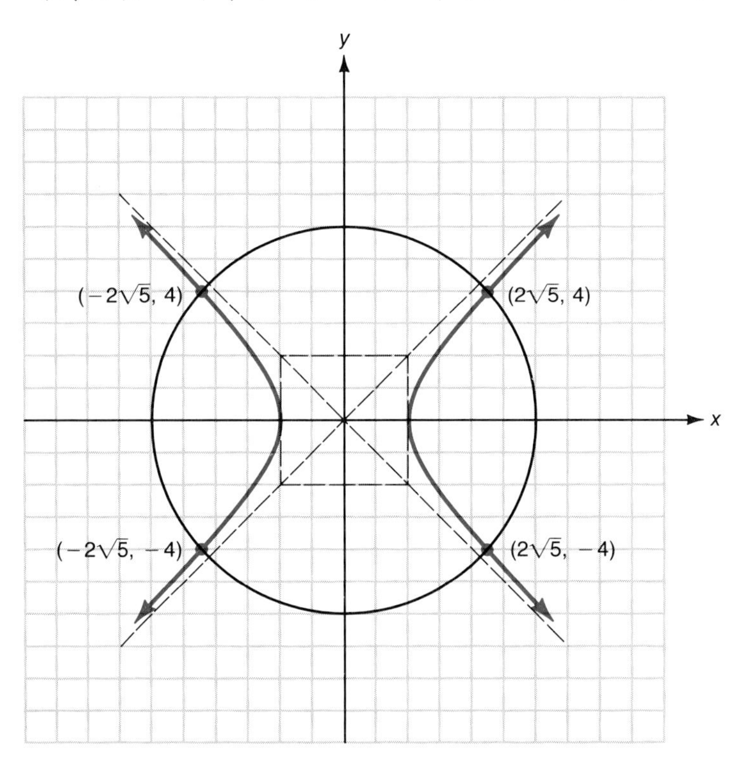


■

**Practice Problems**

Solve each of the following systems algebraically.

1. $\begin{cases} y = x^2 - 4 \\ x - y = 2 \end{cases}$

2. $\begin{cases} x^2 + y^2 = 72 \\ x = y^2 \end{cases}$

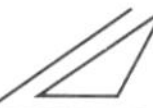

## EXERCISES 10.4

Solve each of the systems of equations in Exercises 1–16. Sketch the graphs.

1. $\begin{cases} y = x^2 + 1 \\ 2x + y = 4 \end{cases}$

2. $\begin{cases} y = 3 - x^2 \\ x + y = -3 \end{cases}$

3. $\begin{cases} y = 2 - x \\ y = (x - 2)^2 \end{cases}$

4. $\begin{cases} x^2 + y^2 = 25 \\ y + x + 5 = 0 \end{cases}$

5. $\begin{cases} x^2 + y^2 = 20 \\ x - y = 2 \end{cases}$

6. $\begin{cases} x^2 - y^2 = 16 \\ 3x + 5y = 0 \end{cases}$

7. $\begin{cases} y = x - 2 \\ x^2 = y^2 + 16 \end{cases}$

8. $\begin{cases} x^2 + 3y^2 = 12 \\ x = 3y \end{cases}$

9. $\begin{cases} x^2 + y^2 = 9 \\ x^2 - y^2 = 9 \end{cases}$

10. $\begin{cases} x^2 + y^2 = 9 \\ x^2 - y + 3 = 0 \end{cases}$

11. $\begin{cases} 4x^2 + y^2 = 25 \\ 3x - y^2 + 3 = 0 \end{cases}$

12. $\begin{cases} x^2 - 4y^2 = 9 \\ x + 2y^2 = 3 \end{cases}$

13. $\begin{cases} x^2 + y^2 + 4x - 2y = 4 \\ x + y = 2 \end{cases}$

14. $\begin{cases} x^2 + y^2 = 9 \\ x^2 + y^2 - 2x - 3 = 0 \end{cases}$

15. $\begin{cases} x^2 - y^2 = 5 \\ x^2 + 4y^2 = 25 \end{cases}$

16. $\begin{cases} 2x^2 - 3y^2 = 6 \\ 2x^2 + y^2 = 22 \end{cases}$

Solve each of the systems in Exercises 17–30.

17. $\begin{cases} x^2 - y^2 = 20 \\ x^2 - 9y = 0 \end{cases}$

18. $\begin{cases} x^2 + 5y^2 = 16 \\ x^2 + y^2 = 4x \end{cases}$

19. $\begin{cases} x^2 + y^2 = 10 \\ x^2 + y^2 - 4y + 2 = 0 \end{cases}$

20. $\begin{cases} x^2 + y^2 = 20 \\ 4x + 8 = y^2 \end{cases}$

21. $\begin{cases} 2x^2 - y^2 = 7 \\ 2x^2 + y^2 = 29 \end{cases}$

22. $\begin{cases} y = x^2 + 2x + 2 \\ 2x + y = 2 \end{cases}$

23. $\begin{cases} 4y + 10x^2 + 7x - 8 = 0 \\ 6x - 8y + 1 = 0 \end{cases}$

24. $\begin{cases} x^2 + y^2 - 4x + 6y + 3 = 0 \\ 2x - y - 2 = 0 \end{cases}$

25. $\begin{cases} x^2 + y^2 - 4y = 16 \\ x - y = 0 \end{cases}$

26. $\begin{cases} 4x^2 + y^2 = 11 \\ y = 4x^2 - 9 \end{cases}$

27. $\begin{cases} x^2 - y^2 - 2y = 22 \\ 2x + 5y + 5 = 0 \end{cases}$

28. $\begin{cases} x^2 + y^2 - 6y = 0 \\ 2x^2 - y^2 + 15 = 0 \end{cases}$

29. $\begin{cases} y = x^2 - 2x + 3 \\ y = -x^2 + 2x + 3 \end{cases}$

30. $\begin{cases} y^2 = x^2 - 5 \\ 4x^2 - y^2 = 32 \end{cases}$

**Answers to Practice Problems** 1. $(-1,-3)$ and $(2,0)$ 2. $(8,2\sqrt{2})$ and $(8,-2\sqrt{2})$

## CHAPTER 10 SUMMARY

### Key Terms and Formulas

**Conic sections** are curves on a plane that are found when the plane intersects a cone. Four such sections are the circle, ellipse, parabola, and hyperbola. The equations representing these curves are called **quadratic equations.** [10.1]

Equations of the form

$$x = ay^2 + by + c \qquad \text{where } a \neq 0$$

or

$$x = a(y - k)^2 + h$$

represent **parabolas.** The parabola opens left (if $a < 0$) or right (if $a > 0$) with vertex at $(h,k)$. The line $y = k$ is the line of symmetry. [10.1]

The formula for the **distance between two points** $(x_2,y_2)$ and $(x_1,y_1)$ is

$$d = \sqrt{(x_2 - x_1)^2 + (y_2 - y_1)^2} \quad [10.1]$$

A **circle** is the set of all points in a plane that are a fixed distance (called the **radius**) from a fixed point (called the **center**). The standard form for the equation of a circle with center at $(h,k)$ and radius $r$ is

$$(x - h)^2 + (y - k)^2 = r^2 \quad [10.1]$$

The standard form for the equation of an ellipse with center at the origin is

$$(1)\ \frac{x^2}{a^2} + \frac{y^2}{b^2} = 1 \qquad \left(\text{or } (2)\ \frac{x^2}{b^2} + \frac{y^2}{a^2} = 1\right) \qquad \text{where } a^2 \geq b^2$$

In equations of form (1), the $x$-intercepts are $(a,0)$ and $(-a,0)$ and the segment of length $2a$ joining these intercepts is called the **major axis.** The **minor axis** is the segment of length $2b$ joining the $y$-intercepts $(0,b)$ and $(0,-b)$. [10.3]

The standard form for the equation of a **hyperbola** with its center at $(0,0)$ is

$$\frac{x^2}{a^2} - \frac{y^2}{b^2} = 1 \qquad \text{or} \qquad \frac{y^2}{b^2} - \frac{x^2}{a^2} = 1 \quad [10.3]$$

The lines $y = \frac{b}{a}x$ and $y = -\frac{b}{a}x$ are **asymptotes** of the hyperbola. [10.3]

### Procedures

Systems of quadratic equations (in this text) can be solved by either substitution or addition. [10.4]

## CHAPTER 10 REVIEW

In each of the Exercises 1–6, find the vertex, the $x$-intercept, the $y$-intercept, and the line of symmetry. Then draw the graph. [10.1]

**1.** $x - 2 = y^2$
**2.** $x = 2y^2 + 1$
**3.** $x + 4 = (y - 1)^2$
**4.** $y + 2 = (x + 1)^2$
**5.** $x = y^2 - 2y - 5$
**6.** $y = 9x^2 - 12x + 4$

In Exercises 7–9, find the distance between the two points. [10.2]

**7.** $(2,3)$, $(-4,1)$
**8.** $(5,-2)$, $(1,6)$
**9.** $(3,3)$, $(-7,5)$

**10.** Show that the triangle determined by the points $A(-2,-2)$, $B(0,2)$, and $C(5,-3)$ is an isosceles triangle. [10.2]

**11.** Find the perimeter of the triangle determined by the points $A(1,3)$, $B(-2,2)$, and $C(4,6)$. [10.2]

**12.** Show that the triangle determined by the points $A(-3,1)$, $B(0,-5)$, and $C(1,3)$ is a right triangle. [10.2]

Find equations for each of the circles in Exercises 13–16. [10.2]

**13.** Center (0,0); $r = 5$

**14.** Center (0,0); $r = 2\sqrt{3}$

**15.** Center (1,2); $r = 3$

**16.** Center (−2,3); $r = \sqrt{7}$

Write each of the equations in Exercises 17–20 in standard form. Find the center and radius of the circle. Then sketch the graph. [10.2]

**17.** $x^2 + y^2 = 4$

**18.** $x^2 = 18 - y^2$

**19.** $x^2 + y^2 + 2x - 3 = 0$

**20.** $x^2 + y^2 - 4x + 6y - 3 = 0$

Write each of the equations in Exercises 21–30 in standard form; then sketch the graph. Write the equations and graph the asymptotes for each hyperbola. [10.3]

**21.** $4x^2 + y^2 = 4$

**22.** $x^2 + 4y^2 = 16$

**23.** $x^2 - 9y^2 = 9$

**24.** $9y^2 - 4x^2 = 36$

**25.** $4x^2 + 25y^2 = 100$

**26.** $16x^2 - 9y^2 = 144$

**27.** $16y^2 - 25x^2 = 400$

**28.** $25x^2 + 9y^2 = 225$

**29.** $x^2 + 3y^2 = 12$

**30.** $2x^2 - y^2 = 18$

Solve each of the systems of equations in Exercises 31–36. [10.4]

**31.** $\begin{cases} x^2 + y^2 = 10 \\ 2x + y = 1 \end{cases}$

**32.** $\begin{cases} 4x^2 + y^2 = 13 \\ x + y = 2 \end{cases}$

**33.** $\begin{cases} 4x^2 + y^2 = 16 \\ y = x^2 - 4 \end{cases}$

**34.** $\begin{cases} x^2 + y^2 = 12 \\ 4x^2 - y^2 = 8 \end{cases}$

**35.** $\begin{cases} x^2 + y^2 + 4x = 15 \\ y^2 = x + 1 \end{cases}$

**36.** $\begin{cases} x^2 + y^2 = 25 \\ x^2 + 4x + y^2 - 21 = 0 \end{cases}$

## CHAPTER 10 TEST

In each of the Exercises 1–4, find the vertex, the $x$-intercept, the $y$-intercept, and the line of symmetry. Then draw the graph.

**1.** $y - 3 = x^2$

**2.** $x = y^2 - 5$

**3.** $x + 4 = y^2 + 3y$

**4.** $y = x^2 - 2x - 8$

**5.** Find the distance between (5,−2) and (−4,1).

**6.** Show that the triangle determined by the points $A(-4,-3)$, $B(0,3)$ and $C(6,-1)$ is a right triangle.

Find the equation for each of the circles in Exercises 7 and 8.

**7.** Center (0,0); $r = 3\sqrt{2}$

**8.** Center (−3,1); $r = 5$

Write each of the equations in Exercises 9 and 10 in standard form. Find the center and radius of the circle; then sketch the graph.

**9.** $x^2 + y^2 = 9$

**10.** $x^2 + y^2 - 2y - 8 = 0$

Write each of the equations in Exercises 11–16 in standard form; then sketch the graph. Write the equations and graph the asymptotes for each hyperbola.

**11.** $9x^2 - 4y^2 = 36$

**12.** $x^2 + 4y^2 = 9$

**13.** $3x^2 + 3y^2 = 12$

**14.** $16y^2 - 9x^2 = 144$

**15.** $x - y^2 + 6y = 7$

**16.** $25x^2 + 4y^2 = 100$

Solve each of the systems of equations in Exercises 17–20.

**17.** $\begin{cases} x^2 + y^2 = 29 \\ x - y = 3 \end{cases}$

**18.** $\begin{cases} x^2 + 2y^2 = 4 \\ x = y^2 - 2 \end{cases}$

**19.** $\begin{cases} x^2 + y^2 = 25 \\ x^2 - y^2 = 7 \end{cases}$

**20.** $\begin{cases} x^2 + y^2 + 2y = 8 \\ x^2 - 2y^2 = 7 \end{cases}$

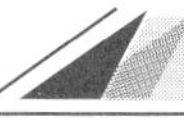

## CUMULATIVE REVIEW (10)

Factor completely Exercises 1–3.

**1.** $2x^3 + 54$

**2.** $2 + 9x^{-1} - 35x^{-2}$

**3.** $x^3 - 4x^2 + 3x - 12$

Perform the indicated operations and simplify in Exercises 4 and 5.

**4.** $2\sqrt{12} + 5\sqrt{108} - 7\sqrt{27}$

**5.** $3\sqrt{48x} - 2\sqrt{75x} + 5\sqrt{24}$

**6.** Find an equation for the line parallel to $5x - 2y = 8$ and passing through $(-2,3)$.

**7.** Find an equation for the line perpendicular to $4x + 3y = 8$ and passing through $(4,-1)$.

**8.** Determine the inverse of the function: $f(x) = \dfrac{1}{x + 2}$

Solve each of the equations in Exercises 9–12.

**9.** $x - 6x^{1/2} + 8 = 0$

**10.** $x = 1 + \sqrt{3x + 15}$

**11.** $6e^{-0.04x} = 23.4$

**12.** $\log(x + 1) + \log(x - 2) = 1$

**13.** Solve the inequality: $\dfrac{x^2 - 4x - 5}{x} \geq 0$

**14.** Solve the system of equations: $\begin{cases} 2x - 3y = 4 \\ 3x - 4y = 7 \end{cases}$

Graph each of the equations in Exercises 15–18.

**15.** $5x + 2y = 8$

**16.** $y = 2x^2 - 4x - 3$

**17.** $4x^2 - y^2 = 16$

**18.** $x^2 + 2x + y^2 - 2y = 4$

**19.** Kevin purchased two shirts and one pair of trousers for a total of \$96.00. To purchase three shirts and two pairs of trousers, it would have cost \$165.50. What was the price of each?

**20.** The population of a city $t$ years from now is estimated to be $25{,}000e^{0.03t}$. In how many years from now will the population be 40,000?

CHAPTER

# 11 MATRICES AND DETERMINANTS

## MATHEMATICAL CHALLENGES

Find all possible sets of four consecutive integers such that the sum of the cubes of the smallest three is the cube of the fourth.

Sheryl is $\frac{1}{2}$ as old as Deirdre will be in 7 years. Fourteen years from now, Sheryl will be $\frac{6}{7}$ as old as Deirdre is now. How old was Deirdre when Sheryl was exactly $\frac{1}{2}$ her age?

Laura can build a brick wall in exactly 9 hours if she works alone. Crystale can build it alone in 10 hours. If they work together their combined rate decreases by 10 bricks an hour and they can finish the wall in exactly 5 hours. How many bricks are in the wall?

## CHAPTER OUTLINE



In Chapter 7, we discussed the methods of graphing, substitution, and addition for solving systems of linear equations. While these methods are certainly important and useful, they become somewhat difficult to apply with large numbers of equations and variables. In this chapter, two new topics, determinants and matrices, are introduced that provide powerful methods for solving systems with any number of variables. These new methods are particularly adaptable to computer programming.

General techniques for evaluating large determinants are left for later courses, and we will restrict our discussion to the evaluation of $2 \times 2$ and $3 \times 3$ determinants. With these determinants we will apply a method, called Cramers' Rule, for solving systems of two equations in two variables and systems of three equations in three variables. Similarly, our discussion of matrices and the Gaussian elimination method for solving systems will be limited in size and scope. The student should understand that we are laying groundwork here and more general applications can be expected in higher level courses in mathematics and economics.

## 11.1 Determinants

**OBJECTIVES**

In this section, you will be learning to:

1. Evaluate second- and third-order determinants.
2. Solve equations involving determinants.

Certain problems in business dealing with minimizing costs and maximizing profits and various problems in mathematics and physics lead to systems of equations or systems of inequalities. In many cases, the solutions to these systems depend on the evaluation of one or more **determinants.**

***Determinant*** A **determinant** is a real number associated with a square array of real numbers and is indicated by enclosing the array between two vertical bars.

Examples of determinants are

$$\begin{vmatrix} 3 & 4 \\ 7 & -2 \end{vmatrix} \qquad \begin{vmatrix} 1 & 6 & -3 \\ 4 & 6 & 5 \\ -1 & -1 & -1 \end{vmatrix} \qquad \begin{vmatrix} 0 & 5 & 6 & 1 \\ 9 & 10 & -2 & 14 \\ 8 & 0 & 0 & 3 \\ 4 & -2 & -1 & 4 \end{vmatrix}$$

**(a)** **(b)** **(c)**

Example (a) is a **second-order,** or 2 by 2 ($2 \times 2$), **determinant** and has two rows and two columns.

column 1 column 2

$$\begin{matrix} \text{row 1} \rightarrow \\ \text{row 2} \rightarrow \end{matrix} \begin{vmatrix} 3 & 4 \\ 7 & -2 \end{vmatrix} \qquad \begin{vmatrix} 3 & 4 \\ 7 & -2 \end{vmatrix}$$

Example (b) is a **third-order,** or 3 by 3 ($3 \times 3$), **determinant** and has three rows and three columns. Example (c) is a **fourth-order,** or 4 by 4 ($4 \times 4$), **determinant** and has four rows and four columns. A determinant may be of any order $n \geq 2$ where $n$ is a positive integer.

**Every determinant has a value.** In this text, we will evaluate second- and third-order determinants only.

***Value of a Second-order Determinant***

$$\begin{vmatrix} a_1 & b_1 \\ a_2 & b_2 \end{vmatrix} = a_1b_2 - a_2b_1$$

**EXAMPLES** Evaluate the following 2 by 2 determinants.

**1.** $\begin{vmatrix} 3 & 4 \\ 7 & -2 \end{vmatrix} = 3(-2) - 7(4) = -6 - 28 = -34$

**2.** $\begin{vmatrix} -5 & -\frac{1}{2} \\ 6 & 3 \end{vmatrix} = -5(3) - 6\left(-\frac{1}{2}\right) = -15 + 3 = -12$

**3.** $\begin{vmatrix} 1 & 7 \\ 2 & 14 \end{vmatrix} = 1(14) - 2(7) = 14 - 14 = 0$

■

***Value of a Third-order Determinant***

$$\begin{vmatrix} a_1 & b_1 & c_1 \\ a_2 & b_2 & c_2 \\ a_3 & b_3 & c_3 \end{vmatrix} = a_1 \begin{vmatrix} b_2 & c_2 \\ b_3 & c_3 \end{vmatrix} - b_1 \begin{vmatrix} a_2 & c_2 \\ a_3 & c_3 \end{vmatrix} + c_1 \begin{vmatrix} a_2 & b_2 \\ a_3 & b_3 \end{vmatrix}$$

In the definition, each $2 \times 2$ determinant is found by mentally crossing out both the row and the column that contains the coefficient.

$$\begin{vmatrix} a_1 & b_1 & c_1 \\ a_2 & b_2 & c_2 \\ a_3 & b_3 & c_3 \end{vmatrix} \rightarrow a_1 \begin{vmatrix} b_2 & c_2 \\ b_3 & c_3 \end{vmatrix}$$

$$\begin{vmatrix} a_1 & b_1 & c_1 \\ a_2 & b_2 & c_2 \\ a_3 & b_3 & c_3 \end{vmatrix} \rightarrow -b_1 \begin{vmatrix} a_2 & c_2 \\ a_3 & c_3 \end{vmatrix}$$

$$\begin{vmatrix} a_1 & b_1 & c_1 \\ a_2 & b_2 & c_2 \\ a_3 & b_3 & c_3 \end{vmatrix} \rightarrow c_1 \begin{vmatrix} a_2 & b_2 \\ a_3 & b_3 \end{vmatrix}$$

(**WARNING:** The negative sign in the middle expression is a critical part of the definition and is a source of error for many students. Be careful.)

**EXAMPLES** Evaluate the following $3 \times 3$ determinants.

**4.** $\begin{vmatrix} 5 & 1 & -4 \\ 2 & 6 & 3 \\ 2 & 2 & 1 \end{vmatrix}$ Mentally delete the shaded region.

$$= 5\begin{vmatrix} 5 & 1 & -4 \\ 2 & 6 & 3 \\ 2 & 2 & 1 \end{vmatrix} - 1\begin{vmatrix} 5 & 1 & -4 \\ 2 & 6 & 3 \\ 2 & 2 & 1 \end{vmatrix} - 4\begin{vmatrix} 5 & 1 & -4 \\ 2 & 6 & 3 \\ 2 & 2 & 1 \end{vmatrix}$$

$$= 5\begin{vmatrix} 6 & 3 \\ 2 & 1 \end{vmatrix} - 1\begin{vmatrix} 2 & 3 \\ 2 & 1 \end{vmatrix} - 4\begin{vmatrix} 2 & 6 \\ 2 & 2 \end{vmatrix}$$

$$= 5(6 \cdot 1 - 2 \cdot 3) - 1(2 \cdot 1 - 2 \cdot 3) - 4(2 \cdot 2 - 2 \cdot 6)$$
$$= 5(6 - 6) - 1(2 - 6) - 4(4 - 12)$$
$$= 5(0) - 1(-4) - 4(-8)$$
$$= 0 + 4 + 32$$
$$= 36$$

**5.** $$\begin{vmatrix} 6 & -2 & 4 \\ 1 & 7 & 0 \\ -3 & 2 & -1 \end{vmatrix} = 6\begin{vmatrix} 7 & 0 \\ 2 & -1 \end{vmatrix} + 2\begin{vmatrix} 1 & 0 \\ -3 & -1 \end{vmatrix} + 4\begin{vmatrix} 1 & 7 \\ -3 & 2 \end{vmatrix}$$
$$= 6(-7 - 0) + 2(-1 + 0) + 4(2 + 21)$$
$$= -42 - 2 + 92$$
$$= 48$$

**6.** Find $x$ if $\begin{vmatrix} 2 & 3 & 0 \\ 6 & x & 5 \\ 1 & -2 & 9 \end{vmatrix} = 53$

**Solution** $$\begin{vmatrix} 2 & 3 & 0 \\ 6 & x & 5 \\ 1 & -2 & 9 \end{vmatrix} = 2\begin{vmatrix} x & 5 \\ -2 & 9 \end{vmatrix} - 3\begin{vmatrix} 6 & 5 \\ 1 & 9 \end{vmatrix} + 0\begin{vmatrix} 6 & x \\ 1 & -2 \end{vmatrix}$$
$$= 2(9x + 10) - 3(54 - 5) + 0$$
$$= 18x + 20 - 162 + 15$$
$$= 18x - 127$$

$$18x - 127 = 53$$
$$18x = 180$$
$$x = 10$$

■

The technique used to evaluate $3 \times 3$ determinants may be used (with appropriate adjustments) to evaluate any $n \times n$ determinant. Also, there are techniques for simplifying determinants and rules for arithmetic with determinants. That is, determinants may be added, subtracted, multiplied, and divided. The general rules governing these operations are discussed in courses in precalculus mathematics, finite mathematics, and linear algebra.

**Practice Problems** Evaluate each of the following determinants.

**1.** $\begin{vmatrix} -3 & 2 \\ 4 & 7 \end{vmatrix}$ **2.** $\begin{vmatrix} 6 & 3 \\ 4 & 2 \end{vmatrix}$ **3.** $\begin{vmatrix} 1 & 4 & 0 \\ 2 & -1 & 5 \\ 0 & 7 & -1 \end{vmatrix}$

## EXERCISES 11.1

Evaluate the determinants in Exercises 1–20.

**1.** $\begin{vmatrix} 2 & 7 \\ 4 & 3 \end{vmatrix}$ **2.** $\begin{vmatrix} 7 & 3 \\ 8 & 5 \end{vmatrix}$ **3.** $\begin{vmatrix} -5 & 2 \\ 4 & 8 \end{vmatrix}$ **4.** $\begin{vmatrix} -6 & 5 \\ 4 & 0 \end{vmatrix}$

**5.** $\begin{vmatrix} 1 & 3 \\ -2 & 5 \end{vmatrix}$ **6.** $\begin{vmatrix} 7 & 2 \\ 3 & -6 \end{vmatrix}$ **7.** $\begin{vmatrix} 6 & 3 \\ -11 & -5 \end{vmatrix}$ **8.** $\begin{vmatrix} 2 & 3 \\ 3 & -4 \end{vmatrix}$

**9.** $\begin{vmatrix} 9 & 4 \\ 4 & 7 \end{vmatrix}$ **10.** $\begin{vmatrix} 3 & -4 \\ 8 & -6 \end{vmatrix}$ **11.** $\begin{vmatrix} 0 & -1 & 2 \\ 3 & 5 & -7 \\ -3 & 4 & 1 \end{vmatrix}$ **12.** $\begin{vmatrix} 1 & 0 & -1 \\ -2 & 3 & 5 \\ 6 & -3 & 4 \end{vmatrix}$

**13.** $\begin{vmatrix} 1 & -1 & 2 \\ -2 & 5 & -7 \\ 6 & 4 & 1 \end{vmatrix}$ **14.** $\begin{vmatrix} 2 & -1 & -3 \\ 5 & 9 & 4 \\ 7 & 6 & -2 \end{vmatrix}$ **15.** $\begin{vmatrix} 2 & 1 & 3 \\ 3 & 4 & 5 \\ 1 & 7 & 2 \end{vmatrix}$ **16.** $\begin{vmatrix} -3 & 2 & 1 \\ 1 & -4 & -1 \\ 2 & 5 & 3 \end{vmatrix}$

**17.** $\begin{vmatrix} 2 & 1 & -1 \\ 4 & 3 & 2 \\ 1 & 5 & 5 \end{vmatrix}$ **18.** $\begin{vmatrix} 6 & 7 & 1 \\ 0 & 3 & 3 \\ 4 & 1 & -5 \end{vmatrix}$ **19.** $\begin{vmatrix} 3 & -1 & -1 \\ 2 & 4 & 1 \\ -1 & 1 & 2 \end{vmatrix}$ **20.** $\begin{vmatrix} 2 & 3 & 2 \\ 1 & -1 & 5 \\ 0 & 5 & 1 \end{vmatrix}$

Solve for $x$ in Exercises 21–25.

**21.** $\begin{vmatrix} 1 & 3 & 4 \\ 2 & x & 3 \\ 1 & 3 & 5 \end{vmatrix} = 1$ **22.** $\begin{vmatrix} -2 & -1 & 1 \\ x & 1 & -1 \\ 4 & 3 & -2 \end{vmatrix} = 7$ **23.** $\begin{vmatrix} 1 & x & x \\ 2 & -2 & 1 \\ -1 & 3 & 2 \end{vmatrix} = 0$

**24.** $\begin{vmatrix} x & x & 1 \\ 1 & 5 & 0 \\ 0 & 1 & -2 \end{vmatrix} = -15$ **25.** $\begin{vmatrix} 3 & 1 & -2 \\ 1 & x & 4 \\ 2 & x & 0 \end{vmatrix} = 38$

The determinant $\begin{vmatrix} x & y & 1 \\ x_1 & y_1 & 1 \\ x_2 & y_2 & 1 \end{vmatrix} = 0$ is an equation of the line passing through $P_1(x_1,y_1)$ and $P_2(x_2,y_2)$. Find an equation for the line determined by the pairs of points given in Exercises 26–28.

**26.** $(3,2)$, $(-1,4)$ **27.** $(-2,1)$, $(5,3)$ **28.** $(4,-4)$, $(0,6)$

The area of the triangle having the vertices $P_1(x_1,y_1)$, $P_2(x_2,y_2)$, and $P_3(x_3,y_3)$ is given by the absolute value of $\frac{1}{2}\begin{vmatrix} x_1 & y_1 & 1 \\ x_2 & y_2 & 1 \\ x_3 & y_3 & 1 \end{vmatrix}$. Find the area of the triangles determined by the sets of vertices given in Exercises 29 and 30.

**29.** $(3,1)$, $(5,2)$, $(1,-1)$ **30.** $(4,0)$, $(7,1)$, $(5,-2)$

**Answers to Practice Problems** 1. $-29$ 2. 0 3. $-26$

## 11.2 Cramer's Rule

**OBJECTIVE**

In this section, you will be learning to solve systems of linear equations by using Cramer's Rule.

For illustration, the following system of linear equations is solved leaving the indicated products and sums of constant coefficients.

Eliminating $y$ gives

$$\begin{cases} [1] & 2x + 3y = -5 \\ [-3] & 4x + y = 5 \end{cases} \qquad \begin{aligned} 1 \cdot 2x + 1 \cdot 3y &= 1(-5) \\ -3 \cdot 4x - 3 \cdot 1y &= -3(5) \\ \hline (1 \cdot 2 - 3 \cdot 4)x &= 1(-5) - 3(5) \\ x &= \frac{1(-5) - 3(5)}{1 \cdot 2 - 3 \cdot 4} \end{aligned}$$

Eliminating $x$ gives

$$\begin{cases} [-4] & 2x + 3y = -5 \\ [2] & 4x + y = 5 \end{cases} \qquad \begin{aligned} -4 \cdot 2x - 4 \cdot 3y &= -4(-5) \\ 2 \cdot 4x + 2 \cdot 1y &= 2(5) \\ \hline (2 \cdot 1 - 4 \cdot 3)y &= 2(5) - 4(-5) \\ y &= \frac{2(5) - 4(-5)}{2 \cdot 1 - 4 \cdot 3} \end{aligned}$$

Notice that the **denominators** for both $x$ and $y$ are the same number. This number is the value of the determinant formed by the coefficients of the variables with the equations in standard form.

$$D = \begin{vmatrix} 2 & 3 \\ 4 & 1 \end{vmatrix} = 2 \cdot 1 - 4 \cdot 3 = -10$$

The numerators are

$$D_x = \begin{vmatrix} -5 & 3 \\ 5 & 1 \end{vmatrix} = 1(-5) - 3(5) = -20$$

and

$$D_y = \begin{vmatrix} 2 & -5 \\ 4 & 5 \end{vmatrix} = 2(5) - 4(-5) = 30$$

So we can write

$$x = \frac{D_x}{D} = \frac{-20}{-10} = 2 \qquad \text{and} \qquad y = \frac{D_y}{D} = \frac{30}{-10} = -3$$

**The determinant $D_x$ is formed as follows:**

1. Form $D$, the determinant of the coefficients, and
2. replace the coefficients of $x$ with the corresponding constants on the right of the equal signs.

**The determinant $D_y$ is formed as follows:**

1. Form $D$, the determinant of the coefficients, and
2. replace the coefficients of $y$ with the corresponding constants on the right of the equal signs.

Using determinants to solve systems of linear equations is called **Cramer's Rule.** We state Cramer's Rule here only for $2 \times 2$ systems and $3 \times 3$ systems, but it applies to all $n \times n$ systems of linear equations.

***Cramer's Rule for 2 × 2 Systems***

For the system $\begin{cases} a_1x + b_1y = k_1 \\ a_2x + b_2y = k_2 \end{cases}$

where $$D = \begin{vmatrix} a_1 & b_1 \\ a_2 & b_2 \end{vmatrix}$$

$$D_x = \begin{vmatrix} k_1 & b_1 \\ k_2 & b_2 \end{vmatrix}$$

and $$D_y = \begin{vmatrix} a_1 & k_1 \\ a_2 & k_2 \end{vmatrix}$$

if $D \neq 0$, then

$$x = \frac{D_x}{D} \quad \text{and} \quad y = \frac{D_y}{D}$$

is a unique solution of the system.

***Cramer's Rule for 3 × 3 Systems***

For the system $\begin{cases} a_1x + b_1y + c_1z = k_1 \\ a_2x + b_2y + c_2z = k_2 \\ a_3x + b_3y + c_3z = k_3 \end{cases}$

where $$D = \begin{vmatrix} a_1 & b_1 & c_1 \\ a_2 & b_2 & c_2 \\ a_3 & b_3 & c_3 \end{vmatrix}$$

$$D_x = \begin{vmatrix} k_1 & b_1 & c_1 \\ k_2 & b_2 & c_2 \\ k_3 & b_3 & c_3 \end{vmatrix}$$

$$D_y = \begin{vmatrix} a_1 & k_1 & c_1 \\ a_2 & k_2 & c_2 \\ a_3 & k_3 & c_3 \end{vmatrix}$$

and $$D_z = \begin{vmatrix} a_1 & b_1 & k_1 \\ a_2 & b_2 & k_2 \\ a_3 & b_3 & k_3 \end{vmatrix}$$

if $D \neq 0$, then

$$x = \frac{D_x}{D}, \quad y = \frac{D_y}{D}, \quad \text{and} \quad z = \frac{D_z}{D}$$

is a unique solution of the system.

**If $D = 0$, there are two possibilities:**

**For the 2 × 2 Case**

1. If either $D_x \neq 0$ or $D_y \neq 0$, the system is **inconsistent and there is no solution.**
2. If both $D_x = 0$ and $D_y = 0$, the system is **dependent and there are an infinite number of solutions.**

**For the 3 × 3 Case**

1. If $D_x \neq 0$ or $D_y \neq 0$ or $D_z \neq 0$, the system **has no solution.**
2. If $D_x = 0$ and $D_y = 0$ and $D_z = 0$, the system **has an infinite number of solutions.**

EXAMPLES Using Cramer's Rule, solve the following systems of linear equations. The solutions are not checked here, but they can be checked by substituting the solutions into all the equations in the system.

1. $\begin{cases} 2x + y = 3 \\ 3x - 2y = 5 \end{cases}$

**Solution**

$$D = \begin{vmatrix} 2 & 1 \\ 3 & -2 \end{vmatrix} = -7, \quad D_x = \begin{vmatrix} 3 & 1 \\ 5 & -2 \end{vmatrix} = -11, \quad D_y = \begin{vmatrix} 2 & 3 \\ 3 & 5 \end{vmatrix} = 1$$

$$x = \frac{D_x}{D} = \frac{-11}{-7} = \frac{11}{7} \qquad y = \frac{D_y}{D} = \frac{1}{-7} = -\frac{1}{7}$$

2. $\begin{cases} 2x + \sqrt{2}y = 3\sqrt{2} \\ -x + 3\sqrt{2}y = 2\sqrt{2} \end{cases}$

**Solution** $D = \begin{vmatrix} 2 & \sqrt{2} \\ -1 & 3\sqrt{2} \end{vmatrix} = 7\sqrt{2}$

$$D_x = \begin{vmatrix} 3\sqrt{2} & \sqrt{2} \\ 2\sqrt{2} & 3\sqrt{2} \end{vmatrix} = 14$$

$$D_y = \begin{vmatrix} 2 & 3\sqrt{2} \\ -1 & 2\sqrt{2} \end{vmatrix} = 7\sqrt{2}$$

$$x = \frac{D_x}{D} = \frac{14}{7\sqrt{2}} = \sqrt{2} \qquad y = \frac{D_y}{D} = \frac{7\sqrt{2}}{7\sqrt{2}} = 1$$

**3.** $\begin{cases} x + 2y + 3z = 3 \\ 4x + 5y + 6z = -1 \\ 7x + 8y + 9z = 0 \end{cases}$

**Solution**

$$D = \begin{vmatrix} 1 & 2 & 3 \\ 4 & 5 & 6 \\ 7 & 8 & 9 \end{vmatrix} = 1\begin{vmatrix} 5 & 6 \\ 8 & 9 \end{vmatrix} - 2\begin{vmatrix} 4 & 6 \\ 7 & 9 \end{vmatrix} + 3\begin{vmatrix} 4 & 5 \\ 7 & 8 \end{vmatrix}$$

$$= 1(-3) - 2(-6) + 3(-3) = 0$$

$$D_x = \begin{vmatrix} 3 & 2 & 3 \\ -1 & 5 & 6 \\ 0 & 8 & 9 \end{vmatrix} = 3\begin{vmatrix} 5 & 6 \\ 8 & 9 \end{vmatrix} - 2\begin{vmatrix} -1 & 6 \\ 0 & 9 \end{vmatrix} + 3\begin{vmatrix} -1 & 5 \\ 0 & 8 \end{vmatrix}$$

$$= 3(-3) - 2(-8) + 3(-9) = -20$$

The system has no solution since $D = 0$ and $D_x \neq 0$.

**4.** $\begin{cases} x + y + 3z = 7 \\ 2x - y - 3z = -4 \\ 5x - 2y = -5 \end{cases}$

**Solution**

$$D = \begin{vmatrix} 1 & 1 & 3 \\ 2 & -1 & -3 \\ 5 & -2 & 0 \end{vmatrix}$$

$$= 1\begin{vmatrix} -1 & -3 \\ -2 & 0 \end{vmatrix} - 1\begin{vmatrix} 2 & -3 \\ 5 & 0 \end{vmatrix} + 3\begin{vmatrix} 2 & -1 \\ 5 & -2 \end{vmatrix} = -18$$

$$D_x = \begin{vmatrix} 7 & 1 & 3 \\ -4 & -1 & -3 \\ -5 & -2 & 0 \end{vmatrix}$$

$$= 7\begin{vmatrix} -1 & -3 \\ -2 & 0 \end{vmatrix} - 1\begin{vmatrix} -4 & -3 \\ -5 & 0 \end{vmatrix} + 3\begin{vmatrix} -4 & -1 \\ -5 & -2 \end{vmatrix} = -18$$

$$D_y = \begin{vmatrix} 1 & 7 & 3 \\ 2 & -4 & -3 \\ 5 & -5 & 0 \end{vmatrix}$$

$$= 1\begin{vmatrix} -4 & -3 \\ -5 & 0 \end{vmatrix} - 7\begin{vmatrix} 2 & -3 \\ 5 & 0 \end{vmatrix} + 3\begin{vmatrix} 2 & -4 \\ 5 & -5 \end{vmatrix} = -90$$

$$D_z = \begin{vmatrix} 1 & 1 & 7 \\ 2 & -1 & -4 \\ 5 & -2 & -5 \end{vmatrix}$$

$$= 1\begin{vmatrix} -1 & -4 \\ -2 & -5 \end{vmatrix} - 1\begin{vmatrix} 2 & -4 \\ 5 & -5 \end{vmatrix} + 7\begin{vmatrix} 2 & -1 \\ 5 & -2 \end{vmatrix} = -6$$

$$x = \frac{-18}{-18} = 1 \qquad y = \frac{-90}{-18} = 5 \qquad z = \frac{-6}{-18} = \frac{1}{3}$$

■

***Practice Problems***

**1.** Solve the following system using Cramer's Rule.

$$\begin{cases} 2x - y = 11 \\ x + y = -2 \end{cases}$$

**2.** Find $D_x$ for the following system.

$$\begin{aligned} x + 2y + z &= 0 \\ 2x + y - 2z &= 5 \\ 3x - y + z &= -3 \end{aligned}$$

## *EXERCISES 11.2*

Use Cramer's Rule to solve the following systems.

**1.** $\begin{cases} 2x - 5y = -7 \\ 3x - 2y = 6 \end{cases}$

**2.** $\begin{cases} 3x + 5y = 17 \\ x + 3y = 15 \end{cases}$

**3.** $\begin{cases} 6x - 4y = 5 \\ 3x + 8y = 0 \end{cases}$

**4.** $\begin{cases} 3x + 4y = 24 \\ 2x + y = 11 \end{cases}$

**5.** $\begin{cases} 3x + y = 1 \\ -9x - 3y = 2 \end{cases}$

**6.** $\begin{cases} 4x + 8y = 12 \\ 3x + 6y = 9 \end{cases}$

**7.** $\begin{cases} 12x + 4y = 3 \\ -10x + 3y = 7 \end{cases}$

**8.** $\begin{cases} 4x - 9y = 2 \\ 8x - 15y = 3 \end{cases}$

**9.** $\begin{cases} 2x + 3y = 4 \\ 3x - 4y = 5 \end{cases}$

**10.** $\begin{cases} 5x + 2y = 7 \\ 2x - 3y = 4 \end{cases}$

**11.** $\begin{cases} 7x + 3y = 9 \\ 4x + 8y = 11 \end{cases}$

**12.** $\begin{cases} 5x - 9y = 3 \\ 11x + 6y = 12 \end{cases}$

**13.** $\begin{cases} 6x - 13y = 21 \\ 5x - 12y = 18 \end{cases}$

**14.** $\begin{cases} 10x + 7y = 15 \\ 13x - 4y = 11 \end{cases}$

**15.** $\begin{cases} 8x - 9y = -14 \\ 15x + 6y = 7 \end{cases}$

**16.** $\begin{cases} 17x - 5y = 21 \\ 4x + 3y = 6 \end{cases}$

**17.** $\begin{cases} 0.8x + 0.3y = 4 \\ 0.9x - 1.2y = 5 \end{cases}$

**18.** $\begin{cases} 0.4x + 0.7y = 3 \\ 0.5x + y = 6 \end{cases}$

**19.** $\begin{cases} 1.6x - 4.5y = 1.5 \\ 0.4x + 1.2y = 3.1 \end{cases}$

**20.** $\begin{cases} 2.3x + 1.8y = 4.6 \\ 0.8x - 1.4y = 3.2 \end{cases}$

**21.** $\begin{cases} x - 2y - z = -7 \\ 2x + y + z = 0 \\ 3x - 5y + 8z = 13 \end{cases}$

**22.** $\begin{cases} 2x + 3y + z = 0 \\ 5x + y - 2z = 9 \\ 10x - 5y + 3z = 4 \end{cases}$

**23.** $\begin{cases} 5x - 4y + z = 17 \\ x + y + z = 4 \\ -10x + 8y - 2z = 11 \end{cases}$

**24.** $\begin{cases} 9x + 10y = 2 \\ 3x + 6z = 4 \\ -3y + 3z = 1 \end{cases}$

**25.** $\begin{cases} 2x - 3y - z = -4 \\ -x + 2y + z = 6 \\ x - y + 2z = 14 \end{cases}$

**26.** $\begin{cases} 2x - 3y - z = 4 \\ x - 2y - z = 1 \\ x - y + 2z = 9 \end{cases}$

**27.** $\begin{cases} 3x + 2y + z = 5 \\ 2x + y - 2z = 4 \\ 5x + 3y - z = 9 \end{cases}$

**28.** $\begin{cases} 8x + 3y + 2z = 15 \\ 3x + 5y + z = -4 \\ 2x + 3y = -7 \end{cases}$

**29.** $\begin{cases} 2x - y + 3z = 1 \\ 5x + 2y - z = 2 \\ x - 2y + 5z = 2 \end{cases}$

**30.** $\begin{cases} 2x + 3y + 2z = -5 \\ 2x - 2y + z = -1 \\ 5x + y + z = 1 \end{cases}$

**Answers to Practice Problems** **1.** $x = 3$, $y = -5$ **2.** $D_x = 0$

# 11.3 Matrices and Gaussian Elimination

**OBJECTIVES**

In this section, you will be learning to:

1. Identify the order of matrices.
2. Identify specified entries in matrices.
3. Transform matrices into triangular form using elementary row operations.
4. Solve systems of linear equations using the Gaussian elimination method.

In this section, we will discuss a method for solving systems of linear equations that involves the use of rectangular arrays of numbers called **matrices** (singular **matrix**). For the definition of a matrix, we introduce double subscript notation that indicates both the row and the column in which a particular number is located. For example, the notation $a_{23}$ (read "$a$ sub two three") indicates that the number is in the second row and third column of a matrix.

***Matrix***

An $m \times n$ (read "$m$ by $n$") matrix is a rectangular array of numbers with $m$ rows and $n$ columns in the form

$$A = \overbrace{\begin{bmatrix} a_{11} & a_{12} & a_{13} & \cdot & \cdot & \cdot & a_{1n} \\ a_{21} & a_{22} & a_{23} & \cdot & \cdot & \cdot & a_{2n} \\ a_{31} & a_{32} & a_{33} & \cdot & \cdot & \cdot & a_{3n} \\ \cdot & \cdot & \cdot & \cdot & \cdot & \cdot & \cdot \\ \cdot & \cdot & \cdot & \cdot & \cdot & \cdot & \cdot \\ a_{m1} & a_{m2} & a_{m3} & \cdot & \cdot & \cdot & a_{mn} \end{bmatrix}}^{n \text{ columns}} \Bigg\} m \text{ rows}$$

Matrix $A$ is said to be of **dimension $m \times n$** (or **order $m \times n$**) and each number $a_{ij}$ is said to be an **element** (or an **entry**) of the matrix. If $m = n$, then the matrix is said to be **square of order $n$.**

(**Note:** In this text, we will deal only with matrices in which each entry is a real number.)

## EXAMPLE

**1.** Identify the order of each matrix and find the indicated entry.

**a.** $A = \begin{bmatrix} 2 & 3 \\ 5 & -1 \end{bmatrix}$; $a_{12}$

**b.** $B = \begin{bmatrix} 4 & -1 & 2 \\ -1 & 1 & 3 \end{bmatrix}$; $b_{22}$

**c.** $C = \begin{bmatrix} 5 \\ 0 \\ 4 \\ 1 \end{bmatrix}$; $c_{21}$

**d.** $D = \begin{bmatrix} 3 & 6 & -2 & 1 \\ 0 & 4 & -1 & 5 \\ 2 & 7 & 8 & 0 \end{bmatrix}$; $d_{32}$

### Solution

**a.** $A$ is a square matrix of order 2; $a_{12} = 3$.
**b.** $B$ is of order $2 \times 3$; $b_{22} = 1$.
**c.** $C$ is of order $4 \times 1$; $c_{21} = 0$.
**d.** $D$ is of order $3 \times 4$; $d_{32} = 7$. ■

Now, consider the following system of linear equations in three variables written

1. in standard form with the constant terms on the right, and
2. with the like variables and constant terms aligned vertically.

$$\begin{aligned} y + z &= 6 \\ x + 5y - 4z &= 4 \\ x - 3y + 5z &= 7 \end{aligned}$$

There are two matrices associated with such a system: the **coefficient matrix** made up of the coefficients of the variables with 0 supplied for any missing variable, and the **augmented matrix** which includes the constant terms.

**Coefficient Matrix**

$$\begin{bmatrix} 0 & 1 & 1 \\ 1 & 5 & -4 \\ 1 & -3 & 5 \end{bmatrix}$$

**Augmented Matrix**

$$\left[\begin{array}{ccc|c} 0 & 1 & 1 & 6 \\ 1 & 5 & -4 & 4 \\ 1 & -3 & 5 & 7 \end{array}\right]$$

The **Gaussian elimination** method (named after the famous German mathematician Carl Friedrich Gauss, 1777–1855) for solving a system of linear equations makes use of the augmented matrix. To use this method, we employ the following three **elementary row operations** on matrices.

### Elementary Row Operations

1. Multiply a row by a nonzero constant.
2. Interchange any two rows.
3. Change a row by adding a multiple of another row to it.

If any elementary row operation is applied to an augmented matrix, the new matrix is said to be **row-equivalent** to the original matrix. The new matrix represents a system that is **equivalent** to the original system. That is, the two systems have the same solutions, if any. Note that the augmented matrix is essentially the system without the variables and the elementary row operations correspond to the technique of solving a system by adding that was discussed in Sections 7.1 and 7.3.

We also need to know that a matrix is in **triangular form** if its entries in the lower left triangular region are all 0s as shown here.

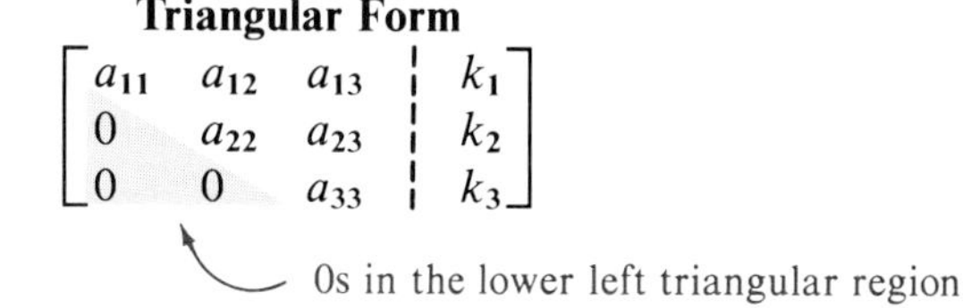


The following is an outline of the **Gaussian elimination method** for solving a system of linear equations. The method is valid for any number of equations with any number of variables and is particularly easy to program for computer use.

**Strategy for Gaussian Elimination**

1. Write the augmented matrix for the system.
2. Use elementary row operations to transform the matrix into triangular form.
3. Solve the corresponding system of equations by substituting back into the system.

If any row has all 0s except for the last entry on the right, then the original system of linear equations has no solution and is **inconsistent.** For example, the second row of the matrix

$$\left[\begin{array}{ccc|c} 1 & 3 & 4 & 1 \\ 0 & 0 & 0 & 3 \\ 0 & 0 & 5 & 10 \end{array}\right]$$

would correspond to the equation $0x + 0y + 0z = 3$ which is impossible.

## EXAMPLES

**2.** Solve the following system of linear equations using the Gaussian elimination method.

$$\begin{aligned} 2x - 3y - \phantom{2}z &= -4 \\ -x + 2y + \phantom{2}z &= 6 \\ x - \phantom{2}y + 2z &= 14 \end{aligned}$$

**Solution**

**Step 1** Write the augmented matrix.

$$\left[\begin{array}{rrr|r} 2 & -3 & -1 & -4 \\ -1 & 2 & 1 & 6 \\ 1 & -1 & 2 & 14 \end{array}\right]$$

**Step 2** Proceed to obtain the triangular form.

$$\left[\begin{array}{rrr|r} 1 & -1 & 2 & 14 \\ -1 & 2 & 1 & 6 \\ 2 & -3 & -1 & -4 \end{array}\right]$$

Get 1 in the upper left corner by exchanging row 1 and row 3. This will always be a first goal because the remaining 0s in the first column are easier to calculate.

$$\begin{array}{l} \\ R2 + R1 \longrightarrow \\ R3 - 2 \cdot R1 \longrightarrow \end{array} \left[\begin{array}{rrr|r} 1 & -1 & 2 & 14 \\ 0 & 1 & 3 & 20 \\ 0 & -1 & -5 & -32 \end{array}\right]$$

Get 0s under 1 in the first column. Add the entries in row 1 to the corresponding entries in row 2. Add $-2$ times the entries in row 1 to the corresponding entries in row 3. Note that the entries in row 1 are not changed.

$$\begin{array}{l} \\ \\ R3 + R2 \longrightarrow \end{array} \left[\begin{array}{rrr|r} 1 & -1 & 2 & 14 \\ 0 & 1 & 3 & 20 \\ 0 & 0 & -2 & -12 \end{array}\right]$$

Add the entries in row 2 to row 3 to arrive at the triangular form.

The matrix represents the system of linear equations

$$\begin{aligned} x - y + 2z &= 14 \\ y + 3z &= 20 \\ -2z &= -12 \end{aligned}$$

Solving the last equation for $z$ we get

$$\begin{aligned} -2z &= -12 \\ z &= 6 \end{aligned}$$

Substituting $z = 6$ into the equation $y + 3z = 20$ gives

$$\begin{aligned} y + 3(6) &= 20 \\ y &= 2 \end{aligned}$$

Substituting $z = 6$ and $y = 2$ into the equation $x - y + 2z = 14$ gives

$$\begin{aligned} x - 2 + 2(6) &= 14 \\ x &= 4 \end{aligned}$$

Thus, the solution is $x = 4$, $y = 2$, and $z = 6$, or we can write the solution in the form of an ordered triple (4,2,6).

**Note:** The triangular form of a matrix is not unique. However, the solutions to the system, if they exist, are unique regardless of the final form of the matrix. For example, multiplying the last row of the final matrix in Example 2 by $-\frac{1}{2}$ will give a different matrix but an equivalent one.

**3.** Solve the following system of linear equations using the Gaussian elimination method.

$$\begin{aligned} w + 2x + y - z &= 7 \\ x - y + 2z &= -7 \\ 2w + y &= 2 \\ w + x - 2y &= -8 \end{aligned}$$

**Solution**

**Step 1** Write the augmented matrix.

$$\left[\begin{array}{cccc|c} 1 & 2 & 1 & -1 & 7 \\ 0 & 1 & -1 & 2 & -7 \\ 2 & 0 & 1 & 0 & 2 \\ 1 & 1 & -2 & 0 & -8 \end{array}\right]$$

**Step 2** Proceed to obtain the triangular form.

$$\begin{array}{l} \\ \\ R3 - 2 \cdot R1 \longrightarrow \\ R4 - R1 \longrightarrow \end{array} \left[\begin{array}{cccc|c} 1 & 2 & 1 & -1 & 7 \\ 0 & 1 & -1 & 2 & -7 \\ 0 & -4 & -1 & 2 & -12 \\ 0 & -1 & -3 & 1 & -15 \end{array}\right]$$

1 is in the upper left corner. Add $-2$ times row 1 to row 3. Add $-1$ times row 1 to row 4. The leading 1 in the first column now has 0s under it.

$$\begin{array}{l} \\ \\ R3 + 4 \cdot R2 \longrightarrow \\ R4 + R2 \longrightarrow \end{array} \left[\begin{array}{cccc|c} 1 & 2 & 1 & -1 & 7 \\ 0 & 1 & -1 & 2 & -7 \\ 0 & 0 & -5 & 10 & -40 \\ 0 & 0 & -4 & 3 & -22 \end{array}\right]$$

Get 0s under the 1 in the second row and second column. Add 4 times row 2 to row 3. Add row 2 to row 4.

$$-\frac{1}{5} \cdot R3 \longrightarrow \left[\begin{array}{cccc|c} 1 & 2 & 1 & -1 & 7 \\ 0 & 1 & -1 & 2 & -7 \\ 0 & 0 & 1 & -2 & 8 \\ 0 & 0 & -4 & 3 & -22 \end{array}\right]$$

Multiply $-\frac{1}{5}$ times row 3.

$$R4 + 4 \cdot R3 \longrightarrow \left[\begin{array}{cccc|c} 1 & 2 & 1 & -1 & 7 \\ 0 & 1 & -1 & 2 & -7 \\ 0 & 0 & 1 & -2 & 8 \\ 0 & 0 & 0 & -5 & 10 \end{array}\right]$$

Add 4 times row 3 to row 4 to obtain the triangular form.

The matrix represents the system of linear equations

$$\begin{aligned} w + 2x + y - z &= 7 \\ x - y + 2z &= -7 \\ y - 2z &= 8 \\ -5z &= 10 \end{aligned}$$

Solving the last equation for $z$ and substituting back in the preceding equations will give the solutions

$$w = -1,\ x = 1,\ y = 4, \text{ and } z = -2$$

To check that no errors have been made in the elementary row operations, the values found for $w$, $x$, $y$, and $z$ can be substituted back into the original system of equations. If they do not satisfy every equation, then an error has been made and the steps must be retraced with a careful check of the arithmetic at each step. ■

**Practice Problem**

1. Solve the following system of equations by using the Gaussian elimination method.

$$\begin{aligned} x - 2y + 3z &= 4 \\ 2x + y \phantom{{}-z} &= 0 \\ 3x + y - z &= -4 \end{aligned}$$

**Answers to Practice Problem** 1. $x = -1,\ y = 2,\ z = 3$

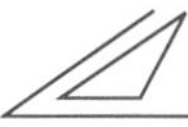

## EXERCISES 11.3

Identify the order of each matrix and find the indicated entry for each matrix in Exercises 1–10.

**1.** $\begin{bmatrix} 2 & 2 \\ 4 & 0 \end{bmatrix}$; $a_{21}$

**2.** $\begin{bmatrix} -3 & 5 \\ 1 & -1 \\ 3 & 2 \end{bmatrix}$; $a_{22}$

**3.** $\begin{bmatrix} -5 \\ 0 \\ 4 \end{bmatrix}$; $a_{31}$

**4.** $[10 \quad -3 \quad 5 \quad 17]$; $a_{13}$

**5.** $\begin{bmatrix} 7 & -2 & 7 & 2 \\ -5 & 3 & 0 & 2 \\ 0 & 4 & 11 & -8 \\ -3 & -9 & 2 & 6 \end{bmatrix}$; $a_{34}$

**6.** $\begin{bmatrix} -8 & 1 & -1 & 6 \\ 2 & 0 & 3 & -3 \\ -4 & -2 & 5 & 13 \end{bmatrix}$; $a_{24}$

**7.** $\begin{bmatrix} 0 & 3 & -1 & 5 \\ -2 & 10 & 9 & -4 \end{bmatrix}$; $a_{22}$

**8.** $\begin{bmatrix} 2 & -9 & 14 \\ -3 & 0 & -8 \\ 17 & -6 & 7 \end{bmatrix}$; $a_{33}$

**9.** $\begin{bmatrix} 22 & 17 & 13 \\ 16 & 21 & 19 \\ 20 & 18 & 23 \\ 31 & 15 & 27 \end{bmatrix}$; $a_{12}$

**10.** $\begin{bmatrix} 12 & 15 \\ 20 & 18 \\ 15 & 11 \\ 22 & 27 \end{bmatrix}$; $a_{41}$

Solve each system of equations in Exercises 11–26 by using the Gaussian elimination method.

**11.** $\begin{aligned} x - 3y + 2z &= 11 \\ -2x + 4y + z &= -3 \\ x - 2y + 3z &= 12 \end{aligned}$

**12.** $\begin{aligned} x + 2y - z &= 6 \\ 3x - y + 2z &= 9 \\ x + y + z &= 6 \end{aligned}$

**13.** $\begin{aligned} x + 2y + 3z &= 4 \\ x - y - z &= 0 \\ 4x - 3y + z &= 5 \end{aligned}$

**14.** $\begin{aligned} x + y - 2z &= -1 \\ 3x + 4y - 2z &= 0 \\ x - y + z &= 4 \end{aligned}$

**15.** $\begin{aligned} x - y - 2z &= 3 \\ x + 2y - z &= 5 \\ 2x - 3y - 2z &= 3 \end{aligned}$

**16.** $\begin{aligned} x - y + 5z &= -6 \\ x + 2z &= 0 \\ 6x + y + 3z &= 0 \end{aligned}$

**17.** $\begin{aligned} x - 3y - z &= -4 \\ 3x - 2y + z &= 1 \\ -2x + y + 2z &= 13 \end{aligned}$

**18.** $\begin{aligned} 2x - y - 5z &= -9 \\ x - 3y + 2z &= 0 \\ 3x + 2y + 10z &= 4 \end{aligned}$

**19.** $\begin{aligned} x - 2y + 3z &= 0 \\ x + y + 4z &= 7 \\ 2x - y + 5z &= 5 \end{aligned}$

**20.** $\begin{aligned} 2x - y + 5z &= -2 \\ 4x + y + 3z &= -2 \\ x + 3y - z &= 6 \end{aligned}$

**21.** $\begin{aligned} 3x + 4z &= 11 \\ x + y &= -2 \\ 2y + z &= -4 \end{aligned}$

**22.** $\begin{aligned} y + z &= 2 \\ x + y &= 5 \\ x + z &= 5 \end{aligned}$

**23.** $\begin{aligned} x - 3y - 3z + 2w &= 5 \\ x + 2y + z &= -2 \\ 2x - 3z - 2w &= 7 \\ y - z - 5w &= 6 \end{aligned}$

**24.** $\begin{aligned} 3x + y - z + 2w &= 6 \\ x - y + 2z - w &= -8 \\ 2y + 5z + w &= 2 \\ x + 3y + 3w &= 14 \end{aligned}$

**25.** $\begin{aligned} x + 3y - z + 2w &= 1 \\ 2x + 5y + 2z - w &= -1 \\ 2x - 6y + 5z &= -8 \\ -x + 2y - z + 2w &= -1 \end{aligned}$

**26.** $\begin{aligned} 4x + y + 3z - 2w &= 13 \\ x - 2y + z - 4w &= -3 \\ x + y + 4z + 2w &= 12 \\ -2x + 3y - z - 3w &= 5 \end{aligned}$

27. The sum of three numbers is 169. The first number is twelve more than the second number. The third number is fifteen less than the sum of the first and second numbers. Find the numbers.

28. Julie bought a pound of bacon, a dozen eggs, and a loaf of bread. The total cost was \$5.65. The eggs cost \$0.50 less than the bacon. The combined cost of the bread and eggs was \$1.07 more than the cost of the bacon. Find the cost of each.

29. A pizzeria sells three sizes of pizzas, small, medium and large. The pizzas sell for \$6.00, \$8.00, and \$9.50, respectively. One evening they sold 68 pizzas and received \$528.00. If they sold twice as many medium-sized pizzas as large-sized pizzas, how many of each size did they sell?

30. An investment firm is responsible for investing \$250,000 from an estate. The money is invested in three acounts paying 6%, 8%, and 11% interest. The amount invested in the 6%-account is \$5000 more than the total invested in the other two accounts. If the total annual interest is \$19,250, how much is invested in each account?

## 11.4 Operations with Matrices

**OBJECTIVES**

In this section, you will be learning to:

1. Determine the corresponding entries of equal matrices.
2. Multiply matrices by scalars.
3. Add and subtract matrices.
4. Multiply matrices.

In Section 11.3, matrices were used to solve systems of linear equations using elementary row operations. In this section, we will discuss the operations of multiplication by numbers (called scalars), matrix addition (and subtraction), and matrix multiplication. Applications using these operations with matrices are abundant in finite mathematics, linear algebra, and economics. We will leave the applications to those courses and concentrate here on developing the operational skills.

Matrices were defined using the double subscript notation $a_{ij}$ to represent the entry in row $i$ and column $j$ of an $m \times n$ matrix $A$. We now write

$$A = [a_{ij}]$$

to represent the entire $m \times n$ matrix $A$. Thus,

$$A = [a_{ij}] = \begin{bmatrix} a_{11} & a_{12} & a_{13} & \cdots & a_{1n} \\ a_{21} & a_{22} & a_{23} & \cdots & a_{2n} \\ a_{31} & a_{32} & a_{33} & \cdots & a_{3n} \\ \cdot & \cdot & \cdot & & \cdot \\ \cdot & \cdot & \cdot & & \cdot \\ \cdot & \cdot & \cdot & & \cdot \\ a_{m1} & a_{m2} & a_{m3} & \cdots & a_{mn} \end{bmatrix}$$

For two matrices to be **equal,** they must have the same dimensions and their corresponding entries must be the same.

**Definition**

If $A = [a_{ij}]$ and $B = [b_{ij}]$ are both matrices of dimension $m \times n$, then

$$A = B \text{ if and only if } a_{ij} = b_{ij}$$

for all $i = 1,2,3, \ldots, m$ and all $j = 1,2,3, \ldots, n$.

**EXAMPLE**

**1.** Find the value for each of the entries $a_{11}$, $a_{12}$, $a_{13}$, $a_{21}$, $a_{22}$, and $a_{23}$, given the matrix equation

$$\begin{bmatrix} a_{11} & a_{12} & a_{13} \\ a_{21} & a_{22} & a_{23} \end{bmatrix} = \begin{bmatrix} 3 & -1 & 0 \\ 2 & 4 & -5 \end{bmatrix}$$

**Solution** The matrices are equal if and only if the corresponding entries are equal. Therefore,

$$a_{11} = 3,\ a_{12} = -1,\ a_{13} = 0,\ a_{21} = 2,\ a_{22} = 4, \text{ and } a_{23} = -5.$$ ■

When working with matrices, numbers that are not entries are called **scalars.** To multiply a matrix by a scalar, we multiply each entry in the matrix by the scalar. In this text, the scalars will be real numbers.

***Multiplication by Scalars***

If $c$ is a scalar and $A = [a_{ij}]$ is a matrix of dimension $m \times n$, then the **scalar multiple** *$cA$ is defined to be the $m \times n$* matrix

$$cA = [ca_{ij}]$$

**EXAMPLE**

**2.** Given $A = \begin{bmatrix} 5 & 6 \\ 8 & -1 \end{bmatrix}$, find the scalar multiples (a) $3A$ and (b) $\frac{1}{2}A$.

**Solution**

**a.** $3A = \begin{bmatrix} 3 \cdot 5 & 3 \cdot 6 \\ 3 \cdot 8 & 3(-1) \end{bmatrix} = \begin{bmatrix} 15 & 18 \\ 24 & -3 \end{bmatrix}$

**b.** $\frac{1}{2}A = \begin{bmatrix} \frac{1}{2} \cdot 5 & \frac{1}{2} \cdot 6 \\ \frac{1}{2} \cdot 8 & \frac{1}{2}(-1) \end{bmatrix} = \begin{bmatrix} \frac{5}{2} & 3 \\ 4 & -\frac{1}{2} \end{bmatrix}$ ■

To add two matrices with the same dimensions, we add their corresponding entries. If two matrices do not have the same dimensions, then they cannot be added.

***Matrix Addition***

If $A = [a_{ij}]$ and $B = [b_{ij}]$ are both of dimension $m \times n$, then their **sum** is defined to be the $m \times n$ matrix

$$A + B = [a_{ij} + b_{ij}]$$

With $-B = (-1)B$, we define matrix subtraction in terms of matrix addition as follows:

$$A - B = A + (-B)$$

We define the $m \times n$ **zero matrix** to be the $m \times n$ matrix in which every entry is 0. Boldface $\mathbf{0}$ is used to indicate each zero matrix. For example,

$$\mathbf{0} = \begin{bmatrix} 0 & 0 & 0 \\ 0 & 0 & 0 \\ 0 & 0 & 0 \end{bmatrix} \text{ for } 3 \times 3 \text{ matrices.}$$

If matrix $A$ and matrix $\mathbf{0}$ are both of dimension $m \times n$, then

$$A + \mathbf{0} = A$$

## EXAMPLE

**3.** Let $A = \begin{bmatrix} 2 & 3 & -1 \\ 5 & 4 & 7 \end{bmatrix}$, $B = \begin{bmatrix} 6 & 8 & 10 \\ -2 & 3 & 1 \end{bmatrix}$, $C = \begin{bmatrix} 1 & -3 \\ 4 & -2 \end{bmatrix}$,

and $D = \begin{bmatrix} -1 & 3 \\ -4 & 2 \end{bmatrix}$.

Find (a) $A + B$, (b) $A - B$, (c) $A + C$, and (d) $C + D$, if possible.

**Solution**

**a.** $A + B = \begin{bmatrix} 2 & 3 & -1 \\ 5 & 4 & 7 \end{bmatrix} + \begin{bmatrix} 6 & 8 & 10 \\ -2 & 3 & 1 \end{bmatrix}$

$= \begin{bmatrix} 2 + 6 & 3 + 8 & -1 + 10 \\ 5 + (-2) & 4 + 3 & 7 + 1 \end{bmatrix} = \begin{bmatrix} 8 & 11 & 9 \\ 3 & 7 & 8 \end{bmatrix}$

**b.** $A - B = \begin{bmatrix} 2 - 6 & 3 - 8 & -1 - 10 \\ 5 - (-2) & 4 - 3 & 7 - 1 \end{bmatrix}$

$= \begin{bmatrix} -4 & -5 & -11 \\ 7 & 1 & 6 \end{bmatrix}$

**c.** $A + C$ is not defined since $A$ and $C$ are not of the same order.

**d.** $C + D = \begin{bmatrix} 1 & -3 \\ 4 & -2 \end{bmatrix} + \begin{bmatrix} -1 & 3 \\ -4 & 2 \end{bmatrix} = \begin{bmatrix} 1 - 1 & -3 + 3 \\ 4 - 4 & -2 + 2 \end{bmatrix}$

$= \begin{bmatrix} 0 & 0 \\ 0 & 0 \end{bmatrix} = \mathbf{0}$ ■

We begin the discussion of **matrix multiplication** by illustrating the process with a $2 \times 3$ matrix $A$ and a $3 \times 2$ matrix $B$. The product is found by multiplying the entries in each row of $A$ times the corresponding entries of each column of $B$ and adding these products to find each entry in $AB$. Study the following step-by-step development carefully.

Let

$$A = \begin{bmatrix} 1 & 2 & 3 \\ 4 & 5 & 6 \end{bmatrix} \quad \text{and} \quad B = \begin{bmatrix} a & b \\ c & d \\ e & f \end{bmatrix}$$

**Step 1** Multiply the entries of **row 1** in $A$ times the corresponding entries of **column 1** in $B$ and add the results.

$$AB = \begin{bmatrix} 1 & 2 & 3 \\ 4 & 5 & 6 \end{bmatrix} \begin{bmatrix} a & b \\ c & d \\ e & f \end{bmatrix} = \begin{bmatrix} 1a + 2c + 3e & ? \\ ? & ? \end{bmatrix}$$

**Step 2** Multiply the entries of **row 1** in $A$ times the corresponding entries of **column 2** in $B$ and add the results.

$$AB = \begin{bmatrix} 1 & 2 & 3 \\ 4 & 5 & 6 \end{bmatrix} \begin{bmatrix} a & b \\ c & d \\ e & f \end{bmatrix} = \begin{bmatrix} 1a + 2c + 3e & 1b + 2d + 3f \\ ? & ? \end{bmatrix}$$

**Steps 3 and 4** Proceed as before using **row 2** in $A$ with **column 1** and **column 2** in $B$.

$$AB = \begin{bmatrix} 1 & 2 & 3 \\ 4 & 5 & 6 \end{bmatrix} \begin{bmatrix} a & b \\ c & d \\ e & f \end{bmatrix} = \begin{bmatrix} 1a + 2c + 3e & 1b + 2d + 3f \\ 4a + 5c + 6e & 4b + 5d + 6f \end{bmatrix}$$

Thus, we see that the product $AB$ of a $2 \times 3$ matrix $A$ with a $3 \times 2$ matrix $B$ is a $2 \times 2$ matrix. Next, we will see that the product $BA$ of the same two matrices is a $3 \times 3$ matrix. In other words, $AB \neq BA$.

$$BA = \begin{bmatrix} a & b \\ c & d \\ e & f \end{bmatrix} \begin{bmatrix} 1 & 2 & 3 \\ 4 & 5 & 6 \end{bmatrix} = \begin{bmatrix} a1 + b4 & a2 + b5 & a3 + b6 \\ c1 + d4 & c2 + d5 & c3 + d6 \\ e1 + f4 & e2 + f5 & e3 + f6 \end{bmatrix}$$

***Matrix Multiplication***

If $A = [a_{ij}]$ is of dimension $m \times n$ and $B = [b_{ij}]$ is of dimension $n \times p$, then the **product** $AB$ is the $m \times p$ matrix

$$AB = [c_{ij}]$$

where $c_{ij}$ is found by adding the products of the corresponding entries in row $i$ of matrix $A$ and column $j$ of matrix $B$ for all $i = 1,2,3, \ldots, m$ and for all $j = 1,2,3, \ldots, p$.

The following diagram illustrates the condition that for the product $AB$ to even exist, the number of columns in $A$ must be the same as the number of rows in $B$:

$$\begin{array}{ccccc} A & \times & B & = & AB \\ m \times n & & n \times p & & m \times p \end{array}$$

same

rows columns

### EXAMPLES

**4.** Find the product $AB$ given

$$A = \begin{bmatrix} 2 & 4 & -1 \\ 3 & -2 & 1 \end{bmatrix} \quad \text{and} \quad B = \begin{bmatrix} 3 & 0 & 1 \\ 2 & 5 & 0 \\ -1 & 3 & 4 \end{bmatrix}$$

**Solution** Since the matrix on the left is of order $2 \times 3$ and the matrix on the right is $3 \times 3$, the product will be of order $2 \times 3$.

$$AB = \begin{bmatrix} 2 & 4 & -1 \\ 3 & -2 & 1 \end{bmatrix} \begin{bmatrix} 3 & 0 & 1 \\ 2 & 5 & 0 \\ -1 & 3 & 4 \end{bmatrix}$$

$$= \begin{bmatrix} 2 \cdot 3 + 4 \cdot 2 + (-1)(-1) & 2 \cdot 0 + 4 \cdot 5 + (-1)(3) & 2 \cdot 1 + 4 \cdot 0 + (-1)(4) \\ 3 \cdot 3 + (-2)(2) + 1(-1) & 3 \cdot 0 + (-2)(5) + 1 \cdot 3 & 3 \cdot 1 + (-2)(0) + 1 \cdot 4 \end{bmatrix}$$

$$= \begin{bmatrix} 15 & 17 & -2 \\ 4 & -7 & 7 \end{bmatrix}$$

**5.** If possible, (a) find the product $CD$ and (b) find the product $DC$ given

$$C = [4 \quad 0 \quad 1] \quad \text{and} \quad D = \begin{bmatrix} 6 & 1 \\ 5 & 2 \\ 0 & 3 \end{bmatrix}$$

**Solution**

**a.** Since $C$ is $1 \times 3$ and $D$ is $3 \times 2$, the product $CD$ is defined and will be $1 \times 2$. After some practice, you should be able to perform the arithmetic operations mentally.

$$CD = [4 \quad 0 \quad 1] \begin{bmatrix} 6 & 1 \\ 5 & 2 \\ 0 & 3 \end{bmatrix} = [24 \quad 7]$$

**b.** Since $D$ is $3 \times 2$ and $C$ is $1 \times 3$, the product $DC$ is not defined. The dimensions (orders) do not fit the requirements for multiplication to be defined. ■

### Practice Problems

**1.** Solve the following matrix equation for $x$, $y$, and $z$.

$$\begin{bmatrix} 5 & x \\ y & z - 1 \end{bmatrix} = \begin{bmatrix} 5 & -4 \\ 6 & 3 \end{bmatrix}$$

**2.** If $A = \begin{bmatrix} 2 & -1 \\ 3 & 1 \end{bmatrix}$ and $B = \begin{bmatrix} 5 & 6 \\ 3 & -1 \end{bmatrix}$, find (a) $A + B$, (b) $AB$, and (c) $BA$.

**Answers to Practice Problems** **1.** $x = -4$, $y = 6$, $z = 4$ **2. (a)** $\begin{bmatrix} 7 & 5 \\ 6 & 0 \end{bmatrix}$ **(b)** $\begin{bmatrix} 7 & 13 \\ 18 & 17 \end{bmatrix}$ **(c)** $\begin{bmatrix} 28 & 1 \\ 3 & -4 \end{bmatrix}$

## EXERCISES 11.4

Solve for $x$, $y$, $z$, and $w$ in each of the matrix equations in Exercises 1–8.

**1.** $\begin{bmatrix} 1 & x \\ y & -4 \end{bmatrix} = \begin{bmatrix} z & 5 \\ 4 & w+1 \end{bmatrix}$

**2.** $\begin{bmatrix} 3x & -1 \\ 6 & y+1 \end{bmatrix} = \begin{bmatrix} 6 & z-2 \\ 2w & 9 \end{bmatrix}$

**3.** $\begin{bmatrix} 2x & 4 & -3 \\ 5 & y & 0 \end{bmatrix} = \begin{bmatrix} -2 & 4 & 3z \\ 5 & -5 & w+2 \end{bmatrix}$

**4.** $\begin{bmatrix} -8 & 6 & x+3 \\ 2 & 9 & 3y \end{bmatrix} = \begin{bmatrix} 4z & 6 & -1 \\ w-5 & 9 & 12 \end{bmatrix}$

**5.** $\begin{bmatrix} -3x & 2 \\ 4 & y+3 \\ 14 & -3 \end{bmatrix} = \begin{bmatrix} 9 & 2 \\ z-1 & -2 \\ 2w & -3 \end{bmatrix}$

**6.** $\begin{bmatrix} 7 & -8 \\ 2x & 5 \\ y-1 & 0 \end{bmatrix} = \begin{bmatrix} 7 & z-4 \\ 10 & 5 \\ -1 & 3w \end{bmatrix}$

**7.** $\begin{bmatrix} -1 & 2 & 2x+1 \\ 3-y & 5 & -3 \\ z+6 & 4 & -7 \end{bmatrix} = \begin{bmatrix} -1 & 2 & 7 \\ -1 & 5 & -3 \\ 2 & 4 & 3w-1 \end{bmatrix}$

**8.** $\begin{bmatrix} 9 & 3 & 6 \\ -2 & -5 & 7 \\ 3x+1 & 4 & 0 \end{bmatrix} = \begin{bmatrix} 2y+3 & 3 & 6 \\ -2 & 3z+1 & 7 \\ 10 & 2w-4 & 0 \end{bmatrix}$

In Exercises 9–34, perform the indicated operations, if possible.

**9.** $2\begin{bmatrix} -1 & 0 \\ 3 & -2 \end{bmatrix}$

**10.** $-3\begin{bmatrix} 2 & -1 & 4 \\ -1 & 0 & -5 \end{bmatrix}$

**11.** $-0.5\begin{bmatrix} 4 & 2 \\ 0 & -1 \\ 3 & -4 \end{bmatrix}$

**12.** $0.2\begin{bmatrix} 5 & 1 & 0 \\ -2 & 4 & -3 \\ 10 & 6 & -8 \end{bmatrix}$

**13.** $\begin{bmatrix} 4 & 5 \\ 0 & 4 \end{bmatrix} + \begin{bmatrix} -2 & 5 \\ 3 & -1 \end{bmatrix}$

**14.** $\begin{bmatrix} 3 \\ -4 \\ 5 \end{bmatrix} + \begin{bmatrix} -3 \\ 4 \\ 5 \end{bmatrix}$

**15.** $\begin{bmatrix} 2 & -1 & 7 \\ 3 & 0 & -2 \end{bmatrix} + \begin{bmatrix} -1 & 5 \\ 6 & 0 \\ -3 & 1 \end{bmatrix}$

**16.** $\begin{bmatrix} 1 & 1 & 1 \\ -2 & 1 & 3 \\ 0 & 0 & 1 \end{bmatrix} + \begin{bmatrix} -4 & -3 & 1 \\ 2 & -1 & 3 \\ 1 & 1 & 0 \end{bmatrix}$

**17.** $\begin{bmatrix} 4 & 1 \\ -2 & 1 \end{bmatrix} - \begin{bmatrix} 2 & -2 \\ 4 & 3 \end{bmatrix}$

**18.** $\begin{bmatrix} -3 & 5 \\ 0 & 6 \end{bmatrix} - \begin{bmatrix} 2 & 3 & -1 \\ 4 & 7 & 0 \\ 1 & 1 & 2 \end{bmatrix}$

**19.** $\begin{bmatrix} 1 & 0 \\ 0 & 1 \end{bmatrix} - \begin{bmatrix} 0.3 & 0.6 \\ 0.2 & 0.4 \end{bmatrix}$

**20.** $\begin{bmatrix} -3 & 2 & 2 \\ -4 & 1 & 0 \end{bmatrix} - \begin{bmatrix} 2 & -1 & 3 \\ 1 & 7 & 2 \end{bmatrix}$

**21.** $\begin{bmatrix} 2 \\ -1 \\ 3 \end{bmatrix} [5 \quad 0 \quad 2]$

**22.** $[1 \quad 6 \quad 5] \begin{bmatrix} -2 \\ 2 \\ -3 \end{bmatrix}$

**23.** $\begin{bmatrix} 1 & 0 \\ 0 & 1 \end{bmatrix} \begin{bmatrix} -6 & 1 \\ -2 & 5 \end{bmatrix}$

**24.** $\begin{bmatrix} 3 & 1 \\ -4 & 5 \end{bmatrix} \begin{bmatrix} 0 & 3 & -2 \\ 1 & 6 & -4 \end{bmatrix}$

**25.** $\begin{bmatrix} 1 & -1 & 3 \\ 1 & 0 & -2 \end{bmatrix} \begin{bmatrix} -3 & 1 \\ -1 & 2 \end{bmatrix}$

**26.** $\begin{bmatrix} 2 & -1 \\ 5 & -3 \end{bmatrix} \begin{bmatrix} 3 & -1 \\ 5 & -2 \end{bmatrix}$

**27.** $\begin{bmatrix} 4 & 9 \\ 3 & 7 \end{bmatrix}\begin{bmatrix} 7 & -9 \\ -3 & 4 \end{bmatrix}$

**28.** $\begin{bmatrix} 2 & 1 & 0 \\ -1 & 4 & 1 \\ 3 & 5 & -2 \end{bmatrix}\begin{bmatrix} 3 & 2 & -2 \\ 0 & 1 & 0 \\ 4 & -3 & 2 \end{bmatrix}$

**29.** $\begin{bmatrix} 1 & -2 & 5 \end{bmatrix}\begin{bmatrix} -1 & 3 & 0 \\ -1 & 2 & 4 \\ -2 & 1 & 4 \end{bmatrix}$

**30.** $\begin{bmatrix} 2 & -1 & 6 \end{bmatrix}\begin{bmatrix} -1 & 2 & 1 \\ -3 & 4 & 2 \\ 0 & -2 & 1 \end{bmatrix}$

**31.** $\begin{bmatrix} 4 & 3 & 0 \\ -1 & 3 & -2 \\ 1 & 1 & 2 \end{bmatrix}\begin{bmatrix} -3 \\ 1 \\ -1 \end{bmatrix}$

**32.** $\begin{bmatrix} 2 & 1 & 3 \\ -2 & 0 & 1 \\ 1 & 1 & 0 \end{bmatrix}\begin{bmatrix} 1 \\ 4 \\ -2 \end{bmatrix}$

**33.** $\begin{bmatrix} 1 & 0 & 2 \\ 1 & 2 & 3 \\ -2 & 4 & 1 \end{bmatrix}\begin{bmatrix} 3 & 2 & 0 \\ 4 & -1 & 2 \\ -1 & 2 & 3 \end{bmatrix}$

**34.** $\begin{bmatrix} 1 & 3 & 0 \\ 0 & -1 & 2 \\ 2 & 2 & 1 \end{bmatrix}\begin{bmatrix} 2 & -1 & 3 \\ -3 & 4 & 1 \\ 0 & -2 & 2 \end{bmatrix}$

For Exercises 35–44 let:

$$A = \begin{bmatrix} 3 & -1 \\ 2 & 1 \end{bmatrix},\ B = \begin{bmatrix} 1 & 4 & -3 \\ 2 & 2 & -1 \end{bmatrix},\ C = \begin{bmatrix} 2 & -1 & 3 \\ 2 & -3 & 5 \\ 0 & 4 & 2 \end{bmatrix},\ \text{and } D = \begin{bmatrix} -2 & 3 & 2 \\ 1 & 0 & 1 \\ -1 & 2 & 0 \end{bmatrix}.$$

Find the indicated matrix, if possible.

**35.** $4B$ **36.** $-2A$ **37.** $A + B$ **38.** $A - B$ **39.** $AB$
**40.** $BA$ **41.** $C + D$ **42.** $3C - 2D$ **43.** $B(C + D)$ **44.** $B(C - D)$

## 11.5 Inverses of Matrices

**OBJECTIVES**

In this section, you will be learning to:

1. Find the inverses of square matrices.
2. Write systems of linear equations in matrix equation form.
3. Solve systems of linear equations using inverses of square matrices.

In this section, we will show how to represent systems of linear equations in the form of matrix equations and to solve these equations using matrix multiplication. To accomplish this, we need to develop several new matrix concepts.

Consider the square matrix of order 3

$$A = \begin{bmatrix} a_{11} & a_{12} & a_{13} \\ a_{21} & a_{22} & a_{23} \\ a_{31} & a_{32} & a_{33} \end{bmatrix}$$

main diagonal

The entries $a_{11}$, $a_{22}$, and $a_{33}$ are called the **main diagonal.** If the main diagonal consists of all 1s with the remaining entries all 0s, the matrix is called the **identity matrix** of order 3, denoted by $I_3$.

$$I_3 = \begin{bmatrix} 1 & 0 & 0 \\ 0 & 1 & 0 \\ 0 & 0 & 1 \end{bmatrix}$$

In general, $I_n$ is used to represent the identity matrix of order $n$. Or, we may simply use $I$ to represent the identity matrix when the order is understood.

In the real number system, 1 is called the multiplicative identity because, for any real number $a$,

$$a \cdot 1 = 1 \cdot a = a$$

Similarly, with matrices, for any square matrix $A$ of order $n$,

$$A \cdot I_n = I_n \cdot A = A$$

For example, matrix multiplication gives the following results:

$$\begin{bmatrix} 1 & 2 & 3 \\ 4 & 5 & 6 \\ 7 & 8 & 9 \end{bmatrix} \begin{bmatrix} 1 & 0 & 0 \\ 0 & 1 & 0 \\ 0 & 0 & 1 \end{bmatrix} = \begin{bmatrix} 1 & 2 & 3 \\ 4 & 5 & 6 \\ 7 & 8 & 9 \end{bmatrix}$$

The next concept that we need for solving matrix equations is that of the **multiplicative inverse** (or **inverse**) of a square matrix. This idea corresponds to that of multiplicative inverses (reciprocals) of real numbers in which their product is 1. For example, $\frac{2}{3}$ and $\frac{3}{2}$ are multiplicative inverses because $\frac{2}{3} \cdot \frac{3}{2} = 1$.

***Inverse of a Matrix***

If $A$ and $B$ are both square matrices of order $n$ and

$$A \cdot B = B \cdot A = I_n$$

then $B$ is the **inverse** of $A$ and $A$ is the **inverse** of $B$.

The inverse of $A$ is denoted by the symbol $A^{-1}$ (read "$A$ inverse"). This notation is similar to the inverse function notation used in Chapter 8 and should not be confused with negative exponents which indicate fractions.

Not every square matrix has an inverse. If a square matrix $A$ has an inverse, then it is said to be **nonsingular;** otherwise, it is said to be **singular.** Before we discuss the method for finding inverses, we give two examples of inverses.

**EXAMPLE**

**1.** Given $A = \begin{bmatrix} 1 & 1 & 0 \\ 3 & 2 & 0 \\ -1 & -1 & 1 \end{bmatrix}$ and $B = \begin{bmatrix} -2 & 1 & 0 \\ 3 & -1 & 0 \\ 1 & 0 & 1 \end{bmatrix}$

show that $A$ and $B$ are inverses of each other.

**Solution** We want to show that $AB = I_3$ and $BA = I_3$.

$$AB = \begin{bmatrix} 1 & 1 & 0 \\ 3 & 2 & 0 \\ -1 & -1 & 1 \end{bmatrix} \begin{bmatrix} -2 & 1 & 0 \\ 3 & -1 & 0 \\ 1 & 0 & 1 \end{bmatrix}$$

$$= \begin{bmatrix} 1(-2) + 1 \cdot 3 + 0 \cdot 1 & 1 \cdot 1 + 1(-1) + 0 \cdot 0 & 1 \cdot 0 + 1 \cdot 0 + 0 \cdot 1 \\ 3(-2) + 2 \cdot 3 + 0 \cdot 1 & 3 \cdot 1 + 2(-1) + 0 \cdot 0 & 3 \cdot 0 + 2 \cdot 0 + 0 \cdot 1 \\ -1(-2) + (-1)3 + 1 \cdot 1 & -1 \cdot 1 + (-1)(-1) + 1 \cdot 0 & -1 \cdot 1 + -1(-1) + 1 \cdot 1 \end{bmatrix}$$

$$= \begin{bmatrix} 1 & 0 & 0 \\ 0 & 1 & 0 \\ 0 & 0 & 1 \end{bmatrix} = I_3$$

$$BA = \begin{bmatrix} -2 & 1 & 0 \\ 3 & -1 & 0 \\ 1 & 0 & 1 \end{bmatrix} \begin{bmatrix} 1 & 1 & 0 \\ 3 & 2 & 0 \\ -1 & -1 & 1 \end{bmatrix}$$

$$= \begin{bmatrix} -2 \cdot 1 + 1 \cdot 3 + 0(-1) & -2 \cdot 1 + 1 \cdot 2 + 0(-1) & -2 \cdot 0 + 1 \cdot 0 + 0 \cdot 1 \\ 3 \cdot 1 + (-1)3 + 0(-1) & 3 \cdot 1 + (-1)2 + 0(-1) & 3 \cdot 0 + (-1)0 + 0 \cdot 1 \\ 1 \cdot 1 + 0 \cdot 3 + 1(-1) & 1 \cdot 1 + 0 \cdot 2 + 1(-1) & 1 \cdot 0 + 0 \cdot 1 + 1 \cdot 1 \end{bmatrix}$$

$$= \begin{bmatrix} 1 & 0 & 0 \\ 0 & 1 & 0 \\ 0 & 0 & 1 \end{bmatrix} = I_3$$

Thus, $B = A^{-1}$ and $A = B^{-1}$. ■

We know that $AB \neq BA$ in general. However, whenever $A$ and $B$ are square matrices and $AB = I$, then it can be proven that $BA = I$ also. That is, from the definition, we know that $A \cdot A^{-1} = A^{-1} \cdot A = I$. Therefore, in Example 1 the calculation of the second product, $BA$, was not necessary.

**EXAMPLE**

**2.** Given $C = \begin{vmatrix} 2 & 5 \\ 4 & 9 \end{vmatrix}$ and $D = \begin{vmatrix} -\frac{9}{2} & \frac{5}{2} \\ 2 & -1 \end{vmatrix}$ show that $D = C^{-1}$.

**Solution** We want to show that $CD = I_2$.

$$CD = \begin{bmatrix} 2 & 5 \\ 4 & 9 \end{bmatrix} \begin{bmatrix} -\frac{9}{2} & \frac{5}{2} \\ 2 & -1 \end{bmatrix}$$

$$= \begin{bmatrix} 2\left(-\frac{9}{2}\right) + 5 \cdot 2 & 2\left(\frac{5}{2}\right) + 5(-1) \\ 4\left(-\frac{9}{2}\right) + 9 \cdot 2 & 4\left(\frac{5}{2}\right) + 9(-1) \end{bmatrix}$$

$$= \begin{bmatrix} 1 & 0 \\ 0 & 1 \end{bmatrix} = I_2$$

Therefore, $D = C^{-1}$. (Also, $C = D^{-1}$.) ■

The following procedure shows how elementary row operations can be used to find the inverse of a matrix, if the inverse exists. The process outlined involves steps somewhat similar to those used in the Gaussian elimination method in that the initial goal is to get 0s below the leading entry in the first column. The proof that this procedure does indeed give the inverse of a matrix is not given in this text.

***To Find the Inverse of a Square Matrix A***

Let $A$ be a square matrix of order $n$.

1. Write the matrix $A$ and the corresponding identity matrix $I_n$ side-by-side in the format $A \mid I_n$.
2. Perform elementary row operations on $A$ and $I_n$ so that $A$ is transformed into $I_n$. In so doing, $I_n$ will have been transformed into $A^{-1}$, if it exists.
3. Check by multiplying to verify that $A \cdot A^{-1} = I_n$.

**Note:** If, in the process just described, any row consists entirely of 0s, then $A^{-1}$ does not exist.

## EXAMPLES

**3.** Find $A^{-1}$ for $A = \begin{bmatrix} 1 & 2 \\ 1 & 3 \end{bmatrix}$.

**Solution** Write $A$ and $I_2$ side-by-side and proceed to transform $A$ into $I_2$.

$$\begin{bmatrix} 1 & 2 \\ 1 & 3 \end{bmatrix} \Bigg| \begin{bmatrix} 1 & 0 \\ 0 & 1 \end{bmatrix}$$

$$\text{R2} - \text{R1} \longrightarrow \begin{bmatrix} 1 & 2 \\ 0 & 1 \end{bmatrix} \Bigg| \begin{bmatrix} 1 & 0 \\ -1 & 1 \end{bmatrix}$$

$$\text{R1} - 2 \cdot \text{R2} \longrightarrow \begin{bmatrix} 1 & 0 \\ 0 & 1 \end{bmatrix} \Bigg| \begin{bmatrix} 3 & -2 \\ -1 & 1 \end{bmatrix}$$

Thus, $A^{-1} = \begin{bmatrix} 3 & -2 \\ -1 & 1 \end{bmatrix}$

Checking will show that $A \cdot A^{-1} = I_2$.

**4.** Find $B^{-1}$ for $B = \begin{bmatrix} 1 & 0 & -1 \\ -2 & 7 & -4 \\ 1 & -1 & 0 \end{bmatrix}$.

**Solution** Write $B$ and $I_3$ side-by-side and proceed to transform $B$ into $I_3$.

$$\begin{bmatrix} 1 & 0 & -1 \\ -2 & 7 & -4 \\ 1 & -1 & 0 \end{bmatrix} \Bigg| \begin{bmatrix} 1 & 0 & 0 \\ 0 & 1 & 0 \\ 0 & 0 & 1 \end{bmatrix}$$

$$\begin{matrix} \\ \text{R2} + 2 \cdot \text{R1} \longrightarrow \\ \text{R3} - \text{R1} \longrightarrow \end{matrix} \begin{bmatrix} 1 & 0 & -1 \\ 0 & 7 & -6 \\ 0 & -1 & 1 \end{bmatrix} \Bigg| \begin{bmatrix} 1 & 0 & 0 \\ 2 & 1 & 0 \\ -1 & 0 & 1 \end{bmatrix}$$

$$\begin{matrix} \\ -\text{R3} \\ \text{R2} \end{matrix} \begin{bmatrix} 1 & 0 & -1 \\ 0 & 1 & -1 \\ 0 & 7 & -6 \end{bmatrix} \Bigg| \begin{bmatrix} 1 & 0 & 0 \\ 1 & 0 & -1 \\ 2 & 1 & 0 \end{bmatrix}$$

$$\begin{matrix} \\ \\ \text{R3} - 7 \cdot \text{R2} \longrightarrow \end{matrix} \begin{bmatrix} 1 & 0 & -1 \\ 0 & 1 & -1 \\ 0 & 0 & 1 \end{bmatrix} \Bigg| \begin{bmatrix} 1 & 0 & 0 \\ 1 & 0 & -1 \\ -5 & 1 & 7 \end{bmatrix}$$

$$\begin{matrix} \text{R1} + \text{R3} \longrightarrow \\ \text{R2} + \text{R3} \longrightarrow \\ \end{matrix} \begin{bmatrix} 1 & 0 & 0 \\ 0 & 1 & 0 \\ 0 & 0 & 1 \end{bmatrix} \Bigg| \begin{bmatrix} -4 & 1 & 7 \\ -4 & 1 & 6 \\ -5 & 1 & 7 \end{bmatrix}$$

Thus,

$$B^{-1} = \begin{bmatrix} -4 & 1 & 7 \\ -4 & 1 & 6 \\ -5 & 1 & 7 \end{bmatrix}$$

Checking will show that $B \cdot B^{-1} = I_3$. ■

Now we will discuss how to represent a system of linear equations in the form of a matrix equation and to use the inverse of the coefficient matrix to solve the system. The idea is to represent the system in the form of a matrix equation such as

$$A \cdot X = K$$

then multiply both sides of the equation by $A^{-1}$.

Giving $$A^{-1} \cdot A \cdot X = A^{-1} \cdot K$$
and $$I \cdot X = A^{-1} \cdot K$$
or $$X = A^{-1} \cdot K$$

For example, we can write the system

$$\begin{aligned} x + 2y &= 3 \\ x + 3y &= 5 \end{aligned} \quad \text{as} \quad \begin{bmatrix} 1 & 2 \\ 1 & 3 \end{bmatrix} \begin{bmatrix} x \\ y \end{bmatrix} = \begin{bmatrix} 3 \\ 5 \end{bmatrix}$$

where $A = \begin{bmatrix} 1 & 2 \\ 1 & 3 \end{bmatrix}$, $X = \begin{bmatrix} x \\ y \end{bmatrix}$, and $K = \begin{bmatrix} 3 \\ 5 \end{bmatrix}$. From Example 3, we know that $A^{-1} = \begin{bmatrix} 3 & -2 \\ -1 & 1 \end{bmatrix}$.

Now, multiplying both sides of the equation by $A^{-1}$ gives

$$\begin{bmatrix} 3 & -2 \\ -1 & 1 \end{bmatrix} \begin{bmatrix} 1 & 2 \\ 1 & 3 \end{bmatrix} \begin{bmatrix} x \\ y \end{bmatrix} = \begin{bmatrix} 3 & -2 \\ -1 & 1 \end{bmatrix} \begin{bmatrix} 3 \\ 5 \end{bmatrix}$$

$$\begin{bmatrix} 1 & 0 \\ 0 & 1 \end{bmatrix} \begin{bmatrix} x \\ y \end{bmatrix} = \begin{bmatrix} 3 \cdot 3 + (-2)5 \\ -1 \cdot 3 + 1 \cdot 5 \end{bmatrix}$$

$$\begin{bmatrix} x \\ y \end{bmatrix} = \begin{bmatrix} -1 \\ 2 \end{bmatrix}$$

Equating the corresponding entries in the last equation gives the solution to the system: $x = -1$ and $y = 2$.

**EXAMPLE**

**5.** Use matrix inversion to solve the system

$$\begin{aligned} x \qquad\quad - z &= 5 \\ -2x + 7y - 4z &= 2 \\ x - y \qquad\; &= 3 \end{aligned}$$

**Solution** First write the system in the form of a matrix equation.

$$\begin{bmatrix} 1 & 0 & -1 \\ -2 & 7 & -4 \\ 1 & -1 & 0 \end{bmatrix} \begin{bmatrix} x \\ y \\ z \end{bmatrix} = \begin{bmatrix} 5 \\ 2 \\ 3 \end{bmatrix}$$

Now multiply both sides of the equation by the inverse of the coefficient matrix, found in Example 4.

$$\begin{bmatrix} -4 & 1 & 7 \\ -4 & 1 & 6 \\ -5 & 1 & 7 \end{bmatrix}\begin{bmatrix} 1 & 0 & -1 \\ -2 & 7 & -4 \\ 1 & -1 & 0 \end{bmatrix}\begin{bmatrix} x \\ y \\ z \end{bmatrix} = \begin{bmatrix} -4 & 1 & 7 \\ -4 & 1 & 6 \\ -5 & 1 & 7 \end{bmatrix}\begin{bmatrix} 5 \\ 2 \\ 3 \end{bmatrix}$$

$$\begin{bmatrix} 1 & 0 & 0 \\ 0 & 1 & 0 \\ 0 & 0 & 1 \end{bmatrix}\begin{bmatrix} x \\ y \\ z \end{bmatrix} = \begin{bmatrix} -4 \cdot 5 + 1 \cdot 2 + 7 \cdot 3 \\ -4 \cdot 5 + 1 \cdot 2 + 6 \cdot 3 \\ -5 \cdot 5 + 1 \cdot 2 + 7 \cdot 3 \end{bmatrix}$$

$$\begin{bmatrix} x \\ y \\ z \end{bmatrix} = \begin{bmatrix} 3 \\ 0 \\ -2 \end{bmatrix}$$

Equating the corresponding entries in the last equation gives the solution to the system: $x = 3$, $y = 0$, and $z = -2$. ■

***Practice Problems***

1. Show that $A = \begin{bmatrix} 5 & 6 \\ -1 & -1 \end{bmatrix}$ and $B = \begin{bmatrix} -1 & -6 \\ 1 & 5 \end{bmatrix}$ are inverses of each other.

2. Find $A^{-1}$ if $A = \begin{bmatrix} 1 & 0 & 2 \\ 4 & 1 & 3 \\ -1 & 0 & -1 \end{bmatrix}$.

## EXERCISES 11.5

In Exercises 1–8, show that the two matrices are inverses of each other.

1. $\begin{bmatrix} 5 & 7 \\ 2 & 3 \end{bmatrix}; \begin{bmatrix} 3 & -7 \\ -2 & 5 \end{bmatrix}$

2. $\begin{bmatrix} 2 & 3 \\ 3 & 4 \end{bmatrix}; \begin{bmatrix} -4 & 3 \\ 3 & -2 \end{bmatrix}$

3. $\begin{bmatrix} 2 & 3 \\ 4 & 7 \end{bmatrix}; \begin{bmatrix} \frac{7}{2} & -\frac{3}{2} \\ -2 & 1 \end{bmatrix}$

4. $\begin{bmatrix} -1 & -2 \\ 3 & 4 \end{bmatrix}; \begin{bmatrix} 2 & 1 \\ -\frac{3}{2} & -\frac{1}{2} \end{bmatrix}$

5. $\begin{bmatrix} 1 & -7 & 1 \\ 2 & 5 & -1 \\ -1 & 1 & 0 \end{bmatrix}; \begin{bmatrix} 1 & 1 & 2 \\ 1 & 1 & 3 \\ 7 & 6 & 19 \end{bmatrix}$

6. $\begin{bmatrix} 1 & -1 & 3 \\ 1 & 0 & 2 \\ 4 & 3 & 6 \end{bmatrix}; \begin{bmatrix} -6 & 15 & -2 \\ 2 & -6 & 1 \\ 3 & -7 & 1 \end{bmatrix}$

7. $\begin{bmatrix} 1 & 0 & 1 \\ 1 & 7 & 4 \\ 1 & -2 & 0 \end{bmatrix}; \begin{bmatrix} -8 & 2 & 7 \\ -4 & 1 & 3 \\ 9 & -2 & -7 \end{bmatrix}$

8. $\begin{bmatrix} 1 & -1 & 1 \\ 1 & 3 & 4 \\ 2 & 1 & 4 \end{bmatrix}; \begin{bmatrix} -8 & -5 & 7 \\ -4 & -2 & 3 \\ 5 & 3 & -4 \end{bmatrix}$

**Answers to Practice Problems** 1. $AB = I$ 2. $A^{-1} = \begin{bmatrix} -1 & 0 & -2 \\ 1 & 1 & 5 \\ 1 & 0 & 1 \end{bmatrix}$

In Exercises 9–20, find the inverse for each matrix.

**9.** $\begin{bmatrix} 1 & -3 \\ -2 & 5 \end{bmatrix}$

**10.** $\begin{bmatrix} 2 & 9 \\ 1 & 4 \end{bmatrix}$

**11.** $\begin{bmatrix} 4 & -2 \\ -3 & 1 \end{bmatrix}$

**12.** $\begin{bmatrix} 5 & 6 \\ 2 & 3 \end{bmatrix}$

**13.** $\begin{bmatrix} 1 & 2 & -3 \\ 4 & 3 & 2 \\ 3 & 2 & 2 \end{bmatrix}$

**14.** $\begin{bmatrix} 1 & 2 & -1 \\ 1 & 1 & 1 \\ 3 & 2 & 4 \end{bmatrix}$

**15.** $\begin{bmatrix} 1 & -1 & 0 \\ 2 & -1 & 2 \\ 1 & 0 & 3 \end{bmatrix}$

**16.** $\begin{bmatrix} 1 & 1 & -6 \\ 1 & -1 & 5 \\ 2 & -1 & 4 \end{bmatrix}$

**17.** $\begin{bmatrix} 1 & 2 & 0 \\ 2 & 4 & -1 \\ 1 & 3 & 3 \end{bmatrix}$

**18.** $\begin{bmatrix} 1 & 5 & 0 \\ 5 & 3 & 3 \\ 2 & 3 & 1 \end{bmatrix}$

**19.** $\begin{bmatrix} 2 & 0 & 3 \\ 5 & 1 & -2 \\ 3 & 0 & 4 \end{bmatrix}$

**20.** $\begin{bmatrix} 2 & -2 & 5 \\ -3 & 2 & 0 \\ -2 & 1 & 3 \end{bmatrix}$

In Exercises 21–30, use matrix inversion to solve the system of equations.

**21.** $\begin{aligned} x + 3y &= 5 \\ 3x + 5y &= 7 \end{aligned}$

**22.** $\begin{aligned} x - 2y &= 8 \\ 4x - 3y &= 22 \end{aligned}$

**23.** $\begin{aligned} 6x - y &= 3 \\ 5x - 2y &= 1 \end{aligned}$

**24.** $\begin{aligned} 2x + 5y &= 16 \\ 3x + 4y &= 10 \end{aligned}$

**25.** $\begin{aligned} x + 3y + 3z &= 6 \\ x + z &= 3 \\ 5x - 2y + 4z &= 14 \end{aligned}$

**26.** $\begin{aligned} x + 3y - 6z &= 2 \\ 2y + 3z &= 7 \\ x + 4y - 5z &= 5 \end{aligned}$

**27.** $\begin{aligned} x + 2y + 2z &= 4 \\ 2y + 3z &= 8 \\ 7x + 9y + 6z &= 7 \end{aligned}$

**28.** $\begin{aligned} x + y + 3z &= -5 \\ 7x - 2y + 5z &= -14 \\ 4x + 5z &= -11 \end{aligned}$

**29.** $\begin{aligned} x + y &= 4 \\ 2x + 4y - z &= 14 \\ 3x + 2y + 2z &= 15 \end{aligned}$

**30.** $\begin{aligned} x - y + 4z &= 30 \\ x + 2y + 5z &= 30 \\ x - 3y + 3z &= 28 \end{aligned}$

## CHAPTER 11 SUMMARY

### *Key Terms and Formulas*

A **determinant** is a real number associated with a square array of real numbers and is indicated by enclosing the array between two vertical bars. [11.1]

The value for a 2 × 2 determinant is

$$\begin{vmatrix} a_1 & b_1 \\ a_2 & b_2 \end{vmatrix} = a_1b_2 - a_2b_1 \quad [11.1]$$

The value for a 3 × 3 determinant is

$$\begin{vmatrix} a_1 & b_1 & c_1 \\ a_2 & b_2 & c_2 \\ a_3 & b_3 & c_3 \end{vmatrix} = a_1 \begin{vmatrix} b_2 & c_2 \\ b_3 & c_3 \end{vmatrix} - b_1 \begin{vmatrix} a_2 & c_2 \\ a_3 & c_3 \end{vmatrix} + c_1 \begin{vmatrix} a_2 & b_2 \\ a_3 & b_3 \end{vmatrix} \quad [11.1]$$

An $m \times n$ (read "$m$ by $n$") **matrix** is a rectangular array of numbers with $m$ rows and $n$ columns in the form

$$A = \overbrace{\begin{bmatrix} a_{11} & a_{12} & a_{13} & \cdots & a_{1n} \\ a_{21} & a_{22} & a_{23} & \cdots & a_{2n} \\ a_{31} & a_{32} & a_{33} & \cdots & a_{3n} \\ \vdots & \vdots & \vdots & & \vdots \\ a_{m1} & a_{m2} & a_{m3} & \cdots & a_{mn} \end{bmatrix}}^{n \text{ columns}} \Bigg\} m \text{ rows}$$

Matrix $A$ is said to be of **dimension $m \times n$** (or **order $m \times n$**) and each number $a_{ij}$ is said to be an **entry** (or an **element**) of the matrix. If $m = n$, then the matrix is said to be **square of order $n$.** [11.3]

We also write $A = [a_{ij}]$. [11.4]

If a system of linear equations is written

1. in standard form with the constant terms on the right, and
2. with the like variables and constant terms aligned vertically, then there are two matrices associated with such a system: the **coefficient matrix** made up of the coefficients of the variables with 0 supplied for any missing variable, and the **augmented matrix** which includes the constant terms. [11.3]

A matrix is in **triangular form** if its entries in the lower left triangular region are all 0s. [11.3]

If $A = [a_{ij}]$ and $B = [b_{ij}]$ are both matrices of dimension $m \times n$, then

$$A = B \text{ if and only if } a_{ij} = b_{ij}$$

for all $i = 1,2,3, \ldots, m$ and all $j = 1,2,3, \ldots, n$. [11.4]

When working with matrices, numbers that are not entries are called **scalars.** [11.4]

The $m \times n$ **zero matrix** is the $m \times n$ matrix in which every entry is 0. [11.4]

The **identity matrix of order *n*,** denoted $I_n$, is a matrix of order $n$ in which the main diagonal consists of all 1s with the remaining entries all 0s. [11.5]

If $A$ and $B$ are both square matrices of order $n$ and

$$A \cdot B = B \cdot A = I_n$$

then $B$ is the **inverse** of $A$ and $A$ is the **inverse** of $B$. The inverse of $A$ is denoted as $A^{-1}$. [11.5]

If a square matrix $A$ has an inverse, then it is said to be **nonsingular;** otherwise, it is said to be **singular.** [11.5]

## Properties and Rules

**Cramer's Rule for 2 × 2 Systems** [11.2]

For the system $\begin{cases} a_1x + b_1y = k_1 \\ a_2x + b_2y = k_2 \end{cases}$

where

$$D = \begin{vmatrix} a_1 & b_1 \\ a_2 & b_2 \end{vmatrix}, \quad D_x = \begin{vmatrix} k_1 & b_1 \\ k_2 & b_2 \end{vmatrix},$$

and

$$D_y = \begin{vmatrix} a_1 & k_1 \\ a_2 & k_2 \end{vmatrix},$$

if $D \neq 0$, then

$$x = \frac{D_x}{D} \quad \text{and} \quad y = \frac{D_y}{D}$$

is a unique solution of the system.

**Cramer's Rule for 3 × 3 Systems** [11.2]

For the system $\begin{cases} a_1x + b_1y + c_1z = k_1 \\ a_2x + b_2y + c_2z = k_2 \\ a_3x + b_3y + c_3z = k_3 \end{cases}$

where

$$D = \begin{vmatrix} a_1 & b_1 & c_1 \\ a_2 & b_2 & c_2 \\ a_3 & b_3 & c_3 \end{vmatrix}, \quad D_x = \begin{vmatrix} k_1 & b_1 & c_1 \\ k_2 & b_2 & c_2 \\ k_3 & b_3 & c_3 \end{vmatrix},$$

$$D_y = \begin{vmatrix} a_1 & k_1 & c_1 \\ a_2 & k_2 & c_2 \\ a_3 & k_3 & c_3 \end{vmatrix}, \quad \text{and} \quad D_z = \begin{vmatrix} a_1 & b_1 & k_1 \\ a_2 & b_2 & k_2 \\ a_3 & b_3 & k_3 \end{vmatrix},$$

if $D \neq 0$, then

$$x = \frac{D_x}{D}, \quad y = \frac{D_y}{D}, \quad \text{and} \quad z = \frac{D_z}{D}$$

is a unique solution of the system.

## Procedures

**Elementary Row Operations** [11.3]

1. Multiply a row by a nonzero constant.
2. Interchange any two rows.
3. Change a row by adding a multiple of another row to it.

**Strategy for Gaussian Elimination** [11.3]

1. Write the augmented matrix for the system.
2. Use elementary row operations to transform the matrix into triangular form.
3. Solve the corresponding system of equations by substituting back into the system.

**Multiplication by Scalars** [11.4]
If $c$ is a scalar and $A = [a_{ij}]$ is a matrix of dimension $m \times n$, then the **scalar multiple** $cA$ is defined to be the $m \times n$ matrix

$$cA = [ca_{ij}]$$

**Matrix Addition** [11.4]
If $A = [a_{ij}]$ and $B = [b_{ij}]$ are both of dimension $m \times n$, then their **sum** is defined to be the $m \times n$ matrix

$$A + B = [a_{ij} + b_{ij}]$$

Matrix subtraction is defined in terms of matrix addition as $A - B = A + (-B)$. [11.4]

**Matrix Multiplication** [11.4]
If $A = [a_{ij}]$ is of dimension $m \times n$ and $B = [b_{ij}]$ is of dimension $n \times p$, then the **product** $AB$ is the $m \times p$ matrix

$$AB = [c_{ij}]$$

where $c_{ij}$ is found by adding the products of the corresponding entries in row $i$ of matrix $A$ and column $j$ of matrix $B$ for all $i = 1,2,3, \ldots, m$ and for all $j = 1,2,3, \ldots, p$.

**To Find the Inverse of a Square Matrix $A$** [11.5]
Let $A$ be a square matrix of order $n$.

1. Write the matrix $A$ and the corresponding identity matrix $I_n$ side-by-side in the format $A \mid I_n$.
2. Perform elementary row operations on $A$ and $I_n$ so that $A$ is transformed into $I_n$. In so doing, $I_n$ will have been transformed into $A^{-1}$, if it exists.
3. Check by multiplying to verify that $A \cdot A^{-1} = I_n$.

To solve a matrix equation in the form

$$A \cdot X = K$$

multiply boths sides of the equation by $A^{-1}$, if it exists. [11.5]

## CHAPTER 11 REVIEW

Evaluate the determinant in Exercises 1–4. [11.1]

**1.** $\begin{vmatrix} 4 & 7 \\ -2 & 3 \end{vmatrix}$ **2.** $\begin{vmatrix} 6 & -5 \\ 1 & 4 \end{vmatrix}$ **3.** $\begin{vmatrix} 3 & 1 & 4 \\ -2 & 0 & -2 \\ 1 & 4 & -1 \end{vmatrix}$ **4.** $\begin{vmatrix} 2 & -1 & 2 \\ 1 & 3 & 6 \\ 4 & 1 & -2 \end{vmatrix}$

Solve for $x$ in Exercises 5 and 6. [11.1]

**5.** $\begin{vmatrix} 2 & -1 & 2 \\ 4 & 2 & 3 \\ 2 & x & x \end{vmatrix} = 36$ **6.** $\begin{vmatrix} 1 & x & 0 \\ -2 & 5 & -1 \\ -1 & x & 4 \end{vmatrix} = 44$

Solve Exercises 7–11 using Cramer's Rule. [11.2]

**7.** $\begin{aligned} 2x + 3y &= -15 \\ 5x + 2y &= 1 \end{aligned}$ **8.** $\begin{aligned} 4x + 3y &= 2 \\ 24x + 48y &= 17 \end{aligned}$ **9.** $\begin{aligned} 1.7x + 2.1y &= 12.7 \\ 3.2x + 5.4y &= 26.8 \end{aligned}$

**10.** $\begin{aligned} x + 2y - 3z &= -1 \\ -2x + 3y + z &= 1 \\ x + 4y - z &= 9 \end{aligned}$ **11.** $\begin{aligned} 4x + 6y + 3z &= 11 \\ 8x - 2y + z &= 12 \\ 6x - 9y + 3z &= 2 \end{aligned}$

Identify the order of each matrix and find the indicated entry for each matrix in Exercises 12–14. [11.3]

**12.** $\begin{bmatrix} 2 & 1 & 0 \\ 3 & 5 & -1 \end{bmatrix}$; $a_{12}$ **13.** $\begin{bmatrix} -3 \\ 7 \\ 1 \end{bmatrix}$; $a_{31}$ **14.** $\begin{bmatrix} -4 & 3 & 2 \\ -1 & 1 & 0 \\ -2 & 0 & 5 \end{bmatrix}$; $a_{23}$

Solve Exercises 15–17 by using the Gaussian elimination method. [11.3]

**15.** $\begin{aligned} 3x - 2y &= 11 \\ 4x - y &= 8 \end{aligned}$

**16.** $\begin{aligned} x + y \quad\;\; &= 1 \\ 2y - z &= -3 \\ 2x + y + z &= 4 \end{aligned}$

**17.** $\begin{aligned} x + 3y + z &= 3 \\ 2x + 5y - 3z &= 1 \\ 2x + y + 2z &= 6 \end{aligned}$

**18.** A large feed lot mixes two rations, I and II, to obtain the ration to be used. Ration I contains 20% crude protein and costs 60 cents per pound. Ration II contains 16% crude protein and costs 80 cents per pound. The daily ration contains a total of 300 pounds of protein and costs $1180. How many pounds of each ration is used? [11.3]

**19.** At a local fruit stand the prices for apples, bananas, and oranges are 69 cents per pound, 45 cents per pound, and 59 cents per pound, respectively. Traci bought 10 pounds of fruit and the cost was $5.88. The total weight of the bananas and oranges was two pounds more than the weight of the apples. How many pounds of each fruit did she buy? [11.3]

Solve for $x$, $y$, $z$, and $w$ in each of the matrix equations in Exercises 20 and 21. [11.4]

**20.** $\begin{bmatrix} x-1 & 4 \\ 3y & 7 \end{bmatrix} = \begin{bmatrix} 5 & z+2 \\ 9 & 2w-1 \end{bmatrix}$

**21.** $\begin{bmatrix} 6 & -2 & 2x \\ 1 & 3y+1 & -5 \\ 8 & 0 & -2 \end{bmatrix} = \begin{bmatrix} 6 & 3z-5 & 10 \\ 1 & -5 & -5 \\ 8 & 4w & -2 \end{bmatrix}$

In Exercises 22–25, perform the indicated operations, if possible. [11.4]

**22.** $\begin{bmatrix} -1 & 5 \\ 4 & 0 \end{bmatrix} + \begin{bmatrix} 7 & -2 \\ 4 & 1 \end{bmatrix}$

**23.** $\begin{bmatrix} 2 & -1 \\ 3 & 0 \\ -2 & 5 \end{bmatrix} - \begin{bmatrix} 1 & 1 \\ 2 & -4 \\ -1 & 3 \end{bmatrix}$

**24.** $\begin{bmatrix} -3 & 0 & 1 \\ 2 & -5 & 4 \end{bmatrix} \begin{bmatrix} 1 & -1 & 2 \\ 0 & 4 & 3 \\ 1 & 6 & -2 \end{bmatrix}$

**25.** $\begin{bmatrix} 3 & 1 & 1 \\ 1 & -2 & 0 \\ 4 & -1 & 6 \end{bmatrix} \begin{bmatrix} 7 \\ -2 \\ 2 \end{bmatrix}$

In Exercises 26–28, let:

$$A = \begin{bmatrix} 1 & 4 \\ 2 & -3 \end{bmatrix}, \quad B = \begin{bmatrix} 0 & -5 \\ 2 & -1 \\ 5 & -3 \end{bmatrix}, \quad \text{and} \quad C = \begin{bmatrix} 1 & 0 & -1 \\ 2 & 3 & -2 \\ -2 & 5 & -4 \end{bmatrix}.$$

Find the indicated matrix, where possible. [11.4]

**26.** $B + A$

**27.** $BA$

**28.** $CB$

In Exercises 29 and 30, show that the two matrices are inverses of each other. [11.5]

**29.** $\begin{bmatrix} -1 & 4 \\ 2 & -9 \end{bmatrix}; \begin{bmatrix} -9 & -4 \\ -2 & -1 \end{bmatrix}$

**30.** $\begin{bmatrix} 1 & 2 & 0 \\ 1 & 1 & 3 \\ 0 & 1 & -2 \end{bmatrix}; \begin{bmatrix} 5 & -4 & -6 \\ -2 & 2 & 3 \\ -1 & 1 & 1 \end{bmatrix}$

In Exercises 31–34, find the inverse for each matrix. [11.5]

**31.** $\begin{bmatrix} -2 & 3 \\ -3 & 5 \end{bmatrix}$

**32.** $\begin{bmatrix} 4 & 5 \\ -3 & -2 \end{bmatrix}$

**33.** $\begin{bmatrix} 1 & 2 & 2 \\ 1 & 1 & 4 \\ 0 & 1 & -1 \end{bmatrix}$

**34.** $\begin{bmatrix} 2 & 1 & 5 \\ 3 & 2 & 4 \\ 1 & 1 & 0 \end{bmatrix}$

In Exercises 35–38, use matrix inversion to solve the system of equations. [11.5]

**35.** $\begin{aligned} 4x + 5y &= -7 \\ 3x - 2y &= 12 \end{aligned}$

**36.** $\begin{aligned} 5x + 2y &= 29 \\ 2x - 3y &= 4 \end{aligned}$

**37.** $\begin{aligned} x - y + z &= 6 \\ 2x + 2y - z &= 2 \\ 2x + y - z &= 3 \end{aligned}$

**38.** $\begin{aligned} x + y - z &= 7 \\ 4x + 2y - z &= 14 \\ x \qquad\;\; + 2z &= -3 \end{aligned}$

## CHAPTER 11 TEST

Evaluate the determinants in Exercises 1 and 2.

**1.** $\begin{vmatrix} 6 & -3 \\ 4 & 5 \end{vmatrix}$

**2.** $\begin{vmatrix} 1 & 3 & 2 \\ 2 & 5 & 1 \\ 0 & 2 & 1 \end{vmatrix}$

Solve for $x$ in Exercises 3 and 4.

**3.** $\begin{vmatrix} 3 & x \\ 5 & 7 \end{vmatrix} = -9$

**4.** $\begin{vmatrix} -2 & 3 & 1 \\ 1 & 3 & x \\ 2 & 3 & x \end{vmatrix} = 14$

Solve Exercises 5 and 6 using Cramer's Rule.

**5.** $\begin{aligned} 3x + 8y &= 14 \\ 2x + 7y &= 22 \end{aligned}$

**6.** $\begin{aligned} x + 2y - z &= 2 \\ x - 4y - 5z &= -7 \\ x + 3y + 4z &= 5 \end{aligned}$

**7.** Identify the order of the matrix and find the indicated entry.

$\begin{bmatrix} 4 & -1 \\ 7 & 2 \\ 0 & -3 \end{bmatrix}; a_{32}$

Solve Exercises 8 and 9 using the Gaussian elimination method.

**8.** $\begin{aligned} 5x - y &= 8 \\ 2x + 3y &= 27 \end{aligned}$

**9.** $\begin{aligned} x + 2y - 3z &= 7 \\ x - y - z &= 6 \\ 2x - 3y - 5z &= 19 \end{aligned}$

**10.** A company manufactures two types of surfboards, the standard model and the pro model. The standard model requires 4.5 hours to complete and costs $82. The pro model requires 5.5 hours to complete and costs $112. How many can be made in 138 hours if the total cost is $2656?

**11.** Solve for $x$, $y$, $z$, and $w$ in the matrix equation:

$$\begin{bmatrix} 7 & x - 3 \\ 2 & -5 \\ 2y + 3 & 9 \end{bmatrix} = \begin{bmatrix} 7 & -2 \\ 2 & 3z + 4 \\ 5 & w + 3 \end{bmatrix}$$

In Exercises 12–15, perform the indicated operation, if possible.

**12.** $\begin{bmatrix} 5 & -3 \\ 6 & 1 \end{bmatrix} + \begin{bmatrix} -1 & 7 \\ -2 & 0 \end{bmatrix}$

**13.** $\begin{bmatrix} 2 & -1 & 5 \\ 1 & 4 & -2 \end{bmatrix} - \begin{bmatrix} 3 & -2 & -1 \\ 5 & 2 & -2 \end{bmatrix}$

**14.** $\begin{bmatrix} 3 & 4 \\ -1 & -2 \\ 5 & 1 \end{bmatrix} \begin{bmatrix} -1 & 3 \\ 6 & 2 \end{bmatrix}$

**15.** $\begin{bmatrix} 1 & 0 & 2 \\ -2 & 3 & -1 \\ 4 & -2 & -3 \end{bmatrix} \cdot \begin{bmatrix} 10 \\ 11 \\ 18 \end{bmatrix}$

**16.** Show that the two matrices are inverses of each other.

$\begin{bmatrix} 1 & 3 & 2 \\ -2 & 3 & -3 \\ 1 & 2 & 2 \end{bmatrix}; \begin{bmatrix} 12 & -2 & -15 \\ 1 & 0 & -1 \\ -7 & 1 & 9 \end{bmatrix}$

In Exercises 17 and 18, find the inverse for each matrix.

**17.** $\begin{bmatrix} 5 & -3 \\ -7 & 4 \end{bmatrix}$

**18.** $\begin{bmatrix} 1 & 0 & 1 \\ 2 & 3 & -2 \\ 4 & 5 & -3 \end{bmatrix}$

In Exercises 19 and 20, use matrix inversion to solve the system of equations.

**19.** $3x - 7y = 23$
$x + 5y = -7$

**20.** $3x - y + 4z = 4$
$x + 2z = 5$
$5x - 2y + 7z = 6$

## CUMULATIVE REVIEW (11)

Find the indicated products in Exercises 1–3.

**1.** $(2x^2 - x - 5)(2x + 3)$

**2.** $(5 + \sqrt{7})(5 - \sqrt{7})$

**3.** $(8 + 3i)(8 - 3i)$

**4.** Divide: $(x^4 + 2x^3 - 3x^2 - 8x - 4) \div (x^2 - 4)$

**5.** Use synthetic division to divide: $(x^3 + 3x^2 - 7x - 6) \div (x - 2)$

Perform the indicated operations in Exercises 6 and 7 and simplify.

**6.** $2\sqrt{18} + 3\sqrt{50}$

**7.** $(\sqrt{12})(\sqrt{54})$

**8.** Solve the equation for $x$ and $y$: $3x + 4i = x - 6 + (y - 3)i$

**9.** Write a quadratic equation with the roots: $x = 5 + \sqrt{2}$ and $x = 5 - \sqrt{2}$.

Factor each polynomial in Exercises 10 and 11.

**10.** $8x^3 - 125$

**11.** $3x^3 + 2x^2 - 9x - 6$

Solve the equations in Exercises 12–14.

**12.** $3x^2 - 29x + 18 = 0$

**13.** $x^2 + 6x + 4 = 0$

**14.** $x^2 + 8x + 25 = 0$

**15.** Evaluate the determinant:
$$\begin{vmatrix} -1 & 5 & 2 \\ 6 & -3 & 1 \\ -2 & 1 & 4 \end{vmatrix}$$

**16.** Solve the system of equations using Cramer's Rule.
$8x + 15y = 339$
$1.25x + 1.80y = 45.90$

**17.** Find the product:
$$\begin{bmatrix} 1 & -2 \\ 4 & 9 \end{bmatrix} \cdot \begin{bmatrix} 1 & 3 & -2 \\ 5 & 0 & -3 \end{bmatrix}$$

**18.** Find the inverse for the matrix:
$$\begin{bmatrix} -5 & 5 & -1 \\ 3 & -2 & 2 \\ 0 & 2 & 3 \end{bmatrix}$$

**19.** A car rental agency rented 20 cars per day if they charged \$32. They found that for each \$3 decrease in price, they could rent 4 additional cars. Find the price if the daily revenue is \$728.

**20.** A ski shop ordered 20 pairs of skis and 12 sets of bindings for a total cost of \$2980. Two weeks later they ordered 8 pairs of skis and 9 sets of bindings and the cost was \$1465. Find the price of one pair of skis and the price of one set of bindings.

CHAPTER

# 12 ROOTS AND POLYNOMIALS EQUATIONS

## MATHEMATICAL CHALLENGES

A class survey found that 25 students watched television on Monday, 20 on Tuesday, and 16 on Wednesday. Of those who watched TV on only one of the days, 11 chose Monday, 7 chose Tuesday, and 6 chose Wednesday. If every student watched at least one of the days and 7 students watched all three days, find the number of students in the class.

The product of the ages of a group of teenagers is 10 584 000. Find the number of teenagers in the group and the sum of their ages.

An automobile went up a hill at a speed of 10 kilometers per hour and returned down the hill at a speed of 20 kilometers per hour. What was the average speed for the round trip?

## CHAPTER OUTLINE



The topics in this chapter come under the heading of theory of equations with the understanding that the equations being investigated are polynomial equations. Polynomial equations arise in the solutions of many types of applications throughout mathematics. For example, we have seen many word problems in this text that lead to the creation and solution of first-degree equations and second-degree (or quadratic) equations. Because there is no simple general technique for solving higher degree equations, we analyze polynomial equations from a variety of viewpoints.

**a.** We develop techniques for estimating real roots in decimal form, rational or irrational.
**b.** We develop techniques for limiting the search for rational roots to a list that meets certain conditions related to the coefficients.
**c.** We find upper and lower bounds for roots so that any search need not exceed these bounds.
**d.** We develop techniques for finding how many positive and how many negative roots an equation can have.
**e.** We find the maximum number of roots an equation can have so that any search can be limited.
**f.** We learn to use any root that we find, by whatever method, to help create new equations of smaller degree with fewer roots to find.

While none of these procedures gives specific roots, as does the quadratic formula for quadratic equations, taken together they provide a strategy for attempting to find roots of polynomial equations which is much better than simply experimenting.

## 12.1 Remainder Theorem and Approximating Roots

**OBJECTIVES**

In this section, you will be learning to:

1. Use synthetic division to evaluate polynomials.
2. Show that roots of polynomial equations lie between two given numbers.

Throughout this chapter, we will be discussing formulas and rules related to polynomials of degree $n > 0$. We will need the following definition of polynomial equations with complex coefficients as a basis for much of the discussion.

**Polynomial Equation**

A polynomial equation of degree $n > 0$ is an equation of the form

$$P(x) = a_nx^n + a_{n-1}x^{n-1} + \cdot \cdot \cdot + a_1x + a_0 = 0$$

where $a_n \neq 0$ and $a_n, a_{n-1}, \ldots, a_1, a_0$ are complex numbers.

The coefficient $a_n$ in the definition is called the **leading coefficient** of $P(x)$.

Consider the polynomial $x^3 - 5x^2 - 8x + 32$. By using function notation, we can write

$$P(x) = x^3 - 5x^2 - 8x + 32$$

which allows us to indicate evaluation of the polynomial for various values of $x$. For example,

$$\begin{aligned} P(2) &= 2^3 - 5 \cdot 2^2 - 8 \cdot 2 + 32 \\ &= 8 - 20 - 16 + 32 = 4 \\ P(3) &= 3^3 - 5 \cdot 3^2 - 8 \cdot 3 + 32 \\ &= 27 - 45 - 24 + 32 = -10 \end{aligned}$$

and

$$\begin{aligned} P(-1) &= (-1)^3 - 5(-1)^2 - 8(-1) + 32 \\ &= -1 - 5 + 8 + 32 = 34 \end{aligned}$$

To evaluate a polynomial for a particular value of the variable, we can always substitute the number directly into the polynomial and simplify as we have just done for $P(2)$, $P(3)$, and $P(-1)$. In this section, we will discuss another technique for evaluating polynomials that is more systematic and involves the division algorithm and synthetic division (see Section 3.4).

The division algorithm can be indicated by writing

$$\frac{P(x)}{D(x)} = Q(x) + \frac{R(x)}{D(x)}$$

or

$$P(x) = D(x) \cdot Q(x) + R(x)$$

where degree of $R(x) <$ degree of $D(x)$ and $P(x)$, $D(x)$, $Q(x)$, and $R(x)$ are polynomials.

With synthetic division, we can divide

$$(x^3 - 5x^2 - 8x + 32) \div (x - 2)$$

as follows:

$$\begin{array}{r|rrrr} +2 & 1 & -5 & -8 & 32 \\ & & 2 & -6 & -28 \\ \hline & 1 & -3 & -14 & 4 \end{array}$$

The constants on the bottom line are the coefficients of the quotient, $Q(x)$, and the remainder, $R(x)$. The last constant on the right is, in this case, the remainder. Thus,

$$\underbrace{x^3 - 5x^2 - 8x + 32}_{P(x)} = \underbrace{(x - 2)}_{D(x)}\underbrace{(x^2 - 3x - 14)}_{Q(x)} + \underbrace{4}_{R(x)}$$

In this form, we can write

$$\begin{aligned} P(2) &= D(2) \cdot Q(2) + R(2) \\ &= (2 - 2)(2^2 - 3 \cdot 2 - 14) + 4 \\ &= 0(-16) + 4 \\ &= 4 \end{aligned}$$

Thus, $P(2) = 4$ and 4 is the remainder, the constant term on the right in the synthetic division process.

In general, if $D(x) = (x - c)$, then $D(x)$ is first-degree and $R(x) = r$, a constant; and the division algorithm takes the form

$$P(x) = (x - c) \cdot Q(x) + r$$

We now state and prove the Remainder Theorem which tells us that the remainder, using synthetic division, always gives the value $P(c)$. (**Note:** This theorem is true even if $c$ is a nonreal complex number. We will discuss this situation in more detail as the chapter progresses.)

***Remainder Theorem*** If a polynomial $P(x)$ of degree $n > 0$ is divided by $(x - c)$, then the remainder $r$ is equal to $P(c)$.

**Proof** By the division algorithm, we have

$$\begin{aligned} P(x) &= (x - c) \cdot Q(x) + r \\ \text{So,} \quad P(c) &= (c - c) \cdot Q(c) + r \\ &= (0)\, Q(c) + r \\ &= 0 + r = r \end{aligned}$$

If $P(c) = 0$, then $x = c$ is a **root** (or **solution**) of the polynomial equation $P(x) = 0$. This means that the remainder $r = 0$ and $(x - c)$ is a factor of the polynomial $P(x)$. (See the Factor Theorem in Section 5.1.)

## EXAMPLES

**1.** For $P(x) = x^3 - 5x^2 - 8x + 32$, use synthetic division and the Remainder Theorem to verify that (a) $P(3) = -10$ and (b) $P(-1) = 34$.

**Solution**

**a.**

$$\begin{array}{r|rrrr} 3 & 1 & -5 & -8 & 32 \\ & & 3 & -6 & -42 \\ \hline & 1 & -2 & -14 & -10 \end{array} \quad \leftarrow P(3) = -10$$

**b.**

$$\begin{array}{r|rrrr} -1 & 1 & -5 & -8 & 32 \\ & & -1 & 6 & 2 \\ \hline & 1 & -6 & -2 & 34 \end{array} \quad \leftarrow P(-1) = 34$$

**2.** Given $P(x) = 2x^3 - 7x^2 + 4x + 5$. Use synthetic division and the Remainder Theorem to find (a) $P(-1)$, (b) $P(1)$, and (c) $P(1.5)$.

**Solution**

**a.**
$$\begin{array}{r|rrrr} -1 & 2 & -7 & 4 & 5 \\ & & -2 & 9 & -13 \\ \hline & 2 & -9 & 13 & -8 \end{array} \quad \leftarrow P(-1) = -8$$

**b.**
$$\begin{array}{r|rrrr} 1 & 2 & -7 & 4 & 5 \\ & & 2 & -5 & -1 \\ \hline & 2 & -5 & -1 & 4 \end{array} \quad \leftarrow P(1) = 4$$

**c.**
$$\begin{array}{r|rrrr} 1.5 & 2 & -7 & 4 & 5 \\ & & 3 & -6 & -3 \\ \hline & 2 & -4 & -2 & 2 \end{array} \quad \leftarrow P(1.5) = 2$$ ■

We have techniques for solving (finding roots) of first-degree equations and second-degree equations (quadratic equations). However, if a polynomial equation, $P(x) = 0$, is third-degree or higher, the roots can be difficult to find. In Example 2, we found that for

$$P(x) = 2x^3 - 7x^2 + 4x + 5$$

$P(-1) = -8 < 0$ and $P(1) = 4 > 0$. That is, $P(x)$ is negative for $x = -1$ and positive for $x = 1$. According to the following **Location Principle,** this means that the polynomial equation

$$2x^3 - 7x^2 + 4x + 5 = 0$$

has at least one root between $-1$ and 1. An important restriction to note is that this principle applies only when the equation has real coefficients.

**Location Principle**

If $P(x) = 0$ is a polynomial equation with real coefficients and $P(a) > 0$ and $P(b) < 0$ [or $P(a) < 0$ and $P(b) > 0$], then there is at least one number $c$ between $a$ and $b$ that is a root of the equation. [That is, $P(c) = 0$.]

## EXAMPLES

**3.** Using synthetic division and the Location Principle, show that there is a root of the equation $x^3 + x^2 - 3x - 3 = 0$ between 1 and 2.

**Solution**
$$\begin{array}{r|rrrr} 1 & 1 & 1 & -3 & -3 \\ & & 1 & 2 & -1 \\ \hline & 1 & 2 & -1 & -4 \end{array} \quad \leftarrow P(1) = -4$$

$$\begin{array}{r|rrrr} 2 & 1 & 1 & -3 & -3 \\ & & 2 & 6 & 6 \\ \hline & 1 & 3 & 3 & 3 \end{array} \quad \leftarrow P(2) = 3$$

Since $P(1)$ is negative and $P(2)$ is positive, there is a root between 1 and 2 (possibly more than one root).

**4.** In Example 3, we know that there is a root between 1 and 2. To approximate this root more closely, we can systematically "close in" on the root by dividing synthetically by numbers between 1 and 2 that are accurate to tenths, accurate to hundredths, and so on. Show that the equation $x^3 + x^2 - 3x - 3 = 0$ has a root between 1.7 and 1.8. (**Hint:** A calculator may be useful for finding decimal products.)

**Solution**

$$\begin{array}{r|rrrrl} 1.7 & 1 & 1 & -3 & -3 & \\ & & 1.7 & 4.59 & 2.703 & \\ \hline & 1 & 2.7 & 1.59 & -0.297 & \leftarrow P(1.7) = -0.297 \end{array}$$

$$\begin{array}{r|rrrrl} 1.8 & 1 & 1 & -3 & -3 & \\ & & 1.8 & 5.04 & 3.672 & \\ \hline & 1 & 2.8 & 2.04 & 0.672 & \leftarrow P(1.8) = 0.672 \end{array}$$

Thus, since $P(1.8) > 0$ and $P(1.7) < 0$, there is a root between 1.7 and 1.8. ■

The search for roots of polynomial equations can be time-consuming and difficult even when using a system such as the Location Principle. For speed and accuracy, we would turn to the computer and use a program to choose the decimal numbers and accomplish the synthetic division.

**Practice Problems**

1. Use synthetic division and the Remainder Theorem to find (a) $P(3)$ and (b) $P(-1)$ if $P(x) = 5x^3 - 3x + 4$.
2. Use synthetic division and the Location Principle to show that the equation $2x^3 + 3x^2 - x - 1 = 0$ has a root between 0.6 and 0.7.

## EXERCISES 12.1

Use synthetic division and the Remainder Theorem to find the indicated values for each of the polynomials in Exercises 1–12.

1. $P(x) = 5x^2 - 7x + 1$ (a) $P(-3)$, (b) $P(1)$
2. $P(x) = 2x^2 + 4x - 3$ (a) $P(2)$, (b) $P(-1)$
3. $P(x) = 3x^2 + 8x + 2$ (a) $P(-4)$, (b) $P(1.5)$
4. $P(x) = 7x^2 - 11x + 4$ (a) $P(1)$, (b) $P(2.5)$
5. $P(x) = 7x^3 + 2x - 1$ (a) $P(-2)$, (b) $P(-3)$

**Answers to Practice Problems** 1. (a) 130 (b) 2 2. $P(0.6) = -0.088 < 0$ and $P(0.7) = 0.456 > 0$. Therefore, by the Location Principle, there is a root between 0.6 and 0.7.

**6.** $P(x) = 6x^3 + 3x^2 - x - 2$ (a) $P(-1)$, (b) $P(2)$
**7.** $P(x) = 2x^3 - 5x^2 + 9$ (a) $P(3)$, (b) $P(1.2)$
**8.** $P(x) = 3x^3 + 7x + 5$ (a) $P(-2)$, (b) $P(1.4)$
**9.** $P(x) = x^4 - 7x^3 - 4x^2 + x + 3$ (a) $P(-3)$, (b) $P(1)$
**10.** $P(x) = 2x^4 - 5x^2 + x + 2$ (a) $P(2)$, (b) $P(-1)$
**11.** $P(x) = 2x^4 - 3x^3 + 5x - 7$ (a) $P(-2)$, (b) $P(2.1)$
**12.** $P(x) = x^4 + 2x^3 - 4x^2 + 3x - 9$ (a) $P(-3)$, (b) $P(-1.5)$

Use synthetic division and the Location Principle to show that there is a root of the equation between the two numbers listed in Exercises 13–25.

**13.** $x^2 + 3x - 6 = 0$; between 1 and 2
**14.** $2x^2 - 4x - 9 = 0$; between $-2$ and $-1$
**15.** $x^3 - 3x - 4 = 0$; between 2 and 3
**16.** $x^3 - 2x^2 - 6x - 2 = 0$; between 3 and 4
**17.** $2x^3 + 2x^2 - 5x + 10 = 0$; between $-3$ and $-2$
**18.** $3x^3 - 6x^2 + 4x - 5 = 0$; between 1 and 2
**19.** $x^4 - 3x^3 + 2x^2 + 6x - 3 = 0$; between $-2$ and $-1$
**20.** $x^4 - 2x^3 + 2x^2 - 4x - 7 = 0$; between 2 and 3
**21.** $x^3 + 5x^2 - 2x - 10 = 0$; between 1.4 and 1.5
**22.** $x^3 + x^2 - 10x - 12 = 0$; between $-1.3$ and $-1.2$
**23.** $x^3 + x^2 - 7x - 3 = 0$; between 2.4 and 2.5
**24.** $x^3 - 3x^2 + 2 = 0$; between 2.7 and 2.8
**25.** $x^3 + 3x^2 - 2x - 4 = 0$; between $-3.3$ and $-3.2$

## 12.2 Rational Roots Theorem

**OBJECTIVES**

In this section, you will be learning to:

1. Apply the Rational Roots Theorem to find the possible rational roots of polynomial equations.
2. Find rational roots using synthetic division.
3. Use reduced equations to help in finding all the roots of polynomial equations.

There is no general approach to finding roots of polynomial equations. We do know how to solve first-degree equations, and we have developed several techniques for solving quadratic equations (factoring, completing the square, and using the quadratic formula). In Section 12.1, we discussed the Location Principle which is particularly helpful for approximating roots of equations of any degree.

In this section, we will discuss a method for finding a list of possible rational roots for polynomial equations with the condition that the coefficients must be integers. The proof of the theorem is omitted. To understand the theorem, we need to know that:

If $\frac{p}{q}$ is a rational number **in lowest terms,** then $p$ and $q$ are integers and $+1$ and $-1$ are the only common factors of $p$ and $q$.

**Rational Roots Theorem** If $\frac{p}{q}$ is a rational number in lowest terms and $\frac{p}{q}$ is a root of the polynomial equation

$$P(x) + a_nx^n + a_{n-1}x^{n-1} + \cdots + a_1x + a_0 + 0$$

where $a_n \neq 0$ and the coefficients are integers, then $p$ is a factor of $a_0$ and $q$ is a factor of $a_n$.

In simplified terms, the theorem says that the only possible rational roots of a polynomial equation with integer coefficients can be listed by dividing all factors of $a_0$ by all factors of $a_n$. The theorem does not say that any of these numbers is a root. However, if none of these rational numbers is a root, then the equation has no rational roots. That is, in that case, all the roots will be irrational or nonreal complex numbers.

Synthetic division can be used to test whether or not a number is a root of a polynomial equation. If $P(x)$ is divided by $(x - a)$ and the remainder is 0, then $a$ is a root. That is, $P(a) = 0$ is another way of saying that $a$ is a root of the equation.

### EXAMPLES

**1.** List all the possible rational roots of the following polynomial equation:

$$3x^3 - 7x^2 - 3x + 2 = 0$$

**Solution** Here, $a_0 = 2$ and $a_n = 3$.
The factors of 2 are $+2, -2, +1$, and $-1$.
The factors of 3 are $+3, -3, +1$, and $-1$.
Thus, the possible rational roots are $\pm\frac{2}{3}, \pm 2, \pm\frac{1}{3}, \pm 1$.

**2.** Use synthetic division to determine which, if any, of the possible rational roots found in Example 1 is a root of that equation.

**Solution** There are eight numbers to try, and at this stage we have no reason to choose one number over the others.

Try $x = \frac{2}{3}$.

$$\begin{array}{c|cccc} \frac{2}{3} & 3 & -7 & -3 & 2 \\ & & 2 & -\frac{10}{3} & -\frac{38}{9} \\ \hline & 3 & -5 & -\frac{19}{3} & -\frac{20}{9} \end{array} \leftarrow P\left(\frac{2}{3}\right) = -\frac{20}{9}$$

So, $x = \frac{2}{3}$ is **not** a root of the equation.

Try $x = -\dfrac{2}{3}$.

$$\begin{array}{r|rrrr} -\frac{2}{3} & 3 & -7 & -3 & 2 \\ & & -2 & 6 & -2 \\ \hline & 3 & -9 & 3 & 0 \end{array} \quad \leftarrow P\left(-\frac{2}{3}\right) = 0$$

So, $x = -\dfrac{2}{3}$ is a root of the equation.

Continuing with this process, we will find that $-\dfrac{2}{3}$ is the only rational root. ■

From the last synthetic division operation in Example 2, we know that $x = -\dfrac{2}{3}$ is a root of the equation and that the remaining coefficients are the coefficients of $Q(x)$ in the division algorithm. Thus, we can write the equation

$$x^3 - 7x^2 - 3x + 2 = 0$$

in the form

$$\left(x + \frac{2}{3}\right)(3x^2 - 9x + 3) = 0$$

The equation

$$3x^2 - 9x + 3 = 0$$

is called the **reduced equation;** and solutions (roots) of the reduced equation are also solutions of the original equation.

**EXAMPLE**

**3.** Find all the solutions of the equation

$$3x^3 - 7x^2 - 3x + 2 = 0$$

**Solution** From Example 2, we know that $x = -\dfrac{2}{3}$ is a root:

$$\begin{array}{r|rrrr} -\frac{2}{3} & 3 & -7 & -3 & 2 \\ & & -2 & 6 & -2 \\ \hline & 3 & -9 & 3 & 0 \end{array} \quad \leftarrow P\left(-\frac{2}{3}\right) = 0$$

Now, using the reduced equation $3x^2 - 9x + 3 = 0$ and the quadratic formula, we have

$$x = \frac{9 \pm \sqrt{(-9)^2 - 4(3)(3)}}{2 \cdot 3} = \frac{9 \pm \sqrt{81 - 36}}{6}$$

$$= \frac{9 \pm \sqrt{45}}{6} = \frac{9 \pm 3\sqrt{5}}{6} = \frac{3 \pm \sqrt{5}}{2}$$

The equation has three real solutions, one rational and two irrational:

$$-\frac{2}{3}, \quad \frac{3+\sqrt{5}}{2}, \quad \text{and} \quad \frac{3-\sqrt{5}}{2}.$$

**4.** Find all the solutions of the polynomial equation

$$x^4 - 7x^3 + 6x^2 + 2x - 12 = 0$$

**Solution** Since $a_n = 1$, the possible rational roots are the factors of $a_0 = -12$: $\pm 1, \pm 2, \pm 3, \pm 4, \pm 6, \pm 12$.

Try $x = 6$.

$$\begin{array}{r|rrrrr} 6 & 1 & -7 & 6 & 2 & -12 \\ & & 6 & -6 & 0 & 12 \\ \hline & 1 & -1 & 0 & 2 & 0 \end{array} \quad \leftarrow P(6) = 0$$

So $x = 6$ is a root. Using the reduced equation $x^3 - x^2 + 2 = 0$, try $x = -1$.

$$\begin{array}{r|rrrr} -1 & 1 & -1 & 0 & 2 \\ & & -1 & 2 & -2 \\ \hline & 1 & -2 & 2 & 0 \end{array} \quad \leftarrow P(-1) = 0$$

Thus, two rational roots are $x = 6$ and $x = -1$. Using the quadratic formula with the reduced equation $x^2 - 2x + 2 = 0$, we find that two other roots are nonreal complex numbers as follows:

$$x = \frac{2 \pm \sqrt{(-2)^2 - 4(1)(2)}}{2 \cdot 1} = \frac{2 \pm \sqrt{4-8}}{2} = \frac{2 \pm \sqrt{-4}}{2}$$

$$= \frac{2 \pm 2i}{2} = 1 \pm i$$

■

There are two real roots and two nonreal roots: $6, -1, 1 + i$, and $1 - i$.

## Practice Problems

**1.** List all the possible rational roots of the polynomial equation

$$2x^3 + 3x^2 - 18x + 8 = 0$$

Find all the roots of the following polynomial equations.

**2.** $2x^3 + 3x^2 - 18x + 8 = 0$ **3.** $x^3 - 2x^2 - 6x + 9 = 0$

**Answers to Practice Problems** **1.** $\pm 1, \pm\frac{1}{2}, \pm 2, \pm 4, \pm 8$ **2.** $x = -4, x = \frac{1}{2}, x = 2$ **3.** $x = 3, x = \frac{-1 \pm \sqrt{13}}{2}$

## EXERCISES 12.2

List the possible rational roots of each of the polynomial equations in Exercises 1–10.

**1.** $x^3 + 7x^2 - 2x + 6 = 0$
**2.** $x^3 - 8x^2 + 3x - 9 = 0$
**3.** $x^4 - x^3 + 3x^2 - x + 8 = 0$
**4.** $x^4 + 5x^3 - 4x + 15 = 0$
**5.** $2x^3 - 6x^2 + x - 7 = 0$
**6.** $3x^3 + 6x^2 - x + 11 = 0$
**7.** $6x^4 - 5x^3 + x^2 - 3x - 5 = 0$
**8.** $4x^4 - x^2 + 4x - 2 = 0$
**9.** $3x^4 + x^3 - 7x^2 + 2x + 12 = 0$
**10.** $5x^4 - 6x^3 + x^2 - 3x + 20 = 0$

Use the Rational Roots Theorem and synthetic division to find the rational roots of each of the polynomial equations in Exercises 11–20.

**11.** $x^3 - x^2 - 13x + 4 = 0$
**12.** $x^3 + x^2 - 4x + 6 = 0$
**13.** $2x^3 - 7x^2 - 14x - 5 = 0$
**14.** $3x^3 + 2x^2 - 19x + 6 = 0$
**15.** $3x^3 + 21x^2 + 16x - 12 = 0$
**16.** $2x^3 + 3x^2 - 17x + 12 = 0$
**17.** $6x^3 - 7x^2 - 18x - 5 = 0$
**18.** $10x^3 + 7x^2 - 14x - 3 = 0$
**19.** $5x^4 + 29x^3 + 34x^2 + 67x - 15 = 0$
**20.** $2x^4 - 3x^3 - 25x^2 + 55x - 21 = 0$

Find all roots of each of the polynomial equations in Exercises 21–30.

**21.** $4x^3 + 13x^2 - 8x - 3 = 0$
**22.** $3x^3 - 13x^2 - 9x + 35 = 0$
**23.** $2x^3 + x^2 + x - 1 = 0$
**24.** $3x^3 + 4x^2 + 7x + 2 = 0$
**25.** $x^3 - 4x^2 + x + 6 = 0$
**26.** $x^3 - 2x^2 - 4x + 8 = 0$
**27.** $7x^3 - 25x^2 + 19x - 4 = 0$
**28.** $3x^3 - 4x^2 - 2x + 1 = 0$
**29.** $2x^4 - x^3 + 5x^2 - 4x - 12 = 0$
**30.** $4x^4 + 8x^3 + 11x^2 - 2x - 3 = 0$

## 12.3 Upper and Lower Bounds for Roots

**OBJECTIVES**

In this section, you will be learning to:

1. Show that all the real roots of polynomial equations lie between two given real numbers.
2. Locate upper and lower bounds for the real roots of polynomial equations.

From Section 12.1, we know how to approximate roots of polynomial equations using synthetic division; and from Section 12.2, we know how to investigate a list of possible rational roots when the coefficients are integers. In this section, we will develop a technique for narrowing the search for roots by finding upper and lower bounds for the roots.

| | |
|---|---|
| ***Upper Bound*** | A real number $c$ is an **upper bound** for the roots of an equation if there are no roots greater than $c$. |
| ***Lower Bound*** | A real number $c$ is a **lower bound** for the roots of an equation if there are no roots less than $c$. |

Consider the polynomial equation $4x^4 + 4x^3 - x^2 - 31x + 15 = 0$. From the Rational Roots Theorem, we know that the possible rational roots are the factors of 15 divided by the factors of 4:

$$\pm\frac{1}{4}, \pm\frac{1}{2}, \pm 1, \pm\frac{3}{4}, \pm\frac{3}{2}, \pm 3, \pm\frac{5}{4}, \pm\frac{5}{2}, \pm 5, \pm\frac{15}{4}, \pm\frac{15}{2}, \pm 15$$

This knowledge is certainly better than blind guessing; but, we might divide as many as 24 times and still not find a root. We could narrow the search dramatically if we could show, for example, that there are no roots greater than 2. If this is the case, then we would not need to try synthetic division with any of the rational values greater than 2; namely,

$$3, \frac{5}{2}, 5, \frac{15}{4}, \frac{15}{2}, \text{ or } 15$$

The following theorem tells us how to use the signs of the coefficients in the last row in synthetic division to find upper and lower bounds for roots of polynomial equations.

**Theorem of Upper and Lower Bounds**

Suppose that $P(x) = 0$ is a polynomial equation with real coefficients, where the leading coefficient $a_n$ is positive, and $P(x)$ is divided by $(x - c)$ using synthetic division.

**a.** If $c > 0$ and all of the coefficients in the last row are positive or 0, then $c$ is an upper bound for the roots of the equation.

**b.** If $c < 0$ and the coefficients in the last row alternate in sign (with 0 treated as positive or negative), then $c$ is a lower bound for the roots of the equation.

(**Note:** If the leading coefficient $a_n$ is negative, then multiply both sides of the equation by $-1$. This will not affect the roots and will give a positive leading coefficient so that the theorem can be used.)

**EXAMPLES**

**1.** Show that $x = 2$ is an upper bound for the roots of the equation

$$4x^4 + 4x^3 - x^2 - 31x + 15 = 0$$

**Solution**

$$\begin{array}{r|rrrrr} 2 & 4 & 4 & -1 & -31 & 15 \\ & & 8 & 24 & 46 & 30 \\ \hline & 4 & 12 & 23 & 15 & 45 \end{array}$$

Since all the coefficients in the last row are positive or 0, from the Upper and Lower Bounds Theorem, we know that $x = 2$ is an upper bound for the roots.

**Note:** To understand why these results are reasonable, using the division algorithm, we could write

$$P(x) = (x - 2)(4x^3 + 12x^2 + 23x + 15) + 45$$

For any $k > 2$, $k - 2$ is positive, and substituting $k$ for $x$ gives

$$P(k) = (k - 2)(4k^3 + 12k^2 + 23k + 15) + 45 > 0$$

Thus, $P(k) \neq 0$ for any $k > 2$ and $k$ could not be a root.

**2.** Determine whether or not the integer $x = 1$ is an upper bound for the roots of the equation in Example 1.

**Solution** Divide synthetically by $(x - 1)$.

$$\begin{array}{r|rrrrr} 1 & 4 & 4 & -1 & -31 & 15 \\ & & 4 & 8 & 7 & -48 \\ \hline & 4 & 8 & 7 & -24 & -33 \end{array}$$

Since the coefficients in the last row are not all positive or 0, we cannot conclude from the theorem that $x = 1$ is an upper bound for the roots. Thus, from Examples 1 and 2, we can conclude that $x = 2$ is the smallest integer that is an upper bound for the roots that can be determined from the theorem. (**Note:** The number $x = 1$ may or may not be an upper bound for the roots. The Upper and Lower Bounds Theorem cannot be used to make the decision. That is, the theorem allows us to determine upper and lower bounds for the roots, but not necessarily the least upper bound nor the greatest lower bound.)

**3.** Find the largest negative integer that can be determined by the Upper and Lower Bounds Theorem that is a lower bound for the roots of the equation.

$$2x^4 - x^3 - 2x^2 - 4x - 40 = 0$$

**Solution** Try $x = -3$. [If this works (that is, the signs alternate), then we try $x = -2$. If it does not work, we try $x = -4$; and so on.]

$$\begin{array}{r|rrrrr} -3 & 2 & -1 & -2 & -4 & -40 \\ & & -6 & 21 & -57 & 123 \\ \hline & 2 & -7 & 19 & -61 & 83 \end{array}$$

The signs of the coefficients in the last row alternate so we know that $x = -3$ is a lower bound for the roots. To determine whether there is a greater integer lower bound, we next try $x = -2$.

$$\begin{array}{r|rrrrr} -2 & 2 & -1 & -2 & -4 & -40 \\ & & -4 & 10 & -16 & 40 \\ \hline & 2 & -5 & 8 & -20 & 0 \end{array}$$

In this case, we have found that $x = -2$ is a root and it also satisfies the theorem. Therefore, it is the greatest negative integer that is a lower bound for the roots of the equation. ■

**Practice Problems**

1. Determine whether or not $x = 2$ is an upper bound for the roots of the polynomial equation

$$4x^4 - 8x^3 + 3x^2 - 6x + 1 = 0$$

2. Find the largest negative integer and the smallest positive integer that can be determined by the Upper and Lower Bounds Theorem to be bounds for the roots of the polynomial equation

$$x^3 - 7x^2 + 3x - 10 = 0$$

3. Show that all real roots of the equation $2x^3 - 5x^2 - x + 6 = 0$ are between $-2$ and 3.

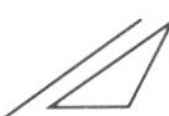

## EXERCISES 12.3

Use the Upper and Lower Bounds Theorem to show that all real roots of each of the polynomial equations in Exercises 1–10 are located between the given integers.

**1.** $x^3 - 5x + 2 = 0$; between $-3$ and 3
**2.** $x^3 + 2x^2 - 5x - 6 = 0$; between $-6$ and 2
**3.** $x^3 - 3x^2 - 16x + 48 = 0$; between $-4$ and 6
**4.** $6x^3 + 17x^2 - 4x - 3 = 0$; between $-4$ and 1
**5.** $x^4 - 4x^3 + 4x^2 - 36x - 45 = 0$; between $-1$ and 5
**6.** $x^4 - 3x^3 - 9x^2 - 3x - 10 = 0$; between $-2$ and 5
**7.** $x^4 + x^3 - 2x + 3 = 0$; between $-3$ and 1
**8.** $2x^4 + 5x^3 - 6x^2 + x - 3 = 0$; between $-4$ and 2
**9.** $x^3 - 8x^2 + 3x - 12 = 0$; between $-1$ and 12
**10.** $2x^3 - 6x^2 + 3x - 8$; between $-1$ and 4

Find the largest negative integer and the smallest positive integer that can be determined by the Upper and Lower Bounds Theorem to be bounds for the roots of each of the polynomial equations in Exercises 11–30.

**11.** $x^3 + 2x^2 + 3x - 4 = 0$
**12.** $x^3 - 5x^2 + 2 = 0$
**13.** $2x^3 + 5x^2 + x + 2 = 0$
**14.** $2x^3 + 3x^2 - 12x - 24 = 0$
**15.** $x^4 + 6x^3 - 5x^2 + 2x - 7 = 0$
**16.** $x^4 - 2x^3 - 11x^2 - 2x - 1 = 0$
**17.** $3x^3 - 7x^2 - 3x + 2 = 0$
**18.** $2x^3 - 6x^2 - 7x - 21 = 0$
**19.** $x^3 - 12x^2 - 11x + 25 = 0$
**20.** $x^3 - 5x^2 - 46x - 36 = 0$
**21.** $x^4 - 20x^2 - 21x - 24 = 0$
**22.** $x^4 - 9x^3 + 15x^2 + 14x + 20 = 0$
**23.** $2x^4 - x^3 - 9x^2 + 21 = 0$
**24.** $x^4 - 5x^3 - 3x^2 - 26x + 8 = 0$
**25.** $4x^4 - 6x^3 + 5x^2 + x - 15 = 0$
**26.** $x^4 + 2x^2 + 3x - 14 = 0$
**27.** $x^5 - x^4 - 10x^3 + 28x^2 - 64x + 16 = 0$
**28.** $x^5 + x^4 - 5x^3 + 3x^2 + 2x - 30 = 0$
**29.** $2x^5 + 2x^4 - 2x^3 - 20x^2 - 21x - 18 = 0$
**30.** $2x^5 - 3x^4 - 7x^3 - 15x^2 + 7x + 14 = 0$

**Answers to Practice Problems** **1.** $x = 2$ is an upper bound. **2.** Synthetic division shows that $x = 7$ is an upper bound and $x = -1$ is a lower bound. **3.** Synthetic division shows that $x = 3$ is an upper bound and $x = -2$ is a lower bound.

## 12.4 Descartes' Rule of Signs

**OBJECTIVES**

In this section, you will be learning to use Descartes' Rule of Signs to determine the possible numbers of positive and negative real roots of polynomial equations.

In this section, we will develop still another technique for narrowing the search for roots of polynomial equations called **Descartes' Rule of Signs.** If a polynomial, $P(x)$, is arranged in order of descending powers, then there is a **variation in sign** if two consecutive terms have opposite signs. For example, the polynomial

$$P(x) = x^4 + 6x^3 - 6x^2 + 6x - 7$$

has three variations in sign. The variations in sign are noted by investigating the signs of the coefficients as follows:

$$\begin{array}{ccccc} +1 & +6 & -6 & +6 & -7 \\ & \diagdown\ \diagup & \diagdown\ \diagup & \diagdown\ \diagup & \\ & 1 & 2 & 3 & \end{array}$$

**Descartes' Rule of Signs**

Suppose that $P(x) = 0$ is a polynomial equation with real coefficients and $a_0 \neq 0$.

**a.** The number of positive real roots is equal to the number of variations in sign of $P(x)$ or to that number decreased by an even integer.

**b.** The number of negative real roots is equal to the number of variations in sign of $P(-x)$ or to that number decreased by an even integer.

The proof of the theorem is left for later courses in mathematics.

### EXAMPLES

**1.** Use Descartes' Rule of Signs to determine the possible numbers of positive real roots and negative real roots for the equation

$$x^4 + 6x^3 - 6x^2 + 6x - 7 = 0$$

**Solution** The polynomial $P(x) = x^4 + 6x^3 - 6x^2 + 6x - 7$ has three changes in sign. Therefore, by Descartes' Rule of Signs, there are either three positive real roots or one positive real root.

For

$$\begin{aligned} P(-x) &= (-x)^4 + 6(-x)^3 - 6(-x)^2 + 6(-x) - 7 \\ &= x^4 - 6x^3 - 6x^2 - 6x - 7 \end{aligned}$$

we see that $P(-x)$ has only one variation in sign. Therefore, the original equation $P(x) = 0$ has exactly one negative real root. Thus, if we were to find a negative real root, it would be the only one, and we would simply be wasting our time if we tried to find any more negative real roots.

**2.** Find all the roots of the equation

$$x^4 + 6x^3 - 6x^2 + 6x - 7 = 0$$

**Solution** From Example 1, we know that there is at least one positive real root. (Possibly as many as three.) Using the Rational Roots Theorem, we first try factors of $a_0 = -7$ since $a_n = 1$.

Try $x = 1$.

$$\begin{array}{r|rrrrr} 1 & 1 & 6 & -6 & 6 & -7 \\ & & 1 & 7 & 1 & 7 \\ \hline & 1 & 7 & 1 & 7 & 0 \end{array}$$

Thus, $x = 1$ is a root. Since the reduced polynomial has no changes in sign, there are no other positive real roots. From Example 1, we know that there is exactly one negative real root and we try $x = -1$ and $x = -7$ (the negative factors of 7) using the reduced polynomial.

$$\begin{array}{r|rrrr} -1 & 1 & 7 & 1 & 7 \\ & & -1 & -6 & 5 \\ \hline & 1 & 6 & -5 & 12 \end{array} \qquad \begin{array}{r|rrrr} -7 & 1 & 7 & 1 & 7 \\ & & -7 & 0 & -7 \\ \hline & 1 & 0 & 1 & 0 \end{array}$$

Thus, $x = -7$ is the negative real root. Furthermore, we now know that there are no other real roots. In fact, using the new reduced equation $x^2 + 1 = 0$, we get

$$\begin{aligned} x^2 + 1 &= 0 \\ (x - i)(x + i) &= 0 \\ x &= \pm i \end{aligned}$$

Thus there are four roots, two real and two nonreal:

$$x = 1,\ x = -7,\ x = i, \text{ and } x = -i$$

We could write $P(x)$ in factored form as

$$\begin{aligned} P(x) &= (x - 1)(x^3 + 7x^2 + x + 7) \\ &= (x - 1)(x + 7)(x^2 + 1) \\ &= (x - 1)(x + 7)(x - i)(x + i) \end{aligned}$$

**3.** Use Descartes' Rule of Signs to determine the possible numbers of positive and negative real roots for the equation

$$x^4 + 3x^3 - 2x^2 - 5x + 3 = 0$$

Then find as many roots as you can using reduced polynomials, upper and lower bounds, the Rational Roots Theorem, and the quadratic formula.

**Solution**

$$P(x) = x^4 + 3x^3 - 2x^2 - 5x + 3 \qquad \text{(two variations in sign)}$$
$$P(-x) = x^4 - 3x^3 - 2x^2 + 5x + 3 \qquad \text{(two variations in sign)}$$

There are either two or no positive real roots and two or no negative real roots.
If there are any rational roots, they are factors of 3. Try $x = 1$.

$$\begin{array}{r|rrrrr} 1 & 1 & 3 & -2 & -5 & 3 \\ & & 1 & 4 & 2 & -3 \\ \hline & 1 & 4 & 2 & -3 & 0 \end{array}$$

So, $x = 1$ is a positive real root and there is one other positive real root, rational or irrational.
Try $x = 3$ using the reduced equation.

$$\begin{array}{r|rrrr} 3 & 1 & 4 & 2 & -3 \\ & & 3 & 21 & 69 \\ \hline & 1 & 7 & 23 & 66 \end{array}$$

$x = 3$ is not a root; so the other positive real root is irrational. Since all the coefficients in the last row are positive, $x = 3$ is an upper bound for the positive real roots.
Now, try $x = -3$.

$$\begin{array}{r|rrrr} -3 & 1 & 4 & 2 & -3 \\ & & -3 & -3 & 3 \\ \hline & 1 & 1 & -1 & 0 \end{array}$$

Thus, $x = -3$ is a root, and there is one other negative real root.
Using the quadratic formula to solve the reduced equation $x^2 + x - 1 = 0$ gives

$$x = \frac{-1 \pm \sqrt{1^2 - 4(1)(-1)}}{2 \cdot 1} = \frac{-1 \pm \sqrt{5}}{2}$$

Thus, the roots are $x = 1$, $x = -3$, $x = \dfrac{-1 + \sqrt{5}}{2}$, and $x = \dfrac{-1 - \sqrt{5}}{2}$.

■

**Practice Problems**

1. Use Descartes' Rule of Signs to determine the possible numbers of positive and negative real roots of the equation

$$2x^4 - 5x^3 - 11x^2 + 20x + 12 = 0$$

Find all the roots of the following polynomial equations.

2. $2x^4 - 5x^3 - 11x^2 + 20x + 12 = 0$
3. $x^4 - 5x^3 + 7x^2 - 7x - 20 = 0$

## EXERCISES 12.4

Use Descartes' Rule of Signs to determine the possible numbers of positive and negative real roots of the following equations.

1. $x^3 + 3x^2 + 4x = 0$
2. $x^3 - 5x - 1 = 0$
3. $x^3 - 2x^2 + 9x - 2 = 0$
4. $x^3 + 15x^2 - 5x - 6 = 0$
5. $x^4 - 5x^3 - 8x + 1 = 0$
6. $x^4 - 3x^2 + 2x - 1 = 0$
7. $2x^4 - 3x^3 + x^2 + 2x - 3 = 0$
8. $6x^4 + 4x^3 - 2x^2 - 5x - 1 = 0$
9. $x^6 - 2x^5 + 3x^3 + 2 = 0$
10. $x^5 + x^3 + x^2 - 2x + 3 = 0$
11. $x^7 + 4x^6 - x^5 + x^3 - 2x - 3 = 0$
12. $2x^7 - 5x^6 + x^4 - 6x^2 - x + 4 = 0$

Use Descartes' Rule of Signs to determine the possible numbers of positive and negative real roots of the following equations. Then find as many roots as you can using reduced polynomials, upper and lower bounds, the Rational Roots Theorem, and the quadratic formula.

13. $2x^3 - 5x^2 + x + 2 = 0$
14. $2x^3 + 8x^2 + 3x + 12 = 0$
15. $3x^3 - 7x^2 - 3x + 2 = 0$
16. $2x^3 + x^2 - 18x - 20 = 0$
17. $6x^4 - 17x^3 - 4x^2 + 21x - 6 = 0$
18. $2x^4 + 11x^3 + 15x^2 + 7x + 1 = 0$
19. $x^4 + x^3 - 11x^2 + 5x + 4 = 0$
20. $2x^4 + 7x^3 - 10x + 4 = 0$
21. $3x^4 + x^3 + 2x^2 + 4x - 40 = 0$
22. $x^4 - 2x^3 - x^2 - 4x - 6 = 0$
23. $3x^4 + 11x^3 + 14x^2 + 6x - 4 = 0$
24. $x^4 - 5x^2 - 10x - 6 = 0$
25. $x^4 - 19x^2 + 3x + 18 = 0$
26. $x^4 - 2x^3 + x^2 - 8x - 12 = 0$
27. $x^5 - x^4 - 2x^3 - x^2 + x + 2 = 0$
28. $4x^5 - 18x^4 + 21x^3 - 14x^2 + 60x - 72 = 0$
29. $2x^5 + 5x^4 + 6x^3 - 14x^2 - 56x - 24 = 0$
30. $3x^5 + 8x^4 - 6x^3 - 38x^2 - 17x + 10 = 0$

**Answers to Practice Problems** 1. There are two positive real roots or none. There are two negative real roots or none. 2. $x = -2$, $x = -\frac{1}{2}$, $x = 2$, $x = 3$ 3. $x = -1$, $x = 4$, $x = 1 + 2i$, $x = 1 - 2i$

# 12.5 The Fundamental Theorem and the Number of Roots

**OBJECTIVES**

In this section, you will be learning to:

1. Find polynomial equations that have given values as roots.
2. Find all the roots of polynomial equations with the help of the Fundamental Theorem of Algebra and the Complex Conjugates Theorem.

In this chapter, we have discussed the nature and location of roots of polynomial equations with the implied assumption that roots do in fact exist for those equations. The important fact that every polynomial equation has at least one root is guaranteed by the **Fundamental Theorem of Algebra.** The theorem is stated here without proof.

***The Fundamental Theorem of Algebra***

If a polynomial $P(x)$ with complex coefficients has degree $n > 0$, then the equation $P(x) = 0$ has at least one root.

The Fundamental Theorem does not tell how many roots a polynomial equation has or how to find them. It merely assures us that at least one root does exist. Thus, while we may have difficulty solving an equation such as $x^3 - 2ix + 4 - 7i = 0$, we do know that at least one root does exist. Many of our previously discussed techniques for solving equations would not apply to this equation since not all the coefficients are real.

Just how many roots are there for a polynomial equation of degree $n > 0$? The following discussion shows the reasoning necessary to answer this question.

Let $P(x) = 0$ be a polynomial equation of degree $n > 0$ with complex coefficients. Then, by the Fundamental Theorem, the equation has at least one root, say $x = r_1$, real or nonreal. Then, $(x - r_1)$ is a factor of $P(x)$, and we can write

$$P(x) = (x - r_1) \cdot Q_1(x)$$

Now, again by the Fundamental Theorem, the reduced equation $Q_1(x) = 0$ has at least one root, say $x = r_2$. Thus, $(x - r_2)$ is a factor of $Q_1(x)$, and we can write

$$P(x) = (x - r_1)(x - r_2) \cdot Q_2(x)$$

Since $P(x)$ is of degree $n > 0$, we can apply the theorem $n$ times and conclude that

$$P(x) = a_n(x - r_1)(x - r_2)(x - r_3) \cdot \cdot \cdot (x - r_n)$$

We have not said that the roots $r_1, r_2, r_3, \ldots, r_n$ are distinct. Some of them may appear in the list more than once. For example, the equation

$$(x - 5)^2(x + 4) = 0$$

has two distinct roots, $x = 5$ and $x = -4$. Since the factor $(x - 5)$ appears twice, we say that $x = 5$ is a **double root** or a **root of multiplicity 2.** For the equation

$$(x - 2)^4(x + 1)^3(x + 5) = 0$$

$x = 2$ is a root of multiplicity 4, $x = -1$ is a root of multiplicity 3, and $x = -5$ is a **simple root** or a **root of multiplicity 1.**

We know that the equation $P(x) = 0$ has the roots $r_1, r_2, \ldots, r_n$, not necessarily distinct. Can there be another root, say $x = c$? If so, $P(c) = 0$ and we can write

$$P(c) = a_n(c - r_1)(c - r_2) \ldots (c - r_n) = 0$$

Since the product is 0 and $a_n \neq 0$, at least one of the factors must be equal to 0. Therefore, $c$ must equal one of the previous roots and there are at most $n$ roots. In fact, we have proved the following theorem concerning the number of roots of a polynomial equation.

***The Number of Roots Theorem***

A polynomial equation $P(x) = 0$ of degree $n > 0$ has at most $n$ distinct roots. The sum of the multiplicities of the roots is $n$.

**EXAMPLE**

**1.** For the equation

$$(x - 2)^4(x + 1)^3(x - 5) = 0$$

find the multiplicity of each root and show that the sum of the multiplicities of the roots is equal to the degree of the equation.

**Solution** The product of all the factors in $P(x)$ would be an 8th degree polynomial so the equation is 8th degree.

The power of each factor indicates that $x = 2$ is a root of multiplicity 4, $x = -1$ is a root of multiplicity 3, and $x = 5$ is a simple root. The sum of the multiplicities is $4 + 3 + 1 = 8$, the degree of the equation. ■

To finalize our discussion of the quantity and nature of the roots of polynomial equations, we state the Complex Conjugates Theorem, without proof, and show how the theorem can be used.

| **The Complex Conjugates Theorem** | If $P(x) = 0$ is a polynomial equation, with real coefficients, and $x = a + bi$ $(b \neq 0)$ is a root, then $x = a - bi$ is also a root. |
|---|---|

The theorem states that nonreal complex roots of polynomial equations with real coefficients always come in conjugate pairs. Thus, if we know one nonreal complex root, we automatically know that its conjugate is also a root, provided that the equation has real coefficients for $P(x)$.

### EXAMPLES

**2.** Find all the roots of the polynomial equation

$$x^4 - 4x^3 + 2x^2 + x + 6 = 0$$

**Solution** Synthetic division gives the following results.

$$\begin{array}{r|rrrrr} 3 & 1 & -4 & 2 & 1 & 6 \\ & & 3 & -3 & -3 & -6 \\ \hline & 1 & -1 & -1 & -2 & 0 \end{array} \qquad \begin{array}{r|rrrr} 2 & 1 & -1 & -1 & -2 \\ & & 2 & 2 & 2 \\ \hline & 1 & 1 & 1 & 0 \end{array}$$

Using the quadratic formula to find the roots of the reduced equation $x^2 + x + 1 = 0$ gives

$$x = \frac{-1 + \sqrt{1^2 - 4(1)(1)}}{2 \cdot 1} = \frac{-1 + \sqrt{-3}}{2} = -\frac{1}{2} + \frac{\sqrt{3}}{2}i$$

Thus, there are four roots:

$$x = 3,\ x = 2,\ x = -\frac{1}{2} + \frac{\sqrt{3}}{2}i \text{ and } x = -\frac{1}{2} - \frac{\sqrt{3}}{2}i$$

and the nonreal complex roots are conjugates of each other as stated in the Complex Conjugates Theorem.

**3.** Find the 4th degree polynomial equation with real coefficients that has $x = 3$ as a double root and the complex root $x = 2 - i$.

**Solution** We know that $(x - 3)$ is a factor twice since $x = 3$ is a double root. Also, since $x = 2 - i$ is a root, $x = 2 + i$ is a root because nonreal complex roots come in conjugate pairs. Thus, we can write the desired equation in the form of a product of factors as follows:

$$(x - 3)^2[x - (2 - i)][x - (2 + i)] = 0$$

Multiplying the factors gives

$$\begin{aligned} &(x^2 - 6x + 9)[x - 2 + i][x - 2 - i] \\ &= (x^2 - 6x + 9)[(x - 2)^2 + 1] \\ &= (x^2 - 6x + 9)[x^2 - 4x + 5] \\ &= x^4 - 10x^3 + 53x^2 - 66x + 45 = 0 \end{aligned}$$

The equation is 4th degree with real coefficients.

**4.** Find a 5th degree polynomial equation with real coefficients that has the simple roots $x = 1$, $x = -1 + i$, and $x = 2 - 3i$.

**Solution** We know that $x = -1 - i$ and $x = 2 + 3i$ are also roots because nonreal complex roots come in conjugate pairs. So, we form an equation using the factors $(x - 1)$, $[x - (-1 + i)]$, $[x - (-1 - i)]$, $[x - (2 + 3i)]$, and $[x - (2 - 3i)]$ and multiply.

The equation, with real coefficients, is

$$\begin{aligned}
&(x - 1)[x - (-1 + i)][x - (-1 - i)][x - (2 + 3i)][x - (2 - 3i)] \\
&= (x - 1)[x + 1 - i][x + 1 + i][x - 2 - 3i][x - 2 + 3i] \\
&= (x - 1)[(x + 1)^2 + 1][(x - 2)^2 + 9] \\
&= (x - 1)[x^2 + 2x + 2][x^2 - 4x + 13] \\
&= x^5 - 3x^4 + 9x^3 + 11x^2 + 8x - 26 = 0
\end{aligned}$$

■

**Practice Problems** Find a polynomial equation of lowest degree with real coefficients that has the given set of numbers as roots.

**1.** $-2, 1, 3$

**2.** $-3, 2 + i$

**3.** $-1$ with multiplicity 3

**4.** Find all the roots, together with their multiplicities, of the polynomial equation

$$x^4 - 8x^3 + 22x^2 - 24x + 9 = 0$$

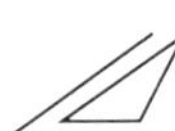

## EXERCISES 12.5

Find a polynomial equation of lowest degree with real coefficients that has the given set of numbers as roots in Exercises 1–12.

**1.** $-1, 3, 4$

**2.** $2, 1, -5$

**3.** $6$, $2$ with multiplicity 2

**4.** $-2$ with multiplicity 3

**5.** $0, -1$, $3$ with multiplicity 2

**6.** $5, 1 + \sqrt{2}, 1 - \sqrt{2}$, $0$ with multiplicity 2

**7.** $-3, 5, 2 - \sqrt{3}, 2 + \sqrt{3}$

**8.** $3, -2, -i\sqrt{2}$

**9.** $2, -1, i\sqrt{7}$

**10.** $1 + i, 2 - \sqrt{2}, 2 + \sqrt{2}$, $1$ with multiplicity 2

**11.** $-4 + i, 1 + \sqrt{5}, 1 - \sqrt{5}$, $-2$ with multiplicity 2

**12.** $4i, 3 + 2i, 3, -3$

For each of the polynomial equations in Exercises 13–30, find all roots together with their multiplicities. Some roots are given.

**13.** $2x^3 - x^2 - 7x + 6 = 0$

**14.** $4x^3 - 21x^2 + 18x + 27 = 0$

**15.** $3x^3 + 4x^2 - 12x - 16 = 0$

**16.** $x^4 - 6x^3 + 13x^2 - 12x + 4 = 0$; 2

**17.** $x^3 + 3x^2 - 5x - 39 = 0$

**18.** $4x^3 - 11x^2 + x + 1 = 0$

**Answers to Practice Problems** **1.** $x^3 - 2x^2 - 5x + 6 = 0$
**2.** $x^3 - x^2 - 7x + 15 = 0$ **3.** $x^3 + 3x^2 + 3x + 1 = 0$
**4.** $x = 3$ with multiplicity 2; $x = 1$ with multiplicity 2

**19.** $x^3 + 2x^2 - 19x + 30 = 0$; $2 - i$

**20.** $x^4 - 6x^3 + 18x^2 - 30x + 25 = 0$; $1 + 2i$

**21.** $x^4 + x^3 - x^2 + 5x + 6 = 0$; $1 - i\sqrt{2}$

**22.** $x^4 - 4x^3 + 16x - 16 = 0$; $2$

**23.** $x^4 - 10x^3 + 54x^2 - 130x + 125 = 0$; $3 - 4i$

**24.** $x^4 - 2x^3 + x^2 - 8x - 12 = 0$; $3$

**25.** $4x^4 - 19x^2 + 3x + 18 = 0$; $-1$

**26.** $x^4 - 5x^2 - 10x - 6 = 0$

**27.** $x^4 - 5x^3 + 11x^2 + 11x - 78 = 0$

**28.** $x^5 - 7x^4 + 20x^3 - 28x^2 + 19x - 5 = 0$

**29.** $x^5 - 5x^4 + 4x^3 + 3x^2 + 9x = 0$

**30.** $x^5 + x^3 - 2x^2 - 12x - 8 = 0$

## CHAPTER 12 SUMMARY

### Key Terms and Formulas

A **polynomial equation** of degree $n > 0$ is an equation of the form

$$P(x) = a_nx^n + a_{n-1}x^{n-1} + \cdots + a_1x + a_0 = 0$$

where $a_n \neq 0$ and $a_n, a_{n-1}, \ldots, a_1, a_0$ are complex numbers. [12.1]

$a_n$ is called the **leading coefficient** of $P(x)$. [12.1]

If $P(c) = 0$, then $x = c$ is a **root** (or **solution**) of the polynomial equation $P(x) = 0$. [12.1]

A real number $c$ is an **upper bound** for the roots of an equation if there are no roots greater than $c$. [12.3]

A real number $c$ is a **lower bound** for the roots of an equation if there are no roots less than $c$. [12.3]

If a polynomial, $P(x)$, is arranged in order of descending powers, then there is a **variation in sign** if two consecutive terms have opposite signs. [12.4]

If $(x - r)$ is a factor of $P(x)$ $m$ times, then $r$ is said to be a root of **multiplicity** $m$ of the equation $P(x) = 0$. [12.5]

### Properties and Rules

**Remainder Theorem** [12.1]
If a polynomial $P(x)$ of degree $n > 0$ is divided by $(x - c)$, then the remainder $r$ is equal to $P(c)$.

**Location Principle** [12.1]
If $P(x) = 0$ is a polynomial equation with real coefficients and $P(a) > 0$ and $P(b) < 0$ [or $P(a) < 0$ and $P(b) > 0$], then there is at least one number $c$ between $a$ and $b$ that is a root of the equation. [That is, $P(c) = 0$.]

**Rational Roots Theorem** [12.2]
If $\frac{p}{q}$ is a rational number in lowest terms and $\frac{p}{q}$ is a root of the polynomial equation

$$P(x) = a_nx^n + a_{n-1}x^{n-1} + \cdots + a_1x + a_0 = 0$$

where $a_n \neq 0$ and the coefficients are integers, then $p$ is a factor of $a_0$ and $q$ is a factor of $a_n$.

**Upper and Lower Bounds Theorem** [12.3]
Suppose that $P(x) = 0$ is a polynomial equation with real coefficients, where the leading coefficient $a_n$ is positive, and $P(x)$ is divided by $(x - c)$ using synthetic division.

**a.** If $c > 0$ and all of the coefficients in the last row are positive or 0, then $c$ is an upper bound for the roots of the equation.

**b.** If $c < 0$ and the coefficients in the last row alternate in sign (with 0 treated as positive or negative), then $c$ is a lower bound for the roots of the equation.

**Descartes' Rule of Signs** [12.4]
Suppose that $P(x) = 0$ is a polynomial equation with real coefficients and $a_0 \neq 0$.

**a.** The number of positive real roots is equal to the number of variations in sign of $P(x)$ or to that number decreased by an even integer.

**b.** The number of negative real roots is equal to the number of variations in sign of $P(-x)$ or to that number decreased by an even integer.

**The Fundamental Theorem of Algebra** [12.5]
If a polynomial $P(x)$ with complex coefficients has degree $n > 0$, then the equation $P(x) = 0$ has at least one root.

**The Number of Roots Theorem** [12.5]
A polynomial equation $P(x) = 0$ of degree $n > 0$ has at most $n$ distinct roots. The sum of the multiplicities of the roots is $n$.

## CHAPTER 12 REVIEW

Use synthetic division and the Remainder Theorem to find the indicated values for each of the polynomials in Exercises 1–3. [12.1]

**1.** $P(x) = x^3 - 2x^2 + 3x - 1$; (a) $P(-1)$, (b) $P(2)$
**2.** $P(x) = x^3 + 5x^2 - 6x - 4$; (a) $P(-2)$, (b) $P(1.5)$
**3.** $P(x) = 2x^4 - 3x^3 + 2x^2 - 7x + 3$; (a) $P(3)$, (b) $P(-1.4)$

Use synthetic division and the Location Principle to show that there is a root of the polynomial equation between the two numbers listed in Exercises 4–6. [12.1]

**4.** $x^2 - 6x + 6 = 0$; between 4 and 5
**5.** $3x^3 - 7x^2 + 9x - 21 = 0$; between 2 and 2.5
**6.** $2x^3 - x^2 - 14x - 12 = 0$; between $-1.3$ and $-1.2$

List the possible rational roots of each of the polynomial equations in Exercises 7–9. [12.2]

**7.** $x^4 - 5x^3 + 2x - 12 = 0$
**8.** $4x^3 - 7x^2 + 3x - 4 = 0$
**9.** $6x^3 - x^2 + 11x + 9 = 0$

Use the Rational Roots Theorem and synthetic division to find the rational roots of each of the polynomial equations in Exercises 10–12. [12.2]

**10.** $3x^3 - 14x^2 + 20x - 8 = 0$
**11.** $5x^3 + 11x^2 - 7x - 4 = 0$
**12.** $2x^3 + 11x^2 - 5x + 6 = 0$

Find all roots of each of the polynomial equations in Exercises 13–16. [12.2]

**13.** $3x^3 - 7x^2 - 11x + 15 = 0$
**14.** $5x^3 - 7x^2 - 3x + 2 = 0$
**15.** $x^3 - 8x^2 + 21x - 20 = 0$
**16.** $2x^4 + 5x^3 - x^2 + 5x - 3 = 0$

Use the Upper and Lower Bounds Theorem to show that all real roots of each of the polynomial equations in Exercises 17–18 are located between the given integers. [12.3]

**17.** $x^3 - 6x^2 - x + 23 = 0$; between $-2$ and 6
**18.** $x^4 - 5x^3 + 3x^2 + 2x - 28 = 0$; between $-2$ and 5

Find the largest negative integer and the smallest positive integer that can be determined by the Upper and Lower Bounds Theorem to be bounds for the roots of each of the polynomial equations in Exercises 19–22. [12.3]

**19.** $2x^3 - 4x^2 - 3x - 7 = 0$
**20.** $2x^3 - 9x^2 - 17x - 10 = 0$
**21.** $x^4 - 6x^3 + 7x^2 + 11x - 16 = 0$
**22.** $5x^4 + 7x^3 - 13x^2 - 6x + 6 = 0$

Use Descartes' Rule of Signs to determine the possible numbers of positive and negative roots of the polynomial equations in Exercises 23–25. [12.4]

**23.** $5x^4 - x^3 + 3x^2 + 6x + 8 = 0$
**24.** $x^3 + 6x^2 + 4x - 2 = 0$
**25.** $7x^4 - 3x^3 - 2x - 3 = 0$

Use Descartes' Rule of Signs to determine the possible numbers of positive and negative roots of the polynomial equations in Exercises 26–29. Then find as many roots as you can using reduced polynomials, upper and lower bounds, the Rational Roots Theorem, and the quadratic formula. [12.4]

**26.** $7x^3 + 24x^2 + 9x - 6 = 0$

**27.** $5x^3 + 8x^2 + 26x - 12 = 0$

**28.** $4x^3 - 13x^2 + 4x + 12 = 0$

**29.** $2x^4 - x^3 + 17x^2 - 9x - 9 = 0$

Find a polynomial equation of lowest degree with real coefficients that has the numbers given in Exercises 30–33 as roots. [12.5]

**30.** $-2, -1, 3$

**31.** $-5i$, $-3$ with multiplicity 2

**32.** $2 - 3i, 1 + \sqrt{2}, 1 - \sqrt{2}$

**33.** $1, 1 - i\sqrt{3}$, 2 with multiplicity 2

For each of the polynomial equations in Exercises 34–37, find all roots together with their multiplicities. Some roots are given. [12.5]

**34.** $x^3 + 9x^2 + 27x + 27 = 0$

**35.** $x^4 - 2x^3 - 3x^2 + 4x + 4 = 0$

**36.** $x^4 - 2x^3 - 4x^2 - 10x + 3 = 0; x = -1 + i\sqrt{2}$

**37.** $x^4 - 4x^3 + 22x^2 - 8x + 40 = 0; x = 2 - 4i$

## CHAPTER 12 TEST

Use synthetic division and the Remainder Theorem to find the indicated values for each of the polynomials in Exercises 1 and 2.

**1.** $P(x) = x^4 + 2x^3 - 3x^2 + 5$; find $P(-2)$.

**2.** $P(x) = 2x^3 + 7x^2 - 4x + 1$; find $P(1.6)$.

Use synthetic division and the Location Principle to show that there is a root of the polynomial equation between the two numbers listed in Exercises 3 and 4.

**3.** $2x^3 - 3x^2 - 14x - 6 = 0$; between $-2$ and $-1.6$

**4.** $4x^3 + 10x^2 - 9 = 0$; between 0.8 and 0.9

**5.** List the possible rational roots of the equation $2x^4 - 11x^3 + 10x^2 - 14 = 0$

Use the Rational Roots Theorem and synthetic division to find the rational roots of each of the polynomial equations in Exercises 6 and 7.

**6.** $5x^3 - 14x^2 - 7x + 12 = 0$

**7.** $9x^3 + 9x^2 - 22x + 8 = 0$

Use the Upper and Lower Bounds Theorem to show that all real roots of each of the polynomial equations in Exercises 8 and 9 are located between the given integers.

**8.** $2x^3 - 3x^2 - 24x - 11 = 0$; between $-3$ and 5

**9.** $2x^3 - 7x^2 - 76x - 18 = 0$; between $-5$ and 9

Find the largest negative integer and the smallest positive integer that can be determined by the Upper and Lower Bounds Theorem to be bounds for the roots of each of the polynomial equations in Exercises 10 and 11.

**10.** $3x^3 + 11x^2 - 36x - 14 = 0$

**11.** $4x^3 - 4x^2 - 23x + 30 = 0$

Use Descartes' Rule of Signs to determine the possible numbers of positive and negative roots of the polynomial equations in Exercises 12 and 13.

**12.** $7x^3 + 4x^2 - 6x - 11 = 0$

**13.** $4x^4 - 2x^3 - 5x^2 + 3x - 2 = 0$

Find a polynomial equation of lowest degree with real coefficients that has the numbers given in Exercises 14 and 15 as roots.

**14.** 5, 2 with multiplicity 2

**15.** $-4, 1 + i\sqrt{5}$, 1 with multiplicity 2

For each of the polynomial equations in Exercises 16–20, find all roots together with their multiplicities. Some roots are given.

**16.** $2x^3 + 5x^2 - 2x - 5 = 0$

**17.** $x^3 - 10x - 12 = 0$

**18.** $2x^4 + x^3 - 11x^2 - 4x + 12 = 0$

**19.** $x^4 + 3x^3 - 5x^2 - 13x + 6 = 0$

**20.** $x^4 - 6x^3 + 19x^2 - 42x + 10 = 0$; $x = 1 - 3i$

## CUMULATIVE REVIEW (12)

Simplify Exercises 1 and 2. Assume that all variables are positive.

**1.** $(3x^{2/3})(5x^{2/5})$

**2.** $\left(\dfrac{x^2y^{-2/3}}{x^{1/3}y^{1/2}}\right)^{1/2}$

Perform the indicated operations in Exercises 3 and 4.

**3.** $\dfrac{x^2 + 2x - 8}{x^2 + x - 12} \div \dfrac{x^2 - x - 2}{x^2 - 3x}$

**4.** $\dfrac{4x}{3x^2 + 4x + 1} - \dfrac{x + 4}{x^2 + 7x + 6}$

Solve the equations in Exercises 5–8.

**5.** $\dfrac{3x + 4}{x - 7} - 1 = \dfrac{2x + 2}{x - 4}$

**6.** $\sqrt{x + 9} = 2 + \sqrt{x - 3}$

**7.** $4e^{-0.7x} = 5$

**8.** $\ln(x^2 + x - 20) - \ln(x + 5) = 1$

**9.** Write an equation for the line passing through the points $(-2,5)$ and $(3,9)$.

**10.** Write an equation for the circle centered at the origin with radius 4.

Draw the graph of each of the equations in Exercises 11 and 12.

**11.** $y = x^2 + 4x + 7$

**12.** $x^2 + 9y^2 = 9$

**13.** Solve the system of inequalities graphically.
$$\begin{cases} 4x + 3y \geq 7 \\ x - 2y \leq 10 \end{cases}$$

**14.** Solve the system of equations.
$$\begin{cases} x - y = 2 \\ y = x^2 - 6x + 10 \end{cases}$$

**15.** Solve the system of equations using the Gaussian elimination method.
$$\begin{cases} x - 2y - z = 13 \\ 2x + y + z = 0 \\ 3x - 5y + 8z = 13 \end{cases}$$

**16.** Solve the system of equations using matrix inversion.
$$\begin{cases} 5x - 3y = 11 \\ 3x + 2y = 18 \end{cases}$$

**17.** Use synthetic division and the Remainder Theorem to find the indicated value for the polynomial: $P(x) = 2x^4 - 3x^3 - 4x^2 + x + 6$.
a. $P(3)$ b. $P(1.5)$

**18.** Find the largest negative integer and the smallest positive integer that can be determined by the Upper and Lower Bounds Theorem to be bounds for the roots of the polynomial equation:
$x^4 - 5x^3 + 2x^2 + x - 12 = 0$

**19.** Find a polynomial equation of lowest degree with real coefficients that has the given numbers as roots.
$3$, $i\sqrt{7}$, $-1$ with multiplicity 2

**20.** Find all roots together with their multiplicities.
$x^4 + 2x^3 + 2x^2 + 16x + 24 = 0$

**21.** If $P$ dollars are invested at a rate, $r$ (expressed as a decimal), and compounded $k$ times a year, the amount, $A$, due at the end of $t$ years is given by $A = P\left(1 + \dfrac{r}{k}\right)^{kt}$ dollars. Find $A$ if \$1200 is invested at 9% compounded quarterly for 5 years.

**22.** An airplane travels 200 mph in calm air. The plane travels 750 miles with the wind and returns against the wind. It takes 2 hours longer to make the return trip. Find the speed of the wind.

## MATHEMATICAL CHALLENGES

Find the tens digit of the following sum:
$1! + 2! + 3! + \cdots + n! + \cdots + 1979!$

If thirty-two teams participate in a standard single-elimination tournament, how many games must be played to decide a champion?

Calculate the sum of this series:
$6 + 3.6 + 2.16 + 1.296 + \cdots$

## CHAPTER OUTLINE



Chapter 13 provides an introduction to a powerful notation using the Greek letter $\Sigma$, capital sigma. With this $\Sigma$-notation and a few basic properties, we will develop some algebraic formulas related to sums of numbers and, in some cases, even infinite sums. The concept of having the sum of an infinite number of numbers equal to some finite number introduces the idea of infinitesimals. Suppose you want to go home today; but first you must leave the classroom through the door and you proceed as follows: you walk half the distance to the door; then you walk half the remaining distance to the door; then you walk half the remaining distance to the door; and so on. Well, you will never make it out the door. You will come close, as close as you want, but. . . . This procedure is purely a mathematical concept that involves an idea called limits. Don't worry. You can get through the door by moving one "foot" at a time.

The other topics in this chapter, permutations, combinations, and the Binomial Theorem, find applications in courses in probability and statistics as well as in more advanced courses in mathematics and computer science.

## 13.1 Sequences

**OBJECTIVES**

In this section, you will be learning to:

1. Write several terms of a sequence given the formula for its general term.
2. Find the formula for the general term of a sequence given several terms.
3. Determine whether sequences are increasing or decreasing.

In mathematics, we are interested in studying lists of numbers that occur in a certain order. For example, the list

$$3, 6, 9, 12, 15, \ldots$$

can be described as the multiples of 3. For any positive integer $n$, the corresponding number in the list is $3n$. Thus, we know that $3 \cdot 6 = 18$ and 18 is the 6th number in the list. The seventh number is 21 since $21 = 3 \cdot 7$. Lists of numbers formed in this way are known as the **terms of a sequence** or simply a **sequence.**

***Infinite Sequence*** An **infinite sequence** (or a **sequence**) is a function that has the positive integers as its domain.

Consider the function

$$f(n) = \frac{1}{2^n} \quad \text{where } n \text{ is any positive integer}$$

For this function,

$$f(1) = \frac{1}{2^1} = \frac{1}{2}$$

$$f(2) = \frac{1}{2^2} = \frac{1}{4}$$

$$f(3) = \frac{1}{2^3} = \frac{1}{8}$$

$$f(4) = \frac{1}{2^4} = \frac{1}{16}$$

$$\vdots$$

$$f(n) = \frac{1}{2^n}$$

$$\vdots$$

Or, using ordered pair notation,

$$f = \left\{\left(1, \frac{1}{2}\right), \left(2, \frac{1}{4}\right), \left(3, \frac{1}{8}\right), \left(4, \frac{1}{16}\right), \ldots, \left(n, \frac{1}{2^n}\right), \ldots\right\}$$

The numbers

$$\frac{1}{2}, \frac{1}{4}, \frac{1}{8}, \frac{1}{16}, \ldots, \frac{1}{2^n}, \ldots$$

are called the **terms** of the sequence. And, since the order corresponds to the positive integers, we customarily indicate a sequence by writing only the terms. A sequence can be indicated with subscript notation as

$$a_1, a_2, a_3, a_4, \ldots, a_n, \ldots$$

The general term $a_n$ is called the ***n*th term** of the sequence. The entire sequence can be denoted by writing the $n$th term in braces as in $\{a_n\}$. Thus,

$$\{a_n\} \quad \text{and} \quad a_1, a_2, a_3, \ldots, a_n, \ldots$$

are both representations of the sequence with

$a_1$ as the first term,
$a_2$ as the second term,
$a_3$ as the third term,
$\vdots$
$a_n$ as the $n$th term,
$\vdots$

## EXAMPLES

**1.** Write the first three terms of the sequence $\left\{\frac{n}{n+1}\right\}$.

**Solution** $a_1 = \frac{1}{1+1} = \frac{1}{2}$

$a_2 = \frac{2}{2+1} = \frac{2}{3}$

$a_3 = \frac{3}{3+1} = \frac{3}{4}$

**2.** If $\{b_n\} = \{2n - 1\}$, find $b_1$, $b_2$, $b_3$, and $b_{50}$.

**Solution** $b_1 = 2 \cdot 1 - 1 = 1$

$b_2 = 2 \cdot 2 - 1 = 3$

$b_3 = 2 \cdot 3 - 1 = 5$

$b_{50} = 2 \cdot 50 - 1 = 99$

**3.** Determine $a_n$ if the first five terms of $\{a_n\}$ are 0, 3, 8, 15, 24.

**Solution** In this case, we simply study the numbers carefully and make an intelligent guess. That is, we keep trying a variety of formulas until we find one that "fits" the given terms. (Questions of this type relating to number patterns are common on intelligence tests.)

The formula is $a_n = n^2 - 1$.
Checking: $a_1 = 1^2 - 1 = 0$, $a_2 = 2^2 - 1 = 3$, $a_3 = 3^2 - 1 = 8$, $a_4 = 4^2 - 1 = 15$, $a_5 = 5^2 - 1 = 24$.

**4.** Write the first five terms of the sequence in which $a_n = \frac{(-1)^n}{n}$.

**Solution** $a_1 = \frac{(-1)^1}{1} = -1$

$a_2 = \frac{(-1)^2}{2} = \frac{1}{2}$

$a_3 = \frac{(-1)^3}{3} = -\frac{1}{3}$

$a_4 = \frac{(-1)^4}{4} = \frac{1}{4}$

$a_5 = \frac{(-1)^5}{5} = -\frac{1}{5}$ ■

Example 4 illustrates a type of sequence known as an **alternating sequence** in which the signs of consecutive terms alternate from positive to negative, and so on. Also, we see that while the domain of a sequence consists of the positive integers, some (or all) of the terms of a sequence may be negative.

The term $a_{n+1}$ is the term following $a_n$ in a sequence and is found by substituting $n + 1$ for $n$ in the formula for the general term. For example,

$$\text{if } a_n = \frac{1}{3n}, \text{ then } a_{n+1} = \frac{1}{3(n+1)} = \frac{1}{3n+3}$$

Similarly,

$$\text{if } b_n = n^2, \text{ then } b_{n+1} = (n+1)^2$$

If the terms of a sequence grow increasingly smaller, then we say that the sequence is **decreasing.** If the terms grow increasingly larger, then we say that the sequence is **increasing.** To determine algebraically whether or not a sequence might be increasing or decreasing, we need the following definition.

***Decreasing Sequence***

A sequence $\{a_n\}$ is

**(a) decreasing** if $a_n > a_{n+1}$ for all $n$

(Each following term is smaller than its preceding term.)

***Increasing Sequence***

**(b) increasing** if $a_n < a_{n+1}$ for all $n$

(Each following term is larger than its preceding term.)

**EXAMPLES** Determine whether each of the following sequences is increasing or decreasing or neither.

**5.** $\{a_n\} = \left\{\frac{1}{2^n}\right\}$

**Solution** $a_n = \frac{1}{2^n}$ and $a_{n+1} = \frac{1}{2^{n+1}}$

Since $\frac{1}{2^n} > \frac{1}{2^{n+1}}$, we have $a_n > a_{n+1}$, and $\{a_n\}$ is decreasing.

**6.** $\{b_n\} = \{n + 3\}$

**Solution** $b_n = n + 3$ and $b_{n+1} = (n + 1) + 3 = n + 4$

Since $n + 3 < n + 4$, we have $b_n < b_{n+1}$, and $\{b_n\}$ is increasing.

**7.** $\{c_n\} = \{(-1)^n\}$

**Solution** $c_n = (-1)^n$ and $c_{n+1} = (-1)^{n+1}$

The value of $c_n$ depends on whether $n$ is even or odd.
If $n$ is even, then $n + 1$ is odd and

$$c_n = (-1)^n = 1 \quad \text{and} \quad c_{n+1} = (-1)^{n+1} = -1 \quad \text{indicating } c_n > c_{n+1}$$

If $n$ is odd, then $n + 1$ is even and

$$c_n = (-1)^n = -1 \quad \text{and} \quad c_{n+1} = (-1)^{n+1} = 1 \quad \text{indicating } c_n < c_{n+1}$$

Therefore, the sequence is neither increasing nor decreasing. ■

**Practice Problems** Write the first three terms of each sequence.

1. $\{n^2\}$ 2. $\{2n + 1\}$ 3. $\left\{\dfrac{1}{n+1}\right\}$

4. Find a formula for the general term for the sequence $-1, 1, 3, 5, 7, \ldots$.

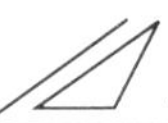

## EXERCISES 13.1

Write the first four terms of each of the sequences in Exercises 1–15.

1. $\{2n - 1\}$
2. $\{4n + 1\}$
3. $\left\{1 + \dfrac{1}{n}\right\}$
4. $\left\{\dfrac{n+3}{n+1}\right\}$
5. $\{n^2 + n\}$
6. $\{n - n^2\}$
7. $\{2^n\}$
8. $\left\{\left(\dfrac{1}{2}\right)^n\right\}$
9. $\{(-1)^n(n^2 + 1)\}$
10. $\left\{(-1)^n\left(\dfrac{n}{n+1}\right)\right\}$
11. $\left\{(-1)^n\left(\dfrac{1}{2n+3}\right)\right\}$
12. $\{(-1)^{n-1}(3^n)\}$
13. $\{2^n - n^2\}$
14. $\left\{\dfrac{n(n-1)}{2}\right\}$
15. $\left\{\dfrac{1 + (-1)^n}{2}\right\}$

Find a formula for the general term for each of the sequences in Exercises 16–25.

16. $2, 5, 8, 11, 14, \ldots$
17. $5, 9, 13, 17, 21, \ldots$
18. $6, 12, 18, 24, 30, \ldots$
19. $1, -3, 5, -7, 9, \ldots$
20. $-3, 7, -11, 15, -19, \ldots$
21. $1, 4, 9, 16, 25, \ldots$
22. $5, 10, 20, 40, 80, \ldots$
23. $\dfrac{1}{3}, \dfrac{1}{4}, \dfrac{1}{5}, \dfrac{1}{6}, \dfrac{1}{7}, \ldots$
24. $\dfrac{1}{2}, \dfrac{1}{4}, \dfrac{1}{8}, \dfrac{1}{16}, \dfrac{1}{32}, \ldots$
25. $2, 5, 10, 17, 26, \ldots$

For each of the sequences in Exercises 26–31, determine whether it is increasing or decreasing. Justify your answer by comparing $a_n$ with $a_{n+1}$.

26. $\{n + 4\}$
27. $\{1 - 2n\}$
28. $\left\{\dfrac{1}{n+3}\right\}$
29. $\left\{\dfrac{1}{3^n}\right\}$
30. $\left\{\dfrac{2n+1}{n}\right\}$
31. $\left\{\dfrac{n}{n+1}\right\}$

Write the sequence described by Exercises 32–35; then solve.

32. A ball is dropped from a height of 250 centimeters. Each time it bounces, it rises to $\dfrac{2}{5}$ of its previous height. How high will it rise after the fourth bounce?

33. A certain automobile costs \$9000 new and depreciates at a rate of $\dfrac{3}{10}$ of its current value each year. What will be its value after 3 years?

**Answers to Practice Problems** **1.** 1, 4, 9 **2.** 3, 5, 7 **3.** $\dfrac{1}{2}, \dfrac{1}{3}, \dfrac{1}{4}$
**4.** $a_n = 2n - 3$

**34.** A culture of bacteria triples every day. If there were 100 bacteria in the original culture, how many would be present after 4 days?

**35.** A university is experiencing a declining enrollment of 3% per year. If the present enrollment is 20,000, what is the projected enrollment after 5 years?

## 13.2 Sigma Notation

**OBJECTIVES**

In this section, you will be learning to:

1. Write sums using $\Sigma$-notation.
2. Find the values of sums written in $\Sigma$-notation.

In Section 13.1 we discussed sequences and the fact that sequences have an infinite number of terms. In this section we will discuss finding the sum of a finite number of terms of a sequence and a special notation called **sigma notation** using the Greek letter capital sigma ($\Sigma$). For example, the sum of the first five terms of the sequence $\{n^2\}$ can be written

$$1^2 + 2^2 + 3^2 + 4^2 + 5^2$$

Using sigma notation, the same sum can be written

$$\sum_{k=1}^{5} k^2 = 1^2 + 2^2 + 3^2 + 4^2 + 5^2$$

The letter $k$ in this notation is called the **index of summation,** the number 5 is the **upper limit of summation,** and the number 1 is the **lower limit of summation.** The understanding is that $k$ takes the integer values from 1 to 5, namely 1, 2, 3, 4, 5.

If the number of terms is large, then three dots are used to indicate missing terms after a pattern has been established with the first three or four terms. For example,

$$\sum_{k=1}^{100} (k-1) = 0 + 1 + 2 + \ldots + 99$$

***Summation Notation***

$$\sum_{k=1}^{n} a_k = a_1 + a_2 + a_3 + \ldots + a_n$$

The **index of summation** $k$ takes the integer values 1, 2, 3, . . . , $n$. $n$ is the **upper limit of summation** and 1 is the **lower limit of summation.**

**EXAMPLES** Write the indicated sums of the terms and find the value of each sum.

**1.** $\sum_{k=1}^{4} k^3$

**Solution** $\sum_{k=1}^{4} k^3 = 1^3 + 2^3 + 3^3 + 4^3 = 100$

**2.** $\sum_{k=5}^{9} (-1)^k k$

**Solution**

$$\sum_{k=5}^{9} (-1)^k k = 5(-1)^5 + 6(-1)^6 + 7(-1)^7 + 8(-1)^8 + 9(-1)^9$$
$$= -5 + 6 - 7 + 8 - 9 = -7$$

**3.** Write the sum $5 + 8 + 11 + 14 + 17 + 20$ in $\Sigma$-notation.

**Solution** After some experimenting, you will find $3k + 2$ as a general term.

Thus,

$$5 + 8 + 11 + 14 + 17 + 20 = \sum_{k=1}^{6} (3k + 2)$$

■

In Sections 13.3 and 13.4, we will develop systematic ways of finding certain types of sums, and we will make use of the following properties of $\Sigma$-notation.

***Properties of $\Sigma$-Notation***

For sequences $\{a_n\}$ and $\{b_n\}$ and any real number $c$:

**I.** $\sum_{k=1}^{n} a_k = \sum_{k=1}^{i} a_k + \sum_{k=i+1}^{n} a_k$

**II.** $\sum_{k=1}^{n} (a_k + b_k) = \sum_{k=1}^{n} a_k + \sum_{k=1}^{n} b_k$

**III.** $\sum_{k=1}^{n} ca_k = c\sum_{k=1}^{n} a_k$

**IV.** $\sum_{k=1}^{n} c = nc$

We can see that these properties "make sense" by applying the associative, commutative, and distributive properties for sums of real numbers.

**I.** $\sum_{k=1}^{n} a_k = a_1 + a_2 + \cdots + a_i + a_{i+1} + a_{i+2} + \cdots + a_n$

$$= (a_1 + a_2 + \cdots + a_i) + (a_{i+1} + a_{i+2} + \cdots + a_n)$$
$$= \sum_{k=1}^{i} a_k + \sum_{k=i+1}^{n} a_k$$

$$\text{II.}\quad \sum_{k=1}^{n} (a_k + b_k) = (a_1 + b_1) + (a_2 + b_2) + \cdots + (a_n + b_n)$$

$$= (a_1 + a_2 + \cdots + a_n) + (b_1 + b_2 + \cdots + b_n)$$

$$= \sum_{k=1}^{n} a_k + \sum_{k=1}^{n} b_k$$

$$\text{III.}\quad \sum_{k=1}^{n} ca_k = ca_1 + ca_2 + \cdots + ca_n$$

$$= c(a_1 + a_2 + \cdots + a_n)$$

$$= c\sum_{k=1}^{n} a_k$$

$$\text{IV.}\quad \sum_{k=1}^{n} c = \underbrace{c + c + c + \cdots + c}_{nc\text{'s}} = nc$$

**EXAMPLES**

**4.** If $\sum_{k=1}^{7} a_k = 40$ and $\sum_{k=1}^{30} a_k = 75$, find $\sum_{k=8}^{30} a_k$.

**Solution** Since

$$\sum_{k=1}^{7} a_k + \sum_{k=8}^{30} a_k = \sum_{k=1}^{30} a_k$$

then

$$40 + \sum_{k=8}^{30} a_k = 75$$

$$\sum_{k=8}^{30} a_k = 35$$

**5.** If $\sum_{k=1}^{50} 3a_k = 600$, find $\sum_{k=1}^{50} a_k$.

**Solution** Since

$$\sum_{k=1}^{50} 3a_k = 3\sum_{k=1}^{50} a_k$$

then

$$3\sum_{k=1}^{50} a_k = 600$$

$$\sum_{k=1}^{50} a_k = 200$$

■

**Practice Problems**

1. Write the indicated sum of the terms and find the value of the sum: $\sum_{k=1}^{4} (k^2 - 1)$
2. Write the sum $10 + 12 + 14 + 16 + 20$ in $\Sigma$-notation.
3. $\sum_{k=1}^{5} a_k = 20$ and $\sum_{k=6}^{10} a_k = 30$. Find $\sum_{k=1}^{10} 2a_k$.

## EXERCISES 13.2

Write Exercises 1–16 in expanded form and evaluate.

1. $\sum_{k=1}^{5} 2k$
2. $\sum_{k=1}^{11} k(k-1)$
3. $\sum_{k=2}^{6} (k+3)$
4. $\sum_{k=9}^{11} (2k+1)$
5. $\sum_{k=2}^{4} \frac{1}{k}$
6. $\sum_{k=1}^{3} \frac{1}{2k}$
7. $\sum_{k=1}^{3} 2^k$
8. $\sum_{k=10}^{15} (-1)^k$
9. $\sum_{k=4}^{8} k^2$
10. $\sum_{k=1}^{4} k^3$
11. $\sum_{k=3}^{6} (9-2k)$
12. $\sum_{k=2}^{7} (4k-1)$
13. $\sum_{k=2}^{5} (-1)^k(k^2+k)$
14. $\sum_{k=1}^{6} (-1)^k(k^2-2)$
15. $\sum_{k=1}^{5} \frac{k}{k+1}$
16. $\sum_{k=3}^{5} (-1)^k\left(\frac{k+1}{k^2}\right)$

Write the sums in Exercises 17–25 in sigma notation.

17. $1 + 3 + 5 + 7 + 9$
18. $16 + 25 + 36 + 49$
19. $-1 + 1 + (-1) + 1 + (-1)$
20. $4 + 7 + 10 + 13 + 16$
21. $\frac{1}{8} - \frac{1}{27} + \frac{1}{64} - \frac{1}{125} + \frac{1}{216}$
22. $\frac{1}{8} + \frac{1}{16} + \frac{1}{32} + \frac{1}{64} + \frac{1}{128}$
23. $\frac{4}{5} + \frac{5}{6} + \frac{6}{7} + \cdots + \frac{15}{16}$
24. $8 + 15 + 24 + 35 + 48$
25. $\frac{6}{25} + \frac{7}{36} + \frac{8}{49} + \frac{9}{64} + \cdots + \frac{13}{144}$

**Answers to Practice Problems** 1. $0 + 3 + 8 + 15 = 26$ 2. $\sum_{k=5}^{9} 2k$ or $\sum_{k=1}^{5} (2k+8)$ 3. 100

Find the indicated sums in Exercises 26–35.

**26.** $\sum_{k=1}^{14} a_k = 18$ and $\sum_{k=1}^{14} b_k = 21$. Find $\sum_{k=1}^{14} (a_k + b_k)$.

**27.** $\sum_{k=1}^{19} a_k = 23$ and $\sum_{k=1}^{19} b_k = 16$. Find $\sum_{k=1}^{19} (a_k - b_k)$.

**28.** $\sum_{k=1}^{15} a_k = 19$. Find $\sum_{k=1}^{15} 3a_k$.

**29.** $\sum_{k=1}^{25} a_k = 63$ and $\sum_{k=1}^{11} a_k = 15$. Find $\sum_{k=12}^{25} a_k$.

**30.** $\sum_{k=1}^{18} a_k = 41$ and $\sum_{k=1}^{18} b_k = 62$. Find $\sum_{k=1}^{18} (3a_k - 2b_k)$.

**31.** $\sum_{k=1}^{21} a_k = -68$ and $\sum_{k=1}^{21} b_k = 39$. Find $\sum_{k=1}^{21} (a_k + 2b_k)$.

**32.** $\sum_{k=1}^{16} a_k = 56$ and $\sum_{k=17}^{40} a_k = 42$. Find $\sum_{k=1}^{40} a_k$.

**33.** $\sum_{k=13}^{29} a_k = 84$ and $\sum_{k=1}^{29} a_k = 143$. Find $\sum_{k=1}^{12} 5a_k$.

**34.** $\sum_{k=1}^{20} b_k = 34$ and $\sum_{k=1}^{20} (2a_k + b_k) = 144$. Find $\sum_{k=1}^{20} a_k$.

**35.** $\sum_{k=1}^{27} a_k = 46$ and $\sum_{k=1}^{10} a_k = 122$. Find $\sum_{k=11}^{27} 2a_k$.

## 13.3 Arithmetic Sequences

**OBJECTIVES**

In this section, you will be learning to:

1. Determine whether or not sequences are arithmetic.
2. Find the general term for arithmetic sequences.
3. Find the sums of the first $n$ terms of arithmetic sequences.

There are many types of sequences studied in higher levels of mathematics. In the next two sections, we will discuss some of the properties of two types: **arithmetic sequences** and **geometric sequences.** In this discussion, we will also use sigma notation and develop formulas for finding sums. For arithmetic sequences, we can find sums of only a finite number of terms. For geometric sequences, we can find sums of a finite number of terms and, in some special cases, we define the sum of an infinite number of terms.

The three sequences

$$3, 5, 7, 9, 11, 13, \ldots$$
$$4, 5, 6, 7, 8, 9, \ldots$$
$$-2, -5, -8, -11, -14, -17, \ldots$$

all have a common characteristic. This characteristic is that **any two consecutive terms have the same difference.**

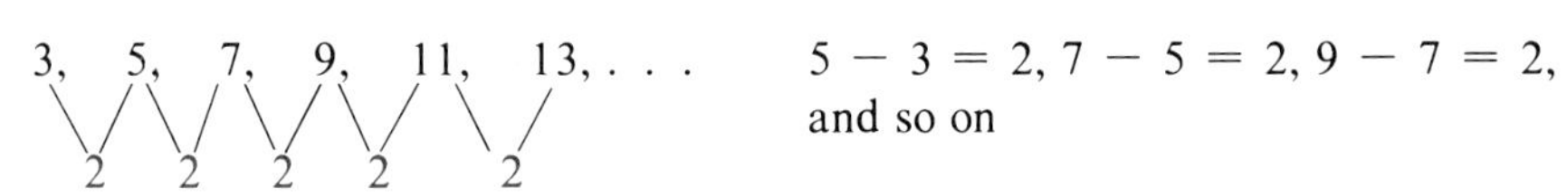


$5 - 3 = 2, 7 - 5 = 2, 9 - 7 = 2,$ and so on

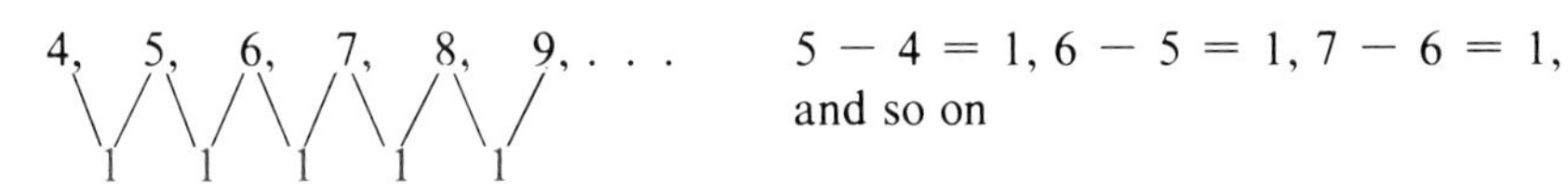


$5 - 4 = 1, 6 - 5 = 1, 7 - 6 = 1,$ and so on

$-2, \quad -5, \quad -8, \quad -11, \quad -14, \quad -17, \ldots$

$-3 \quad -3 \quad -3 \quad -3 \quad -3$

$-5 - (-2) = -3,$
$-8 - (-5) = -3,$
and so on

Such sequences are called **arithmetic sequences** or **arithmetic progressions.**

***Arithmetic Sequence***

A sequence $\{a_n\}$ is called an **arithmetic sequence** (or **arithmetic progression**) if for any natural number $k$,

$$a_{k+1} - a_k = d \qquad \text{where } d \text{ is a constant}$$

$d$ is called the **common difference.**

## EXAMPLES

**1.** Show that the sequence $\{2n - 3\}$ is arithmetic by finding $d$.

**Solution** $a_k = 2k - 3$ and $a_{k+1} = 2(k + 1) - 3 = 2k - 1$

$$a_{k+1} - a_k = (2k - 1) - (2k - 3) = 2k - 1 - 2k + 3 = 2$$

So, $d = 2$, and the sequence $\{2a - 3\}$ is arithmetic.

**2.** Show that the sequence $\{n^2\}$ is not arithmetic.

**Solution** Since $a_3 = 3^2$ and $a_2 = 2^2$ and $a_1 = 1^2$,

$$a_3 - a_2 = 9 - 4 = 5 \quad \text{and} \quad a_2 - a_1 = 4 - 1 = 3$$

Therefore, there is no common difference between consecutive terms, and $\{n^2\}$ is **not arithmetic.** ∎

If the first term is $a_1$ and the common difference is $d$, then the arithmetic sequence can be indicated as follows:

$$\begin{aligned}
a_1 &= a_1 && \text{first term}\\
a_2 &= a_1 + d && \text{second term}\\
a_3 &= a_2 + d = a_1 + 2d && \text{third term}\\
a_4 &= a_3 + d = a_1 + 3d && \text{fourth term}\\
&\vdots\\
a_n &= a_{n-1} + d = a_1 + (n-1)d && n\text{th term}
\end{aligned}$$

Thus,

$$a_n = a_1 + (n-1)d$$

represents the general term or $n$th term of an arithmetic sequence.

**EXAMPLES**

**3.** If in an arithmetic sequence, $a_1 = 5$ and $d = 3$, find $a_{16}$.

**Solution** $a_{16} = a_1 + 15d = 5 + 15 \cdot 3 = 50$

**4.** Find the 20th term of the arithmetic sequence whose first three terms are $-2$, 8, and 18.

**Solution** In this case, $a_1 = -2$ and $a_2 = 8$. Since the sequence is arithmetic, $d = a_2 - a_1 = 8 - (-2) = 10$. To find the 20th term, let $n = 20$ in the formula $a_n = a_1 + (n-1)d$:

$$a_{20} = -2 + (20 - 1)10 = -2 + 190 = 188$$

**5.** Find $a_1$ and $d$ for the arithmetic sequence in which $a_3 = 6$ and $a_{21} = -48$.

**Solution** Using the formula $a_n = a_1 + (n-1)d$ and solving simultaneous equations, we have

$$\begin{aligned} -48 &= a_1 + 20d\\ 6 &= a_1 + 2d \end{aligned} \qquad \begin{aligned} -48 &= a_1 + 20d\\ -6 &= -a_1 - 2d\\ \hline -54 &= 18d\\ -3 &= d \end{aligned} \qquad \begin{aligned} 6 &= a_1 + 2(-3)\\ 12 &= a_1 \end{aligned}$$

So, $a_1 = 12$ and $d = -3$. ■

To find $S = \sum_{k=1}^{6} (4k - 1)$, we can write all the terms and then add them.

$$S = \sum_{k=1}^{6} (4k - 1) = 3 + 7 + 11 + 15 + 19 + 23 = 78$$

Or, we can write the sum as follows, reversing the order:

$$\begin{aligned}
S &= 3 + 7 + 11 + 15 + 19 + 23 \\
S &= 23 + 19 + 15 + 11 + 7 + 3 \\
\hline
2S &= 26 + 26 + 26 + 26 + 26 + 26 \\
2S &= 6 \cdot 26 \\
S &= 78
\end{aligned}$$

This procedure illustrates one technique for finding a formula for the sum of $n$ terms of any arithmetic sequence. Suppose that the $n$ terms are

$$a_1, \quad a_2 = a_1 + d, \quad a_3 = a_1 + 2d, \quad \ldots, \quad a_{n-1} = a_n - d, \quad a_n$$

Thus,

$$\begin{aligned}
S &= a_1 + (a_1 + d) + (a_1 + 2d) + \cdots + (a_n - 2d) + (a_n - d) + a_n \\
S &= a_n + (a_n - d) + (a_n - 2d) + \cdots + (a_1 + 2d) + (a_1 + d) + a_1 \\
\hline
2S &= \underbrace{(a_1 + a_n) + (a_1 + a_n) + (a_1 + a_n) + \cdots + (a_1 + a_n) + (a_1 + a_n) + (a_1 + a_n)}_{(a_1 + a_n) \text{ appears } n \text{ times}}
\end{aligned}$$

$$2S = n(a_1 + a_n)$$

$$S = \frac{n}{2}(a_1 + a_n)$$

The formula can be rewritten using Σ-notation.

The sum of the first $n$ terms of an arithmetic sequence $\{a_n\}$ can be written as

$$\sum_{k=1}^{n} a_k = \frac{n}{2}(a_1 + a_n)$$

A special case of an arithmetic sequence is $\{n\}$ and the corresponding sum

$$\sum_{k=1}^{n} k = 1 + 2 + 3 + \cdots + n$$

In this case, $n =$ the number of terms, $a_1 = 1$, and $a_n = n$, so

$$\sum_{k=1}^{n} k = \frac{n}{2}(1 + n)$$

### EXAMPLES

**6.** $\sum_{k=1}^{75} k = 1 + 2 + 3 + \cdots + 75$

**Solution** This is an example of the special case just discussed where the sequence is $\{n\}$ and the upper limit of summation is $n = 75$.

$$\sum_{k=1}^{75} k = \frac{75}{2}(1 + 75) = \frac{75}{2}(76) = 75(38) = 2850$$

First show that the corresponding sequence is an arithmetic sequence by finding $a_{k+1} - a_k = d$. Then find the indicated sum using the formula.

**7.** $\sum_{k=1}^{50} 3k = 3 + 6 + 9 + \cdots + 150$

**Solution** $a_{k+1} = 3(k + 1) = 3k + 3, \qquad a_k = 3k$

$$a_{k+1} - a_k = (3k + 3) - 3k = 3 = d$$

So, $\{3n\}$ is an arithmetic sequence.

$$\sum_{k=1}^{50} 3k = \frac{50}{2}(3 + 150) \qquad \text{Here, } n = 50, a_1 = 3, \text{ and } a_{50} = 150.$$
$$= 25(153)$$
$$= 3825$$

We can find the same sum using Property III of Section 13.2.

$$\sum_{k=1}^{50} 3k = 3\sum_{k=1}^{50} k \qquad \text{By Property III of Section 13.2.}$$

$$= 3 \cdot \frac{50}{2}(1 + 50) \qquad \text{Here, } n = 50, a_1 = 1, \text{ and } a_{50} = 50.$$

$$= 3 \cdot 25 \cdot 51$$

$$= 3825$$

**8.** $\sum_{k=1}^{70} (-2k + 5) = 3 + 1 + (-1) + (-3) + \cdots + (-135)$

**Solution** $a_{k+1} = -2(k + 1) + 5 = -2k + 3, \qquad a_k = -2k + 5$

$$a_{k+1} - a_k = (-2k + 3) - (-2k + 5) = 3 - 5 = -2 = d$$

So, $\{-2n + 5\}$ is an arithmetic sequence.

$$\sum_{k=1}^{70} (-2k + 5) = \frac{70}{2}[3 + (-135)] \qquad \text{Here, } n = 70, a_1 = 3, \text{ and } a_{70} = -135.$$
$$= 35(-132)$$
$$= -4620$$

We can find the same sum using Properties II, III, and IV of Section 13.2.

$$\begin{aligned}\sum_{k=1}^{70}(-2k+5) &= \sum_{k=1}^{70} -2k + \sum_{k=1}^{70} 5\\ &= -2\sum_{k=1}^{70} k + \sum_{k=1}^{70} 5\\ &= -2\cdot\frac{70}{2}(1+70) + 70\cdot 5\\ &= -2\cdot 35\cdot 71 + 350\\ &= -4970 + 350\\ &= -4620\end{aligned}$$

■

**Practice Problems**

1. Show that the sequence $\{3n + 5\}$ is arithmetic by finding $d$.
2. Find the 40th term of the arithmetic sequence with 1, 6, and 11 as its first three terms.
3. Find $\sum_{k=1}^{50}(3k + 5)$.

## EXERCISES 13.3

Which of the sequences in Exercises 1–10 are arithmetic? Find the common difference and the $n$th term for each arithmetic sequence.

1. 2, 5, 8, 11, . . .
2. $-3, 1, 5, 9, \ldots$
3. 7, 5, 3, 1, . . .
4. 5, 6, 7, 8, . . .
5. 1, 2, 3, 5, 8, . . .
6. 2, 4, 8, 16, . . .
7. $6, 2, -2, -6, \ldots$
8. $4, -1, -6, -11, \ldots$
9. $0, \frac{1}{2}, 1, \frac{3}{2}, \ldots$
10. $2, \frac{7}{3}, \frac{8}{3}, 3, \ldots$

In Exercises 11–20, write the first five terms of the sequence and determine which of the sequences are arithmetic.

11. $\{2n - 1\}$
12. $\{4 - n\}$
13. $\{(-1)^n(3n - 2)\}$
14. $\left\{n + \frac{n}{2}\right\}$
15. $\{5 - 6n\}$
16. $\left\{\frac{1}{n + 1}\right\}$
17. $\left\{7 - \frac{n}{3}\right\}$
18. $\{(-1)^{n-1}(2n + 1)\}$
19. $\left\{\frac{1}{2n}\right\}$
20. $\left\{\frac{2}{3}n - \frac{7}{3}\right\}$

**Answers to Practice Problems** 1. $d = 3$ 2. $a_{40} = 196$ 3. 4075

Find the general term, $a_n$, for each of the arithmetic sequences in Exercises 21–30.

**21.** $a_1 = 1, d = \frac{2}{5}$ **22.** $a_1 = 9, d = -\frac{1}{3}$ **23.** $a_1 = 7, d = -2$ **24.** $a_1 = -3, d = \frac{4}{5}$

**25.** $a_1 = 10, a_3 = 13$ **26.** $a_1 = 6, a_5 = 4$ **27.** $a_{10} = 13, a_{12} = 3$ **28.** $a_5 = 7, a_9 = 19$

**29.** $a_{13} = 60, a_{23} = 75$ **30.** $a_{11} = 54, a_{29} = 180$

In Exercises 31–38, $\{a_n\}$ is an arithmetic sequence.

**31.** $a_1 = 8, a_{11} = 168$. Find $a_{15}$.

**32.** $a_1 = 17, a_9 = -55$. Find $a_{20}$.

**33.** $a_6 = 8, a_4 = 2$. Find $a_{18}$.

**34.** $a_{16} = 12, a_7 = 30$. Find $a_9$.

**35.** $a_{13} = 34, d = 2, a_n = 22$. Find $n$.

**36.** $a_4 = 20, d = 3, a_n = 44$. Find $n$.

**37.** $a_{10} = 41, d = 4, a_n = 77$. Find $n$.

**38.** $a_3 = 15, d = -\frac{3}{2}, a_n = 6$. Find $n$.

Find the indicated sums in Exercises 39–54 using the formula for arithmetic sequences.

**39.** $-2 + 0 + 2 + 4 + \cdots + 24$

**40.** $3 + 6 + 9 + \cdots + 33$

**41.** $1 + 6 + 11 + 16 + \cdots + 46$

**42.** $5 + 9 + 13 + 17 + \cdots + 49$

**43.** $\sum_{k=1}^{9} (3k - 1)$ **44.** $\sum_{k=1}^{12} (4 - 5k)$ **45.** $\sum_{k=1}^{11} (4k - 3)$ **46.** $\sum_{k=1}^{10} (2k + 7)$

**47.** $\sum_{k=1}^{13} \left(\frac{2k}{3} - 1\right)$ **48.** $\sum_{k=1}^{28} (8k - 5)$ **49.** $\sum_{k=7}^{15} \left(k + \frac{k}{3}\right)$ **50.** $\sum_{k=8}^{21} \left(9 - \frac{k}{3}\right)$

**51.** If $\sum_{k=1}^{33} a_k = -12$, find $\sum_{k=1}^{33} (5a_k + 7)$.

**52.** If $\sum_{k=1}^{15} a_k = 60$, find $\sum_{k=1}^{15} (-2a_k - 5)$.

**53.** If $\sum_{k=1}^{100} (-3a_k + 4) = 700$, find $\sum_{k=1}^{100} a_k$.

**54.** If $\sum_{k=1}^{50} (2b_k - 5) = 32$, find $\sum_{k=1}^{50} b_k$.

**55.** On a certain project, a construction company was penalized for taking more than the contractual time to finish the project. The company forfeited \$75 the first day, \$90 the second day, \$105 the third day, and so on. How many additional days were needed if the penalty was \$1215?

**56.** It is estimated that a certain piece of property, now valued at \$48,000, will appreciate as follows: \$1400 the first year, \$1450 the second year, \$1500 the third year, and so on. On this basis, what will be the value of the property after 10 years?

**57.** The rungs of a ladder decrease uniformly in length from 84 cm to 46 cm. What is the total length of the wood in the rungs if there are 25 of them?

**58.** How many blocks are there in a pile if there are 19 in the first layer, 17 in the second layer, 15 in the third layer, and so on, with only 1 block on the top layer?

## 13.4 Geometric Sequences and Series

**OBJECTIVES**

In this section, you will be learning to:

1. Determine whether or not sequences are geometric.
2. Find the general term for geometric sequences.
3. Find specified terms of geometric sequences.
4. Find the sums of the first $n$ terms of geometric sequences.
5. Find the sums of infinite geometric series.

In Section 13.3, we found that arithmetic sequences (or progressions) are characterized by having a common difference between consecutive terms. In this section, we will discuss sequences that have the property that **any two consecutive terms are in the same ratio.** That is, if we divide consecutive terms, the ratio will be the same regardless of which two consecutive terms are divided. Consider the three sequences

$$\frac{1}{2}, \frac{1}{4}, \frac{1}{8}, \frac{1}{16}, \frac{1}{32}, \ldots$$

$$3, 9, 27, 81, 243, \ldots$$

$$-6, 3, -1, \frac{1}{3}, -\frac{1}{9}, \ldots$$

As the following patterns show, each of these sequences has a **common ratio** when consecutive terms are divided.

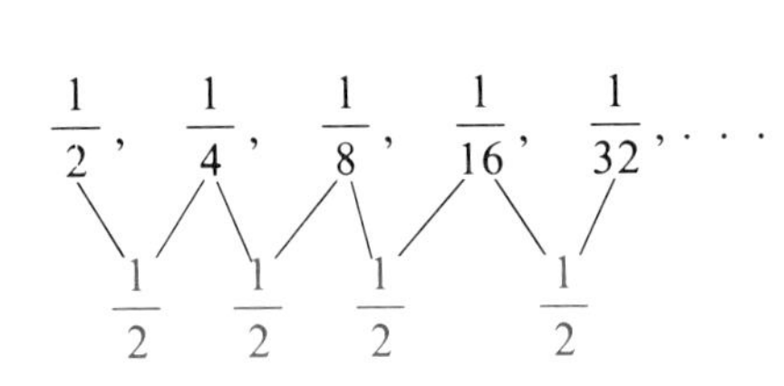

$$\frac{\frac{1}{4}}{\frac{1}{2}} = \frac{1}{2}, \quad \frac{\frac{1}{8}}{\frac{1}{4}} = \frac{1}{2}, \quad \frac{\frac{1}{16}}{\frac{1}{8}} = \frac{1}{2},$$

and so on

$$3, \quad 9, \quad 27, \quad 81, \quad 243, \ldots$$

(ratio between consecutive terms: 3 3 3 3)

$$\frac{9}{3} = 3, \frac{27}{9} = 3, \frac{81}{27} = 3,$$

and so on

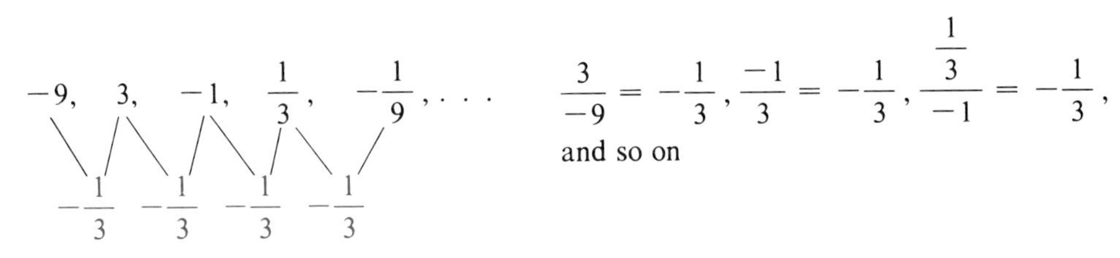

$$\frac{3}{-9} = -\frac{1}{3}, \quad \frac{-1}{3} = -\frac{1}{3}, \quad \frac{\frac{1}{3}}{-1} = -\frac{1}{3},$$

and so on

Such sequences are called **geometric sequences** or **geometric progressions.**

**Geometric Sequence**

A sequence $\{a_n\}$ is called a **geometric sequence** (or **geometric progression**) if for any natural number $k$,

$$\frac{a_{k+1}}{a_k} = r \text{ where } r \text{ is constant and } r \neq 0$$

$r$ is called the **common ratio.**

## EXAMPLES

**1.** Show that the sequence $\left\{\frac{1}{2^n}\right\}$ is geometric by finding $r$.

**Solution** $a_k = \frac{1}{2^k}$ and $a_{k+1} = \frac{1}{2^{k+1}}$

$$\frac{a_{k+1}}{a_k} = \frac{\frac{1}{2^{k+1}}}{\frac{1}{2^k}} = \frac{1}{2^k \cdot 2} \cdot \frac{2^k}{1} = \frac{1}{2} = r$$

**2.** Show that the sequence $\{n^2\}$ is **not** geometric.

**Solution** Since $a_3 = 3^2$ and $a_2 = 2^2$ and $a_1 = 1^2$,

$$\frac{a_3}{a_2} = \frac{3^2}{2^2} = \frac{9}{4} \quad \text{and} \quad \frac{a_2}{a_1} = \frac{2^2}{1^2} = 4$$

Since $\frac{9}{4} \neq 4$, there is no common ratio between consecutive terms, and $\{n^2\}$ is **not geometric.** ■

If the first term is $a_1$ and the common ratio is $r$, then the geometric sequence can be indicated as follows:

$a_1 = a_1 \longrightarrow a_1$ first term

$\frac{a_2}{a_1} = r \longrightarrow a_2 = a_1 r$ second term

$\frac{a_3}{a_2} = r \longrightarrow a_3 = a_2 r = (a_1 r)r = a_1 r^2$ third term

$\frac{a_4}{a_3} = r \longrightarrow a_4 = a_3 r = (a_1 r^2)r = a_1 r^3$ fourth term

$\vdots$ $\vdots$

$\frac{a_n}{a_{n-1}} = r \longrightarrow a_n = a_{n-1} \cdot r = (a_1 r^{n-2})r = a_1 r^{n-1}$ $n$th term

$\vdots$ $\vdots$

Thus,

$$a_n = a_1 r^{n-1}$$

represents the general term or $n$th term of a geometric sequence.

**EXAMPLES**

**3.** If in a geometric sequence, $a_1 = 4$ and $r = -\frac{1}{2}$, find $a_8$.

**Solution** $a_8 = a_1r^7 = 4\left(-\frac{1}{2}\right)^7 = 2^2\left(-\frac{1}{2^7}\right) = -\frac{1}{2^5} = -\frac{1}{32}$

**4.** Find the seventh term of the geometric sequence

$3, \frac{3}{2}, \frac{3}{4}, \ldots$

**Solution** Find $r$ using the formula $r = \frac{a_{k+1}}{a_k}$ with $a_1 = 3$ and $a_2 = \frac{3}{2}$.

$$r = \frac{a_2}{a_1} = \frac{\frac{3}{2}}{3} = \frac{1}{2}$$

Now, the seventh term is

$$a_7 = a_1r^{7-1} = 3\left(\frac{1}{2}\right)^6 = \frac{3}{64}$$

**5.** Find $a_1$ and $r$ for the geometric sequence in which $a_5 = 2$ and $a_7 = 4$.

**Solution** Using the formula $a_n = a_1r^{n-1}$, we get

$$2 = a_1r^4 \qquad \text{and} \qquad 4 = a_1r^6$$

Now, dividing gives

$$\frac{\cancel{a_1}r^6}{\cancel{a_1}r^4} = \frac{4}{2}$$

$$r^2 = 2$$

$$r = \sqrt{2} \qquad \text{or} \qquad r = -\sqrt{2}$$

For $r = \sqrt{2}$,

$$2 = a_1(\sqrt{2})^4$$

$$2 = a_1 \cdot 4$$

$$\frac{1}{2} = a_1$$

For $r = -\sqrt{2}$,

$$2 = a_1(-\sqrt{2})^4$$

$$2 = a_1 \cdot 4$$

$$\frac{1}{2} = a_1$$

There are two geometric sequences with $a_5 = 2$ and $a_7 = 4$. In both cases, $a_1 = \frac{1}{2}$. The two possibilities are

$$a_1 = \frac{1}{2} \quad \text{and} \quad r = \sqrt{2}$$

or

$$a_1 = \frac{1}{2} \quad \text{and} \quad r = -\sqrt{2}$$

■

To find $S = \sum_{k=1}^{6} \frac{1}{3^k}$, we can write all the terms and then add them.

$$\begin{aligned} S = \sum_{k=1}^{6} \frac{1}{3^k} &= \frac{1}{3} + \frac{1}{3^2} + \frac{1}{3^3} + \frac{1}{3^4} + \frac{1}{3^5} + \frac{1}{3^6} \\ &= \frac{3^5 + 3^4 + 3^3 + 3^2 + 3 + 1}{3^6} = \frac{364}{3^6} \end{aligned}$$

Or, we can write the terms and then multiply each term by $\frac{1}{3}$ (the common ratio) and subtract as follows:

$$\begin{aligned} S &= \frac{1}{3} + \frac{1}{3^2} + \frac{1}{3^3} + \frac{1}{3^4} + \frac{1}{3^5} + \frac{1}{3^6} \\ \frac{1}{3}S &= \qquad\ \ \frac{1}{3^2} + \frac{1}{3^3} + \frac{1}{3^4} + \frac{1}{3^5} + \frac{1}{3^6} + \frac{1}{3^7} \\ \hline S - \frac{1}{3}S &= \frac{1}{3} - 0 - 0 - 0 - 0 - 0 - \frac{1}{3^7} \\ \left(1 - \frac{1}{3}\right)S &= \frac{1}{3} - \frac{1}{3^7} \\ S &= \frac{\frac{1}{3} - \frac{1}{3^7}}{1 - \frac{1}{3}} \end{aligned}$$

Following this procedure, we can develop a formula for the sum of $n$ terms of any geometric sequence.

Suppose that the $n$ terms are

$$a_1, \quad a_2 = a_1 r, \quad a_3 = a_1 r^2, \quad \ldots, \quad a_{n-1} = a_1 r^{n-2}, \quad a^n = a_1 r^{n-1}$$

Thus,

$$\begin{aligned} S &= a_1 + a_1 r + a_1 r^2 + \cdots + a_1 r^{n-2} + a_1 r^{n-1} \\ rS &= \phantom{a_1 + {}} a_1 r + a_1 r^2 + a_1 r^3 + \cdots \phantom{{}+ a_1 r^{n-2}} + a_1 r^{n-1} + a_1 r^n \\ \hline S - rS &= a_1 - a_1 r^n \\ (1 - r)S &= a_1(1 - r^n) \\ S &= \frac{a_1(1 - r^n)}{1 - r} \end{aligned}$$

The formula can be rewritten using $\Sigma$-notation.

The sum of the first $n$ terms of a geometric sequence $\{a_n\}$ can be written as

$$S_n = \sum_{k=1}^{n} a_k = \frac{a_1(1 - r^n)}{1 - r} \qquad \text{where } r \neq 1$$

**EXAMPLES** First show that the corresponding sequence is a geometric sequence by finding $\frac{a_{k+1}}{a_k} = r$. Then find the indicated sum using the formula.

**6.** $\displaystyle\sum_{k=1}^{10} \frac{1}{2^k}$

**Solution** $a_k = \dfrac{1}{2^k}$ and $a_{k+1} = \dfrac{1}{2^{k+1}}$

$$\frac{a_{k+1}}{a_k} = \frac{\frac{1}{2^{k+1}}}{\frac{1}{2^k}} = \frac{1}{2^{k+1}} \cdot \frac{2^k}{1} = \frac{1}{2 \cdot \cancel{2^k}} \cdot \frac{\cancel{2^k}}{1} = \frac{1}{2} = r$$

So $\left\{\dfrac{1}{2^n}\right\}$ is a geometric sequence.

$$\sum_{k=1}^{10} \frac{1}{2^k} = \frac{\frac{1}{2}\left(1 - \left(\frac{1}{2}\right)^{10}\right)}{1 - \frac{1}{2}} = \frac{\frac{1}{\cancel{2}}\left(1 - \frac{1}{1024}\right)}{\frac{1}{\cancel{2}}} = \frac{1023}{1024}$$

**7.** $\sum_{k=1}^{5} (-1)^k \cdot 3^{k/2}$

**Solution** $a_k = (-1)^k \cdot 3^{k/2}$ and $a_{k+1} = (-1)^{k+1} \cdot 3^{(k+1)/2}$

$$\frac{a_{k+1}}{a_k} = \frac{(-1)^{k+1} \cdot 3^{(k+1)/2}}{(-1)^k \cdot 3^{k/2}} = \frac{\cancel{(-1)^k}(-1) \cdot \cancel{3^{k/2}} \cdot 3^{1/2}}{\cancel{(-1)^k} \cdot \cancel{3^{k/2}}}$$

$$= (-1) \cdot 3^{1/2} = -\sqrt{3} = r$$

So $\{(-1)^n \cdot 3^{n/2}\}$ is a geometric sequence.

$$\sum_{k=1}^{5} (-1)^k \cdot 3^{k/2} = \frac{(-1) \cdot 3^{1/2} \cdot (1 - (-\sqrt{3})^5)}{1 - (-\sqrt{3})} = \frac{-\sqrt{3}(1 + 9\sqrt{3})}{1 + \sqrt{3}}$$

**8.** The parents of a small child decide to deposit \$1000 annually at the first of each year for 20 years for the child's education. If interest is compounded annually at 8%, what will be the value of the deposits at the end of 20 years? (This type of investment is called an annuity.)

**Solution** The formula for interest compounded annually is $A = P(1 + r)^t$ where $A$ is the amount in the account, $r$ is the annual interest rate (in decimal form), and $t$ is the time (in years).

The first deposit of \$1000 will earn interest for 20 years:

$$A_{20} = 1000(1 + 0.08)^{20} = 1000(1.08)^{20}$$

The second deposit will earn interest for 19 years:

$$A_{19} = 1000(1.08)^{19}$$

$$\vdots$$

The last deposit will earn interest for only 1 year:

$$A_1 = 1000(1.08)^1$$

The accumulated value of all the deposits (plus interest) is the sum of the 20 terms of a geometric sequence:

$$\begin{aligned}
\text{Value} &= A_1 + A_2 + \cdots + A_{20} = \sum_{k=1}^{20} 1000(1.08)^k \\
&= 1000(1.08)^1 + 1000(1.08)^2 + \cdots + 1000(1.08)^{20} \\
&= \frac{1000(1.08)[1 - (1.08)^{20}]}{1 - 1.08} \quad \text{where } a_1 = 1000(1.08) \text{ and } r = 1.08 \\
&= \frac{1080[1 - 4.660957]}{-0.08} \\
&= 49{,}423
\end{aligned}$$

Thus, the accumulated value of the annuity is \$49,423. ■

Now, we will discuss the conditions under which we can define the sum of all the terms of a geometric sequence. In the following definition, the symbol $\infty$ (read "infinity") is used to indicate that the number of terms is unbounded. The symbol $\infty$ does not represent a number.

***Infinite Series***

The indicated sum of all the terms of a sequence is called an **infinite series** (or a **series**). For a sequence $\{a_n\}$, the corresponding series can be written as follows:

$$\sum_{k=1}^{\infty} a_k = a_1 + a_2 + a_3 + \cdots + a_n + \cdots$$

In the case where $|r| < 1$, it can be shown, in higher level mathematics, that $r^n$ approaches 0 as $n$ approaches infinity. We write

$$r^n \longrightarrow 0 \qquad \text{as} \qquad n \longrightarrow \infty$$

Thus, we have the following result:

$$S_n = \frac{a_1(1 - r^n)}{1 - r} \longrightarrow \frac{a_1(1 - 0)}{1 - r} = \frac{a_1}{1 - r} \qquad \text{as } n \longrightarrow \infty$$

***Theorem***

If $\{a_n\}$ is a geometric sequence and $|r| < 1$, then the **sum of the infinite geometric series** is

$$S = \sum_{k=1}^{\infty} a_k = a_1 + a_1r + a_1r^2 + \cdots = \frac{a_1}{1 - r}$$

**EXAMPLES** Find the sum of each of the following geometric series.

**9.** $\displaystyle\sum_{k=1}^{\infty} \left(\frac{2}{3}\right)^{k-1} = \left(\frac{2}{3}\right)^0 + \left(\frac{2}{3}\right)^1 + \left(\frac{2}{3}\right)^2 + \left(\frac{2}{3}\right)^3 + \cdots$

$$= 1 + \frac{2}{3} + \frac{4}{9} + \frac{8}{27} + \cdots$$

**Solution** Here, $a_1 = 1$ and $r = \frac{2}{3}$. Substitution in the formula gives

$$S = \frac{1}{1 - \frac{2}{3}} = \frac{1}{\frac{1}{3}} = 3$$

**10.** $0.33333\ldots = 0.\overline{3}$ (The bar over the 3 indicates a repeating pattern to the decimal.)

**Solution**

$0.33333\ldots = 0.3 + 0.03 + 0.003 + 0.0003 + 0.00003 + \cdots$

Thus, we have a geometric series with $a_1 = 0.3 = \frac{3}{10}$ and $r = 0.1 = \frac{1}{10}$.

Applying the formula gives

$$S = \frac{\frac{3}{10}}{1 - \frac{1}{10}} = \frac{\frac{3}{10}}{\frac{9}{10}} = \frac{3}{10} \cdot \frac{10}{9} = \frac{1}{3}$$

In this way, we can convert an infinite repeating decimal to fraction form:

$$0.33333\ldots = \frac{1}{3}$$

**11.** $5 - 1 + \frac{1}{5} - \frac{1}{25} + \frac{1}{125} - \frac{1}{625} + \cdots$

**Solution** Here, $a_1 = 5$ and $r = -\frac{1}{5}$. A geometric series that alternates in sign will always have a negative value for $r$. Substitution in the formula gives

$$S = \frac{5}{1 - \left(-\frac{1}{5}\right)} = \frac{5}{1 + \frac{1}{5}} = \frac{5}{\frac{6}{5}} = \frac{5}{1} \cdot \frac{5}{6} = \frac{25}{6}$$

■

***Practice Problems***

1. Show that the sequence $\left\{\frac{(-1)^n}{3^n}\right\}$ is geometric by finding $r$.
2. If in a geometric progression $a_1 = 0.1$ and $r = 2$, find $a_6$.
3. Find the sum $\sum_{k=1}^{5} \frac{1}{2^k}$.
4. Represent the decimal $0.\overline{4}$ as a series using $\Sigma$-notation.
5. Find the sum of the series in Problem 4.

**Answers to Practice Problems** **1.** $r = -\frac{1}{3}$ **2.** $a_6 = 3.2$ **3.** $\frac{31}{32}$ **4.** $\sum_{k=1}^{\infty} \frac{4}{10^k}$ **5.** $\frac{4}{9}$

## EXERCISES 13.4

Which of the sequences in Exercises 1–10 are geometric? Find the common ratio for each geometric sequence, and the $n$th term.

**1.** $2, 4, 8, 16 \ldots$

**2.** $\frac{1}{12}, \frac{1}{6}, \frac{1}{3}, \frac{2}{3}, \ldots$

**3.** $3, -\frac{3}{2}, \frac{3}{4}, -\frac{3}{8}, \ldots$

**4.** $5, 9, 13, 17, \ldots$

**5.** $\frac{32}{27}, \frac{4}{9}, \frac{1}{6}, \frac{1}{16}, \ldots$

**6.** $18, 12, 8, \frac{16}{3}, \ldots$

**7.** $\frac{14}{3}, \frac{2}{3}, \frac{2}{15}, \frac{2}{45}, \ldots$

**8.** $1, -\frac{2}{3}, \frac{4}{9}, -\frac{8}{27}, \ldots$

**9.** $48, -12, 3, -\frac{3}{4}, \ldots$

**10.** $4, -8, 12, -16, \ldots$

In Exercises 11–20, write the first four terms of the sequence and determine which of the sequences are geometric sequences.

**11.** $\{(-3)^{n+1}\}$

**12.** $\left\{3\left(\frac{2}{5}\right)^n\right\}$

**13.** $\left\{\frac{2}{3}n\right\}$

**14.** $\left\{(-1)^{n+1}\left(\frac{2}{7}\right)^n\right\}$

**15.** $\left\{2\left(-\frac{4}{5}\right)^n\right\}$

**16.** $\left\{1 + \frac{1}{2^n}\right\}$

**17.** $\{3(2)^{n/2}\}$

**18.** $\left\{\frac{n^2+1}{n}\right\}$

**19.** $\{(-1)^{n-1}(0.3)^n\}$

**20.** $\{6(10)^{1-n}\}$

Find the general term, $a_n$, for each of the geometric sequences in Exercises 21–30.

**21.** $a_1 = 3, r = 2$

**22.** $a_1 = -2, r = \frac{1}{5}$

**23.** $a_1 = \frac{1}{3}, r = -\frac{1}{2}$

**24.** $a_1 = 5, r = \sqrt{2}$

**25.** $a_3 = 2, a_5 = 4, r > 0$

**26.** $a_4 = 19, a_5 = 57$

**27.** $a_2 = 1, a_4 = 9$

**28.** $a_2 = 5, a_5 = \frac{5}{8}$

**29.** $a_3 = -\frac{45}{16}, r = -\frac{3}{4}$

**30.** $a_4 = 54, r = 3$

In Exercises 31–38, $\{a_n\}$ is a geometric sequence.

**31.** $a_1 = -32, a_6 = 1$. Find $a_8$.

**32.** $a_1 = 20, a_6 = \frac{5}{8}$. Find $a_7$.

**33.** $a_1 = 18, a_7 = \frac{128}{81}$. Find $a_5$.

**34.** $a_1 = -3, a_5 = -48$. Find $a_7$.

**35.** $a_3 = \frac{1}{2}, a_7 = \frac{1}{32}$. Find $a_4$.

**36.** $a_5 = 48, a_8 = -384$. Find $a_9$.

**37.** $a_1 = -2, r = \frac{2}{3}, a_n = -\frac{16}{27}$. Find $n$.

**38.** $a_1 = \frac{1}{9}, r = \frac{3}{2}, a_n = \frac{27}{32}$. Find $n$.

In Exercises 39–56, find the indicated sums.

**39.** $3 + 9 + 27 + \cdots + 243$

**40.** $-2 + 4 - 8 + 16$

**41.** $8 + 4 + 2 + \cdots + \frac{1}{64}$

**42.** $3 + 12 + 48 + \cdots + 3072$

**43.** $\sum_{k=1}^{3} -3\left(\frac{3}{4}\right)^k$ **44.** $\sum_{k=1}^{6} \left(-\frac{5}{3}\right)\left(\frac{1}{2}\right)^k$ **45.** $\sum_{k=1}^{5} \left(\frac{2}{3}\right)^k$ **46.** $\sum_{k=1}^{6} \left(\frac{1}{3}\right)^k$

**47.** $\sum_{k=4}^{7} 5\left(\frac{1}{2}\right)^k$ **48.** $\sum_{k=3}^{6} -7\left(\frac{3}{2}\right)^k$ **49.** $\sum_{k=1}^{\infty} \left(\frac{3}{4}\right)^{k-1}$ **50.** $\sum_{k=1}^{\infty} \left(\frac{5}{8}\right)^{k-1}$

**51.** $\sum_{k=1}^{\infty} \left(-\frac{1}{2}\right)^k$ **52.** $\sum_{k=1}^{\infty} \left(-\frac{2}{5}\right)^k$ **53.** $0.\overline{2}$ **54.** $0.\overline{6}$

**55.** $0.\overline{36}$ **56.** $0.\overline{81}$

**57.** Sue deposits \$800 annually at the first of each year for 10 years. If the interest is compounded annually at 9%, what will be the value of the deposits at the end of 10 years?

**58.** If \$1200 is deposited annually at the first of each year for 8 years, what will be the value of the deposits if the interest is compounded annually at 10%?

**59.** An automobile that costs \$8500 new depreciates at a rate of 20% of its value each year. What is its value after 3 years?

**60.** The radiator of a car contains 20 liters of water. Five liters are drained off and replaced by antifreeze. Then, 5 liters of the mixture are drained off and replaced by antifreeze, and so on. This process is continued until six drain-offs and replacements have been made. How much antifreeze is in the final mixture?

**61.** A substance decays at a rate of $\frac{2}{5}$ of its weight per day. How much of the substance will be present after 4 days if initially there are 500 grams?

**62.** A ball rebounds to a height that is $\frac{3}{4}$ of its original height. How high will it rise after the fourth bounce if it is dropped from a height of 24 meters?

## 13.5 The Binomial Theorem

**OBJECTIVES**

In this section, you will be learning to:

1. Calculate factorials.
2. Expand binomials using the Binomial Theorem.
3. Find specified terms in binomial expressions.

The objective in this section is to develop a formula stated as the **Binomial Theorem** (and sometimes called the **binomial expansion**) that will allow you to write products such as

$$(a + b)^3, \qquad (a + b)^7, \qquad (x + 5)^8$$

without having to multiply the binomial factors, such as

$$\begin{aligned}(a + b)^3 &= (a + b)(a + b)(a + b)\\ &= (a^2 + 2ab + b^2)(a + b)\\ &= a^3 + 2a^2b + ab^2 + a^2b + 2ab^2 + b^3\\ &= a^3 + 3a^2b + 3ab^2 + b^3\end{aligned}$$

The idea of **factorial** is basic to the discussion. 6! is read "six factorial" and represents the product of the positive integers from 6 to 1. Thus,

$$6! = 6 \cdot 5 \cdot 4 \cdot 3 \cdot 2 \cdot 1 = 720$$

Also,

$$10! = 10 \cdot 9 \cdot 8 \cdot 7 \cdot 6 \cdot 5 \cdot 4 \cdot 3 \cdot 2 \cdot 1 = 3{,}628{,}800$$

**n Factorial**

For any positive integer $n$,

$$n! = n(n-1)(n-2) \cdot \ldots \cdot 3 \cdot 2 \cdot 1$$

$n!$ is read "$n$ factorial."

To evaluate an expression such as

$$\frac{7!}{6!}$$

do **not** evaluate each factorial. Instead, write the factorials as products and reduce the fraction.

$$\frac{7!}{6!} = \frac{7 \cdot \cancel{6} \cdot \cancel{5} \cdot \cancel{4} \cdot \cancel{3} \cdot \cancel{2} \cdot \cancel{1}}{\cancel{6} \cdot \cancel{5} \cdot \cancel{4} \cdot \cancel{3} \cdot \cancel{2} \cdot \cancel{1}} = 7$$

We can also write $7! = 7 \cdot 6!$. So,

$$\frac{7!}{6!} = \frac{7(6!)}{(6!)} = 7$$

In general,

$$n! = n \cdot (n-1)!$$

Also, for work with formulas involving factorials, zero factorial is defined to be 1.

**0 Factorial**

$0! = 1$

**EXAMPLES** Simplify the following expressions.

**1.** $\dfrac{11!}{8!}$

**Solution** $\dfrac{11!}{8!} = \dfrac{11 \cdot 10 \cdot 9 \cdot (\cancel{8} \cdot \cancel{7} \cdot \cancel{6} \cdot \cancel{5} \cdot \cancel{4} \cdot \cancel{3} \cdot \cancel{2} \cdot \cancel{1})}{(\cancel{8} \cdot \cancel{7} \cdot \cancel{6} \cdot \cancel{5} \cdot \cancel{4} \cdot \cancel{3} \cdot \cancel{2} \cdot \cancel{1})} = 990$

or $\dfrac{11!}{8!} = \dfrac{11 \cdot 10 \cdot 9 \cdot \cancel{8!}}{\cancel{8!}} = 990$

**2.** $\dfrac{n!}{(n-2)!}$

**Solution** $\dfrac{n!}{(n-2)!} = \dfrac{n(n-1)\cancel{(n-2)!}}{\cancel{(n-2)!}} = n(n-1)$

**3.** $\dfrac{30!}{28!2!}$

**Solution** $\dfrac{30!}{28!2!} = \dfrac{\overset{15}{\cancel{30}} \cdot 29 \cdot \cancel{28!}}{\cancel{28!} \cdot \cancel{2} \cdot 1} = 15 \cdot 29 = 435$ ■

The expression in Example 3 can be written in a different notation.

$$\binom{30}{2} = \frac{30!}{28!2!} \quad \text{and} \quad \binom{30}{28} = \frac{30!}{2!28!}$$

$\binom{n}{r}$

$$\binom{n}{r} = \frac{n!}{(n-r)!r!} \quad \text{when } 0 \le r \le n$$

Using the definition with $\binom{n}{n-r}$, we get

$$\binom{n}{n-r} = \frac{n!}{(n-(n-r))!(n-r)!} = \frac{n!}{(n-n+r)!(n-r)!}$$
$$= \frac{n!}{r!(n-r)!} = \binom{n}{r}$$

Thus,

$$\binom{n}{n-r} = \binom{n}{r}$$

**EXAMPLES** Evaluate the following.

**4.** $\binom{8}{2}$

**Solution** $\binom{8}{2} = \frac{8!}{2!6!} = \frac{\overset{4}{\cancel{8}} \cdot 7 \cdot \cancel{6!}}{\cancel{2} \cdot \cancel{6!}} = 28$

**5.** $\binom{17}{0}$

**Solution** $\binom{17}{0} = \frac{17!}{0!17!} = \frac{1}{1} = 1$ ■

The expansions of the binomial $a + b$ from $(a + b)^0$ to $(a + b)^5$ are shown here.

$$\begin{aligned}
(a+b)^0 &= 1\\
(a+b)^1 &= a + b\\
(a+b)^2 &= a^2 + 2ab + b^2\\
(a+b)^3 &= a^3 + 3a^2b + 3ab^2 + b^3\\
(a+b)^4 &= a^4 + 4a^3b + 6a^2b^2 + 4ab^3 + b^4\\
(a+b)^5 &= a^5 + 5a^4b + 10a^3b^2 + 10a^2b^3 + 5ab^4 + b^5
\end{aligned}$$

Three patterns are evident. **First,** in each case, the powers of ***a*** decrease by 1 each term and the powers of ***b*** increase by 1 each term. **Second,** in each term, the sum of the exponents is equal to the exponent on $(a + b)$. The **third** pattern, called **Pascal's Triangle,** is formed from the coefficients. For convenience in memorizing, note the symmetry in each set of coefficients.

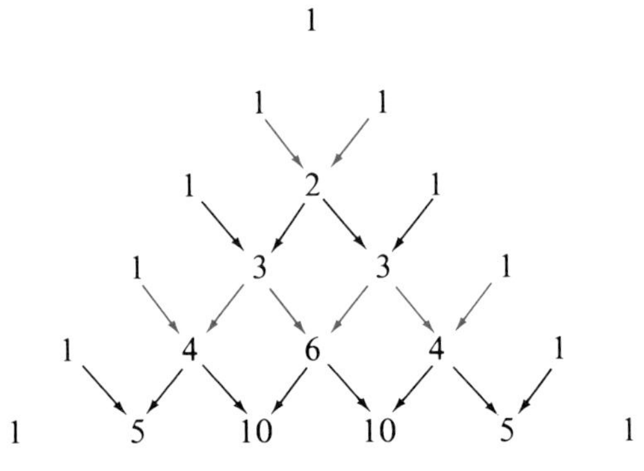


In each case, the first and last coefficients are 1, and the other coefficients are the sums of the two numbers above to the left and above to the right of that coefficient. Thus, for $(a + b)^6$, we have

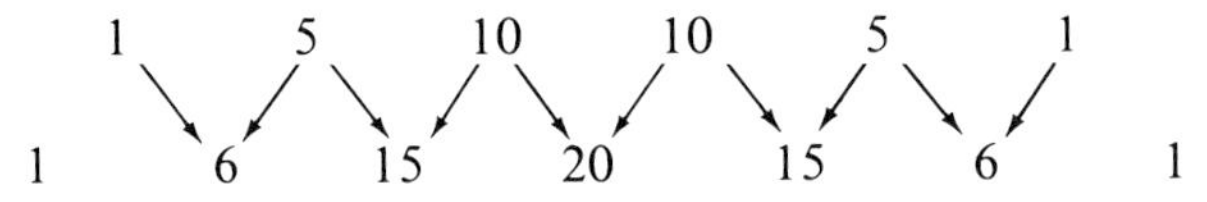


and

$$(a + b)^6 = a^6 + 6a^5b + 15a^4b^2 + 20a^3b^3 + 15a^2b^4 + 6ab^5 + b^6$$

Note that

$$\binom{6}{0} = \frac{6!}{0!6!} = 1 \qquad \binom{6}{1} = \frac{6!}{1!5!} = 6 \qquad \binom{6}{2} = \frac{6!}{2!4!} = 15$$

$$\binom{6}{3} = \frac{6!}{3!3!} = 20 \qquad \binom{6}{4} = \frac{6!}{4!2!} = 15 \qquad \binom{6}{5} = \frac{6!}{5!1!} = 6 \qquad \binom{6}{6} = \frac{6!}{6!0!} = 1$$

So, we can write

$$(a + b)^6 = \binom{6}{0}a^6 + \binom{6}{1}a^5b + \binom{6}{2}a^4b^2 + \binom{6}{3}a^3b^3 + \binom{6}{4}a^2b^4 + \binom{6}{5}ab^5 + \binom{6}{6}b^6$$

This leads to the Binomial Theorem. The proof is beyond the scope of this text.

**The Binomial Theorem**

$$(a + b)^n = \binom{n}{0}a^n + \binom{n}{1}a^{n-1}b + \binom{n}{2}a^{n-2}b^2 + \cdots + \binom{n}{k}a^{n-k}b^k + \cdots + \binom{n}{n}b^n$$

Using $\Sigma$-notation,

$$(a + b)^n = \sum_{k=0}^{n} \binom{n}{k} a^{n-k}b^k$$

Note:

1. **There are $n + 1$ terms in $(a + b)^n$.**
2. **In each term of $(a + b)^n$, the sum of the exponents of $a$ and $b$ is $n$.**

**EXAMPLES**

6. Expand $(x + 3)^5$ by using the Binomial Theorem.

**Solution**

$$\begin{aligned}(x + 3)^5 &= \sum_{k=0}^{5} \binom{5}{k} x^{5-k}3^k \\ &= \binom{5}{0}x^5 + \binom{5}{1}x^4 \cdot 3 + \binom{5}{2}x^3 \cdot 3^2 + \binom{5}{3}x^2 \cdot 3^3 \\ &\quad + \binom{5}{4}x \cdot 3^4 + \binom{5}{5}3^5 \\ &= x^5 + 5 \cdot x^4 \cdot 3 + 10 \cdot x^3 \cdot 9 + 10 \cdot x^2 \cdot 27 \\ &\quad + 5 \cdot x \cdot 81 + 1 \cdot 243 \\ &= x^5 + 15x^4 + 90x^3 + 270x^2 + 405x + 243\end{aligned}$$

7. Expand $(y^2 - 1)^6$ by using the Binomial Theorem.

**Solution**

$$\begin{aligned}(y^2 - 1)^6 &= \sum_{k=0}^{6} \binom{6}{k} (y^2)^{6-k}(-1)^k \\ &= \binom{6}{0}(y^2)^6 + \binom{6}{1}(y^2)^5(-1)^1 + \binom{6}{2}(y^2)^4(-1)^2 \\ &\quad + \binom{6}{3}(y^2)^3(-1)^3 + \binom{6}{4}(y^2)^2(-1)^4 \\ &\quad + \binom{6}{5}(y^2)^1(-1)^5 + \binom{6}{6}(-1)^6 \\ &= 1 \cdot y^{12} + 6 \cdot y^{10}(-1) + 15 \cdot y^8(+1) + 20 \cdot y^6(-1) \\ &\quad + 15 \cdot y^4(+1) + 6 \cdot y^2(-1) + 1(+1) \\ &= y^{12} - 6y^{10} + 15y^8 - 20y^6 + 15y^4 - 6y^2 + 1\end{aligned}$$

**8.** Find the sixth term of the expansion of $\left(2x - \frac{1}{3}\right)^{10}$.

**Solution** Since $\left(2x - \frac{1}{3}\right)^{10} = \sum_{k=0}^{10} \binom{10}{k}(2x)^{10-k}\left(-\frac{1}{3}\right)^k$, and the sum begins with $k = 0$, the sixth term will occur when $k = 5$.

$$\binom{10}{5}(2x)^{10-5}\left(-\frac{1}{3}\right)^5 = \frac{10!}{5!5!}(2x)^5\left(-\frac{1}{3}\right)^5$$

$$= \overset{28}{\cancel{252}} \cdot 32x^5\left(-\frac{1}{\underset{27}{\cancel{243}}}\right) = \frac{-896x^5}{27}$$

The sixth term is $\dfrac{-896x^5}{27}$.

**9.** Find the fourth term of the expansion of $\left(x + \frac{1}{2}y\right)^8$.

**Solution** $\left(x + \frac{1}{2}y\right)^8 = \sum_{k=0}^{8} \binom{8}{k}x^{8-k}\left(\frac{1}{2}y\right)^k$

The fourth term occurs when $k = 3$.

$$\binom{8}{3}x^{8-3}\left(\frac{1}{2}y\right)^3 = \frac{8!}{3!5!} \cdot x^5 \cdot \frac{1}{8}y^3 = 7x^5y^3$$

The fourth term is $7x^5y^3$.

**10.** Using the binomial expansion, approximate $(0.99)^4$ to the nearest thousandth.

**Solution**

$$(0.99)^4 = (1 - 0.01)^4 = \sum_{k=0}^{4} \binom{4}{k}(1)^{4-k}(-0.01)^k$$

$$= \binom{4}{0} \cdot 1^4 + \binom{4}{1} \cdot 1^3 \cdot (-0.01) + \binom{4}{2} \cdot 1^2 \cdot (-0.01)^2$$

$$+ \binom{4}{3} \cdot 1 \cdot (-0.01)^3 + \binom{4}{4} \cdot (-0.01)^4$$

$$= 1 + 4(-0.01) + 6(-0.01)^2 + 4(-0.01)^3 + 1(-0.01)^4$$

$$= 1 - 0.04 + 0.0006 - 0.000004 + \text{(small term)}$$

$$\approx 0.9606$$

$$\approx 0.961 \qquad \text{(to nearest thousandth)}$$

■

**Practice Problems**

1. Simplify: $\frac{10!}{7!}$
2. Evaluate: $\binom{20}{2}$
3. Expand $(x + 2)^4$ by using the Binomial Theorem.
4. Find the third term of the expansion of $(2x - 1)^7$.

## EXERCISES 13.5

Simplify the expressions in Exercises 1–16.

**1.** $\frac{8!}{6!}$ **2.** $\frac{11!}{7!}$ **3.** $\frac{3!8!}{10!}$ **4.** $\frac{5!7!}{8!}$ **5.** $\frac{5!4!}{6!}$ **6.** $\frac{7!4!}{10!}$

**7.** $\frac{n!}{n}$ **8.** $\frac{n!}{(n-3)!}$ **9.** $\frac{(k+3)!}{k!}$ **10.** $\frac{n(n+1)!}{(n+2)!}$ **11.** $\binom{6}{3}$ **12.** $\binom{5}{4}$

**13.** $\binom{7}{3}$ **14.** $\binom{8}{5}$ **15.** $\binom{10}{0}$ **16.** $\binom{6}{2}$

Write the first four terms of the expansions in Exercises 17–28.

**17.** $(x + y)^7$ **18.** $(x + y)^{11}$ **19.** $(x + 1)^9$ **20.** $(x + 1)^{12}$
**21.** $(x + 3)^5$ **22.** $(x - 2)^6$ **23.** $(x + 2y)^6$ **24.** $(x + 3y)^5$
**25.** $(3x - y)^7$ **26.** $(2x - y)^{10}$ **27.** $(x^2 - 4y)^9$ **28.** $(x^2 - 2y)^7$

Using the Binomial Theorem, expand the expressions in Exercises 29–40.

**29.** $(x + y)^6$ **30.** $(x + y)^8$ **31.** $(x - 1)^7$ **32.** $(x - 1)^9$
**33.** $(3x + y)^5$ **34.** $(2x + y)^6$ **35.** $(x + 2y)^4$ **36.** $(x + 3y)^5$
**37.** $(3x - 2y)^4$ **38.** $(5x + 2y)^3$ **39.** $(3x^2 - y)^5$ **40.** $(x^2 + 2y)^4$

Find the specified term in each of the expansions in Exercises 41–46.

**41.** $(x - 2y)^{10}$, fifth term **42.** $(x + 3y)^{12}$, third term **43.** $(2x + 3)^{11}$, fourth term
**44.** $\left(x - \frac{y}{2}\right)^9$, seventh term **45.** $(5x^2 - y^2)^{12}$, tenth term **46.** $(2x^2 + y^2)^{15}$, eleventh term

Approximate the value of each expression in Exercises 47–53 correct to the nearest thousandths.

**47.** $(1.01)^6$ **48.** $(0.96)^8$ **49.** $(0.97)^7$ **50.** $(1.02)^{10}$
**51.** $(2.3)^5$ **52.** $(2.8)^6$ **53.** $(0.98)^8$

**Answers to Practice Problems** 1. 720 2. 190 3. $x^4 + 8x^3 + 48x^2 + 32x + 16$
4. $672x^5$

## 13.6 Permutations

**OBJECTIVES**

In this section, you will be learning to:

1. Evaluate expressions representing permutations.
2. Solve applied problems involving permutations or the Fundamental Principle of Counting.

Many problems in statistics require a systematic approach to counting the number of ways several decisions can be made in succession (or several events can occur in succession). For example, if a seven-person board of education must elect a president and a vice-president from its own membership, how many ways can this be done? The reasoning is as follows:

The presidency can be filled by any one of 7.

After the president is elected, the vice-presidency can be filled by any one of 6.

The number of ways these two decisions can be made in succession is $7 \cdot 6 = 42$.

The procedure for counting the number of independent successive decisions (or events) is based on the **Fundamental Principle of Counting.**

***The Fundamental Principle of Counting***

If an event $E_1$ can occur in $m_1$ ways, an event $E_2$ can occur in $m_2$ ways, . . . , and an event $E_k$ can occur in $m_k$ ways, then the total number of ways that all the events may occur is the product $m_1 \cdot m_2 \cdot \ldots \cdot m_k$.

**EXAMPLE**

**1.** A home contractor offers two basic house plans, each with two possible arrangements for the garage, four color combinations, and three types of landscaping. How many "different" homes can the contractor build?

**Solution** The Fundamental Principle of Counting is used with each of the options considered as a decision or event.

$$\underset{\text{plans}}{\underset{\uparrow}{\mathbf{2}}} \cdot \underset{\text{garages}}{\underset{\uparrow}{\mathbf{2}}} \cdot \underset{\text{colors}}{\underset{\uparrow}{\mathbf{4}}} \cdot \underset{\text{landscaping}}{\underset{\uparrow}{\mathbf{3}}} = 48$$

He can build 48 "different" homes. ■

A concept closely related to the Fundamental Principle of Counting is **permutations.**

***Permutation***

A **permutation** is an arrangement (or ordering) of the elements of a set.

In how many ways can the four letters a, b, c, and d be arranged? That is, how many permutations are there for the four letters a, b, c, and d? All the permutations are listed here.

| | | | |
|---|---|---|---|
| abcd | bacd | cabd | dabc |
| abdc | badc | cadb | dacb |
| acbd | bcad | cbad | dbac |
| acdb | bcda | cbda | dbca |
| adbc | bdac | cdab | dcab |
| adcb | bdca | cdba | dcba |

The number of permutations is the product

$$4 \cdot 3 \cdot 2 \cdot 1 = 4! = 24$$

This is the result we would get using the Fundamental Principle of Counting. However, there are formulas that are helpful in dealing with permutations.

In general **there are $n!$ permutations of $n$ elements.**

Using the eight digits 1, 2, 3, 4, 5, 6, 7, and 8, suppose that we want to form a 5-digit number with no digit repeated. How many numbers can be formed? We can use the fundamental principle of counting in the following way.

| | |
|---|---|
| Leave 5 spaces for the 5 digits. | _ _ _ _ _ |
| 8 digits are "eligible" for the first spot. | $\underline{8} \cdot$ _ _ _ _ |
| Once a digit is chosen for the first spot, there are 7 digits "eligible" for the second spot. | $\underline{8} \cdot \underline{7} \cdot$ _ _ _ |
| Continuing with the same reasoning gives | $\underline{8} \cdot \underline{7} \cdot \underline{6} \cdot \underline{5} \cdot \underline{4}$ |

Thus $8 \cdot 7 \cdot 6 \cdot 5 \cdot 4 = 6720$ different 5-digit numbers can be formed using 8 digits. We say that there are 6720 permutations of 8 digits taken 5 at a time.

$_8P_5 = 8 \cdot 7 \cdot 6 \cdot 5 \cdot 4 = 6720$ The notation $_8P_5$ is read "the number of permutations of 8 elements taken 5 at a time."

The symbol $_nP_r$ denotes the number of permutations of $n$ elements taken $r$ at a time. A formula for $_nP_r$ is

$$_nP_r = n(n-1)(n-2)\cdots(n-r+1)$$

Thus,

$$\begin{aligned} _8P_5 &= 8(8-1)(8-2)(8-3)(8-5+1) \\ &= 8 \cdot 7 \cdot 6 \cdot 5 \cdot 4 = 6720 \end{aligned}$$

As a special case when $r = n$, $n - r + 1 = n - n + 1 = 1$ and the formula is

$$_nP_n = n(n-1)(n-2)\cdots 3\cdot 2\cdot 1 = n!$$

as was stated earlier.

Using factorial notation,

$$_8P_5 = 8\cdot 7\cdot 6\cdot 5\cdot 4 = 8\cdot 7\cdot 6\cdot 5\cdot 4\left(\frac{3\cdot 2\cdot 1}{3\cdot 2\cdot 1}\right)$$

$$= \frac{8\cdot 7\cdot 6\cdot 5\cdot 4\cdot 3\cdot 2\cdot 1}{3\cdot 2\cdot 1} = \frac{8!}{3!}$$

In general,

$$_nP_r = \frac{n!}{(n-r)!}$$

## EXAMPLES

**2.** A sailor has 7 different flags to signal with. How many signals can he send using only 3 of the flags?

**Solution** $_7P_3 = \dfrac{7!}{(7-3)!} = \dfrac{7!}{4!} = 7\cdot 6\cdot 5 = 210$

He can send 210 different signals using 3 flags.

**3.** If the digits 1, 2, 3, 4, 5, and 6 are used to form three-digit numbers, how many numbers can be formed if (a) digits may not be repeated and (b) digits may be repeated?

**Solution**

**a.** $_6P_3 = \dfrac{6!}{(6-3)!} = \dfrac{6!}{3!} = 6\cdot 5\cdot 4 = 120$

**b.** Since any of the digits can be used in more than one position, this part of the problem does not involve permutations. Using the Fundamental Principle of Counting, $6\cdot 6\cdot 6 = 216$.

There are 120 three-digit numbers if digits may not be repeated and 216 three-digit numbers if digits may be repeated. ■

How many permutations are there of the letters in the word BEEHIVE? The three E's are not different from each other (that is, they are not distinct), and changing only these three letters around would account for 3! permutations. Thus, if $N$ represents the number of **distinct permutations,** $N\cdot 3! = 7!$ because there are seven letters involved.

This gives

$$N = \frac{7!}{3!} = 7 \cdot 6 \cdot 5 \cdot 4 = 840$$

The following theorem is stated without proof.

**Theorem**

If in a collection of $n$ elements, $m_1$ are of one kind, $m_2$ are of another kind, . . . , $m_r$ are of still another kind, and

$$n = m_1 + m_2 + \cdots + m_r$$

then the total number of distinct permutations of the $n$ elements is

$$N = \frac{n!}{m_1!m_2! \cdots m_r!}$$

**EXAMPLE**

**4.** Find the number of distinct permutations in the word MISSISSIPPI.

**Solution** There are 4 S's, 4 I's, 2 P's, and 1 M.

$$\frac{11!}{4!4!2!1!} = \frac{11 \cdot 10 \cdot 9 \cdot 8 \cdot 7 \cdot 6 \cdot 5 \cdot 4 \cdot 3 \cdot 2 \cdot 1}{4 \cdot 3 \cdot 2 \cdot 1 \cdot 4 \cdot 3 \cdot 2 \cdot 1 \cdot 2 \cdot 1 \cdot 1} = 34{,}650$$

■

**Practice Problems**

1. Evaluate ${}_6P_6$
2. Evaluate: ${}_6P_2$
3. How many even numbers with four digits can be formed using the digits 1, 2, 3, 5, 7 if no repetitions are allowed?
4. Find the number of distinct permutations in the word SCIENTIFIC.

## EXERCISES 13.6

Evaluate the permutations in Exercises 1–10.

| | | | | |
|---|---|---|---|---|
| **1.** ${}_7P_4$ | **2.** ${}_8P_3$ | **3.** ${}_4P_4$ | **4.** ${}_6P_2$ | **5.** ${}_5P_4$ |
| **6.** ${}_7P_1$ | **7.** ${}_9P_6$ | **8.** ${}_{11}P_9$ | **9.** ${}_{10}P_8$ | **10.** ${}_9P_4$ |

**Answers to Practice Problems** **1.** 720 **2.** 30 **3.** 24 **4.** 302,400

**11.** A football team of eleven players is electing a captain and a most valuable player. In how many ways can this be done if the awards must be given to two players?

**12.** In how many ways can eleven girls be chosen for nine positions in a chorus line?

**13.** There are eight men available for three outfield positions on the baseball team. If each man can play any position, in how many ways can the outfield positions be filled?

**14.** A president, a vice-president, a secretary, and a treasurer are to be selected for an organization of 20 members. In how many ways can these four people be selected?

**15.** A luxury automobile is available in 6 body colors, 3 different vinyl tops, and 4 choices of interior. How many different cars must the dealer stock if he is to have one of each model?

**16.** In a subdivision, there are five basic floor plans. Each floor plan has three different exterior designs and three plans for landscaping the yard. How many different houses could be built?

**17.** How many four-digit numbers can be formed from the digits 1, 2, 5, 6, 7, and 9 if no repetitions are allowed?

**18.** How many **odd** numbers with four digits can be formed from the digits 1, 2, 3, 4, 6, and 7 with no repetitions allowed?

**19.** How many numbers of not more than four digits can be formed from the digits 2, 5, 7, and 9 if no repetitions are allowed?

**20.** How many numbers of not more than four digits can be formed from the digits 2, 5, 7, and 9 if repetitions are allowed?

**21.** A builder recently hired four superintendents and four foremen. He assigned a superintendent and a foreman to each of his four projects. In how many ways can this be done?

**22.** In how many ways can three math books and four English books be arranged on a shelf if:
**a.** they may be placed in any position?
**b.** the math books are together and the English books are together?

**23.** In how many ways can four algebra texts and three geometry texts be arranged on a shelf, keeping the subjects together?

**24.** In Exercise 23, if two of the algebra books are identical and two of the geometry books are identical, in how many ways can the books be arranged, keeping those on each subject together?

**25.** There are eleven flags that are displayed together, one above another. How many signals are possible if four of the flags are blue, two are red, three are yellow, and two are white?

## 13.7 Combinations

**OBJECTIVES**

In this section, you will be learning to:

1. Evaluate expressions representing combinations.
2. Solve applied problems involving combinations.

When we deal with permutations, we are concerned with the order or arrangement of elements. In other situations, such as how many ways a committee can be formed or how many ways a hand of cards can be formed, order is not involved. In these cases, we are interested in the number of **combinations** that can be formed.

**Combination** A **combination** is a collection of elements without regard to the order of the elements.

If $n$ distinct elements are given and a combination of $r$ elements is to be selected, then the total number of combinations of $n$ elements taken $r$ at a time is symbolized ${}_nC_r$. Since each combination of $r$ elements has $r!$ permutations, the product $r! \cdot {}_nC_r$ represents the number of permutations of $n$ elements taken $r$ at a time. Thus,

$$r! \cdot {}_nC_r = {}_nP_r$$

or

$${}_nC_r = \frac{{}_nP_r}{r!} = \frac{n!}{r!(n-r)!}$$

We can illustrate the relationship between combinations and permutations by considering the four letters a, b, c, d and listing both ${}_4C_3$ and ${}_4P_3$.

| | | | | |
|---|---|---|---|---|
| Combinations (${}_4C_3$) | abc | abd | acd | bcd |
| Permutations (${}_4P_3$) | abc | abd | acd | bcd |
| | acb | adb | adc | bdc |
| | bac | bad | cad | cbd |
| | bca | bda | cda | cdb |
| | cab | dab | dac | dbc |
| | cba | dba | dca | dcb |

Notice that each combination of 3 elements has $3! = 6$ corresponding permutations. Thus,

$$3! \cdot {}_4C_3 = 6 \cdot 4 = 24 = {}_4P_3$$

**EXAMPLES**

1. In how many ways can a hand of 5 cards be dealt from a deck of 52 cards?

**Solution** $${}_{52}C_5 = \frac{52!}{5!47!} = \frac{52 \cdot 51 \cdot \overset{5}{\cancel{50}} \cdot 49 \cdot \overset{4}{\cancel{48}}}{\cancel{5} \cdot \cancel{4} \cdot \cancel{3} \cdot \cancel{2} \cdot 1} = 2{,}598{,}960$$

There are 2,598,960 possible hands of 5 cards.

2. A committee of 6 people is to be chosen from a group of 40 members. How many "different" committees are there?

**Solution** $${}_{40}C_6 = \frac{40!}{6!34!} = \frac{\overset{2}{\cancel{40}} \cdot 39 \cdot 38 \cdot 37 \cdot \overset{1}{\cancel{36}} \cdot 35}{\cancel{6} \cdot \cancel{5} \cdot \cancel{4} \cdot \cancel{3} \cdot \cancel{2} \cdot 1} = 3{,}838{,}380$$

There are 3,838,380 "different" committees. ■

These ideas also can be used in conjunction with the Fundamental Principle of Counting, as the following example illustrates.

EXAMPLE

**3.** A Senate committee of 6 members must be chosen with 3 Democrats and 3 Republicans. The eligible members are 10 Democrats and 8 Republicans. How many possible committees are there?

**Solution** $_{10}C_3 = \dfrac{10!}{3!7!} = \dfrac{10 \cdot 9 \cdot 8}{3 \cdot 2 \cdot 1} = 120$ groups of 3 Democrats

$_8C_3 = \dfrac{8!}{3!5!} = \dfrac{8 \cdot 7 \cdot 6}{3 \cdot 2 \cdot 1} = 56$ groups of 3 Republicans

Using the Fundamental Principle of Counting,

$$_{10}C_3 \cdot {}_8C_3 = 120 \cdot 56 = 6720$$

There are 6720 possible committees ■

***Special Note on Notation***

For future studies, the student should be aware of the following commonly used equivalent notations.

For **permutations:** $_nP_r = P_r^n = P(n,r)$.

For **combinations:** $_nC_r = C_r^n = C(n,r)$.

***Practice Problems***

**1.** In how many ways can a jury of 6 men and 6 women be selected from a group of 10 men and 14 women?

**2.** A student senate committee of 5 people is to be chosen from a list of 20 students. In how many ways can this be done?

## *EXERCISES 13.7*

Evaluate the combinations in Exercises 1–10.

**1.** $_7C_3$ **2.** $_8C_4$ **3.** $_4C_4$ **4.** $_8C_5$ **5.** $_9C_6$

**6.** $_5C_3$ **7.** $_6C_1$ **8.** $_5C_5$ **9.** $_{10}C_7$ **10.** $_9C_4$

**11.** A committee of three is selected from the eighteen members of an organization. In how many ways may the committee be chosen?

**12.** You are permitted to answer any ten questions out of thirteen. In how many different ways can you make your ten selections?

**13.** If each girl can play any of the five positions on a basketball team, how many teams can be formed from nine girls?

**14.** A ski club has fifteen members who desire to be on the four-man ski team. How many different ski teams can be formed?

**Answers to Practice Problems** **1.** 630,630 **2.** 15,504

**15.** Twenty people all shake hands with one another. How many handshakes are there?

**16.** How many straight lines are determined by seven distinct points if no three points are collinear? How many triangles are formed?

**17.** Five women and four men are candidates for the debate team. If the team consists of one woman and two men, in how many ways may the team be chosen?

**18.** In how many ways can a committee of two Republicans and three Democrats be selected from a group of seven Republicans and ten Democrats?

**19.** Sandy is packing for a trip. She plans to take five blouses and three skirts. In how many different ways can she make her selections if she has twelve blouses and nine skirts?

**20.** A department store wishes to fill ten positions with four men and six women. In how many ways can these positions be filled if nine men and ten women have applied for the jobs?

**21.** A reading list consists of ten books of fiction and eight books of nonfiction. In how many ways can a student select four fiction and four nonfiction books?

**22.** On a geometry test, there are eight theorems and six constructions to choose from. If you are required to do four theorems and three constructions, in how many ways could you work the test?

**23.** The high school Student Government Committee is composed of three boys and two girls. If there are eight boys and six girls eligible, how many different committees could be formed?

**24.** In a dozen eggs, three are spoiled. In how many ways can you select four eggs and get two spoiled ones?

**25.** On a shelf, there are five English books, six algebra books, and three geometry books. Two English books, three algebra books, and two geometry books are selected. How many different selections are there?

## CHAPTER 13 SUMMARY

### Key Terms and Formulas

A **sequence** is a function that has the set of natural numbers as its domain. [13.1]

A sequence $\{a_n\}$ is
**decreasing** if $a_n > a_{n+1}$ for all $n$
and **increasing** if $a_n < a_{n+1}$ for all $n$. [13.1]

$$\sum_{k=1}^{n} a_k = a_1 + a_2 + a_3 + \cdots + a_n \quad [13.2]$$

**Arithmetic Sequences** [13.3]
An **arithmetic sequence** is a sequence in which

$$a_{k+1} - a_k = d \text{ for any natural number } k.$$

$d$ is called the **common difference.**

The general term (or $n$th term) is $a_n = a_1 + (n-1)d$.

The sum of the first $n$ terms is given by

$$\sum_{k=1}^{n} a_k = \frac{n}{2}(a_1 + a_n)$$

**Geometric Sequences** [13.4]
A **geometric sequence** is a sequence in which

$$\frac{a_{k+1}}{a_k} = r \text{ for any natural number } k$$

$r$ is called the **common ratio.**

The general term (or $n$th term) is $a_n = a_1 r^{n-1}$

The sum of the first $n$ terms is given by

$$S_n = \sum_{k=1}^{n} a_k = \frac{a_1(1 - r^n)}{1 - r} \quad \text{where } r \neq 1$$

An **infinite series** (or just a **series**) is the sum of all the terms of a sequence. We write

$$S = \sum_{k=1}^{\infty} a_k \quad [13.4]$$

If $\{a_n\}$ is a geometric sequence and $|r| < 1$, then

$$S = \sum_{k=1}^{\infty} a_k = \frac{a_1}{1 - r}$$

If $|r| \geq 1$, then there is no sum. [13.4]

***n* Factorial** [13.5]

$$n! = n(n - 1)(n - 2) \cdot \ldots \cdot 3 \cdot 2 \cdot 1$$
$$n! = n \cdot (n - 1)!$$
$$0! = 1$$
$$\binom{n}{r} = \frac{n!}{(n - r)!r!} \quad \text{where } 0 \leq r \leq n$$

**The Binomial Theorem** [13.5]

$$(a + b)^n = \binom{n}{0}a^n + \binom{n}{1}a^{n-1}b + \binom{n}{2}a^{n-2}b^2 + \cdots + \binom{n}{k}a^{n-k}b^k + \cdots + \binom{n}{n}b^n$$

Using $\Sigma$-notation,

$$(a + b)^n = \sum_{k=0}^{n} \binom{n}{k} a^{n-k}b^k \quad [13.2]$$

A **permutation** is an arrangement (or ordering) of the elements of a set. [13.6]

$${}_nP_r = n(n - 1)(n - 2) \cdots (n - r + 1)$$
$${}_nP_r = \frac{n!}{(n - r)!}$$

A **combination** is a collection of elements without regard to the order of the elements. [13.7]

$${}_nC_r = \frac{{}_nP_r}{r!} = \frac{n!}{r!(n - r)!}$$

---

## Properties and Rules

**Properties of $\Sigma$-Notation** [13.2]

For sequences $\{a_n\}$ and $\{b_n\}$ and any real number $c$,

**I.** $$\sum_{k=1}^{n} a_k = \sum_{k=1}^{i} a_k + \sum_{k=i+1}^{n} a_k$$

**II.** $$\sum_{k=1}^{n} (a_k + b_k) = \sum_{k=1}^{n} a_k + \sum_{k=1}^{n} b_k$$

**III.** $$\sum_{k=1}^{n} ca_k = c \sum_{k=1}^{n} a_k$$

**IV.** $$\sum_{k=1}^{n} c = nc$$

**The Fundamental Principle of Counting** [13.6]

If an event $E_1$ can occur in $m_1$ ways, an event $E_2$ can occur in $m_2$ ways, . . . , and an event $E_k$ can occur in $m_k$ ways, then the total number of ways that all the events may occur is the product $m_1 \cdot m_2 \cdot \ldots \cdot m_k$.

If in a collection of $n$ elements, $m_1$ are of one kind, $m_2$ are of another kind, . . . , $m_r$ are of still another kind, and

$$n = m_1 + m_2 + \cdots + m_r$$

then the number of distinct permutations of the $n$ elements is

$$N = \frac{n!}{m_1!m_2! \cdots m_r!} \quad [13.6]$$

## CHAPTER 13 REVIEW

Write the first four terms of each of the sequences in Exercises 1–6. Determine whether the sequence is arithmetic, geometric, or neither. [13.1], [13.3], [13.4]

**1.** $\{5 - 4n\}$ **2.** $\{n^2 + 3\}$ **3.** $\{6(-2)^n\}$

**4.** $\{(-1)^n\}$ **5.** $\{(-1)^n(2n + 5)\}$ **6.** $\left\{\frac{5}{n}\right\}$

Find a general term for a sequence with the first four terms as given in Exercises 7–12. [13.1]

**7.** 6, 11, 16, 21, . . . **8.** 4, 7, 10, 13, . . . **9.** $\frac{1}{6}, \frac{1}{2}, \frac{3}{2}, \frac{9}{2}, \ldots$

**10.** $-5, 7, -9, 11, \ldots$ **11.** $-8, 2, -\frac{1}{2}, \frac{1}{8}, \ldots$ **12.** $2, \frac{3}{2}, \frac{4}{3}, \frac{5}{4}, \ldots$

If $\{a_n\}$ is an arithmetic sequence, find the indicated quantity in Exercises 13–20. [13.3]

**13.** $a_1 = 3, d = 2$. Find $a_5$.

**14.** $a_1 = 6, d = 7$. find $a_8$.

**15.** $a_3 = 7, a_6 = 16$. Find $a_n$.

**16.** $a_4 = 5, a_{10} = 17$. Find $a_n$.

**17.** $a_8 = 1, a_4 = -1$. Find $a_n$.

**18.** $a_2 = 11, a_6 = 27$. Find $\sum_{k=1}^{9} a_k$.

**19.** $a_3 = 19, a_5 = 35$. Find $\sum_{k=1}^{8} a_k$.

**20.** $a_4 = 1, a_8 = 3$. Find $\sum_{k=1}^{12} a_k$.

If $\{a_k\}$ is a geometric sequence, find the indicated quantities in Exercises 21–28. [13.4]

**21.** $a_1 = 5, r = 2$. Find $a_4$.

**22.** $a_1 = \frac{1}{9}, r = -3$. Find $a_5$.

**23.** $a_2 = \frac{2}{3}, a_5 = \frac{16}{3}$. Find $a_n$.

**24.** $a_2 = -\frac{5}{8}, a_5 = \frac{5}{64}$. Find $a_n$.

**25.** $a_3 = 0.8, a_6 = 0.0064$. Find $a_n$.

**26.** $a_2 = -18, a_4 = -162$. Find $\sum_{k=1}^{5} a_k$.

**27.** $a_4 = 4, a_6 = 9, r > 0$. Find $\sum_{k=1}^{5} a_k$.

**28.** $a_2 = \frac{9}{8}, a_4 = \frac{1}{8}$. Find $\sum_{k=1}^{5} a_k$.

Find each of the sums in Exercises 29–38. [13.2]

**29.** $\sum_{k=1}^{10} (3k + 1)$ **30.** $\sum_{k=1}^{17} (2k - 5)$ **31.** $\sum_{k=1}^{8} 7(-2)^k$ **32.** $\sum_{k=1}^{7} \frac{2}{3}(2)^k$

**33.** $\sum_{k=1}^{5} (3^k + 2)$ **34.** $\sum_{k=1}^{4} \left[\left(\frac{1}{4}\right)^k - 3\right]$ **35.** $\sum_{k=1}^{\infty} \left(\frac{3}{5}\right)^k$ **36.** $\sum_{k=1}^{\infty} \left(-\frac{2}{3}\right)^{k-1}$

**37.** $0.\overline{7}$ **38.** $0.\overline{24}$

**39.** If $\sum_{k=1}^{25} a_k = 31$, $\sum_{k=1}^{25} b_k = 44$, find $\sum_{k=1}^{25} (3a_k - 4b_k)$.

**40.** If $\sum_{k=1}^{24} (2a_k + 5) = 210$, find $\sum_{k=1}^{24} a_k$.

Evaluate the expressions in Exercises 41–44. [13.5]

**41.** $\dfrac{9!}{3!4!}$

**42.** $\dfrac{12!}{8!7!}$

**43.** $\dbinom{7}{2}$

**44.** $\dbinom{10}{6}$

Use the Binomial Theorem to expand Exercises 45 and 46. [13.5]

**45.** $(x + 3y)^5$

**46.** $\left(2x - \dfrac{y}{2}\right)^6$

Write the sixth term of the expansions in Exercises 47 and 48. [13.5]

**47.** $(2x - y)^8$

**48.** $\left(\dfrac{x}{2} + 3y\right)^{10}$

**49.** Joan started her job exactly 5 years ago. Her original salary was $12,000 per year. Each year she received a raise of 6% of her current salary. What is her salary after this year's raise? [13.3], [13.4]

**50.** A tank holds 1000 liters of a liquid that readily mixes with water. After 150 liters are drained out, the tank is filled by adding water. Then 150 liters of the mixture are drained out and the tank is filled by adding water. If this process is continued 5 times, how much of the original liquid is left? [13.3], [13.4]

Evaluate the expressions in Exercises 51–56. [13.6], [13.7]

**51.** ${}_6P_3$

**52.** ${}_8P_5$

**53.** ${}_7P_2$

**54.** ${}_9C_4$

**55.** ${}_{10}C_5$

**56.** ${}_8C_6$

**57.** An architect is drawing house plans for a contractor. The contractor requested 4 different floor plans. Each floor plan has 3 different exterior designs. How many different plans must the architect draw? [13.6], [13.7]

**58.** A sailor has 9 different flags to use for signaling. A signal consists of displaying 4 flags in a specific order. How many signals can he send? [13.6], [13.7]

**59.** A swim club has 18 members. A team of 6 is to be selected. How many different teams could be selected? [13.6], [13.7]

**60.** On a geometry test, there are 6 theorems and 5 constructions to choose from. If you are required to do 4 theorems and 2 constructions, in how many ways could you work the test? [13.6], [13.7]

## CHAPTER 13 TEST

**1.** Write the first four terms of the sequence $\left\{\dfrac{1}{3n + 1}\right\}$. Determine whether the sequence is arithmetic, geometric, or neither.

**2.** Find a general term for the sequence $\dfrac{1}{3}, \dfrac{2}{5}, \dfrac{3}{7}, \dfrac{4}{9}, \ldots$

If $\{a_k\}$ is an arithmetic sequence, find the indicated quantity in Exercises 3–5.

**3.** $a_1 = 5, d = 3$. Find $a_8$.

**4.** $a_2 = 4, a_7 = -6$. Find $a_n$.

**5.** $a_4 = 22, a_7 = 37$. Find $\sum_{k=1}^{9} a_k$.

If $\{a_k\}$ is a geometric sequence, find the indicated quantity in Exercises 6–8.

**6.** $a_1 = 8, r = \frac{1}{2}$. Find $a_7$.

**7.** $a_4 = 3, a_6 = 9$. Find $a_n$.

**8.** $a_2 = \frac{1}{3}, a_5 = \frac{1}{24}$. Find $\sum_{k=1}^{6} a_k$.

Find each of the sums in Exercises 9–11.

**9.** $\sum_{k=1}^{8} (3k - 5)$

**10.** $\sum_{k=3}^{6} 2\left(-\frac{1}{3}\right)^k$

**11.** $0.\overline{15}$

**12.** If $\sum_{k=1}^{50} a_k = 88, \sum_{k=1}^{19} a_k = 14$, find $\sum_{k=20}^{50} a_k$.

**13.** An automobile that costs \$10,000 new depreciates at a rate of 20% of its value each year. What is its value after 4 years?

**14.** Evaluate $\binom{11}{5}$.

**15.** Use the Binomial Theorem to expand $(2x - y)^5$.

**16.** Write the fifth term of the expansion $(x + 3y)^8$.

**17.** Find the value of ${}_{10}P_6$.

**18.** Find the value of ${}_{12}C_7$.

**19.** There are ten players available for the three different outfield positions on a baseball team. Assuming each player can play any of the positions, in how many ways can the outfield positions be filled?

**20.** In a box there are eleven apples, three of which are spoiled. In how many ways could you select three good apples and one bad apple?

## CUMULATIVE REVIEW (13)

Simplify each expression in Exercises 1 and 2.

**1.** $10x - [2x + (13 - 4x) - (11 - 3x)] + (2x + 5)$

**2.** $(x^2 + 5x - 2) - (3x^2 - 5x - 3) + (x^2 - 7x + 4)$

Factor completely in Exercises 3–5.

**3.** $64x^3 + 27$

**4.** $6x^2 + 17x - 45$

**5.** $5x^2(2x + 1) - 3x(2x + 1) - 14(2x + 1)$

Perform the indicated operations in Exercises 6–8.

**6.** $\frac{x + 3}{3} + \frac{2x - 1}{5}$

**7.** $\frac{x}{x^2 - 16} - \frac{x + 1}{x^2 - 5x + 4}$

**8.** $\frac{x^2 - 9}{x^4 + 6x^3} \div \frac{x^3 - 2x^2 - 3x}{x^2 + 7x + 6} \cdot \frac{x^2}{x + 3}$

Simplify Exercises 9–12. Assume that all variables are positive.

**9.** $\left(\frac{8x^{-1}y^{1/3}}{x^3y^{-1/3}}\right)^{1/2}$

**10.** $\sqrt[3]{\frac{8x^4}{27y^3}}$

**11.** $\sqrt{12x} - \sqrt{75x} + 2\sqrt{27x}$

**12.** $\frac{1 + 3i}{2 - 5i}$

Solve the inequalities in Exercises 13–15. Graph the solution sets.

**13.** $6(2x - 3) + (x - 5) > 4(x + 1)$

**14.** $|7 - 3x| - 2 \leq 4$

**15.** $8x^2 + 2x - 45 < 0$

Solve the equations in Exercises 16–21.

**16.** $4(x - 7) + 2(3x + 1) = 3x + 2$

**17.** $2x^2 + 4x + 3 = 0$

**18.** $x - \sqrt{x} - 2 = 0$

**19.** $\dfrac{1}{2x} + \dfrac{5}{x + 3} = \dfrac{8}{3x}$

**20.** $2\sqrt{6 - x} = x - 3$

**21.** $5 \ln x = 12$

**22.** Solve for $n$: $P = \dfrac{A}{1 + ni}$

Solve the systems of equations in Exercises 23 and 24.

**23.** $\begin{cases} 9x + 2y = 8 \\ 4x + 3y = -7 \end{cases}$

**24.** $\begin{cases} 3x + y - 2z = 4 \\ x - 4y - 3z = -5 \\ 2x + 2y + z = 3 \end{cases}$

**25.** If $f(x) = 2x - 7$ and $g(x) = x^2 + 1$, find:

**a.** $f^{-1}(x)$

**b.** $f[g(x)]$

**c.** $g(x + 1) - g(x)$

**26.** The following is a graph of $y = f(x)$. Sketch the graph of $y = f(x - 2) - 1$.

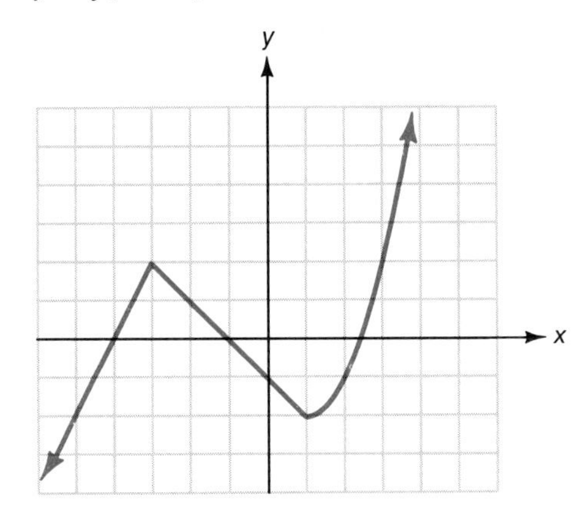


**27.** Solve the system by graphing both equations on the same axes.

$\begin{cases} 4x - 3y = 17 \\ 5x + 2y = 4 \end{cases}$

Graph each of the equations in Exercises 28–30.

**28.** $y = 4x^2 - 8x + 9$

**29.** $\dfrac{x^2}{4} + \dfrac{y^2}{16} = 1$

**30.** $x^2 - 4x + y^2 + 2y = 4$

**31.** If $\{a_n\}$ is an arithmetic sequence where $a_3 = 4$ and $a_8 = -6$,

**a.** find $a_n$.

**b.** find $\sum_{k=1}^{10} a_k$.

**32.** If $\{a_k\}$ is a geometric sequence where $a_1 = 16$ and $r = \dfrac{1}{2}$,

**a.** find $a_n$.

**b.** find $\sum_{k=1}^{6} a_k$.

**33.** Use the Binomial Theorem to expand $(x + 2y)^6$.

Evaluate the expressions in Exercises 34–36.

**34.** $\binom{10}{4}$

**35.** ${}_8P_4$

**36.** ${}_{14}C_5$

**37.** Two cars start together and travel in the same direction, one traveling 3 times as fast as the other. At the end of 3.5 hours, they are 140 miles apart. How fast is each traveling?

**38.** A grocer mixes two kinds of nuts. One costs $1.40 per pound and the other costs $2.60 per pound. If the mixture weighs 20 pounds and costs $1.64 per pound, how many pounds of each kind did he use?

**39.** A rectangular yard is 20 ft by 30 ft. A rectangular swimming pool is to be built leaving a strip of grass of uniform width around the pool. If the area of the grass strip is 184 sq ft, find the dimensions of the pool.

**40.** A radioactive isotope decomposes according to $A = A_0e^{-0.0552t}$, where $t$ is in hours. Determine the time required for one-half of the original amount to decompose.

**Table 1 Squares, Square Roots, and Prime Factors**

| No. | Sq. | Sq. rt. | Prime Factors | No. | Sq. | Sq. rt. | Prime Factors |
|---|---|---|---|---|---|---|---|
| 1 | 1 | 1.000 | | 51 | 2,601 | 7.141 | $3 \cdot 17$ |
| 2 | 4 | 1.414 | 2 | 52 | 2,704 | 7.211 | $2^2 \cdot 13$ |
| 3 | 9 | 1.732 | 3 | 53 | 2,809 | 7.280 | 53 |
| 4 | 16 | 2.000 | $2^2$ | 54 | 2,916 | 7.348 | $2 \cdot 3^3$ |
| 5 | 25 | 2.236 | 5 | 55 | 3,025 | 7.416 | $5 \cdot 11$ |
| 6 | 36 | 2.449 | $2 \cdot 3$ | 56 | 3,136 | 7.483 | $2^3 \cdot 7$ |
| 7 | 49 | 2.646 | 7 | 57 | 3,249 | 7.550 | $3 \cdot 19$ |
| 8 | 64 | 2.828 | $2^3$ | 58 | 3,364 | 7.616 | $2 \cdot 29$ |
| 9 | 81 | 3.000 | $3^2$ | 59 | 3,481 | 7.681 | 59 |
| 10 | 100 | 3.162 | $2 \cdot 5$ | 60 | 3,600 | 7.746 | $2^2 \cdot 3 \cdot 5$ |
| 11 | 121 | 3.317 | 11 | 61 | 3,721 | 7.810 | 61 |
| 12 | 144 | 3.464 | $2^2 \cdot 3$ | 62 | 3,844 | 7.874 | $2 \cdot 31$ |
| 13 | 169 | 3.606 | 13 | 63 | 3,969 | 7.937 | $3^2 \cdot 7$ |
| 14 | 196 | 3.742 | $2 \cdot 7$ | 64 | 4,096 | 8.000 | $2^6$ |
| 15 | 225 | 3.873 | $3 \cdot 5$ | 65 | 4,225 | 8.062 | $5 \cdot 13$ |
| 16 | 256 | 4.000 | $2^4$ | 66 | 4,356 | 8.124 | $2 \cdot 3 \cdot 11$ |
| 17 | 289 | 4.123 | 17 | 67 | 4,489 | 8.185 | 67 |
| 18 | 324 | 4.243 | $2 \cdot 3^2$ | 68 | 4,624 | 8.246 | $2^2 \cdot 17$ |
| 19 | 361 | 4.359 | 19 | 69 | 4,761 | 8.307 | $3 \cdot 23$ |
| 20 | 400 | 4.472 | $2^2 \cdot 5$ | 70 | 4,900 | 8.367 | $2 \cdot 5 \cdot 7$ |
| 21 | 441 | 4.583 | $3 \cdot 7$ | 71 | 5,041 | 8.426 | 71 |
| 22 | 484 | 4.690 | $2 \cdot 11$ | 72 | 5,184 | 8.485 | $2^3 \cdot 3^2$ |
| 23 | 529 | 4.796 | 23 | 73 | 5,329 | 8.544 | 73 |
| 24 | 576 | 4.899 | $2^3 \cdot 3$ | 74 | 5,476 | 8.602 | $2 \cdot 37$ |
| 25 | 625 | 5.000 | $5^2$ | 75 | 5,625 | 8.660 | $3 \cdot 5^2$ |
| 26 | 676 | 5.099 | $2 \cdot 13$ | 76 | 5,776 | 8.718 | $2^2 \cdot 19$ |
| 27 | 729 | 5.196 | $3^3$ | 77 | 5,929 | 8.775 | $7 \cdot 11$ |
| 28 | 784 | 5.292 | $2^2 \cdot 7$ | 78 | 6,084 | 8.832 | $2 \cdot 3 \cdot 13$ |
| 29 | 841 | 5.385 | 29 | 79 | 6,241 | 8.888 | 79 |
| 30 | 900 | 5.477 | $2 \cdot 3 \cdot 5$ | 80 | 6,400 | 8.944 | $2^4 \cdot 5$ |
| 31 | 961 | 5.568 | 31 | 81 | 6,561 | 9.000 | $3^4$ |
| 32 | 1,024 | 5.657 | $2^5$ | 82 | 6,724 | 9.055 | $2 \cdot 41$ |
| 33 | 1,089 | 5.745 | $3 \cdot 11$ | 83 | 6,889 | 9.110 | 83 |
| 34 | 1,156 | 5.831 | $2 \cdot 17$ | 84 | 7,056 | 9.165 | $2^2 \cdot 3 \cdot 7$ |
| 35 | 1,225 | 5.916 | $5 \cdot 7$ | 85 | 7,225 | 9.220 | $5 \cdot 17$ |
| 36 | 1,296 | 6.000 | $2^2 \cdot 3^2$ | 86 | 7,396 | 9.274 | $2 \cdot 43$ |
| 37 | 1,369 | 6.083 | 37 | 87 | 7,569 | 9.327 | $3 \cdot 29$ |
| 38 | 1,444 | 6.164 | $2 \cdot 19$ | 88 | 7,744 | 9.381 | $2^3 \cdot 11$ |
| 39 | 1,521 | 6.245 | $3 \cdot 13$ | 89 | 7,921 | 9.434 | 89 |
| 40 | 1,600 | 6.325 | $2^3 \cdot 5$ | 90 | 8,100 | 9.487 | $2 \cdot 3^2 \cdot 5$ |
| 41 | 1,681 | 6.403 | 41 | 91 | 8,281 | 9.539 | $7 \cdot 13$ |
| 42 | 1,764 | 6.481 | $2 \cdot 3 \cdot 7$ | 92 | 8,464 | 9.592 | $2^2 \cdot 23$ |
| 43 | 1,849 | 6.557 | 43 | 93 | 8,649 | 9.644 | $3 \cdot 31$ |
| 44 | 1,936 | 6.633 | $2^2 \cdot 11$ | 94 | 8,836 | 9.695 | $2 \cdot 47$ |
| 45 | 2,025 | 6.708 | $3^2 \cdot 5$ | 95 | 9,025 | 9.747 | $5 \cdot 19$ |
| 46 | 2,116 | 6.782 | $2 \cdot 23$ | 96 | 9,216 | 9.798 | $2^5 \cdot 3$ |
| 47 | 2,209 | 6.856 | 47 | 97 | 9,409 | 9.849 | 97 |
| 48 | 2,304 | 6.928 | $2^4 \cdot 3$ | 98 | 9,604 | 9.899 | $2 \cdot 7^2$ |
| 49 | 2,401 | 7.000 | $7^2$ | 99 | 9,801 | 9.950 | $3^2 \cdot 11$ |
| 50 | 2,500 | 7.071 | $2 \cdot 5^2$ | 100 | 10,000 | 10.000 | $2^2 \cdot 5^2$ |

**Table 2 Common Logarithms**

| *N* | 0 | 1 | 2 | 3 | 4 | 5 | 6 | 7 | 8 | 9 |
|---|---|---|---|---|---|---|---|---|---|---|
| 1.0 | .0000 | .0043 | .0086 | .0128 | .0170 | .0212 | .0253 | .0294 | .0334 | .0374 |
| 1.1 | .0414 | .0453 | .0492 | .0531 | .0569 | .0607 | .0645 | .0682 | .0719 | .0755 |
| 1.2 | .0792 | .0828 | .0864 | .0899 | .0934 | .0969 | .1004 | .1038 | .1072 | .1106 |
| 1.3 | .1139 | .1173 | .1206 | .1239 | .1271 | .1303 | .1335 | .1367 | .1399 | .1430 |
| 1.4 | .1461 | .1492 | .1523 | .1553 | .1584 | .1614 | .1644 | .1673 | .1703 | .1732 |
| 1.5 | .1761 | .1790 | .1818 | .1847 | .1875 | .1903 | .1931 | .1959 | .1987 | .2014 |
| 1.6 | .2041 | .2068 | .2095 | .2122 | .2148 | .2175 | .2201 | .2227 | .2253 | .2279 |
| 1.7 | .2304 | .2330 | .2355 | .2380 | .2405 | .2430 | .2455 | .2480 | .2504 | .2529 |
| 1.8 | .2553 | .2577 | .2601 | .2625 | .2648 | .2672 | .2695 | .2718 | .2742 | .2765 |
| 1.9 | .2788 | .2810 | .2833 | .2856 | .2878 | .2900 | .2923 | .2945 | .2967 | .2989 |
| 2.0 | .3010 | .3032 | .3054 | .3075 | .3096 | .3118 | .3139 | .3160 | .3181 | .3201 |
| 2.1 | .3222 | .3243 | .3263 | .3284 | .3304 | .3324 | .3345 | .3365 | .3385 | .3404 |
| 2.2 | .3424 | .3444 | .3464 | .3483 | .3502 | .3522 | .3541 | .3560 | .3579 | .3598 |
| 2.3 | .3617 | .3636 | .3655 | .3674 | .3692 | .3711 | .3729 | .3747 | .3766 | .3784 |
| 2.4 | .3802 | .3820 | .3838 | .3856 | .3874 | .3892 | .3909 | .3927 | .3945 | .3962 |
| 2.5 | .3979 | .3997 | .4014 | .4031 | .4048 | .4065 | .4082 | .4099 | .4116 | .4133 |
| 2.6 | .4150 | .4166 | .4183 | .4200 | .4216 | .4232 | .4249 | .4265 | .4281 | .4298 |
| 2.7 | .4314 | .4330 | .4346 | .4362 | .4378 | .4393 | .4409 | .4425 | .4440 | .4456 |
| 2.8 | .4472 | .4487 | .4502 | .4518 | .4533 | .4548 | .4564 | .4579 | .4594 | .4609 |
| 2.9 | .4624 | .4639 | .4654 | .4669 | .4683 | .4698 | .4713 | .4728 | .4742 | .4757 |
| 3.0 | .4771 | .4786 | .4800 | .4814 | .4829 | .4843 | .4857 | .4871 | .4886 | .4900 |
| 3.1 | .4914 | .4928 | .4942 | .4955 | .4969 | .4983 | .4997 | .5011 | .5024 | .5038 |
| 3.2 | .5051 | .5065 | .5079 | .5092 | .5105 | .5119 | .5132 | .5145 | .5159 | .5172 |
| 3.3 | .5185 | .5198 | .5211 | .5224 | .5237 | .5250 | .5263 | .5276 | .5289 | .5302 |
| 3.4 | .5315 | .5328 | .5340 | .5353 | .5366 | .5378 | .5391 | .5403 | .5416 | .5428 |
| 3.5 | .5441 | .5453 | .5465 | .5478 | .5490 | .5502 | .5514 | .5527 | .5539 | .5551 |
| 3.6 | .5563 | .5575 | .5587 | .5599 | .5611 | .5623 | .5635 | .5647 | .5658 | .5670 |
| 3.7 | .5682 | .5694 | .5705 | .5717 | .5729 | .5740 | .5752 | .5763 | .5775 | .5786 |
| 3.8 | .5798 | .5809 | .5821 | .5832 | .5843 | .5855 | .5866 | .5877 | .5888 | .5899 |
| 3.9 | .5911 | .5922 | .5933 | .5944 | .5955 | .5966 | .5977 | .5988 | .5999 | .6010 |
| 4.0 | .6021 | .6031 | .6042 | .6053 | .6064 | .6075 | .6085 | .6096 | .6107 | .6117 |
| 4.1 | .6128 | .6138 | .6149 | .6160 | .6170 | .6180 | .6191 | .6201 | .6212 | .6222 |
| 4.2 | .6232 | .6243 | .6253 | .6263 | .6274 | .6284 | .6294 | .6304 | .6314 | .6325 |
| 4.3 | .6335 | .6345 | .6355 | .6365 | .6375 | .6385 | .6395 | .6405 | .6415 | .6425 |
| 4.4 | .6435 | .6444 | .6454 | .6464 | .6474 | .6484 | .6493 | .6503 | .6513 | .6522 |
| 4.5 | .6532 | .6542 | .6551 | .6561 | .6571 | .6580 | .6590 | .6599 | .6609 | .6618 |
| 4.6 | .6628 | .6637 | .6646 | .6656 | .6665 | .6675 | .6684 | .6693 | .6702 | .6712 |
| 4.7 | .6721 | .6730 | .6739 | .6749 | .6758 | .6767 | .6776 | .6785 | .6794 | .6803 |
| 4.8 | .6812 | .6821 | .6830 | .6839 | .6848 | .6857 | .6866 | .6875 | .6884 | .6893 |
| 4.9 | .6902 | .6911 | .6920 | .6928 | .6937 | .6946 | .6955 | .6964 | .6972 | .6981 |
| 5.0 | .6990 | .6998 | .7007 | .7016 | .7024 | .7033 | .7042 | .7050 | .7059 | .7067 |
| 5.1 | .7076 | .7084 | .7093 | .7101 | .7110 | .7118 | .7126 | .7135 | .7143 | .7152 |
| 5.2 | .7160 | .7168 | .7177 | .7185 | .7193 | .7202 | .7210 | .7218 | .7226 | .7235 |
| 5.3 | .7243 | .7251 | .7259 | .7267 | .7275 | .7284 | .7292 | .7300 | .7308 | .7316 |
| 5.4 | .7324 | .7332 | .7340 | .7348 | .7356 | .7364 | .7372 | .7380 | .7388 | .7396 |

**Table 2** *(Continued)*

| N | 0 | 1 | 2 | 3 | 4 | 5 | 6 | 7 | 8 | 9 |
|---|---|---|---|---|---|---|---|---|---|---|
| 5.5 | .7404 | .7412 | .7419 | .7427 | .7435 | .7443 | .7451 | .7459 | .7466 | .7474 |
| 5.6 | .7482 | .7490 | .7497 | .7505 | .7513 | .7520 | .7528 | .7536 | .7543 | .7551 |
| 5.7 | .7559 | .7566 | .7574 | .7582 | .7589 | .7597 | .7604 | .7612 | .7619 | .7627 |
| 5.8 | .7634 | .7642 | .7649 | .7657 | .7664 | .7672 | .7679 | .7686 | .7694 | .7701 |
| 5.9 | .7709 | .7716 | .7723 | .7731 | .7738 | .7745 | .7752 | .7760 | .7767 | .7774 |
| 6.0 | .7782 | .7789 | .7796 | .7803 | .7810 | .7818 | .7825 | .7832 | .7839 | .7846 |
| 6.1 | .7853 | .7860 | .7868 | .7875 | .7882 | .7889 | .7896 | .7903 | .7910 | .7917 |
| 6.2 | .7924 | .7931 | .7938 | .7945 | .7952 | .7959 | .7966 | .7973 | .7980 | .7987 |
| 6.3 | .7993 | .8000 | .8007 | .8014 | .8021 | .8028 | .8035 | .8041 | .8048 | .8055 |
| 6.4 | .8062 | .8069 | .8075 | .8082 | .8089 | .8096 | .8102 | .8109 | .8116 | .8122 |
| 6.5 | .8129 | .8136 | .8142 | .8149 | .8156 | .8162 | .8169 | .8176 | .8182 | .8189 |
| 6.6 | .8195 | .8202 | .8209 | .8215 | .8222 | .8228 | .8235 | .8241 | .8248 | .8254 |
| 6.7 | .8261 | .8267 | .8274 | .8280 | .8287 | .8293 | .8299 | .8306 | .8312 | .8319 |
| 6.8 | .8325 | .8331 | .8338 | .8344 | .8351 | .8357 | .8363 | .8370 | .8376 | .8382 |
| 6.9 | .8388 | .8395 | .8401 | .8407 | .8414 | .8420 | .8426 | .8432 | .8439 | .8445 |
| 7.0 | .8451 | .8457 | .8463 | .8470 | .8476 | .8482 | .8488 | .8494 | .8500 | .8506 |
| 7.1 | .8513 | .8519 | .8525 | .8531 | .8537 | .8543 | .8549 | .8555 | .8561 | .8567 |
| 7.2 | .8573 | .8579 | .8585 | .8591 | .8597 | .8603 | .8609 | .8615 | .8621 | .8627 |
| 7.3 | .8633 | .8639 | .8645 | .8651 | .8657 | .8663 | .8669 | .8675 | .8681 | .8686 |
| 7.4 | .8692 | .8698 | .8704 | .8710 | .8716 | .8722 | .8727 | .8733 | .8739 | .8745 |
| 7.5 | .8751 | .8756 | .8762 | .8768 | .8774 | .8779 | .8785 | .8791 | .8797 | .8802 |
| 7.6 | .8808 | .8814 | .8820 | .8825 | .8831 | .8837 | .8842 | .8848 | .8854 | .8859 |
| 7.7 | .8865 | .8871 | .8876 | .8882 | .8887 | .8893 | .8899 | .8904 | .8910 | .8915 |
| 7.8 | .8921 | .8927 | .8932 | .8938 | .8943 | .8949 | .8954 | .8960 | .8965 | .8971 |
| 7.9 | .8976 | .8982 | .8987 | .8993 | .8998 | .9004 | .9009 | .9015 | .9020 | .9025 |
| 8.0 | .9031 | .9036 | .9042 | .9047 | .9053 | .9058 | .9063 | .9069 | .9074 | .9079 |
| 8.1 | .9085 | .9090 | .9096 | .9101 | .9106 | .9112 | .9117 | .9122 | .9128 | .9133 |
| 8.2 | .9138 | .9143 | .9149 | .9154 | .9159 | .9165 | .9170 | .9175 | .9180 | .9186 |
| 8.3 | .9191 | .9196 | .9201 | .9206 | .9212 | .9217 | .9222 | .9227 | .9232 | .9238 |
| 8.4 | .9243 | .9249 | .9253 | .9258 | .9263 | .9269 | .9274 | .9279 | .9284 | .9289 |
| 8.5 | .9294 | .9299 | .9304 | .9309 | .9315 | .9320 | .9325 | .9330 | .9335 | .9340 |
| 8.6 | .9345 | .9350 | .9355 | .9360 | .9365 | .9370 | .9375 | .9380 | .9385 | .9390 |
| 8.7 | .9395 | .9400 | .9405 | .9410 | .9415 | .9420 | .9425 | .9430 | .9435 | .9440 |
| 8.8 | .9445 | .9450 | .9455 | .9460 | .9465 | .9469 | .9474 | .9479 | .9484 | .9489 |
| 8.9 | .9494 | .9499 | .9504 | .9509 | .9513 | .9518 | .9523 | .9528 | .9533 | .9538 |
| 9.0 | .9542 | .9547 | .9552 | .9557 | .9562 | .9566 | .9571 | .9576 | .9581 | .9586 |
| 9.1 | .9590 | .9595 | .9600 | .9605 | .9609 | .9614 | .9619 | .9624 | .9628 | .9633 |
| 9.2 | .9638 | .9643 | .9647 | .9652 | .9657 | .9661 | .9666 | .9671 | .9675 | .9680 |
| 9.3 | .9685 | .9689 | .9694 | .9699 | .9703 | .9708 | .9713 | .9717 | .9722 | .9727 |
| 9.4 | .9731 | .9736 | .9741 | .9745 | .9750 | .9754 | .9759 | .9763 | .9768 | .9773 |
| 9.5 | .9777 | .9782 | .9786 | .9791 | .9795 | .9800 | .9805 | .9809 | .9814 | .9818 |
| 9.6 | .9823 | .9827 | .9832 | .9836 | .9841 | .9845 | .9850 | .9854 | .9859 | .9863 |
| 9.7 | .9868 | .9872 | .9877 | .9881 | .9886 | .9890 | .9894 | .9899 | .9903 | .9908 |
| 9.8 | .9912 | .9917 | .9921 | .9926 | .9930 | .9934 | .9939 | .9943 | .9948 | .9952 |
| 9.9 | .9956 | .9961 | .9965 | .9969 | .9974 | .9978 | .9983 | .9987 | .9991 | .9996 |

## Table 3 Natural Logarithms (Base $e$)

| | 0 | 1 | 2 | 3 | 4 | 5 | 6 | 7 | 8 | 9 |
|---|---|---|---|---|---|---|---|---|---|---|
| 1.0 | 0.0000 | 0.0100 | 0.0198 | 0.0296 | 0.0392 | 0.0488 | 0.0583 | 0.0677 | 0.0770 | 0.0862 |
| 1.1 | 0.0953 | 0.1044 | 0.1133 | 0.1222 | 0.1310 | 0.1398 | 0.1484 | 0.1570 | 0.1655 | 0.1740 |
| 1.2 | 0.1823 | 0.1906 | 0.1989 | 0.2070 | 0.2151 | 0.2231 | 0.2311 | 0.2390 | 0.2469 | 0.2546 |
| 1.3 | 0.2624 | 0.2700 | 0.2776 | 0.2852 | 0.2927 | 0.3001 | 0.3075 | 0.3148 | 0.3221 | 0.3293 |
| 1.4 | 0.3365 | 0.3436 | 0.3507 | 0.3577 | 0.3646 | 0.3716 | 0.3784 | 0.3853 | 0.3920 | 0.3988 |
| 1.5 | 0.4055 | 0.4121 | 0.4187 | 0.4253 | 0.4318 | 0.4383 | 0.4447 | 0.4511 | 0.4574 | 0.4637 |
| 1.6 | 0.4700 | 0.4762 | 0.4824 | 0.4886 | 0.4947 | 0.5008 | 0.5068 | 0.5128 | 0.5188 | 0.5247 |
| 1.7 | 0.5306 | 0.5365 | 0.5423 | 0.5481 | 0.5539 | 0.5596 | 0.5653 | 0.5710 | 0.5766 | 0.5822 |
| 1.8 | 0.5878 | 0.5933 | 0.5988 | 0.6043 | 0.6098 | 0.6152 | 0.6206 | 0.6259 | 0.6313 | 0.6366 |
| 1.9 | 0.6419 | 0.6471 | 0.6523 | 0.6575 | 0.6627 | 0.6678 | 0.6729 | 0.6780 | 0.6831 | 0.6881 |
| 2.0 | 0.6931 | 0.6981 | 0.7031 | 0.7080 | 0.7129 | 0.7178 | 0.7227 | 0.7275 | 0.7324 | 0.7372 |
| 2.1 | 0.7419 | 0.7467 | 0.7514 | 0.7561 | 0.7608 | 0.7655 | 0.7701 | 0.7747 | 0.7793 | 0.7839 |
| 2.2 | 0.7885 | 0.7930 | 0.7975 | 0.8020 | 0.8065 | 0.8109 | 0.8154 | 0.8198 | 0.8242 | 0.8286 |
| 2.3 | 0.8329 | 0.8373 | 0.8416 | 0.8459 | 0.8502 | 0.8544 | 0.8587 | 0.8629 | 0.8671 | 0.8713 |
| 2.4 | 0.8755 | 0.8796 | 0.8838 | 0.8879 | 0.8920 | 0.8961 | 0.9002 | 0.9042 | 0.9083 | 0.9123 |
| 2.5 | 0.9163 | 0.9203 | 0.9243 | 0.9282 | 0.9322 | 0.9361 | 0.9400 | 0.9439 | 0.9478 | 0.9517 |
| 2.6 | 0.9555 | 0.9594 | 0.9632 | 0.9670 | 0.9708 | 0.9746 | 0.9783 | 0.9821 | 0.9858 | 0.9895 |
| 2.7 | 0.9933 | 0.9969 | 1.0006 | 1.0043 | 1.0080 | 1.0116 | 1.0152 | 1.0188 | 1.0225 | 1.0260 |
| 2.8 | 1.0296 | 1.0332 | 1.0367 | 1.0403 | 1.0438 | 1.0473 | 1.0508 | 1.0543 | 1.0578 | 1.0613 |
| 2.9 | 1.0647 | 1.0682 | 1.0716 | 1.0750 | 1.0784 | 1.0818 | 1.0852 | 1.0886 | 1.0919 | 1.0953 |
| 3.0 | 1.0986 | 1.1019 | 1.1053 | 1.1086 | 1.1119 | 1.1151 | 1.1184 | 1.1217 | 1.1249 | 1.1282 |
| 3.1 | 1.1314 | 1.1346 | 1.1378 | 1.1410 | 1.1442 | 1.1474 | 1.1506 | 1.1537 | 1.1569 | 1.1600 |
| 3.2 | 1.1632 | 1.1663 | 1.1694 | 1.1725 | 1.1756 | 1.1787 | 1.1817 | 1.1848 | 1.1878 | 1.1909 |
| 3.3 | 1.1939 | 1.1969 | 1.2000 | 1.2030 | 1.2060 | 1.2090 | 1.2119 | 1.2149 | 1.2179 | 1.2208 |
| 3.4 | 1.2238 | 1.2267 | 1.2296 | 1.2326 | 1.2355 | 1.2384 | 1.2413 | 1.2442 | 1.2470 | 1.2499 |
| 3.5 | 1.2528 | 1.2556 | 1.2585 | 1.2613 | 1.2641 | 1.2669 | 1.2698 | 1.2726 | 1.2754 | 1.2782 |
| 3.6 | 1.2809 | 1.2837 | 1.2865 | 1.2892 | 1.2920 | 1.2947 | 1.2975 | 1.3002 | 1.3029 | 1.3056 |
| 3.7 | 1.3083 | 1.3110 | 1.3137 | 1.3164 | 1.3191 | 1.3218 | 1.3244 | 1.3271 | 1.3297 | 1.3324 |
| 3.8 | 1.3350 | 1.3376 | 1.3403 | 1.3429 | 1.3455 | 1.3481 | 1.3507 | 1.3533 | 1.3558 | 1.3584 |
| 3.9 | 1.3610 | 1.3635 | 1.3661 | 1.3686 | 1.3712 | 1.3737 | 1.3762 | 1.3788 | 1.3813 | 1.3838 |
| 4.0 | 1.3863 | 1.3888 | 1.3913 | 1.3938 | 1.3962 | 1.3987 | 1.4012 | 1.4036 | 1.4061 | 1.4085 |
| 4.1 | 1.4110 | 1.4134 | 1.4159 | 1.4183 | 1.4207 | 1.4231 | 1.4255 | 1.4279 | 1.4303 | 1.4327 |
| 4.2 | 1.4351 | 1.4375 | 1.4398 | 1.4422 | 1.4446 | 1.4469 | 1.4493 | 1.4516 | 1.4540 | 1.4563 |
| 4.3 | 1.4586 | 1.4609 | 1.4633 | 1.4656 | 1.4679 | 1.4702 | 1.4725 | 1.4748 | 1.4771 | 1.4793 |
| 4.4 | 1.4816 | 1.4839 | 1.4861 | 1.4884 | 1.4907 | 1.4929 | 1.4951 | 1.4974 | 1.4996 | 1.5019 |
| 4.5 | 1.5041 | 1.5063 | 1.5085 | 1.5107 | 1.5129 | 1.5151 | 1.5173 | 1.5195 | 1.5217 | 1.5239 |
| 4.6 | 1.5261 | 1.5282 | 1.5304 | 1.5326 | 1.5347 | 1.5369 | 1.5390 | 1.5412 | 1.5433 | 1.5454 |
| 4.7 | 1.5476 | 1.5497 | 1.5518 | 1.5539 | 1.5560 | 1.5581 | 1.5602 | 1.5623 | 1.5644 | 1.5665 |
| 4.8 | 1.5686 | 1.5707 | 1.5728 | 1.5748 | 1.5769 | 1.5790 | 1.5810 | 1.5831 | 1.5851 | 1.5872 |
| 4.9 | 1.5892 | 1.5913 | 1.5933 | 1.5953 | 1.5974 | 1.5994 | 1.6014 | 1.6034 | 1.6054 | 1.6074 |
| 5.0 | 1.6094 | 1.6114 | 1.6134 | 1.6154 | 1.6174 | 1.6194 | 1.6214 | 1.6233 | 1.6253 | 1.6273 |
| 5.1 | 1.6292 | 1.6312 | 1.6332 | 1.6351 | 1.6371 | 1.6390 | 1.6409 | 1.6429 | 1.6448 | 1.6467 |
| 5.2 | 1.6487 | 1.6506 | 1.6525 | 1.6544 | 1.6563 | 1.6582 | 1.6601 | 1.6620 | 1.6639 | 1.6658 |
| 5.3 | 1.6677 | 1.6696 | 1.6715 | 1.6734 | 1.6752 | 1.6771 | 1.6790 | 1.6808 | 1.6827 | 1.6845 |
| 5.4 | 1.6864 | 1.6882 | 1.6901 | 1.6919 | 1.6938 | 1.6956 | 1.6974 | 1.6993 | 1.7011 | 1.7029 |

**Table 3** *(Continued)*

| | 0 | 1 | 2 | 3 | 4 | 5 | 6 | 7 | 8 | 9 |
|---|---|---|---|---|---|---|---|---|---|---|
| 5.5 | 1.7047 | 1.7066 | 1.7084 | 1.7102 | 1.7120 | 1.7138 | 1.7156 | 1.7174 | 1.7192 | 1.7210 |
| 5.6 | 1.7228 | 1.7246 | 1.7263 | 1.7281 | 1.7299 | 1.7317 | 1.7334 | 1.7352 | 1.7370 | 1.7387 |
| 5.7 | 1.7405 | 1.7422 | 1.7440 | 1.7457 | 1.7475 | 1.7492 | 1.7509 | 1.7527 | 1.7544 | 1.7561 |
| 5.8 | 1.7579 | 1.7596 | 1.7613 | 1.7630 | 1.7647 | 1.7664 | 1.7681 | 1.7699 | 1.7716 | 1.7733 |
| 5.9 | 1.7750 | 1.7766 | 1.7783 | 1.7800 | 1.7817 | 1.7834 | 1.7851 | 1.7868 | 1.7884 | 1.7901 |
| 6.0 | 1.7918 | 1.7934 | 1.7951 | 1.7967 | 1.7984 | 1.8001 | 1.8017 | 1.8034 | 1.8050 | 1.8066 |
| 6.1 | 1.8083 | 1.8099 | 1.8116 | 1.8132 | 1.8148 | 1.8165 | 1.8181 | 1.8197 | 1.8213 | 1.8229 |
| 6.2 | 1.8245 | 1.8262 | 1.8278 | 1.8294 | 1.8310 | 1.8326 | 1.8342 | 1.8358 | 1.8374 | 1.8390 |
| 6.3 | 1.8405 | 1.8421 | 1.8437 | 1.8453 | 1.8469 | 1.8485 | 1.8500 | 1.8516 | 1.8532 | 1.8547 |
| 6.4 | 1.8563 | 1.8579 | 1.8594 | 1.8610 | 1.8625 | 1.8641 | 1.8656 | 1.8672 | 1.8687 | 1.8703 |
| 6.5 | 1.8718 | 1.8733 | 1.8749 | 1.8764 | 1.8779 | 1.8795 | 1.8810 | 1.8825 | 1.8840 | 1.8856 |
| 6.6 | 1.8871 | 1.8886 | 1.8901 | 1.8916 | 1.8931 | 1.8946 | 1.8961 | 1.8976 | 1.8991 | 1.9006 |
| 6.7 | 1.9021 | 1.9036 | 1.9051 | 1.9066 | 1.9081 | 1.9095 | 1.9110 | 1.9125 | 1.9140 | 1.9155 |
| 6.8 | 1.9169 | 1.9184 | 1.9199 | 1.9213 | 1.9228 | 1.9242 | 1.9257 | 1.9272 | 1.9286 | 1.9301 |
| 6.9 | 1.9315 | 1.9330 | 1.9344 | 1.9359 | 1.9373 | 1.9387 | 1.9402 | 1.9416 | 1.9430 | 1.9445 |
| 7.0 | 1.9459 | 1.9473 | 1.9488 | 1.9502 | 1.9516 | 1.9530 | 1.9544 | 1.9559 | 1.9573 | 1.9587 |
| 7.1 | 1.9601 | 1.9615 | 1.9629 | 1.9643 | 1.9657 | 1.9671 | 1.9685 | 1.9699 | 1.9713 | 1.9727 |
| 7.2 | 1.9741 | 1.9755 | 1.9769 | 1.9782 | 1.9796 | 1.9810 | 1.9824 | 1.9838 | 1.9851 | 1.9865 |
| 7.3 | 1.9879 | 1.9892 | 1.9906 | 1.9920 | 1.9933 | 1.9947 | 1.9961 | 1.9974 | 1.9988 | 2.0001 |
| 7.4 | 2.0015 | 2.0028 | 2.0042 | 2.0055 | 2.0069 | 2.0082 | 2.0096 | 2.0109 | 2.0122 | 2.0136 |
| 7.5 | 2.0149 | 2.0162 | 2.0176 | 2.0189 | 2.0202 | 2.0215 | 2.0229 | 2.0242 | 2.0255 | 2.0268 |
| 7.6 | 2.0281 | 2.0295 | 2.0308 | 2.0321 | 2.0334 | 2.0347 | 2.0360 | 2.0373 | 2.0386 | 2.0399 |
| 7.7 | 2.0412 | 2.0425 | 2.0438 | 2.0451 | 2.0464 | 2.0477 | 2.0490 | 2.0503 | 2.0516 | 2.0528 |
| 7.8 | 2.0541 | 2.0554 | 2.0567 | 2.0580 | 2.0592 | 2.0605 | 2.0618 | 2.0631 | 2.0643 | 2.0656 |
| 7.9 | 2.0669 | 2.0681 | 2.0694 | 2.0707 | 2.0719 | 2.0732 | 2.0744 | 2.0757 | 2.0769 | 2.0782 |
| 8.0 | 2.0794 | 2.0807 | 2.0819 | 2.0832 | 2.0844 | 2.0857 | 2.0869 | 2.0882 | 2.0894 | 2.0906 |
| 8.1 | 2.0919 | 2.0931 | 2.0943 | 2.0956 | 2.0968 | 2.0980 | 2.0992 | 2.1005 | 2.1017 | 2.1029 |
| 8.2 | 2.1041 | 2.1054 | 2.1066 | 2.1078 | 2.1090 | 2.1102 | 2.1114 | 2.1126 | 2.1138 | 2.1150 |
| 8.3 | 2.1163 | 2.1175 | 2.1187 | 2.1199 | 2.1211 | 2.1223 | 2.1235 | 2.1247 | 2.1259 | 2.1270 |
| 8.4 | 2.1282 | 2.1294 | 2.1306 | 2.1318 | 2.1330 | 2.1342 | 2.1353 | 2.1365 | 2.1377 | 2.1389 |
| 8.5 | 2.1401 | 2.1412 | 2.1424 | 2.1436 | 2.1448 | 2.1459 | 2.1471 | 2.1483 | 2.1494 | 2.1506 |
| 8.6 | 2.1518 | 2.1529 | 2.1541 | 2.1552 | 2.1564 | 2.1576 | 2.1587 | 2.1599 | 2.1610 | 2.1622 |
| 8.7 | 2.1633 | 2.1645 | 2.1656 | 2.1668 | 2.1679 | 2.1691 | 2.1702 | 2.1713 | 2.1725 | 2.1736 |
| 8.8 | 2.1748 | 2.1759 | 2.1770 | 2.1782 | 2.1793 | 2.1804 | 2.1815 | 2.1827 | 2.1838 | 2.1849 |
| 8.9 | 2.1861 | 2.1872 | 2.1883 | 2.1894 | 2.1905 | 2.1917 | 2.1928 | 2.1939 | 2.1950 | 2.1961 |
| 9.0 | 2.1972 | 2.1983 | 2.1994 | 2.2006 | 2.2017 | 2.2028 | 2.2039 | 2.2050 | 2.2061 | 2.2072 |
| 9.1 | 2.2083 | 2.2094 | 2.2105 | 2.2116 | 2.2127 | 2.2138 | 2.2148 | 2.2159 | 2.2170 | 2.2181 |
| 9.2 | 2.2192 | 2.2203 | 2.2214 | 2.2225 | 2.2235 | 2.2246 | 2.2257 | 2.2268 | 2.2279 | 2.2289 |
| 9.3 | 2.2300 | 2.2311 | 2.2322 | 2.2332 | 2.2343 | 2.2354 | 2.2364 | 2.2375 | 2.2386 | 2.2396 |
| 9.4 | 2.2407 | 2.2418 | 2.2428 | 2.2439 | 2.2450 | 2.2460 | 2.2471 | 2.2481 | 2.2492 | 2.2502 |
| 9.5 | 2.2513 | 2.2523 | 2.2534 | 2.2544 | 2.2555 | 2.2565 | 2.2576 | 2.2586 | 2.2597 | 2.2607 |
| 9.6 | 2.2618 | 2.2628 | 2.2638 | 2.2649 | 2.2659 | 2.2670 | 2.2680 | 2.2690 | 2.2701 | 2.2711 |
| 9.7 | 2.2721 | 2.2732 | 2.2742 | 2.2752 | 2.2762 | 2.2773 | 2.2783 | 2.2793 | 2.2803 | 2.2814 |
| 9.8 | 2.2824 | 2.2834 | 2.2844 | 2.2854 | 2.2865 | 2.2875 | 2.2885 | 2.2895 | 2.2905 | 2.2915 |
| 9.9 | 2.2925 | 2.2935 | 2.2946 | 2.2956 | 2.2966 | 2.2976 | 2.2986 | 2.2996 | 2.3006 | 2.3016 |

$\ln 10 \approx 2.3026$

## Table 4 Exponential Functions

| $x$ | $e^x$ | $e^{-x}$ | $x$ | $e^x$ | $e^{-x}$ |
|---|---|---|---|---|---|
| 0.00 | 1.0000 | 1.0000 | 1.5 | 4.4817 | 0.2231 |
| 0.01 | 1.0101 | 0.9901 | 1.6 | 4.9530 | 0.2019 |
| 0.02 | 1.0202 | 0.9802 | 1.7 | 5.4739 | 0.1827 |
| 0.03 | 1.0305 | 0.9705 | 1.8 | 6.0496 | 0.1653 |
| 0.04 | 1.0408 | 0.9608 | 1.9 | 6.6859 | 0.1496 |
| 0.05 | 1.0513 | 0.9512 | 2.0 | 7.3891 | 0.1353 |
| 0.06 | 1.0618 | 0.9418 | 2.1 | 8.1662 | 0.1225 |
| 0.07 | 1.0725 | 0.9324 | 2.2 | 9.0250 | 0.1108 |
| 0.08 | 1.0833 | 0.9231 | 2.3 | 9.9742 | 0.1003 |
| 0.09 | 1.0942 | 0.9139 | 2.4 | 11.023 | 0.0907 |
| 0.10 | 1.1052 | 0.9048 | 2.5 | 12.182 | 0.0821 |
| 0.11 | 1.1163 | 0.8958 | 2.6 | 13.464 | 0.0743 |
| 0.12 | 1.1275 | 0.8869 | 2.7 | 14.880 | 0.0672 |
| 0.13 | 1.1388 | 0.8781 | 2.8 | 16.445 | 0.0608 |
| 0.14 | 1.1503 | 0.8694 | 2.9 | 18.174 | 0.0550 |
| 0.15 | 1.1618 | 0.8607 | 3.0 | 20.086 | 0.0498 |
| 0.16 | 1.1735 | 0.8521 | 3.1 | 22.198 | 0.0450 |
| 0.17 | 1.1853 | 0.8437 | 3.2 | 24.533 | 0.0408 |
| 0.18 | 1.1972 | 0.8353 | 3.3 | 27.113 | 0.0369 |
| 0.19 | 1.2092 | 0.8270 | 3.4 | 29.964 | 0.0334 |
| 0.20 | 1.2214 | 0.8187 | 3.5 | 33.115 | 0.0302 |
| 0.21 | 1.2337 | 0.8106 | 3.6 | 36.598 | 0.0273 |
| 0.22 | 1.2461 | 0.8025 | 3.7 | 40.447 | 0.0247 |
| 0.23 | 1.2586 | 0.7945 | 3.8 | 44.701 | 0.0224 |
| 0.24 | 1.2712 | 0.7866 | 3.9 | 49.402 | 0.0202 |
| 0.25 | 1.2840 | 0.7788 | 4.0 | 54.598 | 0.0183 |
| 0.30 | 1.3499 | 0.7408 | 4.1 | 60.340 | 0.0166 |
| 0.35 | 1.4191 | 0.7047 | 4.2 | 66.686 | 0.0150 |
| 0.40 | 1.4918 | 0.6703 | 4.3 | 73.700 | 0.0136 |
| 0.45 | 1.5683 | 0.6376 | 4.4 | 81.451 | 0.0123 |
| 0.50 | 1.6487 | 0.6065 | 4.5 | 90.017 | 0.0111 |
| 0.55 | 1.7333 | 0.5769 | 4.6 | 99.484 | 0.0101 |
| 0.60 | 1.8221 | 0.5488 | 4.7 | 109.95 | 0.0091 |
| 0.65 | 1.9155 | 0.5220 | 4.8 | 121.51 | 0.0082 |
| 0.70 | 2.0138 | 0.4966 | 4.9 | 134.29 | 0.0074 |
| 0.75 | 2.1170 | 0.4724 | 5.0 | 148.41 | 0.0067 |
| 0.80 | 2.2255 | 0.4493 | 5.5 | 244.69 | 0.0041 |
| 0.85 | 2.3396 | 0.4274 | 6.0 | 403.43 | 0.0025 |
| 0.90 | 2.4596 | 0.4066 | 6.5 | 665.14 | 0.0015 |
| 0.95 | 2.5857 | 0.3867 | 7.0 | 1,096.6 | 0.0009 |
| 1.0 | 2.7183 | 0.3679 | 7.5 | 1,808.0 | 0.0006 |
| 1.1 | 3.0042 | 0.3329 | 8.0 | 2,981.0 | 0.0003 |
| 1.2 | 3.3201 | 0.3012 | 8.5 | 4,914.8 | 0.0002 |
| 1.3 | 3.6693 | 0.2725 | 9.0 | 8,103.1 | 0.0001 |
| 1.4 | 4.0552 | 0.2466 | 10.0 | 22,026 | 0.00005 |

# ANSWERS

## CHAPTER 1

### Exercises 1.1, Page 11

**1.** $0, \sqrt{16}, 6$ **3.** $-8, -\sqrt{4}, 0, \sqrt{16}, 6$ **5.** $-8, -\sqrt{4}, -\frac{4}{3}, -1.2, 0, \frac{4}{5}, \sqrt{16}, 4.2, 6$ **7.** Always
**9.** Sometimes **11.** Sometimes **13.** $0.625\overline{0}$ **15.** $-2.3\overline{0}$ **17.** $3.55\overline{0}$ **19.** 0 1 2 3 4 5 6

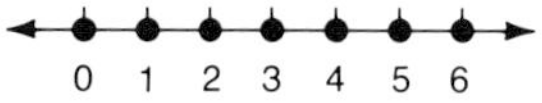

**21.** −4 −3 −2 0 2 **23.** 1 2 3 4 5 **25.** 1 2 3 4 5 6 7 8 9 10 11

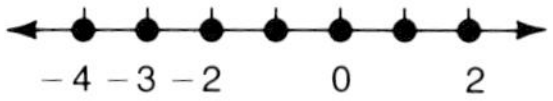

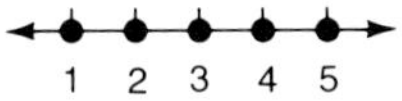

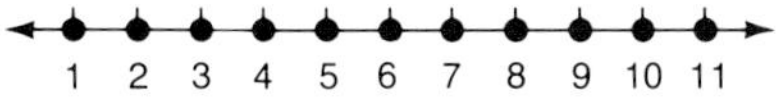

**27.** 3 5 **29.** −1.8 3 **31.** 2 8

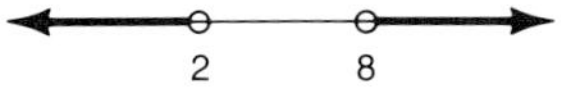

**33.** $-\sqrt{2}$ 0 **35.** −1 0 **37.** −1.6 0 2 3.7

**39.** Closure property for addition **41.** Trichotomy property
**43.** Associative property for addition **45.** Commutative property for multiplication
**47.** Multiplication property of equality **49.** Addition property of equality **51.** Closure property for multiplication
**53.** Distributive property **55.** Identity for multiplication **57.** Transitive property
**59.** Commutative property for addition **61.** Zero factor law **63.** $2(7 + x)$ **65.** $x \cdot 6 + xy$ **67.** $(3x) \cdot z$
**69.** $(x + 2)y + (x + 2)3$

### Exercises 1.2, Page 20

**1.** 7 **3.** $\sqrt{5}$ **5.** $-8$ **7.** $7, -7$ **9.** $2, -2$ **11.** $\frac{4}{5}, -\frac{4}{5}$ **13.** No value **15.** $x \geq 0$ **17.** $-11$

**19.** $-9$ **21.** $\frac{1}{2}$ **23.** $-3$ **25.** 0 **27.** 13 **29.** $-\frac{6}{5}$ **31.** $-6.9$ **33.** 1 **35.** 0 **37.** 20

**39.** $-\frac{13}{6}$ **41.** $-\frac{23}{16}$ **43.** $-\frac{1}{30}$ **45.** 56 **47.** 480 **49.** $\frac{15}{16}$ **51.** $-\frac{1}{28}$ **53.** $-3.92$ **55.** 2
**57.** 13 **59.** Undefined **61.** 0 **63.** $-13.61$ **65.** 0.676 **67.** 5 **69.** 6 **71.** 4 **73.** 5 **75.** 4
**77.** $-11$ **79.** $-7$ **81.** 0.9305 **83.** $-6.4786$ **85.** $-0.819$ **87.** 461.6594

### Exercises 1.3, Page 28

**1.** $4x + 3y$ **3.** $6x + y$ **5.** $3x - 4x^2$ **7.** $6x - 7$ **9.** $5x$ **11.** $6x$ **13.** $-x - 15$ **15.** $5x^2 - 11$
**17.** $x = 3$ **19.** $x = -5$ **21.** $x = 0.6$ **23.** $x = -0.7$ **25.** $x = 3$ **27.** $x = -25$ **29.** $x = 18$

**31.** Any real number **33.** $x = 7$ **35.** $x = 2$ **37.** No solution **39.** $x = \frac{3}{4}$ **41.** $x = 10$ **43.** $x = 5$

**45.** $x = -12$ **47.** $x = 2$ **49.** $x = \frac{3}{2}$ **51.** $x = 4$ **53.** $x = -\frac{3}{2}$ **55.** $x = 9$ **57.** $x = 3$

**59.** $x = 3$ **61.** $x = 8$ or $x = -8$ **63.** No solution **65.** $x = -1$ or $x = -5$ **67.** $x = \frac{9}{2}$ or $x = \frac{7}{2}$

**69.** $x = -\frac{1}{5}$ or $x = 1$ **71.** $x = -16$ or $x = 8$ **73.** $x = -\frac{1}{3}$ or $x = 3$ **75.** $x = -\frac{5}{2}$ or $x = \frac{3}{4}$

**77.** $x = -\frac{20}{7}$ or $x = \frac{4}{5}$ **79.** $x = -20$ or $x = \frac{40}{9}$ **81.** $x = 3.43$ **83.** $x = 1.91$ **85.** $x = -116.12$

## Exercises 1.4, Page 34

**1.** \$1650 **3.** 43° **5.** 13 ft **7.** 61 in. **9.** $196\pi$ sq ft or 615.8 sq ft **11.** $b = P - a - c$ **13.** $m = \frac{f}{a}$

**15.** $w = \frac{A}{L}$ **17.** $n = \frac{R}{p}$ **19.** $p = A - I$ **21.** $m = 2A - n$ **23.** $s = \frac{P}{4}$ **25.** $t = \frac{d}{r}$ **27.** $t = \frac{I}{pr}$

**29.** $b = \frac{P - a}{2}$ **31.** $a = S - Sr$ **33.** $x = \frac{y - b}{m}$ **35.** $r^2 = \frac{A}{4\pi}$ **37.** $M = \frac{(\text{IQ})C}{100}$ **39.** $h = \frac{3V}{\pi r^2}$

**41.** $I = \frac{E}{R}$ **43.** $L = \frac{R}{2A}$ **45.** $b = \frac{2A}{h} - a$ or $b = \frac{2A - ah}{h}$ **47.** $h = \frac{L}{2\pi r}$ **49.** $r = \frac{S - a}{S}$

**51.** $P = \frac{WR - Wr}{2R}$ **53.** $R = \frac{nE - Inr}{I}$ **55.** $a = \frac{Sb - rL}{S - 1}$

## Exercises 1.5, Page 40

**1.** 41 **3.** 9 **5.** 12 **7.** 7 **9.** 18 **11.** 8, 30 **13.** 46 ft by 84 ft **15.** \$1500 **17.** 11 min

**19.** 375 mph, 450 mph **21.** 60 mph, 300 mi **23.** $4\frac{4}{5}$ hr **25.** 36 mph, 60 mph **27.** \$40 **29.** 4500 baskets

**31.** 62,500 pounds **33.** \$86.40; \$72.00 **35.** 150 @ \$3.00, 400 @ \$4.50 **37.** 4.5% on \$10,000, 5.5% on \$6000
**39.** \$7400 @ 5.5%, \$2600 @ 6%

## Exercises 1.6, Page 48

**1.** $(-\infty,1)$ 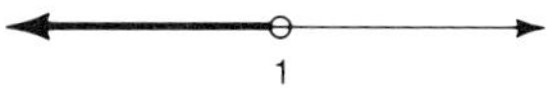


**3.** $(2,+\infty)$ 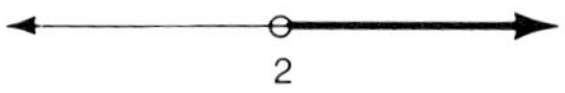


**5.** $\left(\frac{8}{3},+\infty\right)$ 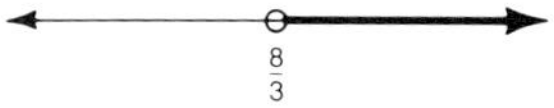


**7.** $[-0.4,+\infty)$ 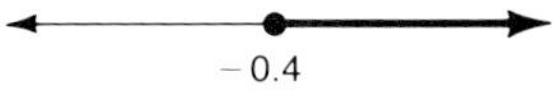


**9.** $(-\infty,2)$ 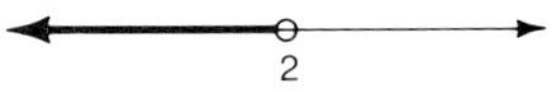


**11.** $(-\infty,-3)$ 

**13.** $(2,+\infty)$ 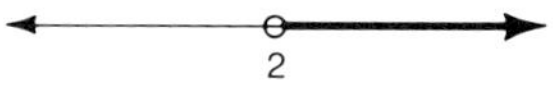


**15.** $(-\infty,5]$ 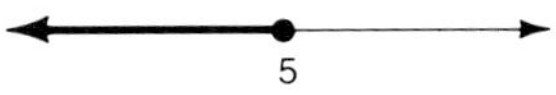


**17.** $\left[-\frac{8}{3},+\infty\right)$ 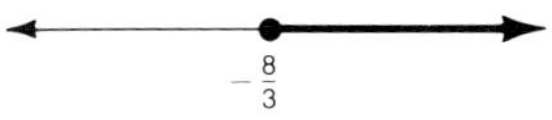


**19.** $\left(-\frac{3}{2},+\infty\right)$ 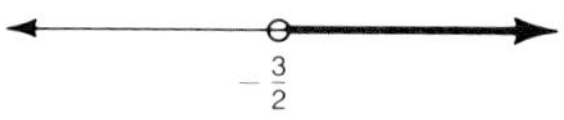

**21.** $[1,+\infty)$ **23.** $(-\infty,-15)$

**25.** $\left(-\frac{9}{4},+\infty\right)$ **27.** $(-\infty,-0.6)$

**29.** $\left(\frac{44}{7},+\infty\right)$ **31.** $(-\infty,-13)$

**33.** $\left(-\infty,\frac{1}{9}\right]$ **35.** $(-\infty,-4]$

**37.** $(-\infty,16)$ **39.** $(-\infty,-9]$

**41.** $(-\infty,6]$ **43.** $[11,+\infty)$

**45.** $(-56,+\infty)$ **47.** $(-\infty,-3]$ or $[3,+\infty)$

**49.** $\left(-\infty,-\frac{7}{2}\right]$ or $\left[\frac{7}{2},+\infty\right)$ **51.** $[-1,9]$

**53.** $(-\infty,+\infty)$ **55.** $\left(-\infty,-\frac{1}{2}\right]$ or $\left[\frac{3}{2},+\infty\right)$

**57.** $\left(-\frac{5}{3},-1\right)$ **59.** $\left(-\infty,-\frac{35}{6}\right)$ or $\left(\frac{7}{2},+\infty\right)$

**61.** $\left(-\infty,\frac{1}{7}\right]$ or $\left[\frac{5}{7},+\infty\right)$ **63.** $\left(\frac{1}{2},\frac{7}{2}\right)$

**65.** $\left(\frac{4}{5},\frac{12}{5}\right)$ **67.** $[1.70,+\infty)$ **69.** $[0.59,0.88]$

## *Chapter 1 Review, Page 52*

**1.** $\sqrt{9}$ **2.** $0, \sqrt{9}$ **3.** $-10, -\sqrt{25}, -1.6, 0, \frac{1}{5}, \sqrt{9}$ **4.** $-10, -\sqrt{25}, 0, \sqrt{9}$ **5.** $-\sqrt{7}, \pi, \sqrt{12}$

**6.** $-10, -\sqrt{25}, -1.6, -\sqrt{7}, 0, \frac{1}{5}, \sqrt{9}, \pi, \sqrt{12}$ **7.** $0.8\overline{3}$ **8.** $1.375\overline{0}$ **9.** $0.8\overline{6}$

**10.** **11.** **12.**

**13.** **14.**

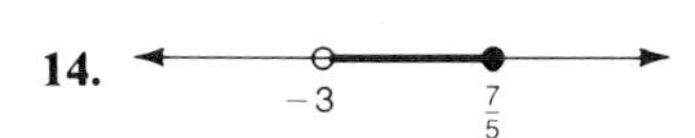

**15.**

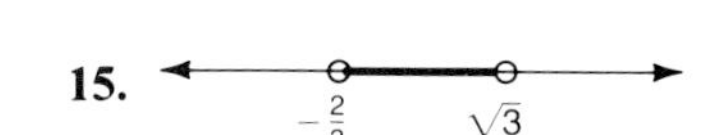

**16.** **17.**

**18.** Commutative property for addition **19.** Distributive property

**20.** Associative property for multiplication **21.** Multiplication property for equality **22.** Inverse for addition **23.** Identity for multiplication **24.** Transitive property **25.** Trichotomy property **26.** $x + (5 + y)$ **27.** $x - 0.3$ **28.** $5x + 5(8z)$ **29.** $x + (x - 7) = 15$ **30.** $y + 4 + 7$ **31.** $x = -3.9$ or $x = 3.9$ **32.** $x = -8$ or $x = 8$ **33.** No value **34.** 4 **35.** $-20$ **36.** 11 **37.** $-8$ **38.** 22 **39.** $-1$ **40.** $\frac{1}{10}$ **41.** 1 **42.** $-0.8$ **43.** $-\frac{1}{12}$ **44.** 84 **45.** $-40$ **46.** $-2$ **47.** $-7.2$ **48.** $-2$ **49.** $\frac{2}{3}$ **50.** Undefined **51.** 7 **52.** 0 **53.** $-126$ **54.** 45 **55.** $-\frac{7}{16}$ **56.** $-2$ **57.** $-35$ **58.** 26 **59.** $-59$ **60.** $-\frac{22}{15}$ **61.** $-2x - 12$ **62.** 0 **63.** $4x$ **64.** $3x^3 - x^2 + 4x - 1$ **65.** $-10x + 4$ **66.** $6x - 6y + 10$ **67.** $x = -3$ **68.** $x = 2$ **69.** $x = -4$ **70.** Any real number **71.** $x = -4$ **72.** $x = -8$ **73.** No solution **74.** $x = -2$ **75.** $x = 2.3$ or $x = -3.3$ **76.** No solution **77.** $x = -9$ or $x = -3$ **78.** $x = -4$ or $x = 16$ **79.** No solution **80.** $x = 2$ or $x = \frac{10}{3}$ **81.** $n = 2A - m$ **82.** $m = \frac{2gk}{v^2}$ **83.** $r = \frac{WR - 2PR}{W}$ **84.** $x = -\frac{1}{3}$ **85.** 7 hr **86.** 10 mi **87.** 18 hr @ \$3.80, 26 hr @ \$4.20 **88.** 85 calves

**89.** $(4, +\infty)$ 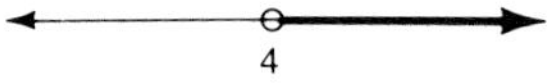


**90.** $(3, +\infty)$ 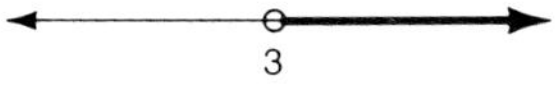


**91.** $[-12.8, +\infty)$ 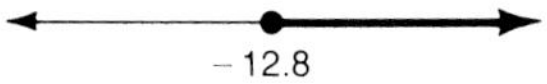


**92.** $(-\infty, 2]$ 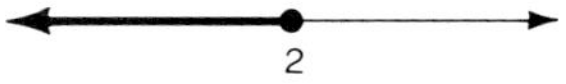


**93.** $(-\infty, -5]$ 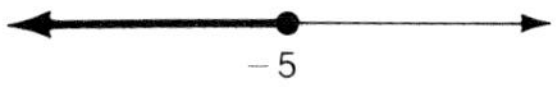


**94.** $(-\infty, -20.4)$ 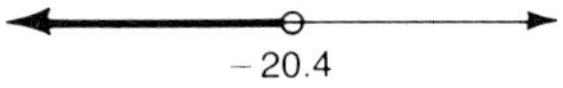


**95.** $(-\infty, 4)$ 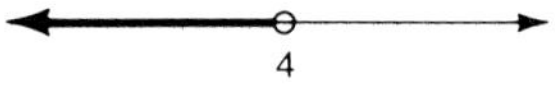


**96.** $(-\infty, 10)$ 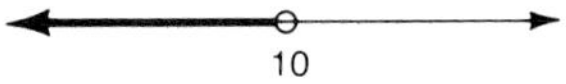


**97.** $\left(-\infty, -\frac{9}{5}\right)$ or $(1, +\infty)$ 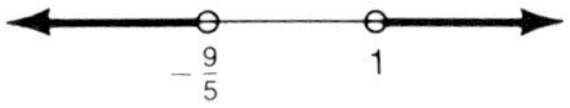


**98.** $\left(-\frac{3}{2}, 3\right)$ 

**99.** $\left(-\frac{8}{3}, \frac{4}{3}\right)$ 

**100.** $(-\infty, +\infty)$ 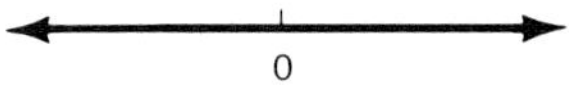


## *Chapter 1 Test, Page 55*

**1.** **(a)** $-2, 0, 2$ **(b)** $-2, -\frac{5}{3}, 0, \frac{1}{2}, 2$ **2.** $0.\overline{384615}$ **3.** 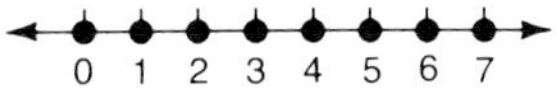


**4.** 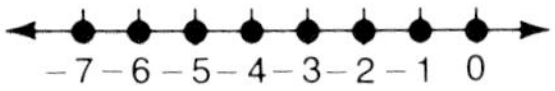


**5.** 

**6.** 

**7.** Distributive property **8.** Associative property for addition **9.** 0 **10.** $2x + 3x = 17$ **11.** 70 **12.** 8 **13.** $-\frac{15}{8}$ **14.** $-\frac{4}{15}$ **15.** $-144$ **16.** 38 **17.** $\frac{25}{6}$ **18.** $6x + 16$ **19.** $7x - 12$ **20.** $x = -1$ **21.** No solution **22.** $x = -6$ **23.** $x = -1.9$ or $x = 0.9$ **24.** $x = \frac{2}{3}$ or $x = \frac{8}{3}$ **25.** $t = \frac{A - p}{pr}$ **26.** $r = \frac{nE - IR}{In}$

**27.** $(-5, +\infty)$ 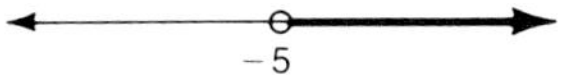


**28.** $\left(-\infty, -\frac{7}{2}\right)$ 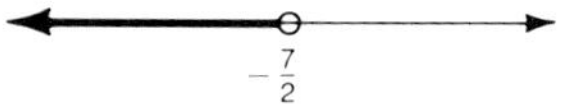

**29.** (2,5) 

**30.** $(-\infty,-0.85)$ or $(1.85,+\infty)$ 

**31.** 4 hr

**32.** 16 packages @ $1.25, 26 packages @ $1.75

## CHAPTER 2

### Exercises 2.1, Page 64

**1.** $(-2)^5 = -32$ **3.** 1 **5.** $-36$ **7.** $\frac{1}{5}$ **9.** $-\frac{1}{64}$ **11.** $x^{12}$ **13.** $x^5$ **15.** $y^{-3} = \frac{1}{y^3}$ **17.** $x^{-2} = \frac{1}{x^2}$
**19.** $x^{-4} = \frac{1}{x^4}$ **21.** $y^5$ **23.** $y^{-4} = \frac{1}{y^4}$ **25.** $x^5$ **27.** $x^{-3} = \frac{1}{x^3}$ **29.** $x^8$ **31.** $x^0 = 1$ **33.** $x^6$
**35.** $-x^{12}$ **37.** $x^{-10} = \frac{1}{x^{10}}$ **39.** $x^{-9} = \frac{1}{x^9}$ **41.** $x^{-4} = \frac{1}{x^4}$ **43.** $x^{-1} = \frac{1}{x}$ **45.** $y^4$ **47.** $y^8$ **49.** $x^3$
**51.** $x^0 = 1$ **53.** $y^{11}$ **55.** $x^3$ **57.** $x^{10}$ **59.** $x^{k+3}$ **61.** $x^{3k+4}$ **63.** $x^k$ **65.** $x^{2k}$ **67.** $x^{2k+1}$
**69.** $x^{k-5}$

### Exercises 2.2, Page 69

**1.** $x^{-4}$ or $\frac{1}{x^4}$ **3.** $4x^8$ **5.** $x^4y^{-6}$ or $\frac{x^4}{y^6}$ **7.** $\frac{a^4}{b^8}$ **9.** $\frac{36b^4}{a^2}$ **11.** $\frac{x^2}{36y^6}$ **13.** $\frac{x^4}{25y^2}$ **15.** $x^{2k}y^k$
**17.** $x^{2k+1}y^{2+n}$ **19.** $x^ny^{k+1}$ **21.** $\frac{x^4}{y}$ **23.** $\frac{9y^7}{x^6}$ **25.** $\frac{a^6}{b^3}$ **27.** $\frac{7}{2x^3y}$ **29.** $\frac{x^{13}}{y^{20}}$ **31.** $\frac{y^6}{x^6}$ **33.** 472,000
**35.** 0.000000128 **37.** 0.00923 **39.** $4.79 \times 10^5$ **41.** $8.71 \times 10^{-7}$ **43.** $4.29 \times 10$ **45.** $1.44 \times 10^5$
**47.** $6 \times 10^{-4}$ **49.** $1.2 \times 10^{-1}$ **51.** $5 \times 10^3$ **53.** $5.6 \times 10^{-2}$ **55.** $1.8 \times 10^{12}$ cm/min; $1.08 \times 10^{14}$ cm/hr
**57.** $4.0678 \times 10^{16}$ m **59.** $1.9926 \times 10^{-26}$ kg

### Exercises 2.3, Page 75

**1.** Binomial **3.** Not a polynomial **5.** Trinomial **7.** Monomial **9.** Not a polynomial
**11.** Not a polynomial **13.** $4x^2 - 3x - 6$ **15.** $-7x - 6$ **17.** $3x^2 - 4y^2$ **19.** $x^2 - x + 10$
**21.** $6x^3 - 7x^2 + 8x + 3$ **23.** $-x^4 + 2x^3 - 3x^2 - x + 5$ **25.** $4x^2 - 4xy + 8y^2$ **27.** $9x^2 + 13xy - 6y^2$
**29.** $12x^3 - 21x^2 - 2x + 12$ **31.** $5x^3 - 2x^2$ **33.** $-2x^2 - 4xy + y^2$ **35.** $8x - 3xy$ **37.** $6x^4 + 10x^3 - 2x^2$
**39.** $x^3z + 4x^2yz - x^2z^2$ **41.** $x^2 - 7x + 10$ **43.** $x^2 + 6x + 8$ **45.** $3y^2 + 17y + 10$ **47.** $2x^2 - 5x + 2$
**49.** $15y^2 - y - 2$ **51.** $14x^2 - 9x - 18$ **53.** $15x^2 - 13x - 44$ **55.** $16x^2 - 24x + 9$
**57.** $49x^2 + 56xy + 16y^2$ **59.** $9x^2 - 25$ **61.** $36x^2 - y^2$ **63.** $147x^5 - 192x$ **65.** $y^3 + 64$
**67.** $y^3 - 2y^2 - 13y - 10$ **69.** $x^3 + 8y^3$ **71.** $x^6 + 4x^3 + 4$ **73.** $24y^4 - 5y^2 - 14$
**75.** $3x^5y - 9x^3y^3 - 54xy^5$ **77.** $x^6 - 8x^3y^3$ **79.** $x^2 + 2x + 1 - y^2$ **81.** $4x^2 + 4x + 1 - 4xy - 2y + y^2$
**83.** $x^2 - 6xy + 9y^2 + 10x - 30y + 25$ **85.** $x^{2k+3} + x^4$ **87.** $x^{2k} - 36$ **89.** $2x^{2k} + x^k - 6$ **91.** 4
**93.** $-1$ **95.** $-22$ **97.** 24 **99.** 5

### Exercises 2.4, Page 83

**1.** $5(x^2+3)$ **3.** $xy(x-2+y)$ **5.** $5x^2y(y+4)$ **7.** $3x^2y(1+7xy+y^2)$ **9.** $(x+11)(x-11)$
**11.** $(3x+5)(3x-5)$ **13.** $(x-5)^2$ **15.** $(3y+2)^2$ **17.** $(x+6)(x+3)$ **19.** $(x-9)(x+3)$
**21.** $(y-7)(y+2)$ **23.** $(x+7)^2$ **25.** $(2x+3)(3x+2)$ **27.** $(2x-5)(x+7)$ **29.** $(4x-5)^2$
**31.** $(5x-3)(7x+6)$ **33.** $4x(x-4)(x+4)$ **35.** $x^2(7x+8)(3x-4)$ **37.** $2x^3y^2(3x^2-14xy-3y^2)$
**39.** $2x(3x^2+2)(2x^2+5)$ **41.** $x(3x^2-4)(3x^2+4)$ **43.** $(2x^2-5y)(x^2+3y)$ **45.** Not factorable
**47.** $(2x^3+y^2)(x^3+4y^2)$ **49.** $(x-5)(x+7)(x-3)$ **51.** $(2x+1)(2x+3)(x-6)$ **53.** $(x+y+3)^2$
**55.** $(x+2y-5)(x+2y+5)$ **57.** $(x+5y+6)(x+5y+2)$ **59.** $(3x-y-1)(3x-y+4)$

### Exercises 2.5, Page 87

**1.** $(x-5)(x^2+5x+25)$ **3.** $(x-2y)(x^2+2xy+4y^2)$ **5.** $(x+6)(x^2-6x+36)$
**7.** $(x+y)(x^2-xy+y^2)$ **9.** $(x-1)(x^2+x+1)$ **11.** $(3x+2)(9x^2-6x+4)$
**13.** $3(x+3)(x^2-3x+9)$ **15.** $(5x-4y)(25x^2+20xy+16y^2)$ **17.** $2(3x-y)(9x^2+3xy+y^2)$
**19.** $y(x+y)(x^2-xy+y^2)$ **21.** $x^2y^2(1-y)(1+y+y^2)$ **23.** $3xy(2x+3y)(4x^2-6xy+9y^2)$
**25.** $(x^2-y^3)(x^4+x^2y^3+y^6)$ **27.** $[3x+(y^2-1)][9x^2-3xy^2+3x+y^4-2y^2+1]$
**29.** $[(x+2)+(y-3)][(x+2)^2-(x+2)(y-3)+(y-3)^2]=[x+y-1][x^2+7x-xy+y^2-8y+19]$
**31.** $[(x+2y)-(y+4)][(x+2y)^2+(x+2y)(y+4)+(y+4)^2]=[x+y-4][x^2+5xy+7y^2+4x+16y+16]$
**33.** $(x-3)(y+4)$ **35.** $(x+4)(y+6)$ **37.** Not factorable **39.** $(4x+3)(y-7)$ **41.** Not factorable
**43.** $(x-y-6)(x-y+6)$ **45.** $(4x+1+y)(4x+1-y)$ **47.** $(x^2+6)(x-5)$
**49.** $(x+12)(x+2)(x-2)$ **51.** $(x+2)(x+2)(x-2)$ **53.** $(y+5)(x+3)(x-3)$
**55.** $(x^k+2)(x^{2k}-2x^k+4)$ **57.** $3x^{3k}(1-2y^k)(1+2y^k+4y^{2k})$ **59.** $4x^{-2}(x+2)$
**61.** $x^{-3}(2x+3y-6)$ **63.** $5x^{-4}(x+2)(x-2)$ **65.** $x^{-3}(x+3)(x+1)$
**67.** $2x^{-2}(x-5)(x+1)$

### Exercises 2.6, Page 92

**1.** $y=\frac{7}{3}$ **3.** 2 **5.** 10 **7.** 36 **9.** 24 **11.** $-\frac{27}{5}$ **13.** 40 **15.** 54 **17.** 32 **19.** 27

**21.** $R=\frac{k\ell}{d^2}$, 15 ohms **23.** $F=kAv^2$, 2025 lb **25.** $E=\frac{km\ell}{A}$, 0.0072 cm **27.** $L=\frac{kwd^2}{\ell}$, 1890 lb

**29.** $F=\frac{km_1m_2}{c^2}$, $9\times10^{-11}$ N

### Chapter 2 Review, Page 94

**1.** $(-3)^7=-2187$ **2.** $\frac{1}{7}$ **3.** $y^3$ **4.** $x$ **5.** $\frac{1}{5}$ **6.** $(-6)^2=36$ **7.** $x^2$ **8.** $x^{-3}$ or $\frac{1}{x^3}$ **9.** $64x^6y^3$

**10.** $49x^{10}y^4$ **11.** $-\frac{1}{8}x^{-9}y^{-6}$ or $-\frac{1}{8x^9y^6}$ **12.** $\frac{36x^4}{y^{10}}$ **13.** $\frac{9y^8}{x^2}$ **14.** $\frac{y^3}{8x^3}$ **15.** $\frac{8x^{10}}{y^{10}}$ **16.** $x^5y^{-1}$ or $\frac{x^5}{y}$

**17.** $x^3y^{-4}$ or $\frac{x^3}{y^4}$ **18.** $x^{-6}y^{-1}$ or $\frac{1}{x^6y}$ **19.** $y$ **20.** $x^5y^{-1}$ or $\frac{x^5}{y}$ **21.** $x^{-4}y^6$ or $\frac{y^6}{x^4}$ **22.** $\frac{1}{6}x^{-3}y^{-1}$ or $\frac{1}{6x^3y}$

**23.** $20x^{-1}y^4$ or $\frac{20y^4}{x}$ **24.** $x^2$ **25.** $\frac{x^3}{y^4}$ **26.** $\frac{y^6}{x^{10}}$ **27.** $\frac{x^4}{4y^3}$ **28.** $3.45\times10^{-7}$ **29.** $6.82\times10^6$

**30.** $1.05 \times 10^{-2}$ **31.** $9 \times 10^{-4}$ **32.** $7x^2 + 10xy - 5y^2$ **33.** $9x^2 + 3xy - 9y^2$ **34.** $-x^3 + 5x^2 - 6$
**35.** $-x^3 + 4x^2 + 5x - 3$ **36.** $-2x^2 - 9xy + 2y^2 - 2y$ **37.** $7x^2y - 7xy^2 + xy + 7y^2$ **38.** $6x^2 - 7xz + z^2$
**39.** $-xz - 8z^2 + 9$ **40.** $x^2 - 9x + 5$ **41.** $-2xy - 2xz + 3yz$ **42.** 22 **43.** 8 **44.** $-3$ **45.** $-6$
**46.** $12x^2 - 13xy - 14y^2$ **47.** $15x^2 + 43xy + 8y^2$ **48.** $6x^4y + 17x^3y^2 - 3x^2y^3$ **49.** $8x^3y^3 - 42x^2y^4 + 27xy^5$
**50.** $8x^3 + 125y^3$ **51.** $49x^4 - 36$ **52.** $2x^{k+2} - x^{k+1} - 55x^k$ **53.** $8x^{2k+2} - 26x^{2k+1} - 45x^{2k}$
**54.** $3x^{2k} + 17x^k - 6$ **55.** $4x^{2k} - 25$ **56.** $(x - 12)(x + 3)$ **57.** $y(x - 7y)(x + 2y)$
**58.** $(5x + 8y)(3x + y)$ **59.** $6y^2(x + 4y)(x - 4y)$ **60.** $(2x + 1)(4x^2 - 2x + 1)$
**61.** $5(2x - 5)(4x^2 + 10x + 25)$ **62.** $(x + 8)(x + 5)$ **63.** $(x - y)(x - 2)$ **64.** $(x + 1)(x + 2)(x - 2)$
**65.** $4xy$ **66.** $(5x^k + 7)(x^k - 2)$ **67.** $x^k(4x - 9)(2x + 5)$ **68.** $(x^2 + y^2)(x^4 - x^2y^2 + y^4)$
**69.** $x^{-3}(3x + 1)^2$ **70.** $x^{-2}(5x + 3)(x - 2)$ **71.** $V = 9.9$ **72.** $Z = 297$ **73.** $W = \frac{7}{15}$
**74.** $F = kAv^2$; $321\frac{3}{7}$ lb **75.** $F = kAv^2$; $178\frac{4}{7}$ lb **76.** $R = kA\sqrt{h}$; $\frac{0.00308}{\sqrt{3}}$ $\text{m}^3$ per sec; about $1.8 \times 10^{-3}$ $\text{m}^3$ per sec

## Chapter 2 Test, Page 96

**1.** $-20x^6y^{-2}$ or $-\frac{20x^6}{y^2}$ **2.** $x^{-1}y^2$ or $\frac{y^2}{x}$ **3.** $y^{-8}$ or $\frac{1}{y^8}$ **4.** $\frac{1}{xy^5}$ **5.** $3.6 \times 10^{-6}$ **6.** $3x^2 + 2x + 13$
**7.** $2x - y + 3$ **8.** $x^2 + 3x - 2$ **9.** $28x^2 + 23x - 15$ **10.** $4x^2 - 28x + 49$ **11.** $64x^2 - 9$
**12.** $6x^3 + 5x^2 - x - 1$ **13.** $x^6 + 10x^3 + 25$ **14.** $x^2 + 2x + 1 - y^2$ **15.** $6xy(5x - 3y + 4)$
**16.** $(7x + 2)(x - 4)$ **17.** $3y(x - 3y)^2$ **18.** $9(x + 3y)(x - 3y)$ **19.** $8x$ **20.** $(x - 2y)(x^2 + 2xy + 4y^2)$
**21.** $3(x + 3y)(x^2 - 3xy + 9y^2)$ **22.** $(x + 2y)(x - 2y + 1)$ **23.** $-55$ **24.** 20 **25.** $z = \frac{400}{9}$
**26.** 1250 ft lb

## Cumulative Review (2), Page 96

**1.** $\frac{5}{6}$ **2.** $\frac{35}{48}$ **3.** $\frac{5}{36}$ **4.** $\frac{8}{5}$ **5.** 6 **6.** $-138$ **7.** $\frac{7}{2}$ **8.** $\frac{49}{40}$ **9.** $2x + 17$ **10.** $12x - 7$
**11.** $x = \frac{9}{10}$ **12.** $x = 21$ **13.** $x = 21$ **14.** $x = -5, x = 10$ **15.** $4x^2 + x - 14$
**16.** $2x^3 + 4x^2 + 13x + 2$ **17.** $-3x^2 + 11x - 4$ **18.** $5x^2 - 4x - 6$ **19.** $5x^2 - 37x - 24$
**20.** $8x^2 + 14x + 3$ **21.** $(2x - 5)(2x + 3)$ **22.** $(3x - 2)(2x - 1)$ **23.** $2x(3x + 1)(x - 4)$
**24.** $(2x + 5)(4x^2 - 10x + 25)$

# CHAPTER 3

## Exercises 3.1, Page 103

**1.** Yes **3.** No **5.** Yes **7.** Yes **9.** No **11.** $-12x^4y$ **13.** $30xy^3$ **15.** $4x^2 + 8x$ **17.** $(x + 1)^2$
**19.** $7x(x + y)$ **21.** $5(x + 2)$ **23.** $-8(x + 2)$ **25.** $7(x^2 - 4x + 16)$ **27.** $-x(4x^2 + 6x + 9)$
**29.** $-x(x^2 + 5x + 25)$ **31.** $\frac{3x}{4y}$; $x \neq 0, y \neq 0$ **33.** $\frac{2x^3}{3y^3}$; $x \neq 0, y \neq 0$ **35.** $\frac{17}{5}$; $x \neq 0, y \neq 0$
**37.** $\frac{2}{5}$; $x \neq -2$ **39.** $\frac{3}{2}$; $x \neq \frac{3}{2}$ **41.** $\frac{1}{x - 3}$; $x \neq 0, 3$ **43.** 7; $x \neq 2$ **45.** $-\frac{3}{4}$; $x \neq 3$

**47.** $\frac{4x + y}{3y + 2}$; $y \neq 0, y \neq \frac{2}{3}$ **49.** $\frac{x}{x - 6}$; $x \neq 0, 6$ **51.** $-\frac{1}{2}$; $x \neq 4$ **53.** $\frac{1}{4x}$; $x \neq 0, -3y$

**55.** $\frac{x - 3}{x + 1}$; $x \neq 2, -1$ **57.** $\frac{3 + 2x}{x(2 - x)}$; $x \neq 0, 2, \frac{1}{4}$ **59.** $\frac{x(3x - 1)}{x - 3}$; $x \neq 3, -\frac{1}{3}$ **61.** $\frac{x + 2}{x^2 + 2x + 4}$; $x \neq 2$

**63.** $\frac{3x - 4}{3 - 2x}$; $x \neq \frac{3}{2}, -6$ **65.** $\frac{x - 3}{y - 2}$; $y \neq -2, 2$ **67.** $\frac{x^2 + 5}{x^2 + 2x + 4}$; $x \neq 2$ **69.** $\frac{1}{3x + 1}$; $x \neq -\frac{1}{3}$

## Exercises 3.2, Page 107

**1.** $\frac{ab}{y}$ **3.** $\frac{8x^2y^3}{15}$ **5.** $\frac{3}{x}$ **7.** $\frac{5}{6}$ **9.** $\frac{x + 3}{x}$ **11.** $\frac{x - 1}{x + 1}$ **13.** $-\frac{1}{x - 8}$ or $\frac{1}{8 - x}$ **15.** $\frac{x - 2}{x}$

**17.** $\frac{4(x + 5)}{x(x + 1)}$ **19.** $\frac{x}{(x + 3)(x - 1)}$ **21.** $-\frac{x + 4}{x(x + 1)}$ **23.** $\frac{x + 2y}{(x - 3y)(x - 2y)}$ **25.** $\frac{x - 1}{x(2x - 1)}$

**27.** $\frac{2x - 1}{2x + 1}$ **29.** $\frac{49y^3}{6x^5}$ **31.** $\frac{6y^7}{x^4}$ **33.** $\frac{x}{12}$ **35.** $\frac{6x + 18}{x^2}$ **37.** $\frac{6}{5x}$ **39.** $\frac{3x + 1}{x + 1}$

**41.** $\frac{-(x - 5)}{(x - 2)(x + 7)}$ or $\frac{5 - x}{(x - 2)(x + 7)}$ **43.** $\frac{x - 2}{2x - 1}$ **45.** $\frac{x + 1}{x - 1}$ **47.** $\frac{1}{x}$ **49.** $\frac{x^2 + 4x + 4}{x^2(2x - 5)}$

**51.** $\frac{x^2 - 3x}{(x - 1)^2}$ **53.** $\frac{x^2 + 5x}{2x + 1}$ **55.** 1 **57.** $x - 3$ **59.** $\frac{xy^2 + y^3}{x^2}$

## Exercises 3.3, Page 115

**1.** 3 **3.** 2 **5.** 1 **7.** $\frac{2}{x - 1}$ **9.** 6 **11.** $\frac{-5}{x - 3}$ **13.** $\frac{1}{2}$ **15.** $\frac{x - 2}{x + 2}$ **17.** $\frac{4x + 5}{2(7x - 2)}$

**19.** $\frac{6x + 15}{(x + 3)(x - 3)}$ **21.** $\frac{x^2 - 2x + 4}{(x + 2)(x - 1)}$ **23.** $\frac{x^2 + 3x + 6}{(x + 3)(x - 3)}$ **25.** $\frac{8x^2 + 13x - 21}{6(x + 3)(x - 3)}$ **27.** $\frac{3x^2 - 20x}{(x + 6)(x - 6)}$

**29.** $\frac{-4x}{x - 7}$ **31.** $\frac{4x^2 - x - 12}{(x + 7)(x - 4)(x - 1)}$ **33.** $\frac{4}{(x + 2)(x + 2)(x - 2)}$ **35.** $\frac{-3x^2 + 17x + 15}{(x - 3)(x - 4)(x + 3)}$

**37.** $\frac{4x - 19}{(7x + 4)(x - 1)(x + 2)}$ **39.** $\frac{-7x - 9}{(4x + 3)(x - 2)}$ **41.** $\frac{4x^2 - 41x - 3}{(x + 4)(x - 4)}$ **43.** $\frac{-2x^2 - 4x + 15}{(x - 4)(x - 2)}$

**45.** $\frac{x^2 - 4x - 6}{(x + 2)(x - 2)(x - 1)}$ **47.** $\frac{6x + 2}{(x - 1)(x + 3)}$ **49.** $\frac{5x + 4}{(x + 1)(x + 3)}$ **51.** $\frac{9x^3 - 19x^2 + 22x + 12}{(3x + 1)(x + 3)(5x + 2)(x - 1)}$

**53.** $\frac{5x + 1}{(y - 3)(x + 1)}$ **55.** $\frac{5xy - 8y + 2}{(y - 4)(y + 1)(x - 2)}$ **57.** $\frac{6x^2 - 12x + 5}{2(2x + 1)(4x^2 - 2x + 1)}$

**59.** $\frac{14x - 43}{(x^2 + 6)(x - 5)(x - 2)}$

## Exercises 3.4, Page 121

**1.** $y^2 - 2y + 3$ **3.** $2x^2 - 3x + 1$ **5.** $10x^3 - 11x^2 + x$ **7.** $-4x^2 + 7x - \frac{5}{2}$ **9.** $y^3 - \frac{7}{2}y^2 - \frac{15}{2}y + 4$

**11.** $3x + 4 + \frac{1}{7x - 1}$ **13.** $x^2 + 2x + 2 + \frac{-12}{2x + 3}$ **15.** $7x^2 + 2x + 1$ **17.** $x^2 + 3x + 2 + \frac{-2}{x - 4}$

**19.** $x^2 + x - 3 + \frac{18}{5x + 3}$ **21.** $2x^2 - 8x + 25 + \frac{-98}{x + 4}$ **23.** $3x^2 + 2x - 5 + \frac{-1}{3x - 2}$ **25.** $3x^2 - 2x + 5$

**27.** $3x + 5 + \frac{x - 1}{x^2 + 2}$ **29.** $x^2 + x - 4 + \frac{-8x + 17}{x^2 + 4}$ **31.** $2x + 3 + \frac{6}{3x^2 - 2x - 1}$

**33.** $3x^2 - 10x + 12 + \frac{-x - 14}{x^2 + x + 1}$ **35.** $x^2 - 2x + 7 + \frac{-17x + 14}{x^2 + 2x - 3}$ **37.** $x^2 + 3x + 9$

**39.** $x^3 - x + \frac{x - 1}{x^2 + 1}$ **41.** $x - 9$ **43.** $x^2 - 4x + 33 + \frac{-265}{x + 8}$ **45.** $4x^2 - 6x + 9 + \frac{-17}{x + 2}$

**47.** $x^2 + 7x + 55 + \frac{388}{x - 7}$ **49.** $2x^2 - 2x + 6 + \frac{-27}{x + 3}$ **51.** $x^2 + 2x + 5 + \frac{17}{x - 3}$

**53.** $x^3 + 2x^2 + 6x + 9 + \frac{23}{x - 2}$ **55.** $x^3 + 4x^2 + 16x + 63 + \frac{255}{x - 4}$ **57.** $x^4 + x^3 + x^2 + x + 1$

**59.** $x^5 - 4x^4 + 16x^3 - 66x^2 + 264x - 1056 + \frac{4228}{x + 4}$

## Exercises 3.5, Page 125

**1.** $\frac{4}{5xy}$ **3.** $\frac{8}{7x^2y}$ **5.** $\frac{2x(x + 3)}{2x - 1}$ **7.** $\frac{7}{2(x + 2)}$ **9.** $\frac{x}{x - 1}$ **11.** $\frac{4x}{3(x + 6)}$ **13.** $\frac{7x}{x + 2}$ **15.** $\frac{x}{6}$

**17.** $\frac{24y + 9x}{18y - 20x}$ **19.** $\frac{xy}{x + y}$ **21.** $\frac{y + x}{y - x}$ **23.** $\frac{2x + 2}{x + 2}$ **25.** $\frac{x^2 - 3x}{x - 4}$ **27.** $\frac{x + 1}{2x - 3}$ **29.** $-\frac{2x + h}{x^2(x + h)^2}$

**31.** $-x^2 + 3xy - 2y^2$ **33.** $\frac{2 - x}{x + 2}$

## Exercises 3.6, Page 131

**1.** $x = 7$ **3.** $x = -\frac{74}{9}$ **5.** $x = \frac{1}{4}$ **7.** $x = \frac{3}{2}$ **9.** $x = 4$ **11.** $x = \frac{10}{3}$ **13.** $x = \frac{1}{4}$

**15.** $x = -\frac{3}{16}$ **17.** $x = -3$ **19.** $x = -39$ **21.** $x = \frac{62}{7}$ **23.** $x = -3$ **25.** $x = 2$

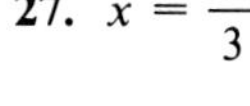

**27.** $x = \frac{2}{3}$ **29.** $x = -2$ **31.** No solution **33.** $x = \frac{7}{11}$ **35.** $x \le -4$ or $x > 0$

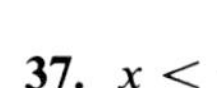

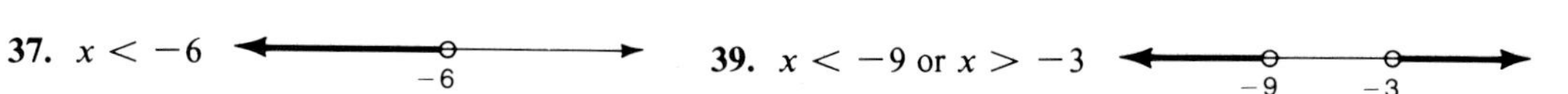


**41.** $2 < x < \frac{5}{2}$ **43.** $x > 7$

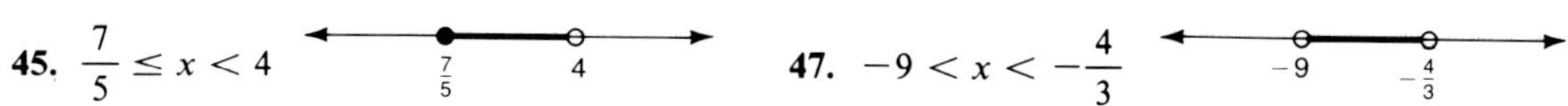


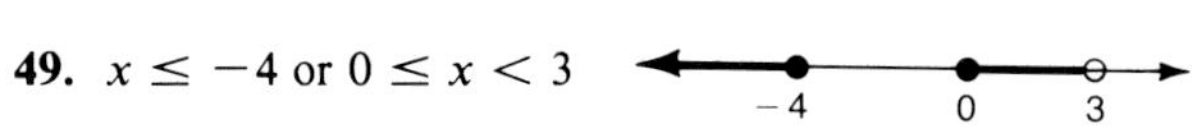

## Exercises 3.7, Page 139

**1.** 10 **3.** 14 **5.** $\frac{12}{7}$ **7.** 15, 9 **9.** 2, 5 **11.** $1500 **13.** 1875 mi **15.** $1\frac{4}{5}$ hr **17.** 6 hr
**19.** 120 mph, 300 mph **21.** 250 mph, 500 mph **23.** 50 mph **25.** 4.5 days, 9 days
**27.** boat—14 mph; current—2 mph

## Chapter 3 Review, Page 141

**1.** $28x^3y^2$ **2.** $5x(3x - 1)$ **3.** $-11x(x + 2)$ **4.** $9y(x^2 - 4x + 16)$ **5.** $\frac{7x^2}{10y^3}$ **6.** $\frac{1}{2x}$ **7.** $\frac{-2x - 3}{x^2 + 2x + 4}$
**8.** $\frac{x + 3}{y - 4}$ **9.** $\frac{3y^3}{20x}$ **10.** $\frac{x^4}{8y^5}$ **11.** $\frac{66}{35x^2}$ **12.** $\frac{5xy - 10x + 3y}{12x^2y^2}$ **13.** $\frac{5x^2}{3}$ **14.** $\frac{1}{(x - 1)(2x + 1)}$
**15.** $\frac{-10x^2 + 19x + 18}{6(x + 2)(x - 2)}$ **16.** $\frac{-3x^2 + 4x + 12}{(x + 1)(x - 1)}$ **17.** $\frac{x + 1}{x^2 + 3x + 9}$ **18.** $\frac{3x^2 - 29x + 52}{(2x - 5)(x + 2)(x - 2)}$ **19.** $\frac{x - 1}{x + 6}$
**20.** $\frac{-6x - 7}{(x - 1)(2x - 7)}$ **21.** $x + 4 + \frac{2}{2x + 3}$ **22.** $2x + 1 + \frac{4}{6x - 5}$ **23.** $x - 8 + \frac{-4x}{x^2 + 4}$
**24.** $x^2 + 3 + \frac{x - 5}{x^2 + x + 2}$ **25.** $x + 15 + \frac{68}{x - 5}$ **26.** $2x^2 + x - 3 + \frac{9}{x + 2}$ **27.** $x^2 + 2x + 12 + \frac{81}{x - 6}$
**28.** $x^2$ **29.** $\frac{x - 1}{(2x - 1)(x + 1)}$ **30.** $\frac{x(x + 2)(x + 3)}{(x + 5)(x - 3)}$ **31.** $x = 10$ **32.** $x = 56$ **33.** $x = 7$
**34.** $x = -\frac{1}{3}$ **35.** $x = 2$ **36.** $x = 2$ **37.** $x = \frac{3}{2}$ **38.** $x = \frac{26}{11}$ **39.** $-\frac{5}{4} < x < 2$ 

**40.** $x < 0$ or $x \geq 1$ 

**41.** $-5 \leq x \leq 1, x > 4$ 

**42.** $\frac{6}{5}$ hr **43.** $\frac{2}{3}$ hr **44.** $37\frac{1}{2}$ mph **45.** 10 mph

## Chapter 3 Test, Page 143

**1.** $\frac{x}{x + 4}$ **2.** $-\frac{x(x - 5)}{4 + x}$ or $\frac{x(5 - x)}{4 + x}$ **3.** $x^2 + x - 2$ **4.** $\frac{x + 3}{x + 4}$ **5.** $\frac{3x - 2}{3x + 2}$ **6.** $\frac{-2x^2 - 13x}{(x + 5)(x + 2)(x - 2)}$
**7.** $\frac{-2x^2 - 7x + 18}{(3x + 2)(x - 4)(x + 1)}$ **8.** $2x^2$ **9.** $\frac{x^2 - 7x + 1}{(x + 3)(x - 3)}$ **10.** $2x - 3 - \frac{x}{2x^2 + 1}$
**11.** $2x^2 + 8x + 43 + \frac{166}{x - 4}$ **12.** $\frac{3x}{2 - x}$ **13.** $x = -\frac{7}{2}$ **14.** $x = -1$
**15.** $x \leq -\frac{5}{2}$ or $x > 3$ 

**16.** $x < -\frac{5}{3}$ or $x > -\frac{1}{2}$ 

**17.** $\frac{2}{7}$ **18.** 4 hr **19.** 42 mph—Carlo's; 57 mph—Mario's **20.** 11 mph

## Cumulative Review (3), Page 144

**1.** $x = -\frac{9}{8}$ **2.** $x = \frac{29}{11}$ **3.** $x = -\frac{4}{3}, x = 4$ **4.** $\left(-\infty, -\frac{1}{4}\right]$

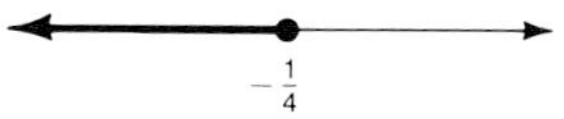

**5.** $\left(-\infty, -\frac{50}{3}\right)$

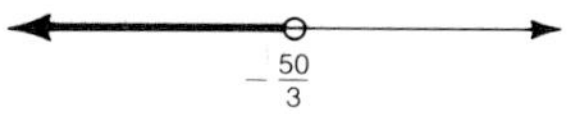

**6.** $\left[-\frac{8}{3}, 4\right]$

**7.** $\frac{1}{9}x^{-4}y^6$ or $\frac{y^6}{9x^4}$ **8.** $\frac{x^6}{y^6}$ **9.** $\frac{3}{2x^4y}$ **10.** 6.8 **11.** $1.05 \times 10^6$ **12.** $(3x + 2)(5x + 4)$

**13.** $(x + 4)(2x^2 + 3)$ **14.** $2(2x - 3)(4x^2 + 6x + 9)$ **15.** $x^{-3}(2x + 1)(x - 3)$ **16.** $4x^2 + 3x + 8 + \frac{15}{x - 2}$

**17.** $\frac{2x - 1}{(x + 1)(x - 1)}$ **18.** $\frac{x^2 + 15x - 26}{(x - 4)(x + 2)(2x + 3)}$ **19.** $\frac{2}{3}$ hr or 40 minutes **20.** 120 sq yd

## CHAPTER 4

### Exercises 4.1, Page 150

**1.** 3 **3.** $\frac{1}{10}$ **5.** $-7$ **7.** $\frac{2}{5}$ **9.** 8 **11.** $-5$ **13.** 4 **15.** $\frac{1}{4}$ **17.** $\frac{5}{4}$ **19.** Undefined

**21.** $-\frac{1}{1000}$ **23.** $\frac{1}{27}$ **25.** $\frac{9}{16}$ **27.** $\frac{1}{4}$ **29.** $-\frac{7}{1000}$ **31.** 64 **33.** $x^2$ **35.** $81x^2$ **37.** $\frac{1}{2}x^{-3/4}$

**39.** $3x^{11/3}$ **41.** $x^{1/15}$ **43.** $x^{7/12}$ **45.** $x^{5/9}$ **47.** $x^{-9/8}$ **49.** 1 **51.** $x^2y^{-6/5}$ or $\frac{x^2}{y^{6/5}}$ **53.** $8x^{3/2}y$ **55.** $x^5$

**57.** $x^{-1}y^{3/2}z^2$ **59.** $\frac{1}{3}x^2y^{-1/2}$ **61.** $-3x^{2/3}yz$ **63.** $x^{-5/6}y^{2/3}$ **65.** $\frac{5}{8}x^{7/5}y^{-23/10}$ or $\frac{5x^{7/5}}{8y^{23/10}}$ **67.** $x^{7/4}y^{-3/4}$ or $\frac{x^{7/4}}{y^{3/4}}$

**69.** $x^{73/30}$

### Exercises 4.2, Page 155

**1.** $\sqrt[5]{x^3}$ **3.** $\sqrt[8]{y^3}$ **5.** $-\sqrt[5]{x^4}$ **7.** $4\sqrt[7]{(xy)^2}$ **9.** $\sqrt[3]{(8y)^4}$ **11.** $\sqrt[3]{x^2} \cdot \sqrt{y}$ or $\sqrt[6]{x^4y^3}$ **13.** $x^{1/3}$ **15.** $x^{1/5}$

**17.** $4y^{2/5}$ **19.** $(xy^2)^{1/3}$ or $x^{1/3}y^{2/3}$ **21.** $(7x^3)^{1/4}$ or $7^{1/4}x^{3/4}$ **23.** $(x^2y^3)^{1/6}$ or $x^{1/3}y^{1/2}$ **25.** $2\sqrt{3}$ **27.** $7\sqrt{2}$

**29.** $-9\sqrt{2}$ **31.** $2\sqrt[3]{2}$ **33.** $3\sqrt[3]{4}$ **35.** $2x\sqrt[3]{x}$ **37.** $2x\sqrt[4]{2x}$ **39.** $\frac{2}{3}\sqrt{3}$ **41.** $\frac{2}{x}\sqrt{x}$ **43.** $-\frac{y}{4x}\sqrt{10x}$

**45.** $\frac{1}{y^2}\sqrt[3]{7xy^2}$ **47.** $\frac{y}{3x^2}\sqrt[4]{6x^2}$ **49.** $-3$ **51.** $-9|x|$ **53.** $-3y^3\sqrt[3]{3}$ **55.** $9y^2$ **57.** $4x^2y^2\sqrt{2}$

**59.** $3y^2\sqrt[3]{4x}$ **61.** $x^{5/6} = \sqrt[6]{x^5}$ **63.** $x^{7/12} = \sqrt[12]{x^7}$ **65.** $\frac{1}{x^{1/30}}$ or $x^{-1/30} = \frac{1}{\sqrt[30]{x}}$ **67.** $x^{1/6} = \sqrt[6]{x}$ **69.** $x^{1/9} = \sqrt[9]{x}$

### Exercises 4.3, Page 159

**1.** $-6\sqrt{2}$ **3.** $5\sqrt{x}$ **5.** $-3\sqrt[3]{7x^2}$ **7.** $10\sqrt{3}$ **9.** $-7\sqrt{2}$ **11.** $20\sqrt{3}$ **13.** $3\sqrt{6} + 7\sqrt{3}$ **15.** $20x\sqrt{5x}$

**17.** $14\sqrt{5} - 3\sqrt{7}$ **19.** $11\sqrt{2x} + 14\sqrt{3x}$ **21.** $4\sqrt[3]{5} - 13\sqrt[3]{2}$ **23.** $x^2\sqrt{2x}$ **25.** $4x^2y\sqrt{x}$

**27.** $3x - 9\sqrt{3x} + 8$ **29.** $24 + 10\sqrt{2x} + 2x$ **31.** 2 **33.** $13 + 4\sqrt{10}$ **35.** $-3$

**37.** $6 + \sqrt{30} - 2\sqrt{3} - \sqrt{10}$ **39.** $5 - \sqrt{33}$ **41.** $49x - 2$ **43.** $9x + 6\sqrt{xy} + y$ **45.** $12x + 7\sqrt{2xy} + 2y$
**47.** $-\frac{3\sqrt{3} + 15}{22}$ **49.** $\frac{3\sqrt{6} - \sqrt{42}}{2}$ **51.** $4\sqrt{5} + 4\sqrt{3}$ **53.** $\frac{-5\sqrt{2} - 4\sqrt{5}}{3}$ **55.** $\frac{59 + 21\sqrt{5}}{44}$
**57.** $8 + 3\sqrt{6}$ **59.** $\frac{x^2 + 4x\sqrt{y} + 4y}{x^2 - 4y}$

### Exercises 4.4, Page 163

**1.** Real part is 4; imaginary part is $-3$. **3.** Real part is $-11$; imaginary part is $\sqrt{2}$. **5.** Real part is $\frac{2}{3}$, imaginary part is $\sqrt{17}$. **7.** Real part is $\frac{4}{5}$; imaginary part is $\frac{7}{5}$. **9.** Real part is $\frac{3}{8}$; imaginary part is 0.
**11.** $7i$ **13.** $-8i$ **15.** $21\sqrt{3}$ **17.** $10i\sqrt{6}$ **19.** $-12i\sqrt{3}$ **21.** $11\sqrt{2}$ **23.** $10i\sqrt{10}$ **25.** $6 + 2i$
**27.** $1 + 7i$ **29.** $2 - 6i$ **31.** $14i$ **33.** $(3 + \sqrt{5}) - 6i$ **35.** $5 + (\sqrt{6} + 1)i$ **37.** $\sqrt{3} - 5$
**39.** $-2 - 5i$ **41.** $11 - 16i$ **43.** 2 **45.** $3 + 4i$ **47.** $x = 6, y = -3$ **49.** $x = -2, y = \sqrt{5}$
**51.** $x = \sqrt{2} - 3, y = 1$ **53.** $x = 2, y = -8$ **55.** $x = 2, y = -6$ **57.** $x = 3, y = 10$
**59.** $x = -\frac{4}{3}, y = -3$

### Exercises 4.5, Page 167

**1.** $16 + 24i$ **3.** $-7\sqrt{2} + 7i$ **5.** $3 + 12i$ **7.** $1 - i\sqrt{3}$ **9.** $5\sqrt{2} + 10i$ **11.** $2 + 8i$ **13.** $-7 - 11i$
**15.** $13 + 0i$ **17.** $-3 - 7i$ **19.** $5 + 12i$ **21.** $5 - i\sqrt{3}$ **23.** $21 + 0i$ **25.** $23 - 10i\sqrt{2}$
**27.** $(2 + \sqrt{10}) + (2\sqrt{2} - \sqrt{5})i$ **29.** $(9 - \sqrt{30}) + (3\sqrt{5} + 3\sqrt{6})i$ **31.** $0 + 3i$ **33.** $0 - \frac{5}{4}i$
**35.** $-\frac{1}{4} + \frac{1}{2}i$ **37.** $-\frac{4}{5} + \frac{8}{5}i$ **39.** $\frac{24}{25} + \frac{18}{25}i$ **41.** $-\frac{1}{13} + \frac{5}{13}i$ **43.** $-\frac{1}{29} - \frac{12}{29}i$ **45.** $-\frac{17}{26} - \frac{7}{26}i$
**47.** $\frac{4 + \sqrt{3}}{4} + \left(\frac{4\sqrt{3} - 1}{4}\right)i$ **49.** $-\frac{1}{7} + \frac{4\sqrt{3}}{7}i$ **51.** $0 + i$ **53.** $-1 + 0i$ **55.** $0 + i$ **57.** $1 + 0i$
**59.** $0 - i$

### Chapter 4 Review, Page 169

**1.** 11 **2.** $\frac{1}{4}$ **3.** $\frac{8}{9}$ **4.** 27 **5.** $-\frac{1}{9}$ **6.** Undefined **7.** $\frac{1}{8}$ **8.** $16x^{2/3}$ **9.** $125x^{3/2}$ **10.** $5x^{7/3}$
**11.** $3x^{1/6}$ **12.** $x^{7/4}$ **13.** $x^{7/6}$ **14.** $\frac{3x}{y^2}$ **15.** $\frac{1}{8}x^{-3}y^{-3}$ or $\frac{1}{8x^3y^3}$ **16.** $x^{-1}y^{1/3}$ or $\frac{y^{1/3}}{x}$ **17.** $x^{5/12}y^{7/12}$
**18.** $\sqrt[3]{x^8}$ **19.** $-\sqrt[4]{x^3}$ **20.** $4\sqrt[3]{xy^2}$ **21.** $7x\sqrt[5]{y^2}$ **22.** $6x^{1/2}$ **23.** $3x^{3/4}$ **24.** $-2x^{2/3}y^{1/3}$ **25.** $11x^{3/5}y^{2/5}$
**26.** $x^{3/4}$ or $\sqrt[4]{x^3}$ **27.** $x^{-7/12}$ or $\frac{1}{\sqrt[12]{x^7}}$ **28.** $x^{13/12}$ or $x\sqrt[12]{x}$ **29.** $5\sqrt{5}$ **30.** 14 **31.** $5\sqrt[3]{2}$ **32.** $4\sqrt[3]{3}$
**33.** $24\sqrt{2}$ **34.** $12\sqrt{5}$ **35.** $3xy\sqrt[3]{2x^2y}$ **36.** $4x^2y^2\sqrt[3]{2xy^2}$ **37.** $8|x|y^2$ **38.** $6\sqrt{3}\,x^2|y|$ **39.** $\frac{\sqrt{2}}{2}$

**40.** $\sqrt{3}$ **41.** $\dfrac{7\sqrt{5}-7\sqrt{2}}{3}$ **42.** $\dfrac{15-3\sqrt{2}}{23}$ **43.** $35\sqrt{2}$ **44.** $-2\sqrt{3}$ **45.** $13\sqrt[3]{2}$ **46.** $4\sqrt[3]{3}$
**47.** $16x\sqrt{2y}$ **48.** $3x\sqrt{2xy}-3y\sqrt{3xy}$ **49.** $8+2\sqrt{10}$ **50.** $\sqrt{15}+\sqrt{10}+6\sqrt{3}+6\sqrt{2}$ **51.** $\sqrt{15}-4$
**52.** $\dfrac{3\sqrt{2}+2\sqrt{3}+\sqrt{6}+2}{4}$ **53.** $2+6i$ **54.** $-6+i$ **55.** $-1-5i$ **56.** $-7-8i$ **57.** $14-5i$
**58.** $11-23i$ **59.** $\dfrac{8}{13}+\dfrac{1}{13}i$ **60.** $-\dfrac{7}{2}+\dfrac{3}{2}i$ **61.** $-\dfrac{8}{17}-\dfrac{19}{17}i$ **62.** $0-i$ **63.** $0+i$ **64.** $1+0i$
**65.** $x=-1, y=3$ **66.** $x=1, y=5$ **67.** $x=\dfrac{2}{3}, y=-5$ **68.** $x=7, y=17$
**69.** $x=3+2\sqrt{3}, y=4$ **70.** $x=\dfrac{7}{13}, y=\dfrac{5}{39}$

## *Chapter 4 Test, Page 170*

**1.** 4 **2.** $\dfrac{1}{27}$ **3.** $4x^{7/6}$ **4.** $7x^{1/4}y^{-1/3}$ or $\dfrac{7x^{1/4}}{y^{1/3}}$ **5.** $8x^{-3}y^{3/2}$ or $\dfrac{8y^{3/2}}{x^3}$ **6.** $\sqrt[3]{4x^2}$ **7.** $8^{1/6}x^{1/3}y^{2/3}$ or $2^{1/2}x^{1/3}y^{2/3}$
**8.** $\sqrt[12]{x^{11}}$ **9.** $4\sqrt{7}$ **10.** $2y\sqrt[3]{6x^2y^2}$ **11.** $\dfrac{y}{4x^2}\sqrt{10x}$ **12.** $17\sqrt{3}$ **13.** $(7x-6)\sqrt{x}$ **14.** $8\sqrt[3]{3}$
**15.** $5-2\sqrt{6}$ **16.** $-(\sqrt{3}+\sqrt{5})$ **17.** $16+4i$ **18.** $-5+16i$ **19.** $23-14i$ **20.** $\dfrac{8}{13}-\dfrac{1}{13}i$
**21.** $x=\dfrac{11}{12}, y=3$ **22.** $0-i$

## *Cumulative Review (4), Page 171*

**1.** $2x^3-4x^2+8x-4$ **2.** $6x^2+19x-7$ **3.** $-5x^2+18x+8$ **4.** $n=\dfrac{s-a}{d}+1$ or $\dfrac{s-a+d}{d}$
**5.** $p=\dfrac{A}{1+rt}$ **6.** $(4x+3)(3x-4)$ **7.** $(7+2x)(4-x)$ **8.** $5(x-4)(x^2+4x+16)$
**9.** $(x+4)(x+1)(x-1)$ **10.** $\dfrac{4x^2-12x-1}{(x+3)(x-2)(x-5)}$ **11.** $\dfrac{4x^2+6x+17}{(2x+1)(2x-1)(x+3)}$ **12.** $\dfrac{x^2-1}{2x-4}$
**13.** $\dfrac{x+6}{x-1}$ **14.** $x=-\dfrac{5}{7}$ **15.** $x=\dfrac{83}{4}$ **16.** $\left(-2,\dfrac{3}{5}\right]$ 

**17.** $\left[-2,\dfrac{1}{3}\right)$ 

**18.** $\dfrac{1}{16}$ **19.** $x^{7/3}$ **20.** $12\sqrt{2}$ **21.** $2x^2y^3\sqrt[3]{2y}$ **22.** $23-14i$
**23.** $\dfrac{-8-19i}{17}$ **24.** 10 mph **25.** 7.5 hours

## CHAPTER 5

### Exercises 5.1, Page 177

**1.** $y^2 - y - 6 = 0$ **3.** $x^2 - \frac{5}{4}x + \frac{3}{8} = 0$ **5.** $x^2 - 7 = 0$ **7.** $x^2 - 2x - 2 = 0$ **9.** $y^2 + 4y - 1 = 0$
**11.** $x^2 + 16 = 0$ **13.** $y^2 + 6 = 0$ **15.** $x^2 - 4x + 5 = 0$ **17.** $x^2 - 2x + 3 = 0$ **19.** $x = -4, x = -9$
**21.** $x = 6, x = 8$ **23.** $x = -5, x = 10$ **25.** $x = -7, x = 7$ **27.** $x = \frac{2}{3}, x = 5$ **29.** $x = -\frac{4}{3}, x = -\frac{1}{2}$
**31.** $x = -5i, x = 5i$ **33.** $z = -\frac{7}{2}i, z = \frac{7}{2}i$ **35.** $z = \frac{1}{2}, z = 3$ **37.** $x = -\frac{3}{4}, x = \frac{2}{3}$
**39.** $x = -\frac{5}{2}, x = \frac{7}{3}$ **41.** $y = -\frac{4}{5}, y = \frac{4}{5}$ **43.** $x = -6, x = 8$ **45.** $x = 2, x = 6$
**47.** $x = -11, x = 3$ **49.** $y = \frac{1}{6}, y = \frac{2}{3}$ **51.** $x = -\frac{2}{9}, x = \frac{6}{7}$ **53.** $x = -3, x = -2, x = 0$
**55.** $x = 0, x = -\frac{5}{2}i, x = \frac{5}{2}i$ **57.** $x = -\frac{5}{2}, x = 0, x = \frac{7}{3}$ **59.** $x = -\frac{3}{2}, x = 0, x = \frac{7}{5}$

### Exercises 5.2, Page 181

**1.** $x^2 - 12x + \underline{36} = (x - 6)^2$ **3.** $x^2 + 6x + 9 = (x + 3)^2$ **5.** $x^2 - 5x + \frac{25}{4} = \left(x - \frac{5}{2}\right)^2$
**7.** $y^2 + y + \frac{1}{4} = \left(y + \frac{1}{2}\right)^2$ **9.** $x^2 + \frac{1}{3}x + \frac{1}{36} = \left(x + \frac{1}{6}\right)^2$ **11.** $x = 5, x = -1$ **13.** $x = -1 \pm \sqrt{5}$
**15.** $x = -3 \pm \sqrt{3}$ **17.** $x = 3 \pm 2i$ **19.** $x = -2 \pm i\sqrt{7}$ **21.** $x = 1, x = -5$ **23.** $y = -1 \pm \sqrt{6}$
**25.** $5 \pm \sqrt{22}$ **27.** $x = 3 \pm i$ **29.** $x = 1, x = 11$ **31.** $z = \frac{-3 \pm \sqrt{29}}{2}$ **33.** $x = \frac{-5 \pm \sqrt{17}}{2}$
**35.** $x = 1 \pm 2i$ **37.** $y = \frac{-3 \pm i\sqrt{3}}{2}$ **39.** $x = 2, x = -3$ **41.** $x = \frac{5 \pm \sqrt{10}}{3}$ **43.** $y = \frac{-5 \pm \sqrt{61}}{6}$
**45.** $x = \frac{-1 \pm i\sqrt{11}}{6}$ **47.** $x = -4, x = -\frac{1}{2}$ **49.** $x = \frac{1 \pm i\sqrt{7}}{4}$

### Exercises 5.3, Page 186

**1.** 68, two real solutions **3.** 0, one real solution **5.** $-8$, two nonreal, complex solutions **7.** 4, two real solutions
**9.** 19,600, two real solutions **11.** $-11$, two nonreal, complex solutions **13.** $k < 16$ **15.** $k = \frac{81}{4}$ **17.** $k > 3$
**19.** $k > -\frac{1}{36}$ **21.** $k = \frac{49}{48}$ **23.** $k > \frac{4}{3}$ **25.** $x = \frac{-3 \pm \sqrt{29}}{2}$ **27.** $x = \frac{5 \pm \sqrt{17}}{2}$ **29.** $x = \frac{-7 \pm \sqrt{33}}{4}$
**31.** $x = 1, x = -\frac{1}{6}$ **33.** $x = \frac{\pm 2\sqrt{6}}{3}$ **35.** $x = \frac{9 \pm \sqrt{65}}{2}, x = 0$ **37.** $x = \frac{-3 \pm \sqrt{5}}{2}, x = 0$
**39.** $x = -1, x = 4$ **41.** $z = \frac{-4 \pm i\sqrt{2}}{2}$ **43.** $x = \pm\sqrt{7}$ **45.** $x = 0, x = -1$ **47.** $x = \frac{7 \pm i\sqrt{51}}{10}$

**49.** $x = -2, x = \frac{5}{3}$ **51.** $x = \pm\frac{3}{2}i$ **53.** $x = \frac{2 \pm \sqrt{3}}{3}$ **55.** $x = \frac{7 \pm i\sqrt{287}}{12}$ **57.** $x = \frac{2 \pm \sqrt{2}}{2}$
**59.** $x = \frac{-7 \pm \sqrt{17}}{4}$

## Exercises 5.4, Page 191

**1.** 6, 13 **3.** 4, 14 **5.** $-6, -11$ **7.** $-14, -12$ or 12, 14 **9.** 10, 11, 12 **11.** $7\sqrt{2}$ cm, $7\sqrt{2}$ cm
**13.** Mel—30 mph; John—40 mph **15.** 4 ft by 10 ft **17.** 32 cm by 12 cm × 4 cm **19.** 40 seats
**21.** 17 in., 6 in. **23.** 3 in. **25.** 2 amp or 8 amp **27.** 45¢ or 35¢ **29.** Shirt—$13; trousers—$22
**31.** 64 mph **33.** Sam—$7\frac{1}{2}$ hr; Bob—$12\frac{1}{2}$ hr **35.** 150 mph **37.** 190 reserved; 150 general
**39.** **(a)** $6\frac{3}{4}$ sec **(b)** 3 sec **(c)** $3\frac{3}{4}$ sec **41.** **(a)** 7 sec **(b)** 2 sec **(c)** 12 sec

## Exercises 5.5, Page 198

**1.** $x = 3$ **3.** $x = 13$ **5.** $x = 3$ **7.** $x = 2$ **9.** $x = -4, x = 1$ **11.** $x = -5, x = \frac{5}{2}$ **13.** $x = -2$
**15.** $x = \pm 5$ **17.** 2 **19.** 4 **21.** 0 **23.** 5 **25.** No solution **27.** 4 **29.** $-1, 3$ **31.** 5 **33.** 2
**35.** 6 **37.** $x = -4$ **39.** $x = 12$

## Exercises 5.6, Page 202

**1.** $x = \pm 2, x = \pm 3$ **3.** $x = \pm 2, x = \pm\sqrt{5}$ **5.** $y = \pm\sqrt{7}, y = \pm 2i$ **7.** $y = \pm\sqrt{5}, y = \pm i\sqrt{5}$
**9.** $z = \frac{1}{6}, z = -\frac{1}{4}$ **11.** No solution **13.** $x = 1, x = 4$ **15.** $x = \frac{1}{8}, x = -8$ **17.** $x = \frac{1}{25}$
**19.** $x = -\frac{1}{3}, x = \frac{3}{8}$ **21.** $x = 0, x = -27, x = -8$ **23.** $x = 1, x = 2$ **25.** $x = -3, x = -\frac{7}{2}$
**27.** $x = 0, x = 8$ **29.** $x = -3, x = \frac{3}{2}$ **31.** $x = \pm\sqrt{1 + i}, x = \pm\sqrt{1 - i}$ **33.** $x = \pm\sqrt{1 + 3i}$,
$x = \pm\sqrt{1 - 3i}$ **35.** $x = \pm\sqrt{2 - i\sqrt{3}}, x = \pm\sqrt{2 + i\sqrt{3}}$ **37.** $x = \pm\frac{1}{5}\sqrt{5}, x = \pm 1$ **39.** $x = \pm\frac{1}{2}\sqrt{6}$,
$x = \pm\frac{1}{3}i$ **41.** $x = -\frac{1}{4}$ **43.** $x = -4, x = 1$ **45.** $x = -\frac{1}{2}$ **47.** $x = -7, x = -\frac{3}{2}$ **49.** $x = \frac{26}{5}$

## Exercises 5.7, Page 206

**1.** $-2 < x < 6$ 

**3.** $x < \frac{2}{3}$ or $x > 5$ 

**5.** $x \leq -7$ or $x \geq \frac{5}{2}$ 

**7.** $-2 \leq x \leq -\frac{1}{3}$ 

**9.** $x < -\frac{4}{3}$ or $0 < x < 5$ **11.** $x = -2$

**13.** $x < -\frac{5}{2}$ or $x > 3$ **15.** $-\frac{1}{4} < x < \frac{3}{2}$

**17.** $x \leq \frac{1}{2}$ or $x \geq 2$ **19.** $-\frac{2}{3} < y < -\frac{1}{2}$ **21.** $z \neq \frac{5}{2}$ **23.** $-\frac{5}{2} \leq x \leq \frac{7}{4}$ **25.** $-1 < x < 0$ or $x > 3$

**27.** $0 < x < 1$ or $x > 4$ **29.** $x < -1$ or $x > 4$ **31.** $x < -2$, $-1 < x < 1$, or $x > 2$

**33.** $-3 \leq y \leq -2$ or $2 \leq y \leq 3$ **35.** $x \leq -4$ or $x \geq 2$ **37.** $-\frac{2}{3} < x < 2$

**39.** $x < -1 - \sqrt{5}$ or $x > -1 + \sqrt{5}$ **41.** $x \leq -3 - \sqrt{2}$ or $x \geq -3 + \sqrt{2}$

**43.** $\frac{-5 - \sqrt{13}}{6} < x < \frac{-5 + \sqrt{13}}{6}$ **45.** $x \leq -\frac{4}{5}$ or $x \geq 0$ **47.** $x < -3$ or $0 < x < 3$

**49.** $1 \leq x < 2$ or $x \geq 3$ **51.** All reals **53.** No solution

## Chapter 5 Review, Page 208

**1.** $x = -\frac{7}{2}, x = \frac{7}{2}$ **2.** $x = -\frac{3}{2}, x = -\frac{1}{4}$ **3.** $x = -\frac{7}{3}, x = 0$ **4.** $x = -3, x = \frac{5}{2}$

**5.** $x = -\frac{3}{5}, x = 0, x = 2$ **6.** $4x^2 - 5x - 6 = 0$ **7.** $x^2 - 2x - 4 = 0$ **8.** $x^2 + 6x + 11 = 0$

**9.** $y^2 + 4y + 20 = 0$ **10.** $x^2 + 10x + \underline{25} = (x + 5)^2$ **11.** $y^2 - 7y + \frac{49}{4} = \left(x - \frac{7}{2}\right)^2$

**12.** $3y^2 + 4y + \frac{4}{3} = 3\left(x + \frac{2}{3}\right)^2$ **13.** 73, two real solutions **14.** 0, one real solution **15.** $-4$, two nonreal, complex solutions **16.** $-144$, two nonreal, complex solutions **17.** $k < \frac{1}{16}$ **18.** $k = \pm 6\sqrt{2}$ **19.** $k > \frac{25}{8}$

**20.** $x = \frac{-3 \pm \sqrt{17}}{2}$ **21.** $x = \frac{-3 \pm \sqrt{29}}{2}$ **22.** $z = \frac{5 \pm i\sqrt{3}}{2}$ **23.** $x = \frac{-2 \pm \sqrt{6}}{2}$ **24.** $x = \frac{3 \pm 2\sqrt{6}}{5}$

**25.** $x = \frac{-7 \pm \sqrt{33}}{4}$ **26.** $y = \frac{3}{4}, y = -\frac{6}{7}$ **27.** $y = \frac{-3 \pm \sqrt{21}}{4}$ **28.** $x = \frac{-5 \pm i\sqrt{23}}{4}$

**29.** $y = \frac{-1 \pm i\sqrt{11}}{6}$ **30.** $y = -\frac{2}{7}, y = \frac{2}{3}$ **31.** $x = 0, x = 4$ **32.** $x = 6$ **33.** $x = \pm\sqrt{6}, x = \pm i\sqrt{2}$

**34.** $x = \frac{9}{4}$ **35.** $x = 4$ **36.** $x = 0, x = 11$ **37.** $x = \frac{1}{2}, x = -7$ **38.** $x \leq -\frac{5}{4}$ or $x \geq 6$

**39.** $-\frac{1}{2} < x < \frac{5}{3}$ **40.** $2 - \sqrt{11} < x < 2 + \sqrt{11}$ **41.** $x \leq \frac{1}{3} - \frac{1}{3}\sqrt{7}$ or $x \geq \frac{1}{3} + \frac{1}{3}\sqrt{7}$

**42.** $x < -3$ or $0 < x < \frac{5}{6}$ **43.** $x < \frac{-1 - \sqrt{33}}{4}$ or $0 < x < \frac{-1 + \sqrt{33}}{4}$ **44.** $x < 0$ or $1 \leq x \leq 3$

**45.** 12, 14, 16 **46.** Base—19 cm; alt.—8 cm **47.** $2\frac{1}{2}$ hr after Jim leaves **48.** \$240 or \$280 **49.** \$22

**50.** 50 mph

## Chapter 5 Test, Page 209

**1.** $x = -\frac{1}{7}, x = 1$ **2.** $x = -\frac{1}{2}, x = 0, x = \frac{1}{2}$ **3.** $x = -5, x = \frac{5}{2}$ **4.** $x^2 - 2x - 1 = 0$

**5.** $x = -2 \pm \sqrt{3}$ **6.** Two real solutions **7.** $k = \pm 2\sqrt{6}$ **8.** $x = \frac{-1 \pm i\sqrt{7}}{4}$ **9.** $x = \frac{3 \pm \sqrt{41}}{4}$

**10.** $x = \frac{2 \pm i\sqrt{2}}{2}$ **11.** $x = 1$ **12.** $x = 0, x = 36$ **13.** $x = \pm 1, x = \pm 3$ **14.** $x = \frac{3}{2}, x = -1$

**15.** $x = 0, x = 1$ **16.** $-2 \leq x < -1$ or $x \geq 2$ **17.** $x < -5$ or $2 < x < 4$ **18.** 7, 18 **19.** 12 in. by 16 in.

**20.** Going—50 mph; returning—40 mph

## Cumulative Review (5), Page 210

**1.** $x = 6$ **2.** $x = \frac{7}{8}$ **3.** $\left[-\frac{19}{2}, +\infty\right)$

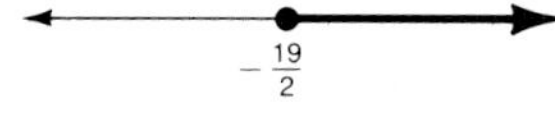

**4.** $\left(-\infty, \frac{7}{5}\right]$ $\frac{7}{5}$

**5.** $y = -\frac{3}{2}x + 3$ **6.** $y = -\frac{3}{2}x + 10$ **7.** $x^{-5}$ or $\frac{1}{x^5}$ **8.** $x^3y^6$ **9.** $9x^{-2}y^{1/2}$ or $\frac{9y^{1/2}}{x^2}$ **10.** $\frac{27}{8}x^2y^{-1}$ or $\frac{27x^2}{8y}$

**11.** $-3x^2y^2\sqrt[3]{y^2}$ **12.** $2x^2y^3\sqrt[4]{2xy^3}$ **13.** $\frac{9}{2}\sqrt{2}$ **14.** $\frac{4}{3}\sqrt{3}$ **15.** $x = -\frac{3}{2}, x = \frac{2}{5}$ **16.** $x = \frac{-7 \pm \sqrt{33}}{8}$

**17.** $x = -1$ **18.** $x = -2, x = \frac{4}{3}$ **19.** \$35 **20.** 20 m by 26 m

## CHAPTER 6

### Exercises 6.1, Page 217

**1.** **(a)** (0,5) **(b)** $\left(\frac{5}{2}, 0\right)$ **(c)** (−2,9) **(d)** (1,3) **3.** **(a)** (0,−4) **(b)** $\left(\frac{4}{3}, 0\right)$ **(c)** (2,2) **(d)** (3,5)

**5.** **(a)** (0,5) **(b)** $\left(\frac{5}{2}, 0\right)$ **(c)** (2,1) **(d)** (−1,7) **7.** **(a)** (0,−3) **(b)** (2,0) **(c)** (−2,−6) **(d)** (4,3)

**9.**

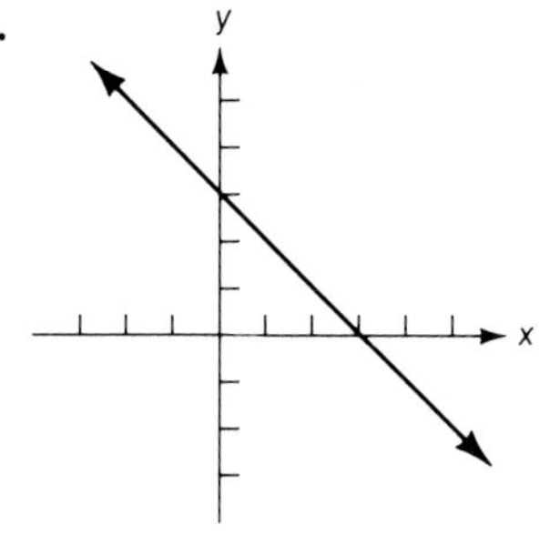


**11.**

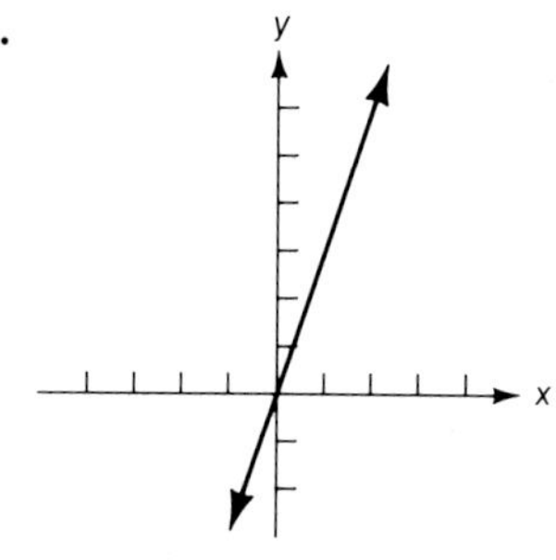


**13.**

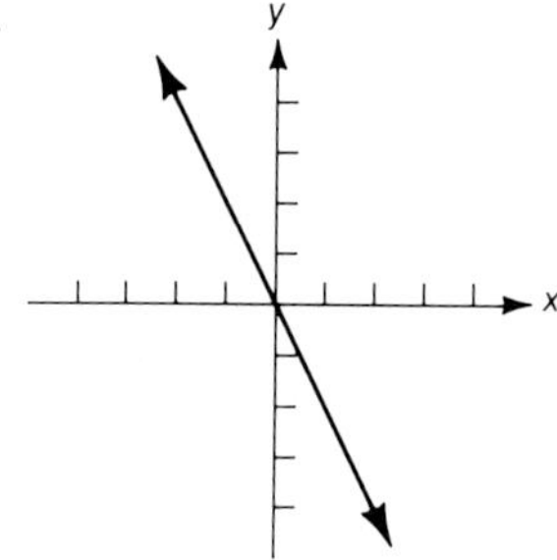

**15.**

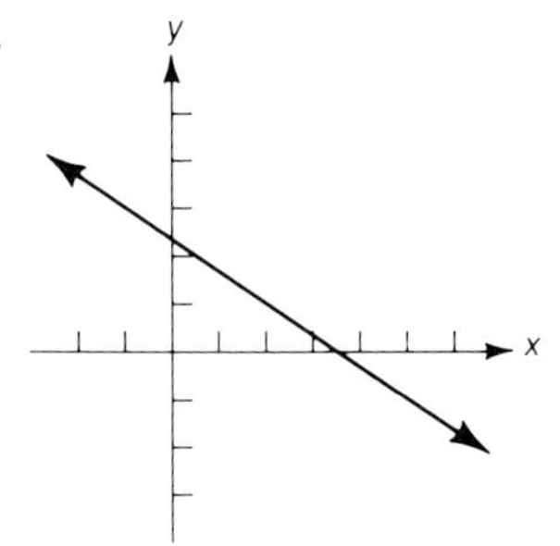


**17.**

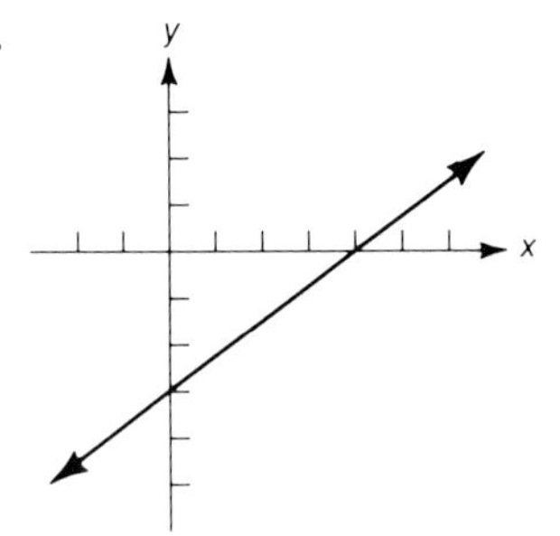


**19.**

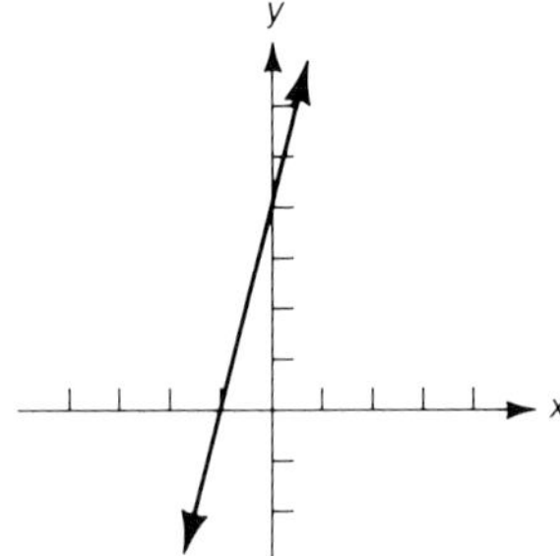


**21.**

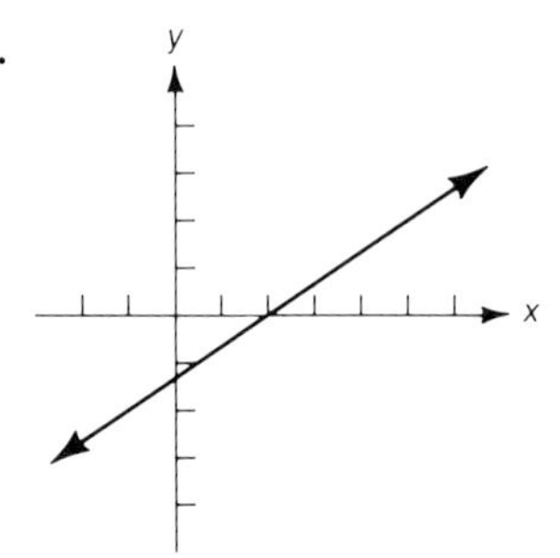


**23.**

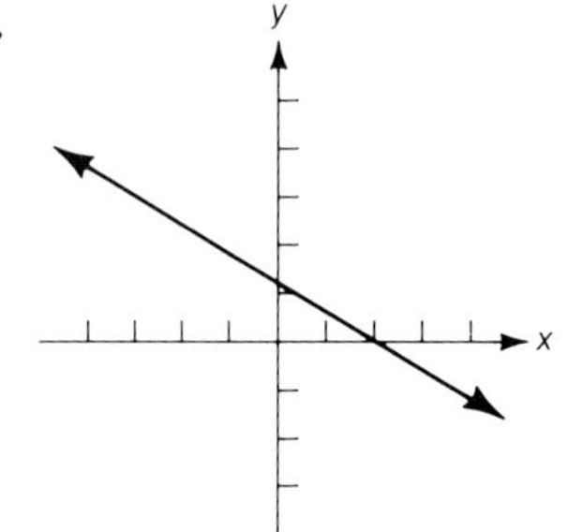


**25.**

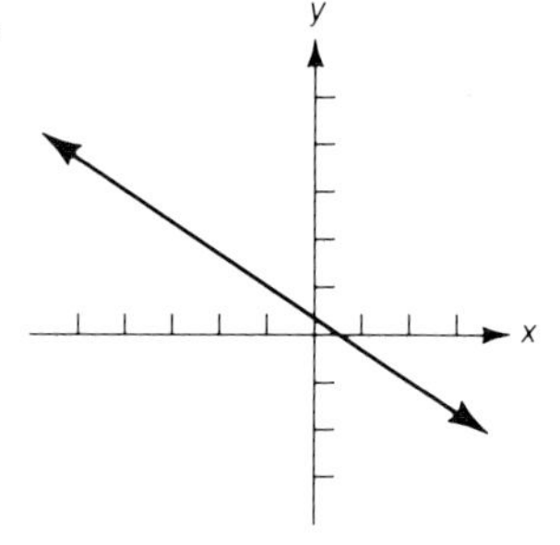


**27.**

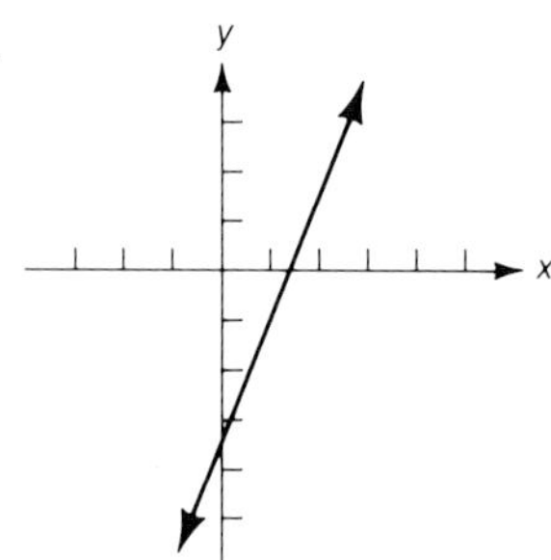


**29.**

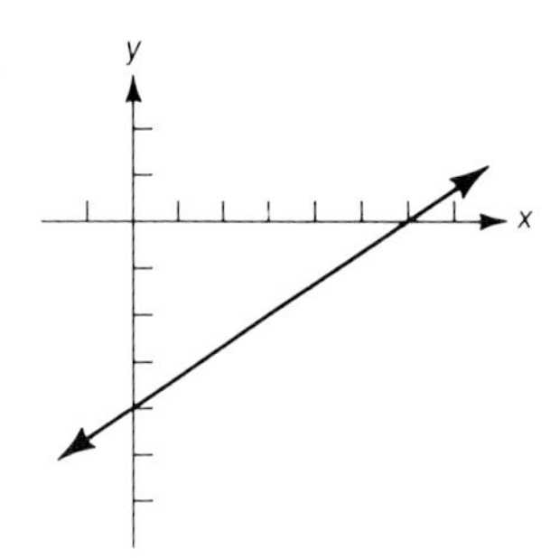


**31.**

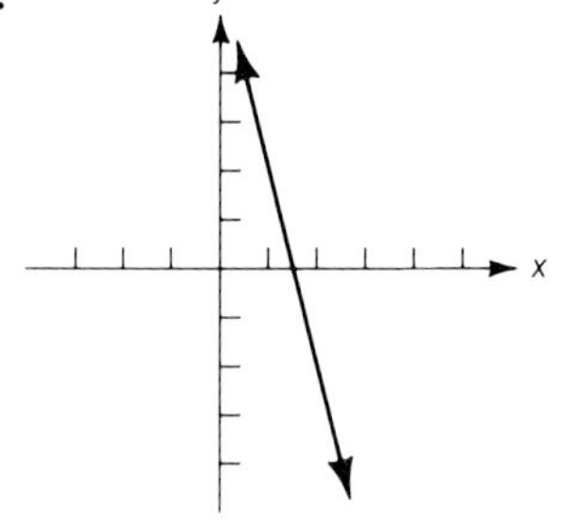


**33.**

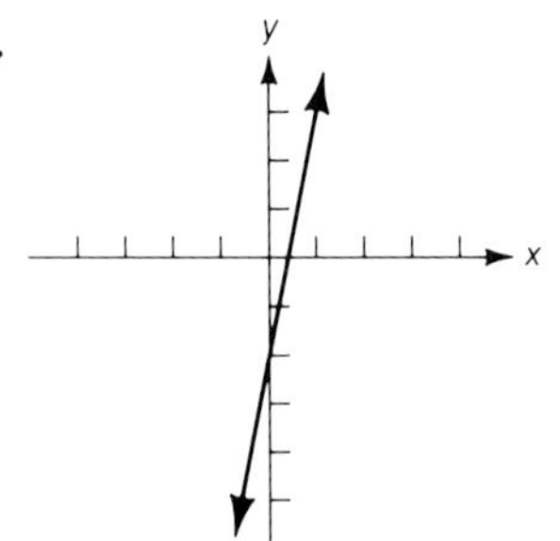

## Exercises 6.2, Page 225

**1.** $m = 5$

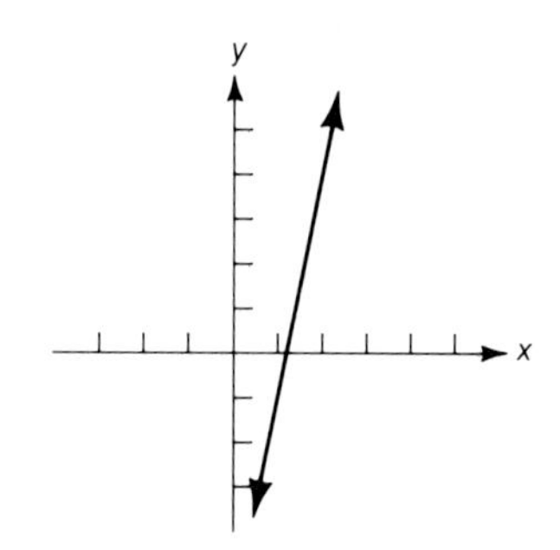


**3.** $m = -\dfrac{8}{7}$

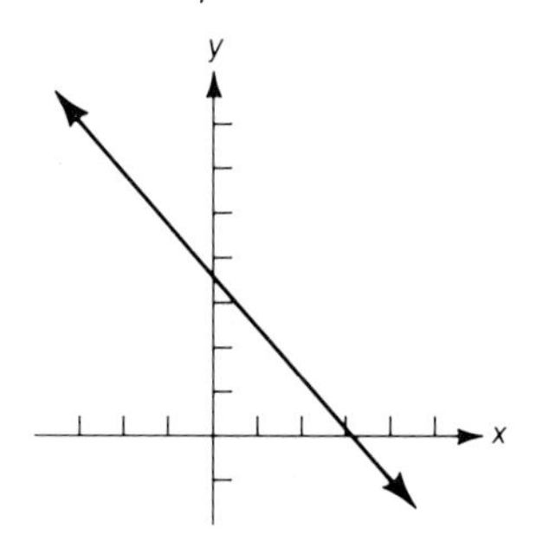


**5.** $m = 0$

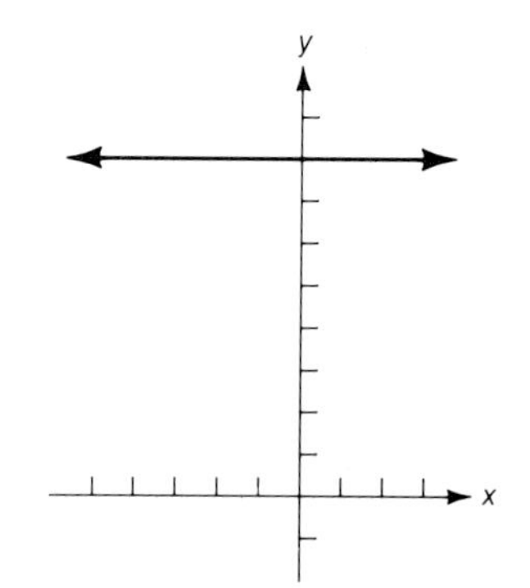


**7.** $m = -\dfrac{3}{10}$

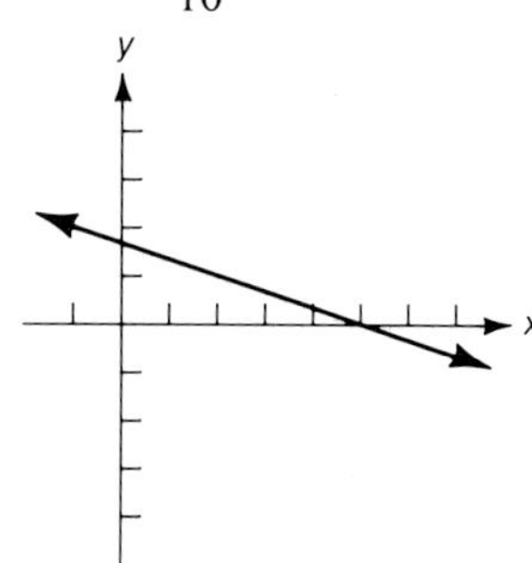


**9.** Slope is undefined.

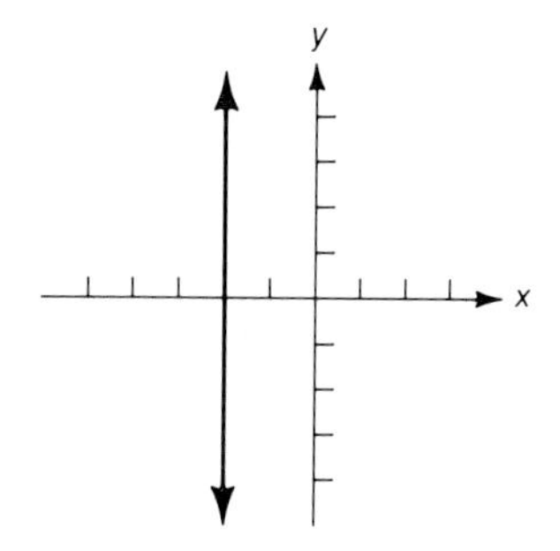


**11.** $m = \dfrac{1}{5}$

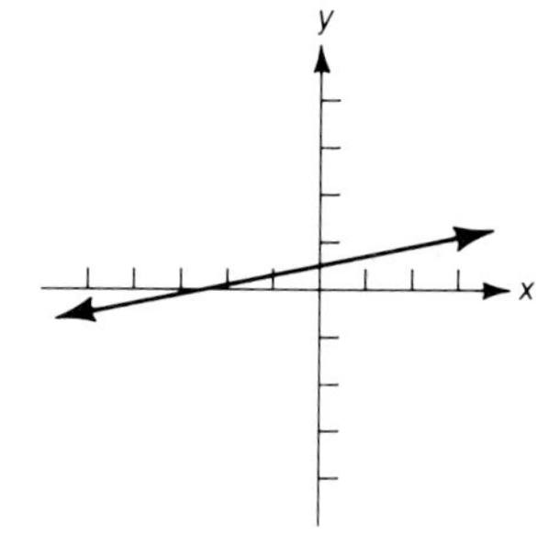


**13.** $y = 2x - 1$
$m = 2, b = -1$

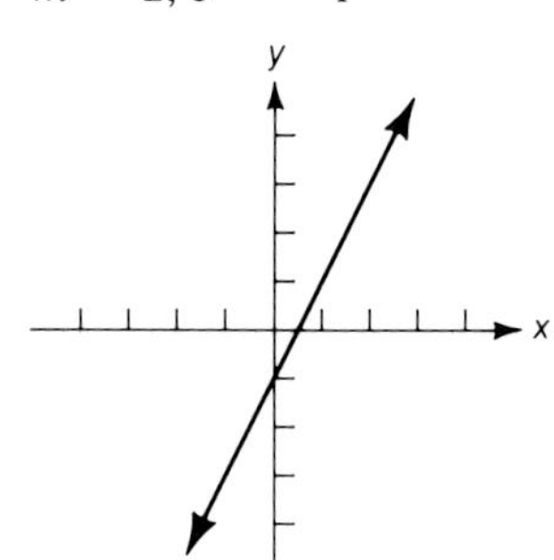


**15.** $y = -4x + 5$
$m = -4, b = 5$

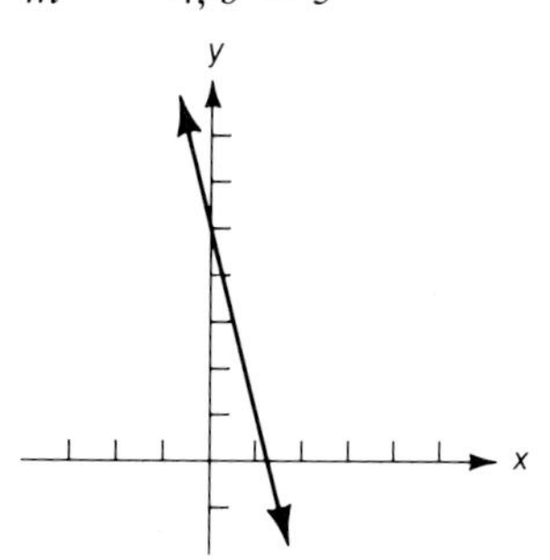


**17.** $y = \dfrac{2}{3}x + 2$

$m = \dfrac{2}{3}, b = 2$

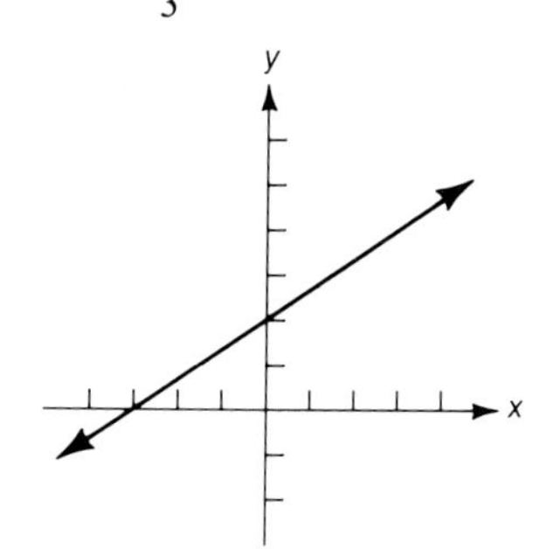

**19.** $y = -x + 5$
$m = -1, b = 5$

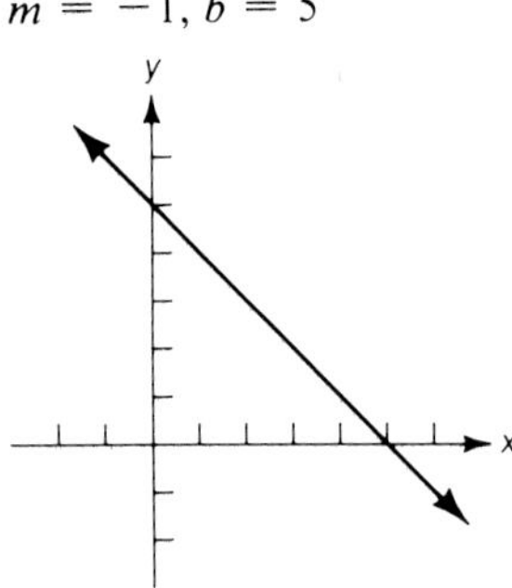


**21.** $y = -\frac{1}{5}x + 2$

$m = -\frac{1}{5}, b = 2$

**23.** $y = 4$
$m = 0, b = 4$

**25.** $y = -4x$
$m = -4, b = 0$

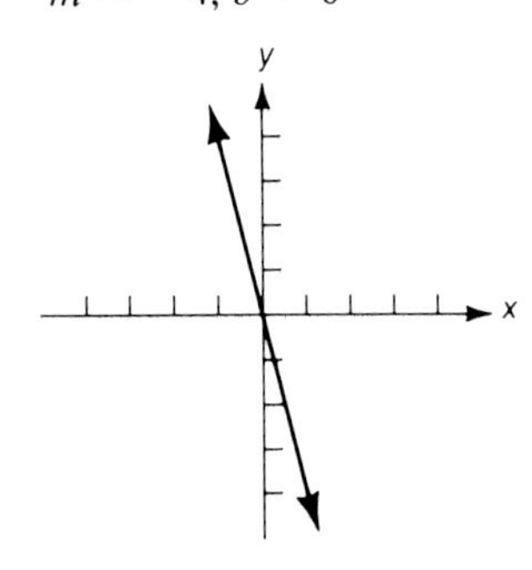


**27.** $y = \frac{2}{3}x - 2$

$m = \frac{2}{3}, b = -2$

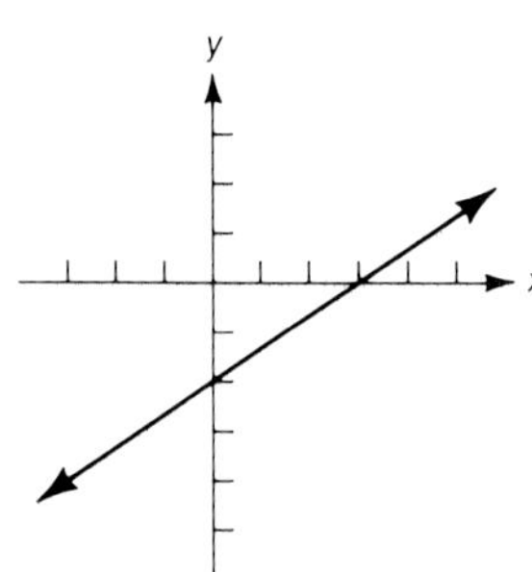


**29.** Cannot be written in slope-intercept form.

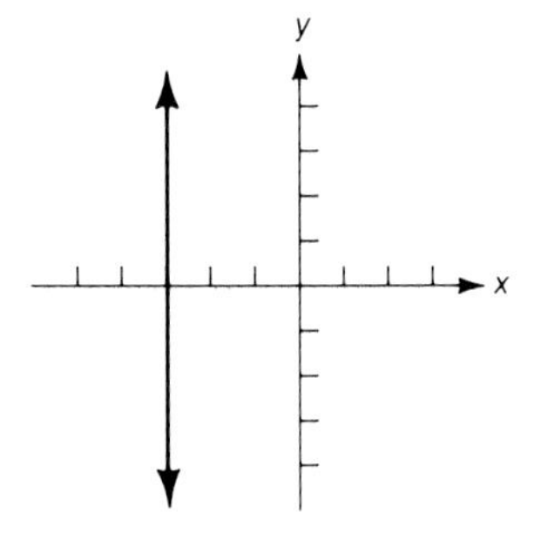


**31.** $y = \frac{5}{6}x - \frac{5}{3}$

$m = \frac{5}{6}, b = -\frac{5}{3}$

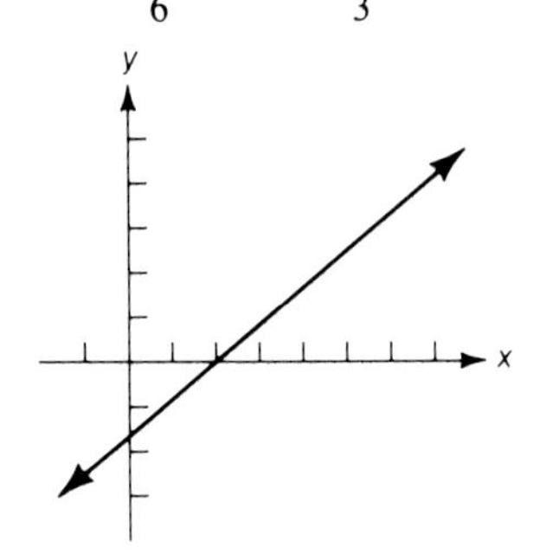


**33.** $y = -\frac{3}{4}x + \frac{5}{4}$

$m = -\frac{3}{4}, b = \frac{5}{4}$

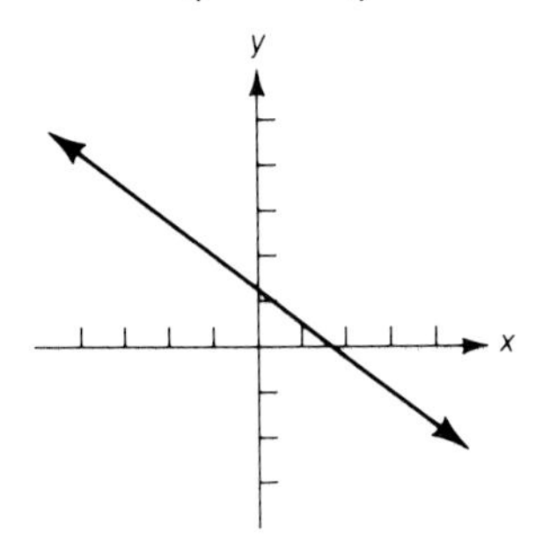


**35.** $y = -\frac{3}{2}x - \frac{7}{4}$

$m = -\frac{3}{2}, b = -\frac{7}{4}$

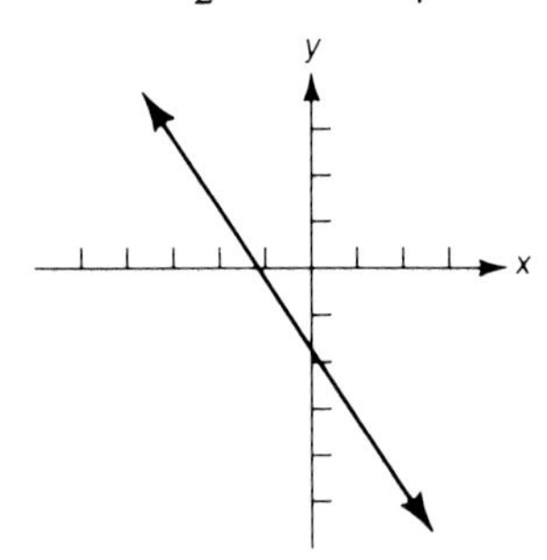

37. $y = \frac{1}{2}x + \frac{2}{3}$

$m = \frac{1}{2}, b = \frac{2}{3}$

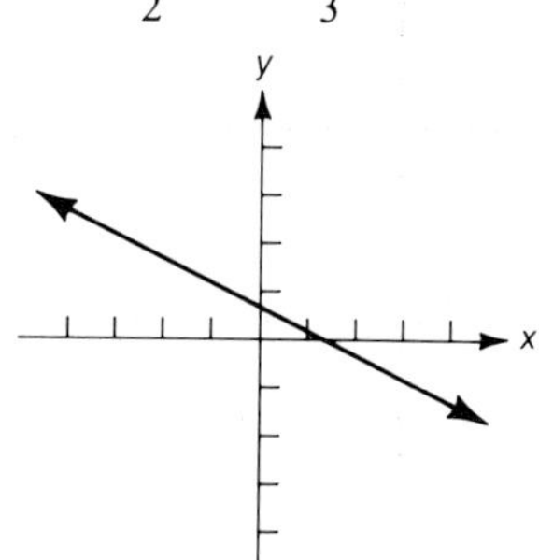


39. $y = \frac{5}{2}x + \frac{5}{2}$

$m = \frac{5}{2}, b = \frac{5}{2}$

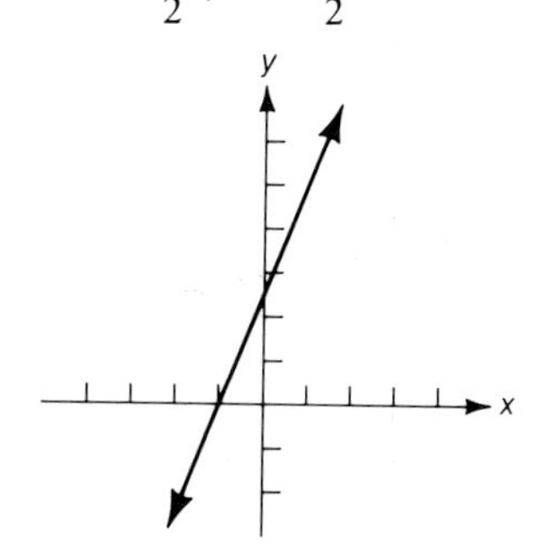


## Exercises 6.3, Page 232

1. $2x + y = -3$

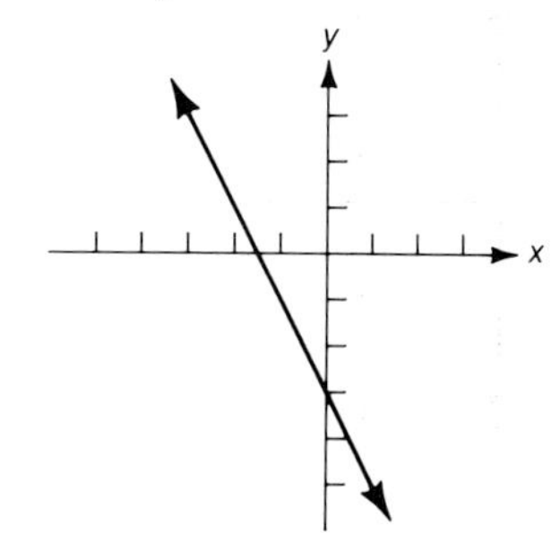


3. $x - 2y = -9$

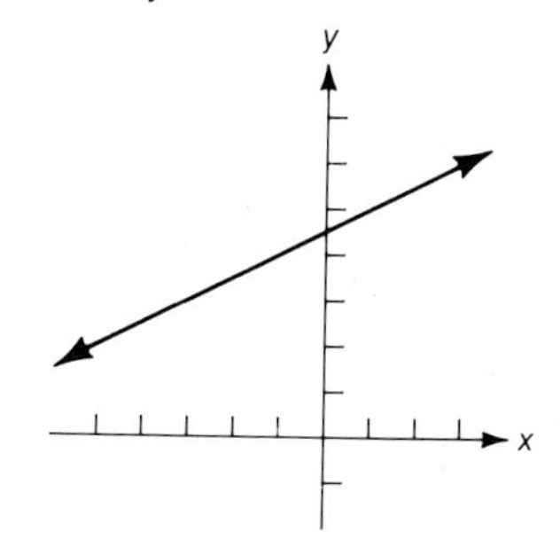


5. $x + 3y = 2$

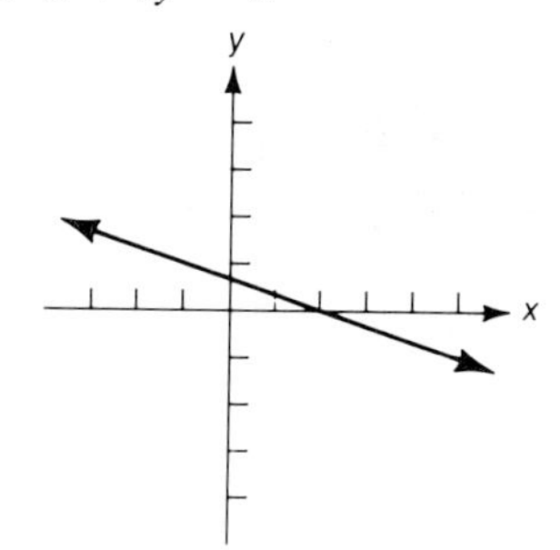


7. $x = 4$

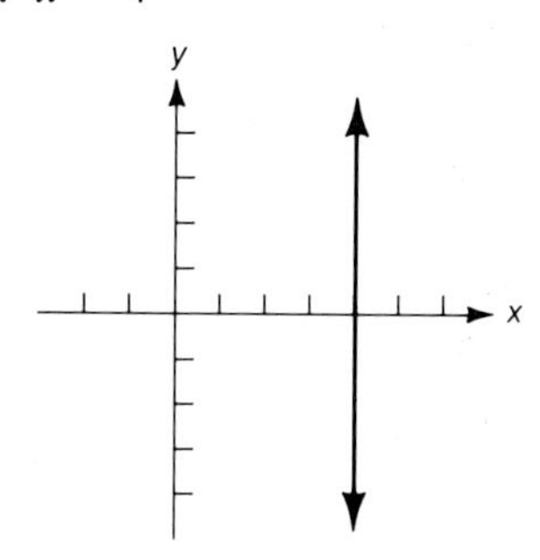


9. $y = 3$

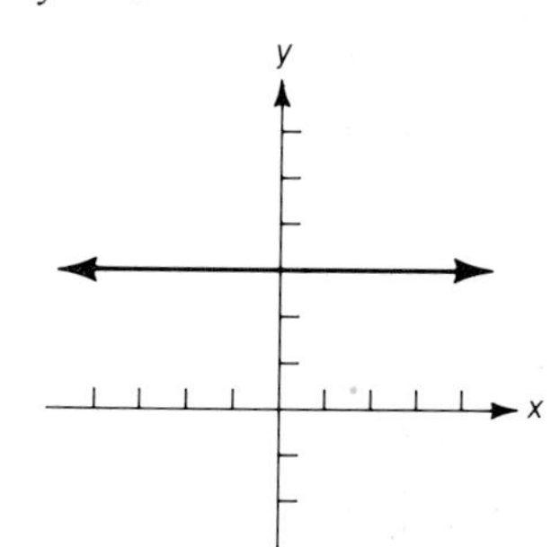


11. $6x + 5y = 23$

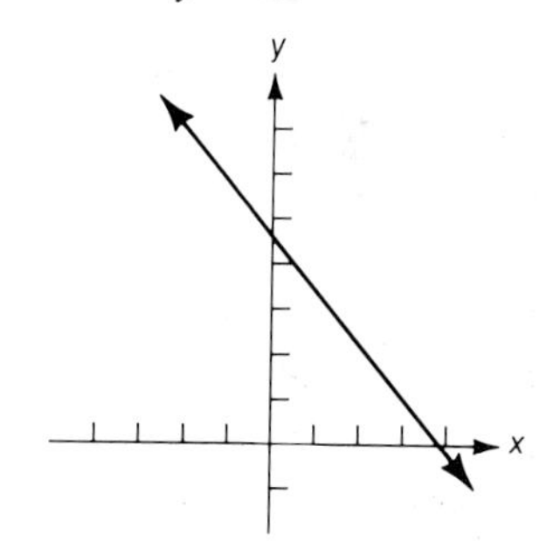

**13.** $2x + 27y = 5$

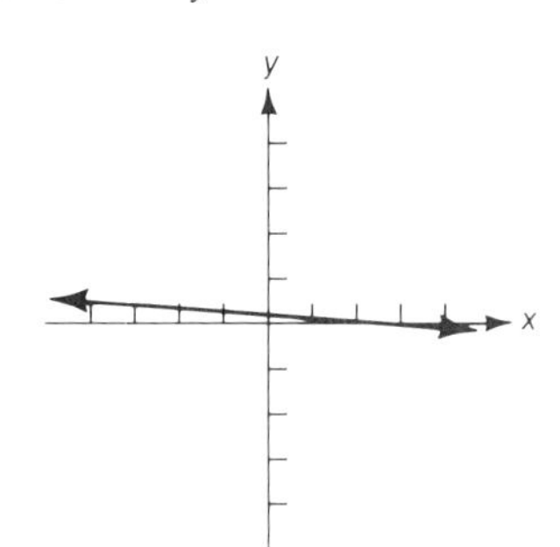


**15.** $4x + 3y = \frac{17}{3}$

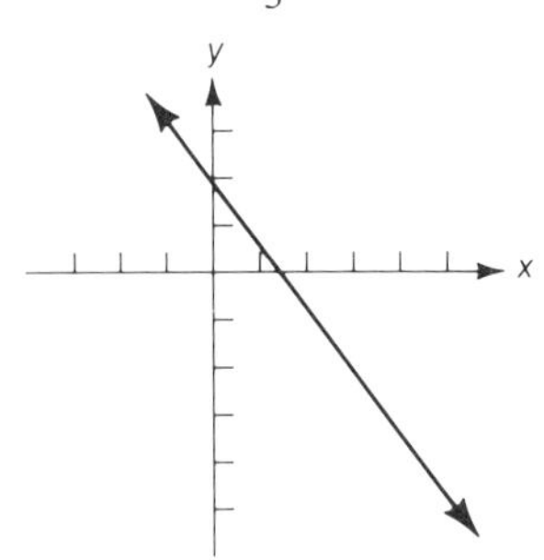


**17.**

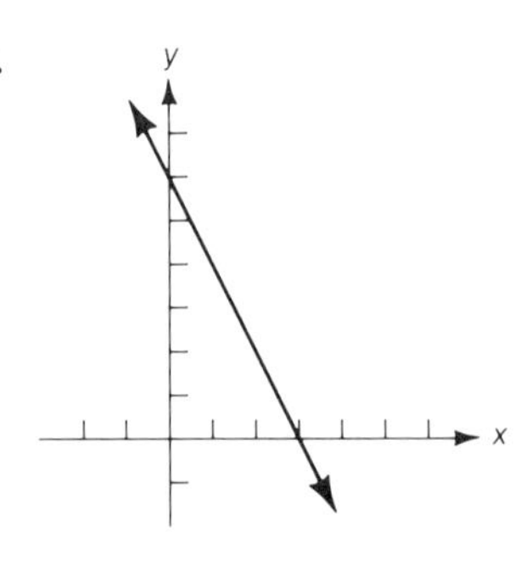


**19.**

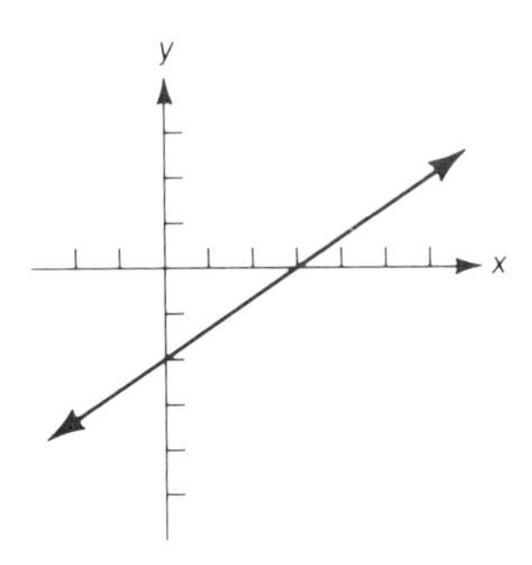


**21.**

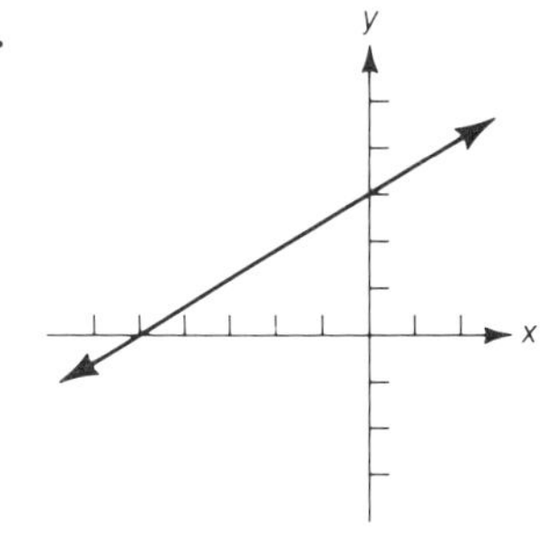


**23.**

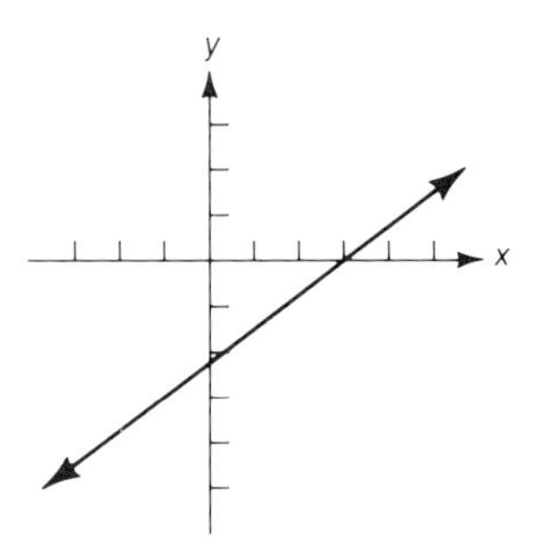


**25.** $x + 2y = 13$ **27.** $5x - y = -2$ **29.** $x = 2$ **31.** $3x + y = 11$ **33.** $2x - 3y = -2$
**35.** $x + 2y = -4$

## *Exercises 6.4, Page 237*

**1.**

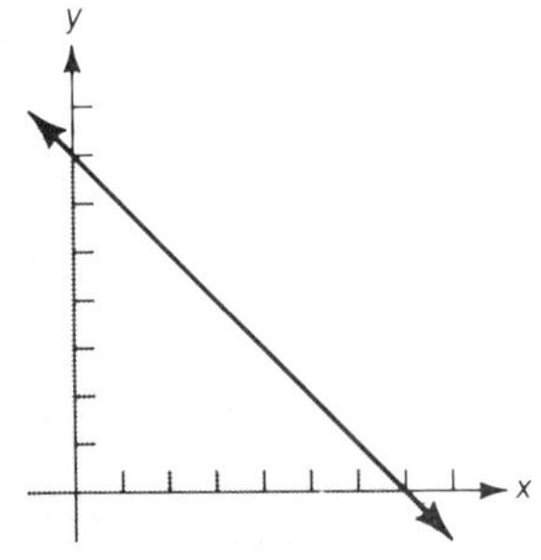


**3.**

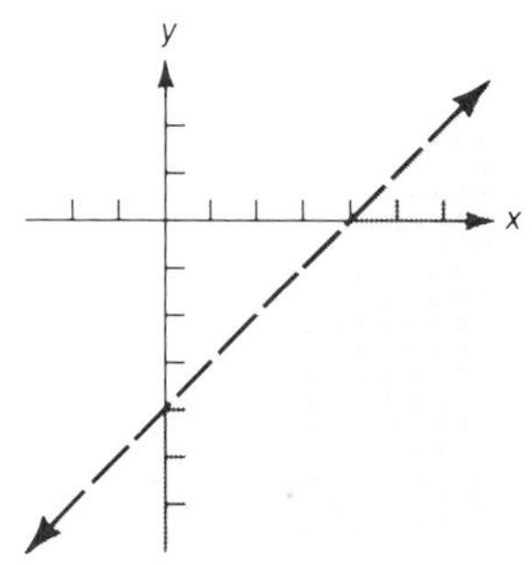


**5.**

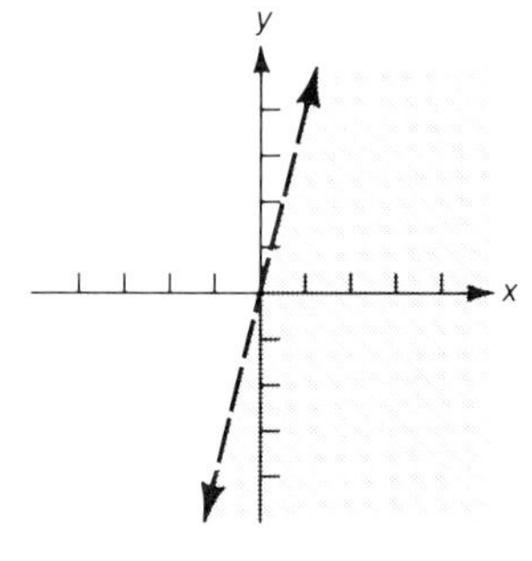


**7.**

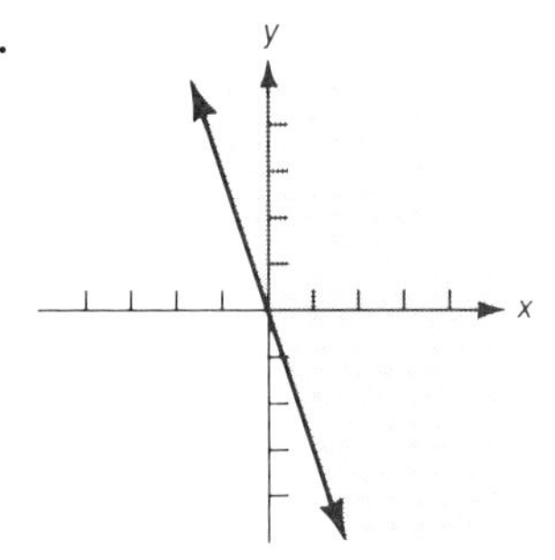


**9.**

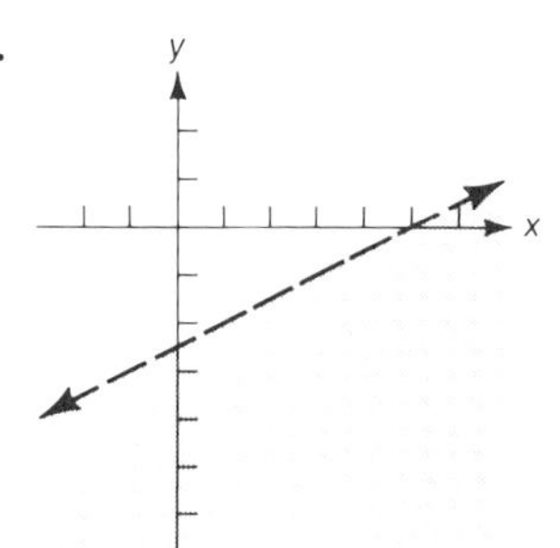


**11.**

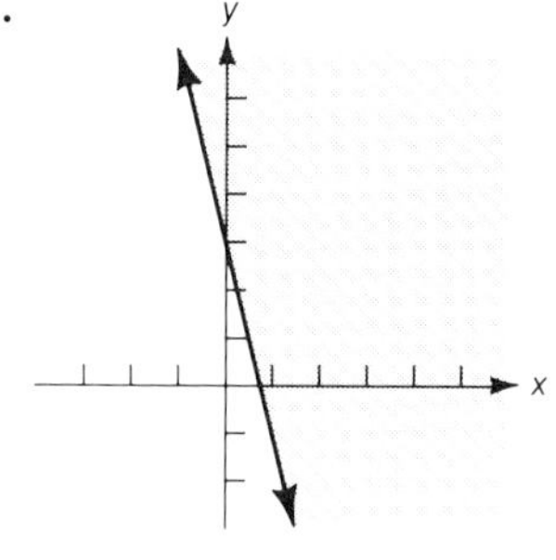

**13.**

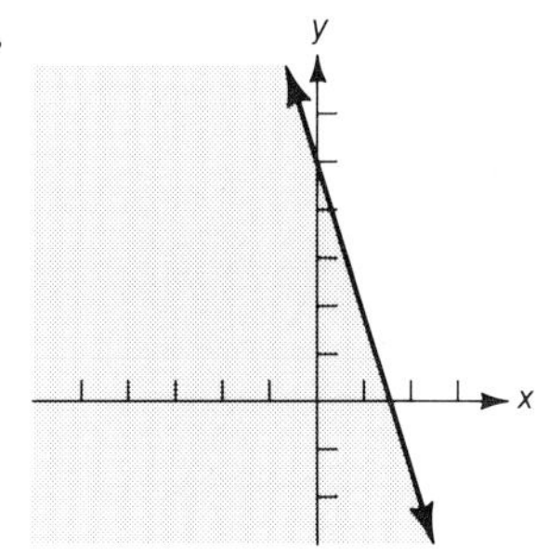


**15.**

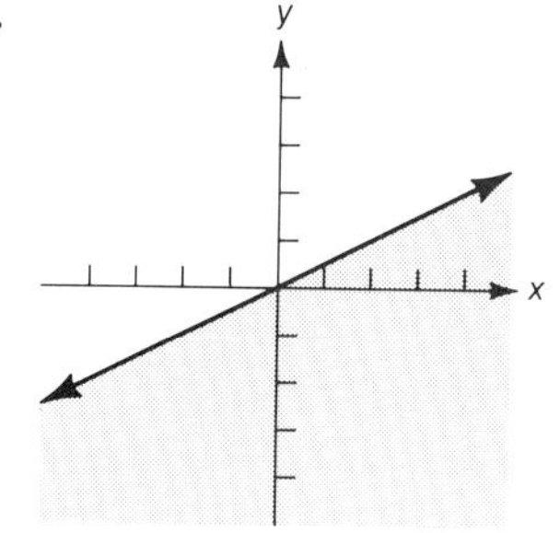


**17.**

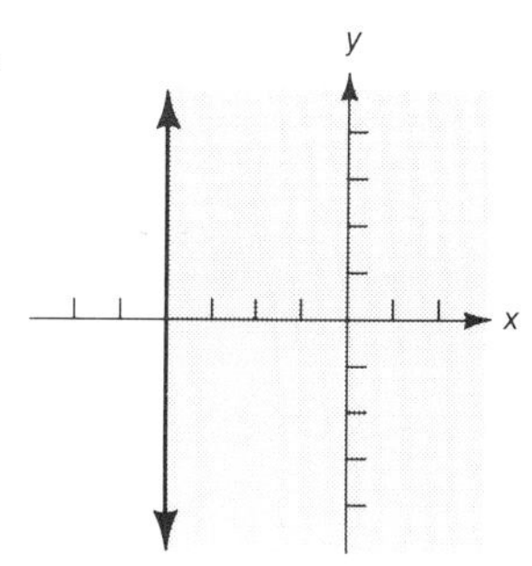


**19.**

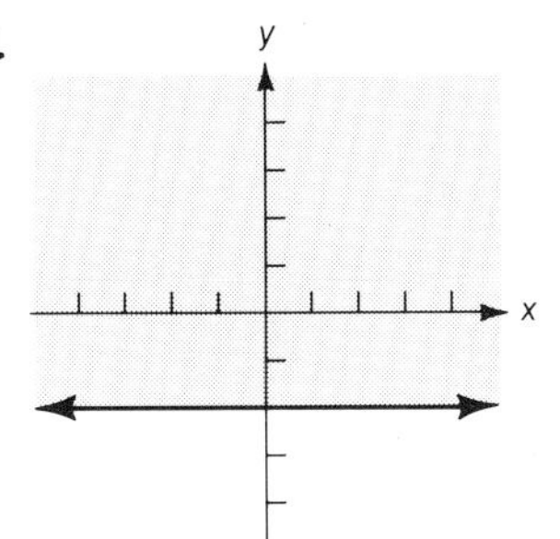


**21.**

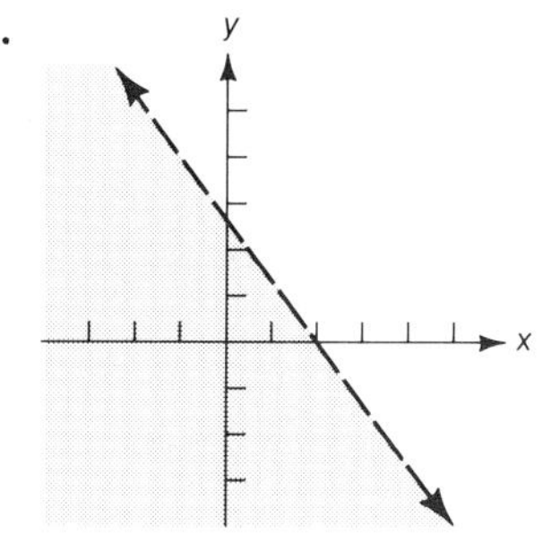


**23.**

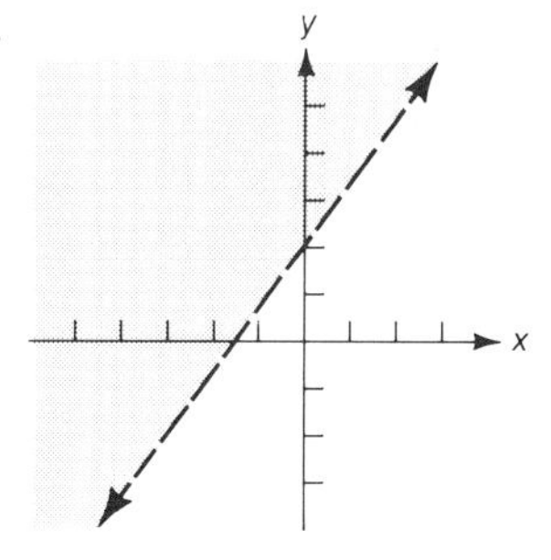


**25.**

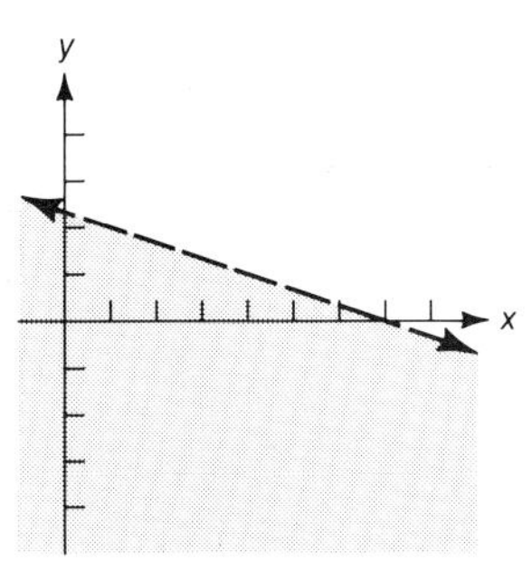


**27.**

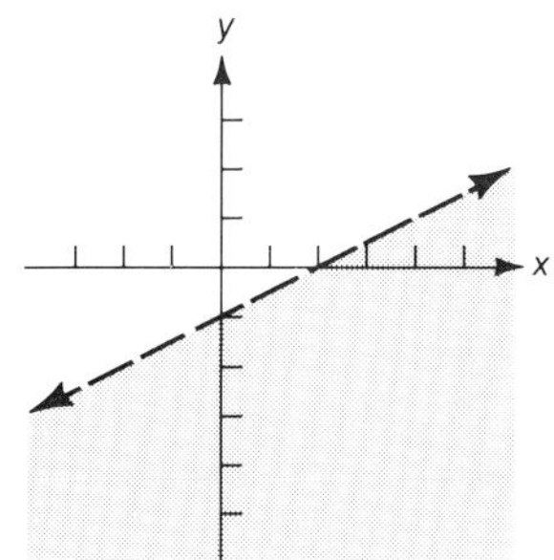


**29.**

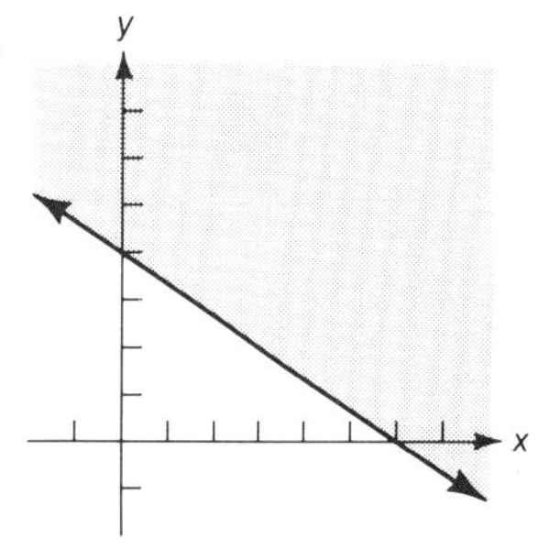


## Chapter 6 Review, Page 238

**1.** **(a)** $(0,-4)$ **(b)** $(2,0)$ **(c)** $(1,-2)$ **(d)** $(1,-2)$ **2.** **(a)** $(0,2)$ **(b)** $(6,0)$ **(c)** $\left(2,\frac{4}{3}\right)$ **(d)** $(9,-1)$
**3.** **(a)** $(0,9)$ **(b)** $(3,0)$ **(c)** $(-2,15)$ **(d)** $\left(\frac{11}{3},-2\right)$ **4.** **(a)** $(0,-6)$ **(b)** $\left(\frac{3}{2},0\right)$ **(c)** $(-2,-14)$
**(d)** $\left(\frac{5}{2},4\right)$

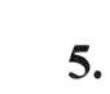

**5.**

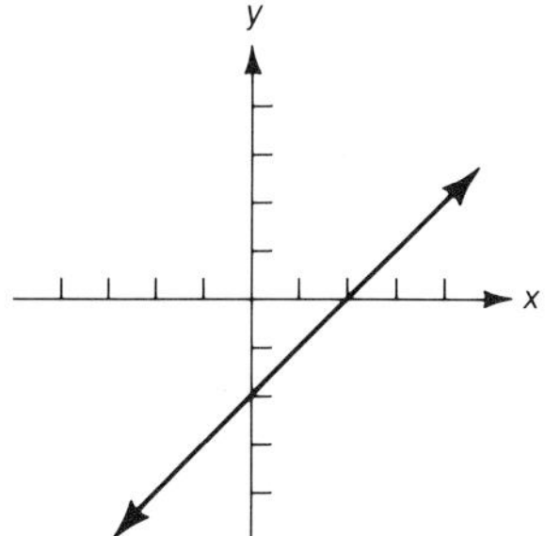


**6.**

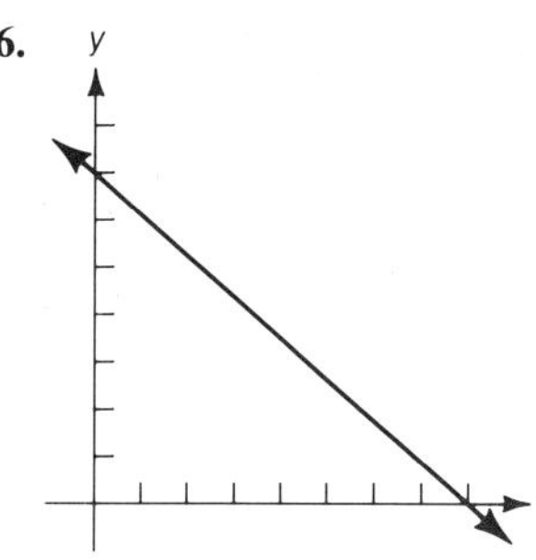


**7.**

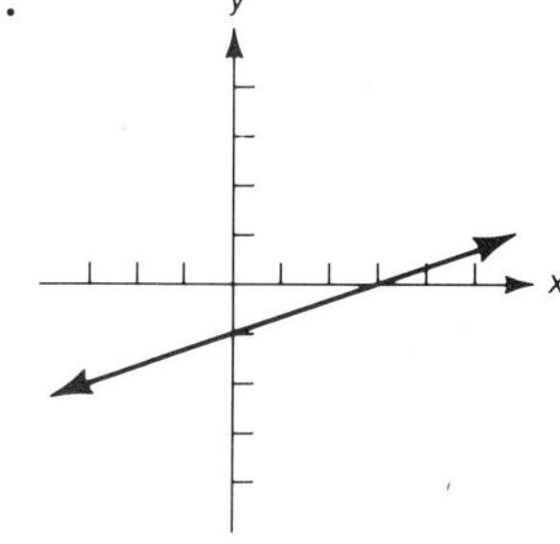

**8.**

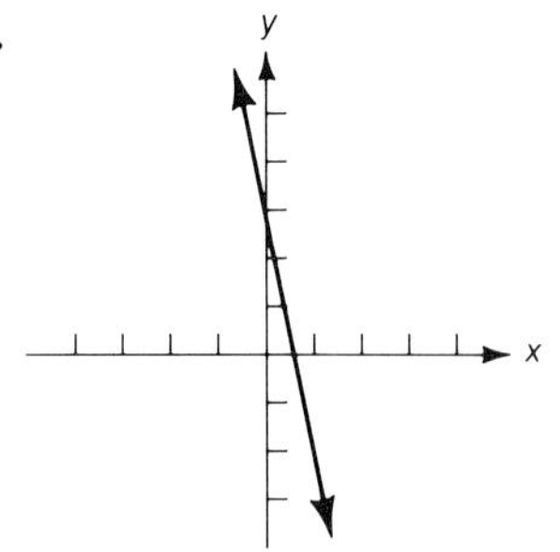


**9.**

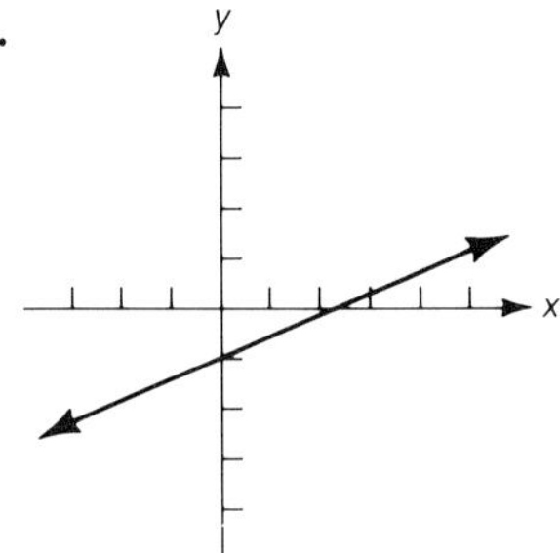


**10.**

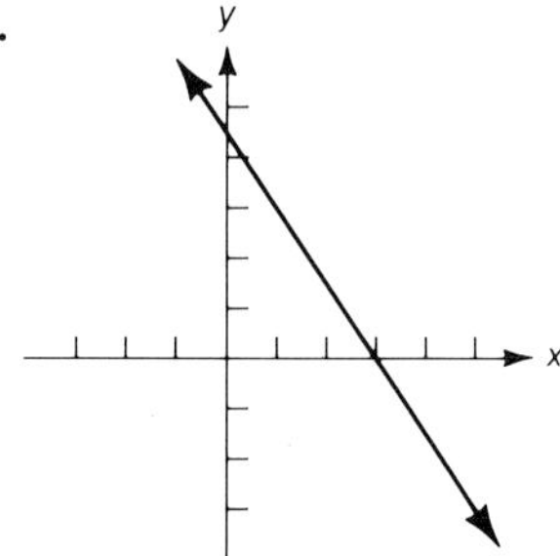


**11.**

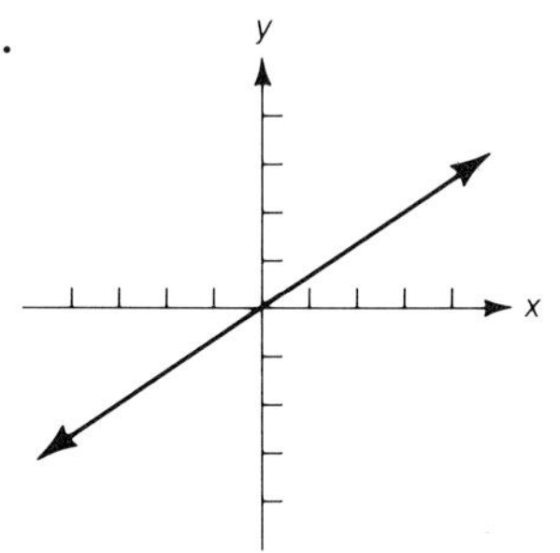


**12.**

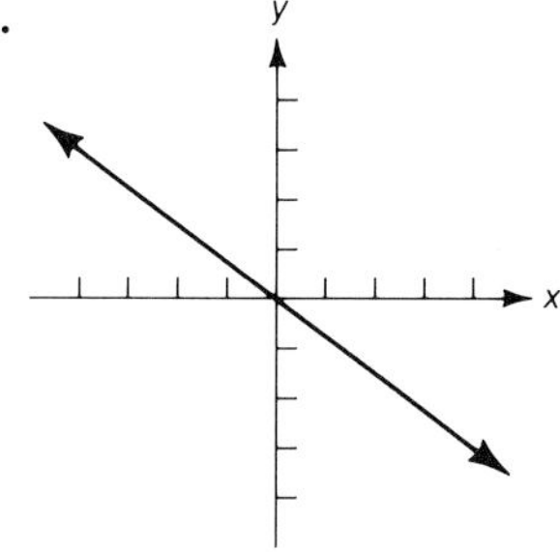


**13.**

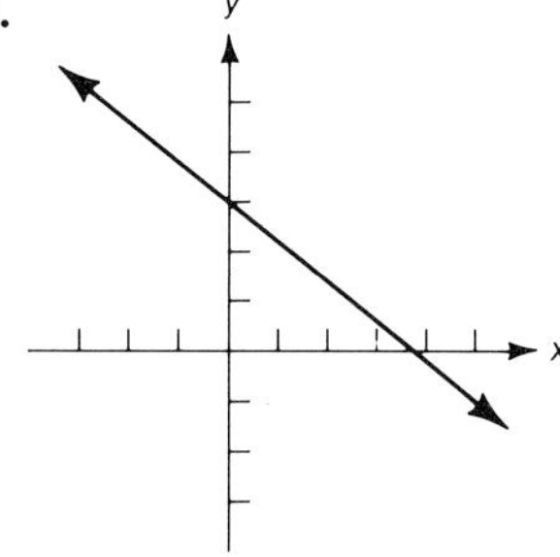


**14.**

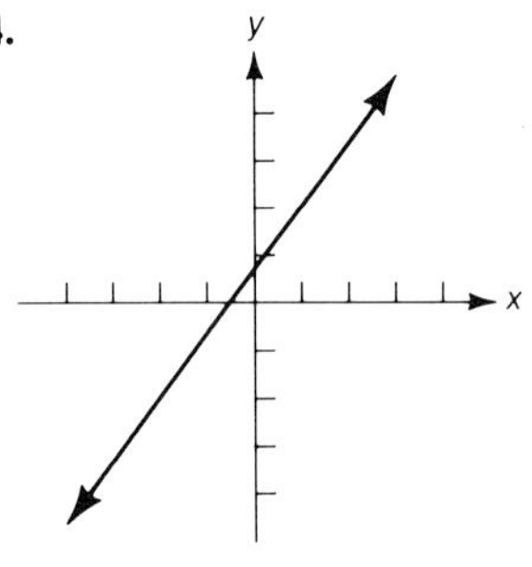


**15.** $m = -\frac{2}{3}$

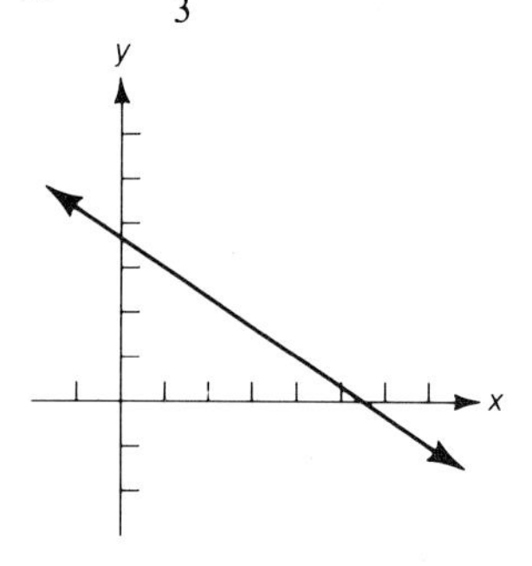


**16.** $m = \frac{5}{7}$

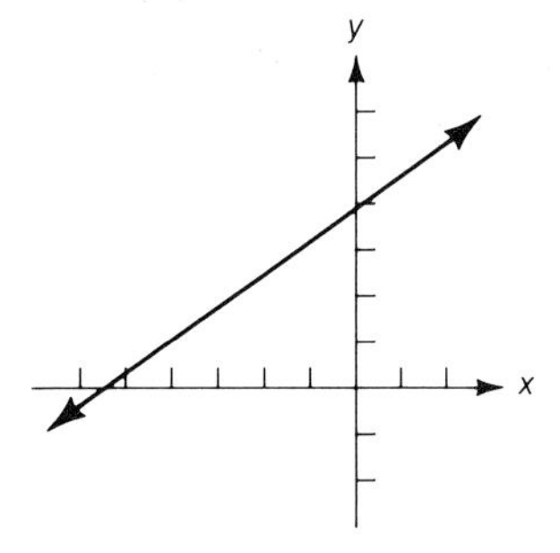


**17.** $m = -\frac{3}{2}$

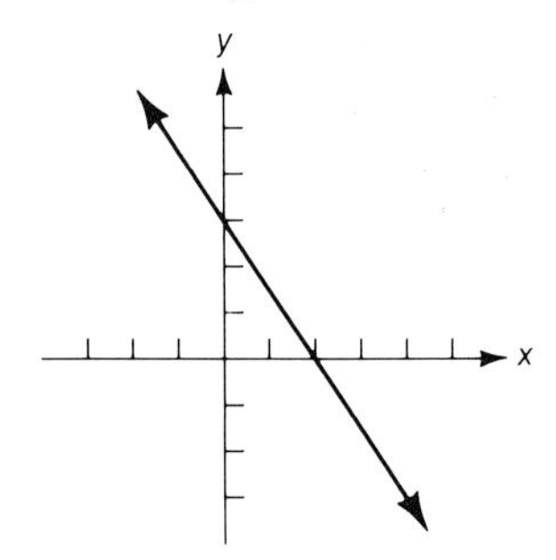


**18.** $m = -\frac{2}{7}$

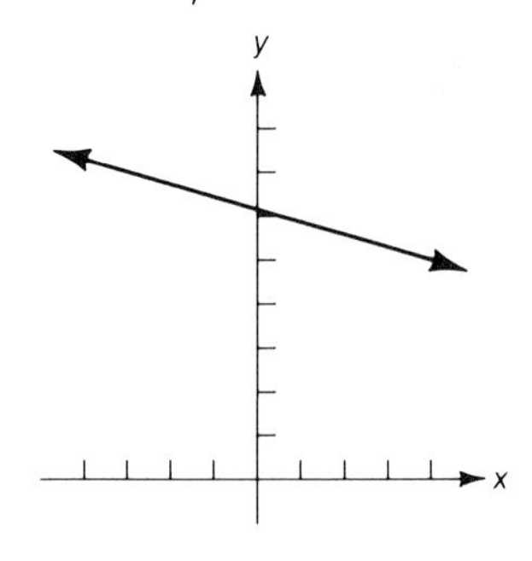


**19.** $m = 6$

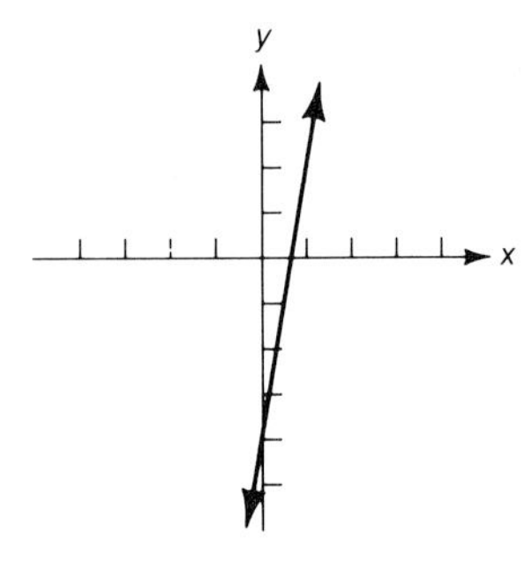

**20.** $m = \frac{2}{7}$

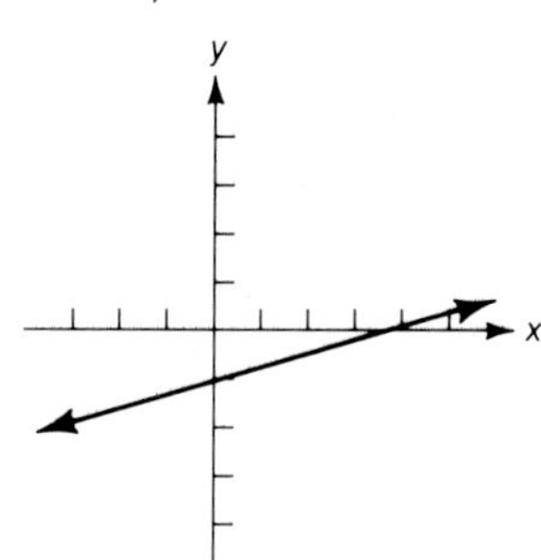


**21.** $y = -\frac{1}{5}x + 2$

$m = -\frac{1}{5}, b = 2$

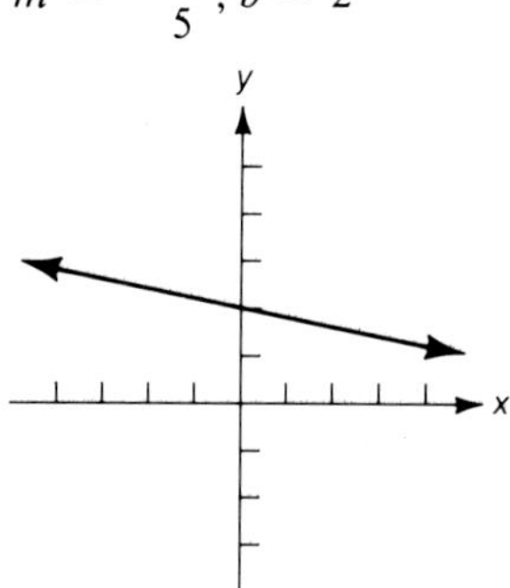


**22.** $y = -3x + 1$
$m = -3, b = 1$

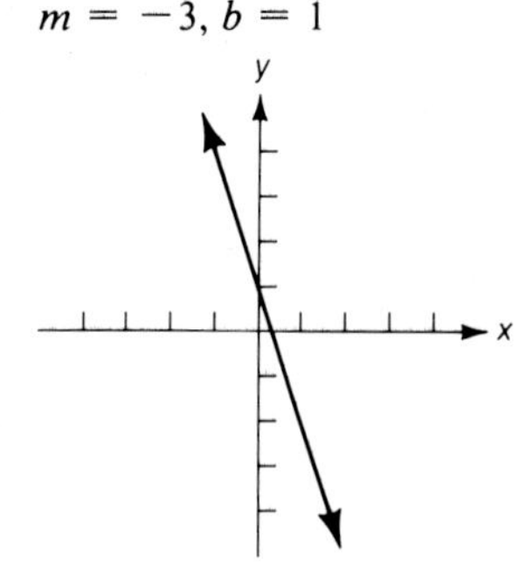


**23.** $y = \frac{3}{7}x - 1$

$m = \frac{3}{7}, b = -1$

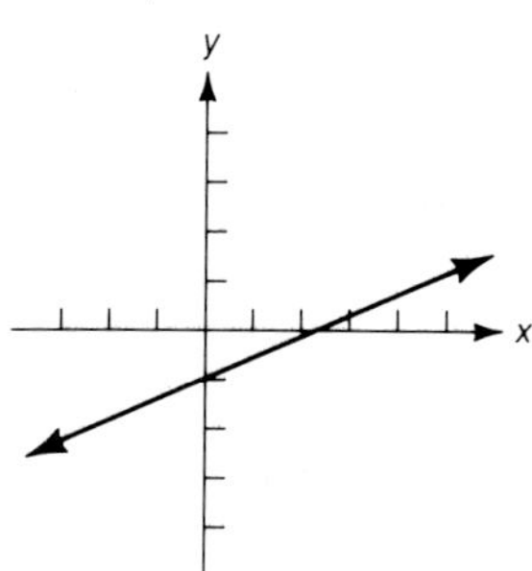


**24.** $y = -\frac{5}{3}x + 2$

$m = -\frac{5}{3}, b = 2$

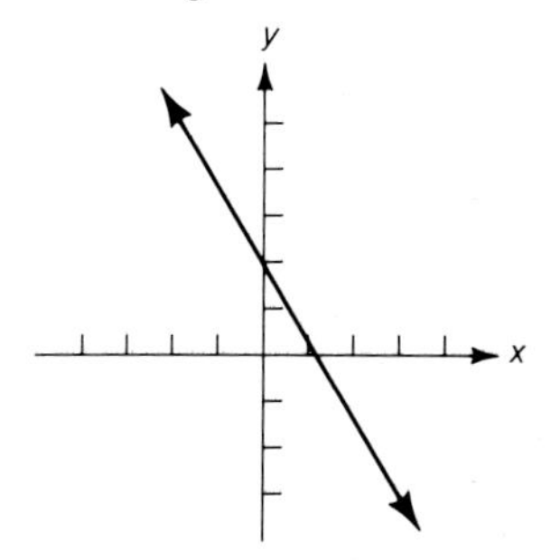


**25.** $y = \frac{9}{4}$

$m = 0, b = \frac{9}{4}$

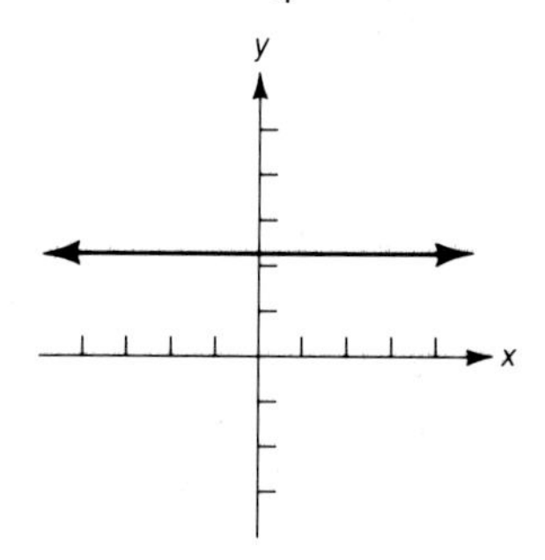


**26.** $y = 2$
$m = 0, b = 2$

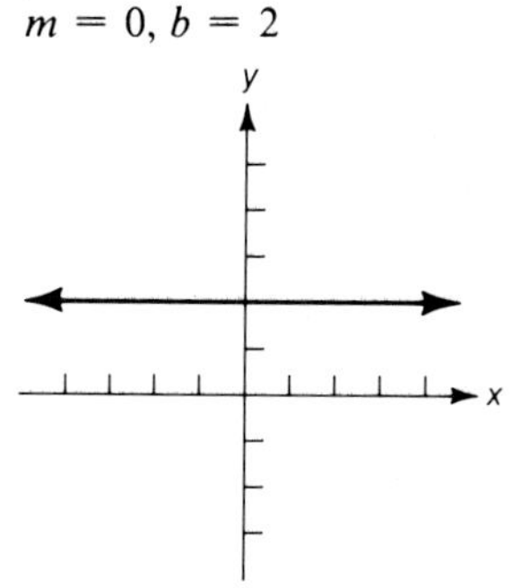


**27.** $y = -\frac{3}{2}x + \frac{5}{2}$

$m = -\frac{3}{2}, b = \frac{5}{2}$

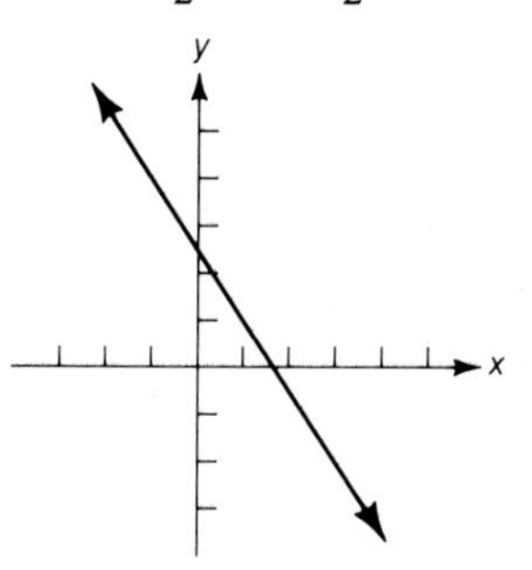


**28.** $y = 2x - \frac{9}{2}$

$m = 2, b = -\frac{9}{2}$

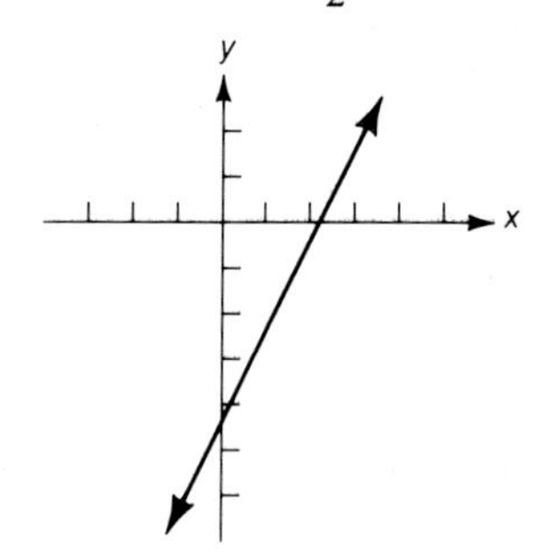

**29.** Cannot be written in slope-intercept form.

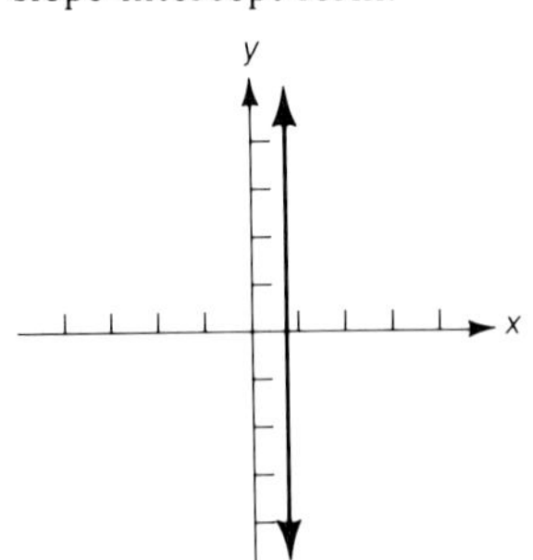


**30.** Cannot be written in slope-intercept form.

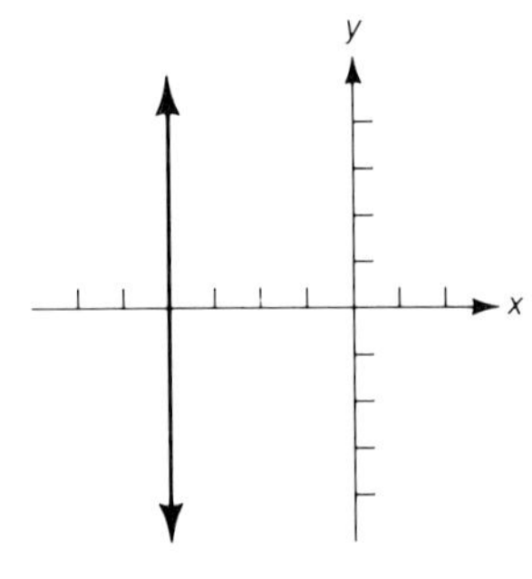


**31.** $y = -\frac{1}{3}x + \frac{9}{2}$

$m = -\frac{1}{3}, b = \frac{9}{2}$

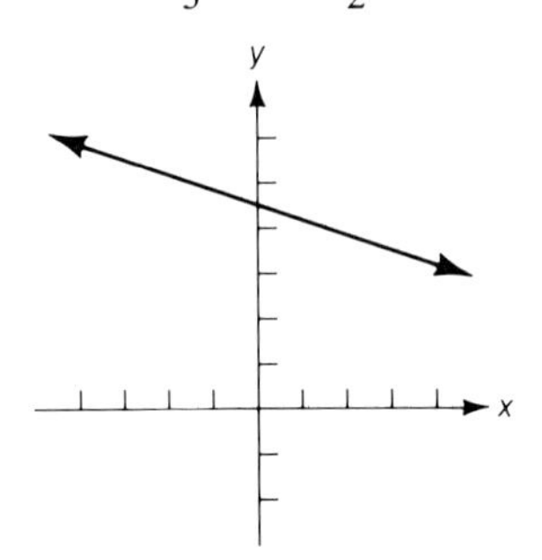


**32.** $y = \frac{1}{4}x$

$m = \frac{1}{4}, b = 0$

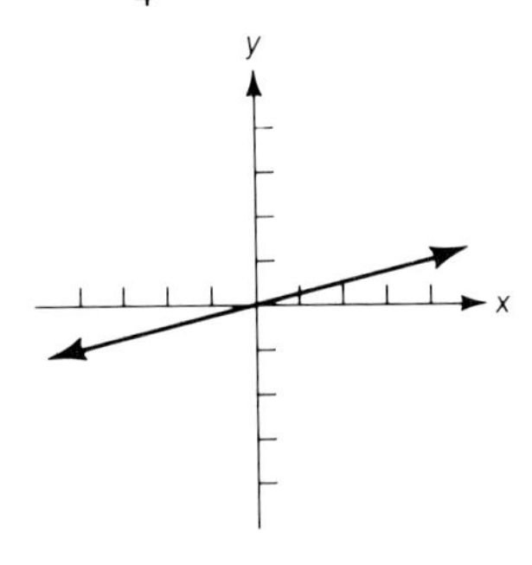


**33.** $x + 4y = -10$ **34.** $y = 1$ **35.** $y = 0$ **36.** $2x - 5y = 17$ **37.** $4x - 3y = -10$ **38.** $2x - y = 0$
**39.** $x = 5$ **40.** $3x + y = -2$ **41.** $4x + 5y = 15$ **42.** $2x + y = 8$ **43.** $3x - y = 13$
**44.** $2x + y = -3$ **45.** $x - 3y = -10$ **46.** $3x - 11y = -12$ **47.** $3x + 2y = 12$ **48.** $x = 1$
**49.** $2x - 5y = 0$ **50.** $y = -5$ **51.** $3x - 4y = 12$ **52.** $5x + 3y = 24$ **53.** $y = 1$ **54.** $x + 3y = -6$

**55.**

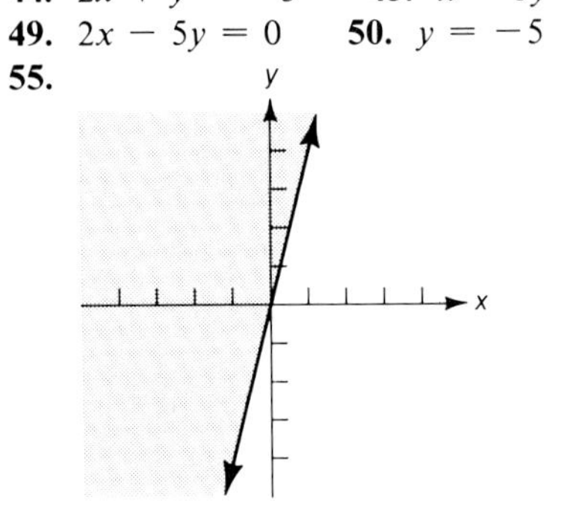


**56.**

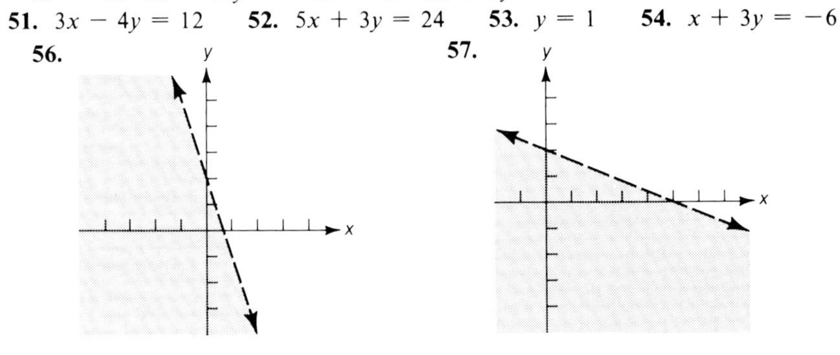


**57.**

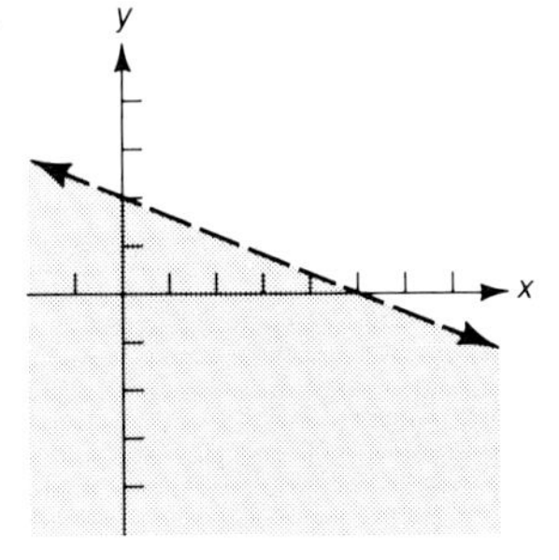

58.

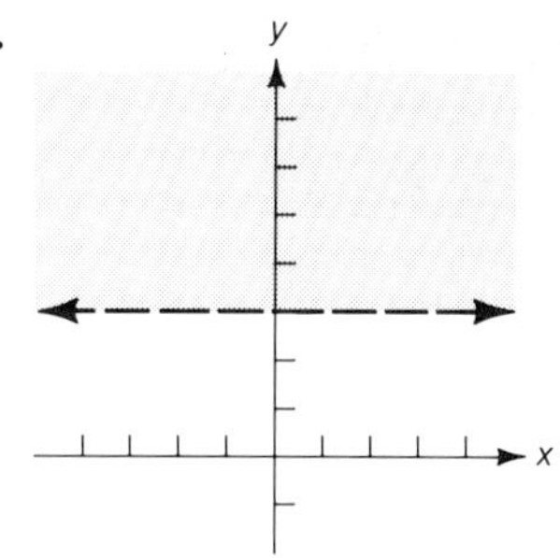


59.

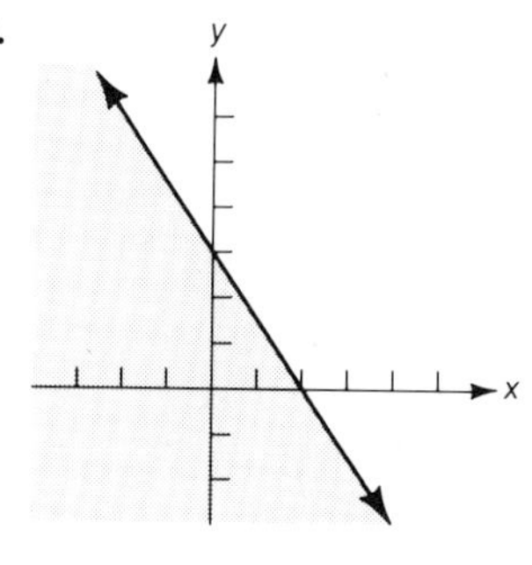


60.

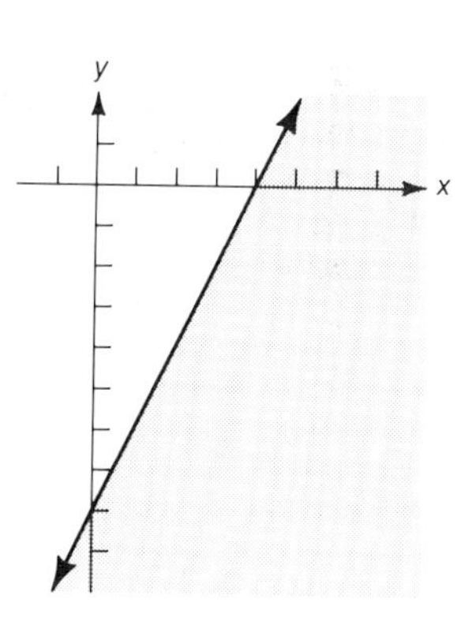


## *Chapter 6 Test, Page 239*

**1.** **(a)** $(0,2)$ **(b)** $\left(\frac{2}{3},0\right)$ **(c)** $(-2,8)$ **(d)** $(3,-7)$ **2.** **(a)** $\left(0,-\frac{6}{5}\right)$ **(b)** $(6,0)$ **(c)** $(11,1)$ **(d)** $(-4,-2)$

**3.**

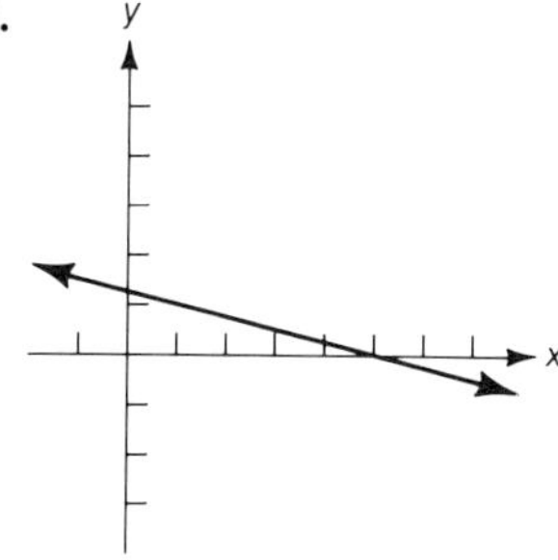


**4.**

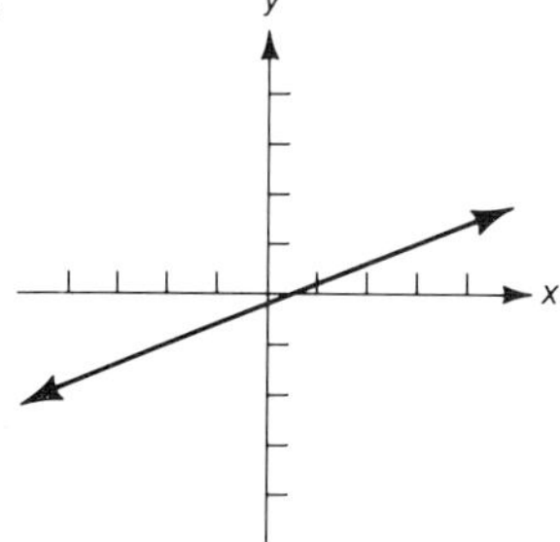


**5.**

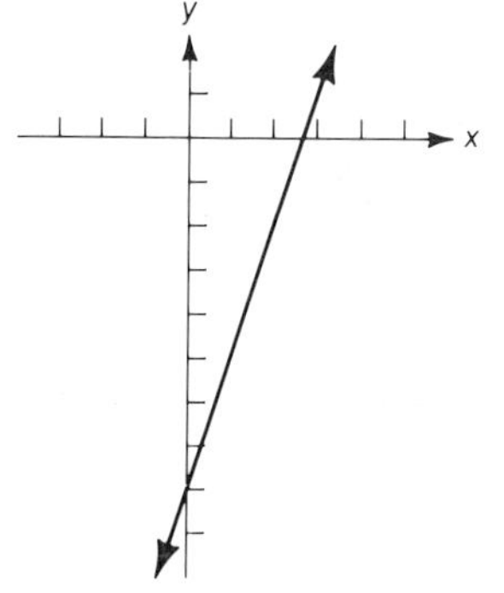


**6.** $m = \frac{9}{8}$

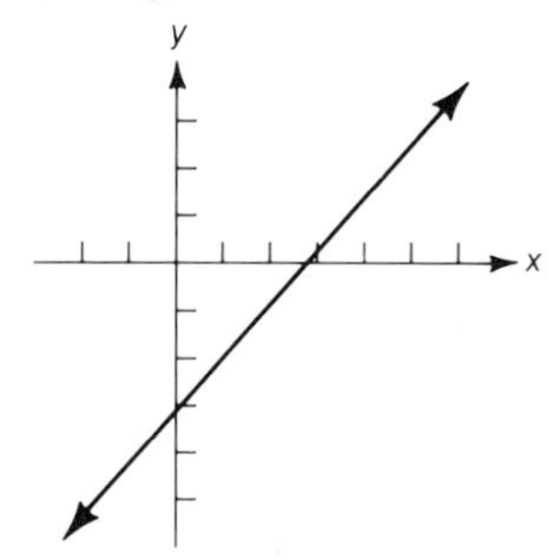


**7.** $m = -\frac{1}{5}$

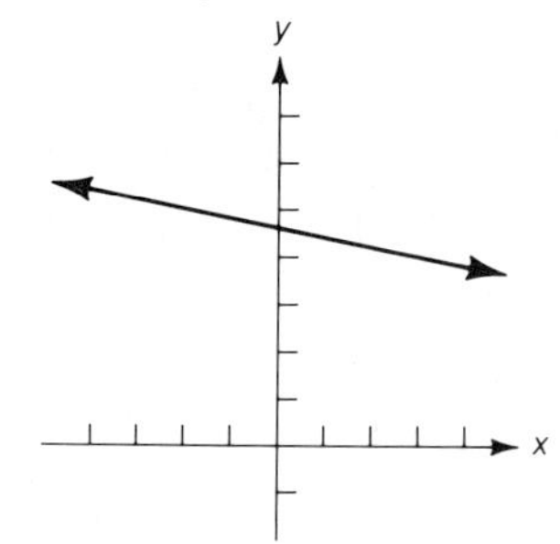


**8.** Slope is undefined.

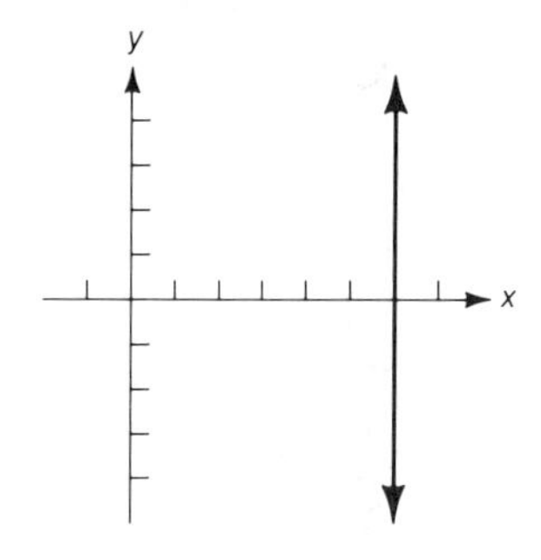

**9.** $y = \frac{1}{3}x - \frac{4}{3}$

$m = \frac{1}{3},\ b = -\frac{4}{3}$

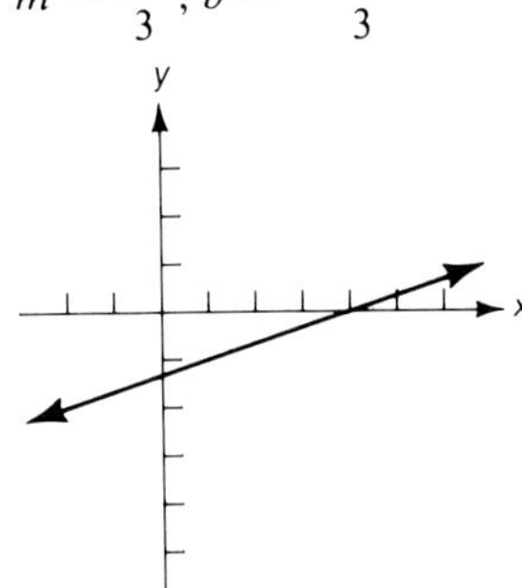

**10.** $y = \frac{4}{3}x - 1$

$m = \frac{4}{3},\ b = -1$

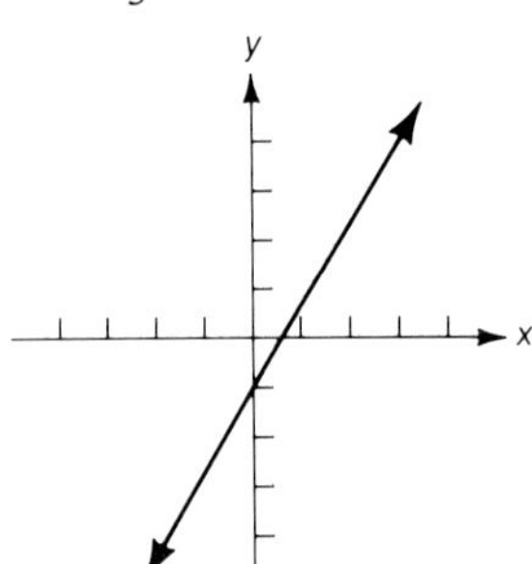

**11.** Cannot be written in slope-intercept form.

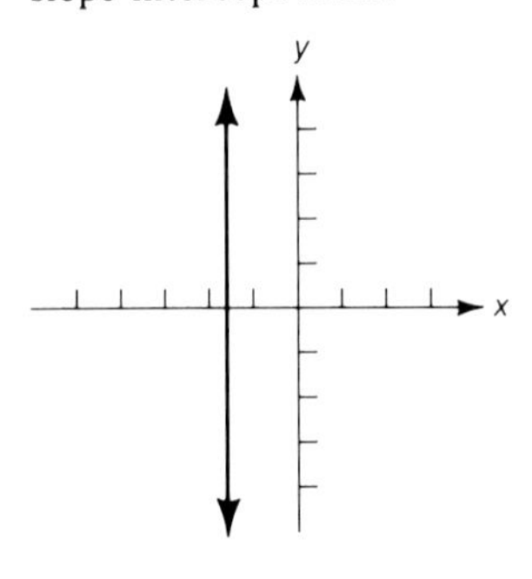

**12.** $5x + 3y = 36$ **13.** $2x - 3y = 0$ **14.** $8x + 7y = 10$ **15.** $8x + 5y = 22$ **16.** $4x + 3y = 15$

**17.** $3x + 2y = 14$ **18.** $y = -2$ **19.**

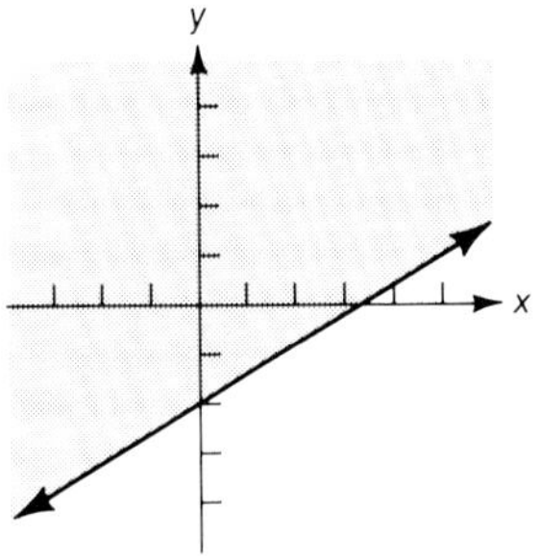

**20.** 

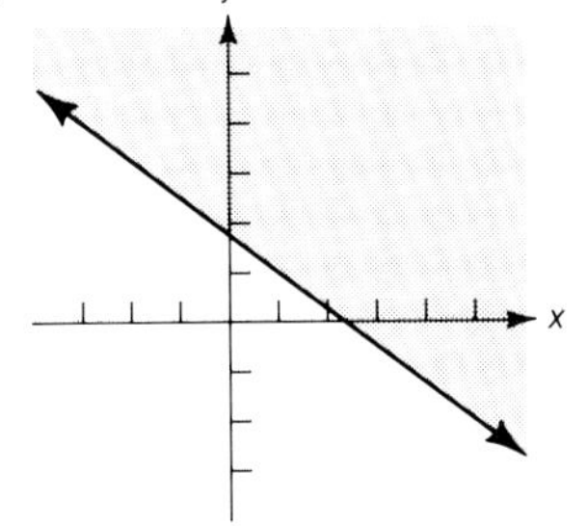

## *Cumulative Review (6), Page 239*

**1.** $x = -\frac{17}{8}$ **2.** $x = -\frac{10}{7}$ **3.** $x \le \frac{16}{7}$ **4.** $x \le \frac{13}{10}$

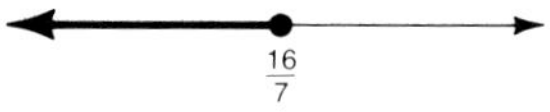

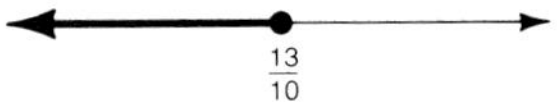

**5.** $4x - 3y - 11$ **6.** $3x - 3y - 9$ **7.** $16x - 17$ **8.** $-x + 24y - 3$ **9.** $y = -6x + 9$

**10.** $x = -\frac{5}{2}y + 4$ **11.** 3, 11 **12.** 11 mph **13.** 750 @ \$1.50; 450 @ \$2.50

**14.**

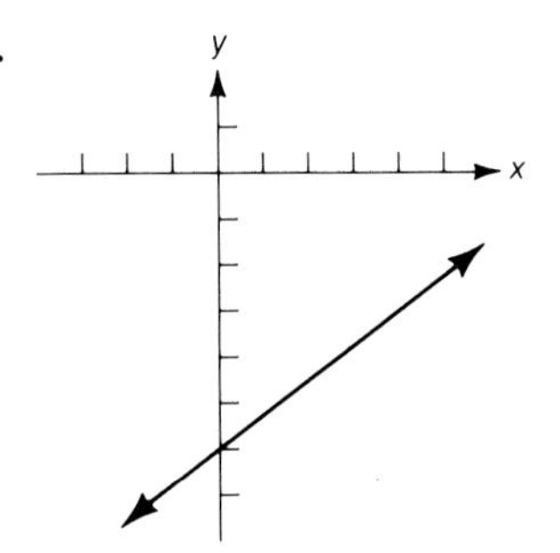

**15.**

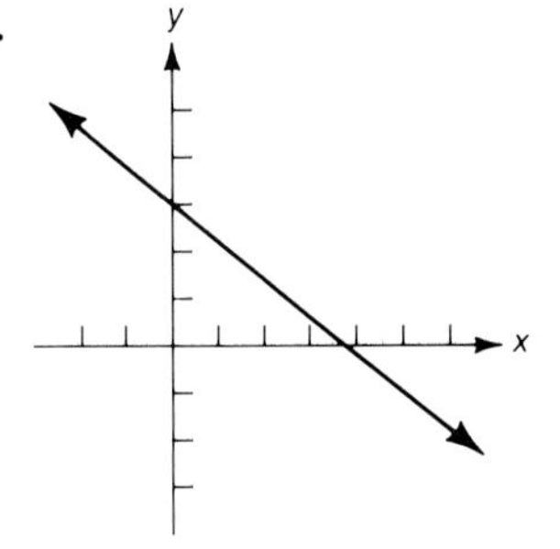

**16.**

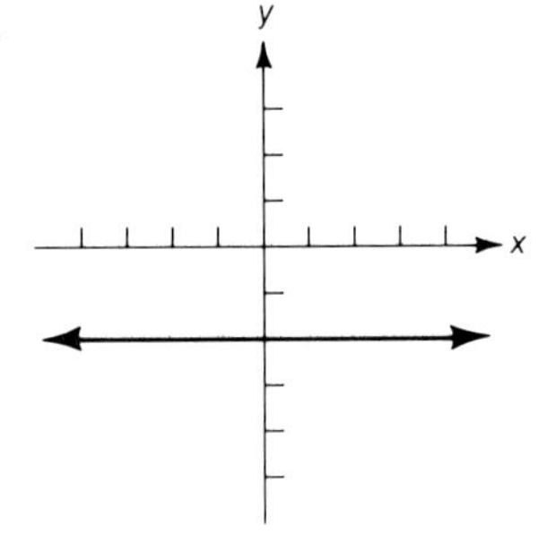

**17.** $4x + 5y = 2$

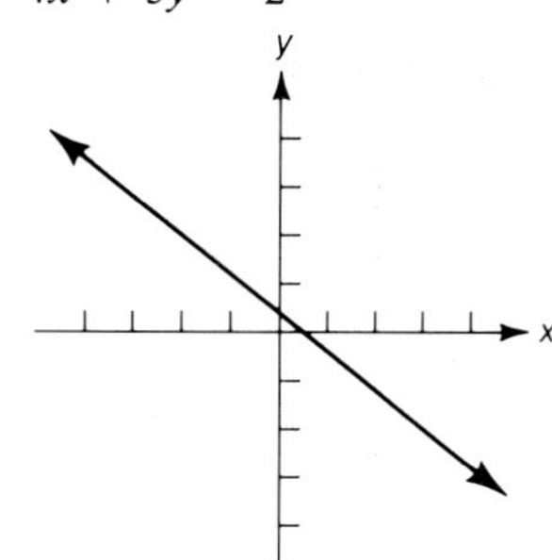

**18.** $5x - 3y = 28$

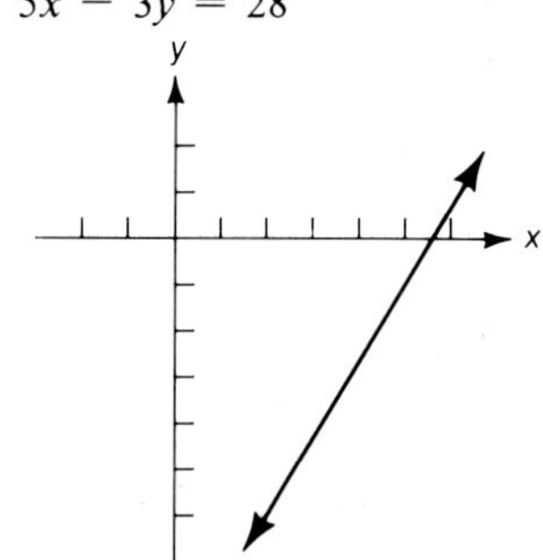

**19.** $3x - 5y = -29$

**20.** $6x - 5y = -28$

## CHAPTER 7

### Exercises 7.1, Page 248

**1.** $x = 5, y = 0$

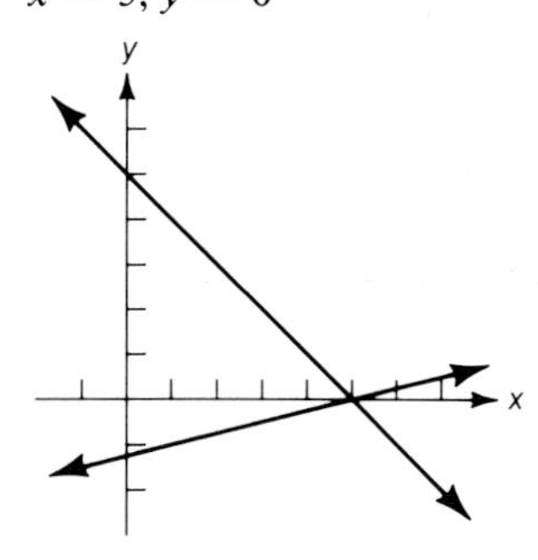

**3.** Inconsistent, no solution

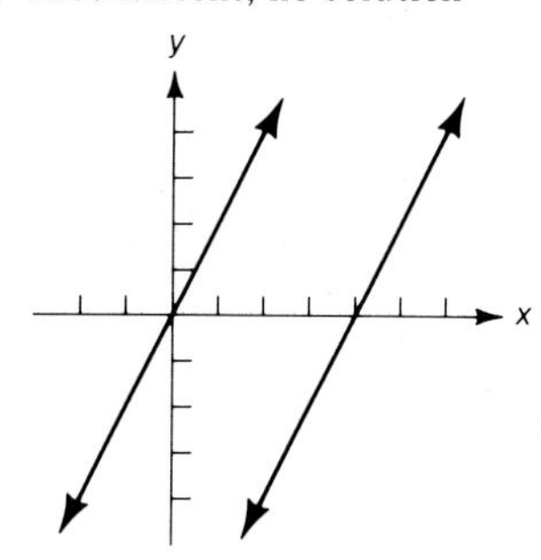

**5.** $x = -6, y = -4$

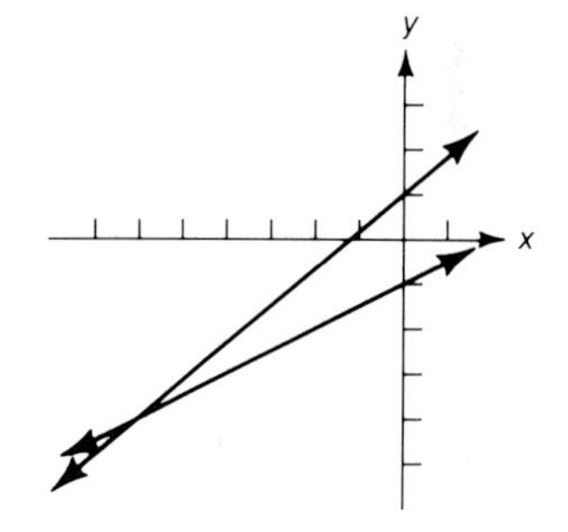

**7.** $x = -1, y = 1$

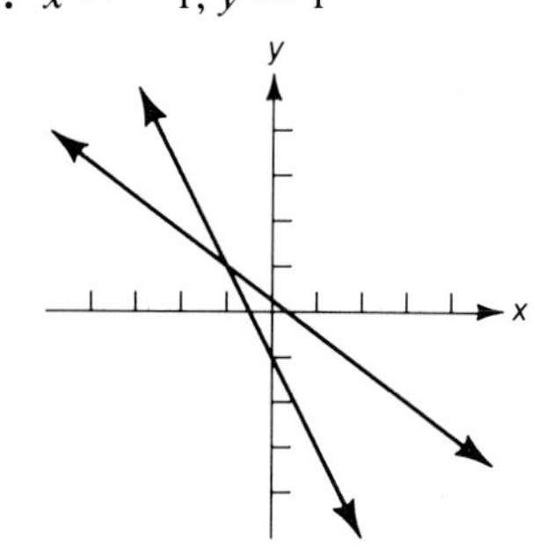

**9.** $x = 3, y = -4$

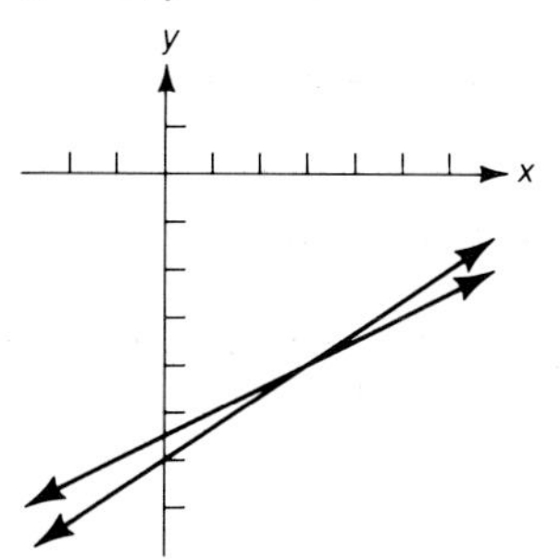

**11.** $x = 2, y = 1$ **13.** $x = -1, y = -3$ **15.** Inconsistent, no solution

**17.** Dependent, infinite number of solutions **19.** $x = 1, y = 6$ **21.** $x = \frac{3}{2}, y = 1$ **23.** $x = 3, y = -\frac{3}{4}$

**25.** Inconsistent, no solution **27.** $x = \frac{1}{2}, y = \frac{1}{4}$ **29.** $x = 5, y = 2$ **31.** $x = 3, y = 3$ **33.** $x = 7, y = 5$

**35.** $x = 4, y = 7$

## Exercises 7.2, Page 251

**1.** 40 liters of 12%, 50 liters of 30% **3.** 240 lb of 5%, 120 lb of 2% **5.** \$3750 certificate, \$8650 in bonds **7.** 325 @ \$3.50, 175 @ \$6.00 **9.** 12 lb @ \$1.96, 18 lb @ \$1.16 **11.** 16 lb @ 35¢, 4 lb @ 65¢ **13.** \$5.50 paperback, \$9.00 hardback **15.** 150 voted "for" **17.** 6:00 P.M. **19.** $a = -2$, $b = 3$ **21.** $a = 5$, $b = 4$ **23.** Labor: \$5.60 per hour; materials: \$4.80 per lb **25.** 9 lb of Ration I; 2 lb of Ration II

## Exercises 7.3, Page 258

**1.** (1,0,1) **3.** (1,2,−1) **5.** Infinite number of solutions **7.** (−2,9,1) **9.** (4,1,1) **11.** (−2,3,1) **13.** No solution **15.** Infinite number of solutions **17.** $\left(\frac{1}{2}, \frac{1}{3}, -1\right)$ **19.** (2,1,−3) **21.** 34, 6, 27 **23.** 11 nickels, 7 dimes, 5 quarters **25.** $a = 1$, $b = 3$, $c = -2$ **27.** 19 cm, 24 cm, 30 cm **29.** Landscaping—\$5500; lot—\$11,000; house—\$45,000

## Exercises 7.4, Page 262

**1.**

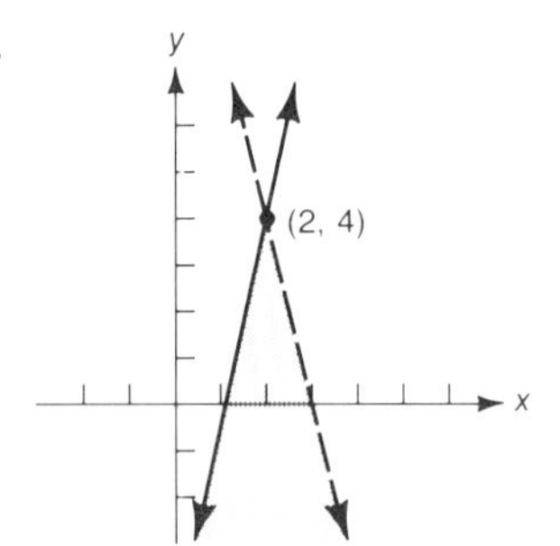


**3.**

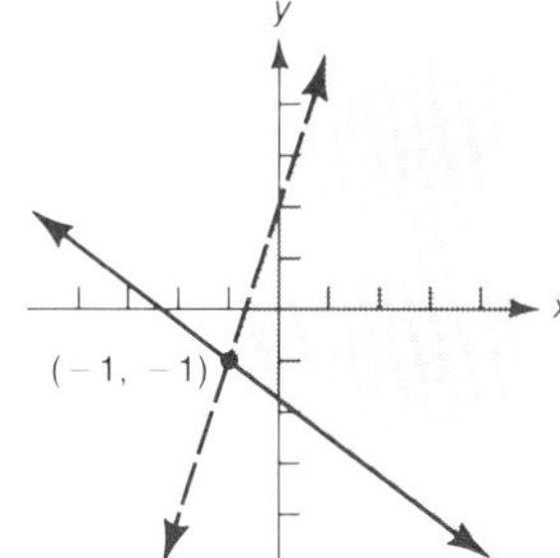


**5.**

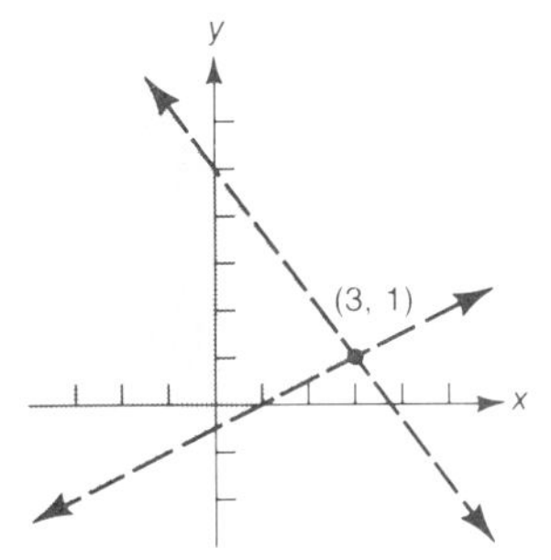


**7.**

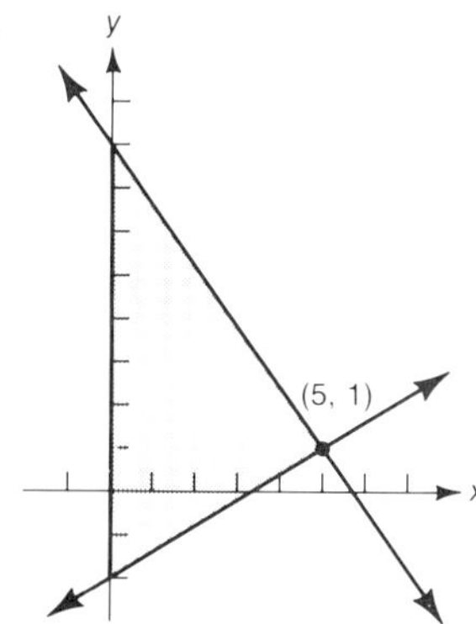


**9.**

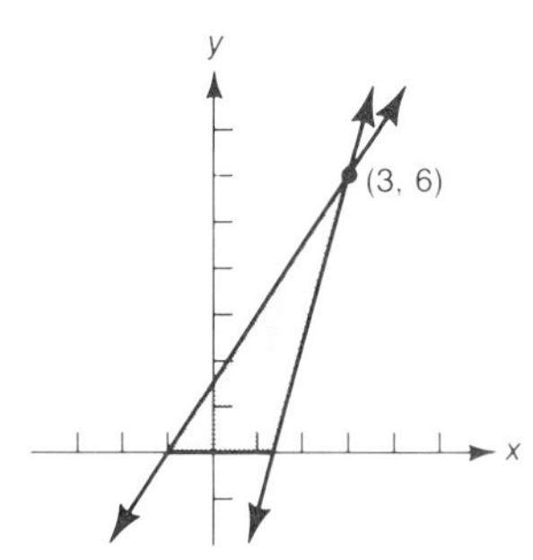


**11.**

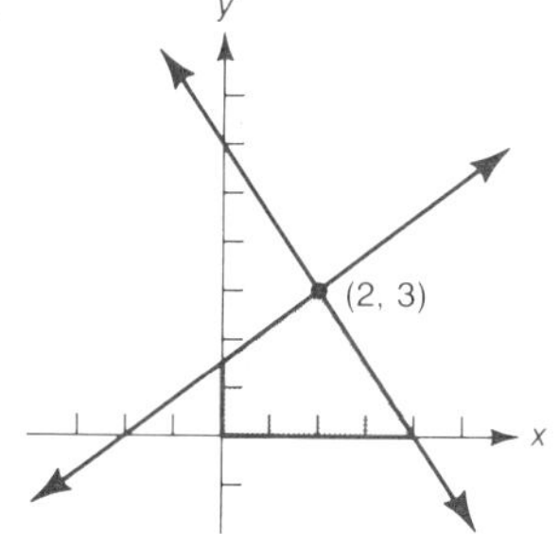

**13.**

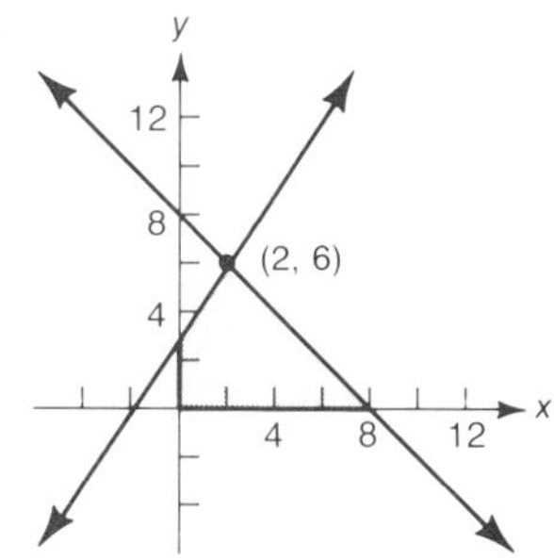


**15.**

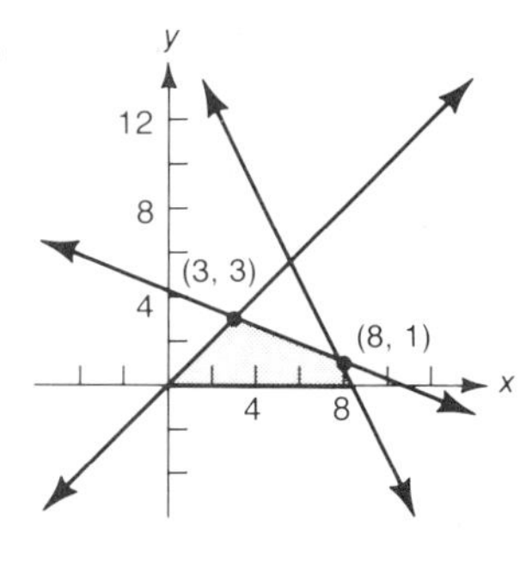


**17.** $F = -3(0) - (0) = 0$

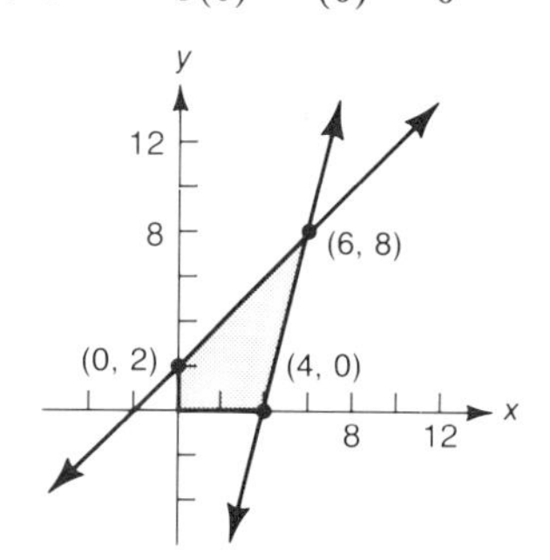


**19.** $F = 7\left(\frac{18}{4}\right) + 14\left(\frac{30}{7}\right) = 78$

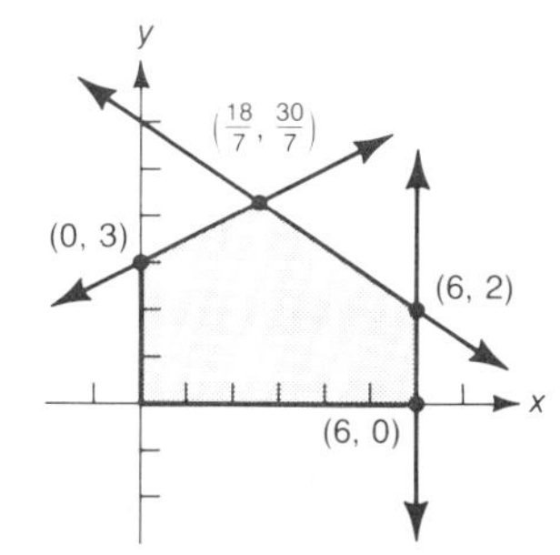


**21.** 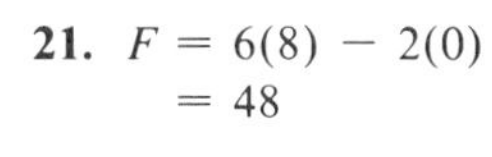
$F = 6(8) - 2(0)$
$= 48$

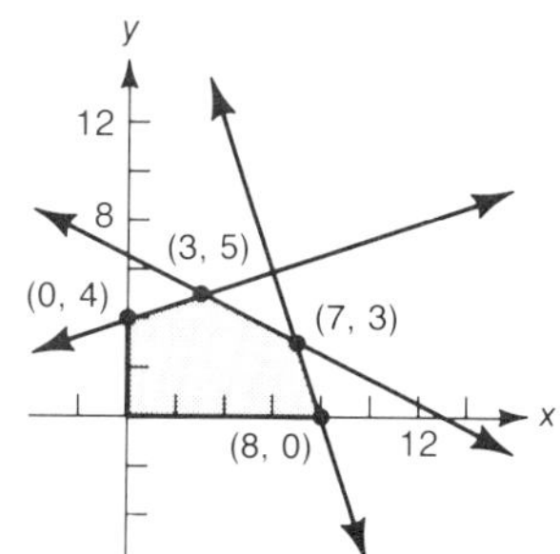


**23.** $x \geq 0, y \geq 0$ $\quad R = 15x + 12y$
$4x + 3y \leq 58$ $\quad = 15(7) + 12(10)$
$8x + 7y \leq 126$ $\quad = \$225$

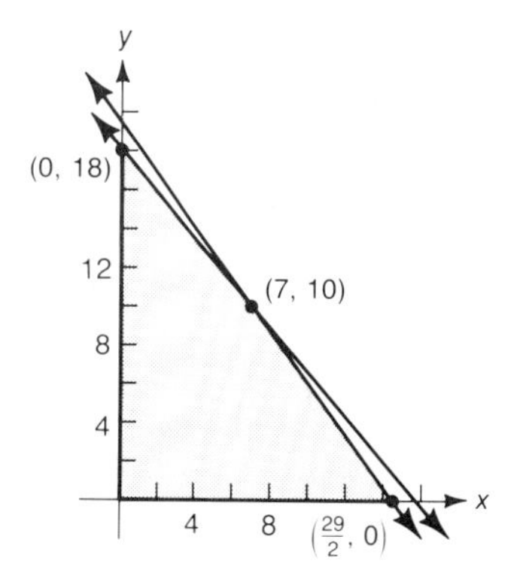


**25.** $x \geq 0, y \geq 0$ $\quad R = 30x + 36y$
$2x + 3y \leq 40$ $\quad = 30(11) + 36(6)$
$6x + 3y \leq 84$ $\quad = \$546$

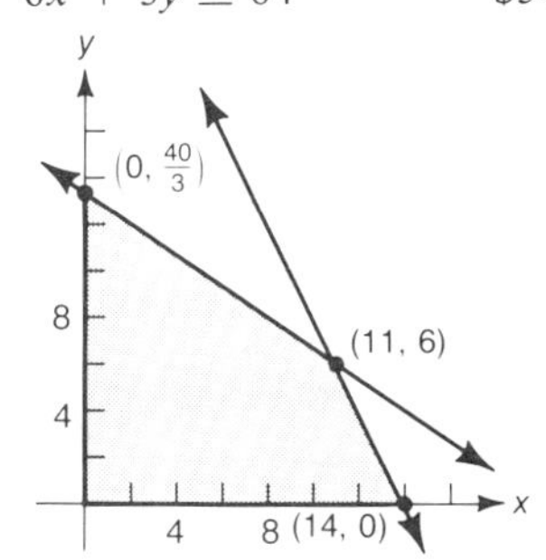

## Chapter 7 Review, Page 264

**1.** $x = 2, y = -2$

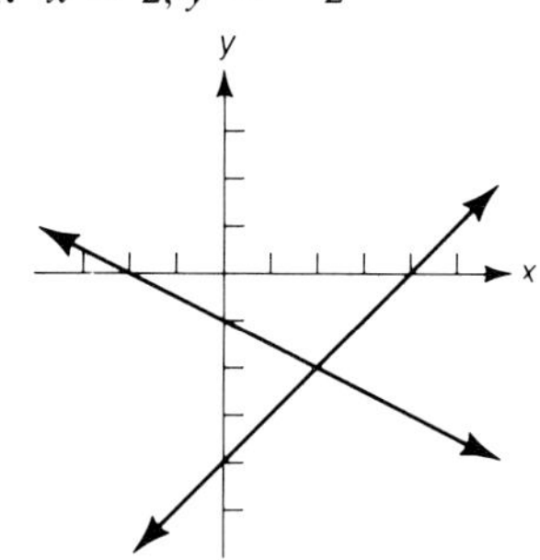


**2.** $x = 3, y = 4$

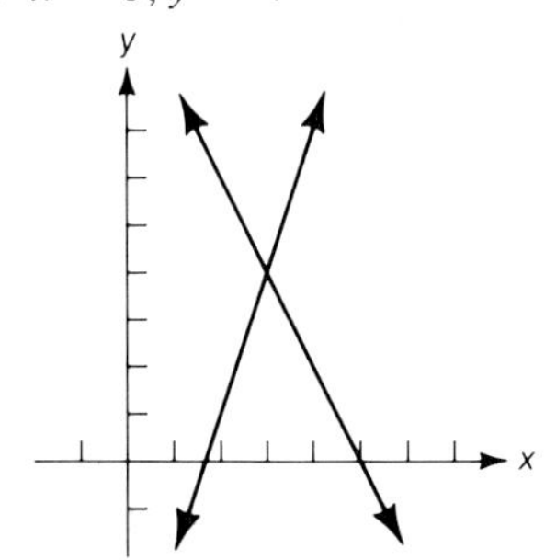


**3.** $x = -2, y = 3$

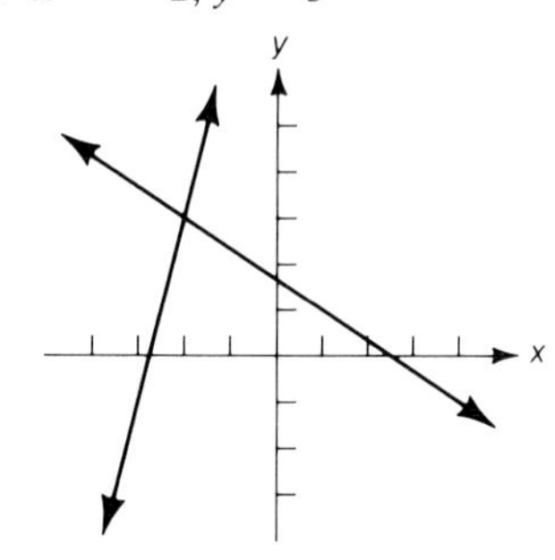


**4.** $x = 2, y = -1$

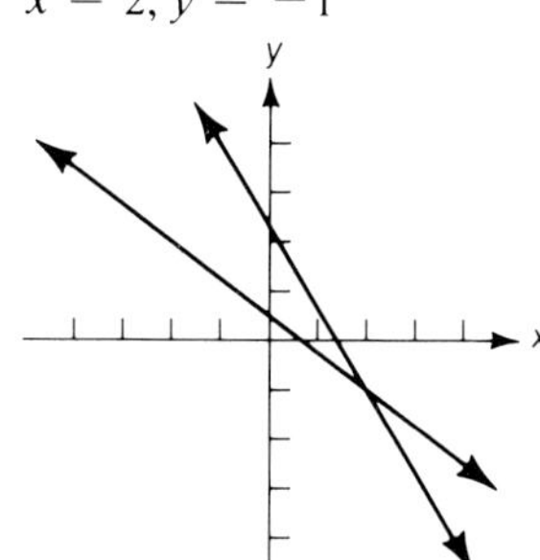


**5.** $x = \frac{1}{2}, y = \frac{3}{2}$ **6.** $x = -2, y = -3$ **7.** $x = \frac{1}{2}, y = 3$
**8.** Inconsistent, no solution **9.** $x = 2, y = 8$ **10.** Dependent, infinite number of solutions **11.** $x = 2, y = \frac{2}{5}, z = -\frac{1}{2}$ **12.** $x = 4, y = 3, z = 2$
**13.** No solution **14.** $x = -1, y = 0, z = 5$ **15.** Infinite number of solutions
**16.** Infinite number of solutions

**17.**

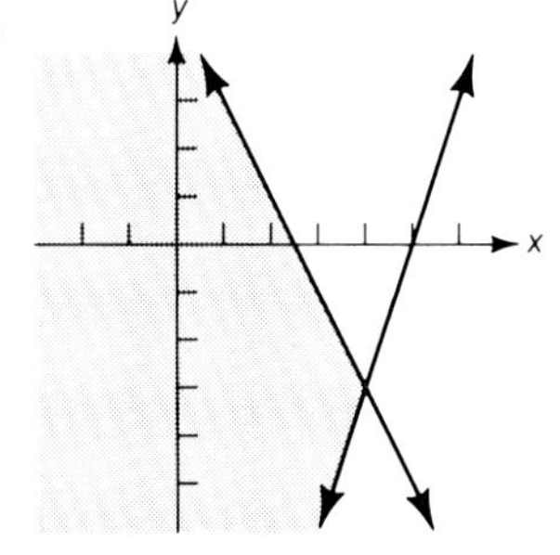


**18.**

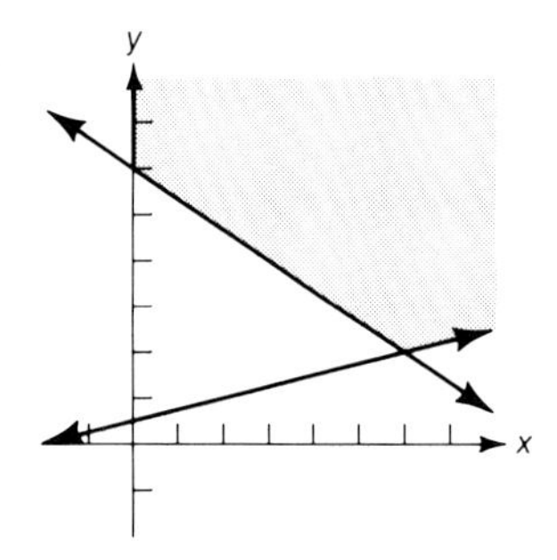


**19.**

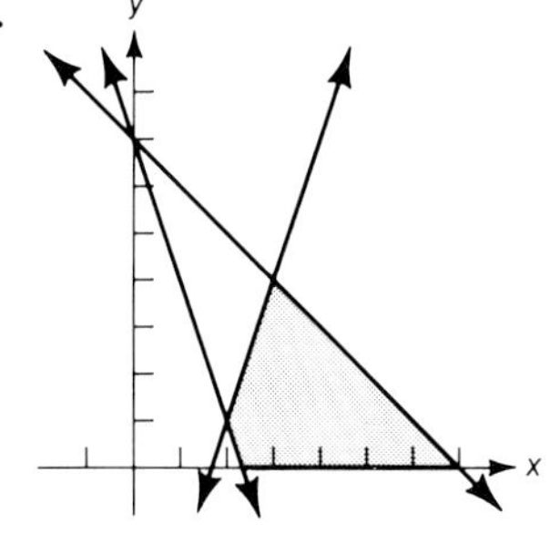


**20.** $F = 13(9) + 15(0)$
$= 117$

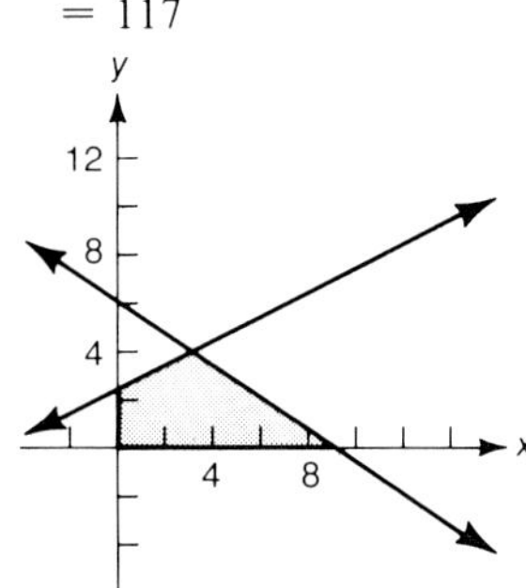


**21.** $F = 8(0) + 20(5)$
$= 100$

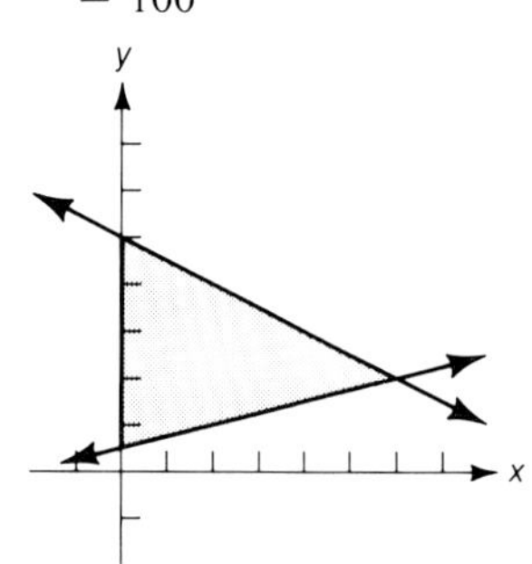


**22.** $F = 18(2) + 24(5)$
$= 156$

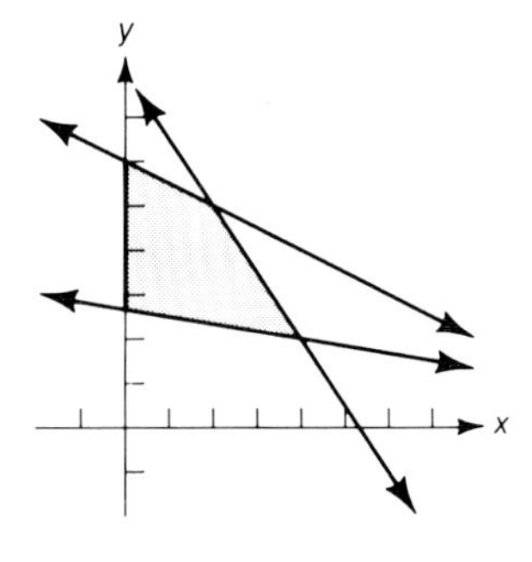

**23.** $a = 8, b = -1$ **24.** Blouse: \$27, skirt: \$32 **25.** 4 nickels, 7 quarters **26.** Grade I: 600 kg, Grade II: 400 kg
**27.** Watch: \$340, ring: \$520 **28.** A: 9 oz, B: 7 oz, C: 5 oz **29.** $a = 2, b = 3, c = 4$

**30.** $P = 25(0) + 17(19)$
$= \$323$

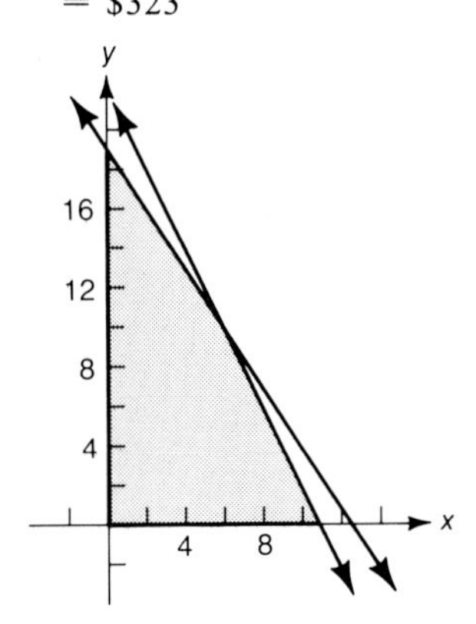


**31.** $P = 104(50) + 130(60)$
$= \$13,000$

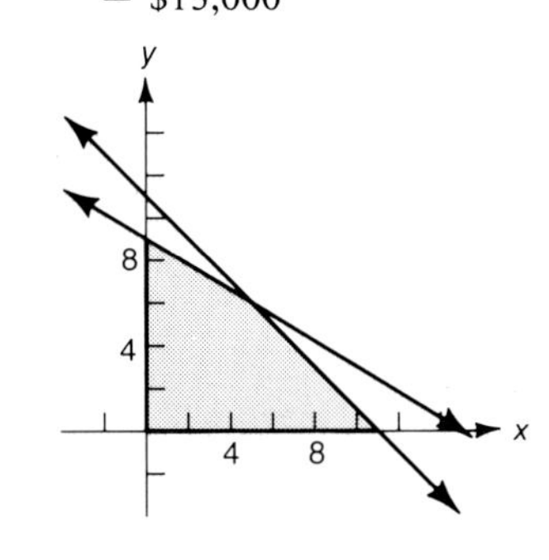


## Chapter 7 Test, Page 265

**1.** $x = 3, y = -1$

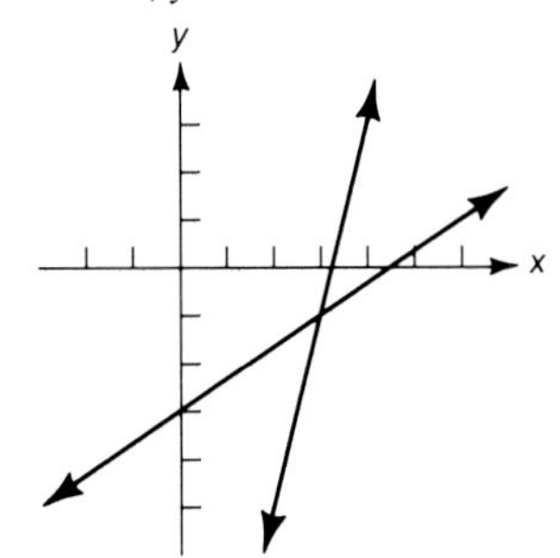


**2.** $x = 4, y = 3$

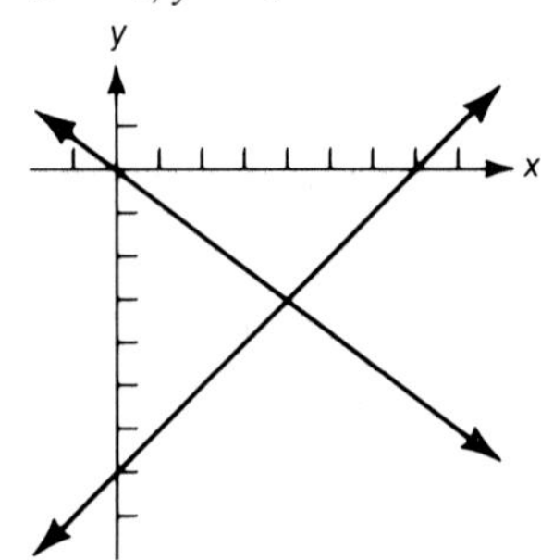


**3.** $x = 7, y = 2$ **4.** Inconsistent, no solution

**5.** $x = \frac{1}{2}, y = \frac{1}{4}$ **6.** $x = 11, y = -7$ **7.** $x = 6, y = 1\frac{1}{2}$
**8.** $a = -3, b = 4$ **9.** 8 ft, 23 ft **10.** Trousers: \$33, shirts: \$23
**11.** $x = 2, y = -1, z = -2$ **12.** $x = -1, y = 1, z = -2$
**13.** Infinite number of solutions **14.** $x = 1, y = -3, z = 2$
**15.** 45 18-cent stamps, 15 20-cent stamps, 40 22-cent stamps

**16.**

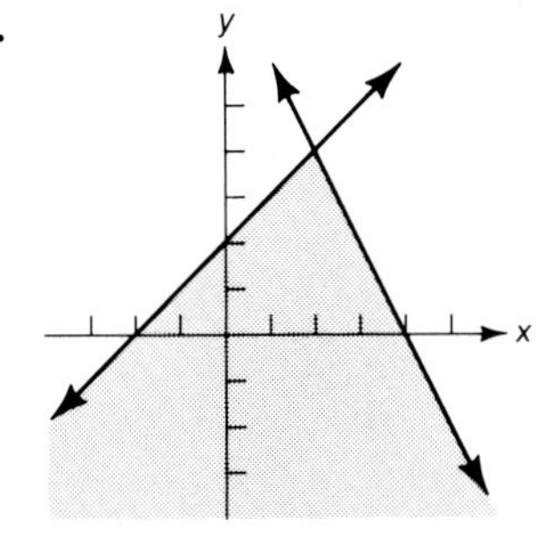


**17.**

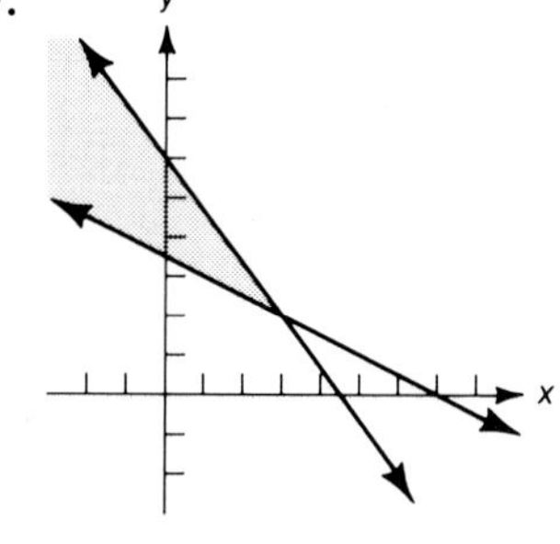


**18.** $F = 12(4) + 9(12)$
$= 156$

y
(4, 12)
10
6
(0, 4)
(8, 0)
4
8
12
x
(0, 0)

**19.** $F = 11(6) + 14(6)$
$= 150$

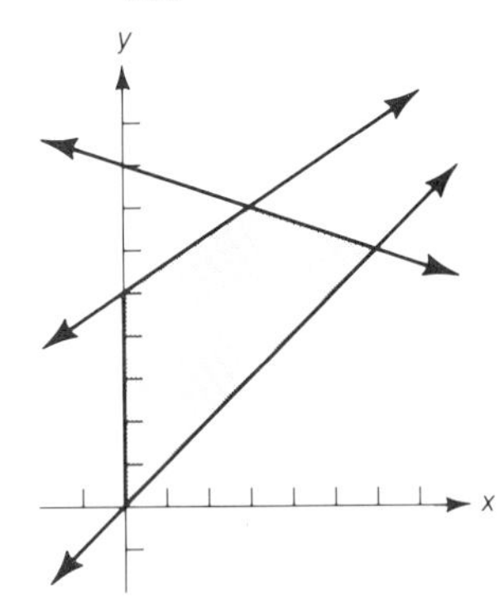

**20.** $R = 12(32) + 16(42.50)$
$= \$1064$

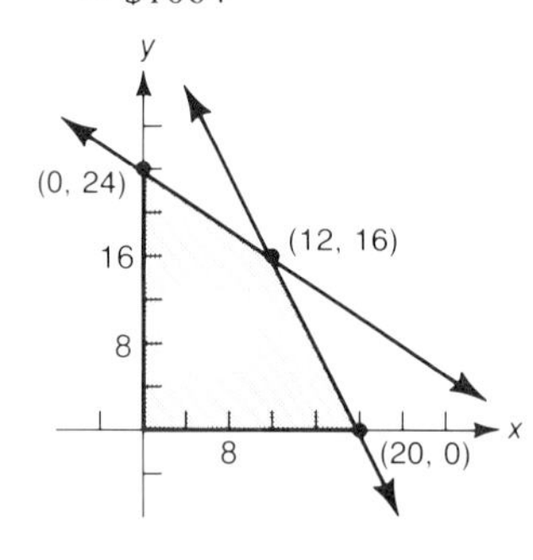


## Cumulative Review (7), Page 267

**1.** 28 **2.** 8 **3.** $(3x + 4)(2x - 5)$ **4.** $2(x - 3)(x^2 + 3x + 9)$ **5.** $(2x + 5)(4x^2 - 10x + 25)$

**6.** $(x + 3)(x + 2)(x - 2)$ **7.** $x^2 + 7x + \underline{\frac{49}{4}} = \left(x + \frac{7}{2}\right)^2$ **8.** $3x^2 - 24x + \underline{48} = 3(x - 4)^2$ **9.** $k = \frac{81}{4}$

**10.** $k < \frac{9}{16}$ **11.** $x = -\frac{3}{2}, x = 5$ **12.** $x = \frac{5 \pm \sqrt{73}}{6}$

**13.**

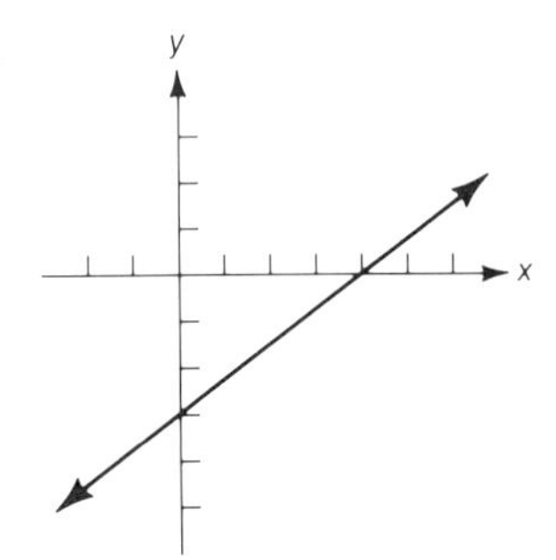

**14.**

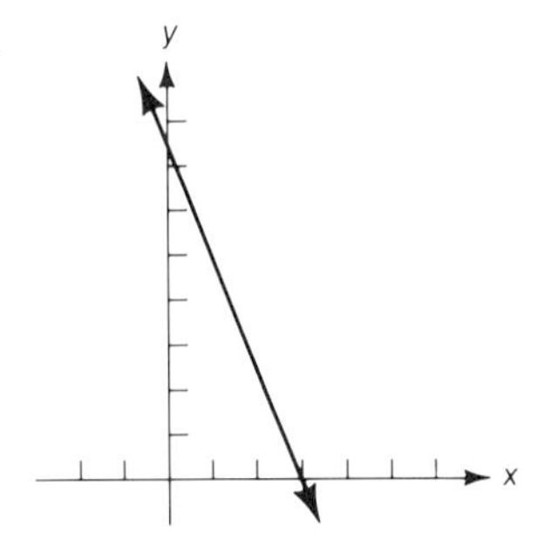

**15.**

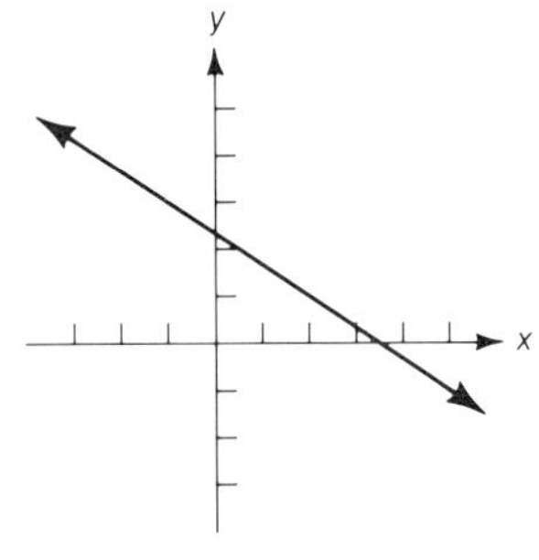

**16.**

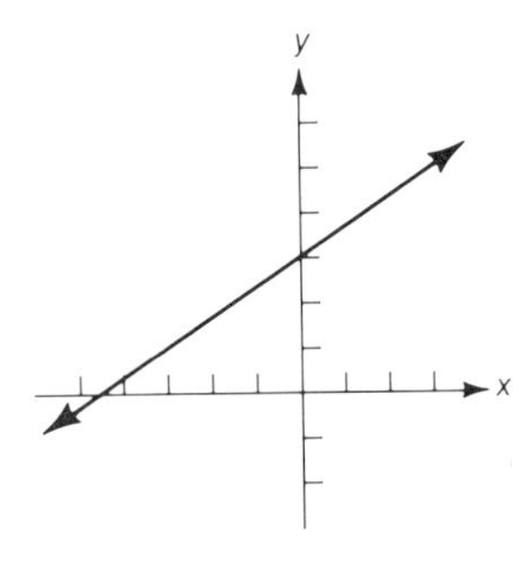

**17.** $x = 1, y = 4$

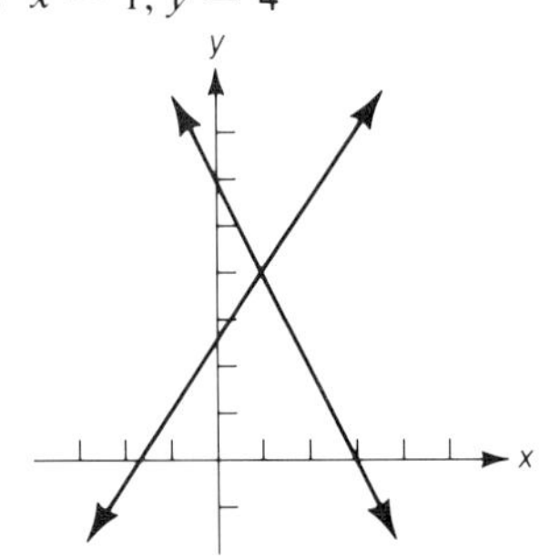

**18.** $x = 1, y = 3$

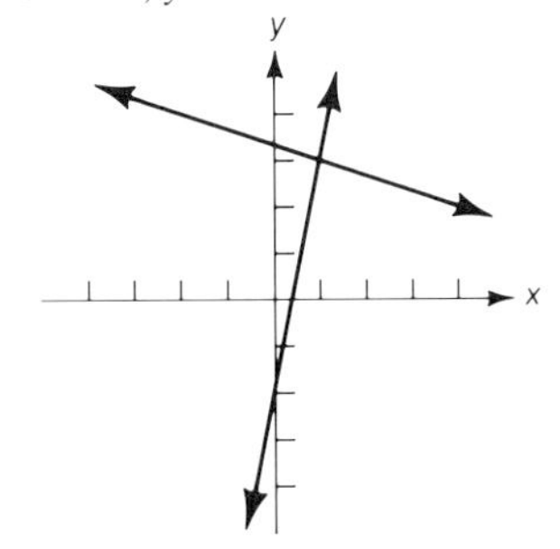

**19.** 3 batches, 2 batches **20.** \$4200 @ 7%; \$2800 @ 8%

## CHAPTER 8

### Exercises 8.1, Page 275

**1.** $\{(-4,0), (-1,4), (1,2), (2,5), (6,-3)\}$; $D = \{-4,-1,1,2,6\}$; $R = \{0,4,2,5,-3\}$; Is a function
**3.** $\{(-5,-4), (-4,-2), (-2,-2), (1,-2), (2,1)\}$; $D = \{-5,-4,-2,1,2\}$; $R = \{-4,-2,1\}$; Is a function
**5.** $\{(-4,1), (-4,-3), (-1,3), (-1,-1), (3,-4)\}$; $D = \{-4,-1,3\}$; $R = \{1,-3,3,-1,-4\}$; Is not a function
**7.** $\{(-5,3), (-5,-5), (0,5), (1,2), (1,-2)\}$; $D = \{-5,0,1\}$; $R = \{3,-5,5,2,-2\}$; Is not a function
**9.** $\{(-5,5), (-3,1), (0,-3), (3,-1), (4,2)\}$; $D = \{-5,-3,0,3,4\}$; $R = \{5,1,-3,-1,2\}$; Is a function

**11.** $D = \{0,1,4,-3,2\}$
$R = \{0,6,-2,5,-1\}$
Is a function

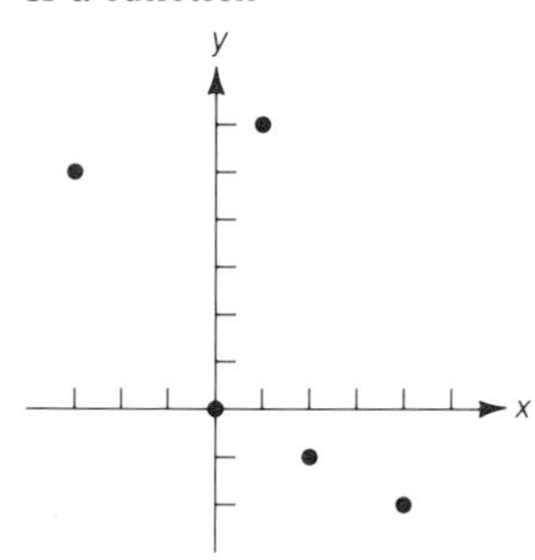

**13.** $D = \{-4,3,1,2\}$
$R = \{4,-2,0,-3,1\}$
Is not a function

**15.** $D = \{-3,-1,0,2,3\}$
$R = \{1,2,4,5\}$
Is a function

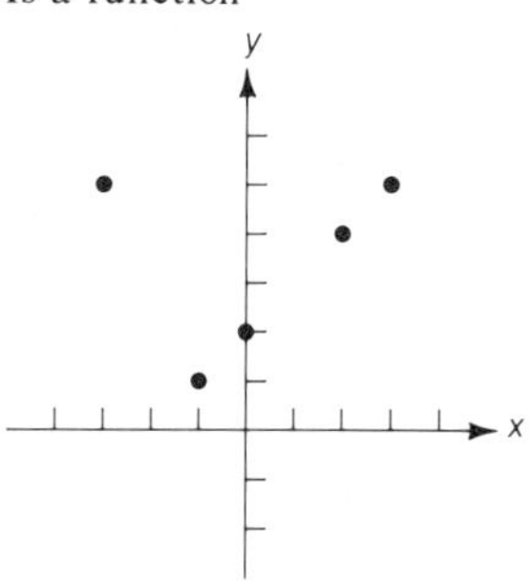

**17.** $D = \{-1,-3,1,0,2\}$
$R = \{-2,-1,0,3,5\}$
Is a function

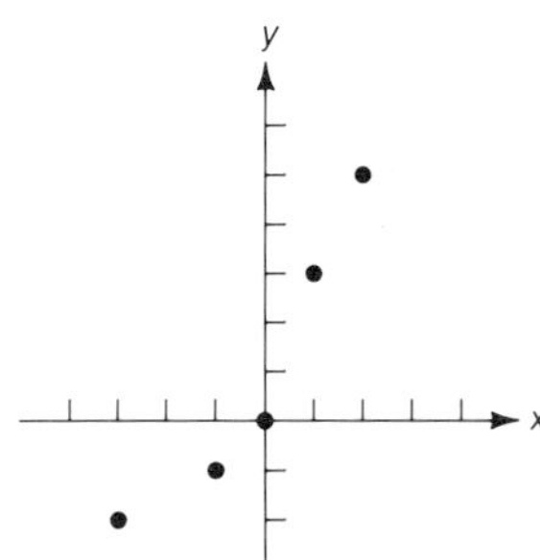

**19.** $D = \{2\}$
$R = \{-4,-2,0,3,5\}$
Is not a function

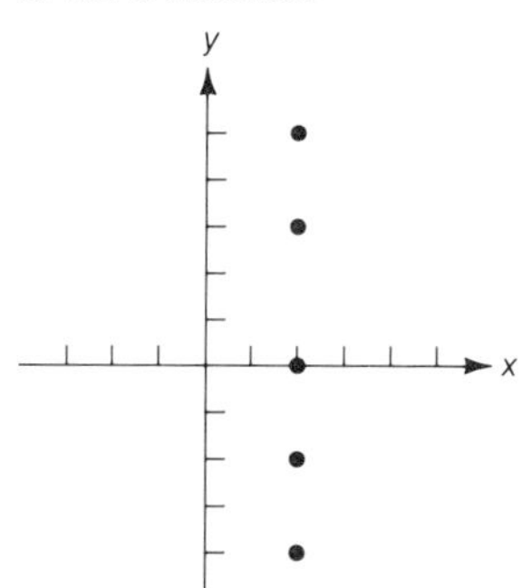

**21.** $D = \{-1\}$
$R = \{-4,-2,0,2,6\}$
Is not a function

**23.** Is a function **25.** Is a function **27.** Is not a function **29.** Is not a function **31.** Is a function

**33.** Is not a function **35.** $\left\{(-9,-26), \left(-\frac{1}{3},0\right), (0,1), \left(\frac{4}{3},5\right) (2,7)\right\}$

**37.** $\{(-2,-11), (-1,-2), (0,1), (1,-2), (2,-11)\}$ **39.** $\left\{(-1,4), \left(0,\frac{3}{2}\right), \left(\frac{1}{2},\frac{13}{10}\right), \left(2,\frac{7}{4}\right), \left(4,\frac{19}{6}\right)\right\}$

**41.** $\{(-2,0), (-1,\sqrt{2}), (0,2), (2,2\sqrt{2}), (3,\sqrt{10}\}$ **43.** $\{x \mid x \neq -3\}$ **45.** $\{x \mid x \neq -2, x \neq 2\}$

**47.** $\left\{x \mid x \geq -\frac{5}{2}\right\}$ **49.** $\{x \mid -3 \leq x \leq 3\}$

## Exercises 8.2, Page 289

**1.** $x = 0$, $(0,-4)$, $\{y \mid y \geq -4\}$ **3.** $x = 0$, $(0,9)$, $\{y \mid y \geq 9\}$ **5.** $x = 0$, $(0,1)$, $\{y \mid y \leq 1\}$

**7.** $x = 0$, $\left(0,\frac{1}{5}\right)$, $\left\{y \mid y \leq \frac{1}{5}\right\}$ **9.** $x = -1$, $(-1,0)$, $\{y \mid y \geq 0\}$ **11.** $x = 4$, $(4,0)$, $\{y \mid y \leq 0\}$

**13.** $x = -3$, $(-3,-2)$, $\{y \mid y \geq -2\}$ **15.** $x = -2$, $(-2,-6)$, $\{y \mid y \geq -6\}$

**17.** $x = \frac{3}{2}$, $\left(\frac{3}{2},\frac{7}{2}\right)$, $\left\{y \mid y \leq \frac{7}{2}\right\}$ **19.** $x = \frac{4}{5}$, $\left(\frac{4}{5},-\frac{11}{5}\right)$, $\left\{y \mid y \geq -\frac{11}{5}\right\}$

**21. (a)**

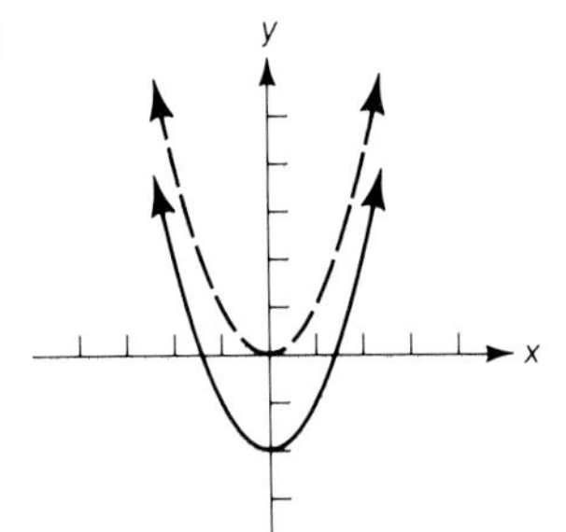

**(b)**

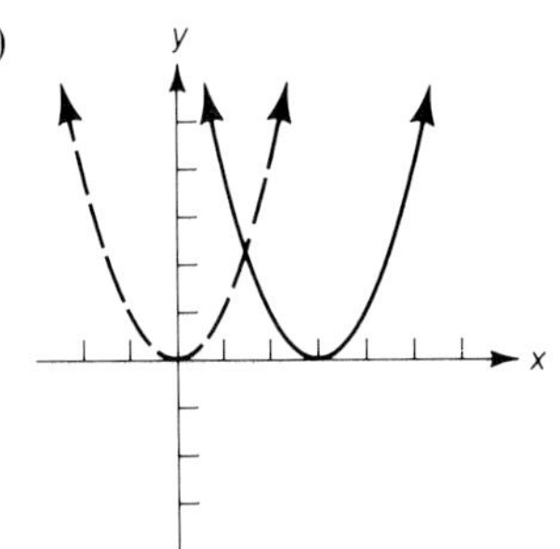


**(c)**

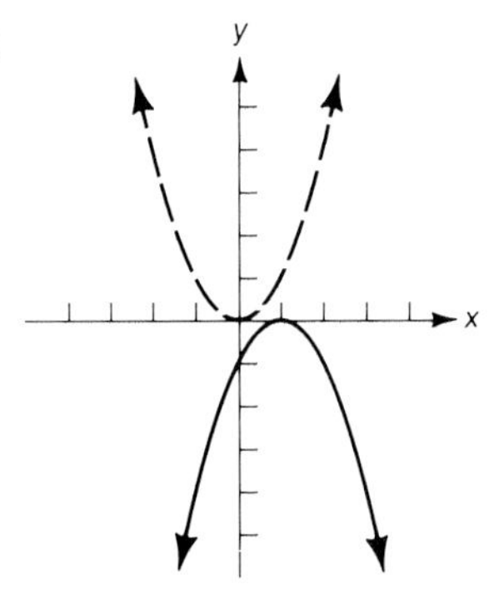


**(d)**

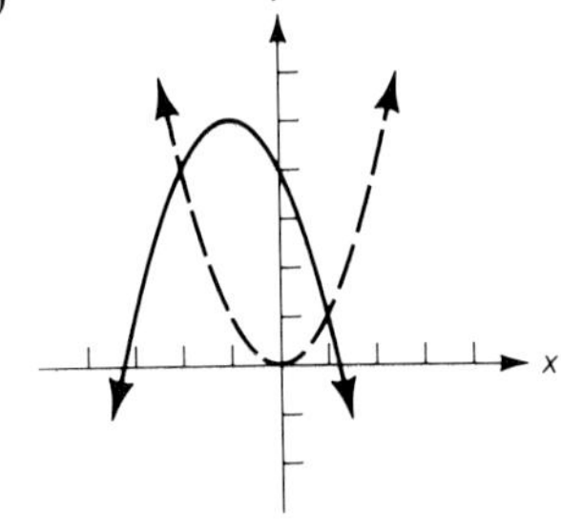


**23. (a)**

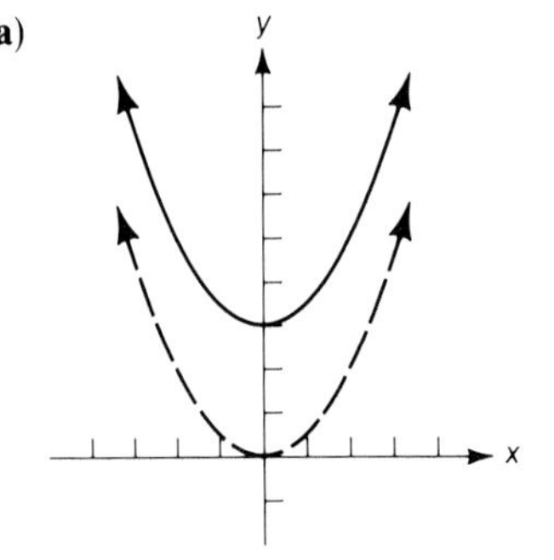


**(b)**

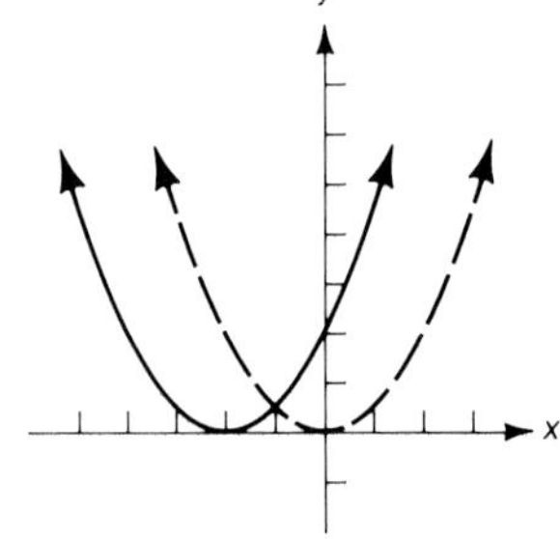


**(c)**

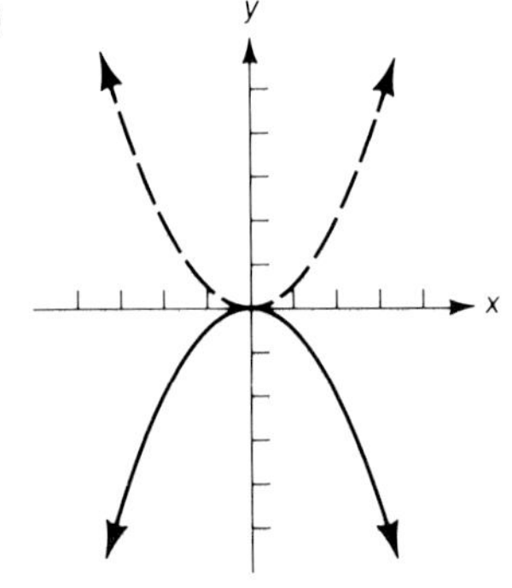


**(d)**

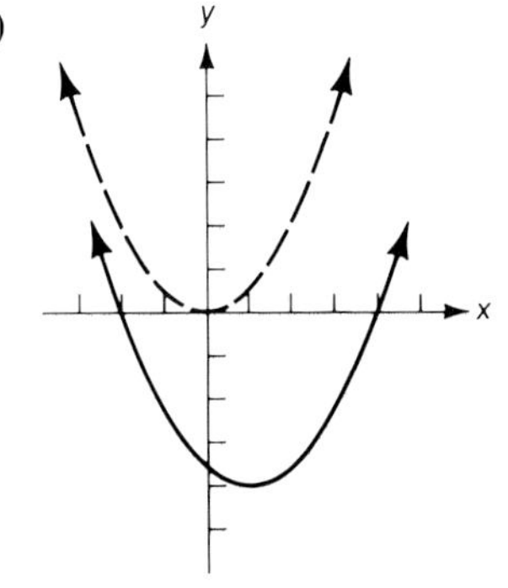

**25.** $y = 2(x - 1)^2$
Vertex $= (1,0)$
$R_f = \{y \mid y \geq 0\}$
Zeros: $x = 1$

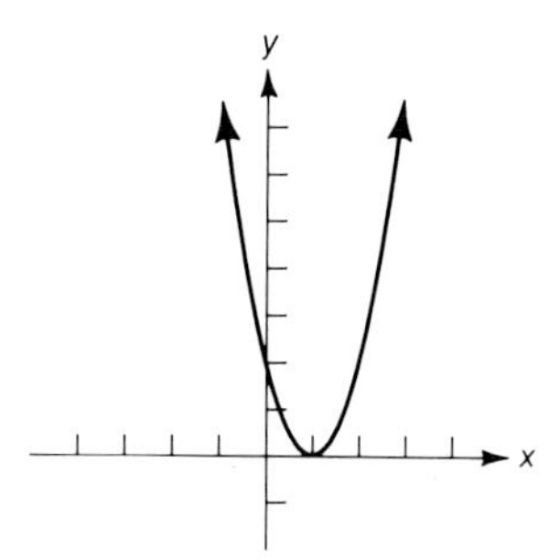


**27.** $y = (x - 1)^2 - 4$
Vertex $= (1,-4)$
$R_f = \{y \mid y \geq -4\}$
Zeros: $x = 3,\ x = -1$

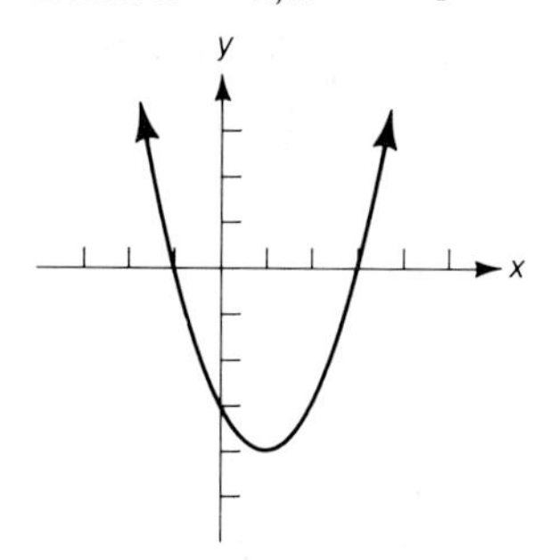


**29.** $y = (x + 3)^2 - 4$
Vertex $= (-3,-4)$
$R_g = \{y \mid y \geq -4\}$
Zeros: $x = -5,\ x = -1$

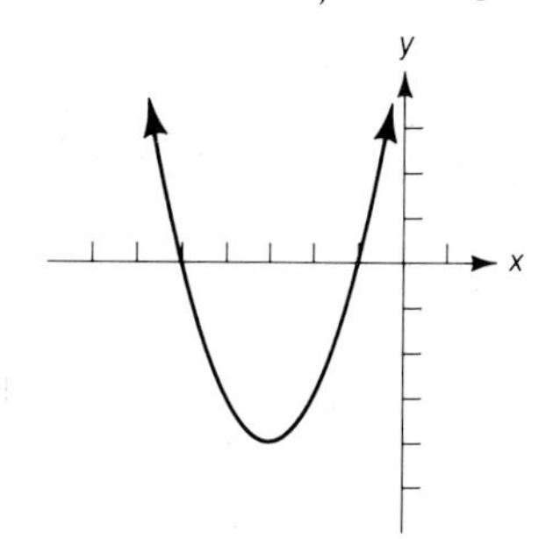


**31.** $y = 2(x - 2)^2 - 3$
Vertex $= (2,-3)$
$R_f = \{y \mid y \geq -3\}$
Zeros: $x = \dfrac{4 \pm \sqrt{6}}{2}$

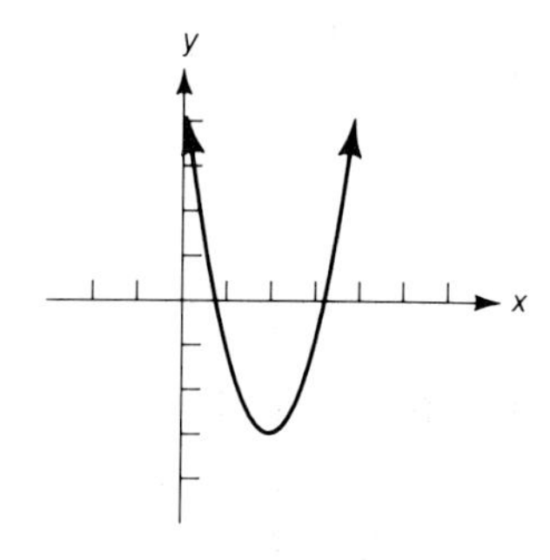


**33.** Vertex $= (-2,3)$
$R_g = \{y \mid y \leq 3\}$
Zeros: $x = -1,\ x = -3$

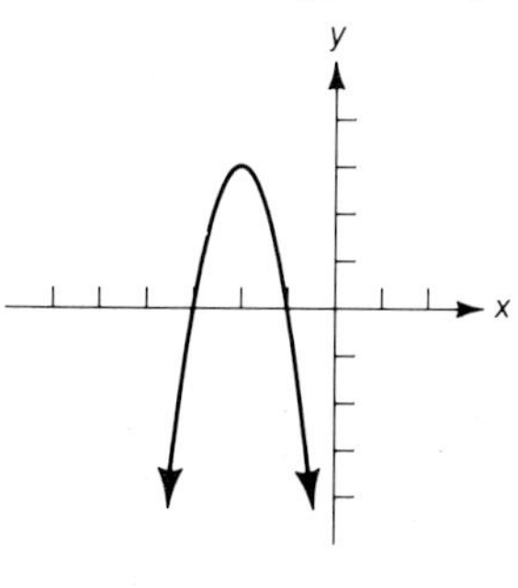


**35.** Vertex $= (1,3)$
$R_f = \{y \mid y \geq 3\}$
Zeros: none

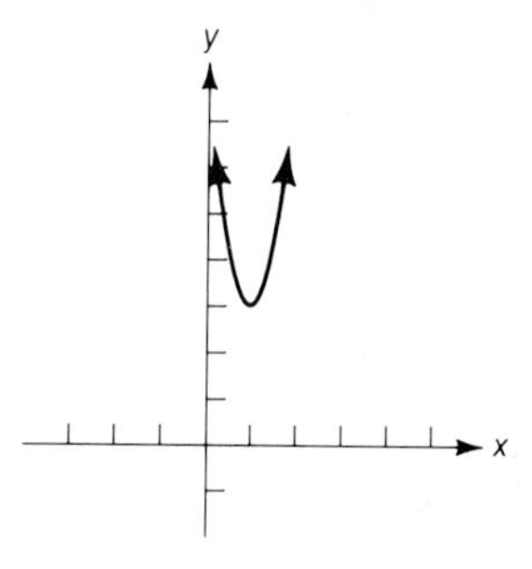


**37.** Vertex $= \left(-\dfrac{5}{2}, \dfrac{17}{4}\right)$
$R_f = \left\{y \mid y < \dfrac{17}{4}\right\}$
Zeros: $x = \dfrac{-5 \pm \sqrt{17}}{2}$

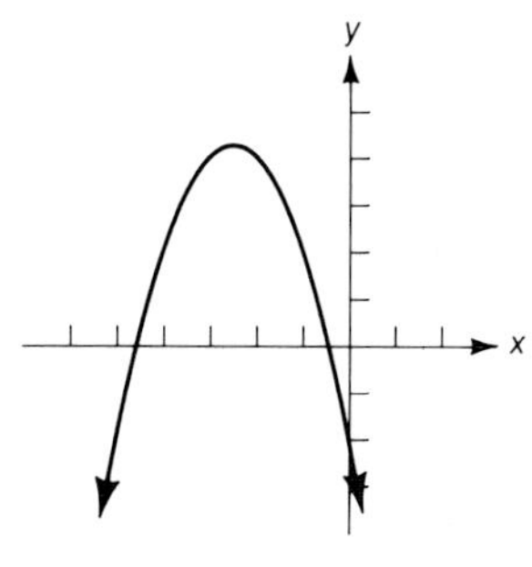


**39.** Vertex $= \left(-\dfrac{7}{4}, -\dfrac{9}{8}\right)$
$R_f = \left\{y \mid y \geq -\dfrac{9}{8}\right\}$
Zeros: $x = -1,\ x = -\dfrac{5}{2}$

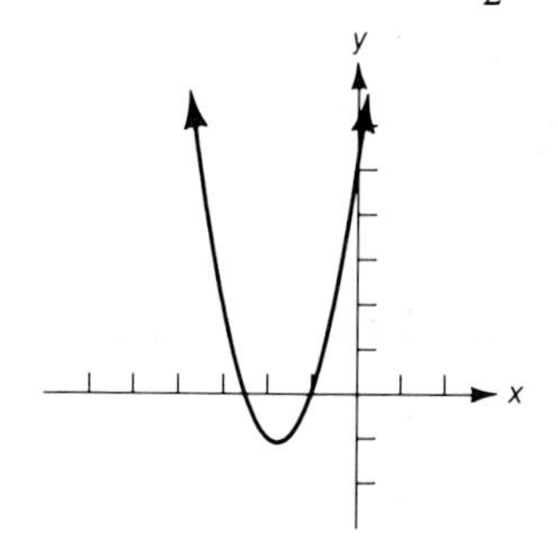


**41.** Vertex $= \left(-\dfrac{7}{6}, -\dfrac{25}{12}\right)$
$R_g = \left\{y \mid y \geq -\dfrac{25}{12}\right\}$
Zeros: $x = -2,\ x = -\dfrac{1}{3}$

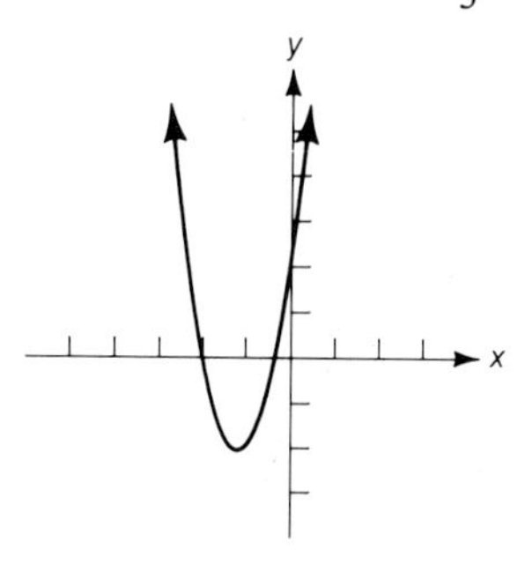


**43.** **(a)** $3\frac{1}{2}$ sec **(b)** 196 ft **45.** **(a)** 4 sec **(b)** 288 ft **47.** \$20 **49.** \$30, \$1800

## Exercises 8.3, Page 299

**1.** **(a)** 12 **(b)** 4 **(c)** $a + 8$ **3.** **(a)** 4 **(b)** 31 **(c)** $x^2 - 4x - 1$ **5.** **(a)** 1 **(b)** $4a + 5$ **(c)** $4x + 4h - 3$ **(d)** 4 **7.** **(a)** 0 **(b)** $a^2 - 6a + 5$ **(c)** $x^2 + 2xh + h^2 - 4$ **(d)** $2x + h$ **9.** **(a)** $-3$ **(b)** $2a^2 - 8a + 5$ **(c)** $2x^2 + 4xh + 2h^2 - 3$ **(d)** $4x + 2h$

**11.**

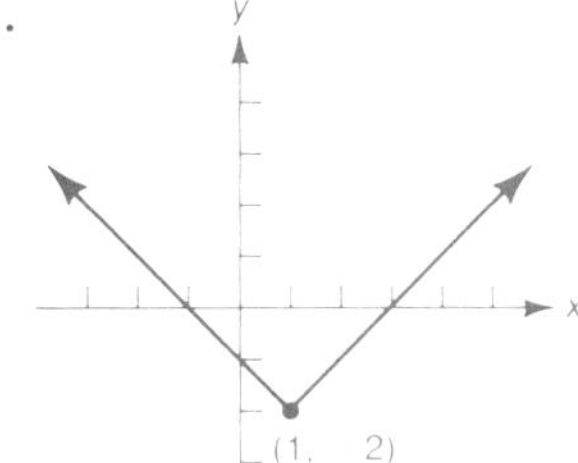


**13.**

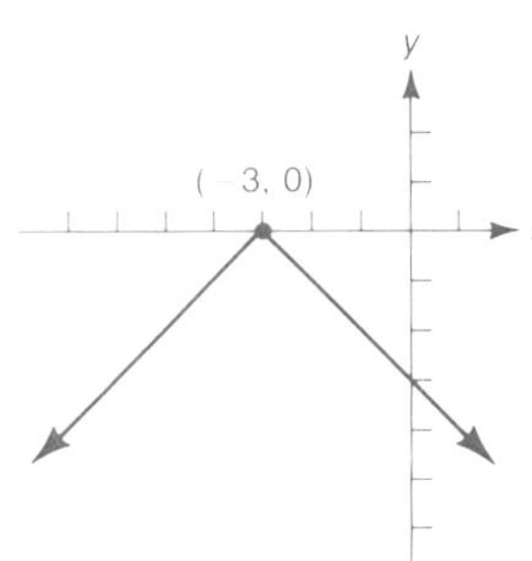


**15.**

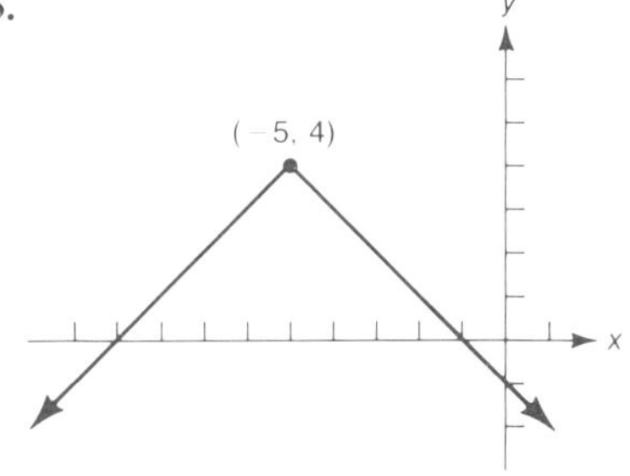


**17.**

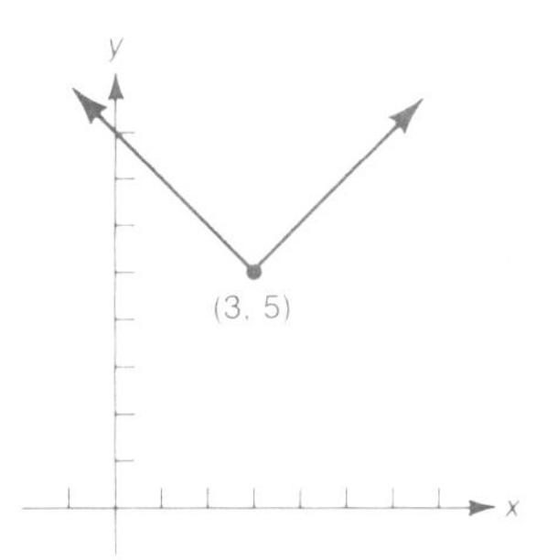


**19.**

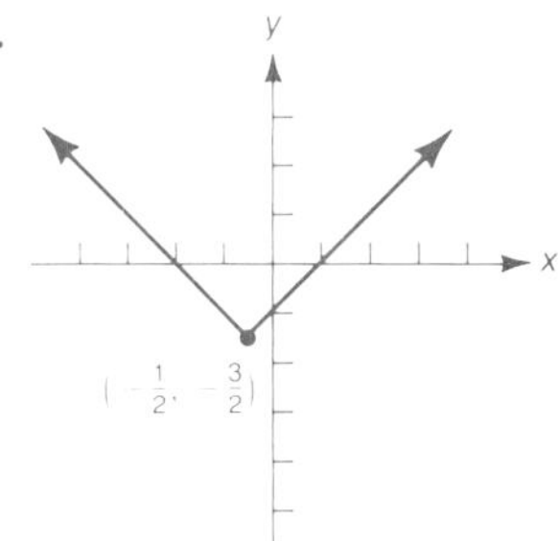


**21.**

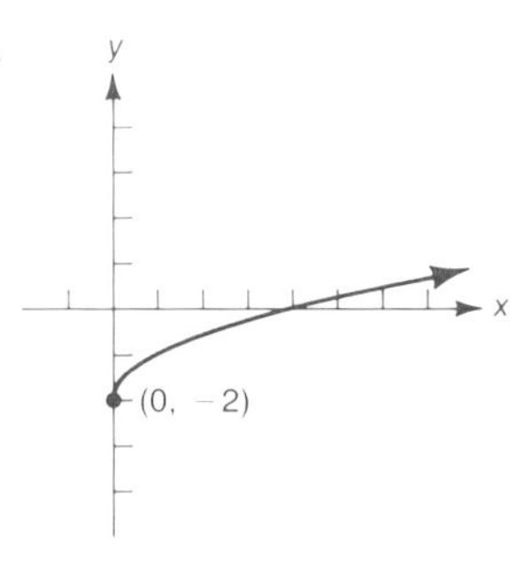


**23.**

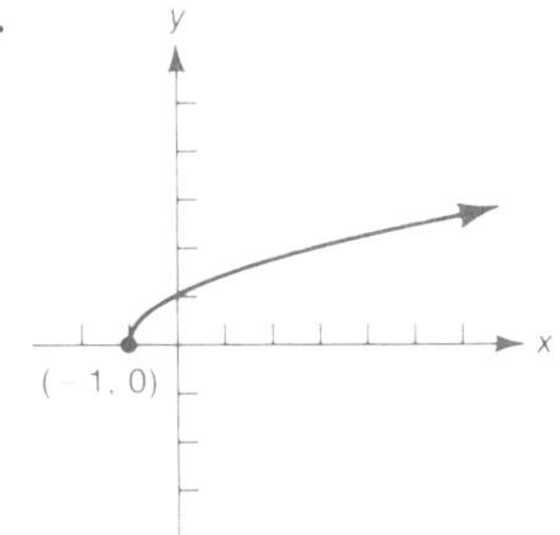


**25.**

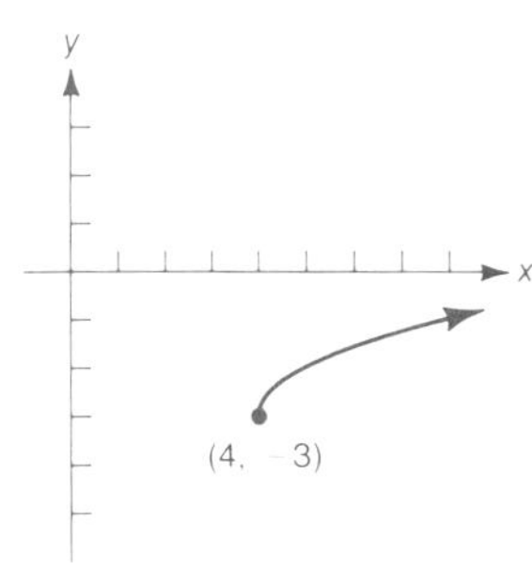


**27.**

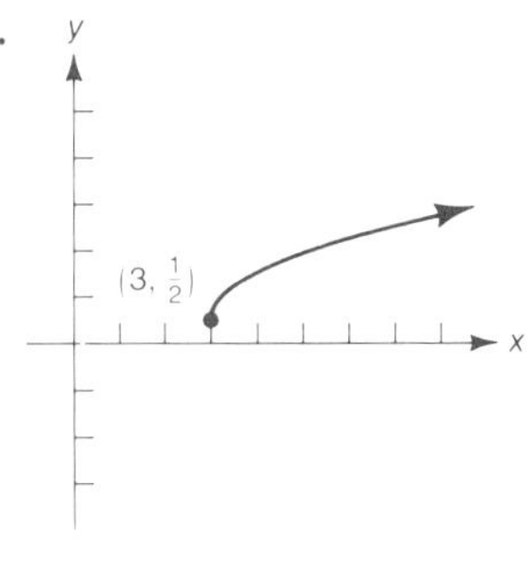


**29.**

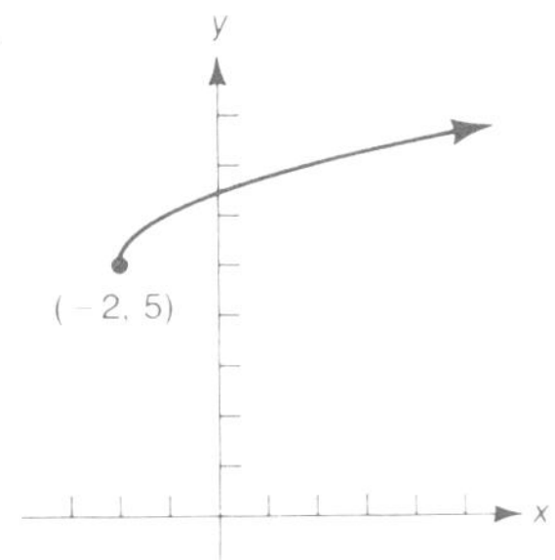


**31.**

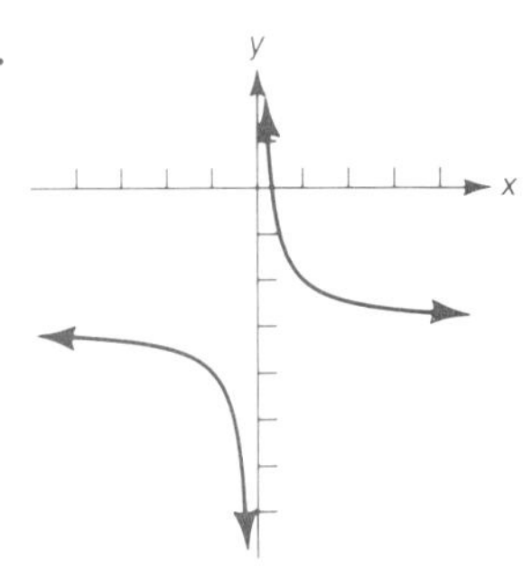


**33.**

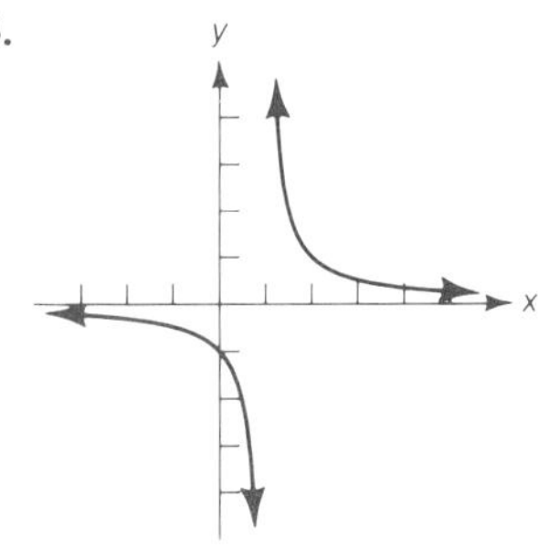

**35.**

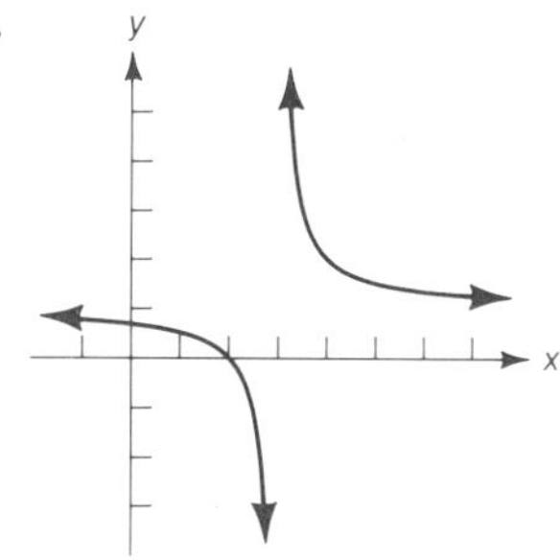


**37.**

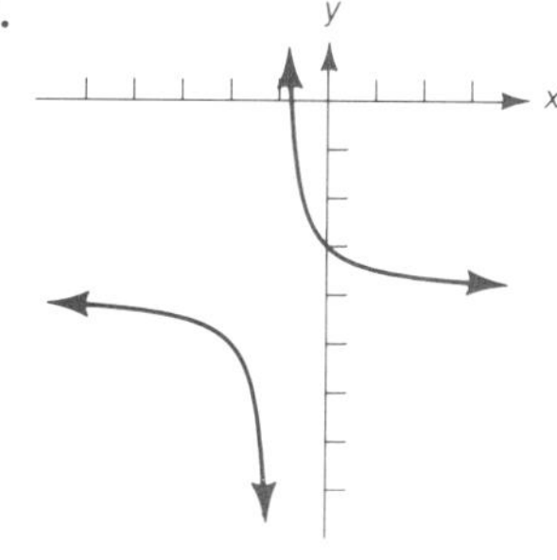


**39.**

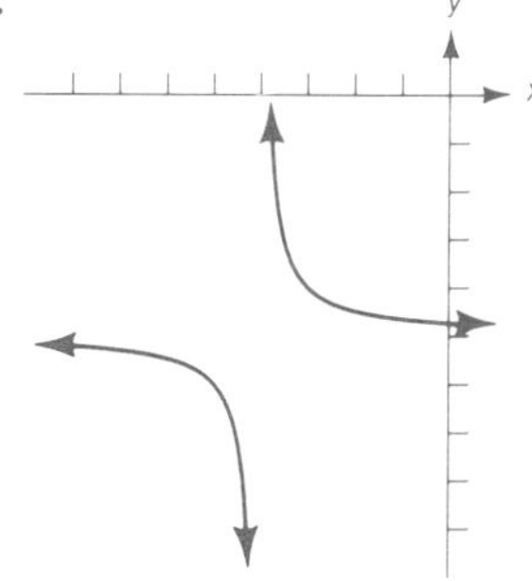


**41.**

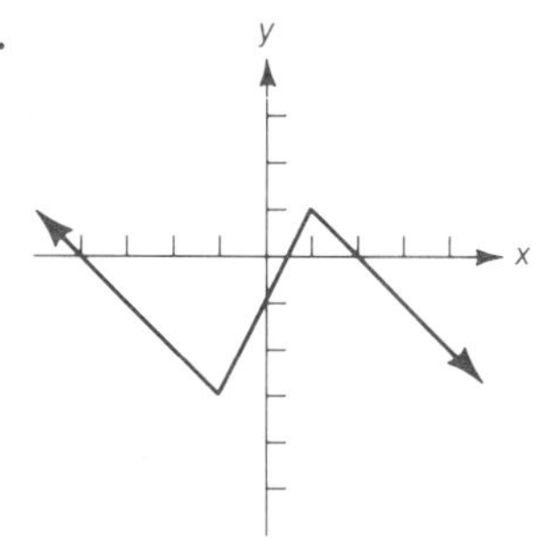


**43.**

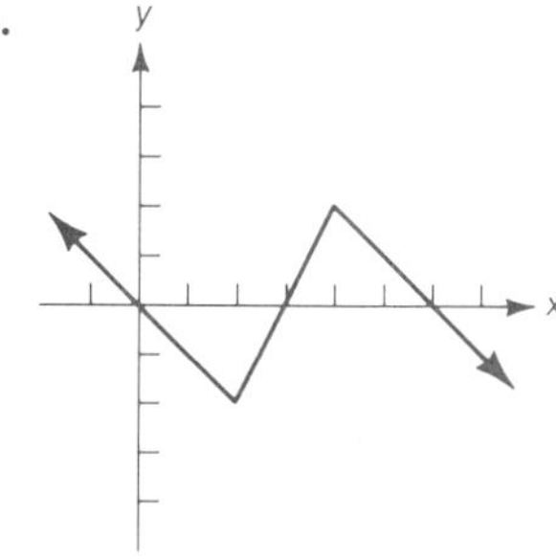


**45.**

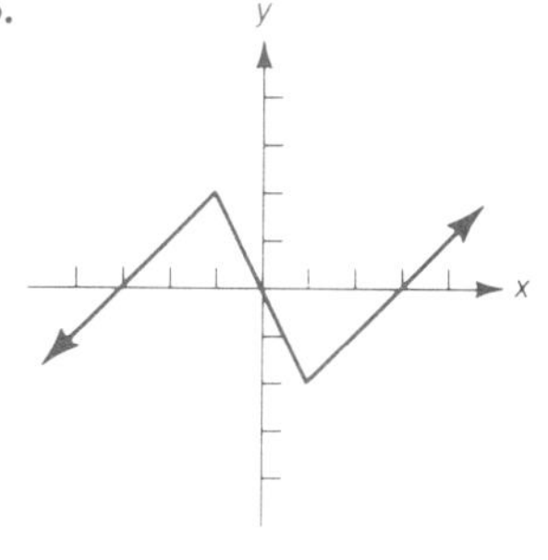


**47.**

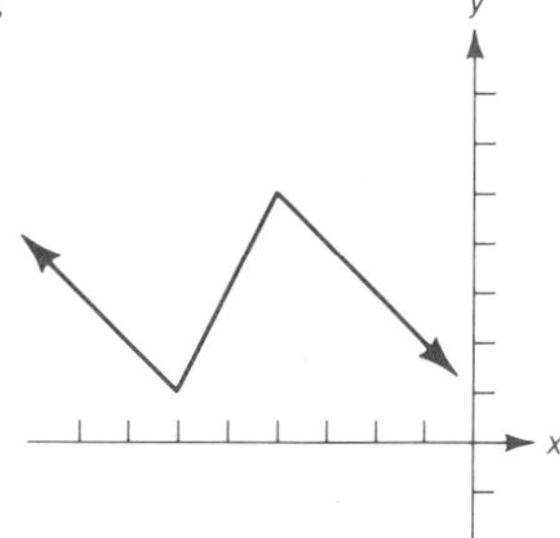


**49.**

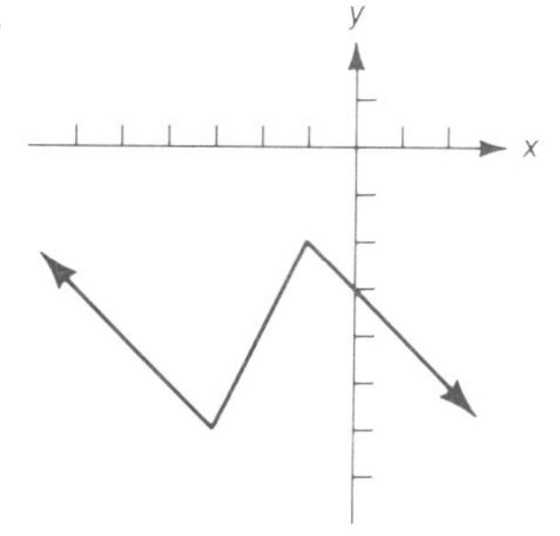


## *Exercises 8.4, Page 308*

**1.** $f[g(x)] = \dfrac{3}{2}x + 11$

$g[f(x)] = \dfrac{3}{2}x + \dfrac{9}{2}$

**3.** $f[g(x)] = 4x^2 + 12x + 9$
$g[f(x)] = 2x^2 + 3$

**5.** $f[g(x)] = \dfrac{1}{5x - 8}$

$g[f(x)] = \dfrac{5}{x} - 8$

**7.** $f[g(x)] = \dfrac{1}{x^2} - 1$

$g[f(x)] = \dfrac{1}{(x - 1)^2}$

**9.** $f[g(x)] = x^3 + 3x^2 + 4x + 3$
$g[f(x)] = x^3 + x + 2$

**11.** $f[g(x)] = \sqrt{x - 2}$
$g[f(x)] = \sqrt{x} - 2$

**13.** $f[g(x)] = \sqrt{x^2} = |x|$
$g[f(x)] = (\sqrt{x})^2 = x$

**15.** $f[g(x)] = \dfrac{1}{\sqrt{x^2 - 4}}$
$g[f(x)] = \left(\dfrac{1}{\sqrt{x}}\right)^2 - 4 = \dfrac{1}{x} - 4$

**17.** $f[g(x)] = x$
$g[f(x)] = x$

**19.** $f[g(x)] = (x - 8)^{3/2}$
$g[f(x)] = \sqrt{x^3 - 8}$

**21.** Is a 1–1 function

**23.** Is not a 1–1 function

**25.** Is a 1–1 function

**27.** Is a 1–1 function

**29.** Is not a 1–1 function

**31.** $f^{-1}(x) = \dfrac{1}{2}x + \dfrac{3}{2}$

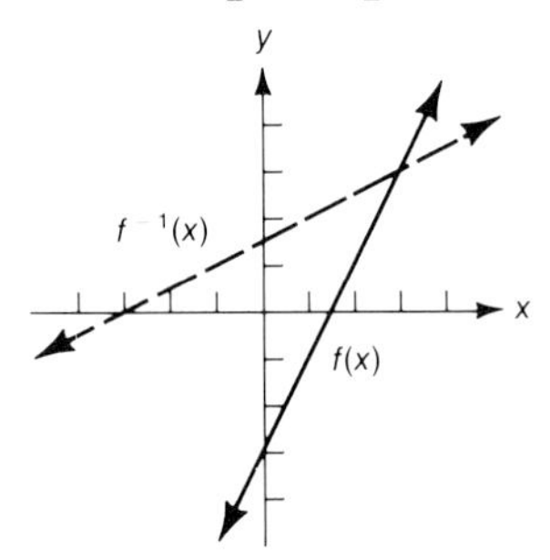


**33.** $g^{-1}(x) = x$

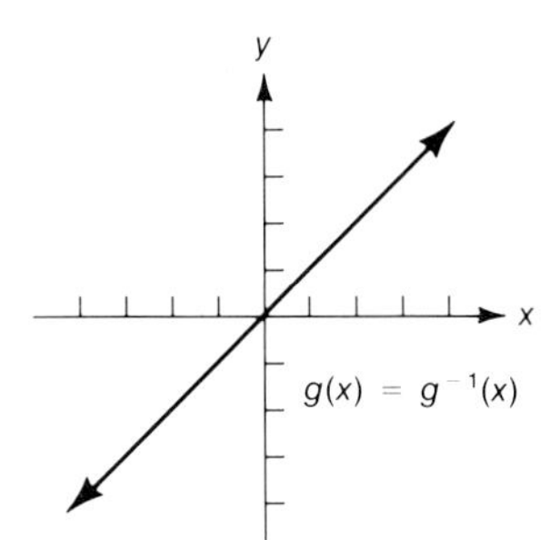


**35.** $f^{-1}(x) = \dfrac{1}{5}x - \dfrac{1}{5}$

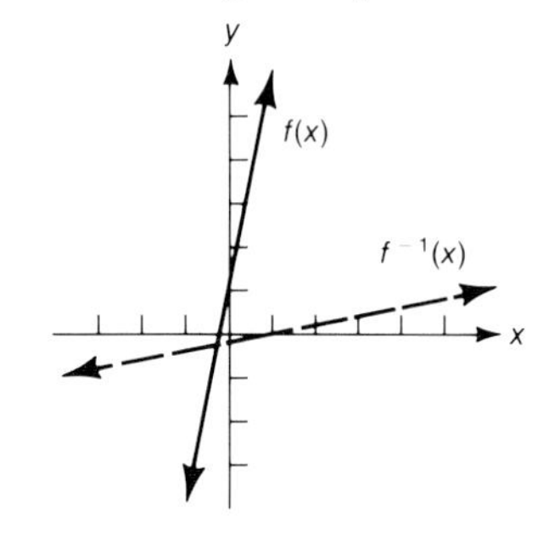


**37.** $f^{-1}(x) = x^{1/3}$

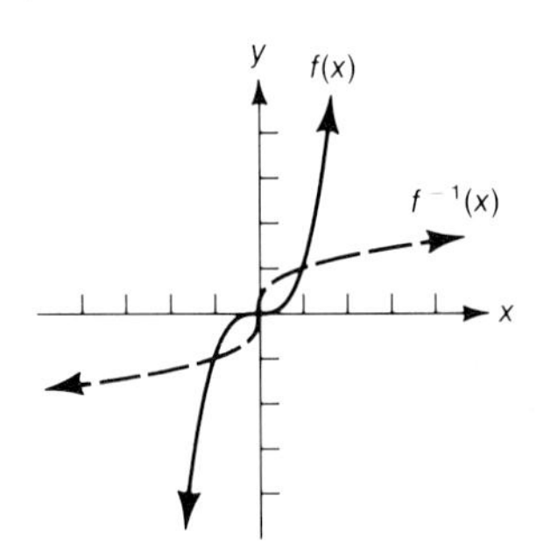


**39.** $f^{-1}(x) = \dfrac{1}{x}$

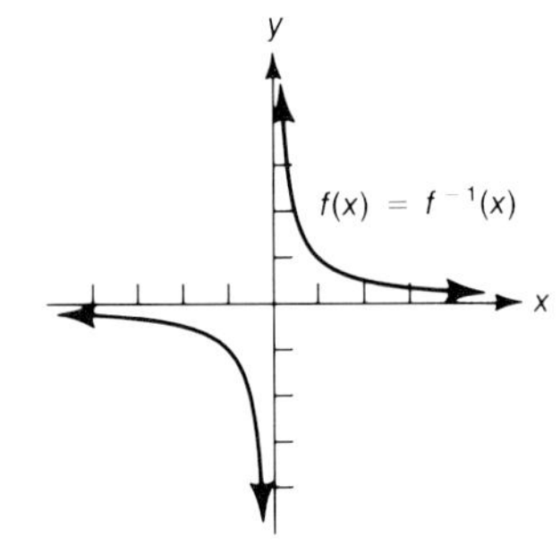


**41.** $f^{-1}(x) = \sqrt{x}$

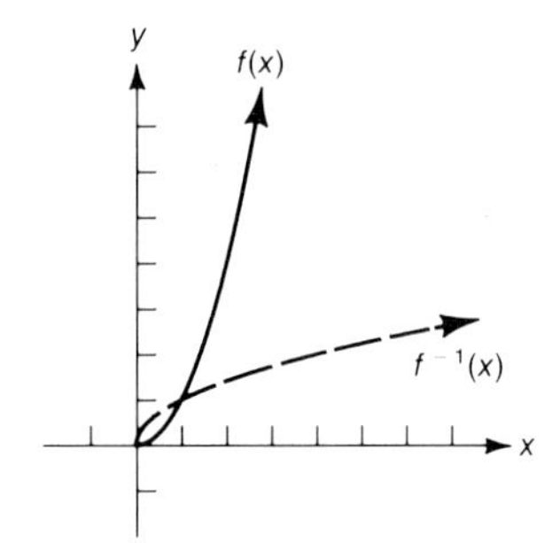


**43.** $g^{-1}(x) = x^2 - 7$
$x \geq 0$

**45.** $f(x) = -\sqrt{x}$ $\quad D_f = \{x \mid x \geq 0\}$ $\quad R_f = \{y \mid y \leq 0\}$
$f^{-1}(x) = x^2$ $\quad D_{f^{-1}} = \{x \mid x \leq 0\}$ $\quad R_{f^{-1}} = \{y \mid y \geq 0\}$

**47.** $f(x) = x^2 + 1$ $\quad D_f = \{x \mid x \geq 0\}$ $\quad R_f = \{y \mid y \geq 1\}$
$f^{-1}(x) = \sqrt{x - 1}$ $\quad D_{f^{-1}} = \{x \mid x \geq 1\}$ $\quad R_{f^{-1}} = \{y \mid y \geq 0\}$

**49.** $f(x) = -x - 2$ $\quad D_f = R_f = D_{f^{-1}} =$ $\quad R_{f^{-1}} = R$
$f^{-1}(x) = -x - 2$

**51.**

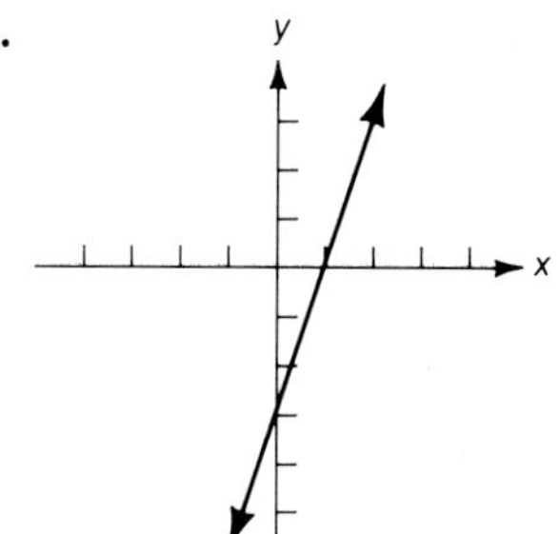


**53.** Is not a 1–1 function **55.** Is not a 1–1 function

**57.**

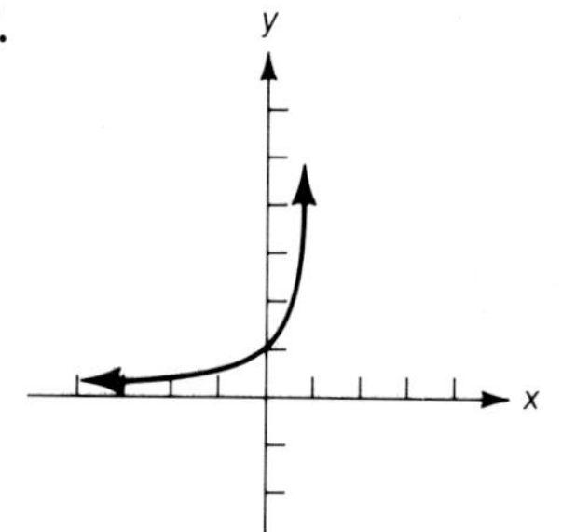


**59.**

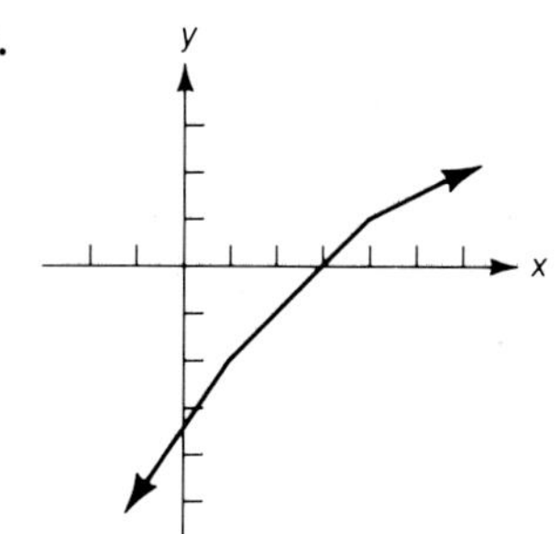


## *Chapter 8 Review, Page 312*

**1.** $\{(-4,-2), (-2,1), (-1,2), (1,2), (3,1)\}$; $D = \{-4,-2,-1,1,3\}$; $R = \{-2,1,2\}$; Is a function
**2.** $\{(-3,-2), (-2,2), (1,1), (1,-2), (4,-1)\}$; $D = \{-3,-2,1,4\}$; $R = \{-2,2,1,-1\}$; Is not a function
**3.** $D = \{0,2,4,1\}$
$R = \{3,1,-1,2,6\}$
Is not a function

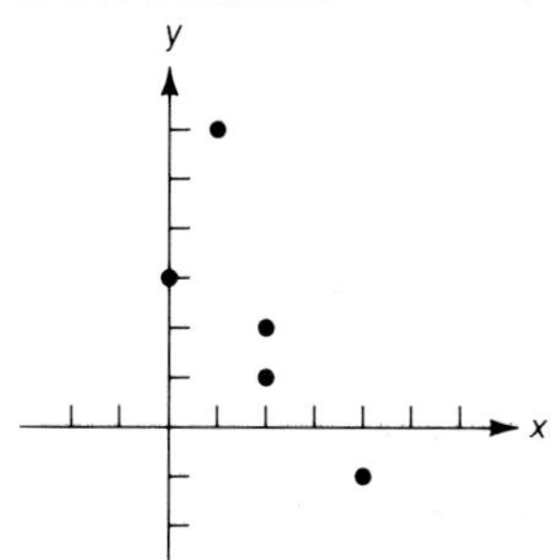


**4.** $D = \{1,2,-1,-2,0\}$
$R = \{-3,5,-1,3\}$
Is a function

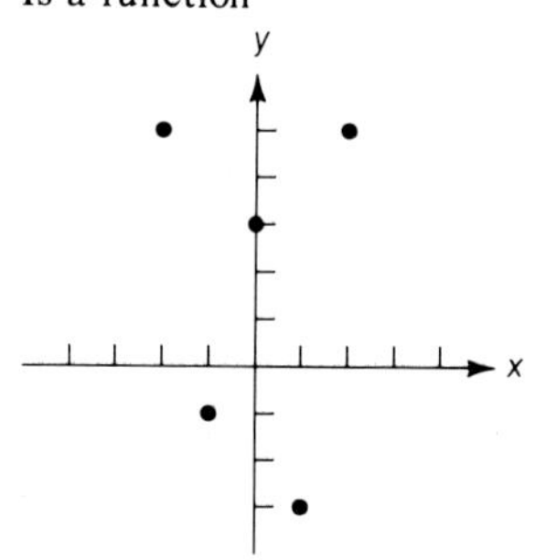


**5.** Is a function
**6.** Is a function
**7.** Is not a function

**8.** Is a function **9.** $\{(-2,13), (-1,8), (0,5), (1,4), (2,5)\}$ **10.** $\{(-2,-28), (-1,-6), (0,0), (1,-4), (2,-12)\}$

**11.** $\left\{\left(-2,\frac{1}{4}\right), \left(-1,\frac{1}{7}\right), (0,0), \left(\frac{1}{2},-\frac{1}{11}\right), (4,-2)\right\}$ **12.**

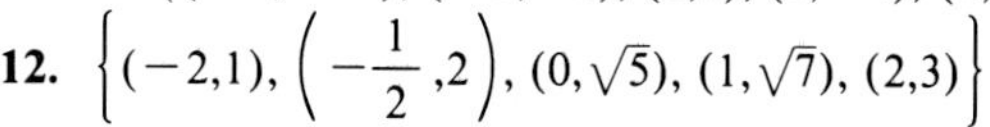
$\left\{(-2,1), \left(-\frac{1}{2},2\right), (0,\sqrt{5}), (1,\sqrt{7}), (2,3)\right\}$

**13.** $\left\{x \mid x \neq -\frac{7}{2}\right\}$ **14.** $\{x \mid x \neq -1, x \neq 3\}$

**15.** $\left\{x \mid x \geq -\frac{4}{3}\right\}$ **16.** $\{x \mid x \leq -1 \text{ or } x \geq 5\}$

**17.** $y = (x + 2)^2 - 2$
Vertex $= (-2, -2)$
Axis: $x = -2$
Range $= \{y \mid y \geq -2\}$
Zeros: $x = -2 \pm \sqrt{2}$

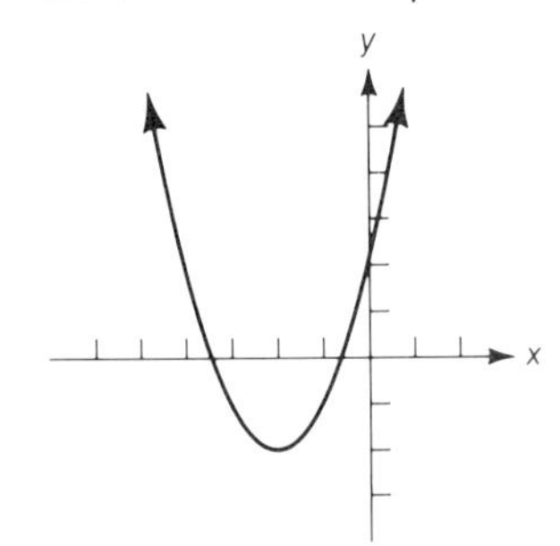


**18.** $y = -2(x - 1)^2 - 3$
Vertex $= (1, -3)$
Axis: $x = 1$
Range $= \{y \mid y \leq -3\}$
Zeros: none

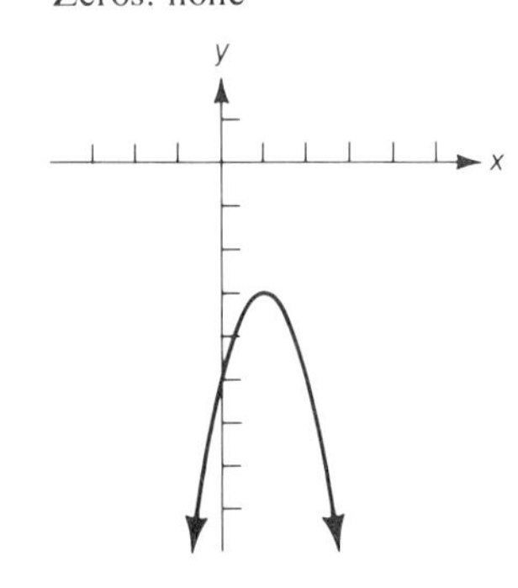


**19.** $y = 2\left(x - \frac{3}{2}\right)^2 - \frac{11}{2}$

Vertex $= \left(\frac{3}{2}, -\frac{11}{2}\right)$

Axis: $x = \frac{3}{2}$

Range $= \left\{y \mid y \geq -\frac{11}{2}\right\}$

Zeros: $x = \frac{3 \pm \sqrt{11}}{2}$

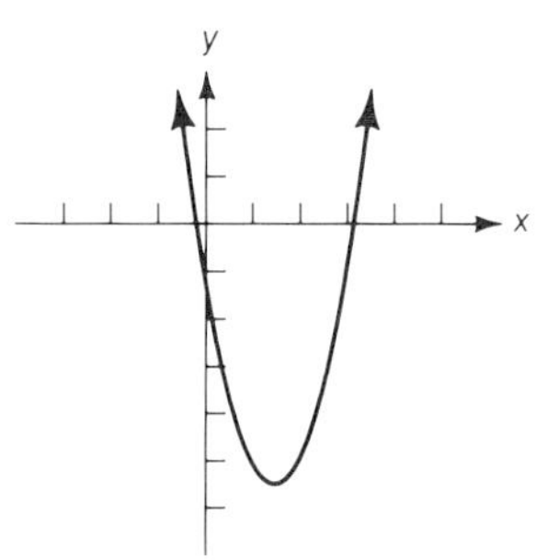


**20.**

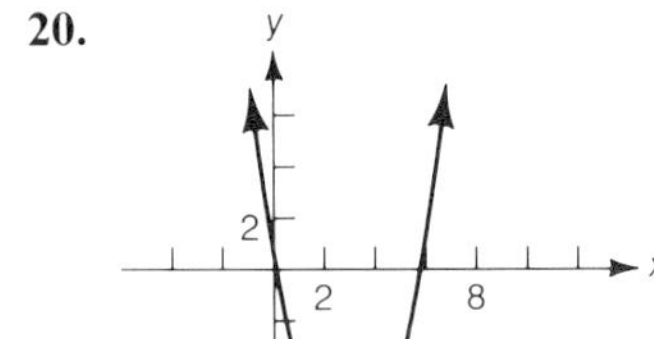


**21.**

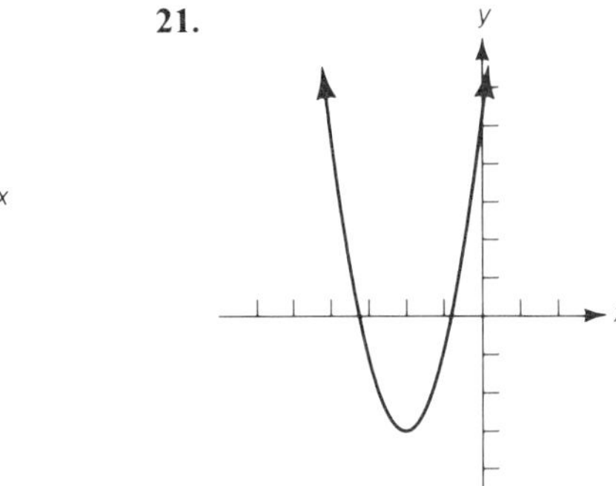


**22.**

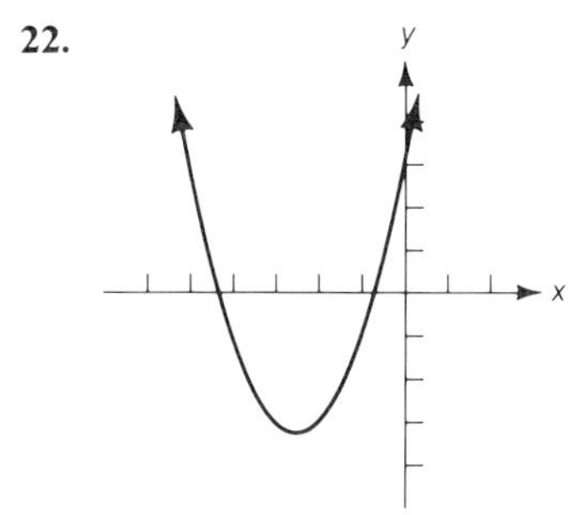


**23.**

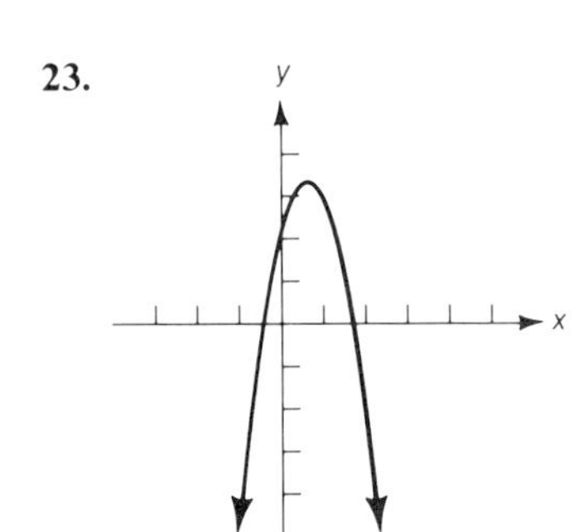


**24.** **(a)** $t = 2\frac{1}{2}$ sec **(b)** 148 ft **25.** \$15 **26.** **(a)** 5 **(b)** $-11$
**(c)** $8x^2 + 12x - 11$ **(d)** $4x + 4h - 7$ **(e)** 4 **27.** **(a)** 13 **(b)** 1
**(c)** $32x^2 - 100x + 76$ **(d)** $2x^2 + 4xh + 2h^2 + 3x + 3h - 1$
**(e)** $4x + 2h + 3$

**28.**

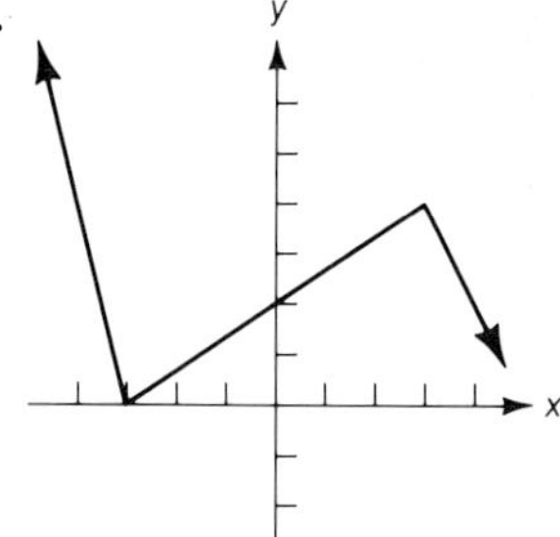

**29.**

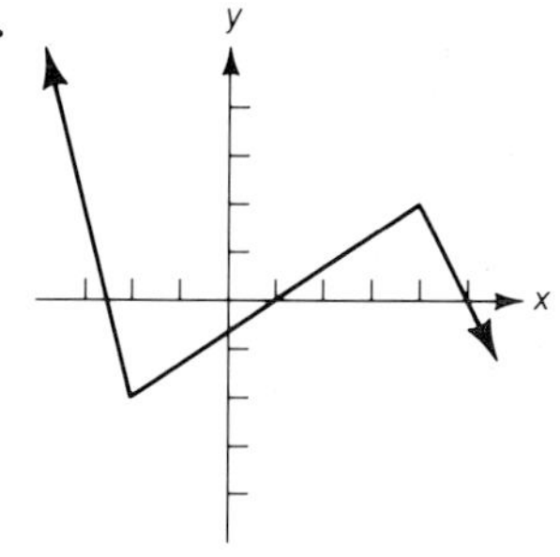

**30.**

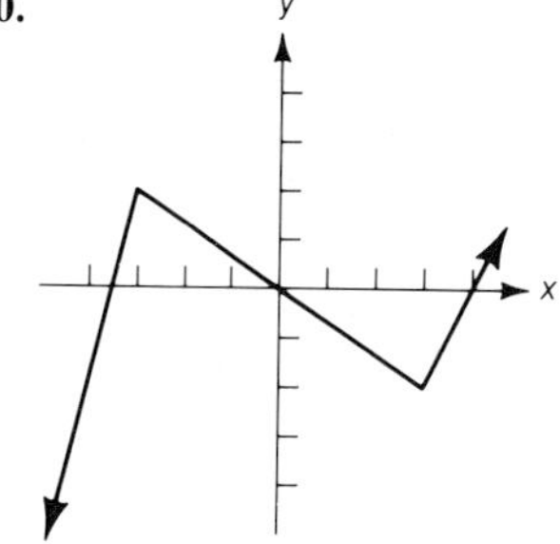

**31.**

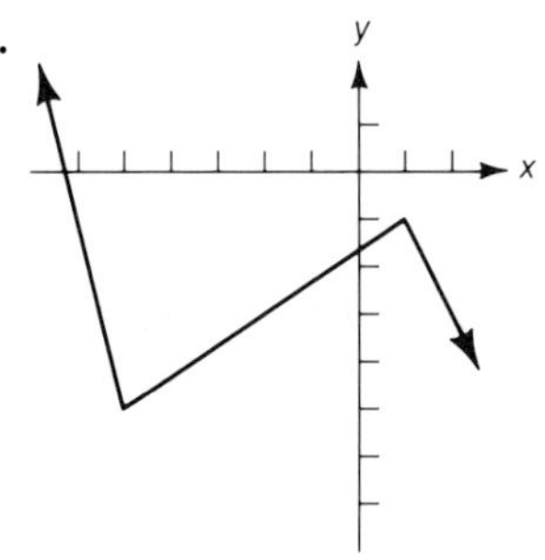

**32.** Is not a function

**33.** Is a function, not a 1–1 function

**34.** Is a 1–1 function

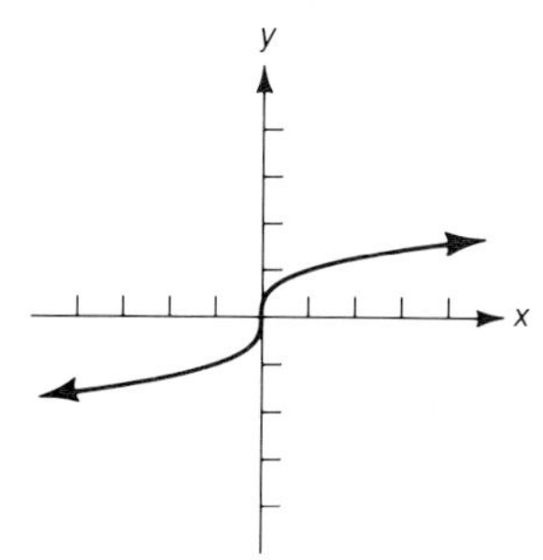

**35.** Is a 1–1 function

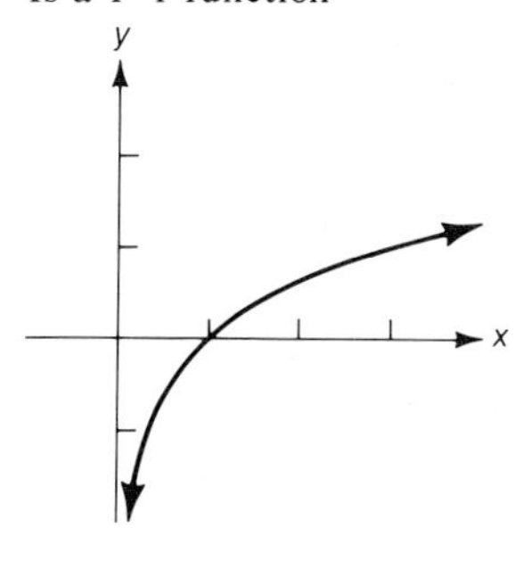

**36.** $f[g(x)] = \frac{1}{2}x + \frac{9}{2}$

**37.** $f[g(x)] = x^2 - 5x + 4$

**38.** $f[g(x)] = \frac{1}{x^2 - 25}$

**39.** $f[g(x)] = (2x - 5)^{3/2}$

**40.** $f^{-1}(x) = \frac{x - 2}{3}$

**40.** $f^{-1}(x) = \frac{x - 2}{3}$

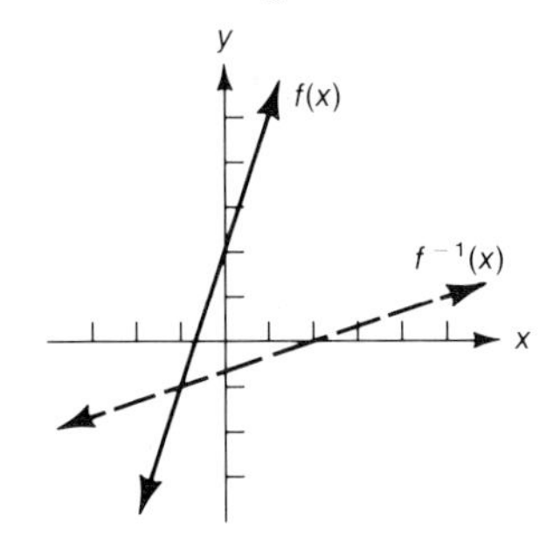


**41.** $g^{-1}(x) = x^{1/3} + 2$

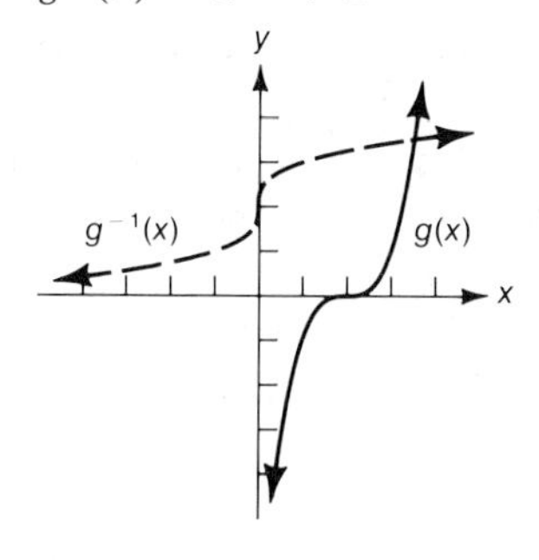


**42.** $f^{-1}(x) = x^2 + 4$
$x \geq 0$

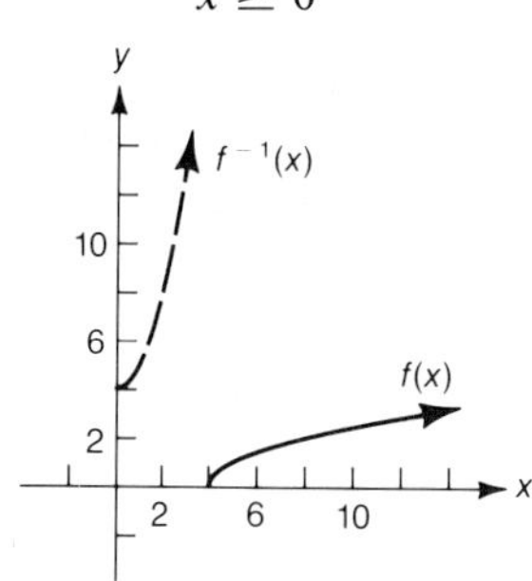

**43.** $f^{-1}(x) = \sqrt{x+4}$

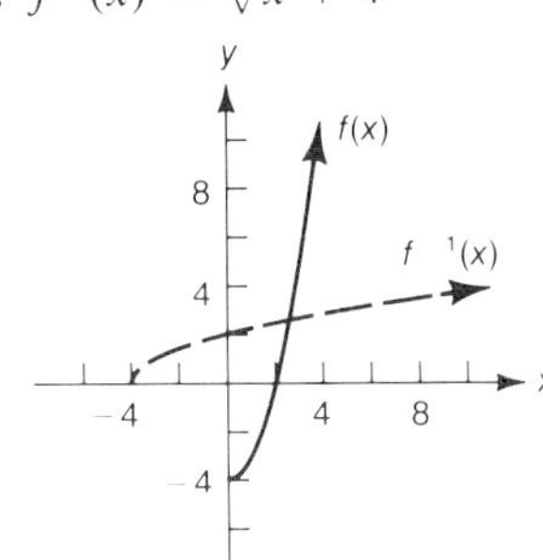


**44.** $f^{-1}(x) = \dfrac{1}{x} - 3$

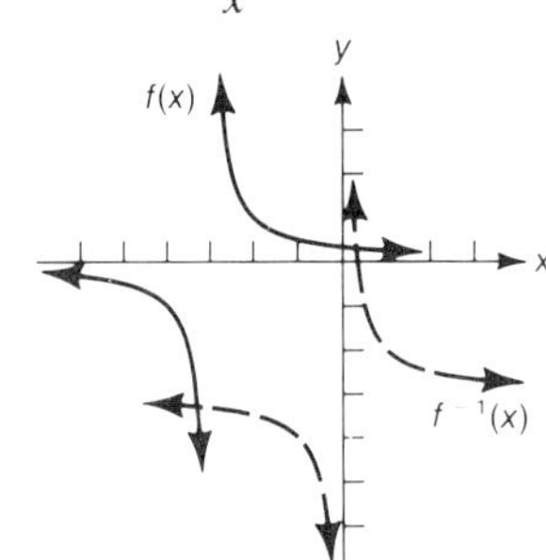


## Chapter 8 Test, Page 315

**1.** $\{(-5,-2), (-3,-2), (-2,4), (1,-1), (2,4)\}$; $D = \{-5,-3,-2,1,2\}$; $R = \{-2,-1,4\}$; Is a function

**2.** $D = \{-2,0,1,5\}$
$R = \{-4,-1,3\}$
Is not a function

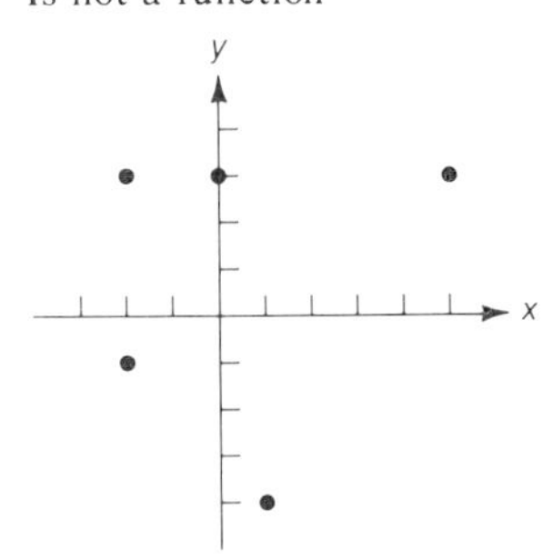


**3.** Is not a function

**4.** Is a function

**5.** $\{(-1,-2), (0,-4), (1,-2), (2,10), (3,38)\}$

**6.** $\left\{\left(-2,-\dfrac{9}{4}\right), (-1,-8), \left(-\dfrac{1}{2},-30\right), \left(\dfrac{1}{2},-26\right), \left(2,-\dfrac{5}{4}\right)\right\}$

**7.** $\left\{x \mid x > \dfrac{4}{5}\right\}$

**8.**

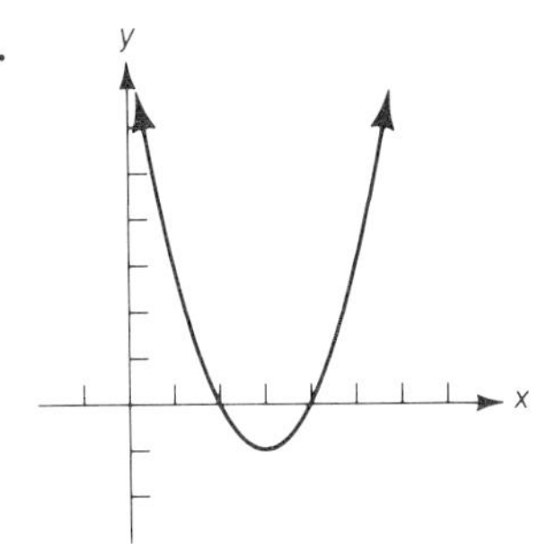


**9.**

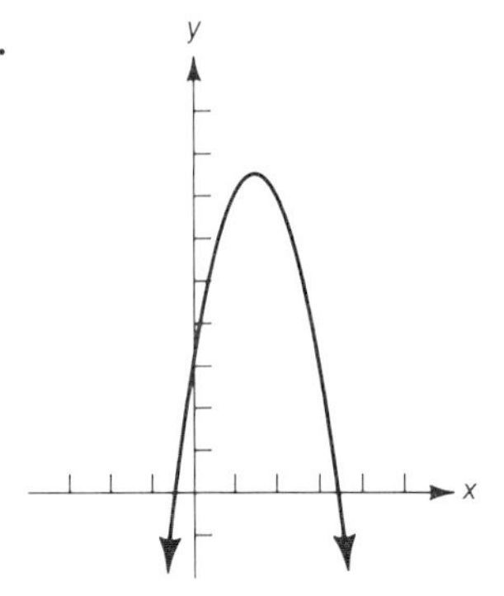


**10.** $y = 2(x-3)^2 + 1$
Vertex: (3,1)
Axis: $x = 3$
Range: $y \geq 1$
Zeros: None

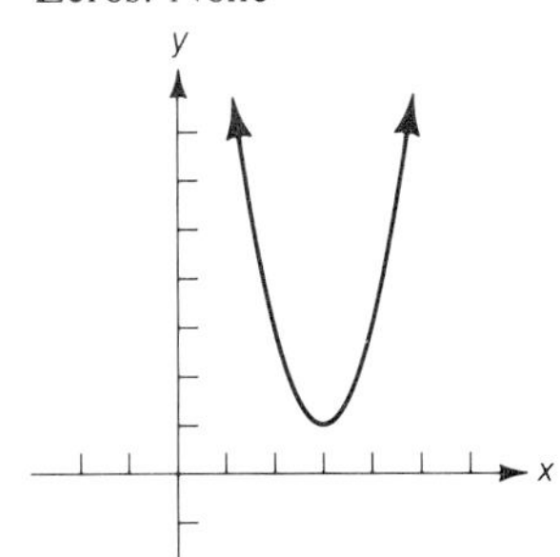

**11.** $-25$ **12.** $5\frac{1}{2}$ in. by $5\frac{1}{2}$ in. **13.** **(a)** 13 **(b)** 13 **(c)** $18x^2 + 12x + 7$ **(d)** $2x^2 + 4xh + 2h^2 + 5$
**(e)** $4x + 2h$ **14.**

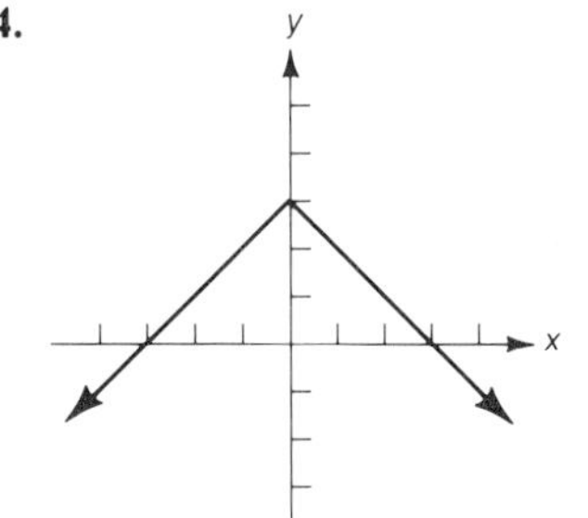

**15.**

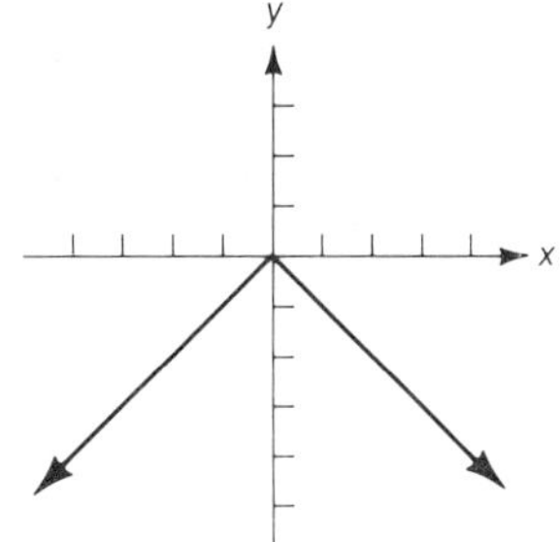

**16.**

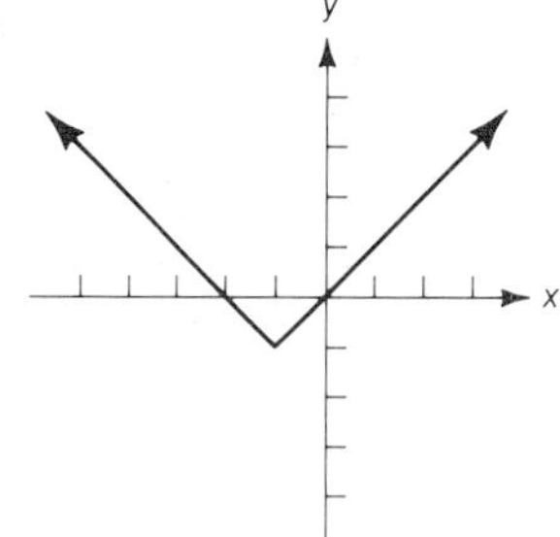

**17.** **(a)** $f(g(x)) = 3 - 2\left(\frac{1}{x+2}\right)^2 = 3 - \frac{2}{x^2 + 4x + 4}$ **(b)** $g(f(x)) = \frac{1}{(3 - 2x^2) + 2} = \frac{1}{5 - 2x^2}$

**18.** $b, c, d$ **19.** $f^{-1}(x) = \frac{2x + 1}{x}$ **20.**

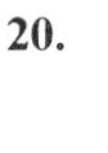

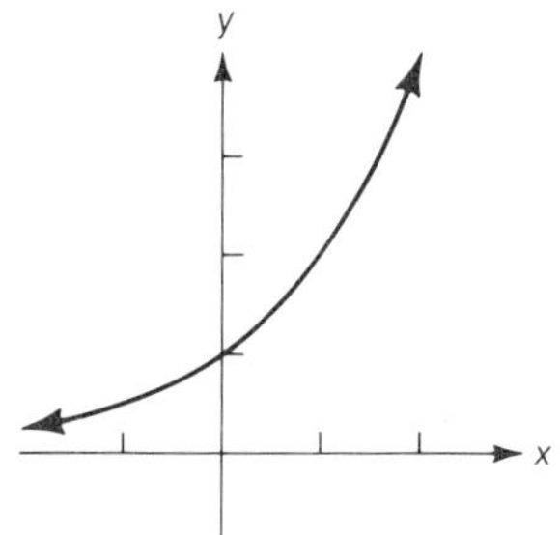

## Cumulative Review (8), Page 317

**1.** 1 **2.** $\frac{9x^8}{4y^8}$ **3.** $5x^{3/4}$ **4.** $\frac{8y^{3/5}}{x}$ **5.** $\sqrt[3]{(7x^3y)^2} = x^2\sqrt[3]{49y^2}$ **6.** $(32x^6y)^{1/3}$

**7.** $\frac{2x^2 + x + 4}{(2x + 3)(x - 4)(3x - 2)}$ **8.** $-1$ **9.** $x = \frac{-1 \pm \sqrt{7}}{3}$ **10.** $x = -9$ **11.** $x = 2$
**12.** $x = -2, y = -3$ **13.**

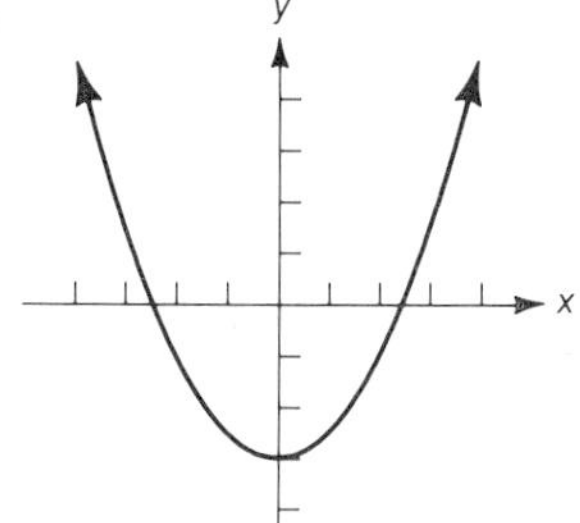

**14.**

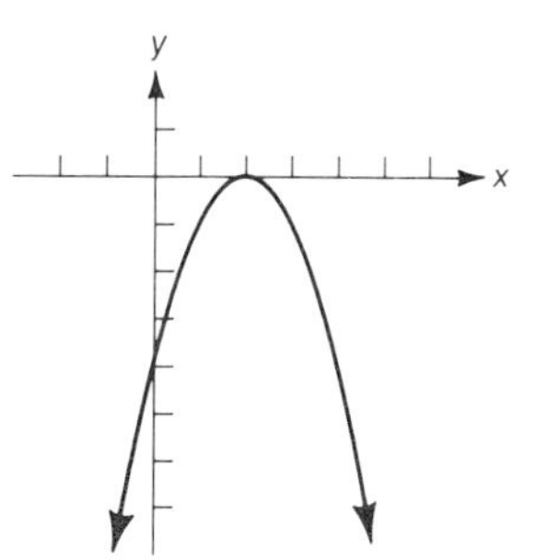

**15.** **(a)** $f(g(x)) = \sqrt{10 - 3x^2}$ **(b)** $g(f(x)) = 10 - 3x$

16.

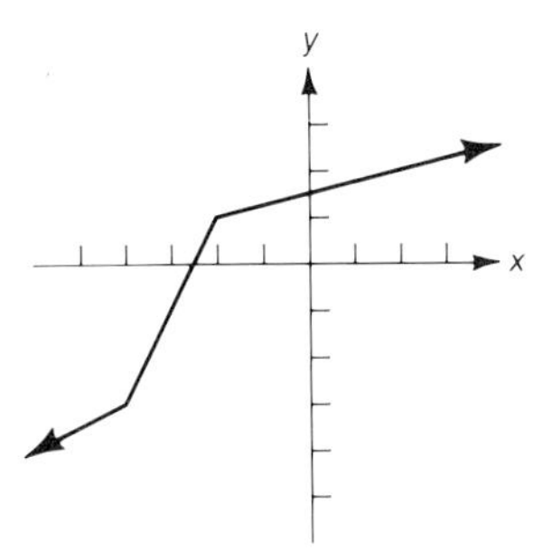

17. $f^{-1}(x) = \frac{2}{5}(x + 3)$

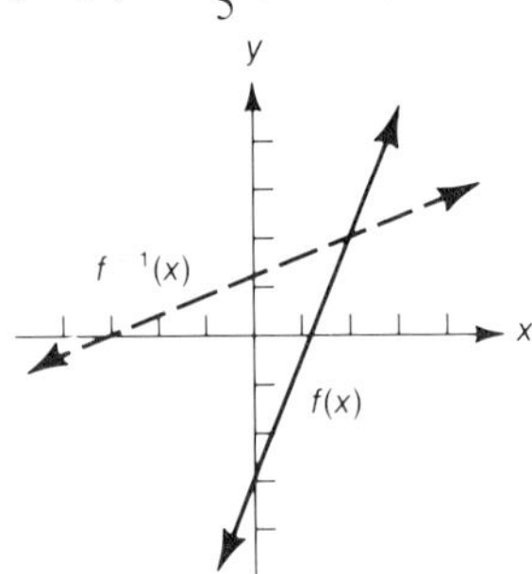


18. $f^{-1}(x) = \sqrt[3]{x - 2}$

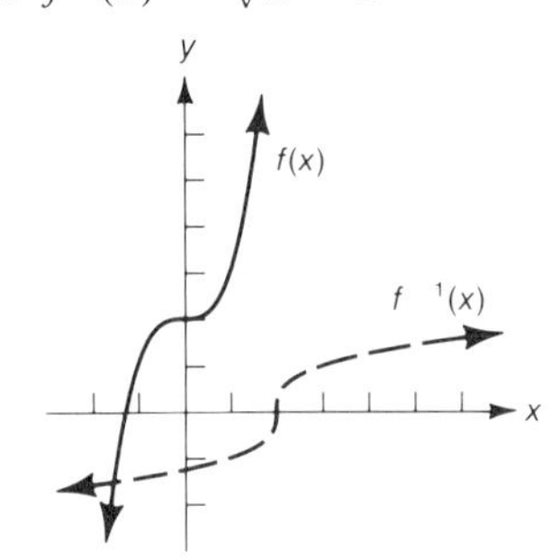


**19.** \$3600 @ 7%; \$5400 @ 8%  **20.** 36

## CHAPTER 9

### Exercises 9.1, Page 330

**1.**

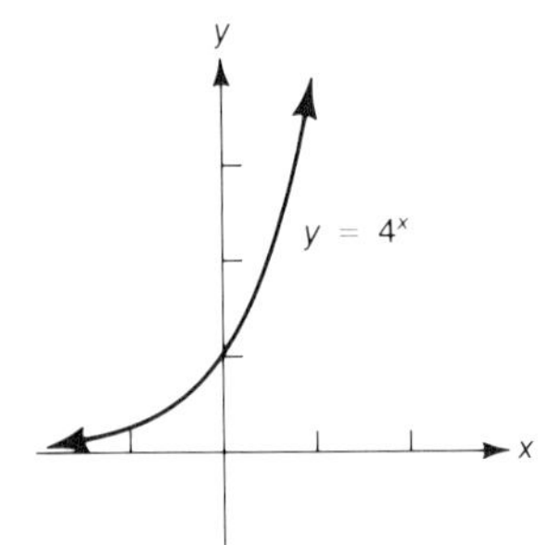


**3.**

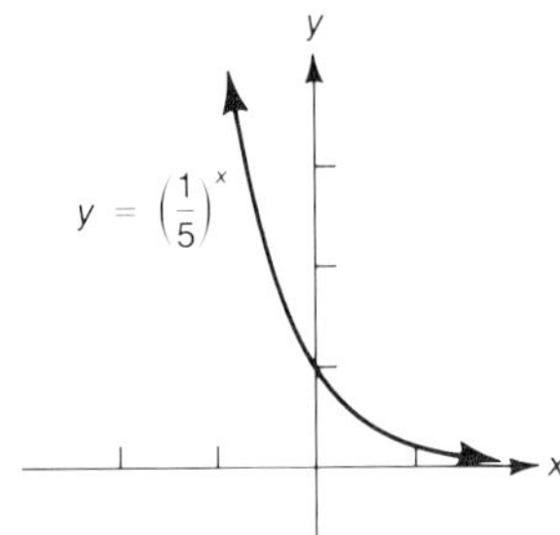


**5.**

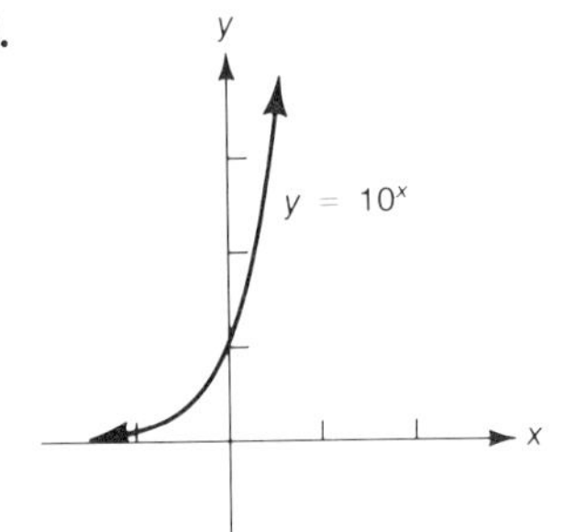


**7.**

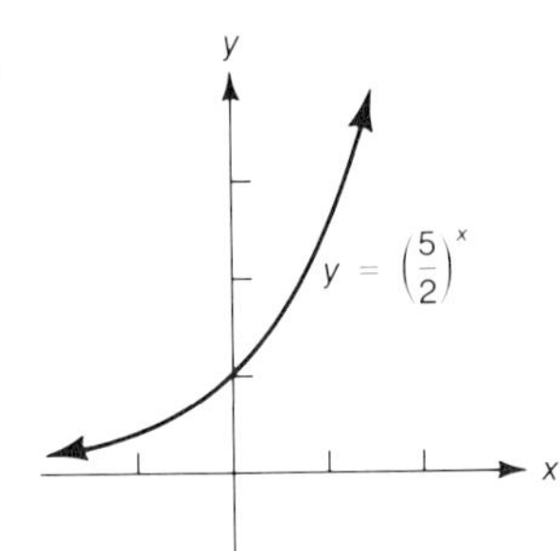


**9.**

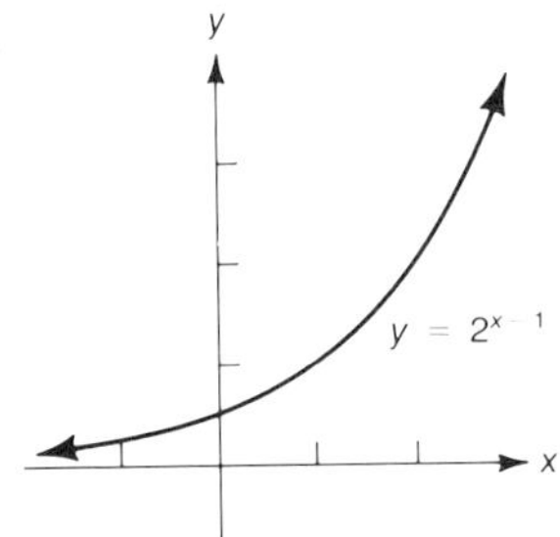


**11.** $x = 11$  **13.** $x = 9$  **15.** $x = \frac{1}{6}$  **17.** $x = \frac{7}{12}$  **19.** $x = \frac{7}{2}$  **21.** $x = -2$  **23.** $x = -3$

**25.** $x = 4, x = -1$  **27.** $x = -1, x = \frac{3}{2}$  **29.** $x = -2, x = 3$  **31.** 25,000 bacteria  **33. (a)** \$5877.31 **(b)** \$5920.98  **(c)** \$5943.79  **(d)** \$5967.04  **(e)** \$5967.30  **35.** \$2220.62  **37.** \$46.88
**39.** 250,000 people  **41.** \$7290  **43.** $\$1200(0.8)^4 = \$491.52$

### Exercises 9.2, Page 339

**1.** $\log_7 49 = 2$ **3.** $\log_5 \frac{1}{25} = -2$ **5.** $\log_2 \frac{1}{32} = -5$ **7.** $\log_{2/3} \frac{4}{9} = 2$ **9.** $\ln 17 = x$ **11.** $\log_{10} 6 = \log_{10} 6$

**13.** $3^2 = 9$ **15.** $9^{1/2} = 3$ **17.** $7^{-1} = \frac{1}{7}$ **19.** $e^{1.74} = N$ **21.** $b^4 = 18$ **23.** $y^x = y^x$ **25.** $x = 16$

**27.** $x = 2$ **29.** $x = \frac{1}{6}$ **31.** $x = 11$ **33.** $x = 32$ **35.** $x = -2$ **37.** $x = 1.52$ **39.** $x = 3$

**41.**

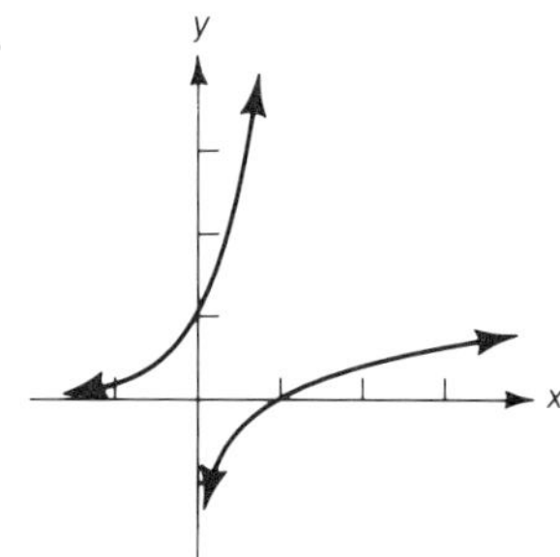


**43.**

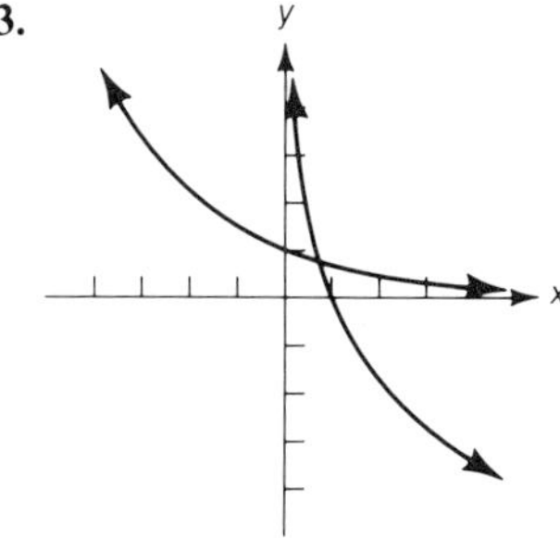


**45.**

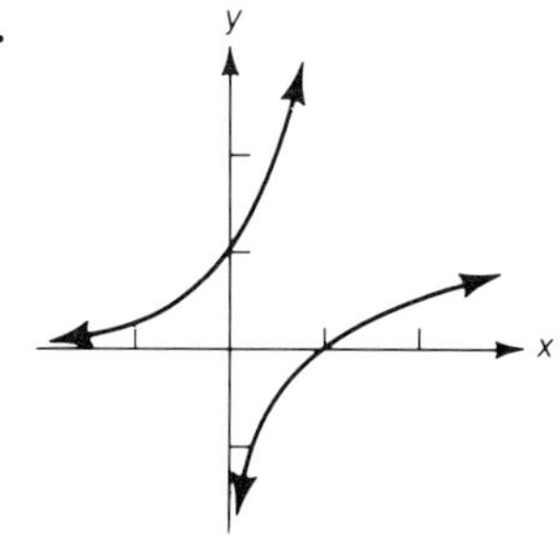


**47.**

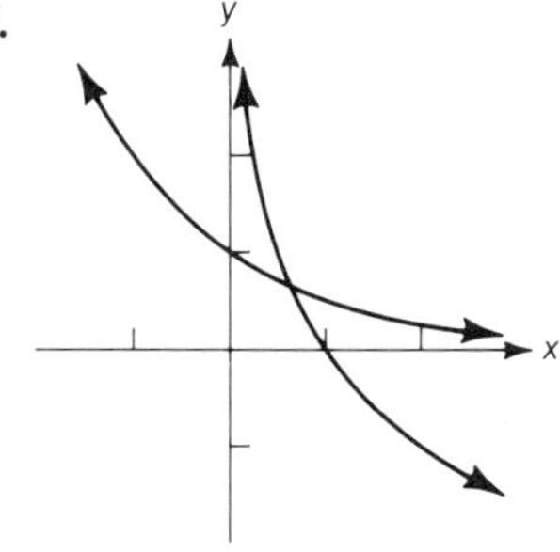


**49.**

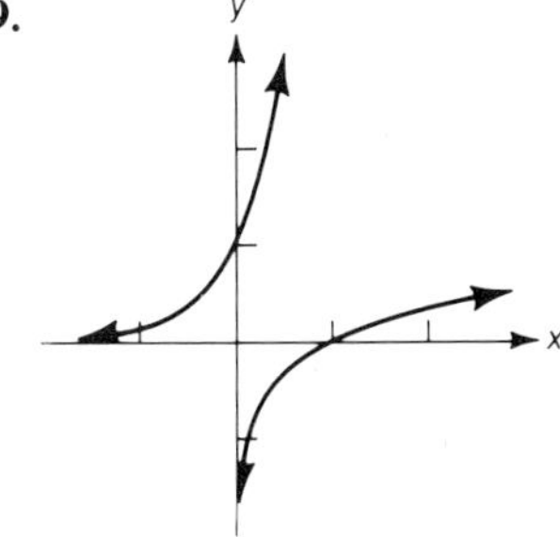


**51.** 2.238046103 **53.** 1.946452265 **55.** $-1.241845378$ **57.** 3.624340933 **59.** Error **61.** $-5.379527422$
**63.** 204.1737945 **65.** 0.019952623 **67.** 0.957194071 **69.** 175.9148375 **71.** 0.000248516
**73.** 1.040290499

### Exercises 9.3, Page 344

**1.** $\log 5 + 4 \log x$ **3.** $\ln 2 - 3 \ln x + \ln y$ **5.** $\log 2 + \log x - 3 \log y$ **7.** $2 \ln x - \ln y - \ln z$

**9.** $-2 \log x - 2 \log y$ **11.** $\frac{1}{3} \log x + \frac{2}{3} \log y$ **13.** $\frac{1}{2} \ln x + \frac{1}{2} \ln y - \frac{1}{2} \ln z$

**15.** $\log 21 + 2 \log x + \frac{2}{3} \log y$ **17.** $-\frac{1}{2} \log x - \frac{5}{2} \log y$ **19.** $-9 \ln x - 6 \ln y + 3 \ln z$ **21.** $\ln \frac{3^2 x}{5} = \ln \frac{9x}{5}$

**23.** $\log \frac{7x^2}{8}$ **25.** $\log x^2 y$ **27.** $\ln \frac{x^3}{y^2}$ **29.** $\ln \sqrt{\frac{x}{y}}$ **31.** $\log \frac{xz}{y}$ **33.** $\log \frac{x}{y^2 z^2}$ **35.** $\log x(2x + 1)$

**37.** $\ln(x - 1)(x + 3)$ **39.** $\log \frac{x^2 - 2x - 3}{x - 3}$ **41.** $\log \frac{x + 6}{2x^2 + 9x - 18}$

## Exercises 9.4, Page 349

**1.** $x = 2$ **3.** $x = 0, x = -\frac{3}{2}$ **5.** $x = 3, x = -1$ **7.** $x \approx 0.7154$ **9.** $x \approx 7.5098$ **11.** $x \approx -1.595$
**13.** $x \approx 24.73$ **15.** $x \approx -65.30$ **17.** $x \approx -1.5193$ **19.** $x = \frac{1}{\log 2} \approx 3.322$ **21.** $x = -\frac{1}{\log 5} \approx -1.431$
**23.** $x = 1$ **25.** $x = \frac{1}{3}\left(2 + \frac{1}{\log 4}\right) \approx 1.220$ **27.** $x = \frac{1}{2}\left(\frac{4}{\log 12} - 7\right) \approx -1.647$ **29.** $x = \frac{\log 9}{\log 7} \approx 1.129$
**31.** $x = 2 + \frac{\log 35}{\log 3} \approx 5.236$ **33.** $x = \frac{1}{2}\left(1 + \frac{\ln 14.8}{\ln 6}\right) \approx 1.252$ **35.** $x = 10^{1.4} \approx 25.12$ **37.** $x \approx 31.6$
**39.** $x = 10^{-4}$ **41.** $x \approx 4.95$ **43.** $x = e^{-1.1} \approx 0.3329$ **45.** $x = 3$ **47.** $x = 100$ **49.** $x = 6$
**51.** $x = 20$ **53.** $x = \frac{25}{2}$ **55.** $x = 1001$ **57.** $x = -2.99$ **59.** $x \approx 1.1353$ **61.** $x \approx 22.1$
**63.** $x = -10, x = 8$ **65.** 2.2619 **67.** 0.3223 **69.** $-0.6279$ **71.** 2.439 **73.** $-1.2222$

## Exercises 9.5, Page 352

**1.** \$4027.51 **3.** 13.9 years **5.** $\frac{3}{10}$ **7.** 1.73 hr **9.** 8166 bees **11.** 12.28 lb per sq. in. **13.** 2350 years
**15.** 2.3 days **17.** 8.75 years **19.** 9 years

## Chapter 9 Review, Page 354

**1.** **2.**

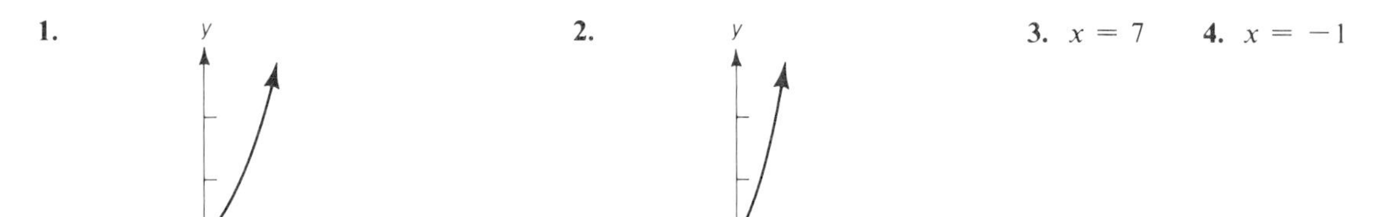


**3.** $x = 7$ **4.** $x = -1$ **5** $x = -10$

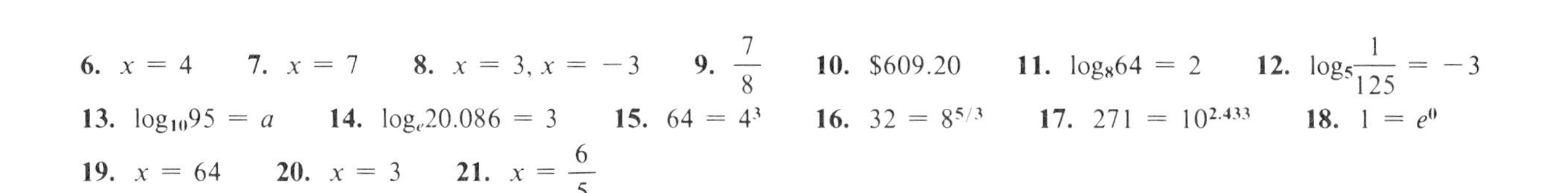

**6.** $x = 4$ **7.** $x = 7$ **8.** $x = 3, x = -3$ **9.** $\frac{7}{8}$ **10.** \$609.20 **11.** $\log_8 64 = 2$ **12.** $\log_5 \frac{1}{125} = -3$
**13.** $\log_{10} 95 = a$ **14.** $\log_e 20.086 = 3$ **15.** $64 = 4^3$ **16.** $32 = 8^{5/3}$ **17.** $271 = 10^{2.433}$ **18.** $1 = e^0$
**19.** $x = 64$ **20.** $x = 3$ **21.** $x = \frac{6}{5}$

**22.**

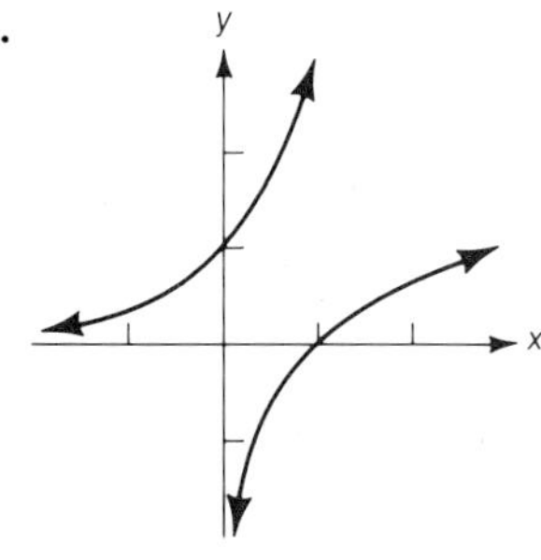

**23.**

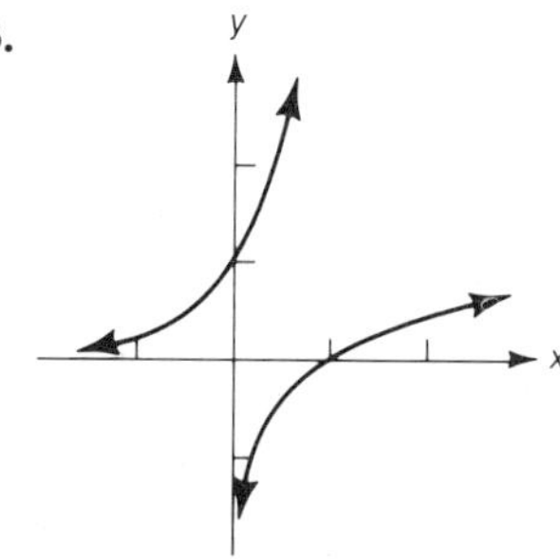

**24.** 2.566 **25.** $-2.730$ **26.** 4.355 **27.** 338.8 **28.** 135.6 **29.** 3.935 **30.** $\log(x^3 \cdot 2x) = \log 2x^4$
**31.** $\ln \frac{\sqrt{x}}{y^2}$ **32.** $\log \frac{x^2 - 4}{x^2}$ **33.** $\ln(x + 3)(2x - 1)$ **34.** $\log x + \frac{1}{2} \log y$ **35.** $\frac{1}{4}[3 \log x + \log y]$
**36.** $\ln 6 + 2 \ln x - \ln y$ **37.** $2 \ln x - \frac{1}{2} \ln 2 - \frac{1}{2} \ln y - \frac{1}{2} \ln z$ **38.** $x = 3$ **39.** $x = -\frac{3}{4}$
**40.** $x = -0.6093$ **41.** $x = 0.3925$ **42.** $x = -0.3562$ **43.** $x = \frac{\ln 2.8}{0.04} \approx 25.74$ **44.** $x = \frac{1}{2}e^{1.6} \approx 2.48$
**45.** $x = 8$ **46.** $x = \sqrt{5}$ **47.** $x \approx 11.3891$ **48.** 17.33 centuries **49.** 4.6 sec **50.** 181 sales

## *Chapter 9 Test, Page 356*

**1.**

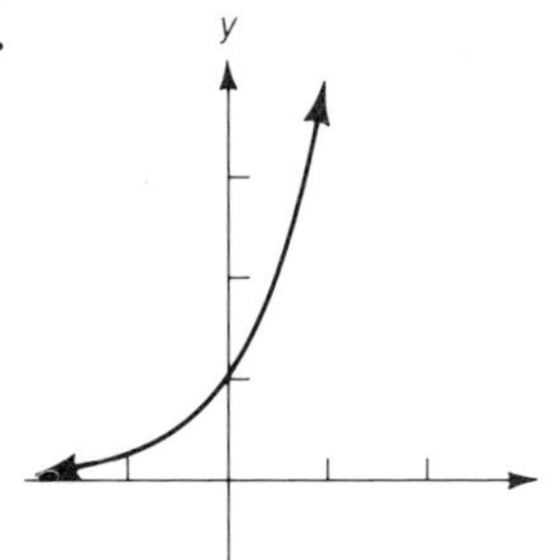

**2.** $x = -4$ **3.** $x = -7$ **4.** 25,981 **5.** $\log_{1/2} 8 = -3$
**6.** $\frac{1}{9} = 3^{-2}$ **7.** $x = 343$ **8.** $x = \frac{3}{2}$

**9.**

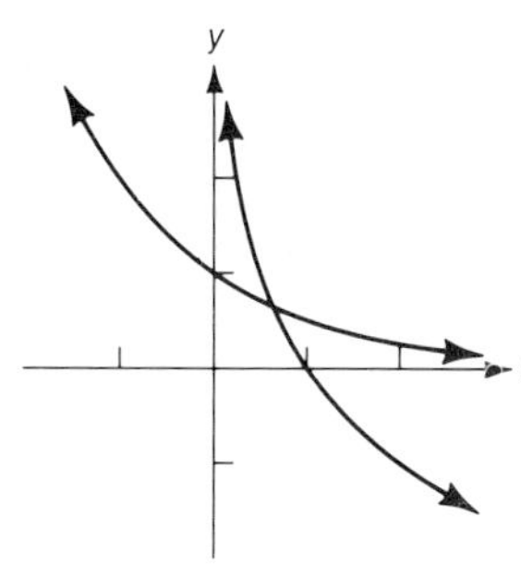

**10.** 2.763 **11.** 6.488 **12.** $\frac{2}{3} \log x - \frac{1}{3} \log y$ **13.** $\ln(x + 5)(x - 4)$
**14.** $x = 0.4518$ **15.** $x = 10.69$ **16.** 1.79 **17.** No solution
**18.** $x = 1 + e^3 \approx 21.09$ **19.** 4.9 years **20.** 134.3 years

## Cumulative Review (9), Page 356

**1.** 1 **2.** $-\dfrac{x+3}{x+1}$ **3.** $\dfrac{2}{x+4}$ **4.** $\dfrac{x-4}{x+1}$ **5.** $\dfrac{x^{1/6}}{y^{1/3}}$ **6.** $\dfrac{x^{1/2}}{2y^2}$ **7.** $x^2y^{3/2}$ **8.** $x^{2/3}y$

**9.** Vertex: $(3, -11)$
Range: $y \geq -11$
Zeros: $x = 3 \pm \sqrt{11}$

**10.** Vertex: $(-2, -5)$
Range: $(y \geq -5)$
Zeros: $x = \dfrac{-4 \pm \sqrt{10}}{2}$

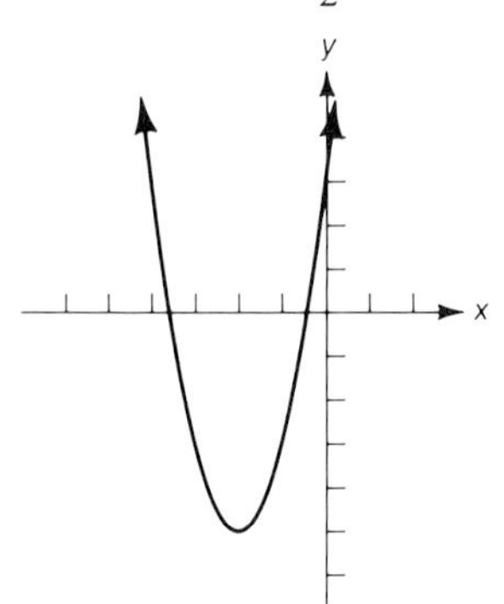


**11.** $x = \dfrac{-5 \pm \sqrt{33}}{2}$ **12.** $x = \pm 2, x = \pm 3$ **13.** $x = 6$ **14.** $x = 4 \pm \sqrt{6}$ **15.** $x = -0.9212$
**16.** $x = 5 + e^{2.5} \approx 17.18$ **17.** $x = 2, y = 5$ **18.** $x = 1, y = -1, z = 2$ **19.** 13 clothbound, 30 paperback
**20.** 10 mph, 4 mph

## CHAPTER 10

### Exercises 10.1, Page 363

**1.** Vertex: $(4,0)$
$y$-intercept: None
Line of symmetry: $y = 0$

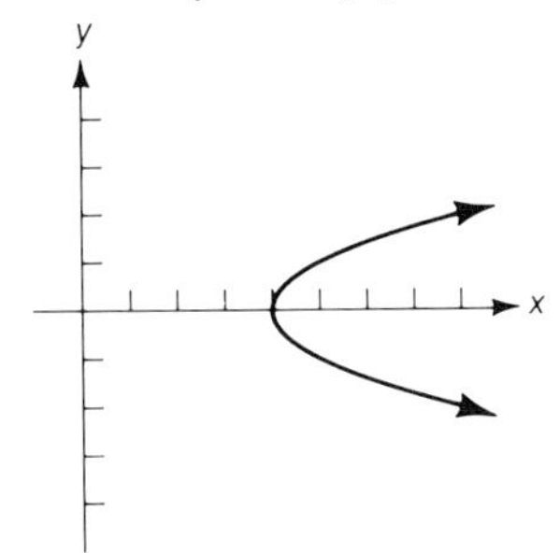


**3.** Vertex: $(-3,0)$
$y$-intercept: $(0, \sqrt{3})$, $(0, -\sqrt{3})$
Line of symmetry: $y = 0$

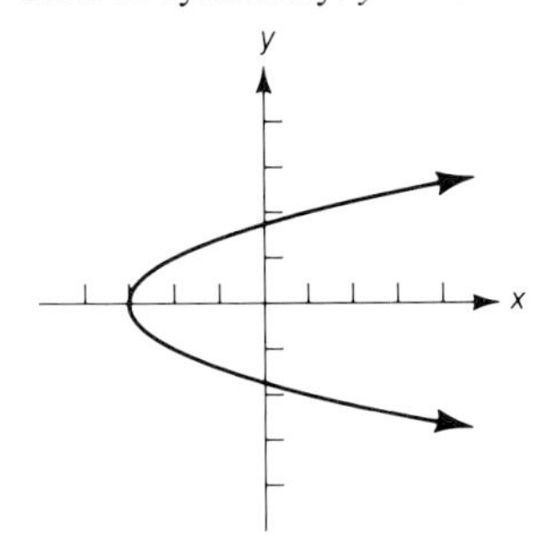


**5.** Vertex: $(3,0)$
$y$-intercept: None
Line of symmetry: $y = 0$

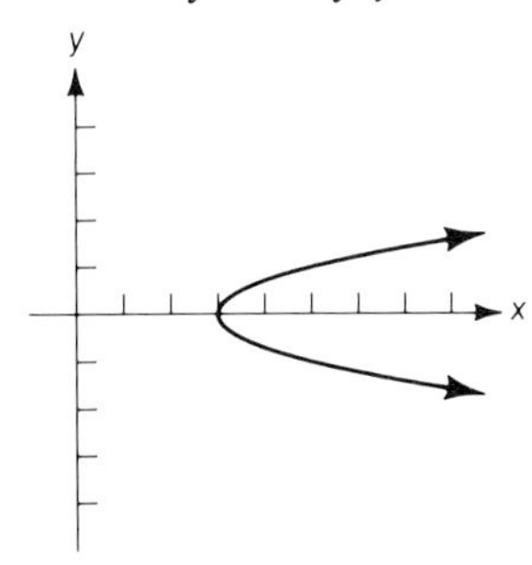

**7.** Vertex: (0,3)
$y$-intercept: (0,3)
Line of symmetry: $y = 3$

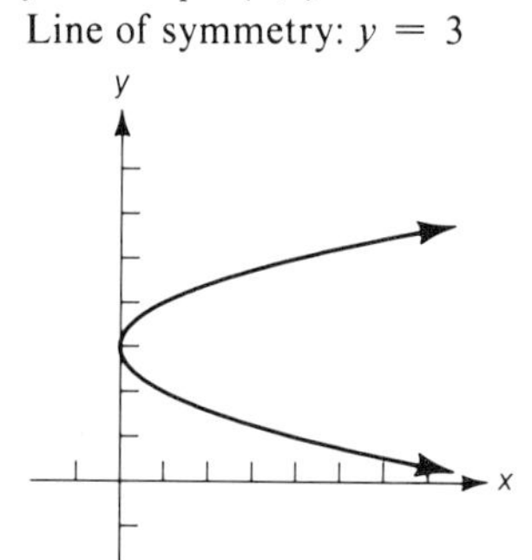

**9.** Vertex: $(4,-2)$
$y$-intercept: None
Line of symmetry: $y = -2$

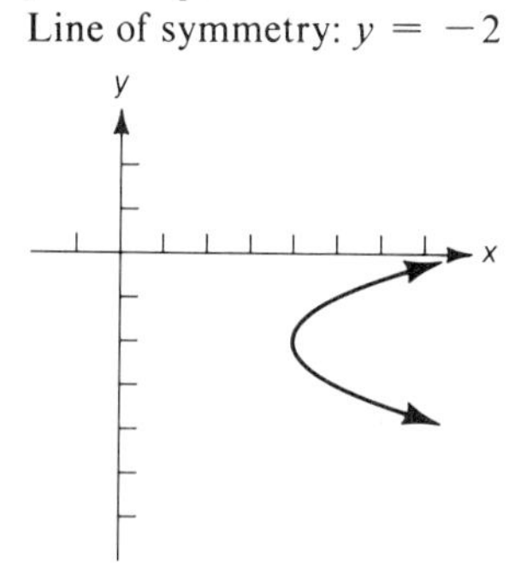

**11.** Vertex: $(-1,1)$
$y$-intercept: (0,0), (0,2)
Line of symmetry: $y = 1$

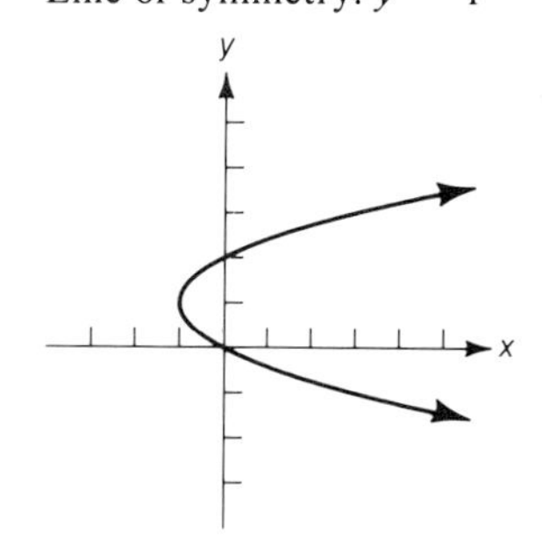

**13.** Vertex: $(0,-2)$
$y$-intercept: $(0,-2)$
Line of symmetry: $y = -2$

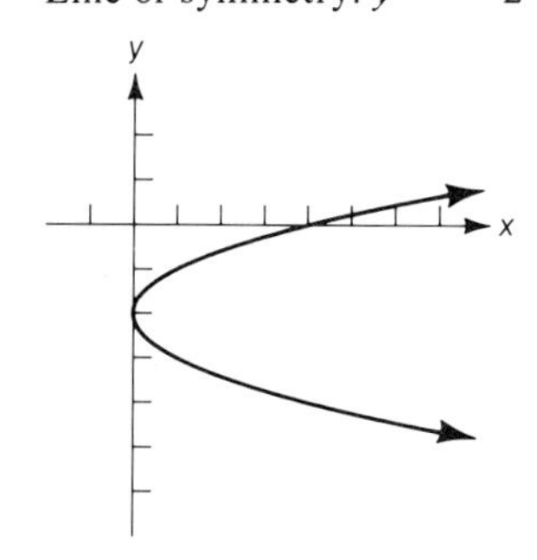

**15.** Vertex: (0,4)
$y$-intercept: (0,4)
Line of symmetry: $y = 4$

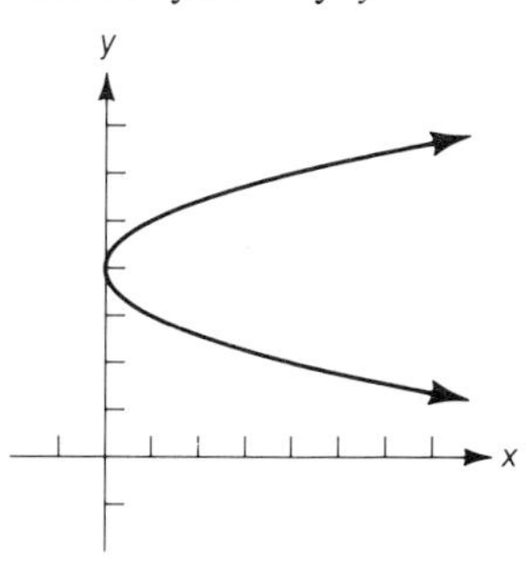

**17.** Vertex: $(-2,9)$
$y$-intercept: (0,5)
Line of symmetry: $x = -2$

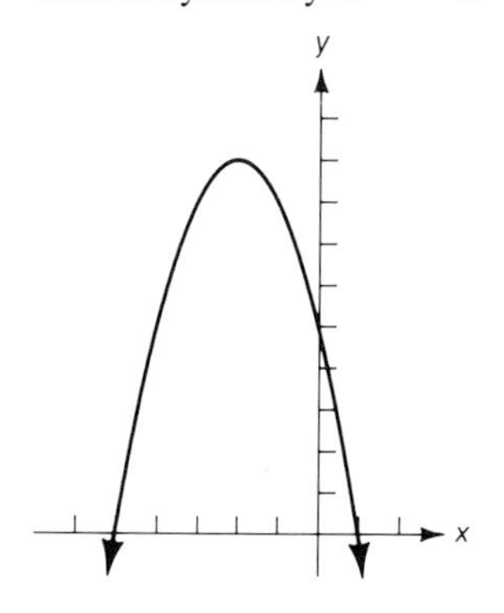

**19.** Vertex: $(-3,-4)$
$y$-intercept: (0,5)
Line of symmetry: $x = -3$

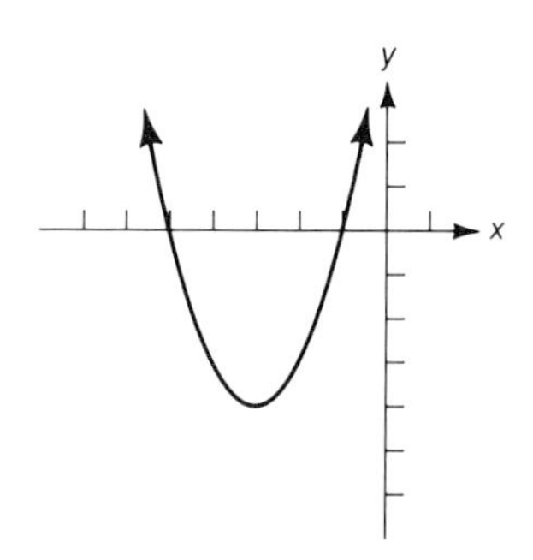

**21.** Vertex: (1,2)
$y$-intercept: (0,1), (0,3)
Line of symmetry: $y = 2$

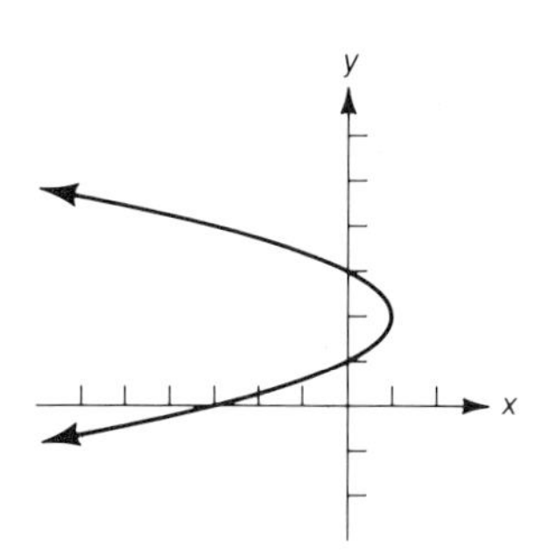

**23.** Vertex: $\left(-\frac{1}{4}, -\frac{9}{8}\right)$
$y$-intercept: $(0,-1)$
Line of symmetry: $x = -\frac{1}{4}$

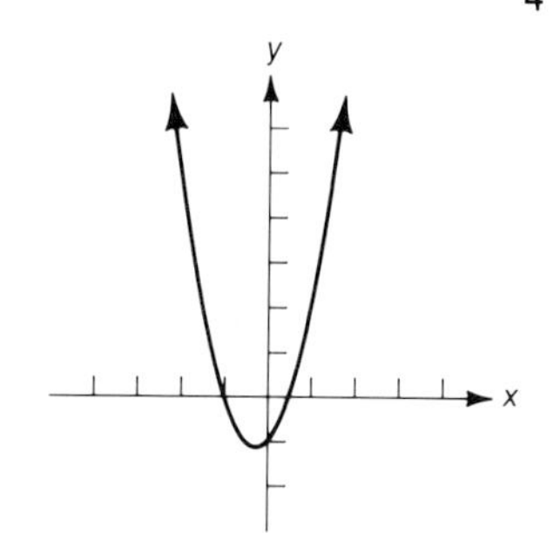

**25.** Vertex: $(-8,-1)$
$y$-intercept: $\left(0, -1 \pm \frac{2}{3}\sqrt{6}\right)$
Line of symmetry: $y = -1$

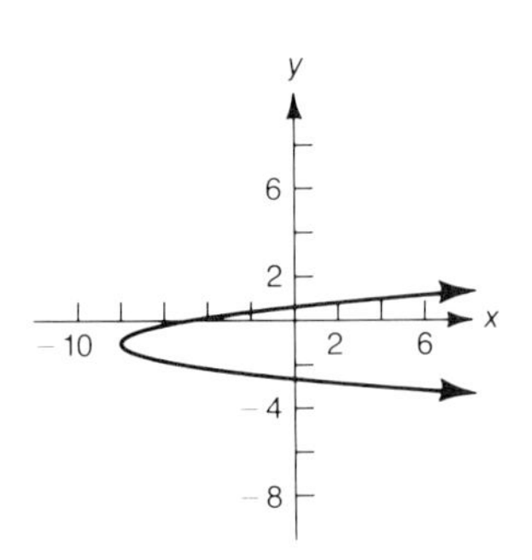


**27.** Vertex: $\left(\frac{9}{8}, \frac{5}{4}\right)$
$y$-intercept: $\left(0, \frac{1}{2}\right)$, $(0,2)$
Line of symmetry: $y = \frac{5}{4}$

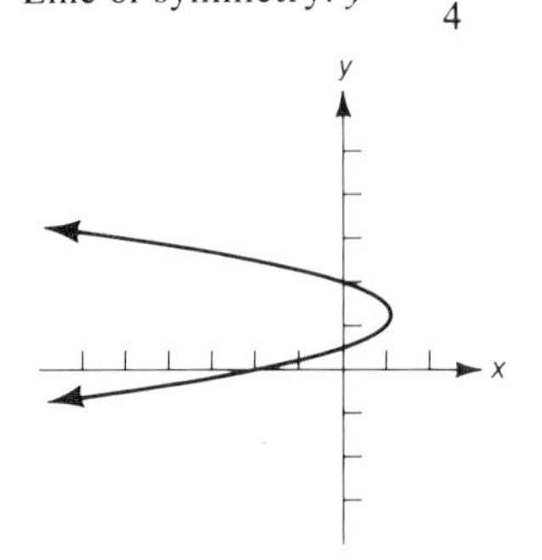


**29.** Vertex: $\left(\frac{3}{2}, 0\right)$
$y$-intercept: $(0,9)$
Line of symmetry: $x = \frac{3}{2}$

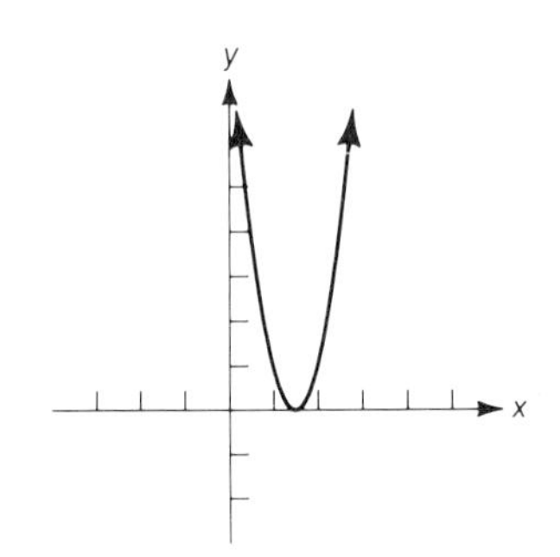


## Exercises 10.2, Page 370

**1.** 5 **3.** 13 **5.** $\sqrt{29}$ **7.** 3 **9.** $\sqrt{13}$ **11.** 17 **13.** $x^2 + y^2 = 16$ **15.** $x^2 + y^2 = 3$
**17.** $x^2 + y^2 = 11$ **19.** $x^2 + y^2 = \frac{4}{9}$ **21.** $x^2 + (y - 2)^2 = 4$ **23.** $(x - 4)^2 + y^2 = 1$
**25.** $(x + 2)^2 + y^2 = 8$ **27.** $(x - 3)^2 + (y - 1)^2 = 36$ **29.** $(x - 3)^2 + (y - 5)^2 = 12$
**31.** $(x - 7)^2 + (y - 4)^2 = 10$

**33.** $x^2 + y^2 = 9$
Center: $(0,0)$, $r = 3$

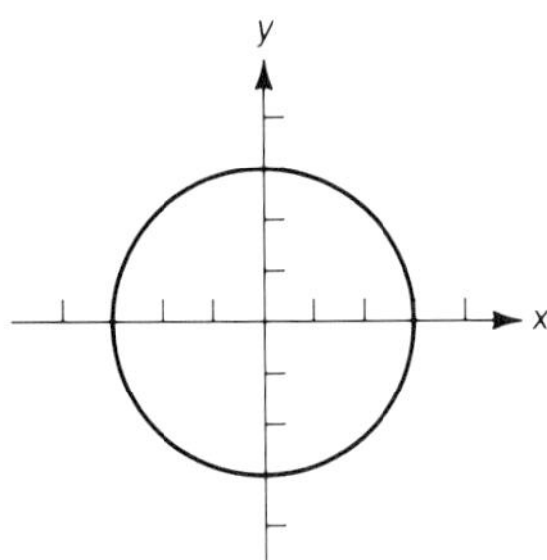


**35.** $x^2 + y^2 = 49$
Center: $(0,0)$, $r = 7$

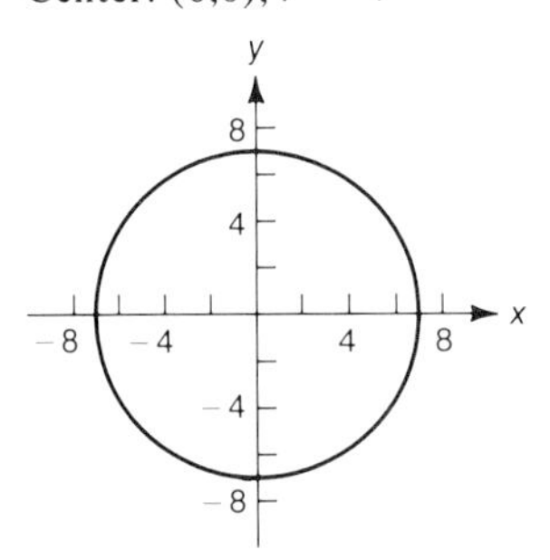


**37.** $x^2 + y^2 = 18$
Center: $(0,0)$, $r = \sqrt{18}$

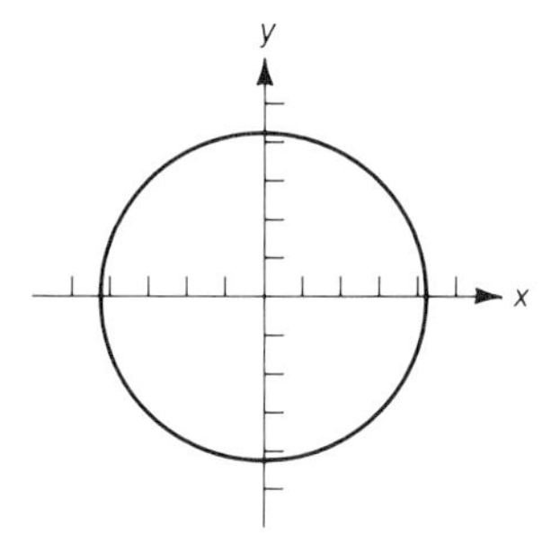


**39.** $(x + 1)^2 + y^2 = 9$
Center: $(-1,0)$, $r = 3$

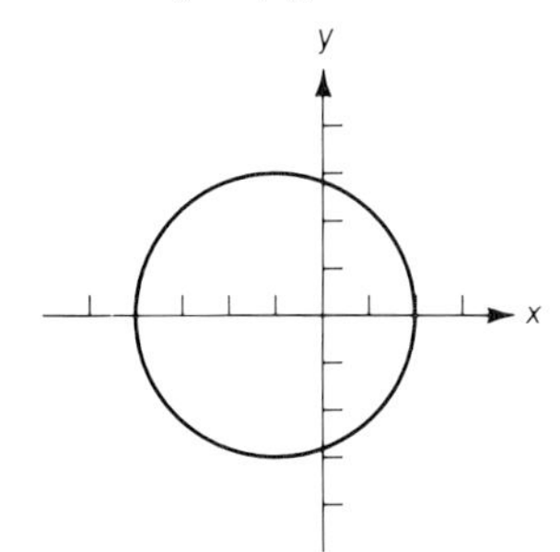


**41.** $x^2 + (y - 2)^2 = 4$
Center: $(0,2)$, $r = 2$

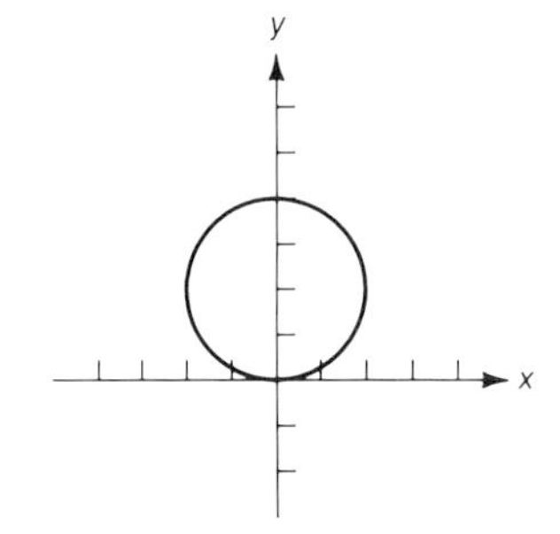


**43.** $(x + 1)^2 + (y + 2)^2 = 16$
Center: $(-1,-2)$, $r = 4$

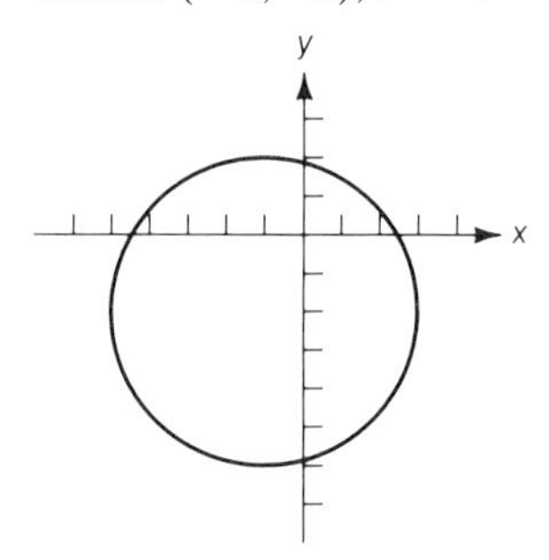

45. $(x + 2)^2 + (y + 2)^2 = 16$
Center: $(-2,-2)$, $r = 4$

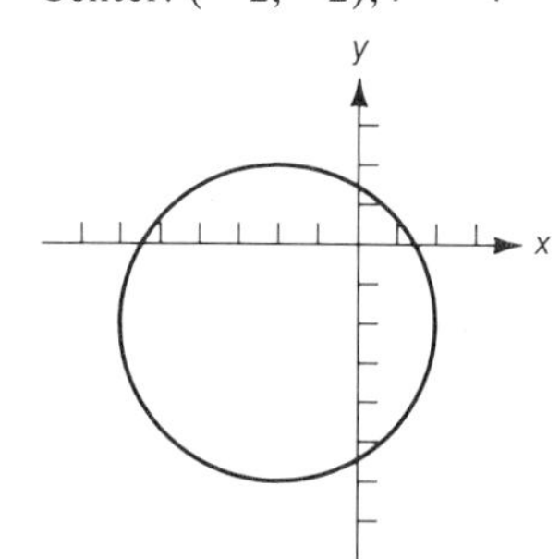


47. $(x - 2)^2 + (y - 3)^2 = 8$
Center: $(2,3)$, $r = \sqrt{8}$

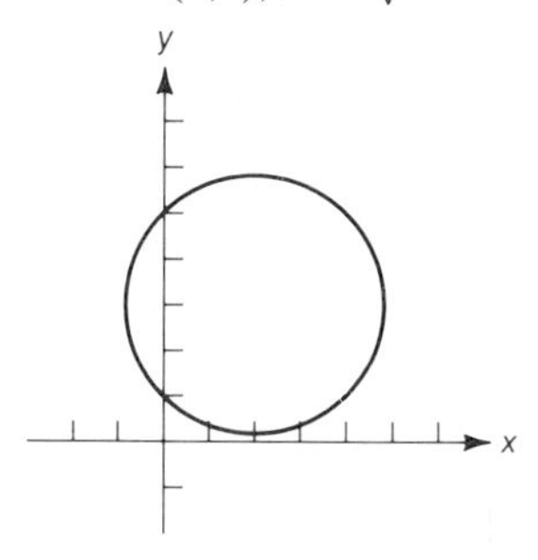


49. $|\overline{AB}|^2 = 45$
$|\overline{AC}|^2 = 65$
$|\overline{BC}|^2 = 20$
$|\overline{AB}|^2 + |\overline{BC}|^2 = |\overline{AC}|^2$

51. $|\overline{AB}| = |\overline{AC}| = 4\sqrt{5}$

53. $|\overline{AB}| = |\overline{AC}| = |\overline{BC}| = 4$

55. $|\overline{AC}| = |\overline{BC}| = \sqrt{61}$

57. $10 + 4\sqrt{5}$

59. $\sqrt{41} + \sqrt{65} + \sqrt{10}$

## *Exercises 10.3, Page 376*

1. $\dfrac{x^2}{36} + \dfrac{y^2}{4} = 1$

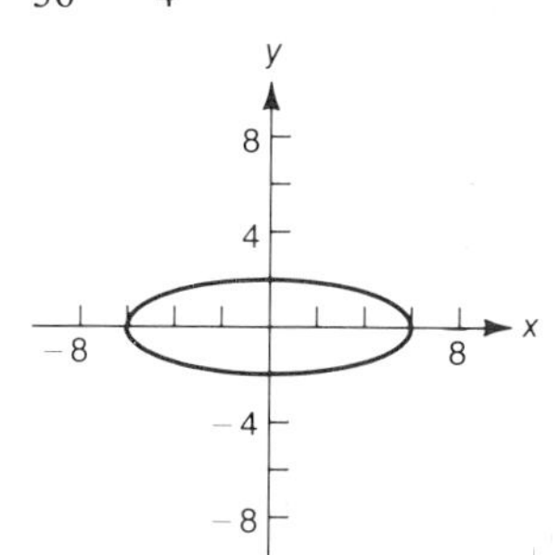


3. $\dfrac{x^2}{25} + \dfrac{y^2}{4} = 1$

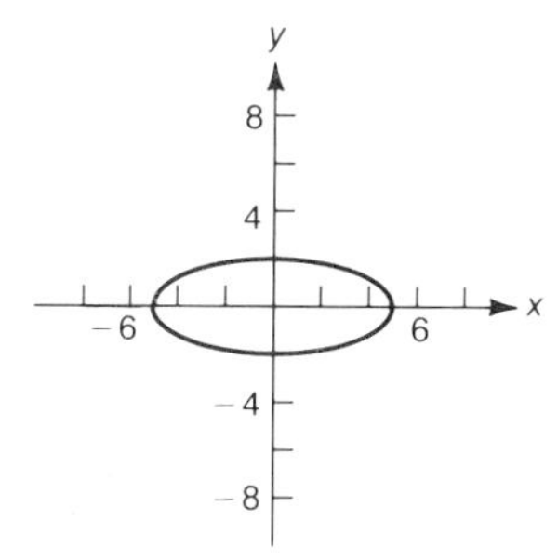


5. $\dfrac{x^2}{1} + \dfrac{y^2}{16} = 1$

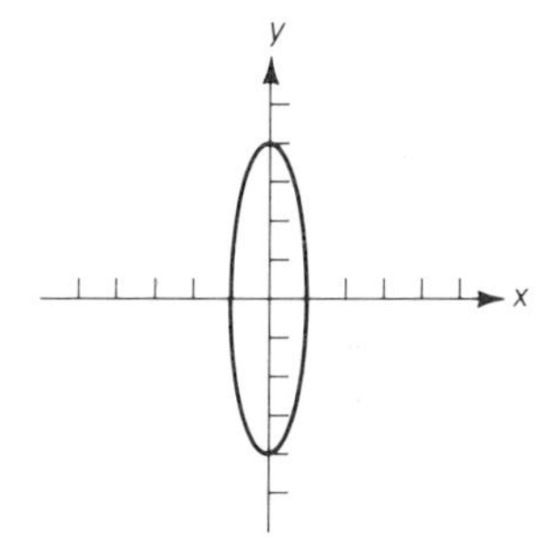


7. $x^2 - y^2 = 1$

$y = x, y = -x$

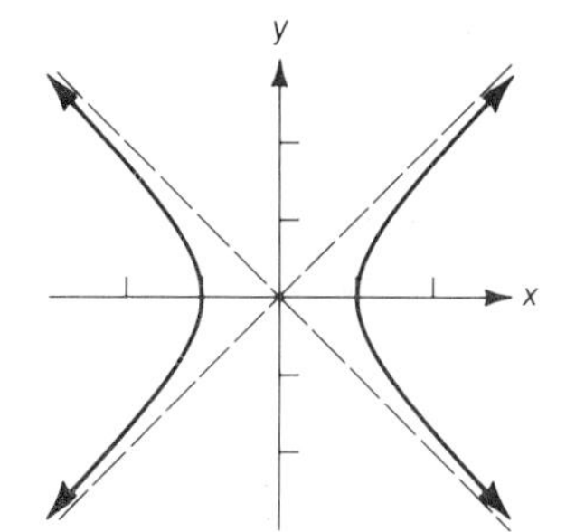


9. $\dfrac{x^2}{1} - \dfrac{y^2}{9} = 1$

$y = 3x, y = -3x$

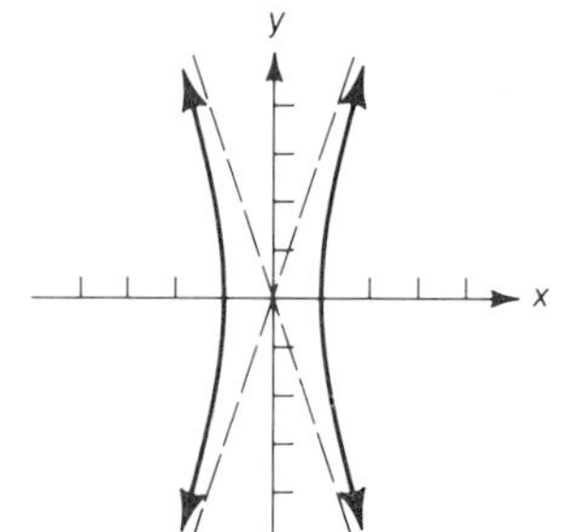


11. $\dfrac{x^2}{9} - \dfrac{y^2}{4} = 1$

$y = \dfrac{2}{3}x, y = -\dfrac{2}{3}x$

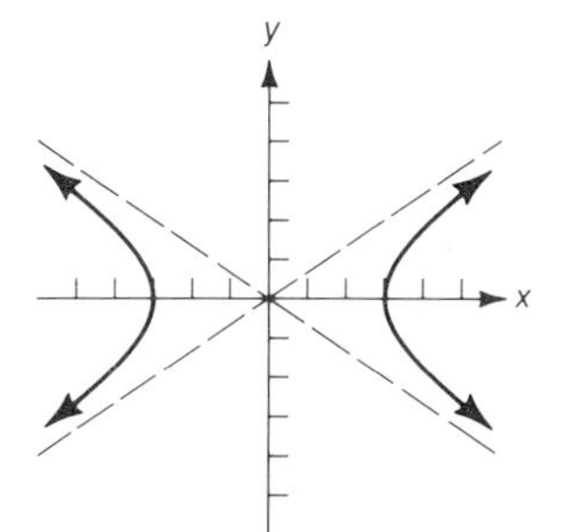

**13.** $\dfrac{x^2}{4} + \dfrac{y^2}{8} = 1$

**15.** $\dfrac{x^2}{20} + \dfrac{y^2}{4} = 1$

**17.** $\dfrac{y^2}{9} - \dfrac{x^2}{9} = 1$

$y = x,\ y = -x$

**19.** $\dfrac{y^2}{8} - \dfrac{x^2}{4} = 1$

$y = \sqrt{2}x,\ y = -\sqrt{2}x$

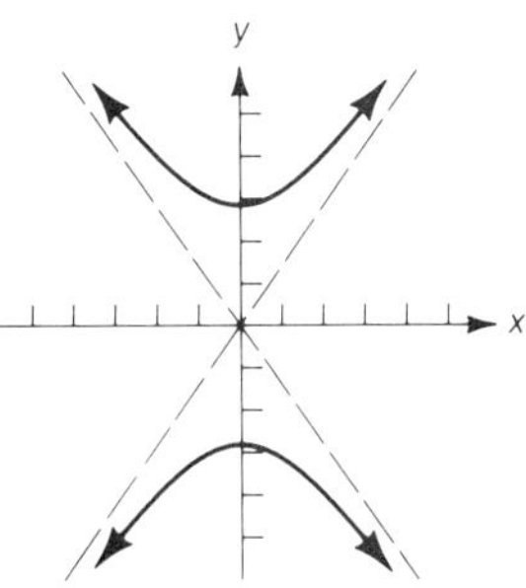


**21.** $\dfrac{y^2}{18} - \dfrac{x^2}{9} = 1$

$y = \sqrt{2}x,\ y = -\sqrt{2}x$

**23.** $\dfrac{x^2}{6} + \dfrac{y^2}{9} = 1$

**25.** $\dfrac{x^2}{5} + \dfrac{y^2}{4} = 1$

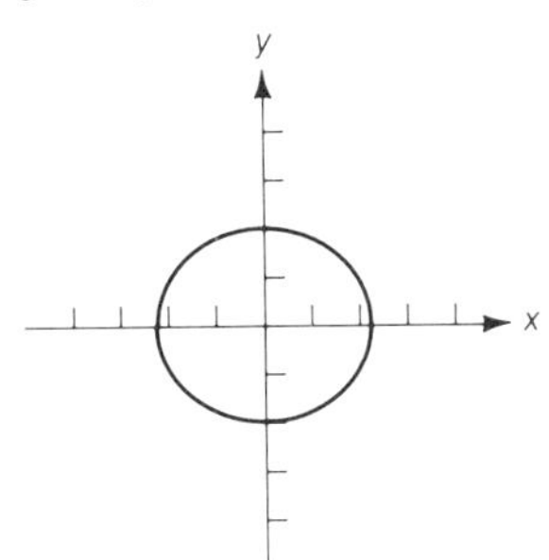


**27.** $\dfrac{x^2}{25} - \dfrac{y^2}{15} = 1$

$y = -\dfrac{\sqrt{15}}{5}x,\ y = \dfrac{\sqrt{15}}{5}x$

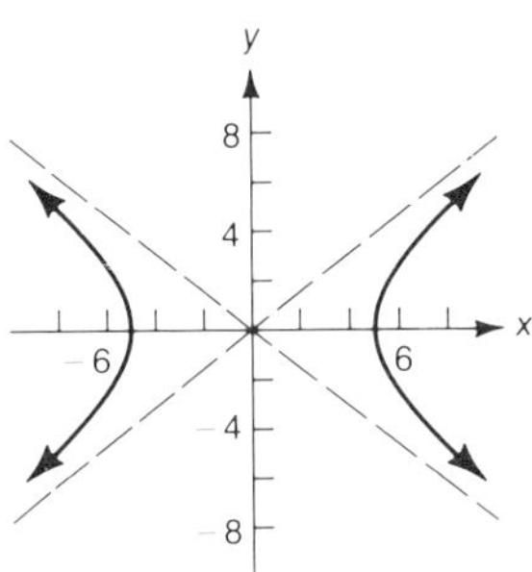


**29.** $\dfrac{y^2}{12} - \dfrac{x^2}{9} = 1$

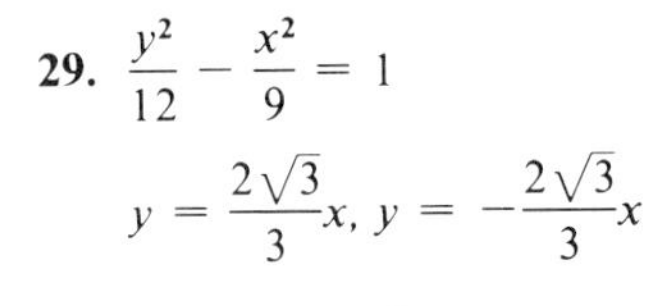

$y = \dfrac{2\sqrt{3}}{3}x,\ y = -\dfrac{2\sqrt{3}}{3}x$

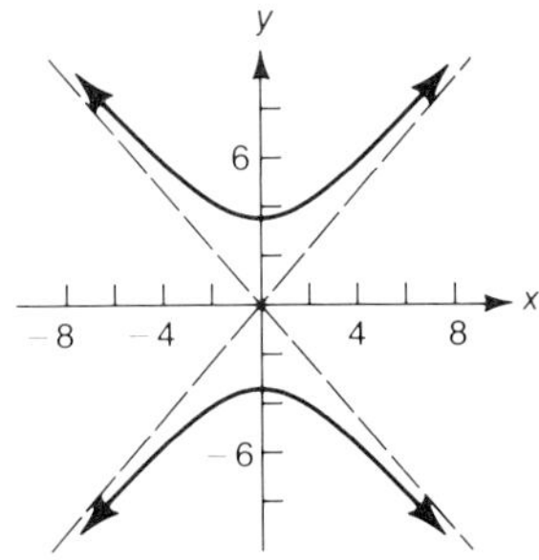

## *Exercises 10.4, Page 381*

**1.** $(-3,10)$, $(1,2)$

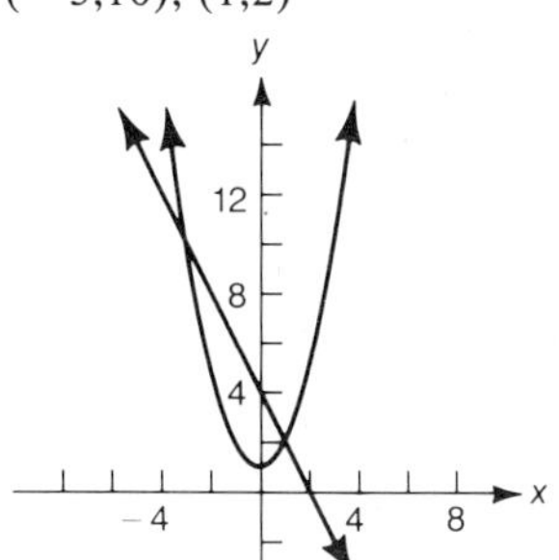


**3.** $(1,1)$, $(2,0)$

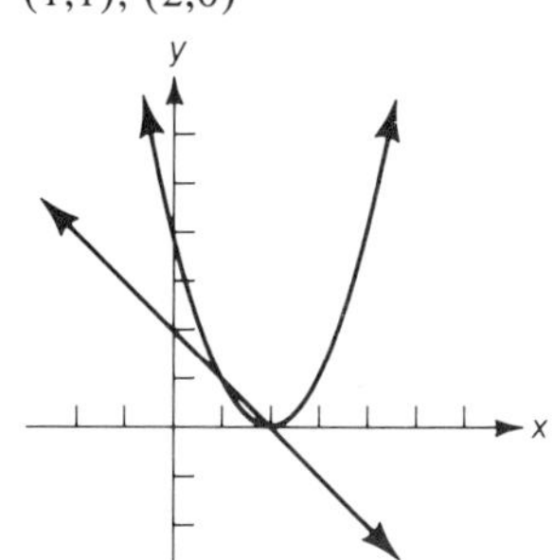


**5.** $(-2,-4)$, $(4,2)$

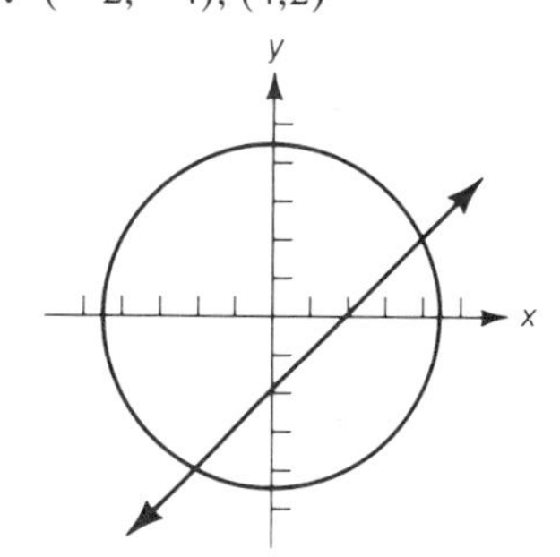


**7.** $(5,3)$

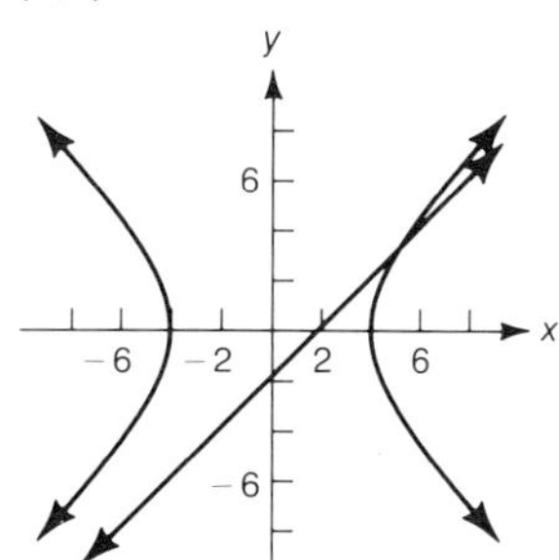


**9.** $(-3,0)$, $(3,0)$

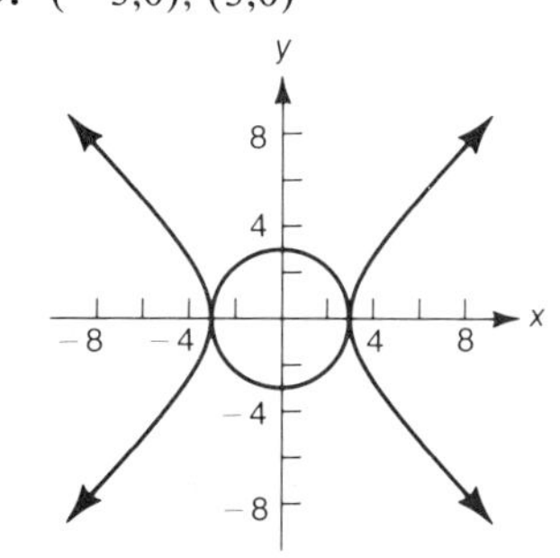


**11.** $(2,3)$, $(2,-3)$

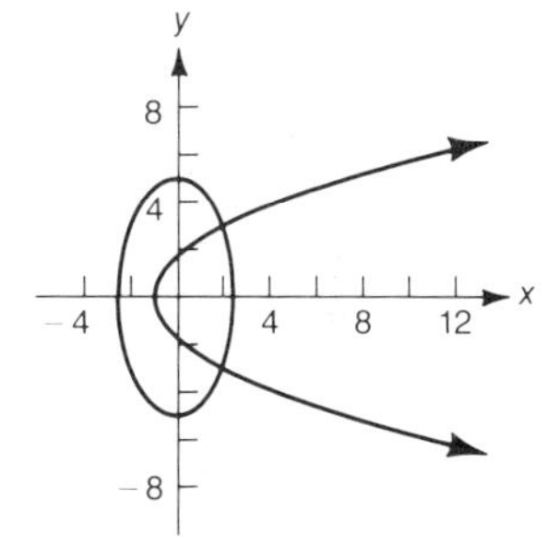


**13.** $(-2,4)$, $(1,1)$

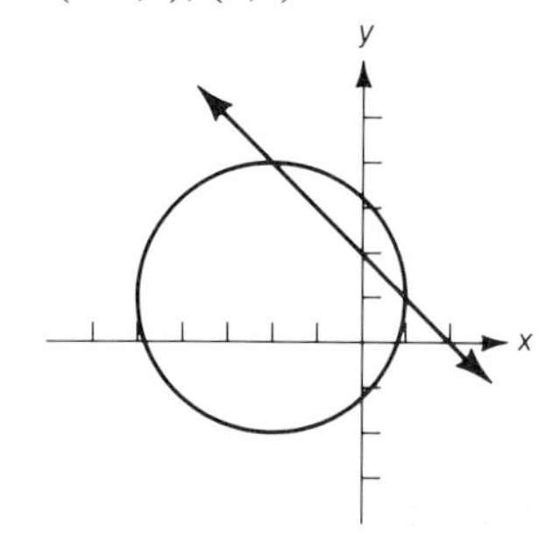


**15.** $(3,2)$, $(-3,-2)$, $(3,-2)$, $(-3,2)$

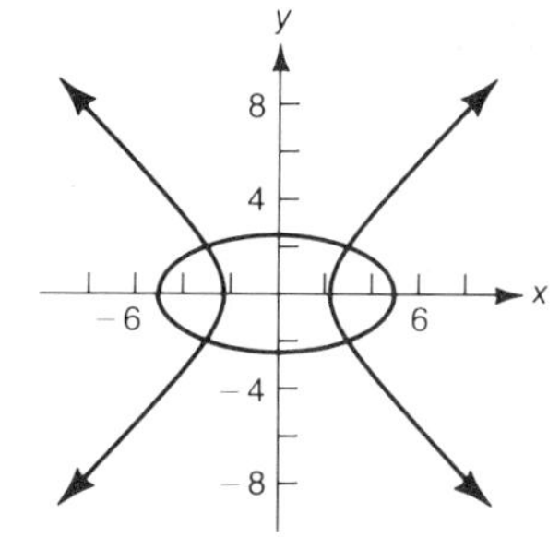


**17.** $(-3\sqrt{5},5)$, $(-6,4)$, $(3\sqrt{5},5)$, $(6,4)$ **19.** $(1,3)$, $(-1,3)$ **21.** $(-3,\sqrt{11})$, $(-3,-\sqrt{11})$, $(3,\sqrt{11})$, $(3,-\sqrt{11})$

**23.** $\left(-\frac{3}{2},-1\right)$, $\left(\frac{1}{2},\frac{1}{2}\right)$ **25.** $(4,4)$, $(-2,-2)$ **27.** $(5,-3)$, $(-5,1)$ **29.** $(0,3)$, $(2,3)$

## Chapter 10 Review, Page 382

**1.** Vertex: (2,0)
$y$-intercept: None
$x$-intercept: (2,0)
Line of symmetry: $y = 0$

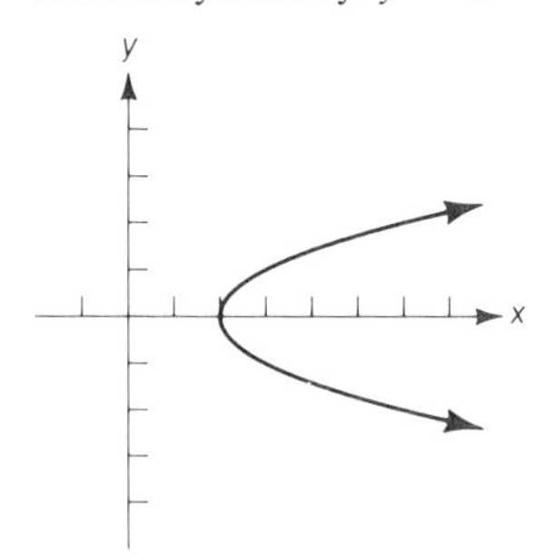

**2.** Vertex: (1,0)
$y$-intercept: None
$x$-intercept: (1,0)
Line of symmetry: $y = 0$

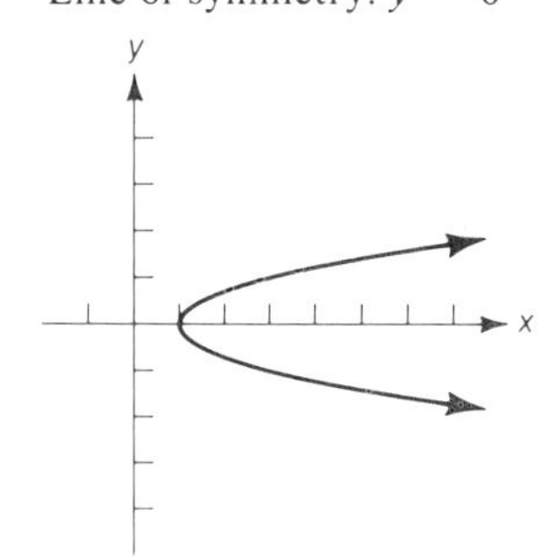

**3.** Vertex: $(-4,1)$
$y$-intercept: (0,3), $(0,-1)$
$x$-intercept: $(-3,0)$
Line of symmetry: $y = 1$

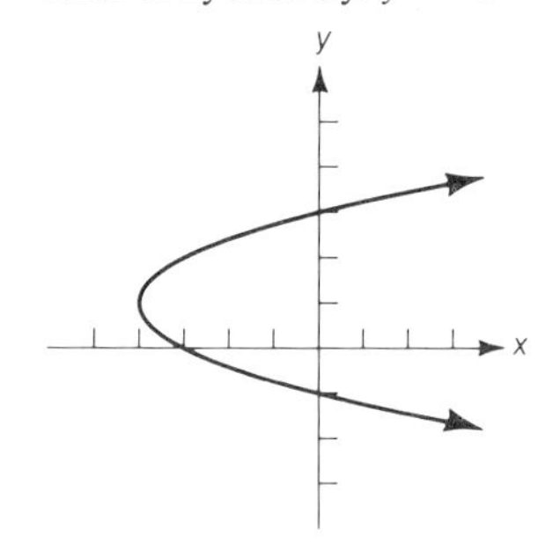

**4.** Vertex: $(-1,-2)$
$y$-intercept: $(0,-1)$
$x$-intercept: $(-1 \pm \sqrt{2},0)$
Line of symmetry: $x = -1$

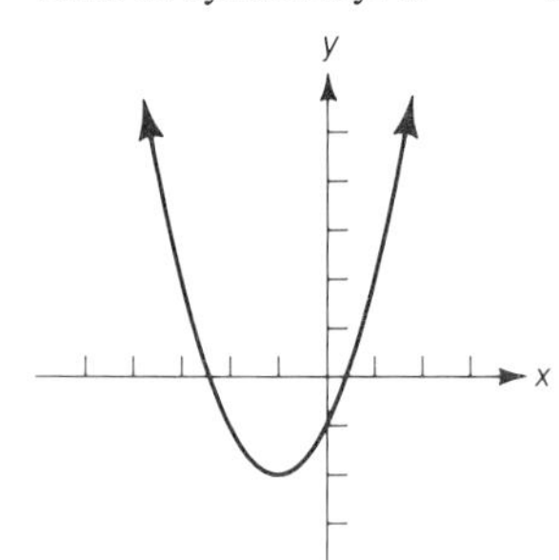

**5.** Vertex: $(-6,1)$
$y$-intercept: $(0,1 \pm \sqrt{6})$
$x$-intercept: $(-5,0)$
Line of symmetry: $y = 1$

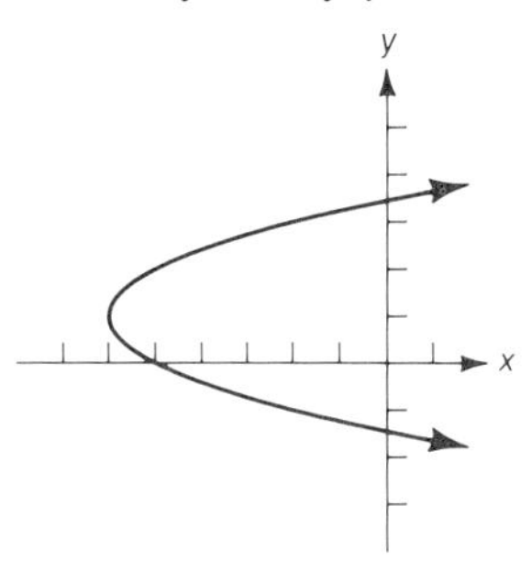

**6.** Vertex: $\left(\frac{2}{3},0\right)$
$y$-intercept: (0,4)
$x$-intercept: $\left(\frac{2}{3},0\right)$
Line of symmetry: $x = \frac{2}{3}$

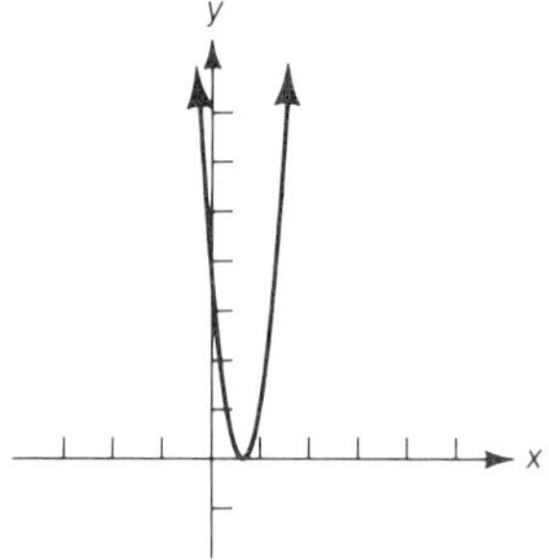

**7.** $2\sqrt{10}$ **8.** $4\sqrt{5}$ **9.** $2\sqrt{26}$ **10.** $\overline{AC} = \sqrt{50}$, $\overline{BC} = \sqrt{50}$ **11.** $\sqrt{10} + 3\sqrt{2} + 2\sqrt{13}$ **12.** $(\overline{AB})^2 = 45$, $(\overline{AC})^2 = 20$, $(\overline{BC})^2 = 65$, $(\overline{AB})^2 + (\overline{AC})^2 = (\overline{BC})^2$

**13.** $x^2 + y^2 = 25$ **14.** $x^2 + y^2 = 12$ **15.** $(x - 1)^2 + (y - 2)^2 = 9$ **16.** $(x + 2)^2 + (y - 3)^2 = 7$

**17.** $x^2 + y^2 = 4$
Center: $(0,0)$, $r = 2$

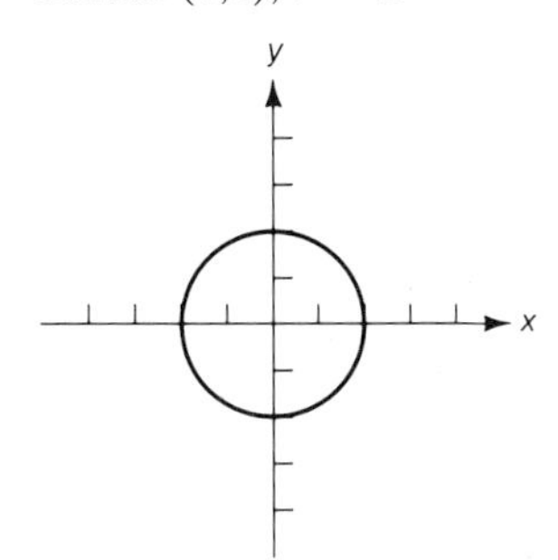


**18.** $x^2 + y^2 = 18$
Center: $(0,0)$, $r = 3\sqrt{2}$

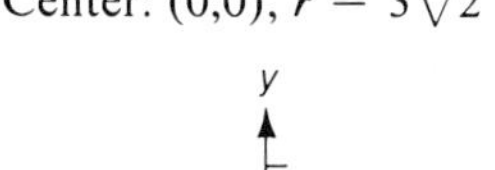
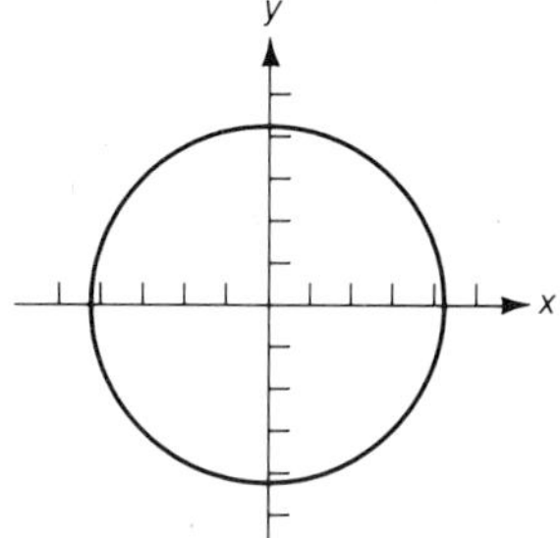


**19.** $(x + 1)^2 + y^2 = 4$
Center: $(-1,0)$, $r = 2$

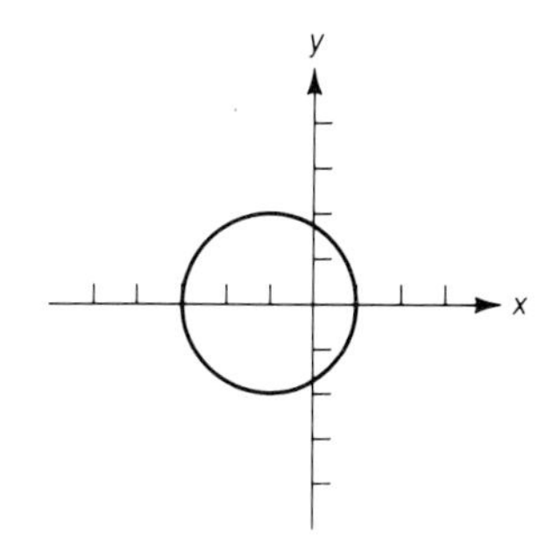


**20.** $(x - 2)^2 + (y + 3)^2 = 16$
Center: $(2,-3)$, $r = 4$

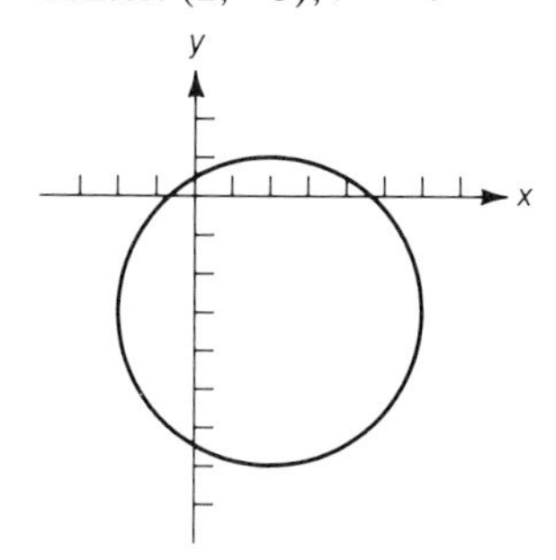


**21.** $\frac{x^2}{1} + \frac{y^2}{4} = 1$

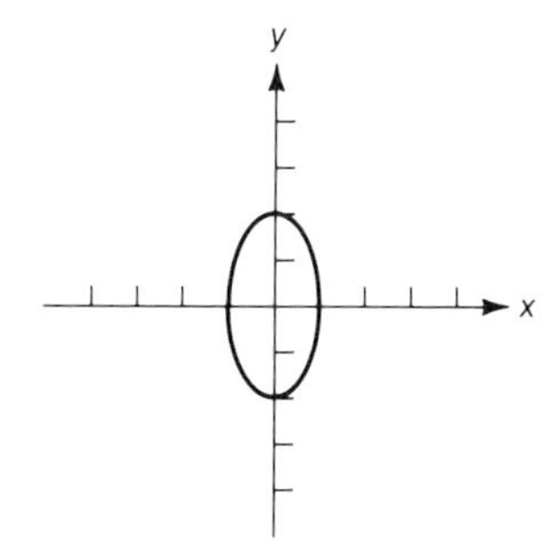


**22.** $\frac{x^2}{16} + \frac{y^2}{4} = 1$

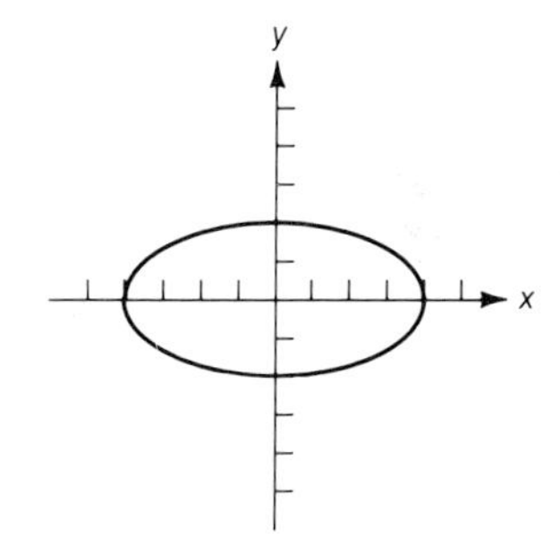


**23.** $\frac{x^2}{9} - \frac{y^2}{1} = 1$

$y = \frac{1}{3}x,\ y = -\frac{1}{3}x$

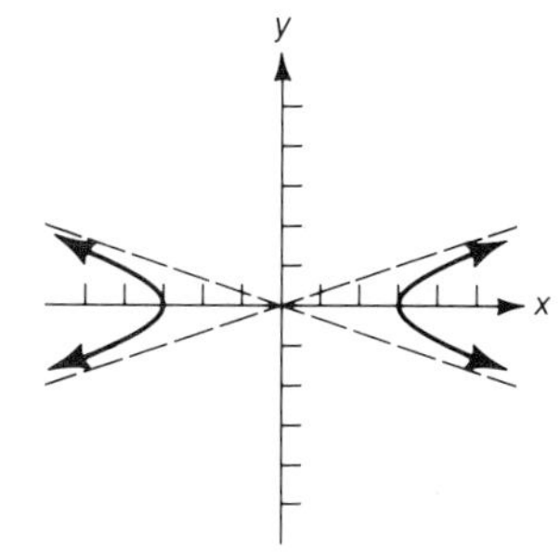


**24.** $\frac{y^2}{4} - \frac{x^2}{9} = 1$

$y = \frac{2}{3}x,\ y = -\frac{2}{3}x$

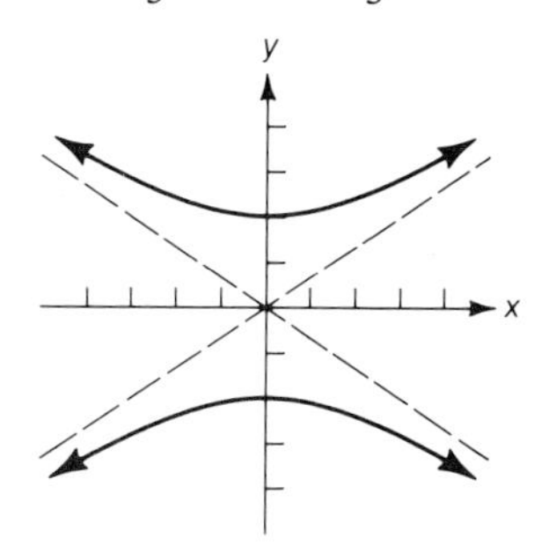


**25.** $\frac{x^2}{25} + \frac{y^2}{4} = 1$

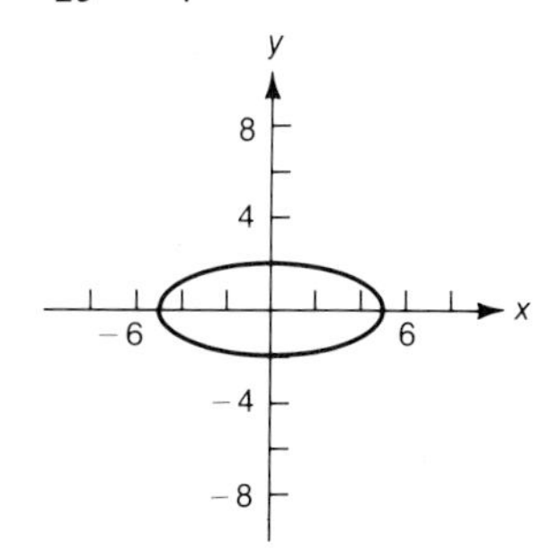

**26.** $\frac{x^2}{9} - \frac{y^2}{16} = 1$

$y = \frac{4}{3}x,\ y = -\frac{4}{3}x$

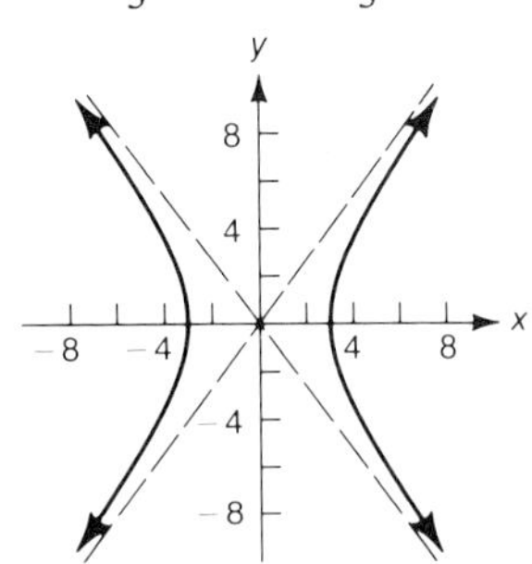

**27.** $\frac{y^2}{25} - \frac{x^2}{16} = 1$

$y = \frac{5}{4}x,\ y = -\frac{5}{4}x$

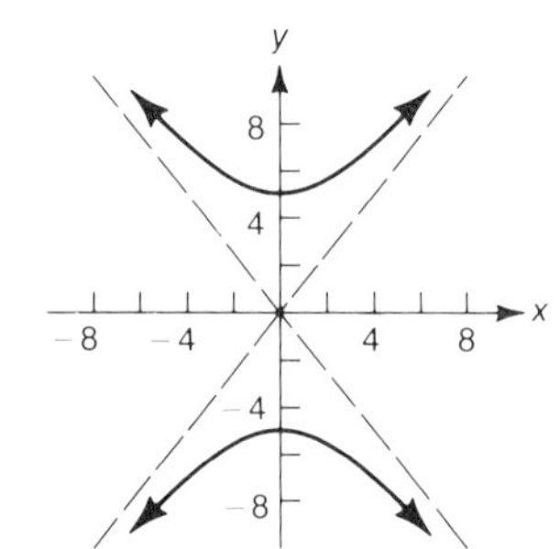

**28.** $\frac{x^2}{9} + \frac{y^2}{25} = 1$

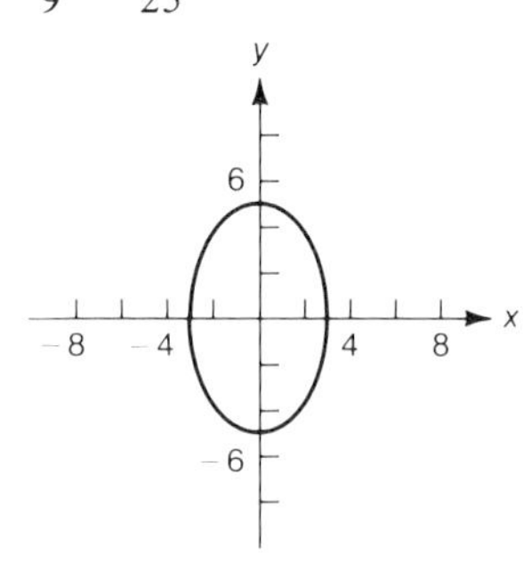

**29.** $\frac{x^2}{12} + \frac{y^2}{4} = 1$

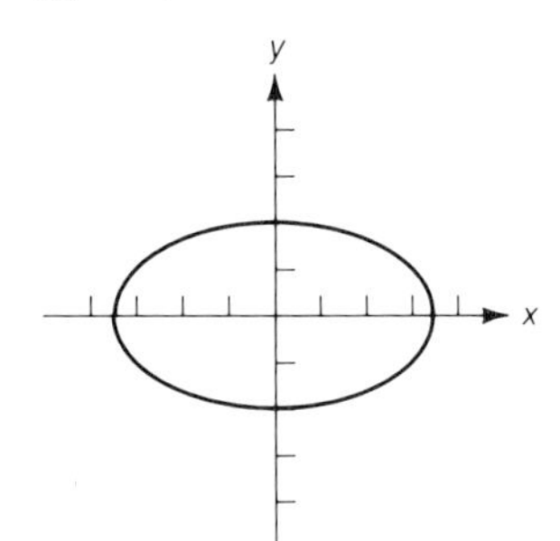

**30.** $\frac{x^2}{9} - \frac{y^2}{18} = 1,\ y = \sqrt{2}x,\ y = -\sqrt{2}x$

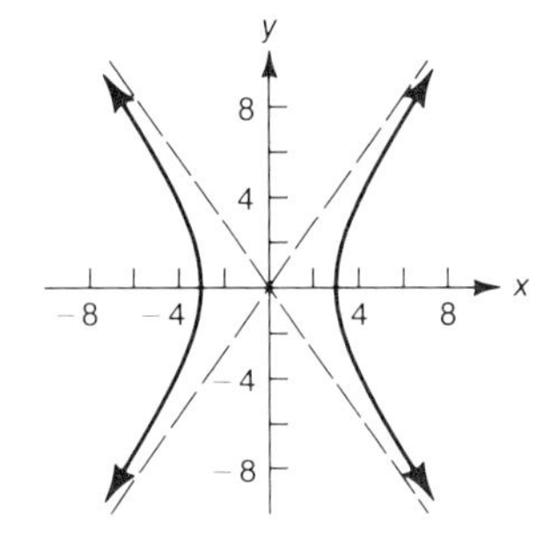

**31.** $(-1,3), \left(\frac{9}{5}, -\frac{13}{5}\right)$

**32.** $(-1,3), \left(\frac{9}{5}, \frac{1}{5}\right)$ **33.** $(2,0), (-2,0), (0,-4)$ **34.** $(2,2\sqrt{2}), (2,-2\sqrt{2}), (-2,2\sqrt{2}), (-2,-2\sqrt{2})$

**35.** $(2,\sqrt{3}), (2,-\sqrt{3})$ **36.** $(-1,2\sqrt{6}), (-1,-2\sqrt{6})$

## Chapter 10 Test, Page 383

**1.** Vertex: $(0,3)$
$y$-intercept: $(0,3)$
$x$-intercept: None
Line of symmetry: $x = 0$

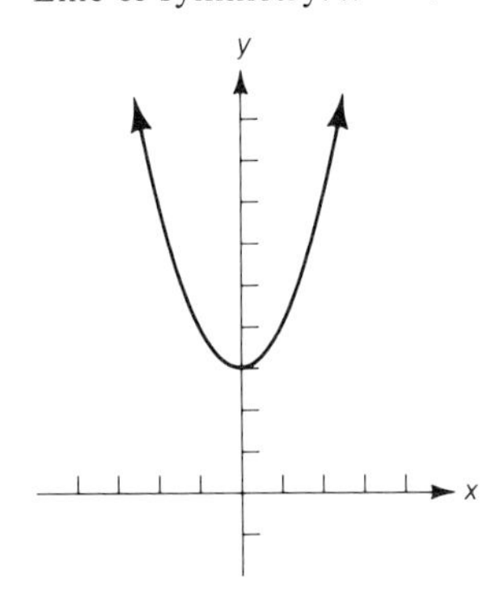

**2.** Vertex: $(-5,0)$
$y$-intercept: $(0,\sqrt{5}), (0,-\sqrt{5})$
$x$-intercept: $(-5,0)$
Line of symmetry: $y = 0$

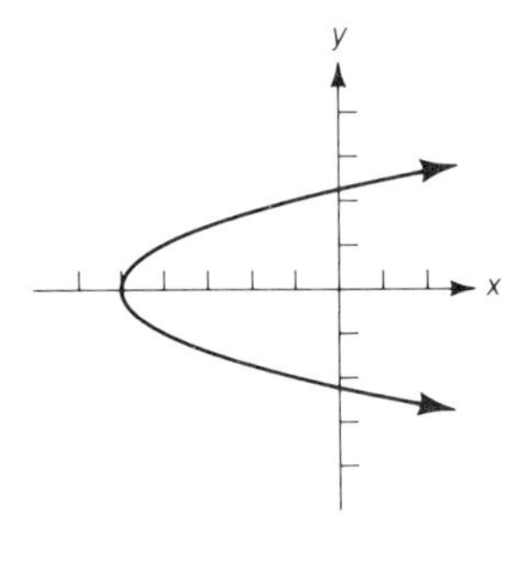

**3.** Vertex: $\left(-\frac{25}{4}, -\frac{3}{2}\right)$
$y$-intercept: $(0,-4), (0,1)$
$x$-intercept: $(-4,0)$
Line of symmetry: $y = -\frac{3}{2}$

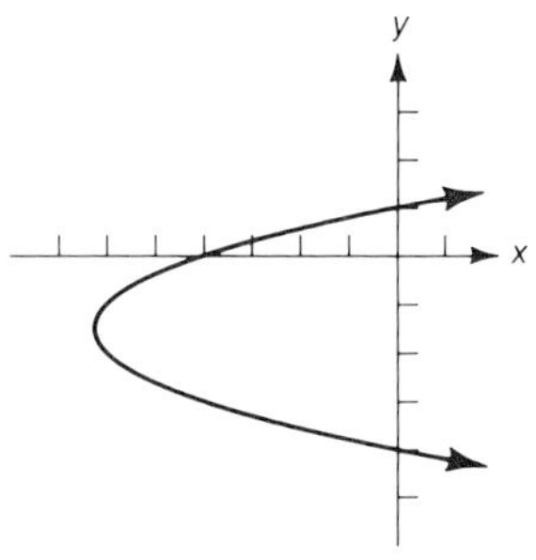

**4.** Vertex: $(1,-9)$
$y$-intercept: $(0,-8)$
$x$-intercept: $(4,0)$, $(-2,0)$
Line of symmetry: $x = 1$

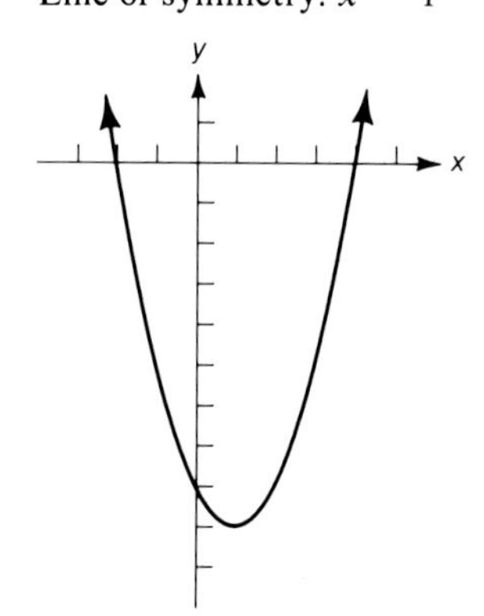


**5.** $3\sqrt{10}$

**6.** $|\overline{AB}|^2 = 52$
$|\overline{AC}|^2 = 104$
$|\overline{BC}|^2 = 52$
$|\overline{AB}|^2 + |\overline{BC}|^2 = |\overline{AC}|^2$

**7.** $x^2 + y^2 = 18$

**8.** $(x + 3)^2 + (y - 1)^2 = 25$

**9.** $x^2 + y^2 = 9$
Center: $(0,0)$, $r = 3$

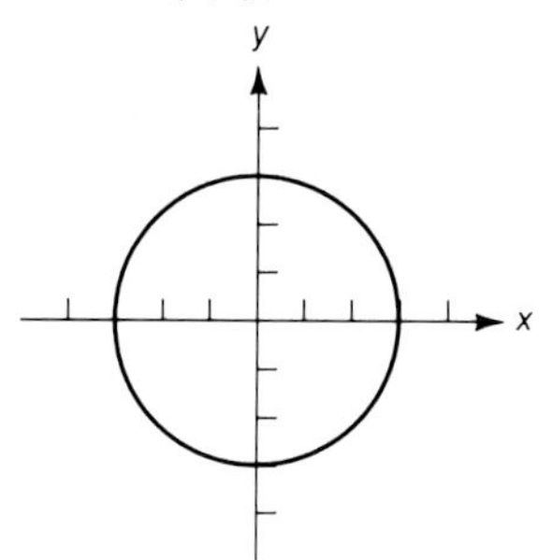


**10.** $x^2 + (y - 1)^2 = 9$
Center: $(0,1)$, $r = 3$

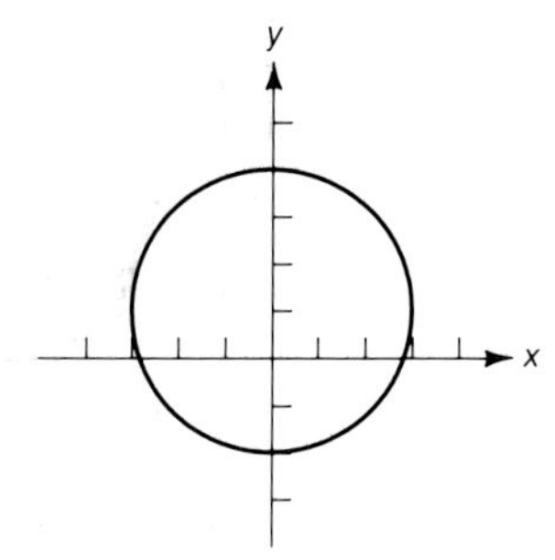


**11.** $\frac{x^2}{4} - \frac{y^2}{9} = 1$

$y = \frac{3}{2}x$, $y = -\frac{3}{2}x$

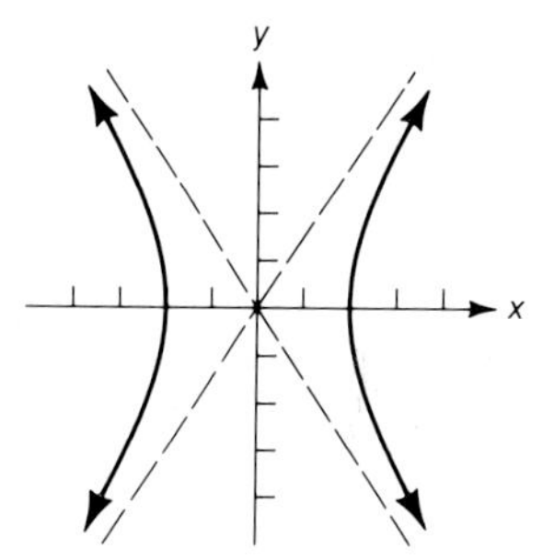


**12.** $\frac{x^2}{9} + \frac{y^2}{\frac{9}{4}} = 1$

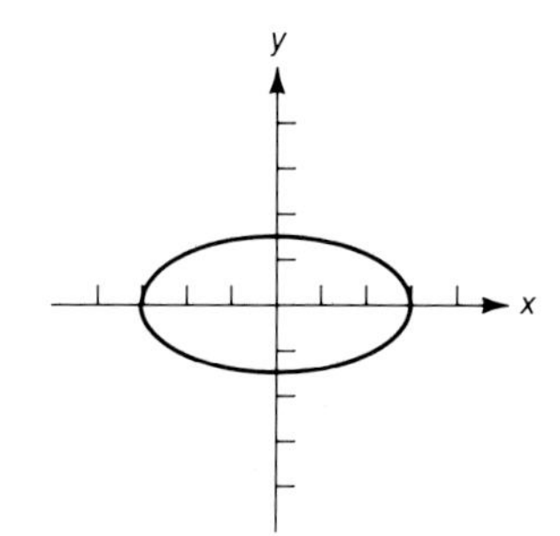


**13.** $x^2 + y^2 = 4$

**14.** $\frac{y^2}{9} - \frac{x^2}{16} = 1$

$y = \frac{3}{4}x$, $y = -\frac{3}{4}x$

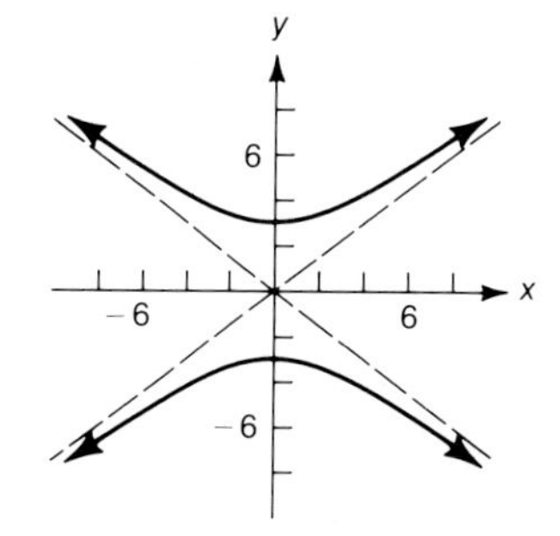

**15.** $x + 2 = (y - 3)^2$

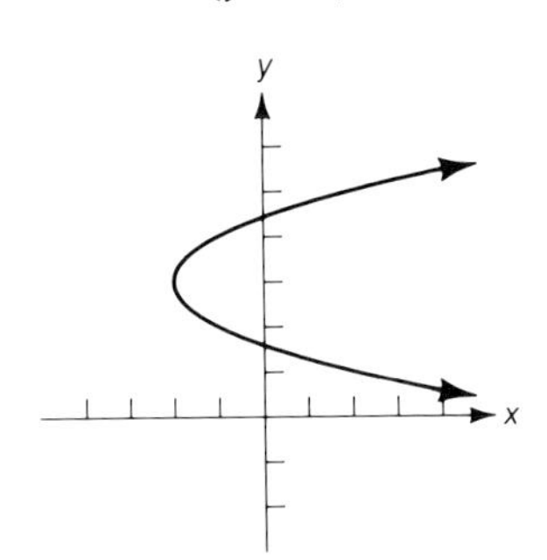


**16.** $\dfrac{x^2}{4} + \dfrac{y^2}{25} = 1$

**17.** $(-2,-5)$, $(5,2)$

**18.** $(0,\sqrt{2})$, $(0,-\sqrt{2})$, $(-2,0)$ **19.** $(4,3)$, $(4,-3)$, $(-4,3)$, $(-4,-3)$

**20.** $(3,-1)$, $(-3,-1)$, $\left(\dfrac{\sqrt{65}}{3},\dfrac{1}{3}\right)$, $\left(-\dfrac{\sqrt{65}}{3},\dfrac{1}{3}\right)$

## *Cumulative Review (10), Page 384*

**1.** $2(x + 3)(x^2 - 3x + 9)$ **2.** $x^{-2}(2x - 5)(x + 7)$ **3.** $(x - 4)(x^2 + 3)$ **4.** $13\sqrt{3}$ **5.** $2\sqrt{3x} + 10\sqrt{6}$

**6.** $5x - 2y = -16$ **7.** $3x - 4y = 16$ **8.** $f^{-1}(x) = \dfrac{1 - 2x}{x}$ **9.** $x = 4$, $x = 16$ **10.** $x = 7$

**11.** $x = -34.02$ **12.** $x = 4$ **13.** $-1 \leq x < 0$ or $x \geq 5$ **14.** $x = 5$, $y = 2$

**15.**

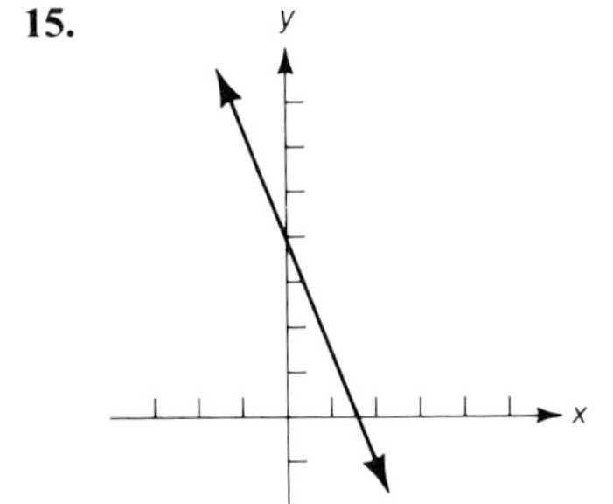


**16.**

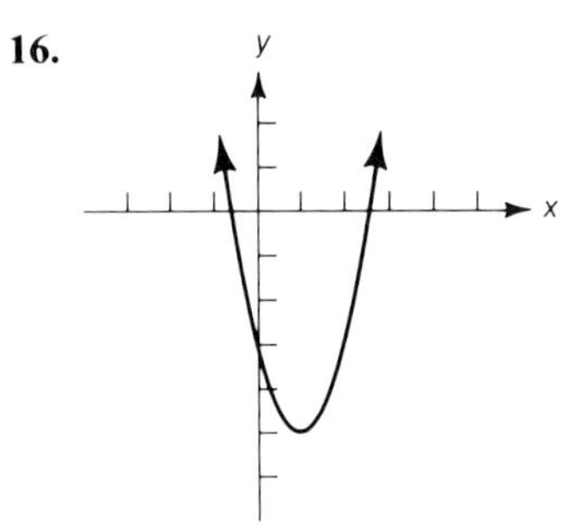


**17.**

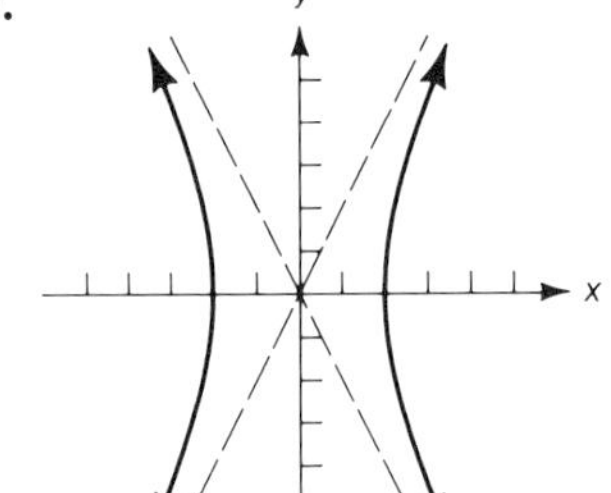


**18.**

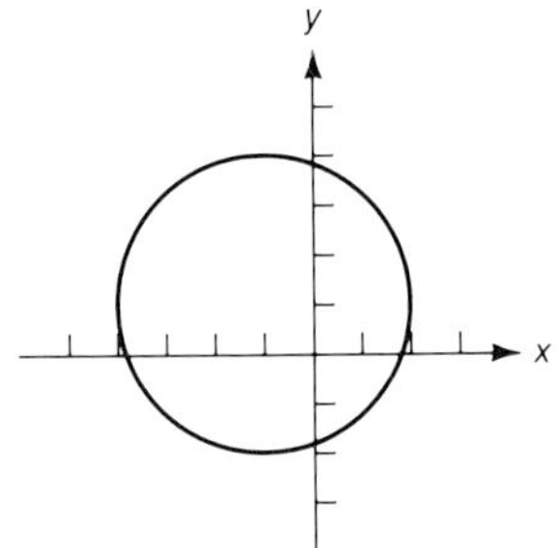


**19.** Shirt: \$26.50, trousers: \$43.00 **20.** $15\frac{2}{3}$ years

# *CHAPTER 11*

## *Exercises 11.1, Page 389*

**1.** $-22$ **3.** $-48$ **5.** 11 **7.** 3 **9.** 47 **11.** 36 **13.** $-3$ **15.** $-4$ **17.** $-25$ **19.** 20

**21.** $x = 7$ **23.** $x = -7$ **25.** $x = -3$ **27.** $2x - 7y + 11 = 0$ **29.** 1

### Exercises 11.2, Page 394

**1.** (4,3) **3.** $\left(\frac{2}{3},-\frac{1}{4}\right)$ **5.** Inconsistent **7.** $\left(-\frac{1}{4},\frac{3}{2}\right)$ **9.** $\left(\frac{31}{17},\frac{2}{17}\right)$ **11.** $\left(\frac{39}{44},\frac{41}{44}\right)$ **13.** $\left(\frac{18}{7},-\frac{3}{7}\right)$

**15.** $\left(-\frac{21}{183},\frac{266}{183}\right)$ **17.** $\left(\frac{210}{41},-\frac{40}{123}\right)$ or (5.12,−0.33) **19.** $\left(\frac{525}{124},\frac{109}{93}\right)$ or (4.23,1.17) **21.** (−2,1,3)

**23.** No solution **25.** (4,2,6) **27.** Infinite number of solutions **29.** $\left(-\frac{2}{3},\frac{11}{3},2\right)$

### Exercises 11.3, Page 401

**1.** $2 \times 2$, $a_{21} = 4$ **3.** $3 \times 1$, $a_{31} = 4$ **5.** $4 \times 4$, $a_{34} = -8$ **7.** $2 \times 4$, $a_{22} = 10$ **9.** $4 \times 3$, $a_{12} = 17$
**11.** $x = -1, y = -2, z = 3$ **13.** $x = 1, y = 0, z = 1$ **15.** $x = 7, y = 0, z = 2$
**17.** $x = -2, y = -1, z = 5$ **19.** $x = 1, y = 2, z = 1$ **21.** $x = 1, y = -3, z = 2$
**23.** $x = 3, y = -3, z = 1, w = -2$ **25.** $x = 1, y = 0, z = -2, w = -1$ **27.** 52, 40, 77
**29.** 20 small, 32 medium, 16 large

### Exercises 11.4, Page 407

**1.** $x = 5, y = 4, z = 1, w = -5$ **3.** $x = -1, y = -5, z = -1, w = -2$ **5.** $x = -3, y = -5, z = 5, w = 7$

**7.** $x = 3, y = 4, z = -4, w = -2$ **9.** $\begin{bmatrix} -2 & 0 \\ 6 & -4 \end{bmatrix}$ **11.** $\begin{bmatrix} -2 & -1 \\ 0 & 0.5 \\ -1.5 & 2 \end{bmatrix}$ **13.** $\begin{bmatrix} 2 & 10 \\ 3 & 3 \end{bmatrix}$ **15.** Undefined

**17.** $\begin{bmatrix} 2 & 3 \\ -6 & -2 \end{bmatrix}$ **19.** $\begin{bmatrix} 0.7 & -0.6 \\ -0.2 & 0.6 \end{bmatrix}$ **21.** $\begin{bmatrix} 10 & 0 & 4 \\ -5 & 0 & -2 \\ 15 & 0 & 6 \end{bmatrix}$ **23.** $\begin{bmatrix} -6 & 1 \\ -2 & 5 \end{bmatrix}$ **25.** Undefined **27.** $\begin{bmatrix} 1 & 0 \\ 0 & 1 \end{bmatrix}$

**29.** $[-9 \quad 4 \quad 12]$ **31.** $\begin{bmatrix} -9 \\ 8 \\ -4 \end{bmatrix}$ **33.** $\begin{bmatrix} 1 & 6 & 6 \\ 8 & 6 & 13 \\ 9 & -6 & 11 \end{bmatrix}$ **35.** $\begin{bmatrix} 4 & 16 & -12 \\ 8 & 8 & -4 \end{bmatrix}$ **37.** Undefined

**39.** $\begin{bmatrix} 1 & 10 & -8 \\ 4 & 10 & -7 \end{bmatrix}$ **41.** $\begin{bmatrix} 0 & 2 & 5 \\ 3 & -3 & 6 \\ -1 & 6 & 2 \end{bmatrix}$ **43.** $\begin{bmatrix} 15 & -28 & 23 \\ 7 & -8 & 20 \end{bmatrix}$

### Exercises 11.5, Page 414

**9.** $\begin{bmatrix} -5 & -3 \\ -2 & -1 \end{bmatrix}$ **11.** $\begin{bmatrix} -\frac{1}{2} & -1 \\ -\frac{3}{2} & -2 \end{bmatrix}$ **13.** $\begin{bmatrix} 2 & -10 & 13 \\ -2 & 11 & -14 \\ -1 & 4 & -5 \end{bmatrix}$ **15.** $\begin{bmatrix} -3 & 3 & -2 \\ -4 & 3 & -2 \\ 1 & -1 & 1 \end{bmatrix}$

**17.** $\begin{bmatrix} 15 & -6 & -2 \\ -7 & 3 & 1 \\ 2 & -1 & 0 \end{bmatrix}$ **19.** $\begin{bmatrix} -4 & 0 & 3 \\ 26 & 1 & -19 \\ 3 & 0 & -2 \end{bmatrix}$ **21.** $x = -1, y = 2$ **23.** $x = \frac{5}{7}, y = \frac{9}{7}$

**25.** $x = 0, y = -1, z = 3$ **27.** $x = -2, y = 1, z = 2$ **29.** $x = -1, y = 5, z = 4$

### Chapter 11 Review, Page 417

**1.** 26 **2.** 29 **3.** −12 **4.** −72 **5.** $x = 5$ **6.** $x = 2.4$ **7.** $x = 3, y = -7$ **8.** $x = \frac{3}{8}, y = \frac{1}{6}$

**9.** $x = 5, y = 2$ **10.** $x = 4, y = 2, z = 3$ **11.** $x = \frac{7}{4}, y = \frac{5}{6}, z = -\frac{1}{3}$ **12.** $2 \times 3, a_{12} = 1$
**13.** $3 \times 1, a_{31} = 1$ **14.** $3 \times 3, a_{23} = 0$ **15.** $x = 1, y = -4$ **16.** $x = 2, y = -1, z = 1$
**17.** $x = 2, y = 0, z = 1$ **18.** 800 lbs of Ration I, 875 lbs of Ration II
**19.** 4 lbs of apples, 3 lbs of bananas, 3 lbs of oranges **20.** $x = 6, y = 3, z = 2, w = 4$
**21.** $x = 5, y = -2, z = 1, w = 0$ **22.** $\begin{bmatrix} 6 & 3 \\ 8 & 1 \end{bmatrix}$ **23.** $\begin{bmatrix} 1 & -2 \\ 1 & 4 \\ -1 & 2 \end{bmatrix}$ **24.** $\begin{bmatrix} -2 & 9 & -8 \\ 6 & 2 & -19 \end{bmatrix}$ **25.** $\begin{bmatrix} 21 \\ 11 \\ 42 \end{bmatrix}$

**26.** Undefined **27.** $\begin{bmatrix} -10 & 15 \\ 0 & 11 \\ -1 & 29 \end{bmatrix}$ **28.** $\begin{bmatrix} -5 & -2 \\ -4 & -7 \\ -10 & 17 \end{bmatrix}$ **31.** $\begin{bmatrix} -5 & 3 \\ -3 & 2 \end{bmatrix}$ **32.** $\begin{bmatrix} -\frac{2}{7} & -\frac{5}{7} \\ \frac{3}{7} & \frac{4}{7} \end{bmatrix}$

**33.** $\begin{bmatrix} 5 & -4 & -6 \\ -1 & 1 & 2 \\ -1 & 1 & 1 \end{bmatrix}$ **34.** $\begin{bmatrix} -4 & 5 & -6 \\ 4 & -5 & 7 \\ 1 & -1 & 1 \end{bmatrix}$ **35.** $x = 2, y = -3$ **36.** $x = 5, y = 2$

**37.** $x = 3, y = -1, z = 2$ **38.** $x = 1, y = 4, z = -2$

## *Chapter 11 Test, Page 419*

**1.** 42 **2.** 5 **3.** $x = 6$ **4.** $x = \frac{17}{3}$ **5.** $x = -\frac{78}{5}, y = \frac{38}{5}$ **6.** $x = -\frac{5}{26}, y = \frac{33}{26}, z = \frac{9}{26}$
**7.** $3 \times 2, a_{32} = -3$ **8.** $x = 3, y = 7$ **9.** $x = 3, y = -1, z = -2$ **10.** Standard: 16, pro: 12
**11.** $x = 1, y = 1, z = -3, w = 6$ **12.** $\begin{bmatrix} 4 & 4 \\ 4 & 1 \end{bmatrix}$ **13.** $\begin{bmatrix} -1 & 1 & 6 \\ -4 & 2 & 0 \end{bmatrix}$ **14.** $\begin{bmatrix} 21 & 17 \\ -11 & -7 \\ 1 & 17 \end{bmatrix}$ **15.** $\begin{bmatrix} 46 \\ -5 \\ -36 \end{bmatrix}$

**16.** The product is $I_3$. **17.** $\begin{bmatrix} -4 & -3 \\ -7 & -5 \end{bmatrix}$ **18.** $\begin{bmatrix} -1 & -5 & 3 \\ 2 & 7 & -4 \\ 2 & 5 & -3 \end{bmatrix}$ **19.** $x = 3, y = -2$ **20.** $x = -1, y = 5, z = 3$

## *Cumulative Review (11), Page 420*

**1.** $4x^3 + 4x^2 - 13x - 15$ **2.** 18 **3.** 73 **4.** $x^2 + 2x + 1$ **5.** $x^2 + 5x + 3$ **6.** $21\sqrt{2}$ **7.** $18\sqrt{2}$
**8.** $x = -3, y = 7$ **9.** $x^2 - 10x + 23 = 0$ **10.** $(2x - 5)(4x^2 + 10x + 25)$ **11.** $(3x + 2)(x^2 - 3)$
**12.** $x = \frac{2}{3}, x = 9$ **13.** $x = -3 - \sqrt{5}, x = -3 + \sqrt{5}$ **14.** $x = -4 - 3i, x = -4 + 3i$
**15.** $-117$ **16.** $x = 18, y = 13$ **17.** $\begin{bmatrix} -9 & 3 & 4 \\ 49 & 12 & -35 \end{bmatrix}$ **18.** $\begin{bmatrix} 10 & 17 & -8 \\ 9 & 15 & -7 \\ -6 & -10 & 5 \end{bmatrix}$ **19.** \$26

**20.** Skis: \$110, bindings: \$65

# *CHAPTER 12*

## *Exercises 12.1, Page 426*

**1.** $P(-3) = 81; P(1) = -1$ **3.** $P(-4) = 18; P(1.5) = 20.75$ **5.** $P(-2) = -61; P(-3) = -196$
**7.** $P(3) = 18; P(1.2) = 5.256$ **9.** $P(-3) = 234; P(1) = -6$ **11.** $P(-2) = 39; P(2.1) = 14.6132$
**13.** $P(1) = -4 < 0$ and $P(2) = 4 > 0$ **15.** $P(2) = -2 < 0$ and $P(3) = 14 > 0$

**17.** $P(-3) = -11 < 0$ and $P(-2) = 12 > 0$ **19.** $P(-2) = 33 > 0$ and $P(-1) = -3 < 0$
**21.** $P(1.4) = -0.256 < 0$ and $P(1.5) = 1.625 > 0$ **23.** $P(2.4) = -0.216 < 0$ and $P(2.5) = 1.375 > 0$
**25.** $P(-3.3) = -0.667 < 0$ and $P(-3.2) = 0.288 > 0$

## Exercises 12.2, Page 431

**1.** $\pm 1, \pm 2, \pm 3, \pm 6$ **3.** $\pm 1, \pm 2, \pm 4, \pm 8$ **5.** $\pm 1, \pm 7, \pm\frac{1}{2}, \pm\frac{7}{2}$

**7.** $\pm 1, \pm 5, \pm\frac{1}{2}, \pm\frac{5}{2}, \pm\frac{1}{3}, \pm\frac{5}{3}, \pm\frac{1}{6}, \pm\frac{5}{6}$ **9.** $\pm 1, \pm 2, \pm 3, \pm 4, \pm 6, \pm 12, \pm\frac{1}{3}, \pm\frac{2}{3}, \pm\frac{4}{3}$

**11.** $x = 4$ **13.** $x = -\frac{1}{2}, x = -1, x = 5$ **15.** $x = -6$ **17.** $x = -1, x = -\frac{1}{3}, x = \frac{5}{2}$

**19.** $x = -5, x = \frac{1}{5}$ **21.** $x = \frac{3}{4}, x = -2 \pm \sqrt{3}$ **23.** $x = \frac{1}{2}, x = \frac{-1 \pm i\sqrt{3}}{2}$ **25.** $x = -1, x = 2, x = 3$

**27.** $x = \frac{4}{7}, x = \frac{3 \pm \sqrt{5}}{2}$ **29.** $x = -1, x = \frac{3}{2}, x = \pm 2i$

## Exercises 12.3, Page 434

**11.** Between $-2$ and 1 **13.** Between $-3$ and 1 **15.** Between $-7$ and 2 **17.** Between $-1$ and 3
**19.** Between $-2$ and 13 **21.** Between $-5$ and 6 **23.** Between $-2$ and 3 **25.** Between $-2$ and 2
**27.** Between $-5$ and 4 **29.** Between $-2$ and 3

## Exercises 12.4, Page 438

**1.** 0 positive roots; 2 negative roots or 0 negative roots
**3.** 3 positive roots or 1 positive root; 0 negative roots
**5.** 2 positive roots or 0 positive roots; 0 negative roots
**7.** 3 positive roots or 1 positive root; 1 negative root
**9.** 2 positive roots or 0 positive roots; 2 negative roots or 0 negative roots
**11.** 3 positive roots or 1 positive root; 4 negative roots, 2 negative roots, or 0 negative roots

**13.** 2 positive roots or 0 positive roots; 1 negative root; $x = 2, x = 1, x = -\frac{1}{2}$

**15.** 2 positive roots or 0 positive roots; 1 negative root; $x = -\frac{2}{3}, x = \frac{3 + \sqrt{5}}{2}, x = \frac{3 - \sqrt{5}}{2}$

**17.** 3 positive roots or 1 positive root; 1 negative root; $x = \frac{1}{3}, x = 1, x = \frac{3 + \sqrt{57}}{4}, x = \frac{3 - \sqrt{57}}{4}$

**19.** 2 positive roots or 0 positive roots; 2 negative roots or 0 negative roots; $x = -4, x = 1, x = 1 + \sqrt{2}, x = 1 - \sqrt{2}$

**21.** 1 positive root; 3 negative roots or 1 negative root; $x = \frac{5}{3}, x = -2, x = 2i, x = -2i$

**23.** 1 positive root; 3 negative roots or 1 negative root; $x = \frac{1}{3}, x = -2, x = -1 + i, x = -1 - i$

**25.** 2 positive roots or 0 positive roots; 2 negative roots or 0 negative roots; no rational roots. 5 is an upper bound, $-5$ is a lower bound.

**27.** 2 positive roots or 0 positive roots; 3 negative roots or 1 negative root; $x = -1, x = 1, x = \frac{-1 + i\sqrt{3}}{2}$, $x = \frac{-1 - i\sqrt{3}}{2}$

**29.** 1 positive root; 4 negative roots; 2 negative roots or 0 negative roots; $x = -2$, $x = -\frac{1}{2}$, $x = 2$, $x = -1 + i\sqrt{5}$, $x = -1 - i\sqrt{5}$

## Exercises 12.5, Page 442

**1.** $x^3 - 6x^2 + 5x + 12 = 0$ **3.** $x^3 - 10x^2 + 28x - 24 = 0$ **5.** $x^4 - 5x^3 + 3x^2 + 9x = 0$
**7.** $x^4 - 6x^3 - 6x^2 + 58x - 15 = 0$ **9.** $x^4 - x^3 + 5x^2 - 7x - 14 = 0$
**11.** $x^6 + 10x^5 + 25x^4 - 54x^3 - 344x^2 - 536x - 272 = 0$ **13.** $x = -2$, $x = 1$, $x = \frac{3}{2}$
**15.** $x = -\frac{4}{3}$, $x = -2$, $x = 2$ **17.** $x = 3$, $x = -3 + 2i$, $x = -3 - 2i$ **19.** $x = -6$, $x = 2 - i$, $x = 2 + i$
**21.** $x = -2$, $x = -1$, $x = 1 - i\sqrt{2}$, $x = 1 + i\sqrt{2}$ **23.** $x = 3 - 4i$, $x = 3 + 4i$, $x = 2 + i$, $x = 2 - i$
**25.** $x = -1$, $x = -2$, $x = \frac{3}{2}$ with multiplicity 2 **27.** $x = -2$, $x = 3$, $x = 2 + 3i$, $x = 2 - 3i$
**29.** $x = 0$, $x = 3$ with multiplicity 2, $x = -\frac{1}{2} + \frac{i\sqrt{3}}{2}$, $x = -\frac{1}{2} - \frac{i\sqrt{3}}{2}$

## Chapter 12 Review, Page 444

**1.** (a) $-7$ (b) 5 **2.** (a) 20 (b) 1.625 **3.** (a) 81 (b) 32.6352 **4.** $P(4) = -2 < 0$; $P(5) = 1 > 0$
**5.** $P(2) = -7 < 0$; $P(2.5) = 4.625 > 0$ **6.** $P(-1.3) = 0.116 > 0$; $P(-1.2) = -0.096 < 0$
**7.** $\pm 1, \pm 2, \pm 3, \pm 4, \pm 6, \pm 12$ **8.** $\pm 1, \pm 2, \pm 4, \pm\frac{1}{2}, \pm\frac{1}{4}$ **9.** $\pm 1, \pm 3, \pm 9, \pm\frac{1}{2}, \pm\frac{3}{2}, \pm\frac{9}{2}, \pm\frac{1}{3}, \pm\frac{1}{6}$
**10.** $x = \frac{2}{3}$, $x = 2$ **11.** $x = \frac{4}{5}$ **12.** $x = -6$ **13.** $x = -\frac{5}{3}$, $x = 3$, $x = 1$ **14.** $x = \frac{2}{5}$, $x = \frac{-1 \pm \sqrt{5}}{2}$
**15.** $x = 4$, $x = 2 \pm i$ **16.** $x = \frac{1}{2}$, $x = -3$, $x = \pm i$ **19.** Between $-1$ and 3 **20.** Between $-2$ and 7
**21.** Between $-2$ and 6 **22.** Between $-3$ and 2
**23.** 2 positive roots or 0 positive roots; 2 negative roots or 0 negative roots
**24.** 1 positive root; 2 negative roots or 0 negative roots **25.** 1 positive root; 1 negative root
**26.** 1 positive root; 2 negative roots or 0 negative roots; $x = -\frac{3}{7}$, $x = \frac{-3 \pm \sqrt{17}}{2}$
**27.** 1 positive root; 2 negative roots or 0 negative roots; $x = \frac{2}{5}$, $x = -1 \pm i\sqrt{5}$
**28.** 2 positive roots or 0 positive roots; 1 negative root; $x = -\frac{3}{4}$, $x = 2$, $x = \frac{1}{2}$
**29.** 3 positive roots or 1 positive root; 1 negative root; $x = -\frac{1}{2}$, $x = 1$, $x = \pm 3i$
**30.** $x^3 - 7x - 6 = 0$ **31.** $x^4 + 6x^3 + 34x^2 + 150x + 225 = 0$ **32.** $x^4 - 6x^3 + 20x^2 - 22x - 13 = 0$
**33.** $x^5 - 7x^4 + 22x^3 - 40x^2 + 40x - 16 = 0$ **34.** $x = -3$ with multiplicity 3
**35.** $x = -1$ with multiplicity 2, $x = 2$ with multiplicity 2 **36.** $x = 2 \pm \sqrt{3}$, $x = -1 \pm i\sqrt{2}$
**37.** $x = 2 \pm 4i$, $x = \pm i\sqrt{2}$

## Chapter 12 Test, Page 445

**1.** $P(-2) = -7$ **2.** $P(1.6) = 20.712$ **3.** $P(-2) = -6 < 0$ and $P(-1.6) = 0.528 > 0$
**4.** $P(0.8) = -0.552 < 0$ and $P(0.9) = 2.016 > 0$ **5.** $1, 2, 7, 14, \frac{1}{2}, \frac{7}{2}$ **6.** $x = -1, x = \frac{5}{4}, x = 3$
**7.** $x = -\frac{4}{3}, x = \frac{1}{3}, x = 2$ **10.** Between $-6$ and $3$ **11.** Between $-3$ and $3$
**12.** 1 positive root, 2 negative roots or 0 negative roots **13.** 3 positive roots or 1 positive root, 1 negative root
**14.** $x^3 - 9x^2 + 24x - 20 = 0$ **15.** $x^5 - 5x^3 + 30x^2 - 50x + 24 = 0$ **16.** $x = -\frac{5}{2}, x = -1, x = 1$
**17.** $x = -2, x = 1 + \sqrt{7}, x = 1 - \sqrt{7}$ **18.** $x = -2, x = -\frac{3}{2}, x = 1, x = 2$
**19.** $x = -3, x = 2, x = -1 + \sqrt{2}, x = -1 - \sqrt{2}$ **20.** $x = 1 - 3i, x = 1 + 3i, x = 2 - \sqrt{3}, x = 2 + \sqrt{3}$

## Cumulative Review (12), Page 446

**1.** $15x^{16/15}$ **2.** $\frac{x^{5/6}}{y^{7/12}}$ **3.** $\frac{x}{x + 1}$ **4.** $\frac{x^2 + 11x - 4}{(3x + 1)(x + 1)(x + 6)}$ **5.** $x = 2$ **6.** $x = 7$ **7.** $x = -0.319$
**8.** $x = 4 + e = 6.718$ **9.** $-4x + 5y = 33$ **10.** $x^2 + y^2 = 16$

**11.**

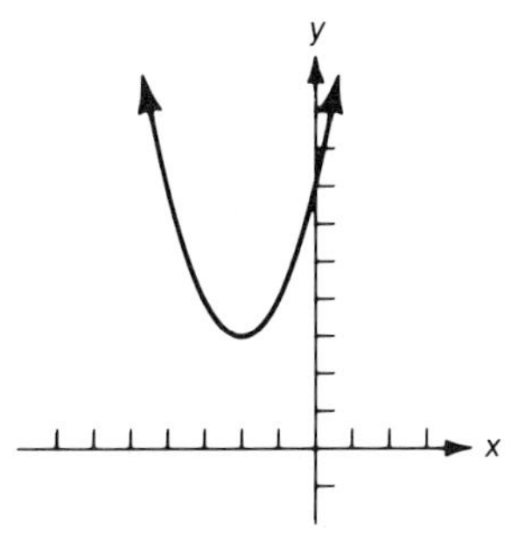


**12.**

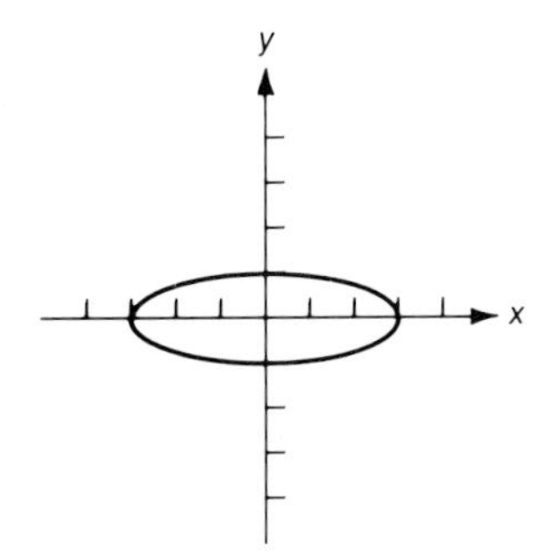


**13.**

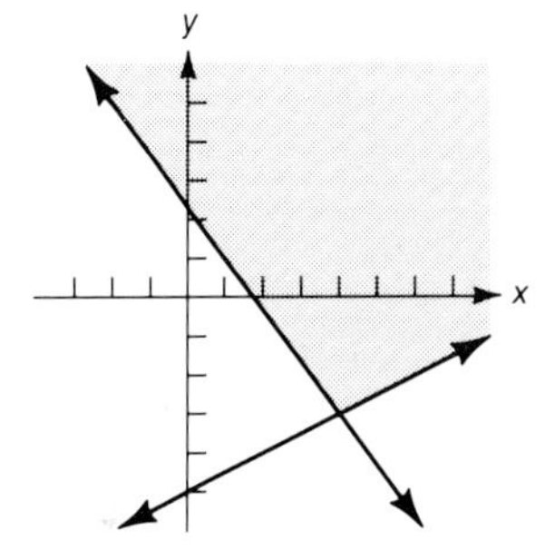


**14.** $x = 4, y = 2$ or $x = 3, y = 1$ **15.** $x = 3, y = -4, z = -2$ **16.** $x = 4, y = 3$ **17.** **(a)** $P(3) = 54$
**(b)** $P(1.5) = -1.5$ **18.** Between $-2$ and $6$ **19.** $x^5 - x^4 + 2x^3 - 10x^2 - 35x - 21 = 0$
**20.** $x = -2$ with multiplicity 2, $x = 1 - i\sqrt{5}, x = 1 + i\sqrt{5}$ **21.** \$1872.61 **22.** 50 mph

## CHAPTER 13

## Exercises 13.1, Page 452

**1.** $1, 3, 5, 7$ **3.** $2, \frac{3}{2}, \frac{4}{3}, \frac{5}{4}$ **5.** $2, 6, 12, 20$ **7.** $2, 4, 8, 16$ **9.** $-2, 5, -10, 17$ **11.** $-\frac{1}{5}, \frac{1}{7}, -\frac{1}{9}, \frac{1}{11}$
**13.** $1, 0, -1, 0$ **15.** $0, 1, 0, 1$ **17.** $\{4n + 1\}$ **19.** $\{(-1)^{n+1}(2n - 1)\}$ **21.** $\{n^2\}$ **23.** $\left\{\frac{1}{n + 2}\right\}$
**25.** $\{n^2 + 1\}$ **27.** Decreasing **29.** Decreasing **31.** Increasing **33.** $\left\{9000 \cdot \left(\frac{7}{10}\right)^n\right\}$; \$3087 when $n = 3$
**35.** $\left\{20{,}000 \cdot \left(\frac{97}{100}\right)^n\right\}$; 17,175 students when $n = 5$

## Exercises 13.2, Page 456

**1.** $2 + 4 + 6 + 8 + 10 = 30$ **3.** $5 + 6 + 7 + 8 + 9 = 35$ **5.** $\frac{1}{2} + \frac{1}{3} + \frac{1}{4} = \frac{13}{12}$ **7.** $2 + 4 + 8 = 14$

**9.** $16 + 25 + 36 + 49 + 64 = 190$ **11.** $3 + 1 + (-1) + (-3) = 0$ **13.** $6 + (-12) + 20 + (-30) = -16$
**15.** $\frac{1}{2} + \frac{2}{3} + \frac{3}{4} + \frac{4}{5} + \frac{5}{6} = \frac{71}{20}$ **17.** $\sum_{k=1}^{5} (2k - 1)$ **19.** $\sum_{k=1}^{5} (-1)^k$ **21.** $\sum_{k=2}^{6} (-1)^k\left(\frac{1}{k^3}\right)$
**23.** $\sum_{k=4}^{15} \frac{k}{k+1}$ **25.** $\sum_{k=5}^{12} \frac{k+1}{k^2}$ **27.** 7 **29.** 48 **31.** 10 **33.** 295 **35.** $-152$

## Exercises 13.3, Page 462

**1.** Arithmetic sequence; $d = 3$, $\{3n - 1\}$ **3.** Arithmetic sequence; $d = -2$, $\{9 - 2n\}$ **5.** Not an arithmetic sequence **7.** Arithmetic sequence; $d = -4$, $\{10 - 4n\}$ **9.** Arithmetic sequence; $d = \frac{1}{2}$, $\left\{\frac{n-1}{2}\right\}$
**11.** 1, 3, 5, 7, 9; arithmetic sequence **13.** $-1, 4, -7, 10, -13$; not an arithmetic sequence **15.** $-1, -7, -13, -19, -25$; arithmetic sequence **17.** $\frac{20}{3}, \frac{19}{3}, 6, \frac{17}{3}, \frac{16}{3}$; arithmetic sequence **19.** $\frac{1}{2}, \frac{1}{4}, \frac{1}{6}, \frac{1}{8}, \frac{1}{10}$; not an arithmetic sequence **21.** $\left\{\frac{3}{5} + \frac{2}{5}n\right\}$ **23.** $\{9 - 2n\}$ **25.** $\left\{\frac{17}{2} + \frac{3}{2}n\right\}$ **27.** $\{63 - 5n\}$
**29.** $\left\{\frac{81}{2} + \frac{3}{2}n\right\}$ **31.** 232 **33.** 44 **35.** 7 **37.** 19 **39.** 154 **41.** 235 **43.** 126 **45.** 231
**47.** $\frac{143}{3}$ **49.** 132 **51.** 171 **53.** $-100$ **55.** 9 days **57.** 1625 cm

## Exercises 13.4, Page 472

**1.** Geometric sequence; $r = 2$, $\{2^n\}$ **3.** Geometric sequence; $r = -\frac{1}{2}$, $\left\{3\left(-\frac{1}{2}\right)^{n-1}\right\}$
**5.** Geometric sequence; $r = \frac{3}{8}$, $\left\{\frac{32}{27}\left(\frac{3}{8}\right)^{n-1}\right\}$ **7.** Not a geometric sequence **9.** Geometric sequence; $r = -\frac{1}{4}$, $\left\{48\left(-\frac{1}{4}\right)^{n-1}\right\}$ **11.** $9, -27, 81, -243$; geometric sequence **13.** $\frac{2}{3}, \frac{4}{3}, 2, \frac{8}{3}$; not a geometric sequence **15.** $-\frac{8}{5}, \frac{32}{25}, -\frac{128}{125}, \frac{512}{625}$; geometric sequence **17.** $3\sqrt{2}, 6, 6\sqrt{2}, 12$; geometric sequence **19.** $0.3, -0.09, 0.027, -0.0081$; geometric sequence **21.** $\{3(2)^{n-1}\}$
**23.** $\left\{\frac{1}{3}\left(-\frac{1}{2}\right)^{n-1}\right\}$ **25.** $\{(\sqrt{2})^{n-1}\}$ **27.** $\left\{\frac{1}{3}(3)^{n-1}\right\}$ **29.** $\left\{-5\left(-\frac{3}{4}\right)^{n-1}\right\}$ **31.** $\frac{1}{4}$ **33.** $\frac{32}{9}$
**35.** $\frac{1}{4}$ **37.** $n = 4$ **39.** 363 **41.** $\frac{1023}{64}$ **43.** $-\frac{333}{64}$ **45.** $\frac{422}{243}$ **47.** $\frac{75}{128}$ **49.** 4 **51.** $-\frac{1}{3}$
**53.** $\frac{2}{9}$ **55.** $\frac{4}{11}$ **57.** \$13,248.23 **59.** \$4352 **61.** $64\frac{4}{5}$ grams

## Exercises 13.5, Page 479

**1.** 56 **3.** $\frac{1}{15}$ **5.** 4 **7.** $(n - 1)!$ **9.** $(k + 3)(k + 2)(k + 1)$ **11.** 20 **13.** 35 **15.** 1
**17.** $x^7 + 7x^6y + 21x^5y^2 + 35x^4y^3$ **19.** $x^9 + 9x^8 + 36x^7 + 84x^6$
**21.** $x^5 + 5x^4(3) + 10x^3(3)^2 + 10x^2(3)^3 = x^5 + 15x^4 + 90x^3 + 270x^2$
**23.** $x^6 + 6x^5(2y) + 15x^4(2y)^2 + 20x^3(2y)^3 = x^6 + 12x^5y + 60x^4y^2 + 160x^3y^3$
**25.** $(3x)^7 + 7(3x)^6(-y) + 21(3x)^5(-y)^2 + 35(3x)^4(-y)^3 = 2187x^7 - 5103x^6y + 5103x^5y^2 - 2835x^4y^3$
**27.** $(x^2)^9 + 9(x^2)^8(-4y) + 36(x^2)^7(-4y)^2 + 84(x^2)^6(-4y)^3 = x^{18} - 36x^{16}y + 576x^{14}y^2 - 5376x^{12}y^3$
**29.** $x^6 + 6x^5y + 15x^4y^2 + 20x^3y^3 + 15x^2y^4 + 6xy^5 + y^6$ **31.** $x^7 - 7x^6 + 21x^5 - 35x^4 + 35x^3 - 21x^2 + 7x - 1$
**33.** $(3x)^5 + 5(3x)^4y + 10(3x)^3y^2 + 10(3x)^2y^3 + 5(3x)y^4 + y^5 = 243x^5 + 405x^4y + 270x^3y^2 + 90x^2y^3 + 15xy^4 + y^5$
**35.** $x^4 + 4x^3(2y) + 6x^2(2y)^2 + 4x(2y)^3 + (2y)^4 = x^4 + 8x^3y + 24x^2y^2 + 32xy^3 + 16y^4$
**37.** $(3x)^4 + 4(3x)^3(-2y) + 6(3x)^2(-2y)^2 + 4(3x)(-2y)^3 + (-2y)^4 = 81x^4 - 216x^3y + 216x^2y^2 - 96xy^3 + 16y^4$
**39.** $(3x^2)^5 + 5(3x^2)^4(-y) + 10(3x^2)^3(-y)^2 + 10(3x^2)^2(-y)^3 + 5(3x^2)(-y)^4 + (-y)^5 = 243x^{10} - 405x^8y + 270x^6y^2 - 90x^4y^3 + 15x^2y^4 - y^5$ **41.** $210x^6(-2y)^4 = 3360x^6y^4$ **43.** $165(2x)^8(3)^3 = 1{,}140{,}480x^8$
**45.** $220(5x^2)^3(-y)^9 = -27{,}500x^6y^9$ **47.** 1.062 **49.** 0.808 **51.** 64.363 **53.** 0.851

## Exercises 13.6, Page 483

**1.** 840 **3.** 24 **5.** 120 **7.** 60,480 **9.** 1,814,400 **11.** 110 **13.** 336 **15.** 72 **17.** 360 **19.** 64
**21.** 576 **23.** 288 **25.** 69,300

## Exercises 13.7, Page 486

**1.** 35 **3.** 1 **5.** 84 **7.** 6 **9.** 120 **11.** 816 **13.** 126 **15.** 190 **17.** 30 **19.** 66,528
**21.** 14,700 **23.** 840 **25.** 600

## Chapter 13 Review, Page 489

**1.** 1, −3, −7, −11; arithmetic sequence **2.** 4, 7, 12, 19; neither **3.** −12, 24, −48, 96; geometric sequence
**4.** −1, 1, −1, 1; geometric sequence **5.** −7, 9, −11, 13; neither **6.** $5, \frac{5}{2}, \frac{5}{3}, \frac{5}{4}$; neither **7.** $\{5n + 1\}$
**8.** $\{3n + 1\}$ **9.** $\left\{\frac{1}{6} \cdot 3^{n-1}\right\}$ **10.** $\{(-1)^n(2n + 3)\}$ **11.** $\left\{(-1)^n 8\left(\frac{1}{4}\right)^{n-1}\right\}$ **12.** $\left\{\frac{n+1}{n}\right\}$ **13.** 11
**14.** 55 **15.** $\{3n - 2\}$ **16.** $\{2n - 3\}$ **17.** $\left\{\frac{1}{2}n - 3\right\}$ **18.** 207 **19.** 248 **20.** 27 **21.** 40 **22.** 9
**23.** $\left\{\frac{1}{3}(2)^{n-1}\right\}$ **24.** $\left\{5\left(-\frac{1}{2}\right)^{n+1}\right\}$ **25.** $\{20(0.2)^{n-1}\}$ **26.** −726 or 366 **27.** $\frac{422}{27}$ **28.** $\frac{121}{24}$ or $-\frac{61}{24}$
**29.** 175 **30.** 221 **31.** 1190 **32.** $\frac{508}{3}$ **33.** 373 **34.** $-11\frac{171}{256}$ **35.** $\frac{3}{2}$ **36.** $\frac{3}{5}$ **37.** $\frac{7}{9}$
**38.** $\frac{8}{33}$ **39.** −83 **40.** 45 **41.** 2520 **42.** $\frac{33}{14}$ **43.** 21 **44.** 210
**45.** $x^5 + \binom{5}{1}x^4(3y) + \binom{5}{2}x^3(3y)^2 + \binom{5}{3}x^2(3y)^3 + \binom{5}{4}x(3y)^4 + \binom{5}{5}(3y)^5$ or $x^5 + 15x^4y + 90x^3y^2 + 270x^2y^3 + 405xy^4 + 243y^5$

**46.** $(2x)^6 + \binom{6}{1}(2x)^5\left(-\frac{y}{2}\right) + \binom{6}{2}(2x)^4\left(-\frac{y}{2}\right)^2 + \binom{6}{3}(2x)^3\left(-\frac{y}{2}\right)^3 + \binom{6}{4}(2x)^2\left(-\frac{y}{2}\right)^4 + \binom{6}{5}(2x)\left(-\frac{y}{2}\right)^5 + \binom{6}{6}\left(-\frac{y}{2}\right)^6$ or $64x^6 - 96x^5y + 60x^4y^2 - 20x^3y^3 + \frac{15}{4}x^2y^4 - \frac{3}{8}xy^5 + \frac{1}{64}y^6$

**47.** $\binom{8}{5}(2x)^3(-y)^5$ or $-448x^3y^5$ **48.** $\binom{10}{5}\left(\frac{x}{2}\right)^5(3y)^5$ or $\frac{15{,}309}{8}x^5y^5$ **49.** 16,058.71 **50.** 443.7 liters

**51.** 120 **52.** 6720 **53.** 42 **54.** 126 **55.** 252 **56.** 28 **57.** 12 **58.** 3024 **59.** 18,564
**60.** 150

## Chapter 13 Test, Page 490

**1.** $\frac{1}{4}, \frac{1}{7}, \frac{1}{10}, \frac{1}{13}, \ldots$; neither **2.** $\left\{\frac{n}{2n+1}\right\}$ **3.** 26 **4.** $a_n = 8 - 2n$ **5.** 243 **6.** $\frac{1}{8}$

**7.** $a_n = (\sqrt{3})^{n-2}$ **8.** $\frac{21}{16}$ **9.** 68 **10.** $-\frac{40}{729}$ **11.** $\frac{5}{33}$ **12.** 74 **13.** \$4096 **14.** 462

**15.** $32x^5 - 80x^4y + 80x^3y^2 - 40x^2y^3 + 10xy^4 - y^5$ **16.** $5670x^4y^4$ **17.** 151,200 **18.** 792 **19.** 720
**20.** 168

## Cumulative Review (13), Page 491

**1.** $11x + 3$ **2.** $-x^2 + 3x + 5$ **3.** $(4x + 3)(16x^2 - 12x + 9)$ **4.** $(3x - 5)(2x + 9)$

**5.** $(2x + 1)(5x + 7)(x - 2)$ **6.** $\frac{11x - 12}{15}$ **7.** $\frac{-6x - 4}{(x + 4)(x - 4)(x - 1)}$ **8.** $\frac{1}{x^2}$ **9.** $\frac{8^{1/2}y^{1/3}}{x^2}$ **10.** $\frac{2x^{4/3}}{3y}$

**11.** $3\sqrt{3x}$ **12.** $\frac{-13 + 11i}{29}$ **13.** $x > 3$ 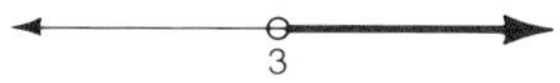

**14.** $\frac{1}{3} \le x \le \frac{13}{3}$ 

**15.** $-\frac{5}{2} < x < \frac{9}{4}$ 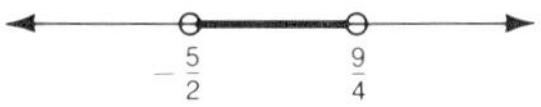

**16.** $x = 4$ **17.** $x = \frac{-2 \pm i\sqrt{2}}{2}$

**18.** $x = 4$ **19.** $x = \frac{39}{17}$ **20.** $x = 5$ **21.** $x = e^{2.4} \approx 11.02$ **22.** $n = \frac{A - P}{Pi}$

**23.** $x = 2, y = -5$ **24.** $x = 0, y = 2, z = -1$ **25.** **(a)** $y = \frac{x + 7}{2}$ **(b)** $2x^2 - 5$ **(c)** $2x + 1$

**26.**

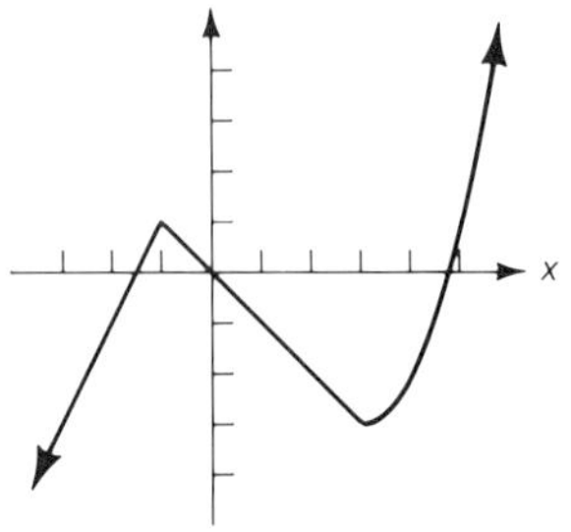


**27.** $x = 2, y = -3$

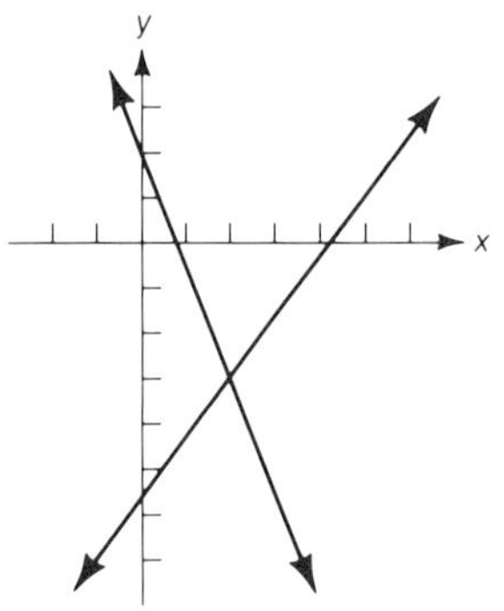


**28.**

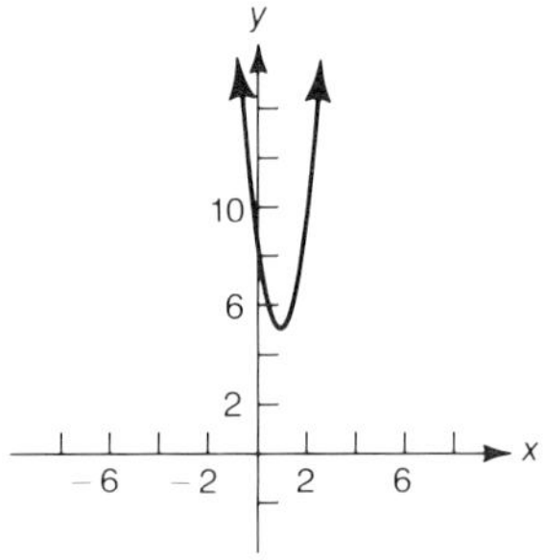

**29.**

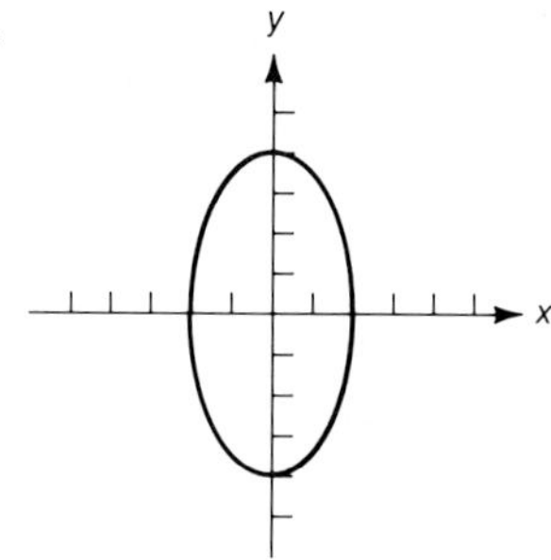


**30.**

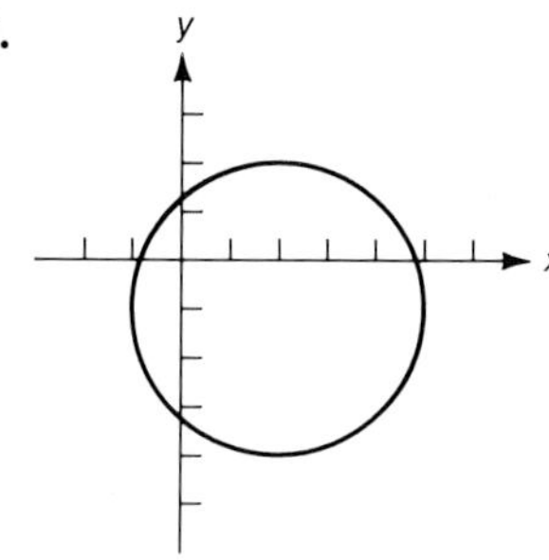


**31.** **(a)** $a_n = 10 - 2n$ **(b)** $-10$ **32.** **(a)** $a_n = 16\left(\frac{1}{2}\right)^{n-1}$ **(b)** $\frac{63}{2}$

**33.** $x^6 + 12x^5y + 60x^4y^2 + 160x^3y^3 + 240x^2y^4 + 192xy^5 + 64y^6$ **34.** 210 **35.** 1680 **36.** 2002

**37.** 20 mph, 60 mph **38.** 16 lb @ \$1.40, 4 lb @ \$2.60 **39.** 16 ft by 26 ft **40.** 12.56 hr

# INDEX

### E



### F



### G



### H



### I

## Q



## R



## S



## T



## U

## V



## W



## X



## Y



## Z



# CREDITS / MATHEMATICAL CHALLENGES

**Page 1** Reprinted with permission from the MATHEMATICS TEACHER; 2/89, 11/88, 3/89, copyright by the National Council of Teachers of Mathematics

**Page 57** Reprinted with permission from the MATHEMATICS TEACHER; 1/89, 12/88, 11/88, copyright by the National Council of Teachers of Mathematics

**Page 98** Reprinted with permission from the MATHEMATICS TEACHER; 1/89, 1/89, 4/89, copyright by the National Council of Teachers of Mathematics

**Page 145** Reprinted with permission from the MATHEMATICS TEACHER; 1/89, 11/88, 3/89, copyright by the National Council of Teachers of Mathematics

**Page 172** Reprinted with permission from the MATHEMATICS TEACHER; 1/89, 11/88, 10/88, copyright by the National Council of Teachers of Mathematics

**Page 211** Reprinted with permission from the MATHEMATICS TEACHER; 12/88, 4/89, 3/89, copyright by the National Council of Teachers of Mathematics

**Page 241** Reprinted with permission from the MATHEMATICS TEACHER; 12/88, 11/88, 12/87, copyright by the National Council of Teachers of Mathematics

**Page 268** Reprinted with permission from the MATHEMATICS TEACHER; 1/89, 2/89, 4/89, copyright by the National Council of Teachers of Mathematics

**Page 319** Reprinted with permission from the MATHEMATICS TEACHER; 1/89, 3/89, 3/88, copyright by the National Council of Teachers of Mathematics

**Page 358** Reprinted with permission from the MATHEMATICS TEACHER; 12/88, 12/88, 3/88, copyright by the National Council of Teachers of Mathematics

**Page 385** Reprinted with permission from the MATHEMATICS TEACHER; 12/88, 3/89, 3/88, copyright by the National Council of Teachers of Mathematics

**Page 421** Reprinted with permission from the MATHEMATICS TEACHER; 3/89, 3/89, 2/89, copyright by the National Council of Teachers of Mathematics

**Page 447** Reprinted with permission from the MATHEMATICS TEACHER; 4/89, 3/89, 3/89, copyright by the National Council of Teachers of Mathematics

# FORMULAS FROM GEOMETRY

## Rectangle

Area: $A = \ell w$

Perimeter: $P = 2\ell + 2w$

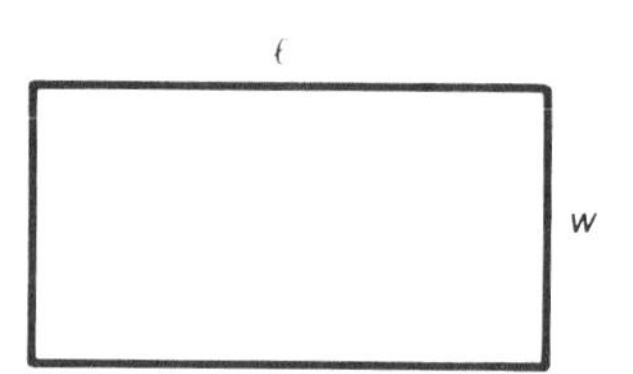


## Rectangular Solid

Volume: $V = \ell wh$

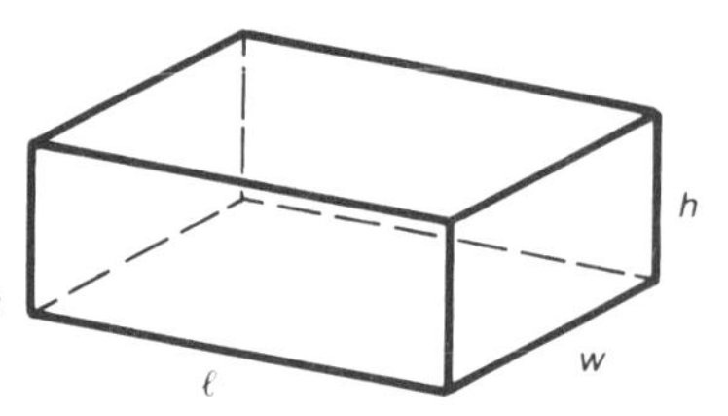


## Square

Area: $A = s^2$

Perimeter: $P = 4s$

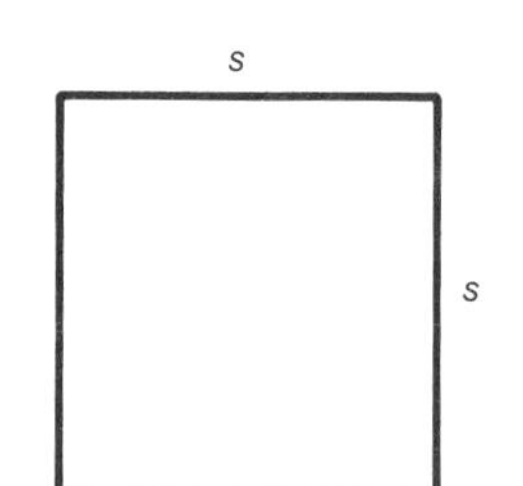


## Cube

Volume: $V = s^3$

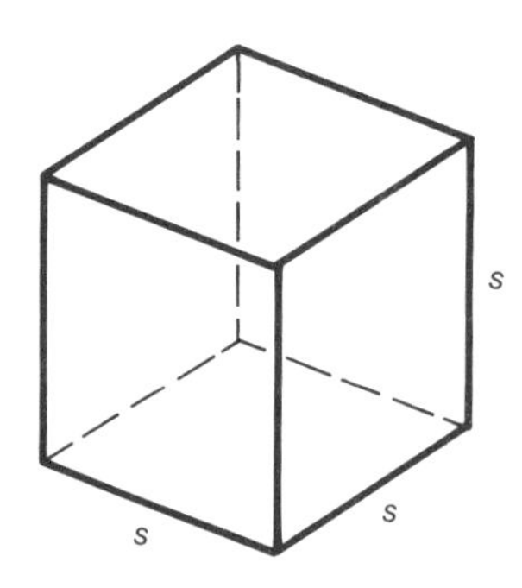


## Triangle

Area: $A = \left(\frac{1}{2}\right)bh$

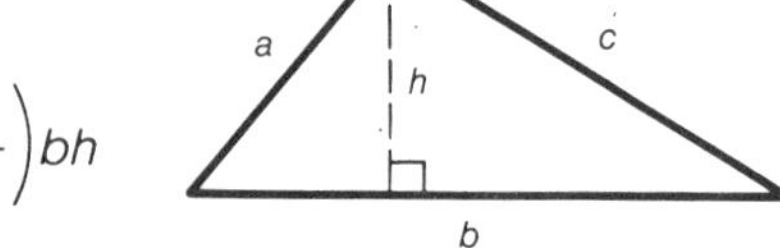


Perimeter: $P = a + b + c$

Angle Measure: $\alpha + \beta + \gamma = 180°$

## Right Circular Cone

Volume: $V = \left(\frac{1}{3}\right)\pi r^2 h$

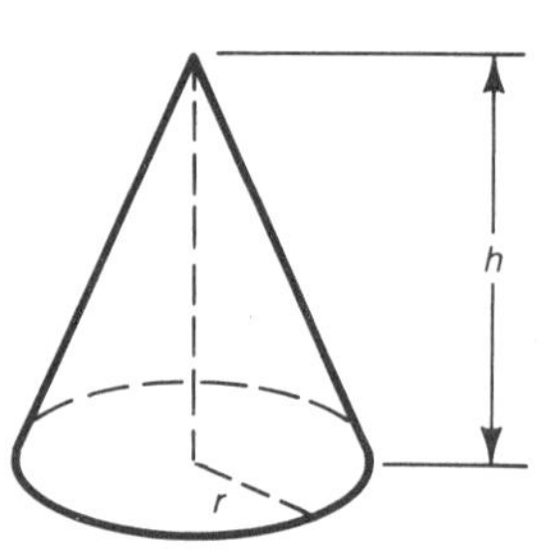


## Circle

Area: $A = \pi r^2$

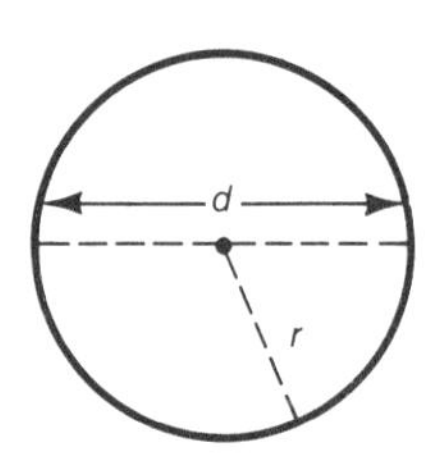


## Right Circular Cylinder

Volume: $V = \pi r^2 h$

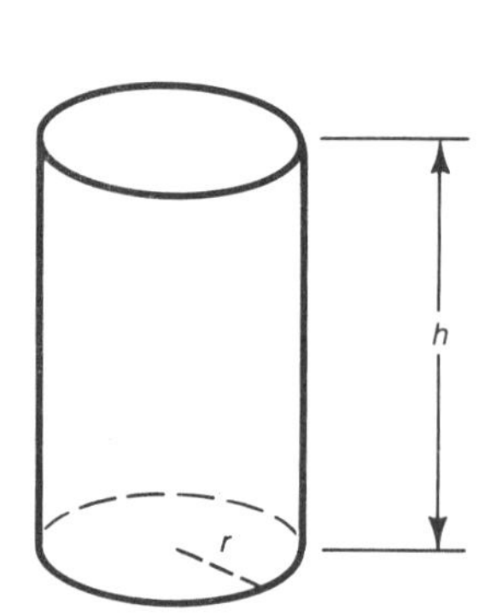